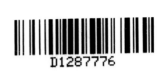

Introduction to Thermodynamics

SERIES IN THERMAL AND TRANSPORT SCIENCES

Gordon J. Van Wylen, Coordinator

FUNDAMENTALS OF CLASSICAL THERMODYNAMICS

Gordon J. Van Wylen and Richard E. Sonntag

FUNDAMENTALS OF STATISTICAL THERMODYNAMICS

Richard E. Sonntag and Gordon J. Van Wylen

FLUID MECHANICS

Arthur G. Hansen

INTRODUCTION TO THERMODYNAMICS: CLASSICAL AND STATISTICAL

Richard E. Sonntag and Gordon J. Van Wylen

Introduction to Thermodynamics: Classical and Statistical

Richard E. Sonntag
University of Michigan
Gordon J. Van Wylen
University of Michigan

JOHN WILEY & SONS, INC.
New York · London · Sydney · Toronto

Library of Congress Catalogue Card Number: 72-129053

ISBN 0-471-81365-6

Printed in the United States of America

10 9 8 7 6 5 4 3 2 1

Preface

In recent years there has been a definite trend to approach the teaching of thermodynamics to engineering and applied science students from both the classical and statistical points of view. In some cases this has been done through separate and distinct courses. In other cases there has been an effort to integrate the two viewpoints in a single or in related courses. This book has been written primarily to provide a textbook that offers considerable flexibility for these latter course arrangements.

Our first goal in writing this book has been to provide a clear and rigorous introduction to both classical and statistical thermodynamics. We have attempted to keep the student's point of view in mind in all our writing, and for this reason have included numerous examples in the text. Thermodynamics is a subject with a wonderful logic and structure. We have attempted to stress this, not only because it is important to a study of thermodynamics, but also because a study of thermodynamics can be a vehicle for a profitable academic experience in logical thinking and intellectual rigor.

Our second goal has been to present the subject matter so that there are several arrangements by which the material can be covered. We have attempted to provide this flexibility in both sections of the book (the classical section, Chapters 1–14, and the statistical section, Chapters 15–20) and in the ways in which material from both sections can be integrated in a given course or in series of courses.

This book is based largely on our two earlier books, *Fundamentals of Classical Thermodynamics*, and *Fundamentals of Statistical Thermodynamics*. We have, however, made a number of changes in certain sections. The control volume analysis of the first and second laws has been extensively revised and significantly simplified. The chapter dealing with mixtures has been revised and concerns only ideal mixtures. There also have been major changes in Chapters 15, 16, and 17, the first three chapters of the statistical section. More than 200 new problems have been included and many of the problems that appeared in earlier books have been revised. A number of problems that require the use of a digital computer have also been incorporated in this book.

Throughout the entire book we have attempted to retain an engineering perspective. Although this emphasis is particularly evident in the examples and the problems, we have also stressed this in the text itself. We believe that the rigorous application of the fundamentals to the solution of problems can provide an excellent vehicle for the integration of theory and practice.

In regard to the symbols used in this text, we were guided by two considerations. First, we have used those symbols commonly used in the general literature of both classical and statistical thermodynamics; second, we have tried to maintain a consistency of symbols throughout the book. In a limited number of cases, we have used a given symbol for more than one purpose. We believe, however, that the context will clarify the meaning of the symbol in these cases.

To emphasize the distinctions between force and mass, the symbols lbf and lbm have been used for force and mass in the English system. Such symbols as lbf/in.^2 have been used for pressure (rather than psi) and $\text{ft}^3\text{/lbm}$ for specific volume (rather than cu ft/lb) in order to stress the fundamental units involved in the various parameters. Concerning the extensive properties, a lower case letter (u, h, s) designates the property per unit mass, an upper case letter (U, H, S), the property for the entire system; a lower case letter with a bar $(\bar{u}, \bar{h}, \bar{s})$, the property per unit mole; and upper case letter with a bar $(\bar{U}, \bar{H}, \bar{S})$, the partial molal property. Following this pattern, we have found it convenient to designate the total heat transfer as Q, the heat transfer per unit mass of the system as q, the total work as W, and the work per unit mass of the system as w.

Further, we represent the rate of flow across a system boundary or control surface by a dot over a given quantity. Thus $\dot{Q}$ represents a rate of heat transfer across the system boundary; $\dot{W}$, the rate at which work crosses the system boundary (that is, the power); and $\dot{m}$, the mass rate of flow across a control surface ($\dot{n}$ is used when the mass rate of flow is expressed in moles per unit time). The rate of heat transfer across a control surface is designated $\dot{Q}_{\text{c.v.}}$. We realize that we have departed from the usual mathematical use of a dotted symbol since in mathematics the dotted symbol typically refers to a derivative with respect to time. However, we have used the dotted symbol only to indicate a flow of heat and work across a system boundary, and heat, work, and mass across a control surface, and we believe that it has contributed to a simple and consistent use of symbols for this book.

We acknowledge with appreciation the suggestions, counsel, and encouragement from many colleagues, both at the University of Michigan and elsewhere. This book has evolved over several years, and during the entire period this assistance has been most helpful to us. A number of different secretaries have aided us immeasurably during this period, and we acknowledge this with many thanks. Students have also aided us. Their perceptive questions have often caused us to rewrite or rethink a given portion of the text. Finally, for each of us the encouragement and patience of our wives and families have been indispensable, and have made this time of writing pleasant and enjoyable in spite of the pressures of the project.

Our hope is that this book will contribute to the effective teaching of thermodynamics to the concerned generation. Your comments, criticism, and suggestions will also be appreciated.

RICHARD E. SONNTAG

Ann Arbor, Michigan, 1970 GORDON J. VAN WYLEN

Organization of the Book

In the classical section of this text, Chapter 1 constitutes a general introduction to thermodynamics, and Chapter 2 deals with definitions and units. Chapter 3 involves the properties of a pure substance, and work and heat are introduced in Chapter 4. Chapters 5 and 6 deal with the first law, Chapter 5 for a system and Chapter 6 from the control volume point of view. These have been arranged so that Chapter 6 can be taken up directly after Chapter 5 or at any subsequent time. Those interested only in the system analysis can omit Chapter 6. A similar arrangement applies to the second law, which is discussed in Chapter 7, with entropy being introduced in Chapter 8. Chapter 9 is concerned with the second law for a control volume, and can be omitted by those interested only in a system analysis. The control volume analyses of Chapters 6 and 9 are a prerequisite for the remaining chapters of the classical section of the book, but are not a prerequisite for statistical thermodynamics.

Chapter 10 deals with thermodynamic cycles, and has been included because many students enjoy a brief coverage of cycles, and find that such a study strengthens their understanding of the first and second laws. This chapter stands quite independently, and is not a prerequisite for any other portion of the book.

Chapter 11 concerns mixtures; Chapter 12 deals with thermodynamic relations; Chapter 13 discusses chemical reactions; and equilibrium is dealt with in Chapter 14. Normally these will be taken up in the order given.

The remainder of the book (Chapters 15 to 20) includes the material on statistical thermodynamics. Chapter 15 is involved with background material on probability and statistics and also the introduction to molecular distributions and models. The subject of statistical mechanics, including simple applications, is treated in Chapter 16. In writing this chapter, we have presumed that the introductory material in Chapter 2 will have been studied previously. Chapter 17 is concerned with thermodynamic properties and laws, and presumes that these topics have been introduced from the classical viewpoint (Chapters 3–5, and 7–8). Chapter 18 deals with an introduction to quantum mechanics, Chapter 19 with evaluation of the properties of gases, and Chapter 20 with solids.

There are, of course, many divergent viewpoints regarding the best order of presentation of classical and statistical material in thermodynamics. During the past several years we have experimented with a number of alternate approaches. Although there are several ways in which the material can be covered, we recommend the following three possibilities.

1. Sequenced—Chapters 1–20 in order. This is particularly useful when different degree programs elect a common first course, with statistical thermodynamics being covered in an optional second course. An alternative to covering all the material in order is to spend the first term on classical thermodynamics (Chapters 1–5 plus 7 and 8, or 1–9 and/or 10 and/or 11, depending on the length and credit hours of the course) and then begin the second term course with Chapter 15.
2. Integrated—Chapters 1–3, 15–16, 4–9, 17 (Sect. 1–3), and then either return to classical thermodynamics or continue with statistical thermodynamics. In this case the control volume is covered fairly early in the course.
3. Integrated—Chapters 1–3, 15–16, 4–5, 7–8, 17 (Sect. 1–3), then return to classical thermodynamics or continue with statistical thermodynamics. This arrangement focuses on an early coverage of statistical thermodynamics, and defers the control volume analysis and the application of classical thermodynamics to the latter portion of the course.

Just as there are many opinions regarding the order of presentation of material, there are also many opinions on the question of how much statistical thermodynamics should be introduced in the undergraduate curriculum. We have provided for a number of alternatives. The first possibility, of course, is classical thermodynamics, Chapters 1–14 only, with a flavor of the molecular viewpoint (for example, the qualitative discussion of specific heat in Section 5.6). Below is an arrangement by which material on statistical thermodynamics could incrementally be added to a given course.

1. Chapters 15 and 16: An introduction to the kinetic theory approach with simple applications.
2. Sections 17.1–17.3: Entropy and an examination of the first and second laws.
3. Sections 17.4, 18.1–18.4, 18.8, 19.1–19.4: Quantum mechanics and its application to the behavior and properties of monatomic gases (Sections 17.5 and 20.3, can be added to include the electron gas).
4. Sections 18.5–18.7, 19.5–19.9: Diatomic and polyatomic gases.
5. Sections 17.5, 18.10: Stronger emphasis on certain fundamental principles.

The remaining topics are independent and can be selected as desired: Section 19.10, photon gas (following 3 above); 19.11, mixtures (related to Chapter 11); 19.12, chemical reactions (related to Chapters 13 and 14); Chapter 20, solids.

R.E.S.
G.J.V.W.

Contents

Symbols

a	acceleration
a, A	specific Helmholtz function and total Helmholtz function
A	area
AF	air-fuel ratio
c	velocity of light
C_p	constant-pressure specific heat
C_v	constant-volume specific heat
C_{po}	zero-pressure constant-pressure specific heat
C_{vo}	zero-pressure constant-volume specific heat
d	electron state symbol
D	total dissociation energy
D	atomic term symbol
D_0	observed dissociation energy
e	electron charge
e, E	specific energy and total energy
f	electron state symbol
f	fugacity
f_i	fugacity of component i in a mixture
F	force
F	atomic term symbol
FA	fuel-air ratio

g	acceleration due to gravity
g, G	specific Gibbs function and total Gibbs function
g_c	a constant that relates force, mass, length, and time
$\mathbf{g}_j$	jth energy level degeneracy
gm	gram
gm mole	gram mole
h	Planck's constant
h, H	specific enthalpy and total enthalpy
i	electrical current
I	irreversibility
I	moment of inertia
$\mathbf{j}$	molecular rotational quantum number
$\mathbf{J}$	atomic total angular momentum number
J	proportionality factor to relate units of work to units of heat
k	specific heat ratio: C_p/C_v
k	Boltzmann constant
$\mathbf{k}_x, \mathbf{k}_y, \mathbf{k}_z$	x-, y-, z-directional translational quantum numbers
K	equilibrium constant
KE	kinetic energy
$\mathbf{l}$	electron azimuthal quantum number
L	length
lbf	pound force
lbm	pound mass
lb mole	pound mole
lw, LW	lost work per unit mass and total lost work
m	mass
$\dot{m}$	mass rate of flow
$\mathbf{m}_l$	electron magnetic quantum number
$\mathbf{m}_s$	electron spin quantum number
m_r	molecular reduced mass
M	molecular weight
mf	mass fraction
$\mathbf{n}$	electron principal quantum number
n	number of moles
n	polytropic exponent
N	number of particles

N_0	Avogadro's number
p	electron state symbol
P	mathematical probability
P	atomic term symbol
P	pressure
P_i	partial pressure of component i in a mixture
PE	potential energy
q, Q	heat transfer per unit mass and total heat transfer
$\dot{Q}$	rate of heat transfer
Q_H, Q_L	heat transfer with high-temperature body and heat transfer with low-temperature body; sign determined from context
r	radius
r_e	molecular equilibrium separation
R	gas constant
$\bar{R}$	universal gas constant
s	electron state symbol
s, S	specific entropy and total entropy
S	atomic term symbol
t	time
T	temperature
u, U	specific internal energy and total internal energy
u_{v0}	zero-point energy per mole
$\mathbf{v}$	molecular vibration quantum number
v, V	specific volume and total volume
vf	volume fraction
V	velocity
w	thermodynamic probability – number of microstates
w, W	work per unit mass and total work
$\dot{W}$	rate of work, or power
w_{rev}	reversible work between two states assuming heat transfer with surroundings
x	quality
x	liquid-phase or solid-phase mole fraction
y	vapor-phase mole fraction

Z	partition function
Z	elevation
Z	compressibility factor
Z	electrical charge

SCRIPT LETTERS

$\mathscr{C}$	number of components
${}_N\mathscr{C}_M$	combinations of N, M at a time
$\mathscr{E}$	electrical potential
$\mathscr{H}$	magnetic field intensity
$\mathscr{M}$	magnetization
$\mathscr{P}$	number of phases
${}_N\mathscr{P}_M$	permutations of N, M at a time
$\mathscr{S}$	surface tension
$\mathscr{T}$	tension
$\mathscr{V}$	variance
$\mathscr{Z}$	nuclear charge

SUBSCRIPTS

c	property at the critical point
c.v.	control volume
e	electronic
e	state of a substance leaving a control volume
f	formation
f	property of saturated liquid
fg	difference in property for saturated vapor and saturated liquid
g	property of saturated vapor
i	energy state
i	state of a substance entering a control volume
i	property of saturated solid
ig	difference in property for saturated vapor and saturated solid
j	energy level
mp	most probable (equilibrium)
n	nuclear
r	rotation

r	reduced property
s	isentropic process
t	translation
v	vibration
0	property of the surroundings

SUPERSCRIPTS

$-$	bar over symbol denotes property on a molal basis (over V, H, S, U, A, G, the bar denotes partial molal property)
$\circ$	property at standard-state condition
$*$	ideal gas
L	liquid-phase
S	solid-phase
V	vapor-phase

1

Some Introductory Comments

In the course of our study of thermodynamics, a number of the examples and problems presented refer to processes that occur in such equipment as a steam power plant, a fuel cell, a vapor compression refrigerator, a thermoelectric cooler, a rocket engine, and an air separation plant. In this introductory chapter a brief description of this equipment is given. There are at least two reasons for including such a chapter. First, many students have had limited contact with such equipment, and the solution of problems will be more significant and relevant when they have some familiarity with the actual processes and the equipment involved. Second, this chapter will provide an introduction to thermodynamics, including the use of certain terms (which will be more formally defined in later chapters), some of the problems for which thermodynamics is relevant, and some accomplishments that have resulted, at least in part, from the application of thermodynamics.

It should be emphasized that thermodynamics is relevant to many other processes than those cited in this chapter. It is basic to the study of materials, chemical reactions, and plasmas. The student should bear in mind that this chapter is only a brief and necessarily very incomplete introduction to the subject of thermodynamics.

1.1 The Simple Steam Power Plant

A schematic diagram of a simple steam power plant is shown in Fig. 1.1. High-pressure superheated steam leaves the boiler, which is also referred to as a steam generator, and enters the turbine. The steam expands in the turbine and in doing so, does work, which enables the turbine to drive the electric generator. The low-pressure steam leaves

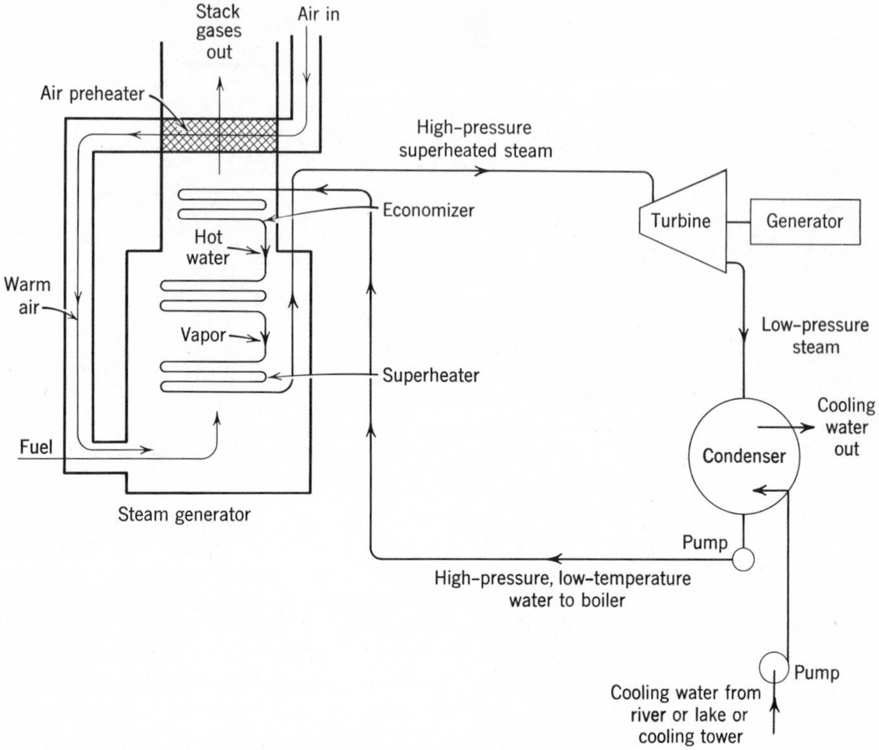

Fig. 1.1 Schematic diagram of a steam power plant.

the turbine and enters the condenser, where heat is transferred from the steam (causing it to condense) to the cooling water. Since large quantities of cooling water are required, power plants are frequently located near rivers or lakes. It is this transfer of heat to the water in lakes and rivers that leads to the thermal pollution problem, which is now being extensively studied. During our study of thermodynamics we shall gain an understanding of why this heat transfer is necessary and ways in which it can be minimized. When the supply of cooling water is limited, a cooling tower may be used. In the cooling tower some of the cooling water evaporates in such a way as to lower the temperature of the water that remains as a liquid.

The pressure of the condensate leaving the condenser is increased in the pump, thus enabling the condensate to flow into the steam generator. In many steam generators an economizer is used. An economizer is simply a heat exchanger in which heat is transferred from the products of combustion (just before they leave the steam generator) to the condensate, with the result that the temperature of the condensate is

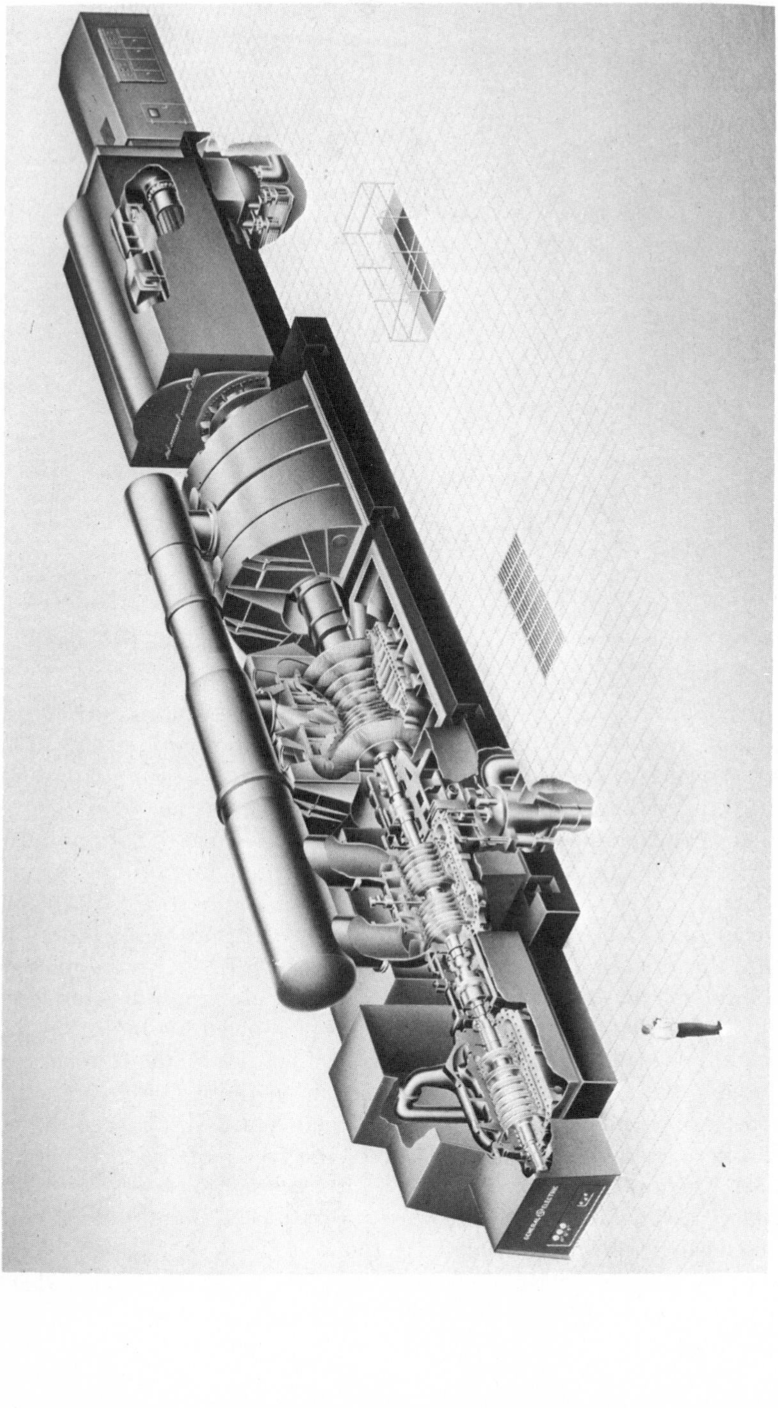

Fig. 1.2 A large steam turbine (Courtesy General Electric Co.).

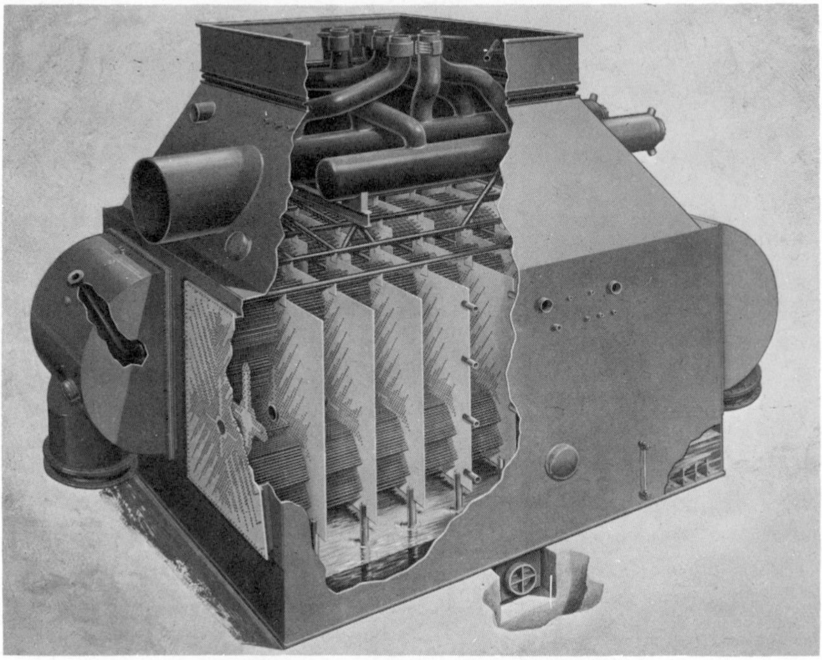

Fig. 1.3 A condenser used in a large power plant (Courtesy Westinghouse Electric Corp.).

increased, but no evaporation takes place. In other sections of the steam generator, heat is transferred from the products of combustion to the water, causing it to evaporate. The temperature at which evaporation occurs is called the saturation temperature. The steam then flows through another heat exchanger known as a superheater, where the temperature of the steam is increased well above the saturation temperature.

In many power plants the air that is used for combustion is preheated in the air preheater by transferring heat from the stack gases as they are leaving the furnace. This air is then mixed with fuel—which might be coal, fuel oil, natural gas, or other combustible material—and combustion takes place in the furnace. As the products of combustion pass through the furnace, heat is transferred to the water in the superheater, the boiler, the economizer, and to the air in the air preheater. The products of combustion from power plants are discharged to the atmosphere, and this constitutes one of the facets of the air pollution problem we now face.

A large power plant will have many other pieces of equipment, some of which will be considered in later chapters.

Figure 1.2 shows a steam turbine and the generator that it drives.

Steam turbines vary in capacity from less than 10 kilowatts to 1,000,000 kilowatts.

Figure 1.3 shows a cutaway view of a condenser. The steam enters at the top and the condensate is collected in the hot well at the bottom while the cooling water flows through the tubes. A large condenser has a tremendous number of tubes, as shown in Fig. 1.3.

Figure 1.4 shows a large steam generator. The flow of air and products of combustion are indicated. The condensate, also called the boiler feedwater, enters at the economizer inlet, and the superheated steam leaves at the superheater outlet.

The number of nuclear power plants in operation is rapidly increas-

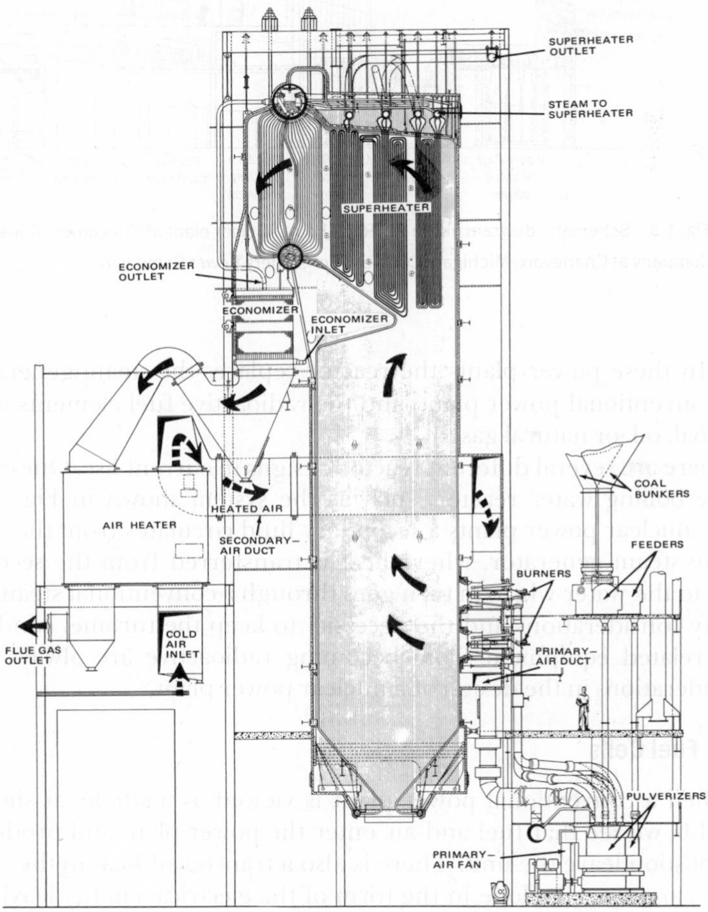

Fig. 1.4 A large steam generator (Courtesy Babcock and Wilcox Co.).

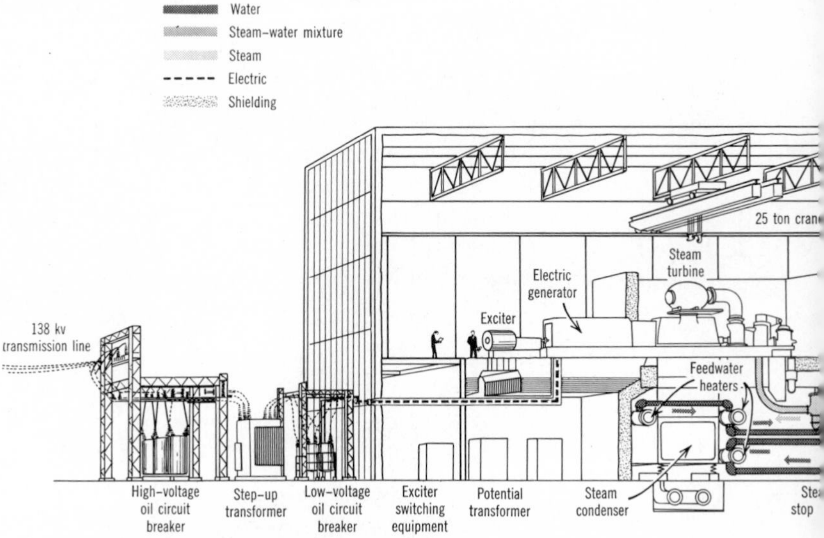

Fig. 1.5 Schematic diagram of the Big Rock Point nuclear plant of Consumers Power. Company at Charlevoix, Michigan (Courtesy Consumers Power Company).

ing. In these power plants the reactor replaces the steam generator of the conventional power plant, and the radioactive fuel elements replace the coal, oil, or natural gas.

There are several different reactor designs in current use. One of these is the boiling water reactor, such as the system shown in Fig. 1.5. In other nuclear power plants a secondary fluid circulates from the reactor to the steam generator, where heat is transferred from the secondary fluid to the water which in turn goes through a conventional steam cycle. Safety considerations and the necessity to keep the turbine, condenser, and related equipment from becoming radioactive are always major considerations in the design of a nuclear power plant.

1.2 Fuel Cells

When a conventional power plant is viewed as a whole, as shown in Fig. 1.6, we see that fuel and air enter the power plant and products of combustion leave the unit. There is also a transfer of heat to the cooling water, and work is done in the form of the electrical energy leaving the power plant. The overall objective of a power plant is to convert the

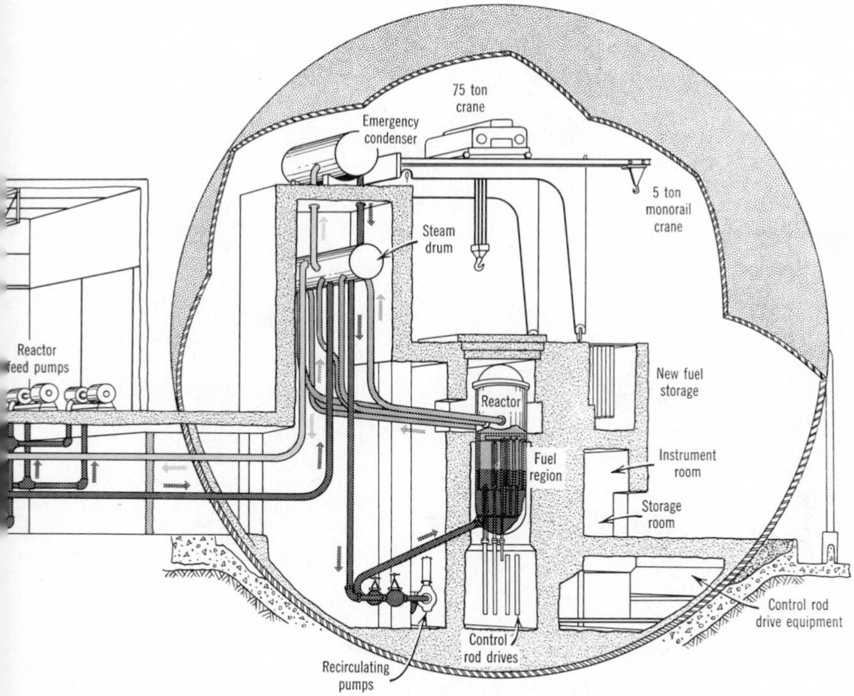

availability (to do work) of the fuel into work (in the form of electrical energy) in the most efficient manner, consistent with such considerations as cost, space, and safety.

We might well ask if all of the equipment in the power plant, such as the steam generator, the turbine, the condenser, and the pump, is necessary. Is it not possible to produce electrical energy from the fuel in a more direct manner?

The fuel cell is a device in which this objective is accomplished. Figure 1.7 shows a schematic arrangement of a fuel cell of the ion-exchange membrane type. In this fuel cell hydrogen and oxygen react to form water. Let us consider the general features of the operation of this type of fuel cell.

The flow of electrons in the external circuit is from anode to cathode. Hydrogen enters at the anode side and oxygen enters at the cathode side. At the surface of the ion-exchange membrane the hydrogen is ionized according to the reaction

$$2H_2 \rightarrow 4H^+ + 4e^-$$

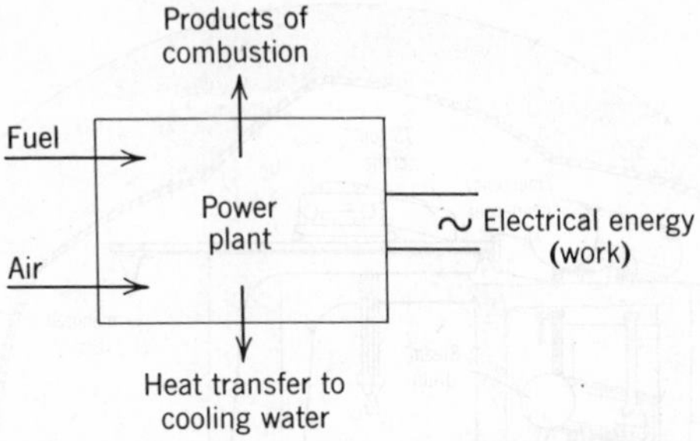

Fig. 1.6 Schematic diagram of a power plant.

The electrons flow through the external circuit and the hydrogen ions flow through the membrane to the cathode, where the following reaction takes place.

$$4H^+ + 4e^- + O_2 \rightarrow 2H_2O$$

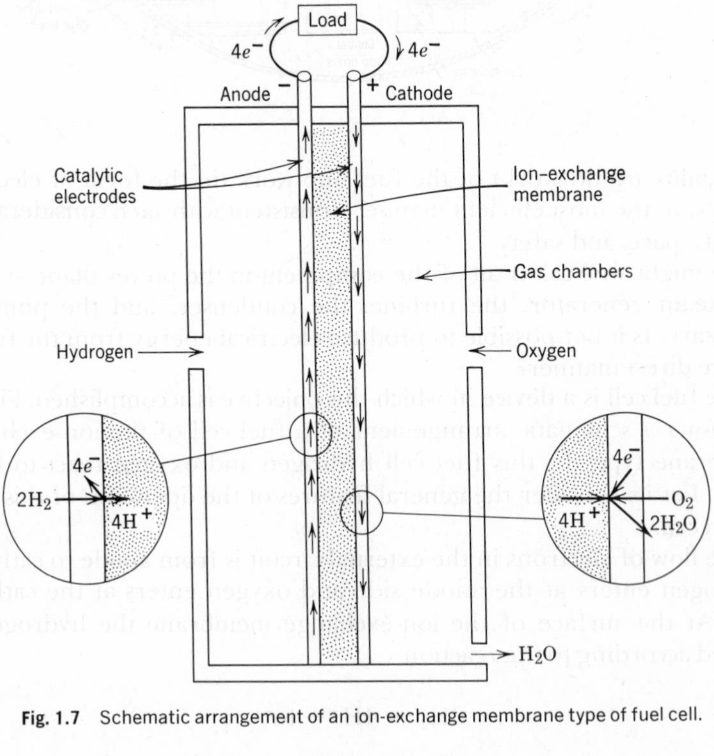

Fig. 1.7 Schematic arrangement of an ion-exchange membrane type of fuel cell.

There is a potential difference between the anode and cathode, and thus there is a flow of electricity through a potential difference which, in thermodynamic terms, is called work. There may also be a transfer of heat between the fuel cell and the surroundings.

At the present time the fuel used in fuel cells is usually either hydrogen or a mixture of gaseous hydrocarbons and hydrogen. The oxidizer is usually oxygen. However, current research is directed toward the development of fuel cells that use hydrocarbon fuels and air. Although the conventional (or nuclear) steam power plant is still used in large-scale power generating systems, and conventional piston engines and gas turbines are still used in most transportation power systems, the fuel cell may eventually become a serious competitor. The fuel cell is already being used to produce power for certain space applications.

Thermodynamics plays a vital role in the analysis, development, and design of all power-producing systems, including reciprocating internal combustion engines and gas turbines. Such considerations as the increase of efficiency, improved design, optimum operating conditions, and alternate methods of power generation involve, among other factors, the careful application of the fundamentals of thermodynamics.

1.3 The Vapor-Compression Refrigeration Cycle

A simple vapor-compression refrigeration cycle is shown schematically in Fig. 1.8. The refrigerant enters the compressor as a slightly super-

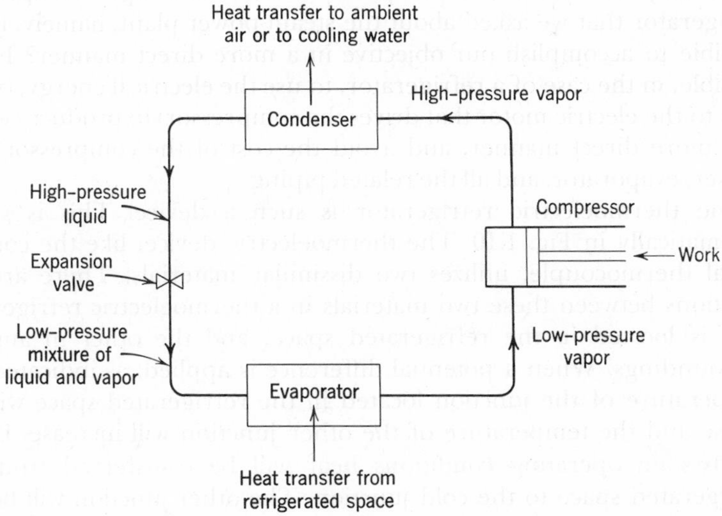

Fig. 1.8 Schematic diagram of a simple refrigeration cycle.

heated vapor at a low pressure. It then leaves the compressor and enters the condenser as a vapor at some elevated pressure, where the refrigerant is condensed as a result of heat transfer to cooling water or to the surroundings. The refrigerant then leaves the condenser as a high-pressure liquid. The pressure of the liquid is decreased as it flows through the expansion valve and, as a result, some of the liquid flashes into vapor. The remaining liquid, now at a low pressure, is vaporized in the evaporator as a result of heat transfer from the refrigerated space. This vapor then enters the compressor.

In a typical home refrigerator the compressor is located in the rear near the bottom of the unit. The compressors are usually hermetically sealed; that is, the motor and compressor are mounted in a sealed housing, and the electric leads for the motor pass through this housing. This is done to prevent leakage of the refrigerant. The condenser is also located at the back of the refrigerator and is so arranged that the air in the room flows past the condenser by natural convection. The expansion valve takes the form of a long capillary tube and the evaporator is located around the outside of the freezing compartment inside the refrigerator.

Figure 1.9 shows a large centrifugal unit that is used to provide refrigeration for an air-conditioning unit. In this unit, water is cooled and then circulated to provide cooling where needed.

1.4 The Thermoelectric Refrigerator

We may well ask the same question about the vapor compression refrigerator that we asked about the steam power plant, namely, isn't it possible to accomplish our objective in a more direct manner? Isn't it possible, in the case of a refrigerator, to use the electrical energy, (which goes to the electric motor that drives the compressor) to produce cooling in a more direct manner, and avoid the cost of the compressor, condenser, evaporator, and all the related piping.

The thermoelectric refrigerator is such a device. This is shown schematically in Fig. 1.10. The thermoelectric device, like the conventional thermocouple, utilizes two dissimilar materials. There are two junctions between these two materials in a thermoelectric refrigerator. One is located in the refrigerated space, and the other in ambient surroundings. When a potential difference is applied, as indicated, the temperature of the junction located in the refrigerated space will decrease and the temperature of the other junction will increase. Under steady-state operating conditions heat will be transferred from the refrigerated space to the cold junction. The other junction will be at a temperature above the ambient, and heat will be transferred from the junction to the surroundings.

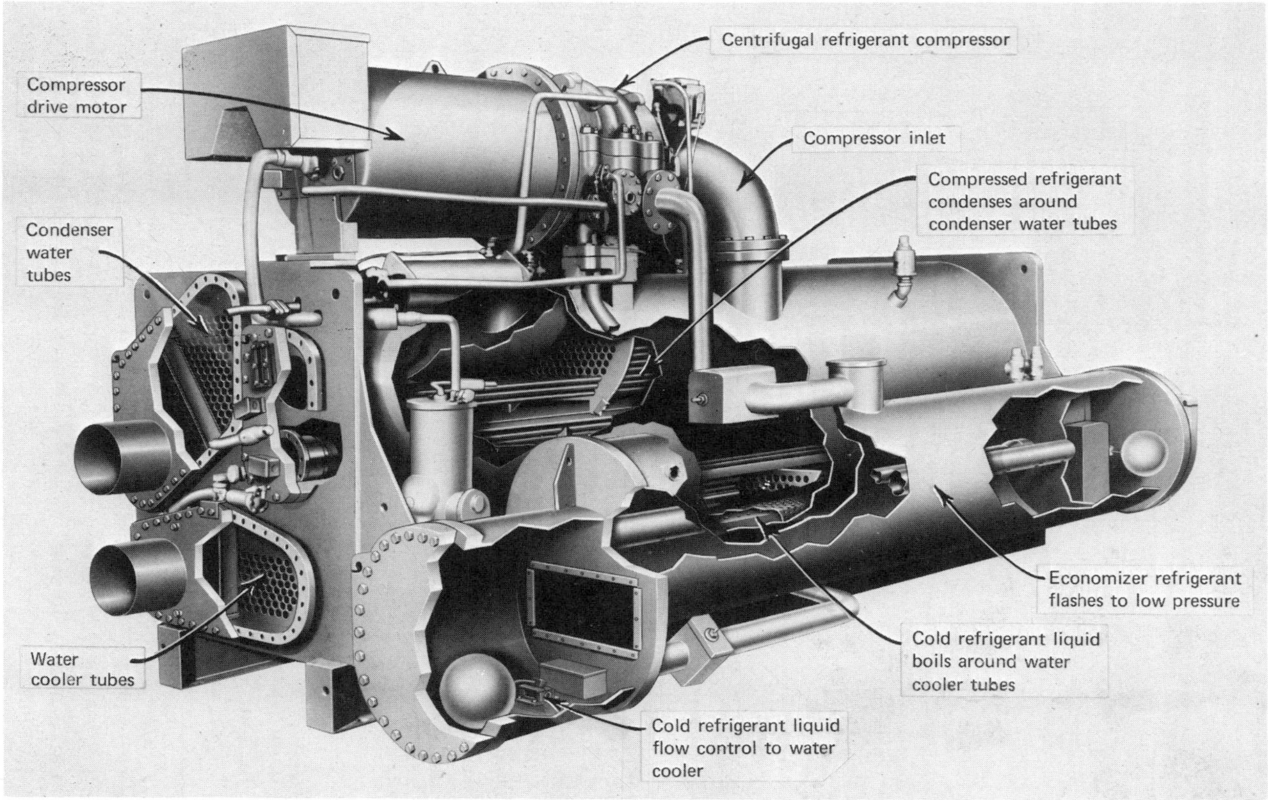

Fig. 1.9 A refrigeration unit for an air-conditioning system (Courtesy Carrier Air Conditioning Co.).

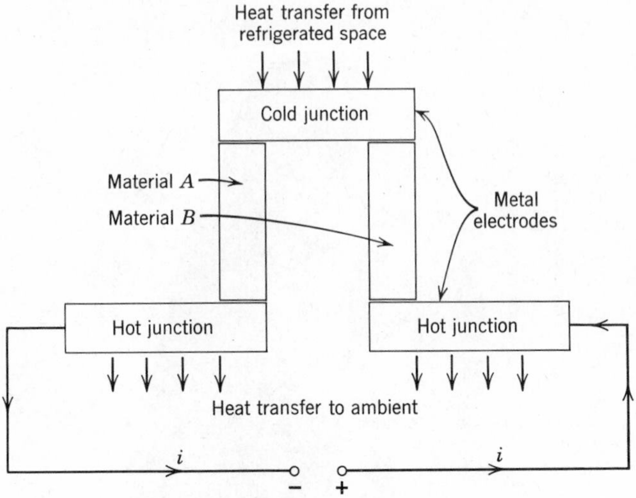

Fig. 1.10 A thermoelectric refrigerator.

It should be emphasized that a thermoelectric device can also be used to generate power by replacing the refrigerated space with a body which is at a temperature above the ambient. Such a system is shown in Fig. 1.11.

The thermoelectric refrigerator cannot yet compete economically with the conventional vapor-compression units. However, in certain special applications the thermoelectric refrigerator is already in use, and in view of the major research and development efforts underway in this field, it is quite possible that the use of thermoelectric refrigerators will be much more extensive in the future.

1.5 The Air Separation Plant

One process of great industrial significance is the air separation plant, in which air is separated into its various components. The oxygen, nitrogen, argon, and rare gases so produced are used extensively in various industrial, research, space, and consumer goods applications. The air separation plant may be considered as an example from two major fields, the chemical process industry and the field of cryogenics. Cryogenics is a term that refers to technology, processes, and research at very low temperatures (in general, below 150°K). In both chemical processing and cryogenics, thermodynamics is basic to an understanding of many phenomena that occur and to the design and development of processes and equipment.

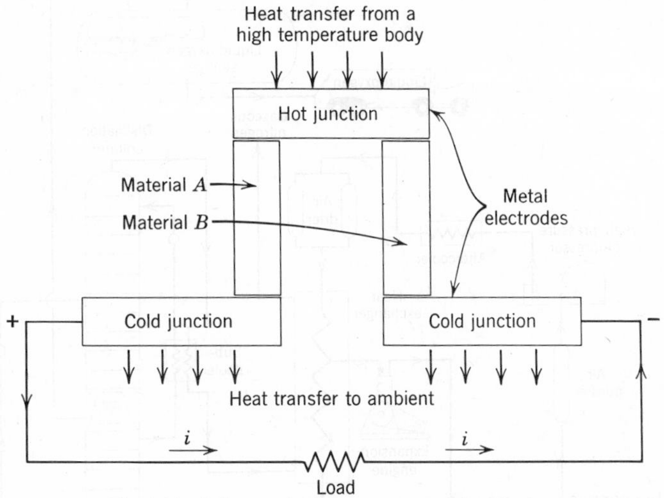

Fig. 1.11 A thermoelectric power generation device.

A number of different designs of air separation plants have been developed. Consider Fig. 1.12, which shows a somewhat simplified sketch of a type of plant that is frequently used. Air from the atmosphere is compressed to a pressure of several hundred pounds per square inch. It is then purified, particularly to remove carbon dioxide (which would plug the flow passages as it solidifies when the air is cooled to its liquefaction temperature). The air is then compressed to a pressure of 2500 to 3000 pounds per square inch, cooled to the ambient temperature in the aftercooler, and dried to remove the water vapor (which would also plug the flow passages as it freezes).

The basic refrigeration in the liquefaction process is provided by two different processes. One involves expansion of the air in the expansion engine. During this process the air does work and as a result the temperature of the air is reduced. The other refrigeration process involves passing the air through a throttle valve that is so designed and so located that there is a substantial drop in the pressure of the air and, associated with this, a substantial drop in the temperature of the air.

As shown in Fig. 1.12, the dry, high pressure air enters a heat exchanger. The air temperature drops as it flows through the heat exchanger. At some intermediate point in the heat exchanger, part of the air is bled off and flows through the expansion engine. The remaining air flows through the rest of the heat exchanger and through the throttle valve. The two streams join, both being at a pressure of 5–10 atmo-

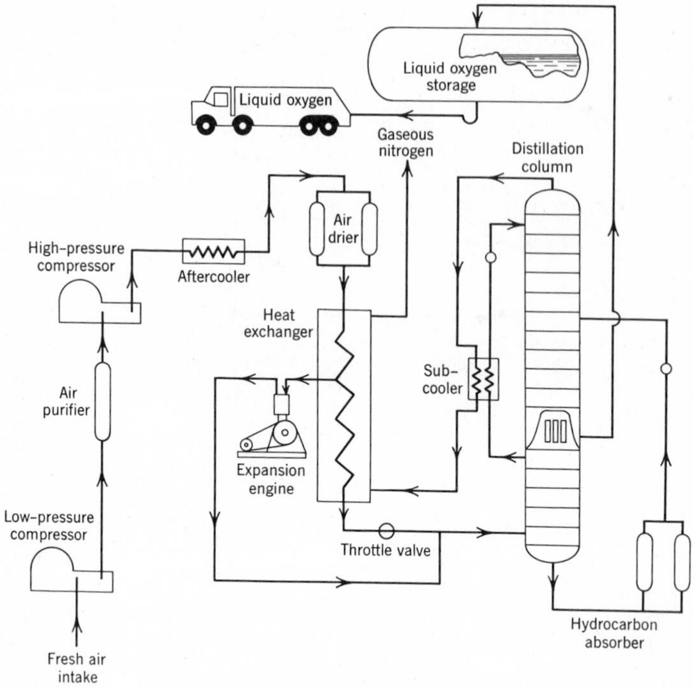

Fig. 1.12 A simplified diagram of a liquid oxygen plant (Courtesy Air Products and Chemicals, Inc.).

spheres, and enter the bottom of the distillation column, which is referred to as the high pressure column. The function of the distillation column is to separate the air into its various components, principally oxygen and nitrogen. Two streams of different composition flow from the high pressure column through throttle valves to the upper column (also called the low-pressure column). One of these is an oxygen-rich liquid that flows from the bottom of the lower column, and the other is a nitrogen-rich stream that flows through the subcooler. The separation is completed in the upper column, with liquid oxygen leaving from the bottom of the upper column and gaseous nitrogen from the top of the column. The nitrogen gas flows through the subcooler and the main heat exchanger. It is the heat transfer to this cold nitrogen gas that causes the cooling of the high pressure air entering the heat exchanger.

Not only is a thermodynamic analysis essential to the design of the system as a whole, but essentially every component of such a system, including the compressors, the expansion engine, the purifiers and driers, and the distillation column, involves thermodynamics. In this separation process we are also concerned with the thermodynamic properties of mixtures and the principles and procedures by which these mixtures can be separated. This is the type of problem encountered in the refining of petroleum and many other chemical processes. It should also be noted that cryogenics is particularly relevant to many aspects of the space program, and a thorough knowledge of thermodynamics is essential for creative and effective work in cryogenics.

1.6 The Chemical Rocket Engine

The advent of missiles and satellites has brought to prominence the use of the rocket engine as a propulsion power plant. Chemical rocket engines may be classified as either liquid propellant or solid propellant, according to the fuel used.

Figure 1.13 shows a simplified schematic diagram of a liquid-propellant

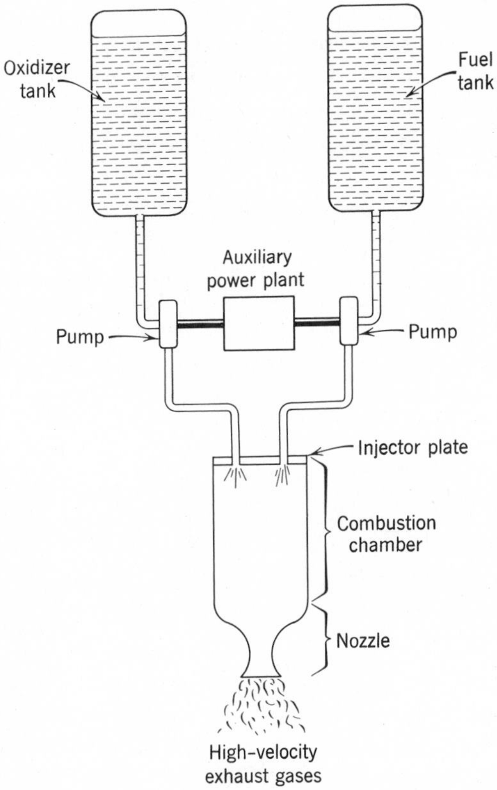

Fig. 1.13 Simplified schematic diagram of a liquid-propellant rocket engine.

rocket. The oxidizer and fuel are pumped through the injector plate into the combustion chamber where combustion takes place at high pressure. The high-pressure, high-temperature products of combustion expand as they flow through the nozzle, and as a result they leave the nozzle with a high velocity. The momentum change associated with this increase in velocity gives rise to the forward thrust on the vehicle.

The oxidizer and fuel must be pumped into the combustion chamber, and some auxiliary power plant is necessary to drive the pumps. In a large rocket this auxiliary power plant must be very reliable and have a relatively high power output, yet it must be light in weight. The oxidizer and fuel tanks occupy the largest part of the volume of an actual rocket, and the range of a rocket is determined largely by the amount of oxidizer and fuel that can be carried. Many different fuels and oxidizers have been considered and tested, and much effort has gone into the development of fuels and oxidizers that will give a higher thrust per unit mass rate of flow of reactants. Liquid oxygen is frequently used as the oxidizer in liquid-propellant rockets.

Much work has also been done on solid-propellant rockets. They have been very successfully used for jet-assisted take-offs of airplanes, military missiles, and space vehicles. They are much simpler in both the basic equipment required for operation and the logistic problems involved in their use.

2

Some Concepts and Definitions

One very excellent definition of thermodynamics is that it is the science of energy and entropy. However, since we have not yet defined those terms, an alternate definition in terms with which we are already familiar is: thermodynamics is the science that deals with heat and work and those properties of substances that bear a relation to heat and work. Like all sciences, the basis of thermodynamics is experimental observation. In thermodynamics these findings have been formalized into certain basic laws, which are known as the first, second, and third laws of thermodynamics. In addition to these, the zeroth law of thermodynamics, which in the logical development of thermodynamics precedes the first law, has been set forth.

In the chapters that follow, we shall present these laws and the thermodynamic properties related to these laws, and apply them to a number of representative examples. The objective of the student should be to gain a thorough understanding of the fundamentals and an ability to apply these fundamentals to thermodynamic problems. The purpose of the examples and problems is to further this two-fold objective. It should be emphasized that it is not necessary for the student to memorize numerous equations, for problems are best solved by the application of the definitions and laws of thermodynamics. In this chapter some concepts and definitions basic to thermodynamics are presented.

2.1 The Thermodynamic System and the Control Volume

A thermodynamic system is defined as a quantity of matter of fixed mass and identity upon which attention is focused for study. Everything external to the system is the surroundings, and the system is separated from the surroundings by the system boundaries. These boundaries may be either movable or fixed.

In Fig. 2.1 the gas in the cylinder is considered the system. If a Bunsen burner is placed under the cylinder, the temperature of the gas will

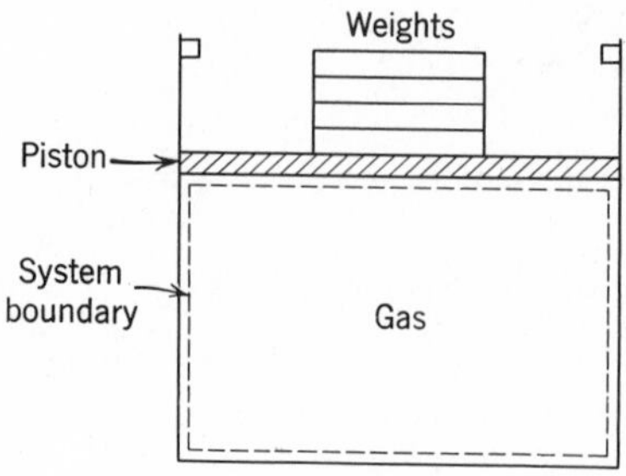

Fig. 2.1 Example of a system.

increase and the piston will rise. As the piston rises, the boundary of the system moves. As we shall see later, heat and work cross the boundary of the system during this process, but the matter that comprises the system can always be identified.

An isolated system is one that is not influenced in any way by the surroundings. This means that no heat or work cross the boundary of the system.

In many cases a thermodynamic analysis must be made of a device, such as an air compressor, that involves a flow of mass into and/or out of the device, as shown schematically in Fig. 2.2. The procedure that is

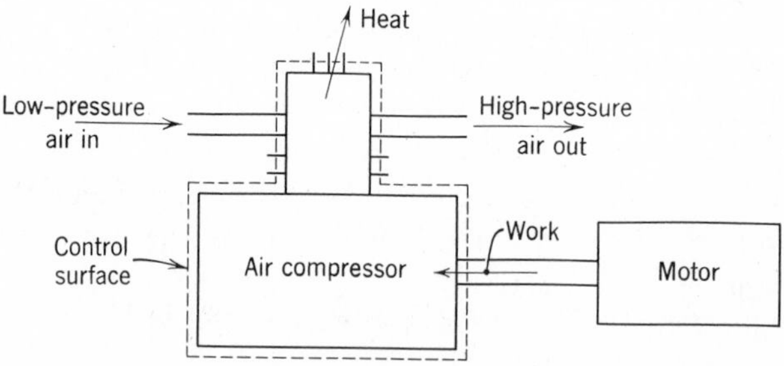

Fig. 2.2 Example of a control volume.

followed in such an analysis is to specify a control volume that surrounds the device under consideration. The surface of this control volume is referred to as a control surface. Mass, as well as heat and work (and momentum) can flow across the control surface.

Thus, a system is defined when dealing with a fixed quantity of mass, and a control volume is specified when an analysis is to be made that involves a flow of mass. The difference in these two approaches is considered in detail in Chapter 6. It should be noted that the terms closed system and open system are sometimes used as the equivalent of the terms system (fixed mass) and control volume (involving a flow of mass). The procedure that will be followed in the presentation of the first and second laws of thermodynamics is first to present these laws for a system, and then to make the necessary transformations to apply them to a control volume.

2.2 Macroscopic vs. Microscopic Point of View

An investigation into the behavior of a system may be undertaken from either a microscopic or macroscopic point of view. Let us briefly consider the problem we have if we attempt to describe a system from a microscopic point of view. Consider a system consisting of a one cubic inch volume of a monatomic gas at atmospheric pressure and temperature. This volume contains approximately 10^{20} atoms. To describe the position of each atom, three coordinates must be specified; to describe the velocity of each atom, three velocity components must be specified.

Thus, to completely describe the behavior of this system from a microscopic point of view, it would be necessary to deal with at least 6×10^{20} equations. Even with a large digital computer, this is a quite hopeless computational task. However, there are two approaches to this problem that reduce the number of equations and variables to a few that can be handled relatively easily in performing computations. One of these approaches is the statistical approach in which, on the basis of statistical considerations and probability theory, we deal with "average" values for all particles under consideration. This is usually done in connection with a model of the atom under consideration. This is the approach used in the disciplines known as kinetic theory and statistical mechanics.

The other approach that reduces the number of variables to a few that can be handled is the macroscopic point of view of classical thermodynamics. As the word macroscopic implies, we are concerned with the gross or average effects of many molecules. Furthermore, these effects can be perceived by our senses and measured by instruments. In so doing, however, what we really perceive and measure is the time-averaged influence of many molecules. For example, consider the pressure a gas exerts on the walls of its container. This pressure results from the change in momentum of the molecules as they collide with the wall. However, from a macroscopic point of view, we are not concerned with the action of the individual molecules but with the time-averaged force on a given area, which can be measured by a pressure gage. In fact, these macroscopic observations are completely independent of our assumptions regarding the nature of matter.

The theory and development of thermodynamics in this book will be presented from both the macroscopic and microscopic points of view. Chapters 3 through 14 cover the principles and a number of applications of classical thermodynamics; chapters 15 through 20 constitute an introduction to the subject of statistical thermodynamics.

A few remarks should be made regarding the continuum. From the macroscopic point of view, we are always concerned with volumes that are very large compared to molecular dimensions, and, therefore, with systems that contain many molecules. Since we are not concerned with the behavior of individual molecules, we can treat the substance as being continuous, disregarding the action of individual molecules, and this is called a continuum. The concept of a continuum, of course, is only a convenient assumption that loses validity when the mean free path of the molecules approaches the order of magnitude of the dimensions of the vessel, as, for example, in high-vacuum technology. In much engineering work the assumption of a continuum is valid and convenient, and goes hand in hand with the macroscopic point of view.

2.3 Properties and State of a Substance

If we consider a given mass of water, we recognize that this water can exist in various forms. If it is a liquid initially, it may become a vapor when it is heated, or solid when it is cooled. Thus we speak of the different phases of a substance. A phase is defined as a quantity of matter that is homogeneous throughout. When more than one phase is present the phases are separated from each other by the phase boundaries. In each phase the substance may exist at various pressures and temperatures or, to use the thermodynamic term, in various states. The state may be identified or described by certain observable, macroscopic properties; some familiar ones are temperature, pressure, and density. In later chapters other properties will be introduced. Each of the properties of a substance in a given state has only one definite value, and these properties always have the same value for a given state, regardless of how the substance arrived at that state. In fact, a property can be defined as any quantity that depends on the state of the system and is independent of the path (i.e., the prior history) by which the system arrived at the given state. Conversely, the state is specified or described by the properties, and later we shall consider the number of independent properties a substance can have, i.e., the minimum number of properties that must be specified in order to fix the state of the substance.

Thermodynamic properties can be divided into two general classes, intensive and extensive properties. An intensive property is independent of the mass; the value of an extensive property varies directly with the mass. Thus, if a quantity of matter in a given state is divided into two equal parts, each part will have the same value of intensive properties as the original, and half the value of the extensive properties. Pressure, temperature, and density are examples of intensive properties. Mass and total volume are examples of extensive properties. Extensive properties per unit mass, such as specific volume, are intensive properties.

Frequently we will refer not only to the properties of a substance but to the properties of a system. When we do so we necessarily imply that the value of the property has significance for the entire system, and this implies what is called equilibrium. For example, if the gas that comprises the system in Fig. 2.1 is in thermal equilibrium, the temperature will be the same throughout the entire system, and we may speak of the temperature as a property of the system. We may also consider mechanical equilibrium and this is related to pressure. If a system is in mechanical equilibrium, there is no tendency for the pressure at any point to change with time as long as the system is isolated from the surroundings. There will be a variation in pressure with elevation, due to the influence of gravitational forces, although under equilibrium conditions there will be no tendency for the pressure at any location to change. However, in many thermodynamic problems this variation in pressure with elevation is so small that it can be neglected. Chemical equilibrium is also important and will be considered in Chapter 14.

When a system is in equilibrium as regards all possible changes of state, we say that the system is in thermodynamic equilibrium.

2.4 Processes and Cycles

Whenever one or more of the properties of a system change we say that a change in state has occurred. For example, when one of the weights on the piston in Fig. 2.3 is removed, the piston rises and a change in state occurs, for the pressure decreases and the specific volume increases. The path of the succession of states through which the system passes is called the process.

Let us consider the equilibrium of a system as it undergoes a change in state. The moment the weight is removed from the piston in Fig. 2.3,

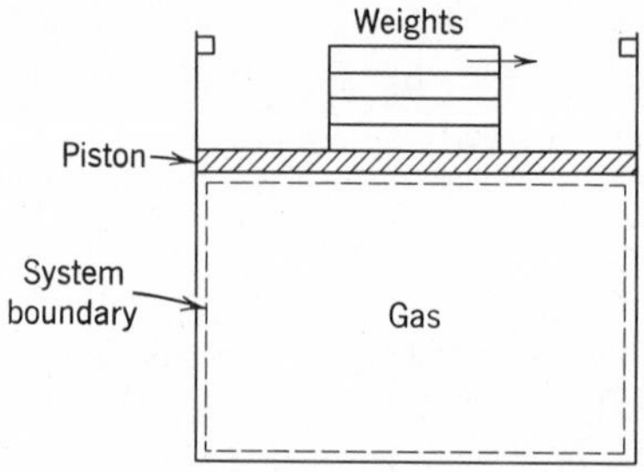

Fig. 2.3 Example of a system that may undergo a quasiequilibrium process.

mechanical equilibrium does not exist and as a result the piston is moved upward until mechanical equilibrium is again restored. The question that arises is this: since the properties describe the state of a system only when it is equilibrium, how can we describe the states of a system during a process if the actual process occurs only when equilibrium does not exist. One step in the answer to this question concerns the definition of an ideal process, which we call a quasiequilibrium process. A quasiequilibrium process is one in which the deviation from thermodynamic equilibrium is infinitesimal, and all the states the system passes through during a quasiequilibrium process may be considered as equilibrium states. Many actual processes closely approach a quasiequilibrium process, and may be so treated with essentially no error. If the weights on the piston in Fig. 2.3 are small and are taken off one by one, the process could be considered quasiequilibrium. On the other hand, if all the weights were removed at once, the piston would rise rapidly until it hit the stops. This would be a nonequilibrium process, and the system would not be in equilibrium at any time during this change of state.

For nonequilibrium processes, we are limited to a description of the system before the process occurs and after the process is completed and equilibrium is restored. We are not able to specify each state through which the system passes, nor the rate at which the process occurs. However, as we shall see later, we are able to describe certain over-all effects which occur during the process.

Several processes are described by the fact that one property remains constant. The prefix iso- is used to describe this. An isothermal process is a constant-temperature process, an isobaric (sometimes called isopiestic) process is a constant-pressure process, and an isometric process is a constant-volume process.

When a system in a given initial state goes through a number of different changes of state or processes and finally returns to its initial state, the system has undergone a cycle. Therefore, at the conclusion of a cycle all the properties have the same value they had at the beginning. Steam (water) that circulates through a steam power plant undergoes a cycle.

A distinction should be made between a thermodynamic cycle, which has just been described, and a mechanical cycle. A four-stroke cycle internal-combustion engine goes through a mechanical cycle once every two revolutions. However, the working fluid does not go through a thermodynamic cycle in the engine, since air and fuel are burned and changed to products of combustion which are exhausted to the atmosphere. In this text the term cycle will refer to a thermodynamic "cycle" unless otherwise designated.

2.5 Units for Mass, Length, Time, and Force

Since we are considering thermodynamic properties from a macroscopic perspective, we are dealing with quantities which can either directly or indirectly be measured and counted. Therefore, the matter of units becomes an important consideration. In the remaining sections of this chapter we will define certain thermodynamic properties and the basic units involved. The relation between force and mass is often a difficult matter for students, and is considered in this section in some detail.

Force, mass, length, and time are related by Newton's second law of motion, which states that the force acting on a body is proportional to the product of the mass and the acceleration in the direction of the force.

$$F \propto ma$$

The concept of time is well established. The basic unit of time is the second, which in the past has been defined in terms of the solar day, the time interval for one complete revolution of the earth relative to the sun. Since this period will vary with the season of the year, an average value over a one-year period is called the mean solar day, and the mean solar second is 1/86,400 of the mean solar day. (The measurement of the earth's rotation is sometimes made relative to a fixed star, in which case the period is called a sidereal day.) In 1967, the International Conference of Weights and Measures adopted a definition of the second in terms of a resonator using a beam of cesium-133 atoms. The time required for 9,192,631,770 cycles of the cesium resonator is now accepted as the definition of the second.

For periods less than one second the terms millisecond (10^{-3} second), microsecond (10^{-6} seconds), and nanosecond (10^{-9} seconds) are frequently used.

The concept of length is also well established. The basic unit of length is the meter, and for many years the accepted standard was the International Prototype Meter, the distance between two marks on a platinum-iridium bar under certain prescribed conditions. This bar is maintained at the International Bureau of Weights and Measures, Sevres, France. In 1960, the International Conference of Weights and Measures adopted a definition of the meter in terms of the wavelength of the orange-red line of krypton 86. The definition of the length of the meter is

$$1 \text{ meter} = 1{,}650{,}763.73 \text{ wavelengths of the orange-red line of Kr-86}$$

The inch is defined in terms of the meter.

$$1 \text{ in.} = 2.540 \text{ cm}$$

The concept of mass is also well understood. It involves the quantity or amount of material under consideration. In the various English systems the unit for mass is the pound mass, designated lbm, which was originally specified as the mass of a certain platinum cylinder in the Tower of London. It has now been defined in terms of the standard kilogram mass as

$$1 \text{ kilogram} = 2.2046 \text{ lbm}.$$

In the English Engineering system of units the concept of force is established as an independent quantity and the unit for force is defined in terms of an experimental procedure as follows. Let the standard pound mass be suspended in the earth's gravitational field at a location where the acceleration due to gravity is 32.1740 ft/sec^2. The force with which the standard pound mass is attracted to the earth (the buoyant effects of the atmosphere on the standard pound mass must also be standardized) is defined as the unit for force and is termed a pound force. Note that we now have arbitrary and independent definitions for force, mass, length, and time. Since these are related by Newton's second law we can write

$$F = \frac{ma}{g_c}$$

where g_c is a constant that relates the units of force, mass, length, and time. For the system of units defined above, namely, the English Engineering System we have

$$1 \text{ lbf} = \frac{1 \text{ lbm} \times 32.174 \text{ ft/sec}^2}{g_c}$$

or

$$g_c = 32.174 \ \frac{\text{lbm-ft}}{\text{lbf-sec}^2}$$

Note that g_c has both a numerical value and dimensions in this system.

To illustrate the use of this equation, let us calculate the force due to gravity on a pound mass at a location where the acceleration due to gravity is 30.0 ft/sec^2.

$$F = \frac{ma}{g_c}$$

$$F = \frac{1 \text{ lbm} \times 30.0 \text{ ft/sec}^2}{32.174 \text{ lbm-ft/lbf-sec}^2} = 0.933 \text{ lbf}$$

Note that the answer is dimensionally correct when the units of g_c as well as the magnitude are used. This is the system of units and the approach that will be used in the majority of this text.

We should briefly discuss three other systems of units. These three systems, namely the Absolute Metric, the Absolute English and the British Gravitational systems, involve a different concept, that of arbitrarily defining three of the four parameters mass, force, length, and time, and defining the fourth in terms of Newton's second law. In the Absolute Metric system (CGS system) the gram is the unit of mass, the second is the unit of time, and the centimeter is the unit of length, and each is arbitrarily and independently defined. The unit of force, the dyne, is defined in terms of Newton's second law.

$$1 \text{ dyne} \equiv 1 \frac{\text{gm-cm}}{\text{sec}^2}$$

Similarly, in the Absolute English system, which is used very little today, the units of mass, length, and time are defined, as in the English Engineering system, as pound mass, the foot, and the second. The unit for force in this system, the poundal, is defined as

$$1 \text{ poundal} \equiv \frac{1 \text{ lbm-ft}}{\text{sec}^2}$$

Note that in these two systems we could have taken the same perspective as in the English Engineering system, and defined force as an independent quantity. For example, we could have defined a dyne as the force acting on a mass of one gram at the location where the acceleration due to gravity is 1 cm/sec² (It might be difficult to locate this exact spot in space and remain there long enough to do the experiment.) We would then introduce g_c as a dimensional constant in Newton's second law.

$$F = \frac{ma}{g_c}$$

$$1 \text{ dyne} = \frac{1 \text{ gm} \times 1 \text{ cm/sec}^2}{g_c}$$

$$g_c = \frac{1 \text{ gm-cm}}{\text{dyne-sec}^2}$$

Similarly, we could have defined the poundal as the force acting on a mass of one pound at a location where the acceleration due to gravity is 1 ft/sec². Then

$$F = \frac{ma}{g_c}$$

$$1 \text{ poundal} = \frac{1 \text{ lbm} \times 1 \text{ ft/sec}^2}{g_c}$$

$$g_c = \frac{1 \text{ lbm-ft}}{\text{poundal-sec}^2}$$

Or, to return to the English Engineering system, we could have defined a pound force (lbf) as

$$1 \text{ lbf} = 32.174 \text{ lbm-ft/sec}^2$$

and have a definition of force parallel to that for the Absolute Metric and Absolute English systems.

There is a fourth system in common use, the British Gravitational system. In this system the independently defined units are force, length, and time. The unit for length is the foot; for time, the second; and for force, the pound force. Although this unit for force is exactly equal to the unit for force in the English Engineering system, we can think of it as an independent quantity in terms of a force required to compress a standard spring a fixed distance. The unit for the mass in this system is the slug and is defined from Newton's second law as

$$1 \text{ slug} \equiv \frac{1 \text{ lbf}}{1 \text{ ft/sec}^2} = \frac{1 \text{ lbf-sec}^2}{\text{ft}}$$

Had we adopted the slug as a fourth independently defined unit of mass as

$$1 \text{ slug} = 32.174 \text{ lbm}$$

we could have written

$$F = \frac{ma}{g_c}$$

$$1 \text{ lbf} = \frac{1 \text{ slug} \times 1 \text{ ft/sec}^2}{g_c}$$

$$g_c = \frac{1 \text{ slug-ft}}{\text{lbf-sec}^2}$$

This matter can be summarized in the tables as follows: If we arbitrarily define mass, length, time, and force as independent quantities, the constant g_c must be introduced into Newton's second law, and for the four systems considered we have

Name of System	Mass	Length	Time	Force	g_c
English Engineering	lbm	ft	sec	lbf	$g_c = 32.174 \frac{\text{lbm-ft}}{\text{lbf-sec}^2}$
not named	slug	ft	sec	lbf	$g_c = \frac{1 \text{ slug-ft}}{\text{lbf-sec}^2}$
not named	lbm	ft	sec	poundal	$g_c = \frac{1 \text{ lbm-ft}}{\text{poundal-sec}^2}$
not named	gm	cm	sec	dyne	$g_c = \frac{1 \text{ gm-cm}}{\text{dyne-sec}^2}$

If we arbitrarily define mass, length, and time as independent quantities we have

Name of System	Mass	Length	Time	Definition of Force
not named	lbm	ft	sec	$1 \text{ lbf} \equiv 32.174 \dfrac{\text{lbm-ft}}{\text{sec}^2}$
Absolute Metric	gm	cm	sec	$1 \text{ dyne} \equiv \dfrac{1 \text{ gm-cm}}{\text{sec}^2}$
Absolute English	lbm	ft	sec	$1 \text{ poundal} \equiv \dfrac{1 \text{ lbm-ft}}{\text{sec}^2}$

If we define force, length, and time as independent quantities we have

Name of System	Force	Length	Time	Definition of Mass
British Gravitational	lbf	ft	sec	$1 \text{ slug} \equiv \dfrac{1 \text{ lbf-sec}^2}{\text{ft}}$

In dealing with the matter of units in equations it is helpful to recall that since

$$1 \text{ lbf} = 32.174 \frac{\text{lbm-ft}}{\text{sec}^2}$$

it follows that

$$1 = 32.174 \frac{\text{lbm-ft}}{\text{lbf-sec}^2}$$

Comparing this with g_c for this system we note that

$$g_c = 32.174 \frac{\text{lbm-ft}}{\text{lbf-sec}^2} = 1$$

Since a pure number can be inserted into an equation at any point, we can, in effect, substitute

$$1 = 32.174 \frac{\text{lbm-ft}}{\text{lbf-sec}^2} = g_c$$

into an equation at any point where it is advantageous to do so from the point of view of units. For example, the viscosity of water (the exact definition of viscosity is not important at this point) at atmospheric pressure and 80 F is given as 1.80×10^{-6} lbf-sec/ft^2. Suppose we wish to know this viscosity in terms of units involving lbm, sec, and ft.

$$\text{Viscosity} = 1.80 \times 10^{-6} \frac{\text{lbf-sec}}{\text{ft}^2} \times 1$$

$$= 1.80 \times 10^{-6} \frac{\text{lbf-sec}}{\text{ft}^2} \times 32.174 \frac{\text{lbm-ft}}{\text{lbf-sec}^2}$$

$$= 57.8 \times 10^{-6} \frac{\text{lbm}}{\text{ft-sec}}$$

The English Engineering system has been adopted in the classical portion of this text and therefore Newton's second law will be written

$$F = \frac{ma}{g_c}$$

and the g_c will be included in equations involving Newton's second law. It should be emphasized that the term pound, the symbol lb (or #) should never be used by itself, since it would not be evident whether pound mass or pound force is being referred to.

It should also be noted that weight always refers to a force. When we say a body weighs so much we mean that this is the force with which it is attracted to the earth (or any other body). The mass of a substance remains constant with elevation, but its weight varies with elevation.

Two other units for mass, namely the pound mole and gram mole are frequently used in thermodynamics. The pound mole, designated lb mole, is the quantity of a substance whose mass in pounds mass is equal to the molecular weight of the substance. Similarly the gram mole, designated gm mole, is the quantity of a substance whose mass in grams is equal to the molecular weight of the substance.

2.6 Energy

One of the very important concepts in a study of thermodynamics is the concept of energy. Energy is a fundamental concept, such as mass or force, and, as is often the case with such concepts, is very difficult to define. Energy has been defined as the capability to produce an effect.

Fortunately the word "energy" and the basic concept that this word represents, are familiar to us in everyday usage and a precise definition is not essential at this point.

It is important to note that energy can be stored within a system and can also be transferred (as heat, for example) from one system to another. In our study of statistical thermodynamics we shall examine, from a molecular viewpoint, the ways in which energy can be stored. However, since it is helpful in a study of classical thermodynamics to have some notion of how this energy is stored, a brief introduction is presented here.

Consider as a system a certain gas that exists at a given pressure and temperature as it is contained within a tank or pressure vessel. When viewed from the molecular point of view, we identify three general forms of energy:

1. Intermolecular potential energy, which is associated with the forces between molecules.
2. Molecular kinetic energy, which is associated with the translational velocity of individual molecules.
3. Intramolecular energy (that within the individual molecules), which is associated with the molecular and atomic structure and related forces.

The first of these, the intermolecular potential energy, depends on the magnitude of the intermolecular forces and the position the molecules have relative to each other at any instant of time. However, it is impossible to determine accurately the magnitude of this energy because we know neither the exact configuration and orientation of the molecules at any time, nor the exact intermolecular potential function. However, there are two situations when we can make good approximations. The first is at low or moderate densities. In this case the molecules are relatively widely spaced, so that only two-molecule or two- and three-molecule interactions contribute to the potential energy. At these low and moderate densities techniques are available for determining, with reasonable accuracy, the potential energy of a system composed of reasonably simple molecules. The second situation is at very low densities; under these conditions the average intermolecular distance between molecules is so large that the potential energy may be assumed to be zero. Consequently, we have in this case a system of independent particles (an ideal gas) and therefore, from a statistical point of view, we are able to concentrate our efforts on evaluating the molecular translational and internal energies.

The translational energy, which depends only on the mass and velocities of the molecules, is determined by using the equations of mechanics – either quantum or classical.

The intramolecular internal energy is more difficult to evaluate because, in general, it may result from a number of contributions. Consider first a simple monatomic gas, such as helium. Each molecule consists of a helium atom. Such an atom possesses electronic energy as a result of both orbital angular momentum of the electrons about the nucleus and angular momentum of the electrons spinning on their axes. The electronic energy is commonly very small compared with the translational energies. (Atoms also possess nuclear energy, which, except in the case of nuclear reactions, is constant. We are not concerned with nuclear energy at this time.) When we consider more complex molecules, such as those comprised of two or three atoms, additional factors much be considered. In addition to having electronic energy, a molecule can rotate about its center of gravity and thus have rotational energy. Further, the atoms may vibrate with respect to each other and have vibrational energy. In some cases there may be an interaction between the rotational and vibrational modes of energy.

In the evaluation of the energy of a molecule, reference is often made to the degree of freedom, f, for these energy modes. For a monatomic molecule, such as helium, $f = 3$, and this represents the 3 directions, x, y, and z, in which the molecule can move. For a diatomic molecule, such as oxygen, $f = 6$. Three of these are the translation of the molecule as a whole in the x, y, and z directions, and two are for rotation. The reason there are only two modes of rotational energy is evident from Fig. 2.4,

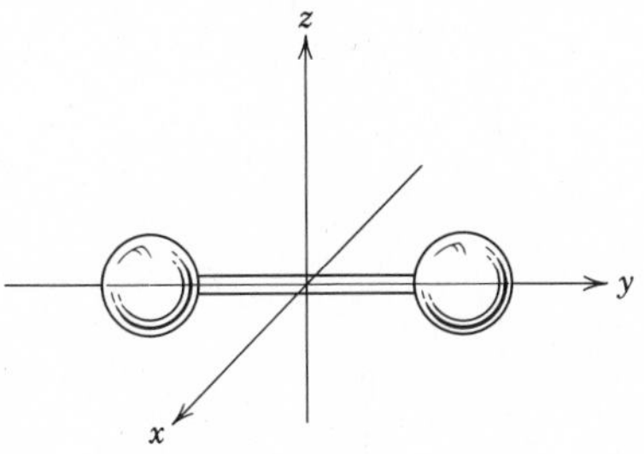

Fig. 2.4 The coordinate system for a diatomic molecule.

where we take the origin of the coordinate system at the center of gravity of the molecule, and the y-axis along the molecule's internuclear axis. The molecule will then have an appreciable moment of inertia about the x-axis and the z-axis, but not about the y-axis. The sixth degree of freedom of the molecule is that of vibration, which relates to stretching of the bond joining the atoms.

For a more complex molecule, such as H_2O, there are additional vibrational degrees of freedom. Figure 2.5 shows a model of the H_2O

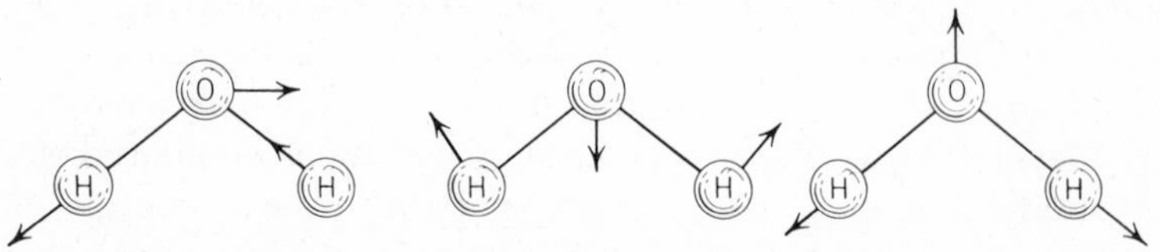

Fig. 2.5 The three principal vibrational modes for the water molecule.

molecule. From this diagram it is evident that there are three vibrational degrees of freedom. It is also possible to have rotational energy about all three axes. Thus, for the H_2O molecule, there are nine degrees of freedom ($f = 9$), 3 translational, 3 rotational, and 3 vibrational.

This general discussion can be summarized by reference to Fig. 2.6.

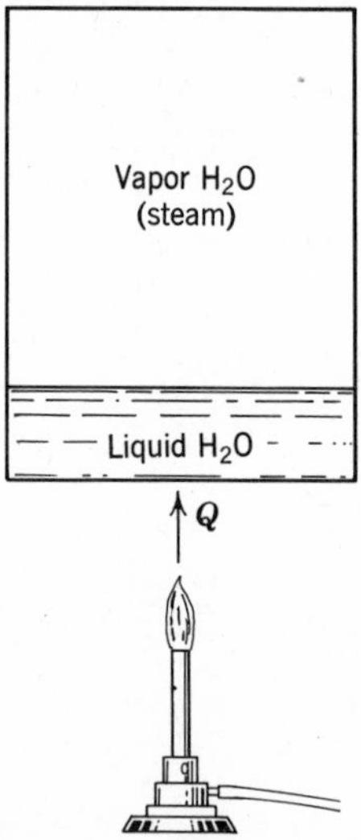

Fig. 2.6 Heat transfer to water.

Let heat be transferred to the water. During this process the temperature of the liquid and vapor (steam) will increase, and eventually all of the liquid will become vapor. From the macroscopic point of view we are concerned only with the energy that is transferred as heat, the change in properties, such as temperature and pressure, and the total amount of energy (relative to some base) that the H_2O contains at any instant of time. Thus questions about the way in which energy is stored in the H_2O do not concern us when we adopt the macroscopic point of view. From a microscopic viewpoint we are concerned about the way in which energy is stored in the molecules. We might be interested in developing a model of the molecule so that we could predict the amount of energy required to change the temperature a given amount. There is an obvious interrelation between the macroscopic or classical and the microscopic or statistical points of view, and this book has been designed to give both and to relate them to each other. Many of these considerations will relate to the matter of energy.

2.7 Specific Volume

The specific volume of a substance is defined as the volume per unit mass, and is given the symbol v. The density of a substance is defined as the mass per unit volume, and is therefore the reciprocal of the specific volume. Density is designated by the symbol ρ. Specific volume and density are intensive properties.

The specific volume of a system in a gravitational field may vary from point to point. For example, considering the atmosphere as a system, the specific volume increases as the elevation increases. Therefore the definition of specific volume involves the specific volume of a substance at a point in a system.

Consider a small volume δV of a system, and let the mass be designated δm. The specific volume is defined by the relation

$$v = \lim_{\delta V \to \delta V'} \frac{\delta V}{\delta m}$$

where $\delta V'$ is the smallest volume for which the system can be considered a continuum.

Thus in a given system we should speak of the specific volume or density at a point in the system, and recognize that this may vary with elevation. However, most of the systems that we consider are relatively small, and the change in specific volume with elevation is not significant. In this case, we can speak of one value of specific volume or density for the entire system.

In this text the specific volume and density will usually be given either on a pound-mass or a pound-mole basis. A bar over the symbol (lower case) will be used to designate the property on a mole basis. Thus, $\bar{v}$ will designate the molal specific volume and $\bar{\rho}$ will designate the molal density. The most common units used in this text for specific volume are ft^3/lbm and ft^3/lb mole; for density the corresponding units are lbm/ft^3 and lb mole/ft^3.

2.8 Pressure

When dealing with liquids and gases we ordinarily speak of pressure; in solids we speak of stresses. The pressure in a fluid at rest at a given point is the same in all directions, and we define pressure as the normal component of force per unit area, resulting from molecules striking the surface of the area. More specifically, if δA is a small area, and $\delta A'$ is the smallest area over which we can consider the fluid a continuum, and δF_n is the component of force normal to δA, we define pressure, P, as

$$P = \lim_{\delta A \to \delta A'} \frac{\delta F_n}{\delta A}$$

The pressure P at a point in a fluid in equilibrium is the same in all directions. In a viscous fluid in motion the variation in the state of stress with orientation becomes an important consideration. These considerations are beyond the scope of this book, and we shall consider pressure only in terms of a fluid in equilibrium.

In general, the unit for pressure that is consistent with the other units used in this text is pounds force per square foot (lbf/ft^2). On the other hand, in common parlance and general experimental work, pressures are often measured in pounds force per square inch (lbf/$in.^2$). Therefore, the student should be careful in numerical calculations to introduce the conversion 144 $in.^2 = 1\ ft^2$, as necessary.

In most thermodynamic investigations we are concerned with absolute pressure. Most pressure and vacuum gages, however, read the difference between the absolute pressure and the atmospheric pressure existing at the gage, and this is referred to as gage pressure. This is shown graphically in Fig. 2.7, and the following examples illustrate the principles involved. Pressures below atmospheric and slightly above atmospheric, and pressure differences (for example, across an orifice in a pipe) are frequently measured with a manometer, which contains water, mercury, alcohol, oil, or other fluids. From the principles of hydrostatics one concludes that for a difference in level of L ft, the pressure difference in pounds per square foot is calculated by the relation

$$\Delta P = \rho \frac{Lg}{g_c}$$

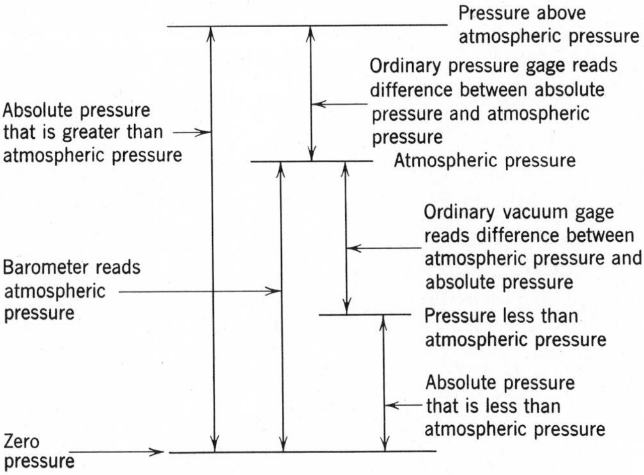

Fig. 2.7 Illustration of terms used in pressure measurement.

where ρ = density of the fluid in lbm/ft^3. Figure 2.8 illustrates such a manometer.

From this relation the student should verify the fact that at ordinary room temperatures

$$1 \text{ in. Hg} = 0.491 \text{ lbf/in.}^2$$

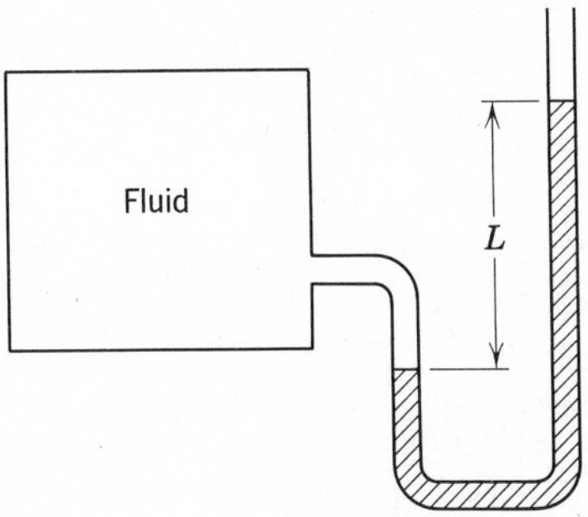

Fig. 2.8 Example of pressure measurement using a column of fluid.

Standard atmospheric pressure is defined as the pressure produced by a column of mercury exactly 760 mm in length, the mercury density being 13.5951 gm/cm^3 and the acceleration due to gravity being standard.

Therefore,

$$\begin{aligned} 1 \text{ std atm} &= 14.6959 \text{ lbf/in.}^2 \\ &= 1.01325 \times 10^6 \text{ dynes/cm}^2 \end{aligned}$$

Another common unit of pressure is the atmosphere, which implies a standard atmosphere. Thus, a pressure of 1000 atmospheres is a pressure of 14,696 lbf/in.^{2}

Extremely low pressures (i.e., a high vacuum) are often measured in microns of mercury (usually only the word micron is used). A micron is one millionth of a meter or 10^{-3} mm. Thus,

$$1 \text{ micron} = 1 \times 10^{-6} \text{ meters} = 1 \times 10^{-3} \text{ mm}$$

$$1 \text{ micron Hg} = 1 \times 10^{-3} \text{ mm Hg} = 1.933 \times 10^{-5} \text{ lbf/in.}^2$$

The term torr (after Evangelista Torricelli, 1608–1647, the pioneer worker in vacuum technology) has been introduced for a pressure of 1 mm Hg.

In order to distinguish between absolute and gage pressures in this text, the term lbf/in.^{2} or lbf/ft^2 will refer to absolute pressure. The gage pressure will be indicated by lbf/in.^{2} gage or lbf/ft^2 gage. It should also be noted that the symbols psia and psig are often used in technical literature to designate absolute and gage pressures, respectively. However, since the matter of units is emphasized in this book, the symbols lbf/in.^{2} and lbf/in.^{2} gage are used.

2.9 Equality of Temperature

Although temperature is a property with which we are all familiar, an exact definition of it is difficult. We are aware of "temperature" first of all as a sense of hotness or coldness when we touch an object. We also learn early in our experience that when a hot body and a cold body are brought into contact, the hot body becomes cooler and the cold body becomes warmer. If these bodies remain in contact for some time, they usually appear to have the same hotness or coldness. However, we also realize that our sense of hotness or coldness is very unreliable. Sometimes very cold bodies may seem hot, and bodies of different materials that are at the same temperature appear to be at different temperatures.

Because of these difficulties in defining temperature, we define equality of temperature. Consider two blocks of copper, one hot and the other cold, each of which is in contact with a mercury-in-glass thermometer. If these two blocks of copper are brought into thermal communication, we then observe that the electrical resistance of the hot block decreases with time and for the cold block it increases with time. After a period of time has elapsed, however, no further changes in resistance are observed. Similarly, when the blocks are first brought in thermal communication, the length of a side of the hot block decreases with time, whereas for the cold block it increases with time. After a period of time, no further change in length of either of the blocks is perceived. Also, the mercury column of the thermometer in the hot block drops at first and in the cold block it rises, but after a period of time no further changes in height are observed. We may say, therefore, that two bodies have equality of temperature when no change in any observable property occurs when they are in thermal communication.

2.10 The Zeroth Law of Thermodynamics

Now consider the same two blocks of copper, and also another thermometer. Let one block of copper be brought into contact with the thermometer until equality of temperature is established, and then removed. Then let the second block of copper be brought into contact with the thermometer, and suppose that no change in the mercury level of the thermometer occurs during this operation with the second block. Then we can say that both blocks are in thermal equilibrium with the given thermometer.

The zeroth law of thermodynamics states that when two bodies have equality of temperature with a third body, they in turn have equality of temperature with each other. This seems very obvious to us because we are so familiar with this experiment. However, since this fact is not derivable from other laws, and since in the logical presentation of thermodynamics it precedes the first and second laws of thermodynamics, it has been called the zeroth law of thermodynamics. This law is really the basis of temperature measurement, for numbers can be placed on the mercury thermometer, and every time a body has equality of temperature with the thermometer, we can say that the body has the temperature we read on the thermometer. The problem remains, however, of relating temperatures that we might read on different mercury thermometers, or that we obtain when using different temperature-measuring devices, such as thermocouples and resistance thermometers. This suggests the need for a standard scale for temperature measurements.

2.11 Temperature Scales

There are two commonly used scales for measuring temperature, namely the Fahrenheit (after Gabriel Fahrenheit, 1686–1736) and Celsius scales. The Celsius scale was formerly called the Centigrade scale, but is now designated the Celsius scale, after Anders Celsius (1701–1744) the Swedish astronomer who devised this scale.

Until 1954 each of these scales was based on two fixed, easily duplicated points, the ice point and the steam point. The temperature of the ice point is defined as the temperature of a mixture of ice and water which is in equilibrium with saturated air at a pressure of 1 atm. The temperature of the steam point is the temperature of water and steam which are in equilibrium at a pressure of 1 atm. On the Fahrenheit scale these two points are assigned the numbers 32 and 212, respectively, and on the Celsius scale the respective points are numbered 0 and 100. The basis for numbers on the Fahrenheit scale has an interesting background. In searching for an easily reproducible point, Fahrenheit selected the temture of the human body and assigned it the number 96. He assigned the number 0 to the temperature of a certain mixture of salt, ice, and salt solution. On this scale the ice point was approximately 32. When this scale was slightly revised and fixed in terms of the ice point and steam point, the normal temperature of the human body was found to be 98.6 F.

In this text the letters F and C will denote the Fahrenheit and Celsius scales, respectively. The usual symbol (°) for degree will not be used, but will rather be implied with the symbol F or C. The symbol T will refer to temperature on all temperature scales.

At the Tenth Conference on Weights and Measures in 1954, the Celsius scale was redefined in terms of a single fixed point and the ideal-gas temperature scale. The single fixed point is the triple point of water (the state in which the solid, liquid, and vapor phases of water exist together in equilibrium). The magnitude of the degree is defined in terms of the ideal-gas temperature scale, which is discussed in Chapter 7. The essential features of this new scale are a single fixed point and a definition of the magnitude of the degree. The triple point of water is assigned the value 0.01 C. On this scale the steam point is experimentally found to be 100.00 C. Thus, there is essential agreement between the old and new temperature scales.

It should be noted that we have not yet considered an absolute scale of temperature. The possibility of such a scale arises from the second law of thermodynamics and is discussed in Chapter 7. On the basis of the second law of thermodynamics a temperature scale which is independent of any thermometric substance can be defined. This absolute scale is usually referred to as the thermodynamic scale of temperature. However, it is very complicated to use this scale directly, and therefore a more practical scale, the International Practical Temperature Scale, which closely represents the thermodynamic scale, has been adopted.

The absolute scale related to the Celsius scale is referred to as the Kelvin scale (after William Thomson, 1824–1907, who is also known as Lord Kelvin), and is designated K. The relation between these scales is

$$°K = °C + 273.15$$

The absolute scale related to the Fahrenheit scale is referred to as the Rankine scale and is designated R. The relation between these scales is

$$°R = °F + 459.67$$

2.12 The International Practical Temperature Scale

In 1968 the International Committee on Weights and Measures adopted a revised International Practical Temperature Scale, IPTS-68, which is described below. This scale, similar to earlier ones of 1927 and 1948, has been extended in range and made to conform more closely to the thermodynamic temperature scale. It is based on a number of fixed and easily reproducible points that are assigned definite numerical values of temperature, and on specified formulas relating temperature to the readings on certain temperature-measuring instruments for the purpose of interpolation between the fixed points. The primary fixed points and a summary of the interpolation techniques are listed here for the sake of completeness, although the student will have limited need for it at this time.

The primary fixed-point temperatures in degrees Celsius are as follows:

1. Triple point (equilibrium between solid, liquid and vapor phases) of equilibrium-hydrogen. −259.34
2. Boiling point (equilibrium between liquid and vapor phases) of equilibrium-hydrogen at 250 mm Hg pressure. −256.108
3. Normal boiling point (1 atm. pressure) of equilibrium-hydrogen. −252.87
4. Normal boiling point of neon. −246.048
5. Triple point of oxygen. −218.789
6. Normal boiling point of oxygen. −182.962

7. Triple point of water. 0.01
8. Normal boiling point of water. 100
9. Normal freezing point (equilibrium between solid and liquid phases at 1 atm. pressure) of zinc. 419.58
10. Normal freezing point of silver. 961.93
11. Normal freezing point of gold. 1064.43

The means available for measurement and interpolation lead to a division of the temperature scale into four general ranges.

1. The range from −259.34 C to 0 C is based on measurements on a platinum resistance thermometer, with temperature expressed in terms of a 20th degree reference function equation. This range is subdivided into four parts. In each, the difference between measured resistance ratios of a particular thermometer and the reference function at the fixed points are used to determine the constants in a specified polynomial interpolation equation.
2. The range from 0 C to 630.74 C (the normal freezing point of antimony, a secondary fixed point) is also based on a platinum resistance thermometer, with constants in a polynomial interpolating equation determined by calibration at the three fixed points in this range.
3. The range from 630.74 C to 1064.43 C is based on measurements on a standard platinum vs. rhodium-platinum thermocouple, and a three-term equation expressing EMF as a function of temperature. The constants are determined by a platinum resistance thermometer measurement at the antimony point and by calibration at the two primary fixed points in this range.
4. The range above 1064.43 C is based on measurements of the intensity of visible-spectrum radiation compared with that of the same wavelength at the gold point, and on Planck's equation for black body radiation.

PROBLEMS

2.1 A 1-kg mass is accelerated with a force of 10 lbf. Calculate the acceleration in ft/sec^2 and cm/sec^2.

2.2 The "standard" acceleration due to gravity is 32.174 ft/sec^2. Calculate the force, in dynes, due to "standard" gravity acting on a mass of 1 gm.

2.3 With what force is a mass of 10 slugs attracted to the earth at a point where the gravitational acceleration is 30.2 ft/sec^2? What is its weight in lbf? What is its mass in lbm?

2.4 A pound mass is "weighed" with a beam balance at a point where $g = 31.0$ ft/sec^2. What reading would be expected? If it is weighed with a spring scale that reads correctly for standard gravity, what reading would be obtained?

2.5 A body of fixed mass is "weighed" at an elevation of 20,000 ft ($g = 32.11$ ft/sec^2) by a spring balance which was calibrated at sea level. The reading on the spring balance is 9.3 pounds. What is the mass of the body?

2.6 A piston has an area of 1 ft^2. What mass must the piston have if it exerts a pressure of 10 lbf/in.^{2} above atmospheric pressure on the gas enclosed in the cylinder? Assume standard gravitational acceleration.

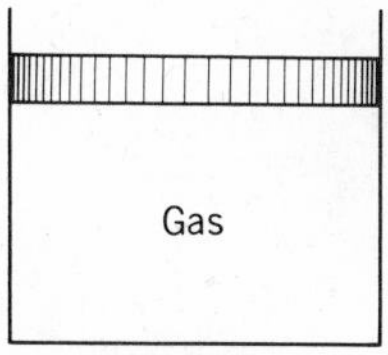

Fig. 2.9 Sketch for Problem 2.6.

2.7 Suppose that in an orbiting space station, an artificial gravity of 5 ft/sec^2 is induced by rotating the station. How much would a 150 lbm man weigh inside?

2.8 Verify the fact that 1 lbf $= 4.448 \times 10^5$ dynes.

2.9 A pressure gage reads 40.7 lbf/in.^{2}, and the barometer reads 29.2 in. Hg. Calculate the absolute pressure in lbf/in.^{2} and atm.

2.10 A manometer contains a fluid having a density of 51.0 lbm/ft^3. The difference in height of the two columns is 20 in. What pressure difference is indicated in lbf/in.^{2}? What would the height difference be if this same pressure difference had been measured by a manometer containing mercury (density of 13.60 gm/cm^3)?

2.11 A mercury manometer which is used to measure a vacuum reads 29.3 in., and the barometer reads 29.7 in. Hg. Determine the pressure in lbf/in.^{2} and in microns.

2.12 In an experimental airplane flying at 40,000 ft ($g = 32.05$ ft/sec^2), the air flow in a piece of apparatus is measured by using a mercury manometer. The difference in the level is 30 in. At sea level and the same temperature, mercury has a density of 13.60 gm/cm^3. Determine the pressure drop across the orifice in lbf/in.^{2}.

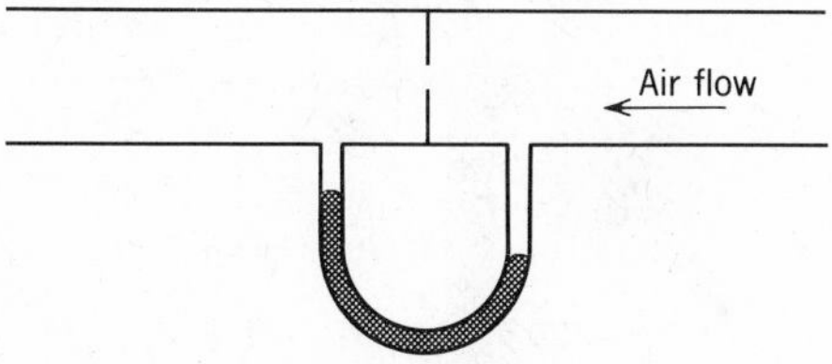

Fig. 2.10 Sketch for Problem 2.12.

2.13 A mercury column is used to measure the pressure difference of 30 lbf/in.^{2} in a piece of apparatus which is located out of doors. The minimum temperature in the winter is 0 F and the maximum temperature in the summer is 100 F. What will be the difference in the height of the mercury column in the summer as compared to the winter when measuring this pressure difference of 30 lbf/in.^{2}? Assume standard gravitational acceleration. The following data are given for the density of mercury:

T °C	Density
−10	13.6198 gm/cm^3
0	13.5951
10	13.5704
20	13.5458
30	13.5213

2.14 In space simulation chambers very low pressures are achieved by cryopumping. This involves maintaining certain surfaces at very low temperatures (as low as 5 K). Essentially all of the gas present (except helium) will freeze on these surfaces. Pressures of 1×10^{-8} torr and lower are achieved in space chambers by this technique. What is this pressure of 10^{-8} torr in atmospheres, lbf/in.^{2}, and dynes/cm^2?

2.15 A cylinder containing a gas is fitted with a piston having a mass of 150 lbm. The cross-sectional area of the piston is 60 in.^{2}. The atmospheric pressure is 14.2 lbf/in.^{2}, and the acceleration due to gravity at this location is 30.9 ft/sec^2. What is the absolute pressure of the gas?

2.16 The level of the water in an enclosed water tank is 100 feet above the ground. The pressure in the air space above the water is 17 lbf/in.^{2}. The average density of the water is 62.4 lbm/ft^3. What is the pressure of the water at the ground level?

2.17 A gas is contained in two cylinders A and B, connected by a piston of two different diameters, as shown in Fig. 2.11. The mass of the piston is 20 lbm and the gas pressure inside cylinder A is 30 lbf/in.^{2}. Claculate the pressure in cylinder B.

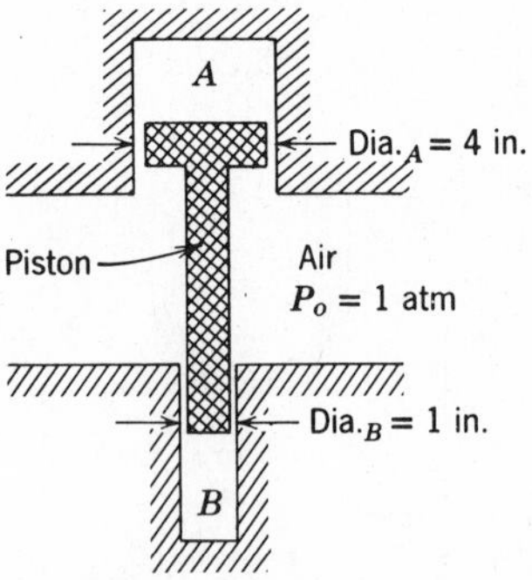

Fig. 2.11 Sketch for Problem 2.17.

3

Properties of a Pure Substance

In the previous chapter we considered three familiar properties of a substance, namely, specific volume, pressure, and temperature. We now turn our attention to pure substances and consider some of the phases in which a pure substance may exist, the number of independent properties a pure substance may have, and methods of presenting thermodynamic properties.

3.1 The Pure Substance

A pure substance is one that has a homogeneous and invariable chemical composition. It may exist in more than one phase, but the chemical composition is the same in all phases. Thus, liquid water, a mixture of liquid water and water vapor (steam), or a mixture of ice and liquid water are all pure substances, for every phase has the same chemical composition. On the other hand, a mixture of liquid air and gaseous air is not a pure substance, since the composition of the liquid phase is different from that of the vapor phase.

Sometimes a mixture of gases, such as air, is considered a pure substance as long as there is no change of phase. Strictly speaking, this is not true, but rather, as we shall see later, we should say that a mixture of gases such as air exhibits some of the characteristics of a pure substance as long as there is no change of phase.

In this text the emphasis will be on those substances which may be called simple compressible substances. By this we understand that surface effects, magnetic effects, and electrical effects are not significant when dealing with these substances. On the other hand changes in volume, such as those associated with the expansion of a gas in a cylinder, are most important. However, reference will be made to other substances in which surface, magnetic, or electrical effects are important. We will refer to a system consisting of a simple compressible substance as a simple compressible system.

3.2 Vapor-Liquid-Solid Phase Equilibrium in a Pure Substance

Consider as a system 1 lbm of water contained in the piston-cylinder arrangement of Fig. 3.1*a*. Suppose that the piston and weight maintain a

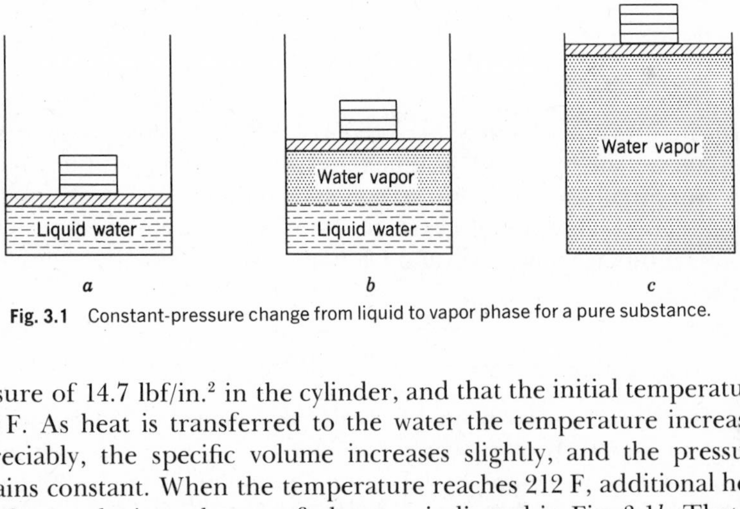

Fig. 3.1 Constant-pressure change from liquid to vapor phase for a pure substance.

pressure of 14.7 lbf/in.^{2} in the cylinder, and that the initial temperature is 60 F. As heat is transferred to the water the temperature increases appreciably, the specific volume increases slightly, and the pressure remains constant. When the temperature reaches 212 F, additional heat transfer results in a change of phase, as indicated in Fig. 3.1*b*. That is, some of the liquid becomes vapor, and during this process both the temperature and pressure remain constant, but the specific volume increases considerably. When the last drop of liquid has vaporized, further transfer of heat results in an increase in both temperature and specific volume of the vapor, Fig. 3.1*c*.

The term saturation temperature designates the temperature at which vaporization takes place at a given pressure, and this pressure is called the saturation pressure for the given temperature. Thus for water at 212 F the saturation pressure is 14.7 lbf/in.^{2}, and for water at 14.7 lbf/in.^{2} the saturation temperature is 212 F. For a pure substance there is a definite relation between saturation pressure and saturation temperature, a typical curve being shown in Fig. 3.2. This is called the vapor-pressure curve.

If a substance exists as liquid at the saturation temperature and pressure, it is called saturated liquid. If the temperature of the liquid is lower than the saturation temperature for the existing pressure, it is called either a subcooled liquid (implying that the temperature is lower than the saturation temperature for the given pressure) or a compressed liquid (implying that the pressure is greater than the saturation pressure for the given temperature). Either term may be used, but the latter term will be used in this text.

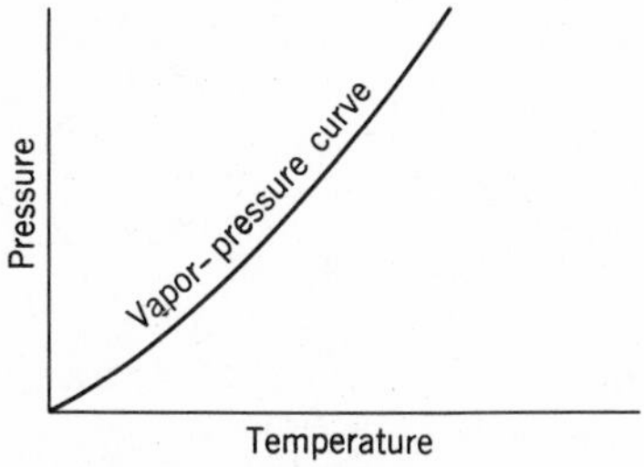

Fig. 3.2 Vapor-pressure curve of a pure substance.

When a substance exists as part liquid and part vapor at the saturation temperature, its quality is defined as the ratio of the mass of vapor to the total mass. Thus, in Fig. 3.1*b*, if the mass of the vapor is 0.2 lbm and the mass of the liquid is 0.8 lbm, the quality is 0.2 or 20 per cent. The quality may be considered as an intensive property, and it has the symbol x. Quality has meaning only when the substance is in a saturated state, i.e., at saturation pressure and temperature.

If a substance exists as vapor at the saturation temperature, it is called saturated vapor. (Sometimes the term dry saturated vapor is used to emphasize that the quality is 100 per cent.) When the vapor is at a temperature greater than the saturation temperature, it is said to exist as superheated vapor. The pressure and temperature of superheated vapor are independent properties, since the temperature may increase while the pressure remains constant. Actually, the substances we call gases are highly superheated vapors.

Consider Fig. 3.1 again, and let us plot on the temperature-volume diagram of Fig. 3.3 the constant-pressure line that represents the states through which the water passes as it is heated from the initial state of 14.7 lbf/in.^{2} and 60 F. Let state A represent the initial state, B the saturated-liquid state (212 F), and line AB the process in which the liquid is heated from the initial temperature to the saturation temperature. Point C is the saturated vapor state, and line BC is the constant-temperature process in which the change of phase from liquid to vapor occurs. Line CD represents the process in which the steam is superheated at constant pressure. Temperature and volume both increase during this process.

Now let the process take place at a constant pressure of 100 lbf/in.^{2},

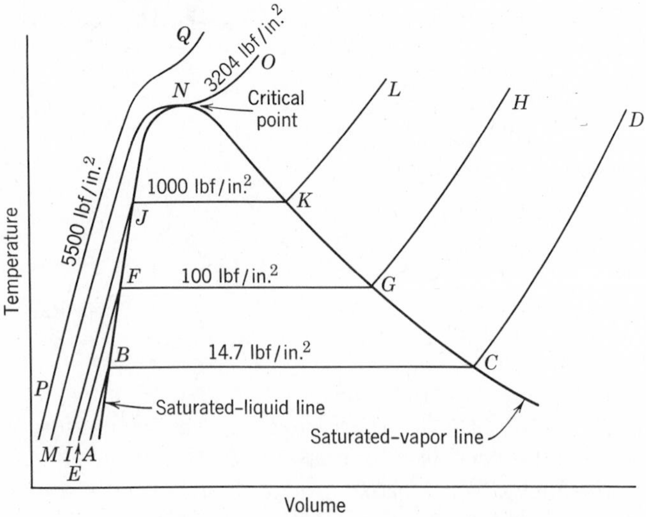

Fig. 3.3 Temperature-volume diagram for water showing liquid and vapor phases (not to scale).

beginning from an initial temperature of 60 F. Point E represents the initial state, the specific volume being slightly less than at 14.7 lbf/in.^{2} and 60 F. Vaporization now begins at point F, where the temperature is 327.9 F. Point G is the saturated-vapor state, and line GH the constant-pressure process in which the steam is superheated.

In a similar manner, a constant pressure of 1000 lbf/in.^{2} is represented by line $IJKL$, the saturation temperature being 544.7 F.

At a pressure of 3203.6 lbf/in.^{2}, represented by line MNO, we find, however, that there is no constant-temperature vaporization process. Rather, point N is a point of inflection with a zero slope. This point is called the critical point, and at the critical point the saturated-liquid and saturated-vapor states are identical. The temperature, pressure, and specific volume at the critical point are called the critical temperature, critical pressure, and critical volume. The critical-point data for some substances are given in Table 3.1, and more extensive data are given in Table B. 5 in the Appendix.

A constant-pressure process at a pressure greater than the critical pressure is represented by line PQ. If water at 5500 lbf/in.^{2}, 60 F is heated in a constant-pressure process in a cylinder such as shown in Fig. 3.1, there will never be two phases present, and the state shown in Fig. 3.1b, will never exist. Rather, there will be a continuous change in density and at all times there will be only one phase present. The question then arises as to when do we have a liquid and when do we have a vapor? The

Table 3.1
Some Critical Point Data

	Critical Temperature (°F)	Critical Pressure (lbf/in.^{2})	Critical Volume (ft^3/lbm)
Water	705	3203.6	0.0505
Carbon dioxide	88	1071	0.0348
Oxygen	−203	735	0.0364
Hydrogen	−400	188	0.534

answer is that this is not a valid question at supercritical pressures. Instead, we simply term the substance a fluid. However, rather arbitrarily at temperatures below the critical temperature we usually refer to it as a compressed liquid and at temperatures above the critical temperature as a superheated vapor. It should be emphasized, however, that at pressures above the critical pressure we never have a liquid and vapor phase of a pure substance existing in equilibrium.

In Fig. 3.3 line *NJFB* represents the saturated-liquid line and line *NKGC* represents the saturated-vapor line.

Let us consider another experiment with the piston-cylinder arrangement. Suppose that the cylinder contains 1 lbm of ice at 0 F, 14.7 lbf/in.^{2}. When heat is transferred to the ice, the pressure remains constant, the specific volume increases slightly, and the temperature increases until it reaches 32 F, at which point the ice melts while the temperature remains constant. In this state the ice is called saturated solid. For most substances the specific volume increases during this melting process but for water the specific volume of the liquid is less than the specific volume of the solid. When all of the ice has melted, a further heat transfer causes an increase in temperature of the liquid.

If the initial pressure of the ice at 0 F is 0.0505 lbf/in.^{2}, heat transfer to the ice first results in an increase in temperature to 20 F. At this point, however, the ice would pass directly from the solid phase to the vapor phase in the process known as sublimation. Further heat transfer would result in superheating of the vapor.

Finally consider an initial pressure of the ice of 0.0887 lbf/in.^{2} and a temperature of 0 F. As a result of heat transfer let the temperature increase until it reaches 32.02 F (0.01 C). At this point, however, further heat transfer may result in some of the ice becoming vapor and some becoming liquid, for at this point it is possible to have the three phases in equilibrium. This is called the triple point, which is defined as the state in which three phases may all be present in equilibrium. The pressure and temperature at the triple point for a number of substances is given in Table 3.2.

Table 3.2
Some Solid-Liquid-Vapor Triple Point Data

	Temperature °F	Pressure atm
Hydrogen (normal)	−435	0.071
Nitrogen	−346	0.1237
Oxygen	−362	0.00150
Mercury	−38	0.0000000013
Water	32	0.00603
Zinc	786	0.05
Silver	1760	0.0001
Copper	1981	0.00000078

This whole matter is best summarized by the diagram of Fig. 3.4, which shows how the solid, liquid and vapor phases may exist together in equilibrium. Along the sublimation line the solid and vapor phases are in equilibrium, along the fusion line the solid and liquid phases are in equilibrium, and along the vaporization line the liquid and vapor phases are in equilibrium. The only point at which all three phases may exist in

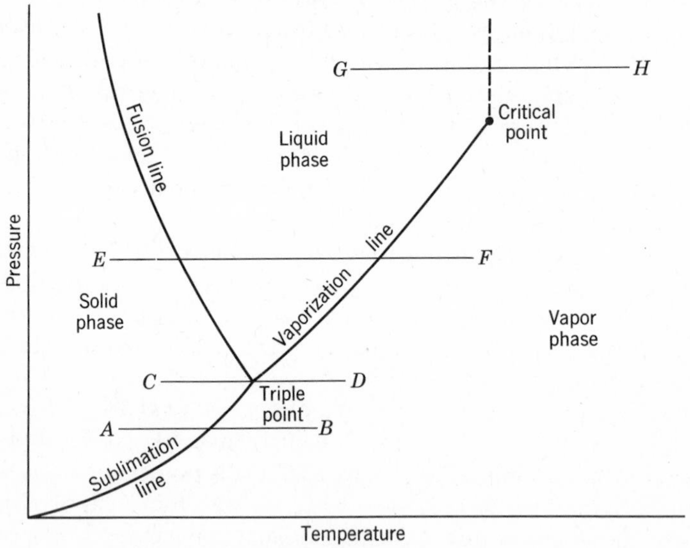

Fig. 3.4 Pressure-temperature diagram for a substance such as water.

equilibrium is the triple point. The vaporization line ends at the critical point because there is no distinct change from the liquid phase to the vapor phase above the critical point.

Consider a solid in state *A*, Fig. 3.4. When the temperature is increased while the pressure (which is less than the triple point pressure) is constant, the substance passes directly from the solid to the vapor phase. Along the constant-pressure line *EF*, the substance first passes from the solid to the liquid phase at one temperature, and then from the liquid to the vapor phase at a higher temperature. Constant-pressure line *CD* passes through the triple point, and it is only at the triple point that the three phases may exist together in equilibrium. At a pressure above the critical pressure, such as *GH*, there is no sharp distinction between the liquid and vapor phases.

Although we have made these comments with rather specific reference to water (only because of our familiarity with water) all pure substances exhibit the same general behavior. However, the triple point temperature and critical temperature vary greatly from one substance to another. For example, the critical temperature of helium, as given in Table B.5 is 9.5 R. Therefore, the absolute temperature of helium at ambient conditions is over 50 times greater than the critical temperature. On the other hand, water has a critical temperature of 705.4 F (1165 R) and at ambient conditions the temperature of water is less than half the critical temperature. Most metals have a much higher critical temperature than water. In considering the behavior of a substance in a given state, it is often helpful to think of this state in relation to the critical state or triple point. For example, if the pressure is greater than the critical pressure, it is impossible to have a liquid and a vapor phase in equilibrium. Or, to consider another example, the states at which vacuum melting a given metal is possible can be ascertained by a consideration of the properties at the triple point. In the case of iron at a pressure just above 0.00005 atm (the triple point pressure), iron would melt at a temperature of about 1535 C (the triple point temperature).

It should also be pointed out that a pure substance can exist in a number of different solid phases. A transition from one solid phase to another is called an allotropic transformation. Figure 3.5 is a pressure-temperature diagram for iron that shows three solid phases, the liquid phase, and the vapor phase. Figure 3.6 shows a number of solid phases for water. It is evident that a pure substance can have a number of triple points, but only one triple point involves solid, liquid, and vapor equilibrium. Other triple points for a pure substance can involve two solid phases and a liquid phase, two solid phases and a vapor phase, or three solid phases.

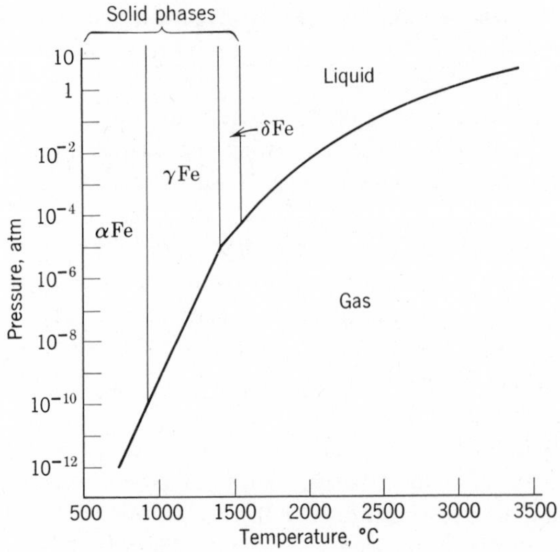

Fig. 3.5 Estimated pressure-temperature diagram for iron (from *Phase Diagrams in Metallurgy*, by F. N. Rhines, copyright 1956, McGraw-Hill Book Company; used by permission).

3.3 Independent Properties of a Pure Substance

One important reason for introducing the concept of a pure substance is that the state of a simple compressible pure substance (i.e., a pure substance in the absence of motion, gravity, and surface, magnetic or electrical effects) is defined by two independent properties. This means, for example, that if the specific volume and temperature of superheated steam are specified, the state of the steam is determined.

To understand the significance of the term independent property, consider the saturated-liquid and saturated-vapor states of a pure substance. These two states have the same pressure and the same temperature, but are definitely not the same state. In a saturation state, therefore, pressure and temperature are not independent properties. Two independent properties such as pressure and specific volume, or pressure and quality, are required to specify a saturation state of a pure substance.

The reason for mentioning previously that a mixture of gases, such as air, has the same characteristics as a pure substance as long as only one phase is present, concerns precisely this point. The state of air, which is a mixture of gases of definite composition, is determined by specifying two properties as long as it remains in the gaseous phase, and in this regard air can be treated as a pure substance.

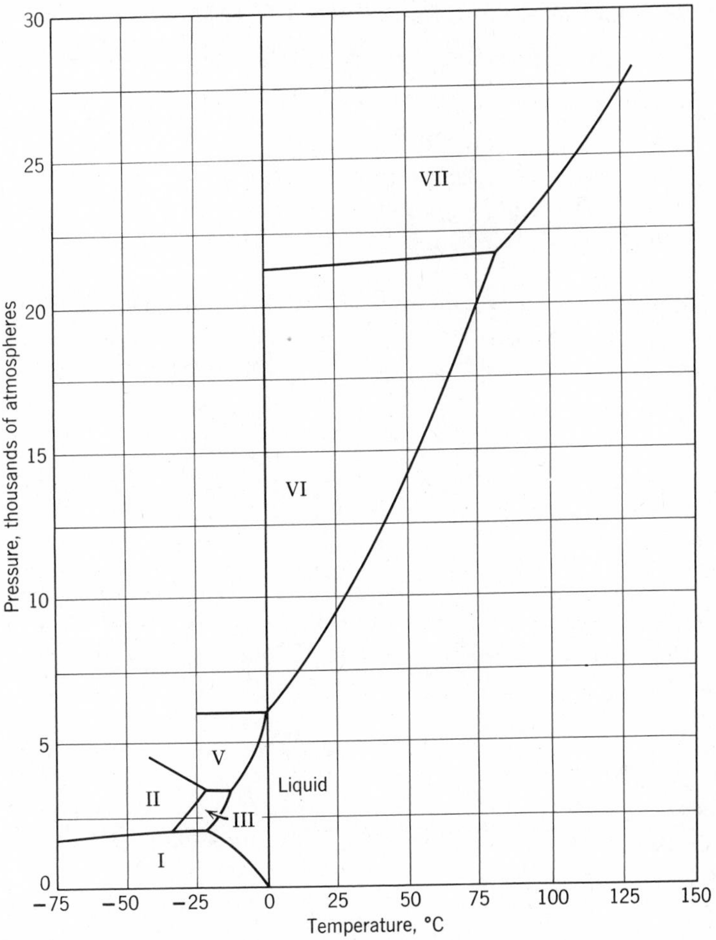

Fig. 3.6 Phase diagram of water (Adapted from the *American Institute of Physics Handbook*, 2nd Ed., 1963, McGraw-Hill).

3.4 Equations of State for the Vapor Phase of a Simple Compressible Substance

From experimental observations it has been established that the P-v-T behavior of gases at low density is closely given by the following equation of state.

$$P\bar{v} = \bar{R}T \tag{3.1}$$

where $\bar{R}$ is the universal gas constant. The value for $\bar{R}$ depends on the units chosen for P, $\bar{v}$, and T. The most frequently used units in this text are:

$$\bar{R} = 1545 \text{ ft-lbf/lb mole R}$$
$$\bar{R} = 1.987 \text{ Btu/lb mole R}$$
$$\bar{R} = 1.987 \text{ cal/gm mole K}$$
$$\bar{R} = 0.08206 \text{ atm-liters/gm mole K}$$

Dividing Eq. 3.1 by M, the molecular weight, we have the equation of state on a unit mass basis.

$$\frac{P\bar{v}}{M} = \frac{\bar{R}T}{M} \tag{3.2}$$

or

$$Pv = RT$$

where

$$R = \frac{\bar{R}}{M} \tag{3.3}$$

R is a constant for a particular gas. The value of R for a number of substances is given in Table B.6 of the Appendix. It follows from Eqs. 3.1 and 3.2 that this equation of state can be as written in terms of the total volume.

$$PV = n\bar{R}T \tag{3.4}$$
$$PV = mRT$$

It should also be noted that Eq. 3.4 can alternately be written in the form

$$\frac{P_1V_1}{T_1} = \frac{P_2V_2}{T_2} \tag{3.5}$$

That is, gases at low density closely follow the well-known Boyle's and Charles' laws. Boyle and Charles, of course, based their statements on experimental observations. (Strictly speaking, neither of these statements should be called a law, since they are only approximately true and even then only under conditions of low density.)

The equation of state given by Eq. 3.1 (or Eq. 3.2) is referred to as the ideal gas equation of state. At very low density all gases and vapors approach ideal gas behavior, with the P-v-T relationship being given by the ideal gas equation of state. At higher densities the behavior may deviate substantially from the ideal gas equation of state.

Because of its simplicity, the ideal gas equation of state is very convenient to use in thermodynamic calculations. However, two questions can appropriately be raised. The ideal gas equation of state is a good approximation at low density. But what constitutes low density? Or, expressed in other words, over what range of density will the ideal gas equation of state hold with accuracy? The second question is, how much does an actual gas at a given pressure and temperature deviate from ideal gas behavior?

To answer both questions we introduce the concept of the compressibility factor, Z, which is defined by the relation

$$Z = \frac{P\bar{v}}{\bar{R}T}$$

or

$$P\bar{v} = Z\bar{R}T \tag{3.6}$$

Note that for an ideal gas, $Z = 1$, and the deviation of Z from unity is a measure of the deviation of the actual relation from the ideal gas equation of state.

Figure 3.7 shows a skeleton compressibility chart for nitrogen. From this chart we make three observations. The first is that at all temperatures, $Z \rightarrow 1$ as $P \rightarrow 0$. That is, as the pressure approaches zero, the

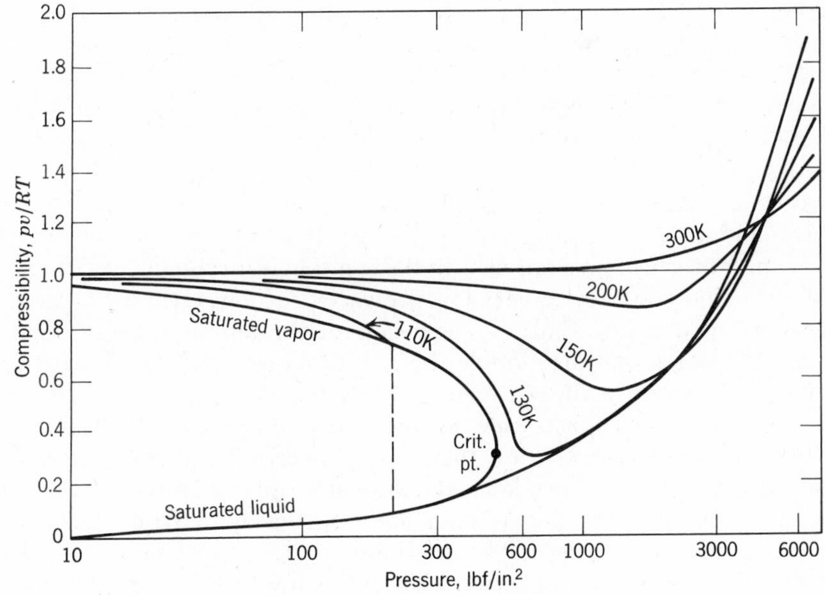

Fig. 3.7 Compressibility of nitrogen.

P-v-T behavior closely approaches that predicted by the ideal gas equation of state. Note also that at temperatures of 300 K and above (i.e., room temperature and above) the compressibility factor is near unity up to pressures well above 1000 lbf/in.^{2}. This means that the ideal gas equation of state can be used for nitrogen (and, as it happens, also air) over this range with considerable accuracy.

Now suppose we reduce the temperature from 300 K while keeping the pressure constant at 600 lbf/in.^{2}. The density will increase and we note a sharp decrease below unity in the value of the compressibility factor. Values of $Z < 1$ mean that the actual density is greater than would be predicted by ideal gas behavior. The physical explanation of this is as follows: As the temperature is reduced from 300 K while pressure remains constant at 600 lbf/in.^{2}, the molecules are brought closer together. In this range of intermolecular distances, and at this pressure and temperature, there is an attractive force between the molecules. The lower the temperature the greater is this intermolecular attractive force. This attractive force between the molecules means that the density is greater than would be predicted by the ideal gas behavior, which assumes no intermolecular forces. Note also from the compressibility chart that at very high densities, for pressures above4000 lbf/in.^{2}, the compressibility factor is always greater than unity. In this range the intermolecular distances are very small, and there is a repulsive force between the molecules. This tends to make the density less than would otherwise be expected.

The precise nature of intermolecular forces is a rather complex matter. These forces are a function of the temperature as well as the density. The preceding discussion should be considered as a qualitative analysis to assist in gaining some understanding of the ideal gas equation of state and how the P-v-T behavior of actual gases deviate from this equation.

From a practical point of view in the solution of problems, two things should be noted. First, at very low pressures, ideal gas behavior can be assumed with good accuracy, regardless of the temperature. Second, at temperatures that are double the critical temperature or above (the critical temperature of nitrogen is 126 K) ideal gas behavior can be assumed with good accuracy to pressures of at least 1000 lbf/in.^{2}. When the temperature is less than twice the critical temperature, and the pressure above a very low value, say atmospheric pressure, we are in the superheated vapor region, and the deviation from ideal gas behavior may be considerable. In this region it is preferable to use tables of thermodynamic properties or charts for a particular substance. These tables are considered in the following section. The concept of the generalized compressibility chart is introduced in Chapter 12.

In order to have an equation of state that accurately represents the P-v-T behavior for a particular gas over the entire superheated vapor range, more complicated equations of state have been developed. Several different forms of these have been used. To illustrate the nature and complexity of these equations we refer to one of the best known, namely, the Beattie-Bridgeman equation of state. This equation is

$$P = \frac{\bar{R}T(1-\epsilon)}{\bar{v}^2}(\bar{v}+B) - \frac{A}{\bar{v}^2} \tag{3.7}$$

where $A = A_0(1 - a/\bar{v})$, $B = B_0(1 - b/\bar{v})$, $\epsilon = c/\bar{v}T^3$, and A_0, a, B_0, b, and c are constants for different gases. The values of these constants for various substances are given in Table 3.3.

Table 3.3
Constants of the Beattie-Bridgeman Equation of State (Pressure in atmospheres; specific volume in liters per gram mole; temperature in degrees Kelvin: $\bar{R} = 0.08206$ atm-liters/gm mole K)

Gas	A_0	a	B_0	b	$10^{-4}c$
Helium	0.0216	0.05984	0.01400	0.0	0.0040
Argon	1.2907	0.02328	0.03931	0.0	5.99
Hydrogen	0.1975	−0.00506	0.02096	−0.04359	0.0504
Nitrogen	1.3445	0.02617	0.05046	−0.00691	4.20
Oxygen	1.4911	0.02562	0.04624	0.004208	4.80
Air	1.3012	0.01931	0.04611	−0.001101	4.34
Carbon dioxide	5.0065	0.07132	0.10476	0.07235	66.00

The matter of equations of state will be discussed further in Chapter 12. The observation to be made here in particular is that an equation of state that accurately describes the relation between pressure, temperature, and specific volume is rather cumbersome and the solution requires considerable time. When using a large digital computer, it is often most convenient to determine the thermodynamic properties in a given state from such equations. However, in hand calculations it is much more convenient to tabulate values of pressure, temperature, specific volume, and other thermodynamic properties for various substances. The Appendix includes summary tables and graphs of the thermodynamic properties of water, ammonia, Freon-12, nitrogen, oxygen, and Freon-13. The tables of the properties of water are usually referred to as the "steam tables" and are extracted from "Steam Tables" by Keenan, Keyes, Hill and Moore. The method for compiling the

P-v-T data for such a table is to find an equation of state that accurately fits the experimental data, and then to solve the equation of state for the values listed in the table.

Example 3.1

What is the mass of air contained in a room 20 ft × 30 ft × 12 ft if the pressure is 14.7 lbf/in.^{2}, and the temperature is 80 F? Assume air to be an ideal gas.

By using Eq. 3.4, and the value of R from Table B.5

$$m = \frac{PV}{RT} = \frac{14.7 \times 144 \text{ lbf/ft}^2 \times 7200 \text{ ft}^3}{53.34 \text{ ft-lbf/lbm R} \times 540 \text{ R}} = 529 \text{ lbm}$$

Example 3.2

A tank has a volume of 15 ft^3 and contains 20 lbm of an ideal gas having a molecular weight of 24. The temperature is 80 F. What is the pressure?

The gas constant is first determined:

$$R = \frac{\bar{R}}{M} = \frac{1545 \text{ ft-lbf/lb mole R}}{24 \text{ lbm/lb mole}} = 64.4 \text{ ft-lbf/lbm R}$$

We now solve for P.

$$P = \frac{mRT}{V} = \frac{20 \text{ lbm} \times 64.4 \text{ ft-lbf/lbm R} \times 540 \text{ R}}{144 \text{ in.}^2/\text{ft}^2 \times 15 \text{ ft}^3} = 321 \text{ lbf/in.}^2$$

3.5 Tables of Thermodynamic Properties

Tables of thermodynamic properties of many substances are available, and in general all these have the same form. In this section we shall refer to the steam tables. The steam tables are selected both as a vehicle for presenting thermodynamic tables and because steam is used extensively in power plants and industrial processes. Once the steam tables are understood, other thermodynamic tables can be readily used.

Several different versions of steam tables have been published. Two new tables have recently been published in the United States. In 1967 The American Society of Mechanical Engineers published a volume entitled "Thermodynamic and Transport Properties of Steam," commonly referred to as the 1967 ASME steam tables. "Steam Tables," by Keenan, Keyes, Hill, and Moore was published in 1969. This is a revision of a very extensively used volume by Keenan and Keyes, which was published in 1936. The Appendix includes a summary of the 1969 edition of "Steam Tables," and reference to these tables is made throughout this text.

In Table B.1.1, the first column after the temperature gives the corresponding saturation pressure in pounds force per square inch. The next two columns give specific volume in cubic feet per pound mass. The first of these gives the specific volume of the saturated liquid, v_f; the second column gives the specific volume of saturated vapor, v_g. The difference between these two, $v_g - v_f$, represents the increase in specific volume when the state changes from staurated liquid to saturated vapor, and is designated v_{fg}.

The specific volume of a substance having a given quality can be found by utilizing the definition of quality. Quality has already been defined as the ratio of the mass of vapor to total mass of liquid plus vapor when a substance is in a saturation state. Let us consider a mass of 1 lbm having a quality x. The specific volume is the sum of the volume of the liquid and the volume of the vapor. The volume of the liquid is $(1-x)v_f$, and the volume of the vapor is xv_g. Therefore the specific volume v is

$$v = xv_g + (1-x)v_f \tag{3.8}$$

Since $v_f + v_{fg} = v_g$, Eq. 3.8 can also be written in the following forms

$$v = v_f + xv_{fg} \tag{3.9}$$

$$v = v_g - (1-x)v_{fg} \tag{3.10}$$

As an example, let us calculate the specific volume of saturated steam at 500 F having a quality of 70 per cent. Using Eq. 3.8,

$$v = 0.3(0.0204) + 0.7(0.6761) = 0.4794 \text{ ft}^3/\text{lbm}$$

In Table B.1.2, the first column after the pressure lists the saturation temperature for each pressure. The next columns list specific volume in a manner similar to Table B.1.1. When necessary, v_{fg} can readily be found by subtracting v_f from v_g.

Table 3 of the steam tables, which is summarized in Table B.1.3 in the Appendix, gives the properties of superheated vapor. In the superheat region, pressure and temperature are independent properties, and therefore, for each pressure a large number of temperatures is given, and for each temperature four thermodynamic properties are listed, the first one being specific volume. Thus, the specific volume of steam at a pressure of 100 lbf/in.^{2} and 500 F is 5.587 ft^3/lbm.

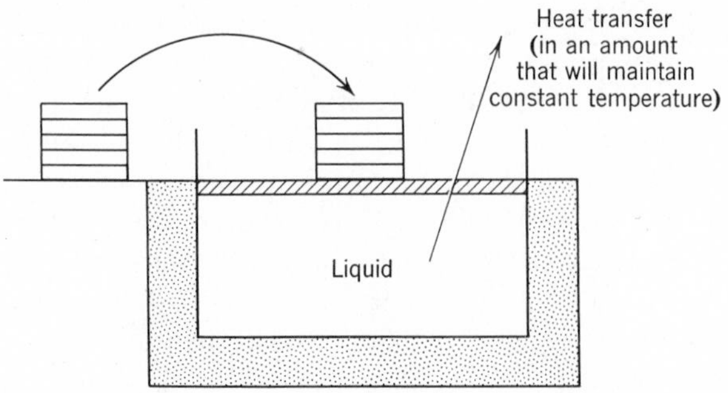

Fig. 3.8 Illustration of compressed liquid state.

Table 4 of the steam tables, summarized in Table B.1.4 in the Appendix, gives the properties of the compressed liquid. To demonstrate the use of this table, consider a piston and a cylinder (as shown in Fig. 3.8) that contains 1 lbm of saturated liquid at 200 F. Its properties are given in Table B.1.1, and we note that the pressure is 11.53 lbf/in.^{2}, and the specific volume is 0.01663 ft^3/lbm. Suppose the pressure is increased to 1000 lbf/in.^{2} while the temperature is held constant at 200 F by the necessary transfer of heat, Q. Since water is slightly compressible, we would expect a slight decrease in specific volume during this process. Table B.1.4 gives this specific volume as 0.01658 ft^3/lbm. Note that this is only a slight decrease, and only a small error would be made if one assumed that the volume of a compressed liquid was equal to the specific volume of the saturated liquid at the same temperature. In many cases this is the most convenient procedure, particularly in those cases when compressed liquid data are not available.

Furthermore, since specific volume does change rapidly with temperature, care should be exercised in interpolation over the wide temperature ranges in Table B.1.4. (In some cases it may be more accurate to use the saturated liquid data from Table B.1.1, and interpolate differences between Table B.1.1, the saturated liquid data, and Table B.1.4, the compressed liquid data.)

Table 6 of the steam tables, which is summarized in Table B.1.5 in the Appendix, gives the properties of saturated solid and saturated vapor that are in equilibrium. The first column gives the temperature, and the second column gives the corresponding saturation pressure. As would be expected, all these pressures are less than the triple-point pressure. The next two columns give the specific volume of the saturated solid and saturated vapor (note that the tabulated value is $v_g \times 10^{-3}$).

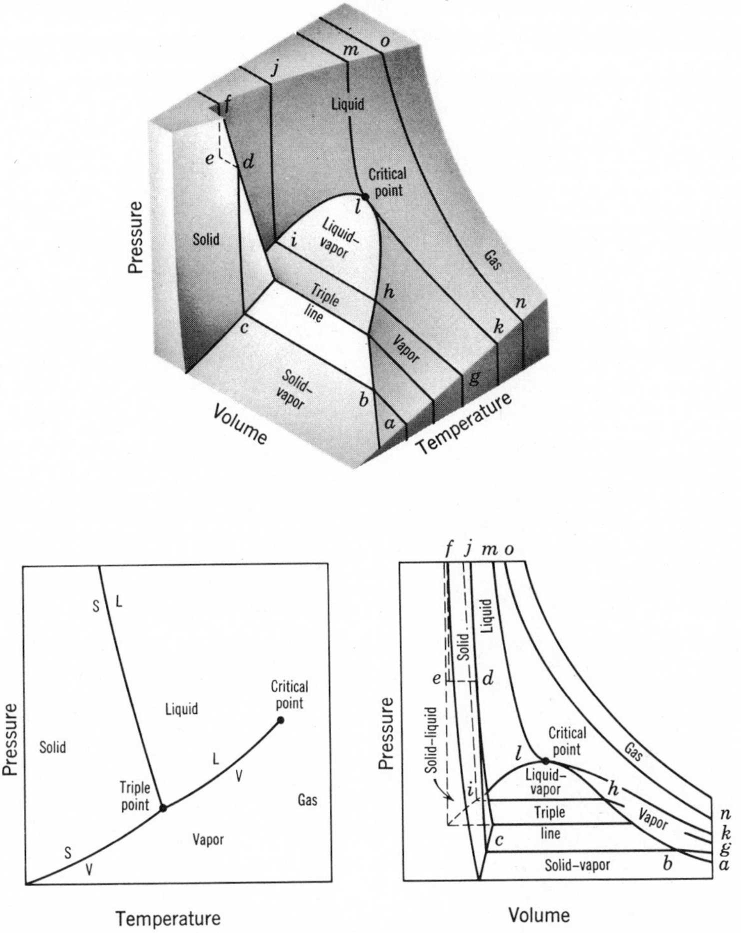

Fig. 3.9 Pressure-volume-temperature surface for a substance that expands on freezing.

3.6 Thermodynamic Surfaces

The matter discussed in this chapter can be well summarized by a consideration of a pressure-specific volume-temperature surface. Two such surfaces are shown in Figs. 3.9 and 3.10. Figure 3.9 shows a substance such as water in which the specific volume increases during freezing, and Fig. 3.10 shows a substance in which the specific volume decreases during freezing.

In these diagrams the pressure, specific volume, and temperature are

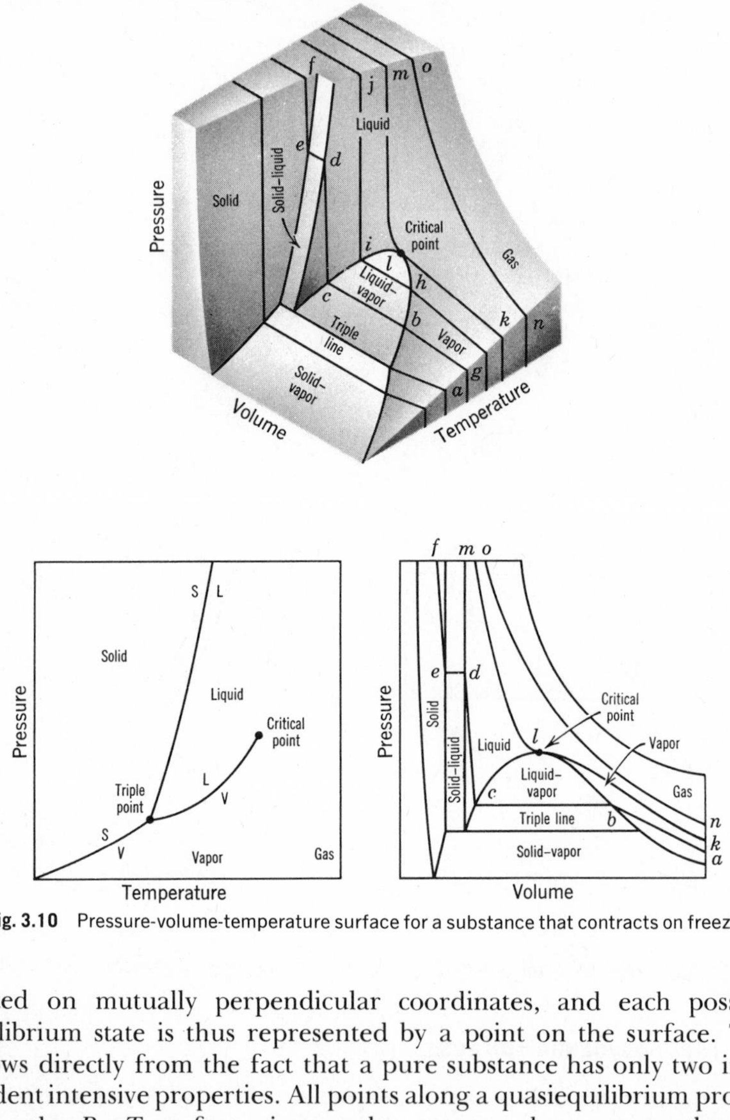

Fig. 3.10 Pressure-volume-temperature surface for a substance that contracts on freezing.

plotted on mutually perpendicular coordinates, and each possible equilibrium state is thus represented by a point on the surface. This follows directly from the fact that a pure substance has only two independent intensive properties. All points along a quasiequilibrium process lie on the *P-v-T* surface, since such a process always passes through equilibrium states.

The regions of the surface that represent a single phase, namely, the solid, liquid, and vapor phases, are indicated, these surfaces being curved. The two-phase regions, namely, the solid-liquid, solid-vapor, and liquid-vapor regions, are ruled surfaces. By this we understand that they are made up of straight lines parallel to the specific volume axis. This, of course, follows from the fact that in the two-phase region, lines of constant pressure are also lines of constant temperature, though the specific volume may change. The triple point actually appears as the triple line on the P-v-T surface, since the pressure and temperature of the triple point are fixed, but the specific volume may vary, depending on the proportion of each phase.

It is also of interest to note the pressure-temperature and pressure-volume projections of these surfaces. We have already considered the pressure-temperature diagram for a substance such as water. It is on this diagram that we observe the triple point. Various lines of constant temperature are shown on the pressure-volume diagram, and the corresponding constant-temperature sections are lettered identically on the P-v-T surface. The critical isotherm has a point of inflection at the critical point.

One notices that with a substance such as water, which expands on freezing, the freezing temperature decreases with an increase in pressure. With a substance that contracts on freezing, the freezing temperature increases as the pressure increases. Thus, as the pressure of vapor is increased along the constant-temperature line *abcdef* in Fig. 3.9, a sub-the substance that contracts on freezing, the corresponding constant-temperature line, Fig. 3.10, indicates that as the pressure on the vapor is increased, it first becomes liquid and then solid.

Example 3.3

A vessel having a volume of 10 ft^3 contains 3.0 lbm of a liquid water and water vapor mixture in equilibrium at a pressure of 100 lbf/in.^{2}. Calculate

1. The volume and mass of liquid.
2. The volume and mass of vapor.

The specific volume is calculated first:

$$v = \frac{10.0}{3.0} = 3.333 \text{ ft}^3/\text{lbm}$$

From the steam tables (Appendix Table B.1.2)

$$v_{fg} = 4.434 - 0.01774 = 4.416$$

The quality can now be calculated, using Eq. 3.10

$$3.333 = 4.434 - (1-x)4.416$$

$$(1-x) = \frac{1.101}{4.416} = 0.249$$

$$x = 0.751$$

Therefore the mass of liquid is

$$3(0.249) = 0.747 \text{ lbm}$$

The mass of vapor is

$$3(0.751) = 2.253 \text{ lbm}$$

The volume of liquid is

$$V_{\text{liq}} = m_{\text{liq}} v_f = 0.747(0.01774) = 0.0133 \text{ ft}^3$$

The volume of the vapor is

$$V_{\text{vap}} = m_{\text{vap}} v_g = 2.253(4.434) = 9.99 \text{ ft}^3$$

Example 3.4

A pressure vessel contains saturated Freon-12 vapor at 60 F. Heat is transferred to the system until the temperature reaches 200 F. What is the final pressure?

Since the volume does not change during this process, the specific volume also remains constant. From the Freon-12 Tables, Table B.3

$$v_1 = v_2 = 0.558 \text{ ft}^3/\text{lbm}$$

We know one other property in the final state, namely the temperature, $T_2 = 200$ F, and therefore the final state is determined.

Since v_g at 200 F is less than 0.558 ft^2/lbm, it is evident that in the final state the Freon-12 is superheated vapor. By interpolating between the 90 lbf/in.^{2} and 100 lbf/in.^{2} columns of Table B.3, we find that

$$P_2 = 97.8 \text{ lbf/in.}^2$$

PROBLEMS

3.1 A spherical balloon has a radius of 10 ft. The atmospheric pressure is 14.7 lbf/in.^{2} and the temperature is 60 F.

(*a*) Calculate the mass and the number of moles of air this balloon displaces.

(*b*) If the balloon is filled with helium at 14.7 lbf/in.^{2}, 60 F, what is the mass and the number of moles of helium?

3.2 The mass of a certain ideal gas in a given container is 0.13 lbm. The pressure is 0.5 atm, the temperature is 60 F, and the volume of the gas is 3 ft^3. Determine the molecular weight of the gas.

3.3 Air is contained in a vertical cylinder fitted with a frictionless piston and a set of stops, as shown in Fig. 3.11. The cross-sectional area of the piston is

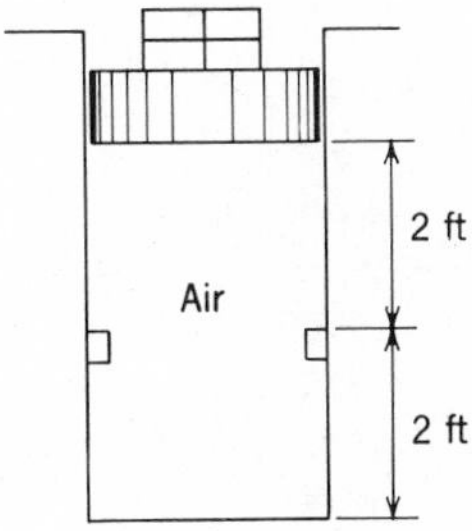

Fig. 3.11 Sketch for Problem 3.3.

0.5 ft^2, and the air is initially at 30 lbf/in.^{2}, 800 F. The air is then cooled as a result of heat transfer to the surroundings.

(*a*) What is the temperature of the air inside when the piston reaches the stops?

(*b*) If the cooling is continued until the temperature reaches 70 F, what is the pressure inside the cylinder at this state?

3.4 A vacuum pump is used to pump a vacuum over a bath of liquid helium. The volume rate of flow into the vacuum pump is 3000 cubic feet per minute. The pressure at the vacuum pump inlet is 0.1 torr and the temperature is -10 F. What mass of helium enters the pump per minute?

3.5 In studying gas behavior from the microscopic viewpoint, it is convenient to write the ideal gas equation of state in terms of the Boltzmann constant k, where

$$k = \bar{R}/N_0$$

(*a*) Calculate the value of the Boltzmann constant, and compare with the value listed in Appendix B.12.

(*b*) In high temperature plasmas, free electrons can be considered to behave as an ideal gas. Calculate the number density N/V of an electron gas at 10,000 K, 0.01 atm.

3.6 A rigid vessel A is connected to a spherical elastic balloon B as shown in Fig. 3.12. Both contain air at the ambient temperature, 80 F. The volume

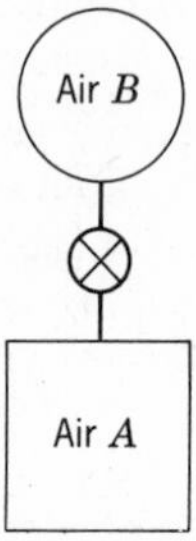

Fig. 3.12 Sketch for Problem 3.6.

of vessel A is 1 ft³ and the initial pressure is 40 lbf/in.². The initial diameter of the balloon is 1 ft and the pressure inside is 15 lbf/in.². The valve connecting A and B is now opened, and remains open. It may be assumed that the pressure inside the balloon is directly proportional to its diameter, and also that the final temperature of the air is uniform throughout at 80 F. Determine the final pressure in the system and the final volume of the balloon.

3.7 Is it reasonable to assume that at the given states the substance behaves as an ideal gas:
  (*a*) Nitrogen at 90 F, 500 lbf/in.².
  (*b*) Carbon dioxide at 90 F, 500 lbf/in.².
  (*c*) Water at 1800 F, 500 lbf/in.².
  (*d*) Water at 120 F, 1 lbf/in.².
  (*e*) Water at 90 F, 1 lbf/in.².

3.8 Determine whether water at each of the following states is a compressed liquid, a superheated vapor, or a mixture of saturated liquid and vapor: 250 F, 20 lbf/in.²; 50 lbf/in.², 8 ft³/lbm; 300 F, 8 ft³/lbm; 30 lbf/in.², 230 F; 400 F, 0.02 ft³/lbm; 0.5 lbf/in.², 60 F.

3.9 Plot the following vapor-pressure curves (saturation pressure vs. saturation temperature):
  (*a*) Water on Cartesian coordinates, −40 F to 60 F.
  (*b*) Water on Cartesian coordinates, 0 to 3500 lbf/in.².
  (*c*) Water, Freon-12, and ammonia, on semilog paper (pressure on log scale), 1.0 to 1000 lbf/in.², −50 F to 300 F.

3.10 Calculate the following specific volumes:
  (*a*) Ammonia, 80 F, 80% quality.

(*b*) Freon-12, 120 F, 15% quality.
(*c*) Water, 1000 lbf/in.^{2}, 98% quality.
(*d*) Nitrogen, −300 F, 40% quality.

3.11 Determine the quality (if saturated) or temperature (if superheated) of the following substances in the given states:
(*a*) Ammonia, 80 F, 1.43 ft^3/lbm; 80 lbf/in.^{2}, 4.75 ft^3/lbm.
(*b*) Freon-12, 50 lbf/in.^{2}, 0.6 ft^3/lbm; 50 lbf/in.^{2}, 0.960 ft^3/lbm.
(*c*) Water, 80 F, 20 ft^3/lbm; 1000 lbf/in.^{2}, 0.4 ft^3/lbm.
(*d*) Nitrogen, 100 lbf/in.^{2}, 0.9 ft^3/lbm; 1 atm, 3 ft^3/lbm.

3.12 Plot a pressure-specific volume diagram on log log paper (3 × 5 cycles) for water showing the following lines:
(*a*) Saturated liquid.
(*b*) Saturated vapor.
(*c*) The following constant-temperature lines (including the compressed-liquid region): 300 F, 500 F, 700 F, 800 F, 1000 F.
(*d*) The following lines of constant quality: 10%, 50%, 90%.

3.13 Plot a pressure-specific volume diagram on log log paper (2 × 3 cycles) for Freon-12, showing the following lines:
(*a*) Saturated liquid.
(*b*) Saturated vapor.
(*c*) The following constant-temperature lines: 0 F, 100 F, 230 F, 300 F.
(*d*) The following constant-quality lines: 10%, 50%, 90%.

3.14 The radiator of a heating system has a volume of 2 ft^3 and contains saturated vapor at 20 lbf/in.^{2}. The valves are then closed on the radiator, and as a result of heat transfer to the room the pressure drops to 15 lbf/in.^{2}. Calculate:
(*a*) The total mass of steam in the radiator.
(*b*) The volume and mass of liquid in the final state.
(*c*) The volume and mass of vapor in the final state.

3.15 Steam at the critical state is contained in a rigid vessel. Heat is transferred from the steam until the pressure is 300 lbf/in.^{2}. Calculate the final quality.

3.16 For a certain experiment, Freon-12 vapor is contained in a sealed glass tube at 80 F. It is desired to know the pressure at this condition, but there is no means for measuring it, since the tube is sealed. However, if the tube is cooled to 50 F, small droplets of liquid are observed on the glass walls. What is the pressure inside at 80 F?

3.17 The rigid vessel shown in Fig. 3.13 contains saturated water at 14.7 lbf/in.^{2}. Determine the proportions by volume of liquid and vapor at 14.7 lbf/in.^{2} necessary to make the water pass through the critical state when heated.

3.18 A rigid vessel of 0.25 ft^3 volume contains 10 lbm of water (liquid plus vapor) at 100 F. The vessel is then slowly heated. Will the liquid level inside eventually rise to the top or drop to the bottom of the vessel? What if the vessel contains 1 lbm instead of 10 lbm?

3.19 A vessel fitted with a sight glass contains Freon-12 at 80 F. Liquid is with-

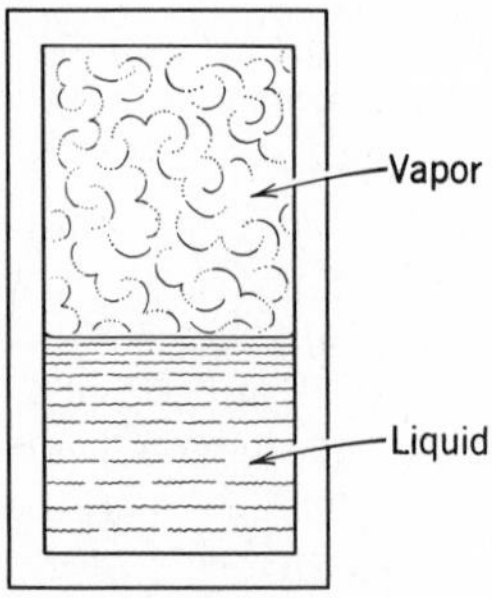

Fig. 3.13 Sketch for Problem 3.17.

drawn from the bottom at a slow rate, and the temperature remains constant during the process. If the area of the vessel is 50 in.^{2} and the level drops 6 in., determine the mass of Freon-12 withdrawn.

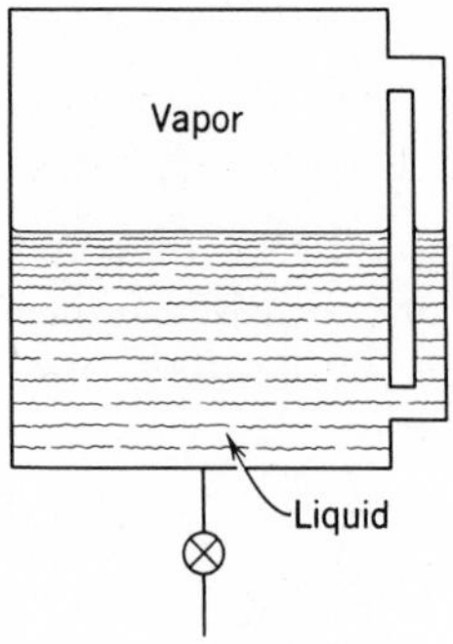

Fig. 3.14 Sketch for Problem 3.19.

3.20 A boiler feed pump delivers 300,000 lbm of water per hour at 560 F, 3000 lbf/in.^{2}. What is the volume rate of flow in ft^3/min? What would be the per cent error if the properties of saturated liquid water at 560 F were used in the calculation?

3.21 Consider compressed liquid water at 100 F. What pressure is required to decrease the specific volume by 1% from its saturated liquid value?

3.22 Liquid nitrogen at a temperature of −240 F exists in a container, and both the liquid and vapor phases are present. The volume of the container is 3 ft^3, and it is determined that the mass of nitrogen in the container is 44.5 lbm. What is the mass of liquid and the mass of vapor present in the container?

3.23 A refrigeration system is to be charged with Freon-12. The system, which has a volume of 0.85 ft^3, is first evacuated, and then slowly charged with Freon-12. The temperature of the Freon-12 remains constant at the ambient temperature of 80 F.

(*a*) What will be the mass of Freon-12 in the system when the pressure reaches 35 lbf/in.^{2}?

(*b*) What will be the mass of Freon-12 in the system when the system is filled with saturated vapor?

(*c*) What fraction of the Freon-12 will exist as a liquid when 3 lbm of Freon-12 have been placed in the system?

3.24 One lbm of H_2O exists at the triple point. The volume of the liquid phase is equal to the volume of the solid phase, and the volume of the vapor phase is equal to 10^4 times the volume of the liquid phase. What is the mass of H_2O in each phase?

3.25 Compare the specific volume of nitrogen at 1000 lbf/in.^{2}, -210 F as given in the nitrogen tables, B.4, with the value calculated from the Beattie–Bridgeman equation of state and with the value calculated from the ideal gas equation of state? Which of the three values is the best?

3.26 Tank *A* (Fig. 3.15) has a volume of 0.1 ft^3 and contains Freon-12 at 80 F, 10% liquid and 90% vapor by volume, while tank *B* is evacuated. The valve is then opened, and the tanks eventually come to the same pressure, which is found to be 30 lbf/in.^{2}. During this process, heat is transferred such that the Freon remains at 80 F. What is the volume of tank *B*?

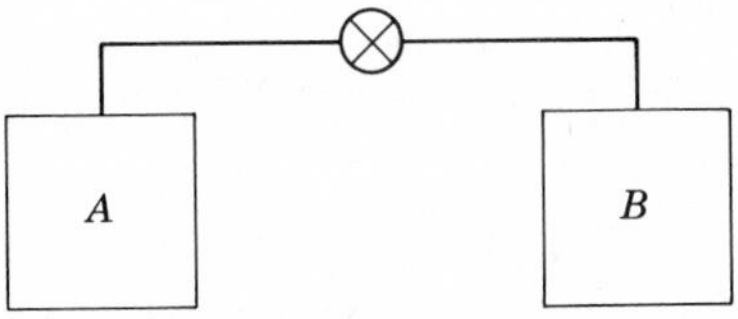

Fig. 3.15 Sketch for Problem 3.26.

3.27 A tank contains Freon-12 at 100 F. The volume of the tank is 2 ft^3, and initially the volume of the liquid in the tank is equal to the volume of the vapor. Additional Freon-12 is forced into the tank until the mass of Freon-12 in the tank reaches 100 lbm. What is the final volume of liquid in the tank, assuming that the temperature is maintained at 100 F? How much mass enters the tank?

3.28 A closed tank contains vapor and liquid H_2O in equilibrium at 500 F. The distance from the bottom of the tank to the liquid level is 20 ft. What is the pressure reading at the bottom of the tank as compared to the pressure reading at the top of the tank?

3.29 A container of liquid nitrogen at 28.12 lbf/in.^{2} pressure has a cross-sectional area of 40 in.^{2}. As the result of heat transfer to the liquid nitrogen, some of the nitrogen evaporates and in one hour, the level drops one inch. The vapor that leaves the insulated container passes through a heater and leaves at 20 lbf/in.^{2}, 0 F. Calculate the volume rate of flow out of the heater in ft^3/hr, assuming ideal gas behavior, and compare this with the result obtained when using the nitrogen tables, Table B.4.

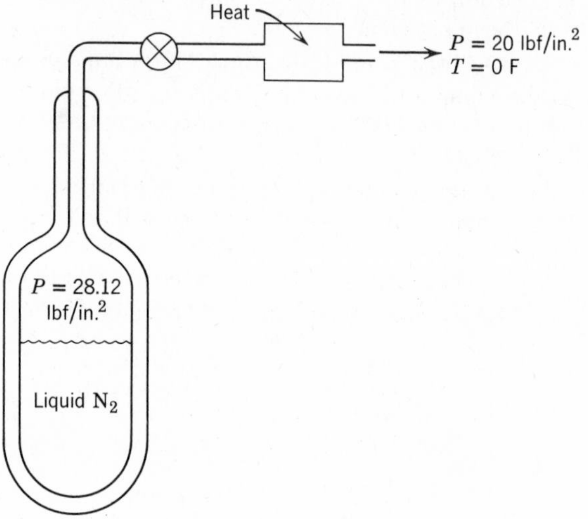

Fig. 3.16 Sketch for Problem 3.29.

3.30 Water is contained in a cylinder fitted with a frictionless piston, as shown in Fig. 3.17. The mass of water is 1 lbm and the area of the piston is 2 ft^2. At the initial state the water is at 230 F, with the quality of 90%, and the spring just touches the piston, but exerts no force on it. Now, heat is trans-

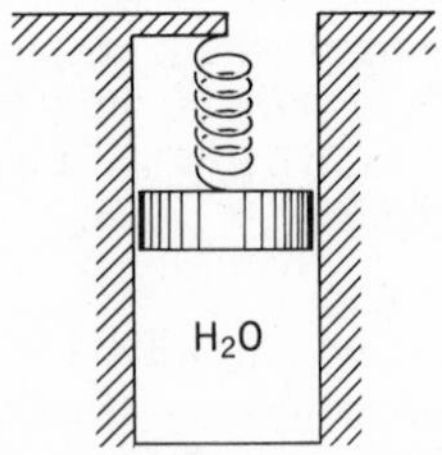

Fig. 3.17 Sketch for Problem 3.30.

ferred to the water, and the piston begins to rise. During this process, the resisting force of the spring is proportional to the distance moved, with a force of 50 lbf/in. Calculate the pressure in the cylinder when the temperature reaches 320 F.

3.31 Write a computer program to solve the following problem. For any specified substance, it is desired to calculate the pressure according to the Beattie–Bridgeman equation of state for any given set (or sets) of temperature and specific volume, and to compare the result with the ideal gas equation of state.

3.32 Write a computer program to solve the following problem. For any specified substance, it is desired to solve the Beattie–Bridgeman equation of state for specific volume at any given set (or sets) of pressure and temperature, and to compare the result with the ideal gas equation of state.

4

Work and Heat

The concept of energy was introduced in Sec. 2.6. At that point we noted that energy can be transferred from one system to another. The two forms in which this transfer can take place are work and heat. The definitions of heat and work, and the difference between them, should be clearly understood, because the correct thermodynamic analysis of many problems depends on an accurate distinction between heat and work. These definitions and distinctions constitute the subject matter of this chapter.

4.1 Definition of Work

Work is usually defined as a force F acting through a displacement x, the displacement being in the direction of the force. That is,

$$W = \int_1^2 F \cdot dx \tag{4.1}$$

This is a very useful relationship because it enables us to find the work required to raise a weight, to stretch a wire, or to move a charged particle through a magnetic field.

However, when treating thermodynamics from a macroscopic point of view, it is advantageous to tie in the definition of work with the concepts of systems, properties, and processes. We therefore define work as follows: work is done by a system if the sole effect on the surroundings (everything external to the system) could be the raising of a weight. Notice that the raising of a weight is in effect a force acting through a distance. Notice, also, that our definition does not state that a weight was actually raised, or that a force actually acted through a given distance but that the sole effect external to the system could be the raising of a weight. Work done *by* a system is considered positive and work done *on* a system is considered negative. The symbol W designates the work done by a system.

In general, we will speak of work as a form of energy. No attempt will be made to give a rigorous definition of energy. Rather, since the concept is familiar, the term energy will be used as appropriate, and various forms of energy will be identified. Work is the form of energy that fulfills the definition given above.

Let us illustrate this definition of work with a few examples. Consider as a system the battery and motor of Fig. 4.1*a* and let the motor drive a

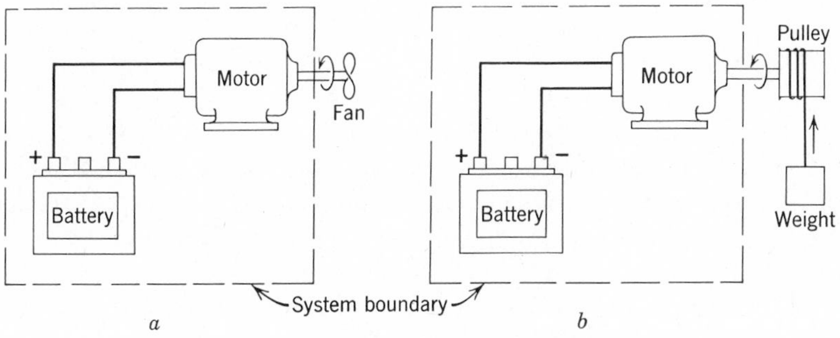

Fig. 4.1 Example of work done at the boundary of a system.

fan. Does work cross the boundary of the system? To answer this question using the definition of work given above, let the fan be replaced with a pulley and weight arrangement shown in Fig.4.1*b*. As the motor turns, the weight is raised, and the sole effect external to the system is the raising of a weight. Thus, for our original system of Fig. 4.1*a*, we conclude that work is crossing the boundary of the system since the sole effect external to the system could be the raising of a weight.

Let the boundaries of the system be changed now to include only the battery shown in Fig. 4.2. Again we ask the question, does work cross the boundary of the system? In answering this question, we will be answering a more general question; namely, does the flow of electrical energy across the boundary of a system constitute work?

The only limiting factor in having the sole external effect the raising of a weight is the inefficiency of the motor. However, as we design a more efficient motor, with lower bearing and electrical losses, we recognize that we can approach a certain limit, which does meet the requirement of having the only external effect the raising of a weight. Therefore, we can conclude that when there is a flow of electricity across the boundary of a system, as in Fig.4.2, it is work with which we are concerned.

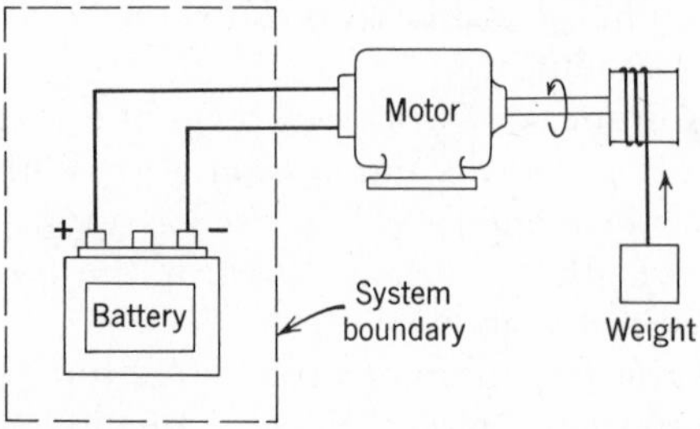

Fig. 4.2 Example of work crossing the boundary of a system because of a flow of an electric current across the system boundary.

4.2 Units for Work

As already noted, we consider work done *by* a system, such as that done by a gas expanding against a piston as positive, and work done *on* a system, such as that done by a piston compressing a gas, as negative. Thus, positive work means that energy leaves the system and negative work means that energy is added to the system.

Our definition of work involves the raising of a weight. The unit of work should therefore be defined in terms of raising a unit weight a given distance at a given location. Let us define our unit of work as the work required to raise a mass of 1 lbm a distance of 1 ft at a location where the acceleration due to gravity is the standard value, 32.174 ft/sec^2. This is exactly equivalent to saying that our unit of work is a force of 1 lbf acting through a distance of 1 ft. This unit for work is called the foot-pound force.

Similarly, in the metric system the unit of work is the erg. An erg is the work done by a force of 1 dyne acting through a distance of 1 cm. This is a very small unit, and for engineering use, a joule is the common unit of work in the metric system. One joule is 10^7 ergs.

Another unit for work that has come into common use as a result of developments in nuclear physics is the electron volt, abbreviated ev. An electron volt is the work required to move an electron through a potential difference of 1 volt. The relation between the electron volt and other units for work is:

$$1 \text{ ev} = 1.602 \times 10^{-12} \text{ erg} = 1.18 \times 10^{-19} \text{ ft-lbf}$$

One million electron volts, abbreviated mev, is also commonly used.

$$1 \text{ mev} = 1.602 \times 10^{-6} \text{ erg} = 1.18 \times 10^{-13} \text{ ft-lbf}$$

Power is the time rate of doing work, and is designated by the symbol $\dot{W}$.

$$\dot{W} \equiv \frac{\delta W}{dt}$$

One familiar unit of power is the horsepower (hp).

$$1 \text{ hp} = 33{,}000 \text{ ft-lbf/min.}$$

Another familiar unit is the kilowatt (kw). Actually the kilowatt is defined in terms of electrical units, but for our purposes we can define the kilowatt as follows:

$$1 \text{ kw} = 44{,}240 \text{ ft-lbf/min.}$$

It follows that

$$1 \text{ hp} = 0.746 \text{ kw}$$

It is often convenient to speak of the work per unit mass of the system. This quantity is designated w and is defined

$$w \equiv \frac{W}{m}$$

These definitions of power lead to two other units of work, the horsepower-hour (hp-hr) and kilowatt-hour (kw-hr). The horsepower-hour is the work done in 1 hr when the power is 1 hp. Similarly, one kilowatt-hour is the work done in 1 hr when the rate of work is 1 kw.

$$1 \text{ hp-hr} = 33{,}000 \times 60 = 1.98 \times 10^6 \text{ ft-lbf} = 2545 \text{ Btu}$$
$$1 \text{ kw-hr} = 44{,}240 \times 60 = 2.654 \times 10^6 \text{ ft-lbf} = 3412 \text{ Btu}$$

4.3 Work Done at the Moving Boundary of a Simple Compressible System in a Quasiequilibrium Process

We have already noted that there are a variety of ways in which work can be done on or by a system. These include work done by a rotating shaft, electrical work, and the work done by the movement of the system boundary, such as the work done in moving the piston in a cylinder. In this section we will consider in some detail the work done at the moving boundary of a simple compressible system during a quasiequilibrium process.

Consider as a system the gas contained in a cylinder and piston, as in Fig.4.3. Let one of the small weights be removed from the piston, causing the piston to move upward a distance dL. We can consider this a quasi-equilibrium process and calculate the amount of work W done by the system during this process. The total force on the piston is PA, where P is the pressure of the gas and A is the area of the piston. Therefore, the work δW is

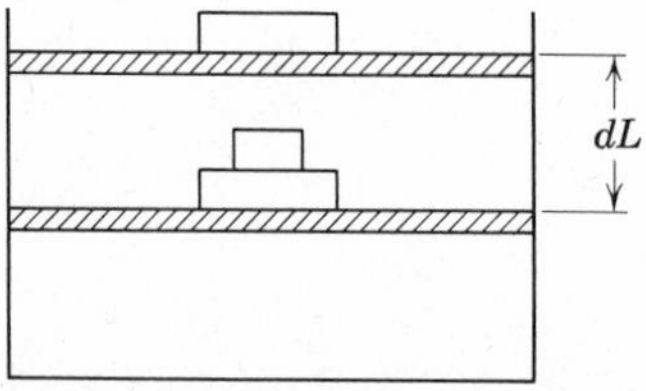

Fig. 4.3 Example of work done at the moving boundary of a system in a quasi-equilibrium process.

$$\delta W = PA\,dL$$

But $A\,dL = dV$, the change in volume of the gas. Therefore,

$$\delta W = P\,dV \tag{4.2}$$

The work done at the moving boundary during a given quasiequilibrium process can be found by integrating Eq.4.2. However, this integration can be performed only if we know the relationship between P and V during this process. This relationship might be expressed in the form of an equation, or it might be shown in the form of a graph.

Let us consider a graphical solution first, using as an example a compression process such as that which occurs during the compression of air in a cylinder, Fig. 4.4. At the beginning of the process the piston is at

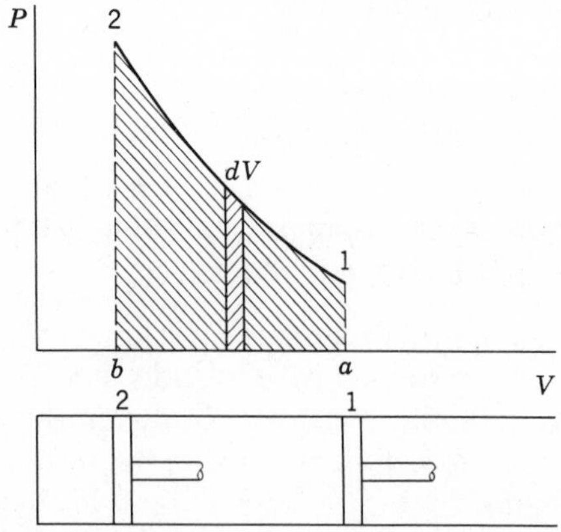

Fig. 4.4 Use of pressure-volume diagram to show work done at the moving boundary of a system in a quasiequilibrium process.

position 1, the pressure being relatively low. This state is represented on a pressure-volume diagram (usually referred to as a *P-V* diagram) as shown. At the conclusion of the process the piston is in position 2, and the corresponding state of the gas is shown at point 2 on the *P-V* diagram. Let us assume that this compression was a quasiequilibrium process, and that during the process the system passed through the states shown by the line connecting states 1 and 2 on the *P-V* diagram. The assumption of a quasiequilibrium process is essential here because each point on line 1-2 represents a definite state, and these states will correspond to the actual state of the system only if the deviation from equilibrium is infinitesimal. The work done on the air during this compression process can be found by integrating Eq.4.2.

$$ {}_1W_2 = \int_1^2 \delta W = \int_1^2 P\,dV \tag{4.3} $$

The symbol ${}_1W_2$ is to be interpreted as the work done during the process from state 1 to state 2. It is clear from examining the *P-V* diagram that the work done during this process, namely,

$$ \int_1^2 P\,dV $$

is represented by the area under the curve 1-2, area *a*-1-2-*b*-*a*. In this example the volume decreased, and the area *a*-1-2-*b*-*a* represents work done on the system. If the process had proceeded from state 2 to state 1 along the same path, the same area would represent work done by the system.

Further consideration of a *P-V* diagram, Fig.4.5, leads to another important conclusion. It is possible to go from state 1 to state 2 along

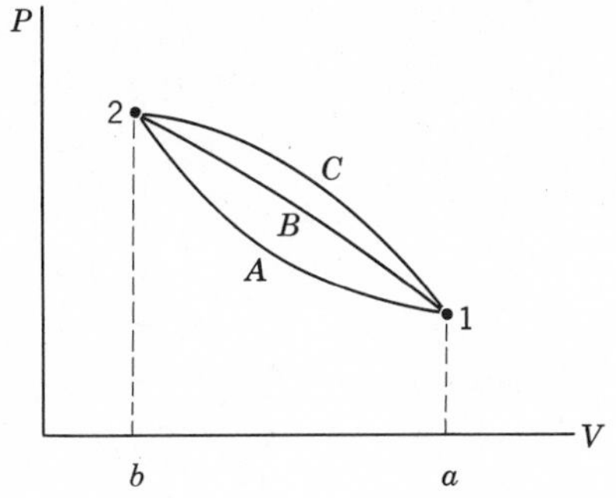

Fig. 4.5 Various quasiequilibrium processes between two given states, indicating that work is a path function.

many different quasiequilibrium paths, such as A, B, or C. Since the area underneath each curve represents the work for each process, it is evident that the amount of work involved in each case is a function not only of the end states of the process, but in addition is dependent on the path that is followed in going from one state to another. For this reason work is called a path function, or in mathematical parlance, δW is an inexact differential.

This leads to a brief consideration of point and path functions or, to use another term, exact and inexact differentials. Thermodynamic properties are point functions, a name that arises from the fact that for a given point on a diagram (such as Fig.4.5) or surface (such as Fig. 3.9), the state is fixed, and thus there is a definite value of each property corresponding to this point. The differentials of point functions are exact differentials, and the integration is simply

$$\int_1^2 dV = V_2 - V_1$$

Thus, we can speak of the volume in state 2 and the volume in state 1, and the change in volume depends only on the initial and final states.

Work on the other hand, is a path function, for, as has been indicated, the work done in a quasiequilibrium process between two given states depends on the path followed. The differentials of path functions are inexact differentials, and the symbol δ will be used in this text to designate inexact differentials (in contrast to d for exact differentials). Thus, for work we would write

$$\int_1^2 \delta W = {}_1W_2$$

It would be more precise to use the notation $({}_1W_{2,A})$ which would indicate the work done during the change from state 1 to 2 along path A. However, implied in the notation ${}_1W_2$ is that the process between states 1 and 2 has been specified. It should be noted, we never speak about the work in the system in state 1 or state 2, and thus we would never write $W_2 - W_1$.

Example 4.1

Consider as a system the gas contained in the cylinder shown in Fig. 4.6, which is fitted with a piston on which a number of small weights are placed. The initial pressure is 20 lbf/in.^{2} and the initial volume of the gas is 1 ft^3.

(*a*) Let a Bunsen burner be placed under the cylinder, and let the volume of the gas increase to 3 ft³ while the pressure remains constant. Calculate the work done by the system during this process.

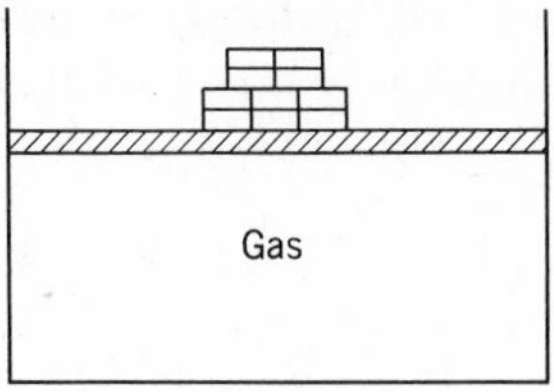

Fig. 4.6 Sketch for Example 4.1.

$$ {}_1W_2 = \int_1^2 P\,dV $$

Since the pressure is constant, we conclude from Eq. 4.3,

$$ {}_1W_2 = P\int_1^2 dV = P(V_2 - V_1) $$

$$ {}_1W_2 = 20\ \text{lbf/in.}^2 \times 144\ \text{in.}^2/\text{ft}^2 \times (3\text{–}1)\,\text{ft}^3 = 5760\ \text{ft-lbf} $$

(*b*) Consider the same system and initial conditions, but at the same time that the Bunsen burner is under the cylinder and the piston is rising, let weights be removed from the piston at such a rate that, during the process, the relation between pressure and volume is given by the expression $PV = \text{constant} = P_1V_1 = P_2V_2$. Let the final volume again be 3 ft³.

We first determine the final pressure.

$$ P_2 = \frac{P_1V_1}{V_2} = 20 \times \frac{1}{3} = 6.67\ \text{lbf/in.}^2 $$

Again we use Eq. 4.3 to calculate the work.

$$ {}_1W_2 = \int_1^2 P\,dV $$

We can substitute $P = \text{constant}/V = P_1V_1/V$ into this equation.

$$ {}_1W_2 = P_1V_1\int_1^2 \frac{dV}{V} = P_1V_1 \ln \frac{V_2}{V_1} $$

$$ {}_1W_2 = 20\ \text{lbf/in.}^2 \times 144\ \text{in.}^2/\text{ft}^2 \times 1\ \text{ft}^3 \times \ln 3 = 3164\ \text{ft-lbf} $$

(*c*) Consider the same system, but during the heat transfer let the weights be removed at such a rate that the expression $PV^{1.3} = \text{constant}$

describes the relation between pressure and volume during the process. Again the final volume is 3 ft^3. Calculate the work.

Let us first solve this problem for the general case of PV^n = constant:

$$PV^n = \text{constant} = P_1V_1{}^n = P_2V_2{}^n$$

$$P = \frac{\text{constant}}{V^n} = \frac{P_1V_1{}^n}{V^n} = \frac{P_2V_2{}^n}{V^n}$$

$$_1W_2 = \int_1^2 P\,dV = \text{constant}\int_1^2 \frac{dV}{V^n} = \text{constant}\left[\frac{V^{-n+1}}{-n+1}\right]_1^2$$

$$= \frac{\text{constant}}{1-n}(V_2^{1-n} - V_1^{1-n}) = \frac{P_2V_2{}^nV_2^{1-n} - P_1V_1{}^nV_1^{1-n}}{1-n}$$

$$_1W_2 = \frac{P_2V_2 - P_1V_1}{1-n}$$

For our problem

$$_1W_2 = \frac{P_2V_2 - P_1V_1}{1-1.3} = \frac{(4.80\times 144\times 3) - (20\times 144\times 1)}{1-1.3}$$

$$= 2688 \text{ ft-lbf}$$

(*d*) Consider the system and initial state given in the first three examples, but let the piston be held by a pin so that the volume remains constant. In addition, let heat be transferred from the system until the pressure drops to 10 lbf/in.^{2}. Calculate the work.

Since $\delta W = P\,dV$ for a quasiequilibrium process, the work is zero, because in this case there is no change in volume.

The process for each of four examples is shown on the *P-V* diagram of Fig. 4.7. Process 1–2*a* is a constant-pressure process, and area 1–2*a*–*f*–*e*–1 represents the work. Similarly, line 1–2*b* represents the process in which PV = constant, line 1–2*c* the process in which $PV^{1.3}$ = constant, and line 1–2*d* represents the constant-volume process. The student should compare the relative areas under each curve with the numerical results obtained above.

4.4 Some Other Systems Involving Work at a Moving Boundary

In the preceding section we considered the work done at the moving boundary of a simple compressible system during a quasiequilibrium

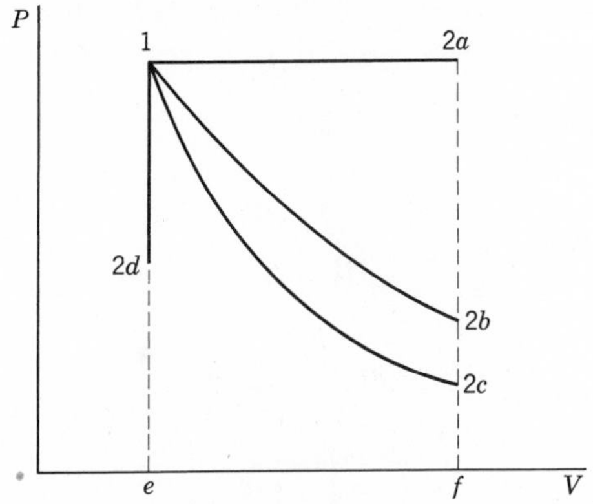

Fig. 4.7 Pressure-volume diagram showing work done in the various processes of Example 4.1.

process. There are other types of systems that involve work at a moving boundary, and in this section we shall briefly consider two such systems, a stretched wire and a surface film.

Consider as a system a stretched wire that is under a given tension $\mathscr{T}$. When the length of the wire changes by the amount dL, the work done by the system is

$$\delta W = -\mathscr{T}\, dL \tag{4.4}$$

The minus sign is necessary because work is done by the system when dL is negative. This can be integrated to give

$${}_1W_2 = -\int_1^2 \mathscr{T}\, dL \tag{4.5}$$

The integration can be performed either graphically or analytically if the relation between $\mathscr{T}$ and L is known. The stretched wire is a simple example of the type of problem in solid body mechanics that involves the calculation of work.

Example 4.2

A metallic wire of initial length L_0 is stretched. Assuming elastic behavior, determine the work done in terms of the modulus of elasticity and the strain.

Let σ = stress, e = strain, and E = modulus of elasticity.

$$\sigma = \frac{\mathscr{T}}{A} = Ee$$

Therefore

$$\mathcal{T} = AEe$$

From the definition of strain,

$$de = \frac{dL}{L_0}$$

Therefore,

$$\delta W = -\mathcal{T}\,dL = -AEeL_0 de$$

$$W = -AEL_0 \int_{e=0}^{e} e\,de = \frac{AEL_0}{2}(e)^2$$

Now consider a system that consists of a liquid film having a surface tension $\mathcal{S}$. A schematic arrangement of such a film is shown in Fig. 4.8,

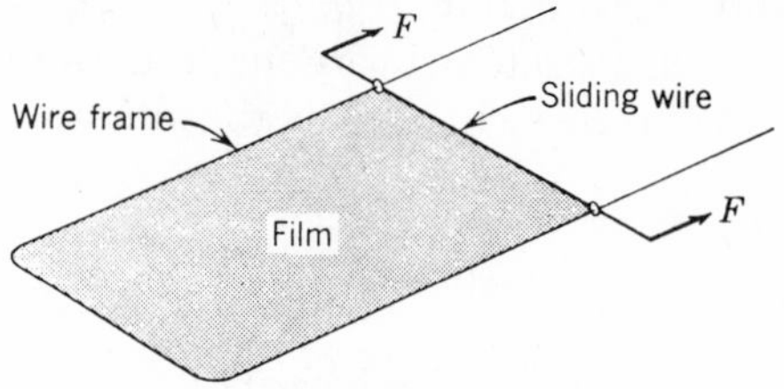

Fig. 4.8 Schematic arrangement showing work done on a surface film.

where a film is maintained on a wire frame, one side of which can be moved. When the area of the film is changed, for example by sliding the movable wire along the frame, work is done on or by the film. When the area changes by an amount dA, the work done by the system is

$$\delta W = -\mathcal{S}\,dA \tag{4.6}$$

For finite changes

$${}_1W_2 = -\int_1^2 \mathcal{S}\,dA \tag{4.7}$$

4.5 Systems That Involve Other Modes of Work

There are systems that involve other modes of work, and in this section we shall consider two of these, namely, systems involving magnetic and systems involving electrical modes of work. We shall consider a quasi-equilibrium process for these systems, and present expressions for the work done during such a process.

In order to visualize how work can be accomplished by magnetic effects, let us briefly describe magnetic cooling, or adiabatic demagnetization, which is a process used to produce temperatures well below 1 K. A temperature of 1.0 K can be produced by pumping a vacuum over a bath of liquid helium (helium has the lowest normal boiling point of any substance, namely 4.2 K at one atmosphere pressure). An apparatus in which the magnetic cooling is accomplished is shown schematically in Fig. 4.9. The paramagnetic salt is the magnetic substance in which

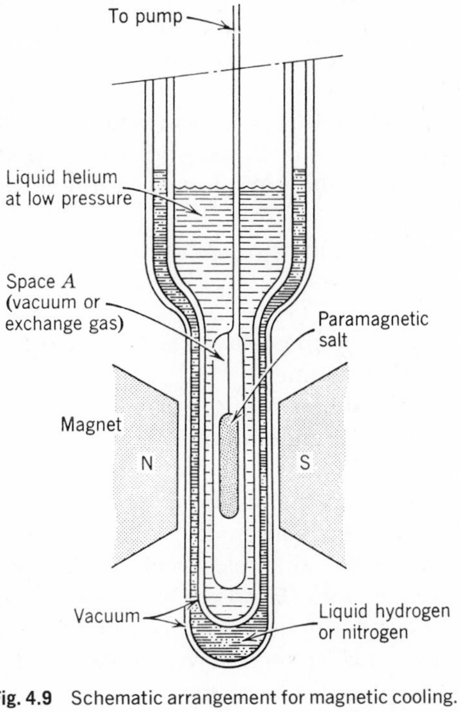

Fig. 4.9 Schematic arrangement for magnetic cooling.

temperatures well below 1 K are achieved. When the magnetic field is slowly increased, work is done on the paramagnetic salt. From a microscopic point of view this work is associated with the fact that in the presence of the magnetic field the ions in the salt tend to align themselves with their magnetic axes in the direction of the field. As a result of this work done on the salt, the temperature of the salt tends to increase. However, at this point in the experiment space A is filled with low pressure helium gas, and heat is transferred from the paramagnetic salt to the liquid helium, which is maintained at about 1 K. When the magnetic field is at full strength, and the paramagnetic salt is at the temperature of the liquid helium, space A is evacuated, thus insulating the paramagnetic salt. The magnetic field is now reduced to zero, and in this process work is done by the paramagnetic salt, and its temperature drops sharply. This entire process may be compared to the compression of a gas that is initially at ambient pressure and temperature. As the result of the compression, the temperature of the gas tends to increase. However, the high pressure gas can be cooled to the ambient temperature. If this gas is now isolated from the surroundings and allowed to expand and do work (against a piston for example) the temperature of the gas will decrease below the ambient temperature during the expansion process.

The basic parameters in this process are the intensity of the magnetic field and the magnetization. It may be shown that in a reversible quasiequilibrium process, the work done on a simple magnetic substance is

$$\delta W = -\mu_0 \mathscr{H} d(V\mathscr{M}) \tag{4.8}$$

where:

μ_0 = permeability of free space
V = volume
$\mathscr{H}$ = intensity of the magnetic field
$\mathscr{M}$ = magnetization

The minus sign indicates that as the magnetization increases, work is done on the simple magnetic substance.

We have already noted that electrical energy flowing across the boundary of a system is work. However, we can gain further insight into such a process by considering a system in which the only work mode is electrical. As an example of such a system we can think of a charged condenser, an electrolytic cell, or the type of fuel cell described in Chapter 1. Consider a quasiequilibrium process for such a system, and during this process let the potential difference be $\mathscr{E}$ and the amount of electrical energy that flows into the system be dZ. For this quasiequilibrium process the work is given by the relation

$$\delta W = -\mathscr{E} dZ \tag{4.9}$$

Since the current, i, equals dZ/dt (where $t =$ time) we can also write

$$\delta W = -\mathscr{E} i dt$$

$$_1W_2 = -\int_1^2 \mathscr{E} i dt \qquad (4.10)$$

Equation 4.10 may also be written as a rate equation for work (the power).

$$\frac{\delta W}{dt} = -\mathscr{E} i \qquad (4.11)$$

This leads to the definition of a unit of power, the watt. A watt is the power developed by a current of 1 ampere flowing through a potential of 1 volt. This is consistent with our earlier observation that electrical energy is work.

4.6 Some Concluding Remarks Regarding Work

The similarity between the expressions for work in the two processes mentioned in Section 4.5 and the three processes involving a moving boundary should be noted. In each of these quasiequilibrium processes the work is given by the integral of the product of an intensive property and the change of an extensive property. These are summarized below:

Simple compressible system $\quad {}_1W_2 = \int_1^2 P dV$

Stretched wire $\quad {}_1W_2 = -\int_1^2 \mathscr{T} dL$

Surface film $\quad {}_1W_2 = -\int_1^2 \mathscr{S} dA \qquad (4.12)$

System involving magnetic work only $\quad {}_1W_2 = -\int_1^2 \mu_0 \mathscr{H} d(V\mathscr{M})$

System involving electrical work only $\quad {}_1W_2 = -\int_1^2 \mathscr{E} dZ$

Although we will deal primarily with systems involving one mode of work, it is quite possible to have more than one work mode involved in a given process. Thus we could write

$$\delta W = PdV - \mathcal{T}dL - \mathcal{S}dA - \mu_0 \mathcal{H} d(V\mathcal{M}) - \mathcal{E}dZ + \cdots \tag{4.13}$$

where the dotted lines represent other products of an intensive property and the derivative of a related extensive property.

It should also be noted that there are many other forms of work which can be identified in processes that are not quasiequilibrium processes. An example of these is the work done by shearing forces in a process involving friction in a viscous fluid or the work done by a rotating shaft that crosses the system boundary.

The identification of work is an important aspect of many thermodynamic problems. We have already noted that work can be identified only at the boundaries of the system. For example, consider Fig. 4.10,

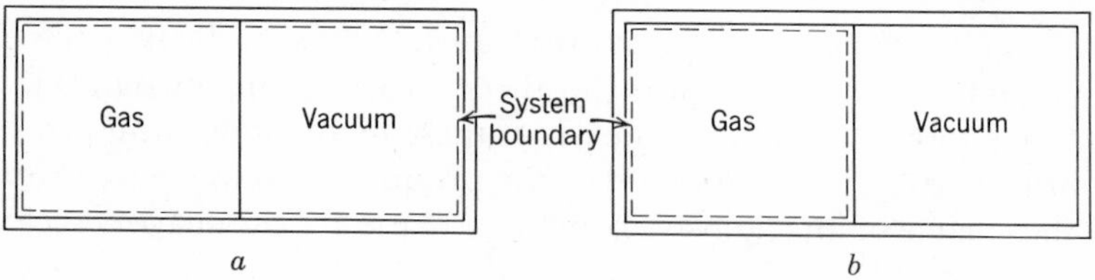

Fig. 4.10 Example of a process involving a change of volume for which the work is zero.

which shows a gas separated from the vacuum by a membrane. Let the membrane rupture and the gas fill the entire volume. Neglecting any work associated with the rupturing of the membrane, we can ask if there is work involved in the process. If we take as our system the gas and the vacuum space, we readily conclude that there is no work involved, since no work can be identified at the system boundary. It we take the gas as a system we do have a change of volume, and we might be tempted to calculate the work from the integral

$$\int_1^2 PdV$$

However, this is not a quasiequilibrium process, and therefore the work cannot be calculated from this relation. Rather, since there is no resistance at the system boundary as the volume increases we conclude that for this system there is no work involved in this process.

Another example can be cited with the aid of Fig. 4.11. In Fig. 4.11*a* the system consists of the container plus the gas. Work crosses the boundary of the system at the point where the system boundary intersects the shaft, and can be associated with the shearing forces in the rotating shaft. In Fig. 4.11*b* the system includes shaft and weight as well as the gas

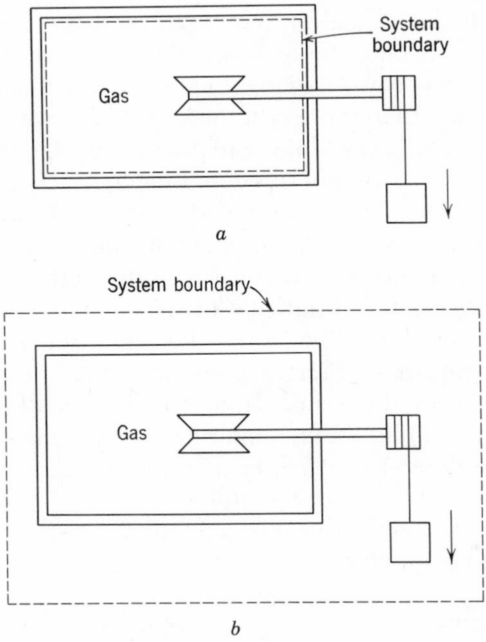

Fig. 4.11 Example showing how selection of the system determines whether work is involved in a process.

and the container. In this case there is no work crossing the system boundary as the weight moves downward. As we will see in the next chapter, we can identify a change of potential energy within the system, but this should not be confused with work crossing the system boundary.

4.7 Definition of Heat

The thermodynamic definition of heat is somewhat different from the everyday understanding of the word. Therefore, it is essential to understand clearly the definition of heat given here, because it is involved in so many thermodynamic problems.

If a block of hot copper is placed in a beaker of cold water, we know from experience that the block of copper cools down and the water warms up until the copper and water reach the same temperature. What causes this decrease in the temperature of the copper and the increase in the temperature of the water? We say that it is the result of the transfer of energy from the copper block to the water. It is out of such a transfer of energy that we arrive at a definition of heat.

Heat is defined as the form of energy that is transferred across the boundary of a system at a given temperature to another system (or the surroundings) at a lower temperature by virtue of the temperature difference between the two systems. That is, heat is transferred from the system at the higher to the system at the lower temperature, and the heat transfer occurs solely because of the temperature difference between the two systems. Another aspect of this definition of heat is that a body never contains heat. Rather heat can be identified only as it crosses the boundary. Thus, heat is a transient phenomenon. If we consider the hot block of copper as one system and the cold water in the beaker as another system, we recognize that originally neither system contains any heat (they do contain energy, of course). When the copper is placed in the water and the two are in thermal communication, heat is transferred from the copper to the water, until equilibrium of temperature is established. At that point we no longer have heat transfer, since there is no temperature difference. Neither of the systems contain heat at the conclusion of the process. It also follows that heat is identified at the boundary of the system, for heat is defined as energy being transferred across the system boundary.

4.8 Units of Heat

We must have units for heat, as for all other quantities in thermodynamics. Consider as a system 1 lbm water at 59.5 F, and let a block of hot copper be placed in the water. Let the block of copper have such a mass and such a temperature that when thermal equilibrium is established the temperature of the water is 60.5 F. We define as our unit of heat the quantity of heat transferred from the copper to the water, and call the unit of heat the British thermal unit, which is abbreviated Btu. More specifically, this is called the 60-degree Btu, which may be defined as the quantity of heat required to raise 1 lbm of water from 59.5 F to 60.5 F.[1]

Similarly, a calorie can be identified as the amount of heat required to raise the temperature of 1 gram of water from 14.5 C to 15.5 C.

Further, heat transferred *to* a system is considered to be positive, and

[1]Actually the Btu as used today is defined in terms of electrical units. This point is explained in the next chapter.

heat transferred *from* a system, negative. Thus, positive heat represents energy transferred to a system, and negative heat represents energy transferred from a system. The symbol Q is used to represent heat.

A process in which there is no heat transfer ($Q = 0$) is called an adiabatic process.

From a mathematical perspective, heat, like work, is a path function and is recognized as an inexact differential. That is, the amount of heat transferred when a system undergoes a change of state from state 1 to state 2 depends on the path that the system follows during the change of state. Since heat is an inexact differential, the differential is written δQ. On integrating, we write

$$\int_1^2 \delta Q = {}_1Q_2$$

In words, ${}_1Q_2$ is the heat transferred during the given process between state 1 and state 2.

The rate at which heat is transferred to a system is designated by the symbol $\dot{Q}$.

$$\dot{Q} \equiv \frac{\delta Q}{dt}$$

It is also convenient to speak of the heat transfer per unit mass of the system, q, which is defined as

$$q \equiv \frac{Q}{m}$$

4.9 Comparison of Heat and Work

At this point it is evident that there are many similarities between heat and work, and these are summarized here.

1. Heat and work are both transient phenomena. Systems never possess heat or work, but either or both cross the system boundary when a system undergoes a change of state.
2. Both heat and work are boundary phenomena. Both are observed only at the boundaries of the system, and both represent energy crossing the boundary of the system.
3. Both heat and work are path functions and inexact differentials.

The definitions of heat and work in this chapter have been from a macroscopic viewpoint, and are based on the difference that can be produced in the surroundings when heat crosses the boundary of a system, as compared to work. In Chapter 17 we shall examine heat and work from a molecular viewpoint. We will find that from this viewpoint, a given amount of energy transferred to a system as work can produce a different effect within the system than the same amount of energy transferred as heat. Thus, from a molecular point of view, the definitions of work and heat need not be related to the surroundings, but will be made on the basis of what occurs within the system.

It should also be noted that in our sign convention, $+Q$ represents heat transferred *to* the system, and thus is energy added to the system, and $+W$ represents work done *by* the system and thus represents energy leaving the system.

A final illustration may be helpful to indicate the difference between heat and work. Figure 4.12 shows a gas contained in a rigid vessel.

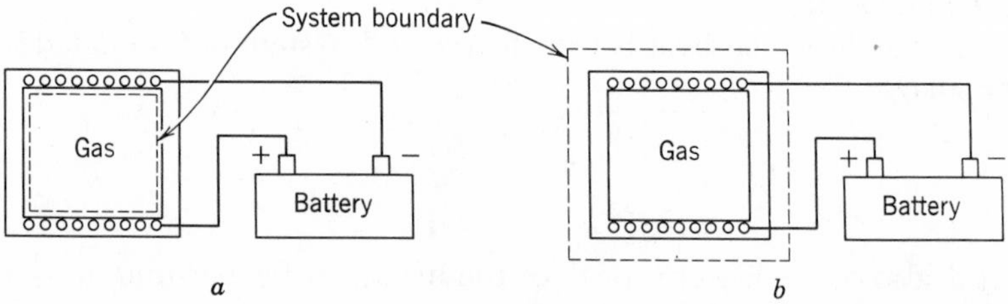

Fig. 4.12 An example showing the difference between heat and work.

Resistance coils are wound around the outside of the vessel. When current flows through the resistance coils, the temperature of the gas increases. Which crosses the boundary of the system, heat or work?

In Fig. 4.12*a* we consider only the gas as the system. In this case the energy crosses the boundary of the system because the temperature of the walls is higher than the temperature of the gas. Therefore, we recognize that heat crosses the boundary of the system.

In Fig. 4.12*b* the system includes the vessel and the resistance heater. Electricity crosses the boundary of the system, and as indicated earlier, this is work.

PROBLEMS

4.1 A cylinder fitted with a piston contains 5 lbm of saturated water vapor at a pressure of 120 lbf/in.^{2}. The steam is heated until the temperature is 500 F. During this process the pressure remains constant. Calculate the work done by the steam during the process.

4.2 One-tenth lbm of oxygen is contained in a cylinder fitted with a piston. The initial conditions are 20 lbf/in.^{2}, 70 F. Weights are then added to the piston, and the O_2 is slowly compressed isothermally until the final pressure is 65 lbf/in.^{2}. Calculate the work done during this process.

4.3 A cylinder in which the piston is restrained by a spring contains 1 ft^3 of air at a pressure of 15 lbf/in.^{2}, which just balances the atmospheric pressure of 15 lbf/in.^{2}. Assume that the weight of the piston is negligible. In this initial state, the spring exerts no force on the piston. The gas is then heated until the volume is doubled. The final pressure of the gas is 50 lbf/in.^{2}, and during the process the spring exerts a force which is proportional to the displacement of the piston from the initial position.

(*a*) Show this process on a *P-V* diagram.

(*b*) Considering the gas as the system, calculate the total work done by the system.

(*c*) Of the total work, how much is done against the atmosphere? How much against the spring?

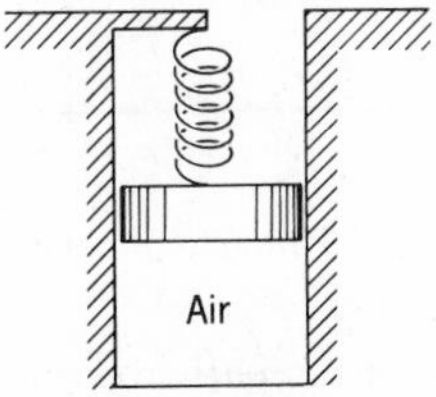

Fig. 4.13 Sketch for Problem 4.3.

4.4 The cylinder-piston arrangement shown in Fig. 4.14 contains carbon dioxide at 40 lbf/in.^{2}, 300 F, at which point the volume is 3 ft^3. Weights are then removed at such a rate that the gas expands according to the relation

$$PV^{1.2} = \text{constant}$$

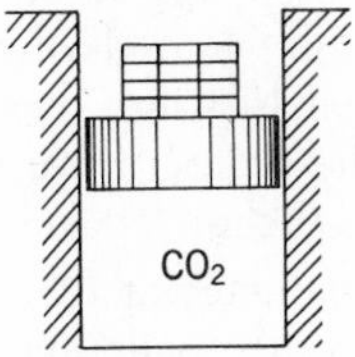

Fig. 4.14 Sketch for Problem 4.4.

until the final temperature is 200 F. Determine the work done during this process.

4.5 A balloon which is initially flat is inflated by filling it with air from a tank of compressed air. The final volume of the balloon is 80 ft^3. The barometer reads 29.2 in. Hg. Consider the tank, the balloon, and the connecting pipe as a system. Determine the work for this process.

4.6 The vertical cylinder shown in Fig. 4.15 contains 0.185 lbm of H_2O at 100 F.

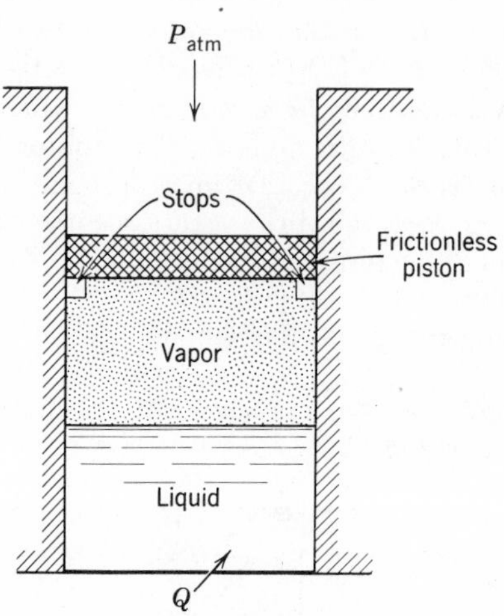

Fig. 4.15 Sketch for Problem 4.6.

The initial volume enclosed beneath the piston is 0.65 ft³. The piston has an area of 60 in.² and a mass of 125 lbm. Initially the piston rests on the stops as shown. The atmospheric pressure is 14.0 lbf/in.² and the gravitational acceleration is 30.9 ft/sec². Heat is then transferred to the steam until the cylinder contains saturated vapor.

(*a*) What is the temperature of the H_2O when the piston first rises from the stops?

(*b*) How much work is done by the steam during the entire process?

(*c*) Show the process on a *T-V* diagram.

4.7 The gas space above the water in a closed tank contains nitrogen at 80 F, 15 lbf/in.². The tank has a total volume of 100 ft³ and contains 1000 lbm of water at 80 F. An additional 1000 lbm of water is now slowly forced into the tank. Assuming that the temperature remains constant, calculate the final pressure of the N_2, and the work done on the N_2 during the process.

4.8 A spherical balloon has a diameter of 10 in., and contains air at a pressure of 20 lbf/in.². The diameter of the balloon increases to 12 in. due to heating, and during this process the pressure is proportional to the diameter. Calculate the work done by the air during this process.

4.9 A spherical balloon having a radius of 20 feet is to be filled with helium from a bank of high pressure gas cylinders that contain helium at 2000 lbf/in.², 80 F. The balloon is initially flat, and the atmospheric pressure is 29.86 in. Hg.

(*a*) How much work is done against the atmosphere as the balloon is inflated? Assume no stretching of the material from which the balloon is made and that the pressure in the balloon is essentially equal to the atmospheric pressure.

(*b*) What is the required volume of the high pressure cylinders, if the final pressure in the cylinders is the same as that in the balloon?

4.10 Ammonia is compressed in a cylinder by a piston. The initial temperature is 100 F, the initial pressure is 60 lbf/in.^{2}, and the final pressure is 180 lbf/in.^{2}. The following data are available for this process:

Pressure, lbf/in.^{2}	Volume, in.^{3}
60	80.0
80	67.5
100	60.0
120	52.5
140	45.0
160	37.5
180	32.5

(*a*) Determine the work for the process considering the ammonia as the system.

(*b*) What is the final temperature of the ammonia?

4.11 During static tests of rocket motors a compressed gas is often used to force the propellants into the combustion chamber. Consider the arrangement shown in Fig. 4.16. Compressed air is used to expel the liquid propellant from the propellant tank. The initial pressure of the air is 3000 lbf/in.^{2} and the initial temperature is 100 F. The propellant has a density of 70 lbm/ft^3

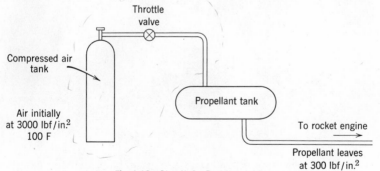

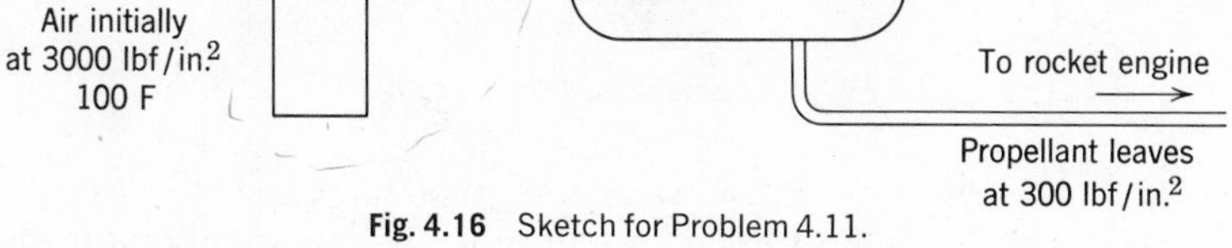

Fig. 4.16 Sketch for Problem 4.11.

and the propellant tank is filled to capacity and contains 2000 lbm of propellant. The propellant leaves at constant pressure of 300 lbf/in.^{2}. Considering the air as the system, determine the work done by the air in forcing the propellant from the propellant tank.

4.12 Saturated water vapor at 400 F is contained in a cylinder fitted with a piston. The initial volume of the steam is 0.3 ft^3. The steam then expands in a quasiequilibrium, isothermal process until the final pressure is 20 lbf/in.^{2}, and in so doing does work against the piston.

(*a*) Determine the work done during this process.

(*b*) How much error would be made by assuming the steam to behave as an ideal gas?

4.13 Tank *A* has a volume of 10 ft^3 and contains argon at 35 lbf/in.^{2}, 90 F. Cylinder *B* contains a frictionless piston of a mass such that a pressure of 20 lbf/in.^{2} inside the cylinder is required to raise the piston. The valve connecting the two is now opened, allowing argon to flow into the cylinder. Eventually, the argon is at a uniform state of 90 F, 20 lbf/in.^{2} throughout. Calculate the work done by the argon during the process.

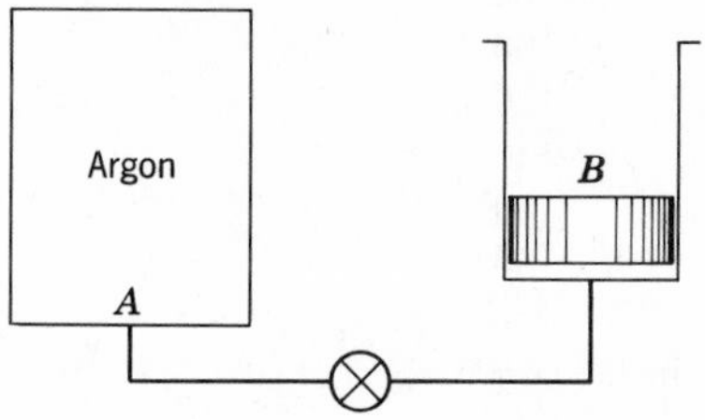

Fig. 4.17 Sketch for Problem 4.13.

4.14 A vertical cylinder fitted with a piston contains 0.1 ft^3 of Freon-12 at 80 F and a quality of 90%. The piston has a mass of 200 lbm, cross-sectional area of 10 in.^{2}, and is held in place by a pin, as shown in Fig. 4.18. The ambient pressure is 15 lbf/in.^{2}.

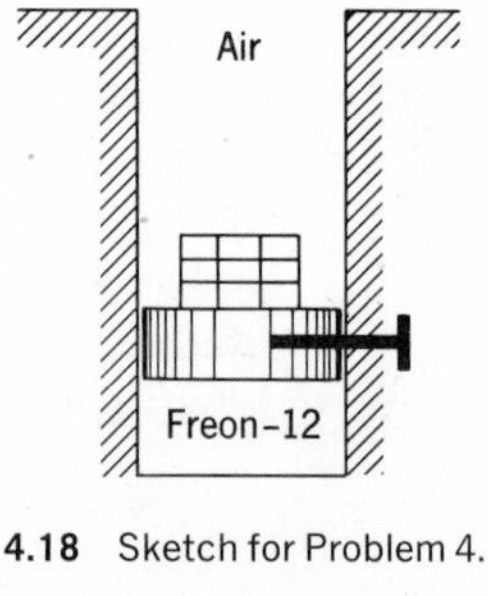

Fig. 4.18 Sketch for Problem 4.14.

The pin is now removed, allowing the piston to move. After a period of time, the system comes to equilibrium, with the final temperature being 80 F.

(*a*) Determine the final pressure and volume of the Freon-12.

(*b*) Calculate the work done by the Freon-12 during this process, and explain what this work is done against.

4.15 The cylinder indicated in Fig. 4.19 is fitted with a piston which is restrained

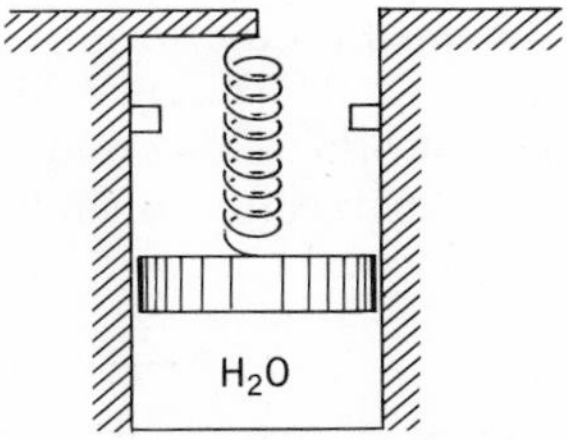

Fig. 4.19 Sketch for Problem 4.15.

by a spring so arranged that for zero volume in the cylinder the spring is fully extended. The spring force is proportional to the spring displacement and the weight of the piston is negligible. The enclosed volume in the cylinder is 4.5 ft^3 when the piston encounters the stops. The cylinder contains 10 lbm of water initially at 50 lbf/in.^{2}, 1% quality and the water is then heated until it exists as saturated vapor. Show the process on a *P-V* diagram which includes the saturation region and determine:

(*a*) The final pressure.

(*b*) The work.

4.16 A metallic wire of initial length L_0 is stretched in the elastic region. Determine the work done in terms of the modulus of elasticity and the strain.

4.17 A sealed vessel having the shape of a rectangular prism with the area of the base A and height L_2, and of negligible mass, is initially floating on a liquid of density ρ, Fig. 4.20. Derive an expression in terms of the given

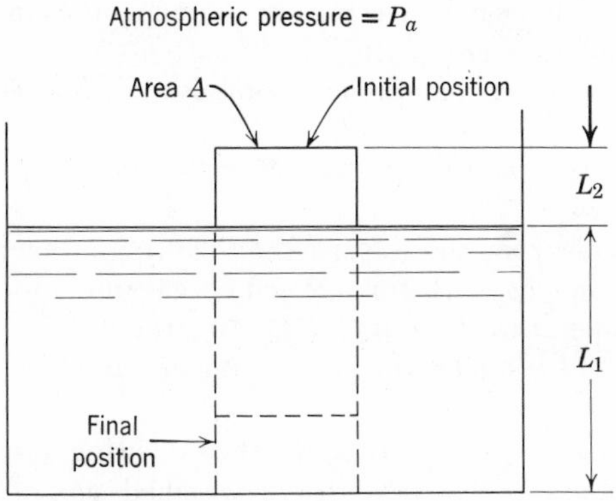

Fig. 4.20 Sketch for Problem 4.17.

variables for the work required to move the vessel to the bottom of a very large tank in which depth of the liquid is L_1.

4.18 Repeat Problem 4.17 assuming that the sealed tank has a mass m.

4.19 At 20 C methanol has a surface tension of 22.6 dynes/cm. Suppose that a film of methanol is maintained on the wire frame as shown in Fig. 4.21, one side of which can be moved. The original dimensions of the wire frame are as shown. Determine the work done (consider the film to be the system), when the wire is moved 1 cm in the direction indicated.

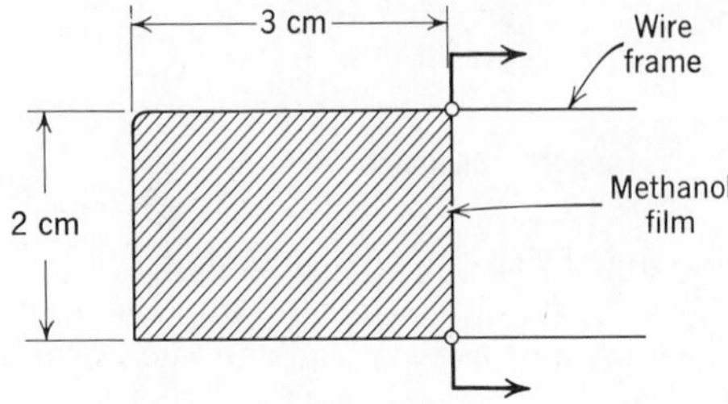

Fig. 4.21 Sketch for Problem 4.19.

4.20 A storage battery is well insulated (thermally) while it is being charged. The charging voltage is 12.3 v and the current is 24.0 amp. Considering the storage battery as the system, what are the heat transfer and work in a 15 min period?

4.21 A room is heated with steam radiators on a winter day. Examine the following systems regarding heat transfer (including sign):

(*a*) The radiator.

(*b*) The room.

(*c*) The radiator and the room.

4.22 Consider a hot-air heating system for a home and examine the following systems for heat transfer:

(*a*) The combustion chamber and combustion gas side of the heat transfer area.

(*b*) The furnace as a whole including the hot and cold air ducts and chimney.

4.23 Write a computer program to solve the following problem. Determine the boundary movement work for a specified gas undergoing an isothermal process at temperature T from v_1 to v_2 using the Beattie–Bridgeman equation of state, and compare the result with that found assuming ideal gas behavior.

4.24 Write a computer program to solve the following problem. Determine the boundary movement work for a specified substance undergoing a process, given a set of data (values of pressure and corresponding volume during the process).

5

The First Law of Thermodynamics

Having completed our consideration of basic definitions and concepts we are ready to proceed to a discussion of the first law of thermodynamics. Often this law is called the law of the conservation of energy, and as we shall see later, this is essentially true. Our procedure will be to state this law first for a system undergoing a cycle, and then for a change of state of a system. The law of the conservation of matter will also be considered in this chapter.

5.1 The First Law of Thermodynamics for a System Undergoing a Cycle

The first law of thermodynamics states that during any cycle a system undergoes, the cyclic integral of the heat is proportional to the cyclic integral of the work.

To illustrate this law, consider as a system the gas in the container shown in Fig. 5.1. Let this system go through a cycle that is comprised of two processes. In the first process work is done on the system by the paddle that turns as the weight is lowered. Let the system then be returned to its initial state by transferring heat from the system until the cycle has been completed.

We have already noted that we can measure the work in foot-pounds force and the heat in Btu. Let measurements of work and heat be made during such a cycle for a wide variety of systems and for various amounts of work and heat. When the amount of work and heat are compared, we

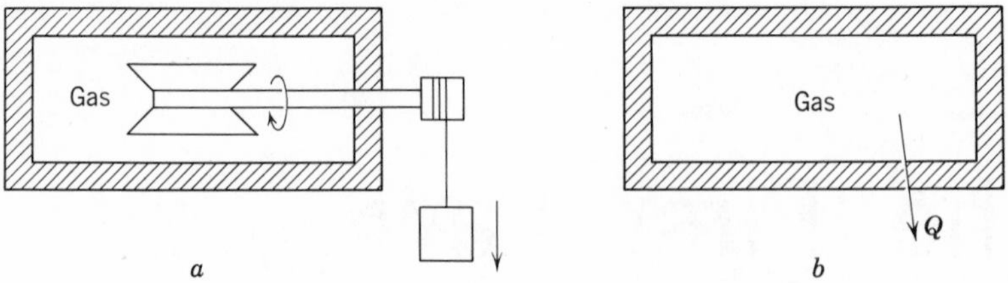

Fig. 5.1 Example of a system undergoing a cycle.

find that these two are always proportional. Observations such as this have led to the formulation of the first law of thermodynamics, which in equation form is written

$$J\oint \delta Q = \oint \delta W \tag{5.1}$$

The symbol $\oint \delta Q$, which is called the cyclic integral of the heat transfer, represents the net heat transfer during the cycle, and $\oint \delta W$, the cyclic integral of the work, represents the net work during the cycle. J is a proportionality factor. Equation 5.1 states that the cyclic integral of the work is proportional to the cyclic integral of the heat transfer.

The basis of every law of nature is experimental evidence, and this is true also of the first law of thermodynamics. Every experiment that has been conducted thus far has verified the first law either directly or indirectly. It has never been disproved, and has been verified by many different experiments.

Thus far we have used the unit foot-pounds for work and Btu for heat. We might also use other units, such as joules for work and calories for heat. The magnitude of the proportionality constant J in Eq. 5.1 will depend, of course, on the units used for heat and work. James P. Joule (1818–1889) did the first accurate work in the 1840's on measurement of the proportionality factor J. The generally accepted value for J, based on the 60-degree Btu and the foot-pound force, is that $J = 778.2$ ft-lbf/Btu.

However, at the International Steam Table Conference in 1929, the Btu was, in effect, defined in terms of the foot-pound force by the following relation:

$$1 \text{ Btu} = 778.26 \text{ ft-lbf}$$

This definition fixes the magnitude of the Btu in terms of previously defined units, and is called the International British thermal unit. Throughout the rest of the text this will be the British thermal unit with which we are concerned whenever the term is used. Thus, when using the international Btu, the question is not to accurately determine the value of J for the 60 F Btu, but rather how many International Btu's are required to raise 1 lbm of water from 59.5 F to 60.5 F.

For much engineering work, however, the accuracy of other data does not warrant more accuracy than the relation 778 ft-lbf = 1 Btu, and, in general, this relation will be used in the examples given in this text.

It is also evident from this discussion that heat and work can be expressed in the same units. Thus, it is perfectly correct to speak of Btu of work and ft-lbf of heat, or calories of work and joules of heat. Therefore we can write Eq. 5.1 without the proportionality factor J.

$$\oint \delta Q = \oint \delta W \tag{5.2}$$

The implication of the equation thus written is that heat and work are expressed in the same units, and in fact can be expressed in any of the units that are used for energy. In this text, therefore, the proportionality factor J will not be written into equations, but the student should realize that each equation must have consistent units throughout.

We also note that

$$1 \text{ hp} = 33{,}000 \text{ ft-lbf/min} = 42.4 \text{ Btu/min} = 2545 \text{ Btu/hr}$$
$$1 \text{ kw} = 44{,}240 \text{ ft-lbf/min} = 56.9 \text{ Btu/min} = 3412 \text{ Btu/hr}$$

5.2 The First Law of Thermodynamics for a Change in State of a System

Equation 5.2 states the first law of thermodynamics for a system during a cycle. Many times, however, we are concerned with a process rather than a cycle, and we now consider the first law of thermodynamics for a system that undergoes a change of state. This can be done by introducing a new property, the energy, which is given the symbol E. Consider a system that undergoes a cycle, changing from state 1 to state 2 by process A, and returning from state 2 to state 1 by process B. This cycle is shown in Fig. 5.2 on a pressure (or other intensive property)-volume (or other extensive property) diagram. From the first law of thermodynamics, Eq. 5.2,

$$\oint \delta Q = \oint \delta W$$

Considering the two separate processes we have

$$\int_{1A}^{2A} \delta Q + \int_{2B}^{1B} \delta Q = \int_{1A}^{2A} \delta W + \int_{2B}^{1B} \delta W$$

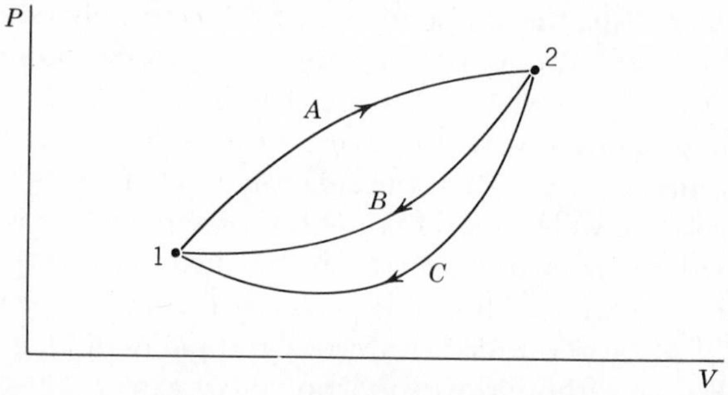

Fig. 5.2 Demonstration of the existence of thermodynamic property *E*.

Now consider another cycle, the system changing from state 1 to state 2 by process *A*, as before, and returning to state 1 by process *C*. For this cycle we can write

$$\int_{1A}^{2A} \delta Q + \int_{2C}^{1C} \delta Q = \int_{1A}^{2A} \delta W + \int_{2C}^{1C} \delta W$$

Subtracting the second of these equations from the first, we have

$$\int_{2B}^{1B} \delta Q - \int_{2C}^{1C} \delta Q = \int_{2B}^{1B} \delta W - \int_{2C}^{1C} \delta W$$

or, by rearranging

$$\int_{2B}^{1B} (\delta Q - \delta W) = \int_{2C}^{1C} (\delta Q - \delta W) \tag{5.3}$$

Since *B* and *C* represent arbitrary processes between states 1 and 2, we conclude that the quantity $(\delta Q - \delta W)$ is the same for all processes between state 1 and state 2. Therefore, $(\delta Q - \delta W)$ depends only on the initial and final states and not on the path followed between the two states. We conclude that this is a point function, and therefore is the differential of a property of the system. This property is the energy of the system and is given the symbol *E*. Thus, we can write

$$\delta Q - \delta W = dE$$
$$\delta Q = dE + \delta W \tag{5.4}$$

Note that since *E* is a property, its derivative is written *dE*. When Eq. 5.4 is integrated from an initial state 1 to the final state 2, we have

$${}_1Q_2 = E_2 - E_1 + {}_1W_2 \tag{5.5}$$

where ${}_1Q_2$ is the heat transferred to the system during the process from state 1 to state 2, E_1 and E_2 are the initial and final values of the energy E of the system, and ${}_1W_2$ is the work done by the system during the process.

The physical significance of the property E is that it represents all the energy of the system in the given state. This energy might be present in a variety of forms, such as the kinetic or potential energy of the system as a whole with respect to the chosen coordinate frame, energy associated with the motion and position of the molecules, energy associated with the structure of the atom, chemical energy such as is present in a storage battery, energy present in a charged condenser, or in any of a number of other forms.

In the study of thermodynamics it is convenient to consider the bulk kinetic and potential energy separately and then to consider all the other energy of the system in a single property which we call the internal energy, and to which we give the symbol U. Thus, we would write

$$E = \text{Internal energy} + \text{Kinetic energy} + \text{Potential energy}$$

or

$$E = U + \text{KE} + \text{PE}$$

The reason for doing this is that the kinetic and potential energy of the system are associated with the coordinate frame that we select and can be specified by the macroscopic parameters of mass, velocity, and elevation. The internal energy U includes all other forms of energy of the system and is associated with the thermodynamic state of the system. In other words, U includes those three general forms of energy discussed in Section 2.6, and constitutes the total energy stored by a system at rest and not subject to external forces.

Since the terms comprising E are point functions, we can write

$$dE = dU + d(\text{KE}) + d(\text{PE}) \tag{5.6}$$

The first law of thermodynamics for a change of state of a system may therefore be written

$$\delta Q = dU + d(\text{KE}) + d(\text{PE}) + \delta W \tag{5.7}$$

In words this equation states that as a system undergoes a change of state, energy may cross the boundary as either heat or work, and each may be positive or negative. The net change in the energy of the system will be exactly equal to the net energy that crosses the boundary of the system. The energy of the system may change in any of three ways, namely, by a change in internal energy, kinetic energy, or potential energy. Since the mass of a system is fixed, we can also say that a quantity of matter may have energy in three forms, internal energy, kinetic energy, or potential energy.

This section will be concluded by deriving an expression for the kinetic and potential energy of a system. Consider first a system that is initially at rest relative to the earth, which is taken as the coordinate frame, and let this system be acted upon by an external horizontal force F which moves the system a distance dx in the direction of the force. Thus there is no change in potential energy. Let there be no heat transfer and no change in internal energy. Then, from the first law, Eq. 5.7,

$$\delta W = -F\,dx = -d\text{KE}$$

But

$$F = \frac{ma}{g_c} = \frac{m}{g_c}\frac{d\mathsf{V}}{dt} = \frac{m}{g_c}\frac{dx}{dt}\frac{d\mathsf{V}}{dx} = \frac{m}{g_c}\mathsf{V}\frac{d\mathsf{V}}{dx}$$

Then

$$d\text{KE} = F dx = \frac{m}{g_c}\mathsf{V}\,d\mathsf{V}$$

Integrating, we obtain

$$\int_{\text{KE}=0}^{\text{KE}} d\text{KE} = \int_{\mathsf{V}=0}^{\mathsf{V}} \frac{m}{g_c}\mathsf{V}\,d\mathsf{V}$$

$$\text{KE} = \frac{1}{2}\frac{m\mathsf{V}^2}{g_c} \tag{5.8}$$

An expression for potential energy can be found in a similar manner. Consider a system that is initially at rest and at the elevation of some reference level. Let this system be acted upon by a vertical force F which is of such magnitude that it raises (in elevation) the system with constant velocity an amount dZ. Let the acceleration due to gravity at this point be g. From the first law, Eq. 5.7

$$\delta W = -F dZ = -d\text{PE}$$

$$F = \frac{ma}{g_c} = \frac{mg}{g_c}$$

Then

$$d\text{PE} = F dZ = \frac{mg}{g_c}dZ$$

Integrating

$$\int_{\text{PE}_1}^{\text{PE}_2} d\text{PE} = m\int_{Z_1}^{Z_2} \frac{g}{g_c}dZ$$

Assuming that g does not vary with Z (which is a very reasonable assumption for moderate changes in elevation)

$$PE_2 - PE_1 = \frac{mg}{g_c}(Z_2 - Z_1) \tag{5.9}$$

Substituting these expressions for kinetic and potential energy into Eq. 5.6 we have

$$dE = dU + \frac{m}{g_c}\mathsf{V}\,d\mathsf{V} + \frac{mg}{g_c}dZ$$

Integrating for a change of state from state 1 to state 2 with constant g we have

$$E_2 - E_1 = U_2 - U_1 + \frac{m\mathsf{V}_2^2}{2g_c} - \frac{m\mathsf{V}_1^2}{2g_c} + \frac{mg}{g_c}Z_2 - \frac{mg}{g_c}Z_1$$

Similarly, substituting these expressions for kinetic and potential energy into Eq. 5.7 we have

$$\delta Q = dU + \frac{d(m\mathsf{V}^2)}{2g_c} + d\left(\frac{mgZ}{g_c}\right) + \delta W \tag{5.10}$$

In the integrated form this equation is, assuming g is a constant,

$${}_1Q_2 = U_2 - U_1 + \frac{m(\mathsf{V}_2^2 - \mathsf{V}_1^2)}{2g_c} + \frac{mg}{g_c}(Z_2 - Z_1) + {}_1W_2 \tag{5.11}$$

Three observations should be made regarding this equation. The first is that the property E, the energy of the system, was found to exist, and we were able to write the first law for a change of state using Eq. 5.5. However, rather than deal with this property E, we find it more convenient to consider the internal energy and the kinetic and potential energies of the system. In general, this will be the procedure followed in the rest of this book.

The second observation is that Eqs. 5.10 and 5.11 are in effect a statement of the conservation of energy. The net change of the energy of the system is always equal to the net transfer of energy across the system boundary as heat and work. This is somewhat analogous to a joint checking account which a man might have with his wife. There are two ways in which deposits and withdrawals can be made, either by the man or by his wife, and the balance will always reflect the net amount of the transaction. Similarly, there are two ways in which energy can cross the boundary of a system, either as heat or work, and the energy of the system will change by the exact amount of the net energy crossing the system boundary. The concept of energy and the law of the conservation of energy are basic to thermodynamics.

The third observation is that Eqs. 5.10 and 5.11 can give only changes in internal energy, kinetic energy, and potential energy. We can learn nothing about absolute values of these quantities from these equations. If we wish to assign values to internal energy, kinetic energy, and potential energy we must assume reference states and assign a value to the quantity in this reference state. The kinetic energy of a body with zero velocity relative to the earth is assumed to be zero. Similarly, the value of the potential energy is assumed to be zero when the body is at some reference elevation. With internal energy, therefore, we must also have a reference state if we wish to assign values of this property. This matter is considered in the following section.

5.3 Internal Energy—A Thermodynamic Property

Internal energy is an extensive property, since it depends upon the mass of the system. Similarly, the kinetic and potential energies are extensive properties.

The symbol U designates the internal energy of a given mass of a substance. Following the convention used with other extensive properties the symbol u designates the internal energy per unit mass. We could speak of u as the specific internal energy, as we do in the case of specific volume. However, since the context will usually make it clear whether u or U is referred to, we will simply use the term internal energy to refer to both internal energy per unit mass and the total internal energy.

In Chapter 3 it was noted that in the absence of the effects of motion, gravity, surface effects, electricity, and magnetism, the state of a pure substance is specified by two independent properties. It is very significant that with these restrictions, the internal energy may be one of the independent properties of a pure substance. This means, for example, that if we specify the pressure and internal energy (with reference to an arbitrary base) of superheated steam, the temperature is also specified.

Thus, in a table of thermodynamic properties such as the steam tables, the value of internal energy can be tabulated along with other thermodynamic properties. Tables 1 and 2 of the steam tables by Keenan et al (Appendix Tables B.1.1 and B.1.2) list the internal energy for saturated states. Included are the internal energy of saturated liquid u_f, the internal energy of saturated vapor u_g, and the difference between the internal energy of saturated liquid and saturated vapor u_{fg}. The values are given in relation to an arbitrarily assumed reference state, which will be discussed later. The internal energy of saturated steam of a given quality is calculated in the same way specific volume is calculated. The relations are

$$u = (1-x)u_f + xu_g$$
$$u = u_f + xu_{fg}$$
$$u = u_g - (1-x)u_{fg}$$

As an example, the specific internal energy of saturated steam having a pressure of 80 lbf/in.^{2} and a quality of 95 per cent can be calculated as follows:

$$u = u_g - (1-x)u_{fg}$$
$$u = 1102.6 - 0.05(820.6) = 1061.6 \text{ Btu/lbm}$$

Example 5.1

A tank containing a fluid is stirred by a paddle wheel. The work input to the paddle wheel is 5090 Btu. The heat transfer from the tank is 1500 Btu. Considering the tank and the fluid as the system, determine the change in the internal energy of the system. The first law of thermodynamics is (Eq. 5.11)

$${}_1Q_2 = U_2 - U_1 + m\frac{(\mathsf{V}_2^2 - \mathsf{V}_1^2)}{2g_c} + \frac{mg}{g_c}(Z_2 - Z_1) + {}_1W_2$$

Since there is no change in kinetic and potential energy, this reduces to

$${}_1Q_2 = U_2 - U_1 + {}_1W_2$$
$$-1500 = U_2 - U_1 - 5090$$
$$U_2 - U_1 = 3590 \text{ Btu}$$

Example 5.2

Consider a system composed of a stone having a mass of 10 lbm and a bucket containing 100 lbm of water. Initially the stone is 77.8 ft above the water and the stone and water are at the same temperature. The stone then falls into the water.

Determine ΔU, ΔKE, ΔPE, Q, and W for the following changes of state.

1. The stone is about to enter the water.
2. The stone has just come to rest in the bucket.

3. Heat has been transferred to the surroundings in such an amount that the stone and water are at the same temperature they were initially.

The first law of thermodynamics is

$$ {}_1Q_2 = U_2 - U_1 + m\frac{(\mathsf{V}_2^{\,2} - \mathsf{V}_1^{\,2})}{2g_c} + \frac{mg}{g_c}(Z_2 - Z_1) + {}_1W_2 $$

1. The stone is about to enter the water: assuming no heat transfer to or from the stone as it falls, we conclude that during the change from the initial state to the state that exists at the moment the stone enters the water,

$$ {}_1Q_2 = 0 \qquad {}_1W_2 = 0 \qquad \Delta U = 0 $$

Therefore the first law reduces to

$$ -\Delta \text{KE} = \Delta \text{PE} = \frac{mg}{g_c}(Z_2 - Z_1) $$

$$ = \frac{10 \text{ lbm} \times 32.17 \text{ ft/sec}^2}{32.17 \text{ lbm-ft/lbf-sec}^2} \times (-77.8) \text{ ft} = -778 \text{ ft-lbf} $$

$$ = -1 \text{ Btu} $$

That is, ΔKE = 1 Btu and ΔPE = -1 Btu.

2. Just after the stone comes to rest in the bucket:

$$ {}_1Q_2 = 0 \qquad {}_1W_2 = 0 \qquad \Delta \text{KE} = 0 $$

Then

$$ \Delta \text{PE} = -\Delta U = \frac{mg}{g_c}(Z_2 - Z_1) = -1 \text{ Btu} $$

$$ \Delta U = 1 \text{ Btu} \qquad \Delta \text{PE} = -1 \text{ Btu} $$

3. After enough heat has been transferred so that the stone and water are at the same temperature they were initially, we conclude that $\Delta U = 0$. Therefore in this case

$$ \Delta U = 0 \qquad \Delta \text{KE} = 0 \qquad {}_1W_2 = 0 $$

$$ {}_1Q_2 = \Delta \text{PE} = \frac{mg}{g_c}(Z_2 - Z_1) = -1 \text{ Btu} $$

Example 5.3

A vessel having a volume of 100 ft^3 contains 1 ft^3 of saturated liquid water and 99 ft^3 of saturated water vapor at 14.7 lbf/in.^{2}. Heat is transferred until the vessel is filled with saturated vapor. Determine the heat transfer for this process.

Consider the total mass within the vessel as our system. Therefore the first law for this process is

$$Q = U_2 - U_1 + m\frac{(\mathsf{V}_2{}^2 - \mathsf{V}_1{}^2)}{2g_c} + \frac{mg}{g_c}(Z_2 - Z_1) + {}_1W_2$$

Since changes in kinetic and potential energy are not involved, this reduces to

$${}_1Q_2 = U_2 - U_1 + {}_1W_2$$

Further, the work for this process is zero, and therefore

$${}_1Q_2 = U_2 - U_1$$

The thermodynamic properties can be found in Table 2 of the steam tables. The initial internal energy U_1 is the sum of the initial internal energy of the liquid and the vapor.

$$U_1 = m_{1\text{liq}}u_{1\text{liq}} + m_{1\text{vap}}u_{1\text{vap}}$$

$$m_{1\text{liq}} = \frac{V_{\text{liq}}}{v_f} = \frac{1}{0.01672} = 59.81 \text{ lbm}$$

$$m_{1\text{vap}} = \frac{V_{\text{vap}}}{v_g} = \frac{99}{26.80} = 3.69 \text{ lbm}$$

$$U_1 = 59.8(180.1) + 3.69(1077.6) = 14{,}740 \text{ Btu}$$

To determine u_2 we need to know two thermodynamic properties, since this determines the final state. The properties we know are the quality, $x = 100\%$, and v_2, the final specific volume, which can readily be determined.

$$m = m_{1\text{liq}} + m_{1\text{vap}} = 59.81 + 3.69 = 63.50 \text{ lbm}$$

$$v_2 = \frac{V}{m} = \frac{100}{63.50} = 1.575 \text{ ft}^3/\text{lbm}$$

In Table 2 of the steam tables we find that at a pressure of 294 lbf/in.^{2} $v_g = 1.575$. The final pressure of the steam is therefore 294 lbf/in.^{2}. Then,

$$u_2 = 1117.0 \text{ Btu/lbm}$$
$$U_2 = mu_2 = 63.50(1117.0) = 70{,}930 \text{ Btu}$$
$${}_1Q_2 = U_2 - U_1 = 70{,}930 - 14{,}740 = 56{,}190 \text{ Btu}$$

The internal energy of superheated vapor steam is listed as a function of temperature and pressure in Table B.1.3. Similarly, Tables B.1.4 and B.1.5 list values in the compressed liquid and the saturated solid-vapor regions, respectively. In summary, all the values of internal energy are listed in the tables in the same manner as are values of specific volume.

5.4 The Thermodynamic Property Enthalpy

In analyzing specific types of processes, we frequently encounter certain combinations of thermodynamic properties, which are therefore also properties of the substance undergoing the change of state. To demonstrate one such situation in which this occurs, let us consider a system undergoing a quasiequilibrium constant-pressure process, as shown in Fig. 5.3. We further assume that there are no changes in kinetic

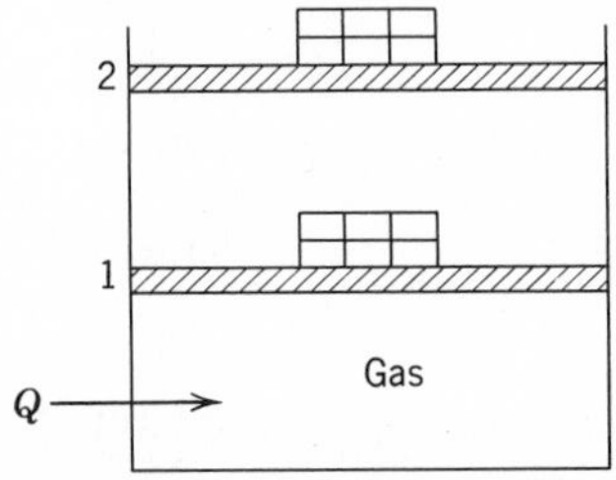

Fig. 5.3 The constant-pressure quasiequilibrium process.

or potential energy and also that the only work done during the process is that associated with the boundary movement. Taking the gas as our system, and applying the first law, Eq. 5.11,

$${}_1Q_2 = U_2 - U_1 + {}_1W_2$$

The work can be calculated from the relation

$${}_1W_2 = \int_1^2 P\,dV$$

Since the pressure is constant

$$ {}_1W_2 = P \int_1^2 dV = P(V_2 - V_1) $$

Therefore

$$ \begin{aligned} {}_1Q_2 &= U_2 - U_1 + P_2V_2 - P_1V_1 \\ &= (U_2 + P_2V_2) - (U_1 + P_1V_1) \end{aligned} $$

We find that in this very restricted case, the heat transfer during the process is given in terms of the change in the quantity $U + PV$ between the initial and final states. Inasmuch as all of these quantities are thermodynamic properties, functions only of the state of the system, their combination must also have these same characteristics. Therefore, we find it convenient to define a new extensive property, called the enthalpy,

$$ H = U + PV \tag{5.12} $$

or, per unit mass,

$$ h = u + Pv \tag{5.13} $$

As in the case of internal energy, we could speak of specific enthalpy, h, and total enthalpy, H. However, we will simply refer to both as enthalpy, since the context will make it clear which is referred to.

In order to add the internal energy and pressure volume product, the units for both must be the same. In this text the usual unit for enthalpy and internal energy is Btu/lbm. When pressure is in lbf/ft^2 and specific volume in ft^3/lbm, the factor $J = 778$ ft-lbf/Btu must be introduced in the denominator (i.e., Pv/J) in order to express the Pv term in Btu/lbm. However, the factor J will not be carried along in the equations because the only essential consideration is that consistent units be used, and many other units than Btu/lbm can be used.

As a result of our definition, the heat transfer in a constant-pressure quasiequilibrium process is equal to the change in enthalpy, which includes both the change in internal energy and the work for this particular process. This is by no means a general result. It is valid for this special case only because the work done during the process is equal to the difference in the PV product for the final and initial states. This would not be true if the pressure had not remained constant during the process.

The significance and use of enthalpy is not restricted to the special process described above. Other cases in which this same combination of properties $u + Pv$ appear will be developed later, notably in Chapter 6, in which we discuss control volume analyses. Our reason for introducing enthalpy at this time is that while the steam tables list values for internal energy, most other tables and charts of thermodynamic properties give values for enthalpy but not for the internal energy. In these cases it is necessary to calculate the internal energy at a state using the tabulated values and Eq. 5.13,

$$u = h - Pv$$

Students often become confused about the validity of this calculation when analyzing system processes that do not occur at constant pressure, for which enthalpy has no physical significance. We must keep in mind that enthalpy, being a property, is a state or point function, and its use in calculating internal energy at the same state is not related to, or dependent on any process that may be taking place.

Tabular values of enthalpy, such as those included in Appendix Tables B.1 through B.4, are all relative to some arbitrarily selected base. In the steam tables, the internal energy of saturated liquid at 32.02 F is the reference state and is given a value of zero. For refrigerants, such as ammonia and Freon-12, the reference state is saturated liquid at −40 F, the enthalpy in this reference state being assigned the value of zero. Thus, it is possible to have negative values of enthalpy, as is the case for saturated solid H_2O in Table B.1.5 of the Appendix. It should be pointed out that when enthalpy and internal energy are given values relative to the same reference state, as is the case in essentially all thermodynamic tables, the difference between internal energy and enthalpy at the reference state is equal to Pv. Since the specific volume of the liquid is very small, this product is negligible as far as the significant figures of the tables are concerned, but the principle should be kept in mind, as in certain cases it is significant.

In the superheat region of most thermodynamic tables, values of the specific internal energy u are not given. As mentioned above, these can be readily calculated from the relation $u = h - Pv$, though it is important to keep the units in mind. As an example, let us calculate the internal energy u of superheated Freon-12 at 100 lbf/in.^{2}, 200 F.

$$u = h - Pv$$

$$u = 105.63 - \frac{100 \times 144 \times 0.5441}{778} = 95.55 \text{ Btu/lbm}$$

The enthalpy of a substance in a saturation state and having a given quality is found in the same way that the specific volume and internal energy were found. The enthalpy of saturated liquid has the symbol h_f, saturated vapor h_g, and the increase in enthalpy during vaporization, h_{fg}. For a saturation state, the enthalpy can be calculated by one of the following relations:

$$h = (1-x)h_f + xh_g$$
$$h = h_f + xh_{fg}$$
$$h = h_g - (1-x)h_{fg}$$

The enthalpy of compressed water may be found from Table B.1.4, while for other substances for which compressed liquid tables are not available, the enthalpy is taken as that of saturated liquid at the same temperature.

Example 5.4

A cylinder fitted with a piston has a volume of 2 ft^3 and contains 0.5 lbm of steam at 60 lbf/in.^{2}. Heat is transferred to the steam until the temperature is 500 F, while the pressure remains constant.

Determine the heat transfer and the work for this process.

For this system changes in kinetic and potential energy are not significant. Therefore

$${}_1Q_2 = m(u_2 - u_1) + {}_1W_2$$

$${}_1W_2 = \int_1^2 P\,dV = P\int_1^2 dV = P(V_2 - V_1) = m(P_2v_2 - P_1v_1)$$

Therefore

$${}_1Q_2 = m(u_2 - u_1) + m(P_2v_2 - P_1v_1) = m(h_2 - h_1)$$

$$v_1 = \frac{V_1}{m} = \frac{2}{0.5} = 4.0 = 7.177 - (1-x_1)7.160$$

$$(1-x_1) = \frac{3.177}{7.160} = 0.443$$

$$h_1 = h_g - (1-x_1)h_{fg}$$
$$= 1178.0 - 0.443 \times 915.8 = 772.0$$
$$h_2 = 1283.0$$
$${}_1Q_2 = 0.5(1283 - 772) = 255.5 \text{ Btu}$$

$${}_1W_2 = mP(v_2 - v_1) = \frac{0.5 \times 60 \times 144}{778}(9.403 - 4.0)$$

$$= 30.0 \text{ Btu}$$

Therefore

$$U_2 - U_1 = {}_1Q_2 - {}_1W_2 = 255.5 - 30.0 = 225.5 \text{ Btu}$$

The heat transfer could also have been found from u_1 and u_2.

$$\begin{aligned} u_1 &= u_g - (1 - x_1)u_{fg} \\ &= 1098.3 - 0.443 \times 836.3 = 727.6 \\ u_2 &= 1178.6 \end{aligned}$$

and

$${}_1Q_2 = 0.5(1178.6 - 727.6) + 30.0 = 255.5 \text{ Btu}$$

5.5 The Constant-Volume and Constant-Pressure Specific Heats

The constant-volume specific heat and the constant-pressure specific heat are useful functions for thermodynamic calculations, and particularly for gases, as will be discussed in the next section.

The constant-volume specific heat C_v is defined by the relation

$$C_v \equiv \left(\frac{\partial u}{\partial T}\right)_v \tag{5.14}$$

The constant-pressure specific heat C_p is defined by the relation

$$C_p \equiv \left(\frac{\partial h}{\partial T}\right)_P \tag{5.15}$$

Note that each of these quantities is defined in terms of properties, and therefore the constant-volume and constant-pressure specific heats are thermodynamic properties of a substance. These definitions assume constant composition, and also that there are no surface, electrical, or magnetic effects.

The term heat capacity is often used instead of specific heat. However, neither term adequately conveys the nature of the thermodynamic property involved. For example, consider the two identical systems of Fig. 5.4. In the first, 100 Btu of heat is transferred to the system, and in the second, 100 Btu of work is done on the system. Thus, the change of internal energy is the same in each, and therefore the final state and the final temperature are the same in each. In accordance with Eq. 5.14, therefore, exactly the same value for the average constant-volume specific heat would be found for the substance for the two processes. However, if the specific heat is defined in terms of the heat transfer, different values will be obtained for the specific heat, since in the first

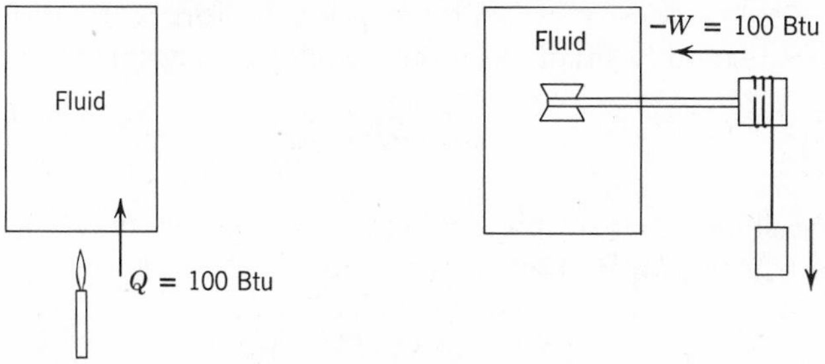

Fig. 5.4 Sketch showing two ways in which a given ΔU may be achieved.

case the heat transfer is finite and in the second case it is zero. Thus, although the terms "specific heat" and "heat capacity" are not the most appropriate, it is always important to keep in mind that these are thermodynamic properties, and are defined by Eqs. 5.14 and 5.15. When other effects are involved, additional specific heats can be defined, such as specific heat at constant magnetization for a system involving magnetic effects.

Example 5.5

Estimate the constant-pressure specific heat of steam at 100 lbf/in.², 400 F.

If we consider a change of state at constant pressure, Eq. 5.15 may be written

$$C_p \approx \left(\frac{\Delta h}{\Delta T}\right)_P$$

From the steam tables

$$\text{at } 100 \text{ lbf/in.}^2, 360 \text{ F}, h = 1205.9$$
$$\text{at } 100 \text{ lbf/in.}^2, 440 \text{ F}, h = 1248.5$$

Since we are interested in C_p at 100 lbf/in.², 400 F,

$$C_p \approx \frac{42.6}{80} = 0.533 \text{ Btu/lbm F}$$

5.6 The Internal Energy, Enthalpy, and Specific Heat of Ideal Gases

At this point certain comments about the internal energy, enthalpy, and the constant-pressure and constant-volume specific heats of an ideal gas should be made. An ideal gas has been defined in Chapter 3 as a gas at sufficiently low density so that intermolecular forces and the associated energy are negligibly small. As a result, an ideal gas has the equation of state

$$Pv = RT$$

It can be shown that for an ideal gas, the internal energy is a function of the temperature only. That is, for an ideal gas

$$u = f(T) \tag{5.16}$$

This means that an ideal gas at a given temperature has a certain definite specific internal energy u, regardless of the pressure. This will be demonstrated mathematically using the methods of classical thermodynamics in Chapter 12 and from the molecular viewpoint in Chapter 17.

In 1843 Joule demonstrated this fact when he conducted the following experiment, which is one of the classical experiments in thermodynamics. Two pressure vessels (Fig. 5.5), connected by a pipe and valve, were

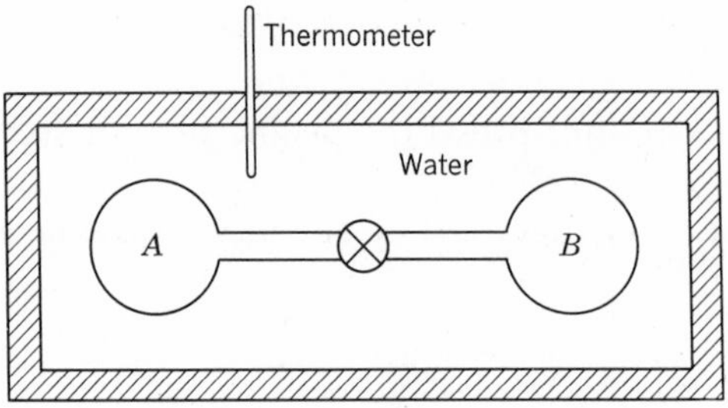

Fig. 5.5 Apparatus for conducting Joule's experiment.

immersed in a bath of water. Initially vessel A contained air at 22 atm pressure and vessel B was highly evacuated. When thermal equilibrium was attained the valve was opened, allowing the pressures in A and B to equalize. No change in the temperature of the bath was detected during or after this process. Because there was no change in the temperature of the bath, Joule concluded that there had been no heat transferred to the air. Since the work was also zero, he concluded from the first law of thermodynamics that there was no change in the internal energy of the gas. Since the pressure and volume changed during this process, one concludes that internal energy is not a function of pressure and volume. Because air does not conform exactly to the definition of an ideal gas, a small change in temperature will be detected when very accurate measurements are made in Joule's experiment.

The relation between the internal energy u and the temperature can be established by using the definition of constant-volume specific heat given by Eq. 5.14.

$$C_v = \left(\frac{\partial u}{\partial T}\right)_v$$

Since the internal energy of an ideal gas is not a function of volume, for an ideal gas we can write

$$C_{vo} = \frac{du}{dT}$$

$$du = C_{vo}\, dT$$

where the subscript o denotes the specific heat of an ideal gas. For a given mass m

$$dU = mC_{vo}\, dT \tag{5.18}$$

From the definition of enthalpy and the equation of state of an ideal gas, it follows that

$$h = u + Pv = u + RT \tag{5.19}$$

Since R is a constant and u is a function of temperature only, it follows that the enthalpy, h, of an ideal gas is also a function of temperature only. That is,

$$h = f(T) \tag{5.20}$$

The relation between enthalpy and temperature is found from the constant-pressure specific heat as defined by Eq. 5.15.

$$C_p = \left(\frac{\partial h}{\partial T}\right)_P$$

Since the enthalpy of an ideal gas is a function of the temperature only, and is independent of the pressure, it follows that

$$C_{po} = \frac{dh}{dT}$$

$$dh = C_{po}\, dT \tag{5.21}$$

For a given mass m,

$$dH = mC_{po}\, dT \tag{5.22}$$

The consequences of Eqs. 5.17 and 5.21 are demonstrated by Fig. 5.6, which shows two lines of constant temperature. Since internal energy and enthalpy are functions of temperature only, these lines of constant temperature are also lines of constant internal energy and constant

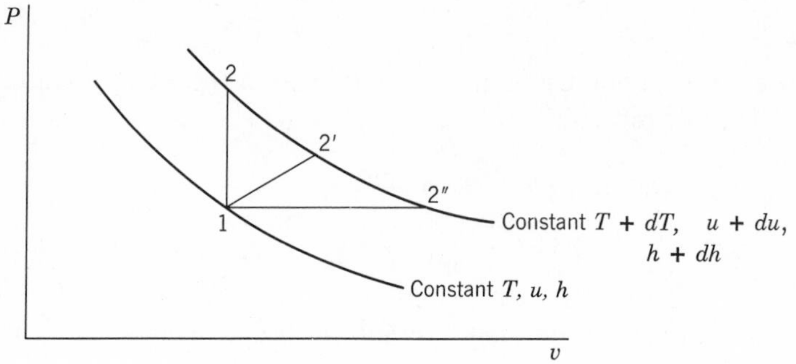

Fig. 5.6 Pressure-volume diagram for an ideal gas.

enthalpy. From state 1 the high temperature can be reached by a variety of paths, and in each case the final state is different. However, regardless of the path, the change in internal energy is the same, as is the change in enthalpy, for lines of constant temperature are also lines of constant u and constant h.

Because the internal energy and enthalpy of an ideal gas are functions of temperature only, it also follows that the constant-volume and constant-pressure specific heats are also functions of temperature only. That is

$$C_{vo} = f(T); \qquad C_{po} = f(T) \tag{5.23}$$

Since all gases approach ideal gas behavior as the pressure approaches zero, the ideal-gas specific heat for a given substance is often called the zero-pressure specific heat, and the zero-pressure constant-pressure specific heat is given the symbol C_{po}. The zero-pressure constant-volume specific heat is given the symbol C_{vo}. $\bar{C}_{po}$ as a function of temperature for a number of different substances is shown in Fig. 5.7. These values are determined by using the techniques of statistical thermodynamics, and will be described in detail in Chapter 19. However, a qualitative discussion at this point should provide some insight into this behavior, and is also beneficial in determining under what conditions an assumption of constant specific heat is justified.

As was discussed in Section 2.6, the energy possessed by molecules may be stored in several forms. The translational and rotational energies increase linearly with temperature, which means that those contributions to the specific heat are not temperature dependent. Contributions from vibrational and electronic modes, on the other hand, are temperature dependent (the electronic usually being very small). From Fig. 5.7, it is evident that the specific heat of a diatomic gas (such as H_2,CO,O_2) increases with an increase in temperature, due primarily to the vibration. A polyatomic gas (such as CO_2, H_2O) shows a much greater increase

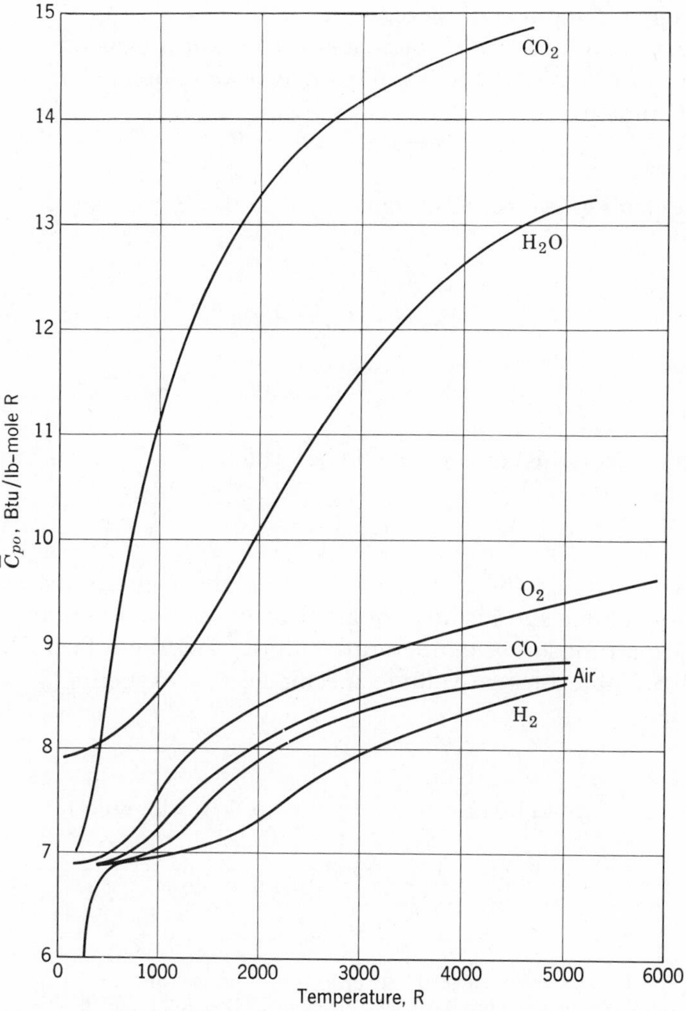

Fig. 5.7 Constant-pressure specific heats for a number of gases at zero pressure.

in specific heat as the temperature increases, and this is due to the additional vibrational modes of a polyatomic molecule. A monatomic gas (such as helium, argon, or neon), possessing only translational and electronic energies, shows little or no variation of specific heat over a wide range of temperatures.

The results of the specific heat calculations from statistical thermodynamics do not lend themselves to convenient mathematical forms. As a result, empirical equations expressing $\overline{C}_{po}$ as a function of temperature have been developed for many substances, and Table B.7 in the Appendix gives a number of these.[1]

A very important relation between the constant-pressure and constant-volume specific heats of an ideal gas may be developed from the definition of enthalpy.

$$h = u + Pv = u + RT$$

Differentiating, and substituting Eqs. 5.17 and 5.21, we have

$$dh = du + RdT$$

$$C_{po}dT = C_{vo}dT + RdT$$

Therefore

$$C_{po} - C_{vo} = R \tag{5.24}$$

On a mole basis this equation would be written

$$\overline{C}_{po} - \overline{C}_{vo} = \overline{R} \tag{5.25}$$

This tells us that the difference between the constant-pressure and constant-volume specific heats of an ideal gas is always constant, though both are a function of temperature. Thus, using the value of $\overline{C}_{po}$ from Table B.7, the constant-volume specific heat of oxygen, $\overline{C}_{vo}$, would be written

$$\overline{C}_{vo} = \overline{C}_{po} - \overline{R}$$

$$\overline{C}_{vo} = 11.515 - \frac{172}{\sqrt{T}} + \frac{1530}{T} - 1.987 \text{ Btu/lb mole-R}$$

$$= 9.528 - \frac{172}{\sqrt{T}} + \frac{1530}{T} \text{ Btu/lb mole-R}$$

[1]The equations given in Table B.7 involve two or three terms. A number of more accurate four-constant equations have been developed. K. A. Kobe and associates have presented a very complete summary in *Petroleum Refiner*, Jan. 1949 through Nov. 1954.

The equations from Table B. 7 can be substituted into Eqs. 5.17 and 5.21, and the change in internal energy and enthalpy then found by integration.

Example 5.6

Calculate the change of enthalpy as 1 lbm of oxygen is heated from 500 R to 2000 R, assuming ideal gas behavior.

Using the specific-heat equation from Table B. 7, and integrating Eq. 5.21, we have

$$\bar{h}_2 - \bar{h}_1 = \int_{T_1}^{T_2} \bar{C}_{po} dT = \int_{T_1}^{T_2} \left(11.515\, dT - \frac{172\, dT}{T^{1/2}} + 1530 \frac{dT}{T}\right)$$

$$\bar{h}_2 - \bar{h}_1 = 11.515(T_2 - T_1) - 172 \times 2(T_2^{1/2} - T_1^{1/2}) + 1530 \ln \frac{T_2}{T_1}$$

$$\bar{h}_{2000} - \bar{h}_{500} = 17{,}280 - 7690 + 2120 = 11{,}710 \text{ Btu/lb mole}$$

$$h_{2000} - h_{500} = \frac{\bar{h}_{2000} - \bar{h}_{500}}{M} = \frac{11{,}710}{32} = 366 \text{ Btu/lbm}$$

The average specific heat for any process is defined by the relation

$$C_{p(\text{av})} = \frac{\int_{T_1}^{T_2} C_p dT}{T_2 - T_1} \tag{5.26}$$

Thus, the average specific heat for Example 5.6 is

$$C_{p(\text{av})} = \frac{366 \text{ Btu/lbm}}{(2000 - 500) \text{ R}} = 0.244 \text{ Btu/lbm R}$$

The integration of specific heat equations can be very conveniently done on a digital computer. For this reason it is important to have accurate specific heat equations available. On the other hand, when a digitial computer is not used, there is great value in having tables that give the zero-pressure enthalpy and internal energy of ideal gases as a function of temperature. Values of enthalpy as a function of temperature for a number of ideal gases are given in Table B. 9, and for air in Table B. 8.

Often, sufficiently accurate results are obtained by assuming that the specific heat is a constant. In this case it follows that

$$h_2 - h_1 = C_{po}(T_2 - T_1); \; u_2 - u_1 = C_{vo}(T_2 - T_1)$$

Example 5.7

A cylinder fitted with a piston has an initial volume of 2 ft^3 and contains nitrogen at 20 lbf/in.^{2}, 80 F. The piston is moved, compressing the nitrogen until the pressure is 160 lbf/in.^{2} and the temperature is 300 F. During this compression process heat is transferred from the nitrogen, and the work done on the nitrogen is 9.15 Btu. Determine the amount of this heat transfer.

We consider the nitrogen as our system, and assume that the nitrogen is an ideal gas. Changes in the kinetic and potential energy are negligible.

First law: ${}_1Q_2 = m(u_2 - u_1) + {}_1W_2$
Property relation: $Pv = RT$; $u = f(T)$

The mass of nitrogen is found from the equation of state with the value of R from Table B. 6.

$$m = \frac{PV}{RT} = \frac{20 \times 144 \times 2}{55.15 \times 540} = 0.1868 \text{ lbm}$$

Assuming constant specific heat as given in Table B. 6,

$$\begin{aligned} {}_1Q_2 &= mC_{vo}(T_2 - T_1) + {}_1W_2 \\ &= 0.1868(0.177)(300 - 80) - 9.15 \\ &= 7.25 - 9.15 = -1.90 \text{ Btu.} \end{aligned}$$

5.7 The First Law as a Rate Equation

We frequently find it desirable to use the first law as a rate equation that expresses either the instantaneous or average rate at which energy crosses the system boundary as heat and work and the rate at which the energy of the system changes. In so doing we are departing from a strictly classical point of view, because basically classical thermodynamics deals with systems that are in equilibrium, and time is not a relevant parameter for systems that are in equilibrium. However, since these rate equations are developed from the concepts of classical thermodynamics, and are used in many applications of thermodynamics, they are included in this book. This rate form of the first law will be used in the development of the first law for the control volume in Chapter 6, and in this form the first law finds extensive applications in thermodynamics, fluid mechanics and heat transfer.

Consider a time interval δt during which an amount of heat δQ crosses the system boundary, an amount of work δW is done by the system, the internal energy change is ΔU, the kinetic energy change is ΔKE, and the potential energy change is ΔPE. From the first law we can write

$$\delta Q = \Delta U + \Delta \text{KE} + \Delta \text{PE} + \delta W$$

Dividing by δt we have the average rate of energy transfer as heat and work and increase of the energy of the system.

$$\frac{\delta Q}{\delta t} = \frac{\Delta U}{\delta t} + \frac{\Delta \text{KE}}{\delta t} + \frac{\Delta \text{PE}}{\delta t} + \frac{\delta W}{\delta t}$$

Taking the limit for each of these quantities as δt approaches zero we have

$$\lim_{\delta t \to 0} \frac{\delta Q}{\delta t} = \dot{Q}, \quad \text{the heat transfer rate}$$

$$\lim_{\delta t \to 0} \frac{\delta W}{\delta t} = \dot{W}, \quad \text{the power}$$

$$\lim_{\delta t \to 0} \frac{\Delta U}{\delta t} = \frac{dU}{dt}; \qquad \lim_{\delta t \to 0} \frac{\Delta(\text{KE})}{\delta t} = \frac{d(\text{KE})}{dt}; \qquad \lim_{\delta t \to 0} \frac{\Delta(\text{PE})}{\delta t} = \frac{d(\text{PE})}{dt}$$

Therefore, the rate equation form of the first law is

$$\dot{Q} = \frac{dU}{dt} + \frac{d(\text{KE})}{dt} + \frac{d(\text{PE})}{dt} + \dot{W} \tag{5.27}$$

We could also write this in the form

$$\dot{Q} = \frac{dE}{dt} + \dot{W} \tag{5.28}$$

Example 5.8

During the charging of a storage battery the current is 20 amp, and the voltage is 12.8 v. The rate of heat transfer from the battery is 25 Btu/hr. At what rate is the internal energy increasing?

Since changes in kinetic and potential energy are not significant, the first law can be written as a rate equation in the form, Eq. 5.27,

$$\dot{Q} = \frac{dU}{dt} + W$$

$$\dot{W} = -\mathscr{E} i = -20 \times 12.8 = -256 \text{ watts}$$

$$1 \text{ watt} = 3.412 \text{ Btu/hr}$$

Therefore,

$$\dot{W} = -256 \text{ watts} \times 3.412 \frac{\text{Btu}}{\text{watt hr}} = -873 \text{ Btu/hr}$$

$$\frac{dU}{dt} = \dot{Q} - \dot{W} = -25 \frac{\text{Btu}}{\text{hr}} - (-873) \frac{\text{Btu}}{\text{hr}}$$

$$= 848 \text{ Btu/hr}$$

5.8 Conservation of Mass

In the previous section we have considered the first law of thermodynamics for a system undergoing a change of state. A system is defined as a a fixed quantity of mass. The question now arises, does the mass of the system change when the energy of a system changes? If it does, then our definition of a system as a fixed quantity of mass is no longer valid when the energy of the system changes.

We know from relativistic considerations that mass and energy are related by the well-known equation

$$E = mc^2 \tag{5.29}$$

where c = velocity of light, and E = energy.

We conclude from this equation that the mass of a system does change when its energy changes. Let us calculate the magnitude of this change of mass for a typical problem, and determine whether this change in mass is significant.

Consider as a system a rigid vessel that contains a 1-lbm stoichiometric mixture of a hydrocarbon fuel (such as gasoline) and air. From our knowledge of combustion, we know that after combustion takes place it will be necessary to transfer about 1250 Btu from the system in order to restore the system to its initial temperature. From the first law

$$_1Q_2 = U_2 - U_1 + {_1W_2}$$

we conclude, since $_1W_2 = 0$ and $_1Q_2 = -1250$ Btu, that the internal energy of the system decreases by 1250 Btu during the heat transfer process. Let us now calculate the decrease in mass during this process using Eq. 5.29.

The velocity of light, c, is 9.83×10^8 ft/sec. Therefore

$$1250 \text{ (Btu)} = \frac{m \text{ (lbm)} \times 9.83^2 \times 10^8 \times 10^8 \text{ ft}^2/\text{sec}^2 \times \frac{1}{778} \text{ Btu/ft-lbf}}{32.17 \text{ lbm-ft/lbf-sec}^2}$$

$$m = 3.24 \times 10^{-11} \text{ lbm}$$

Therefore, when the energy of the system decreases by 1250 Btu, the decrease in mass is 3.24×10^{-11} lbm.

A change in mass of this magnitude cannot be detected by even our most accurate chemical balance. And, certainly, a fractional change in mass of this magnitude is beyond the accuracy required in essentially all engineering calculations. Therefore, if we use the laws of conservation of mass and conservation of energy as separate laws, we will not introduce significant error into most thermodynamic problems, and our definition of a system as having a fixed mass can be used even though the energy of the system changes.

PROBLEMS

5.1 At a toy fair there are four types of power automobiles. *A* is battery operated; *B* has an electric motor that utilizes a-c current; *C* has a spring that can be wound with a key; and *D* has a charged capsule of CO_2 gas and operates like a rocket. Examine each of the automobiles as it operates for heat, work, and changes in internal energy. In each case consider the entire car as the system.

5.2 A thermoelectric refrigerator is used to maintain a constant temperature in a cold space, as indicated in Fig. 5.8. The thermoelectric device operates

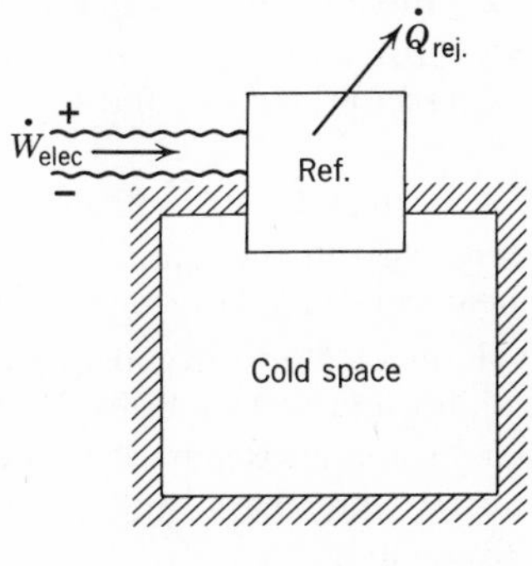

Fig. 5.8 Sketch for Problem 5.2.

on a current of 30 amp. at a voltage of 24 v, and rejects heat to the surroundings at the rate of 5000 Btu/hr. If the temperature in the cold space remains constant, what is the rate of heat leak into the cold space?

5.3 The average heat transfer from a person to the surroundings when he is not actively working is about 400 Btu/hr. Suppose that in an auditorium containing 1000 people the ventilation system fails.

(*a*) How much does the internal energy of the air in the auditorium increase during the first 15 minutes after the ventilation system fails?

(*b*) Considering the auditorium and all the people as a system, and assuming no heat transfer to the surroundings, how much does the internal energy of the system change? How do you account for the fact that the temperature of the air increases?

5.4 A sealed "bomb" containing certain chemicals is placed in a tank of water which is open to the atmosphere. When the chemicals react, heat is transferred from the bomb to the water, causing the temperature of the water to rise. A stirring device is used to circulate the water, and the power input to the rod driving the stirrer is 0.04 hp. In a 15-min period the heat transfer from the bomb is 1000 Btu and the heat transfer from the tank to the surrounding air is 50 Btu. Assuming no evaporation of water, determine the increase in the internal energy of the water.

5.5 A radiator of a steam heating system has a volume of 0.7 ft^3. When the radiator is filled with dry saturated steam at a pressure of 20 lbf/in.^{2} all valves to the radiator are closed. How much heat will have been transferred to the room when the pressure of the steam is 10 lbf/in.^{2}?

5.6 A pressure vessel having a volume of 5 ft^3 contains steam at the critical point. Heat is removed until the pressure is 200 lbf/in.^{2}. Determine the heat transferred from the steam.

5.7 A steam boiler has a total volume of 80 ft^3. The boiler initially contains 60 ft^3 of liquid water and 20 ft^3 of vapor in equilibrium at 14.7 lbf/in.^{2}. The boiler is fired up and heat is transferred to the water and steam in the boiler. Somehow, the valves on the inlet and discharge of the boiler are both left closed. The relief valve lifts when the pressure reaches 800 lbf/in.^{2}. How much heat was transferred to the water and steam in the boiler before the relief valve lifted?

5.8 A rigid vessel having a volume of 20 ft^3 is filled with ammonia at 90 lbf/in.^{2}, 160 F. Heat is transferred from the ammonia until it exists as saturated vapor. Calculate the heat transferred during this process.

5.9 A sealed tube has a volume of 2 in.^{3} and contains a certain fraction of liquid and vapor H_2O in equilibrium at 14.7 lbf/in.^{2}. The fraction of liquid and vapor is such that when heated the steam will pass through the critical point. Calculate the heat transfer when the steam is heated from the initial state at 14.7 lbf/in.^{2} to the critical state.

5.10 A rigid vessel of 5 ft^3 volume contains water at 200 F, 50% quality. The vessel is then cooled to 0 F. Calculate the heat transfer during the process.

5.11 A dewar vessel having a total volume of 100 gal contains liquid nitrogen at 1 atm pressure. The vessel is filled with 90% liquid and 10% vapor by volume. The dewar vessel is accidentally sealed off, so that as heat is transferred to the liquid nitrogen across the vacuum space, the pressure will increase. It is anticipated that the inner vessel will rupture when the pres-

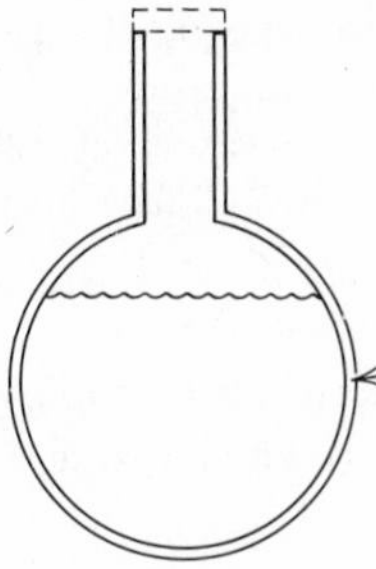

Fig. 5.9 Sketch for Problem 5.11.

sure reaches 60 lbf/in.^{2}. How long will it take to reach this pressure if the heat leak into the dewar vessel is 90 Btu/hr?

5.12 Steam which is contained in a cylinder expands against a piston. Following are the conditions before and after expansion.

Before Expansion		After Expansion	
Pressure:	160 lbf/in.^{2}	Pressure:	20 lbf/in.^{2}
Temperature:	500 F	Volume:	0.6 ft^3
Volume:	0.1 ft^3		

The heat transfer during expansion = −0.8 Btu. Calculate the work done during this process.

5.13 A tank is to be charged with ammonia. It is initially evacuated. A charging bottle filled with saturated liquid ammonia at 80 F is then connected to the tank, and the ammonia flows into the tank. The charging bottle remains connected and open to the tank, and the final temperature is 80 F. The charging bottle has a volume of 0.5 ft^3, and the tank has volume of 15 ft^3. What is the heat transfer to the ammonia?

5.14 Five lbm of water at 60 F is contained in a vertical cylinder by a frictionless piston of a mass such that the pressure of the water is 100 lbf/in.^{2}. Heat is

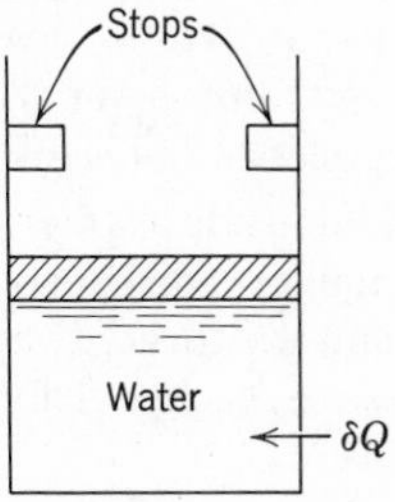

Fig. 5.10 Sketch for Problem 5.14.

transferred slowly to the water, causing the piston to rise until it reaches the stops, at which point the volume inside the cylinder is 15 ft^3. More heat is transferred to the water until it exists as saturated vapor.

(*a*) Find the final pressure in the cylinder and the heat transfer and work done during the process.

(*b*) Show this process on a *T-V* diagram.

5.15 Capsule *A* contains Freon-12 at the critical point and has a volume of 0.1 ft^3. Volume *B* is initially evacuated. The capsule ruptures and the Freon fills the volume. The volume of $A+B$ is 1.0 ft^3. After a certain amount of heat transfer takes place, the pressure is 100 lbf/in.^{2}.

Determine the amount of heat transferred to the Freon-12.

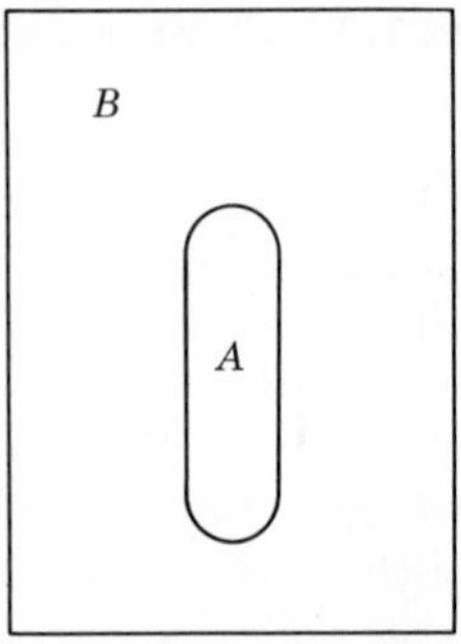

Fig. 5.11 Sketch for Problem 5.15.

5.16 An insulated cylinder fitted with a piston contains Freon-12 at 80 F with a quality of 90%. The volume at this state is 1 ft^3. The piston is then allowed to move, and the Freon expands until it exists as saturated vapor. During this process, the Freon does 3.9 Btu of work against the piston. Determine the final temperature if the process is adiabatic.

5.17 Two insulated tanks, *A* and *B*, are connected by a valve. Tank *A* has a volume of 20 ft^3 and contains steam at 20 lbf/in.^{2}, 400 F. Tank *B* has a volume of 10 ft^3 and contains steam at 80 lbf/in.^{2} with a quality of 90%. The valve is then opened, and the two tanks come to a uniform state. If there is no heat transfer during the process, what is the final pressure?

5.18 One lbm of steam is confined inside a spherical elastic membrane or balloon which supports an internal pressure proportional to its diameter. The initial condition of the steam is saturated vapor at 220 F. Heat is transferred to the steam until the pressure reaches 20 lbf/in.^{2}. Determine:

(*a*) The final temperature.

(*b*) The heat transfer.

5.19 Tank *A* (Fig. 5.12) contains 1 lbm saturated Freon-12 vapor at a temperature of 80 F. The valve is then opened slightly and Freon flows slowly into the cylinder *B*. The mass of the piston is such that the pressure of the Freon-12 in cylinder *B* is 20 lbf/in.². The process ends when the pressure in

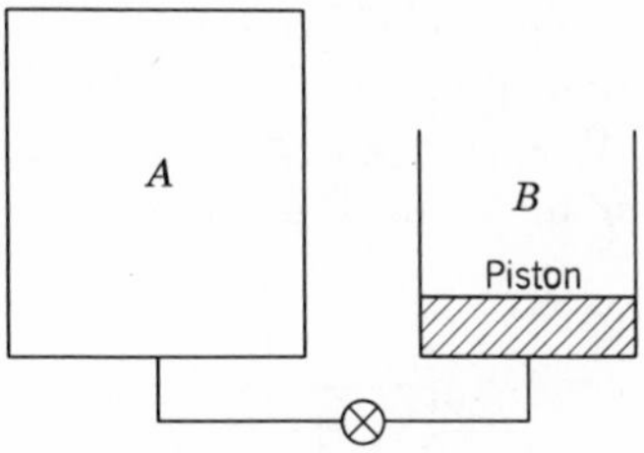

Fig. 5.12 Sketch for Problem 5.19.

tank *A* has dropped to 20 lbf/in.². During this process, heat is transferred to the Freon-12 so that the temperature remains constant at 80 F. Calculate the heat transfer during this process.

5.20 An insulated cylinder containing water has a piston held by a pin, as shown in Fig. 5.13. The water is initially saturated vapor at 150 F, and the volume

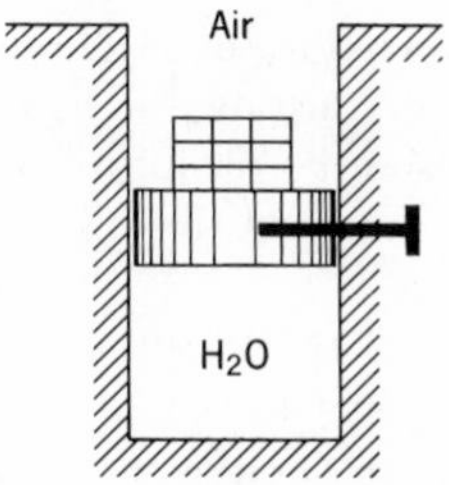

Fig. 5.13 Sketch for Problem 5.20.

is 0.2 ft³. The piston and weights have a total mass of 25 lbm, the piston area is 5 in.², and atmospheric pressure is 15 lbf/in.². The pin is now released, allowing the piston to move. Determine the final state of the water, assuming the process to be adiabatic.

5.21 A cylinder fitted with a piston contains 5 lbm of H_2O at a pressure of 200 lbf/in.² and a quality of 80%. The piston is restrained by a spring which is so arranged that for zero volume in the cylinder the spring is fully extended. The spring force is proportional to the spring displacement. The weight of the piston may be neglected so that the force on the spring is exactly balanced by the pressure forces on the H_2O.

Heat is transferred to the H_2O until its volume is 150% of the initial volume.

(*a*) What is the final pressure?

(*b*) What is the quality (if saturated) or temperature (if superheated) in the final state?

(*c*) Draw a *P-V* diagram and determine the work.

(*d*) Determine the heat transfer.

5.22 A cylinder contains 0.2 lbm of saturated vapor water at 220 F, as shown in Fig. 5.14. At this state the spring is touching the piston but exerts no force

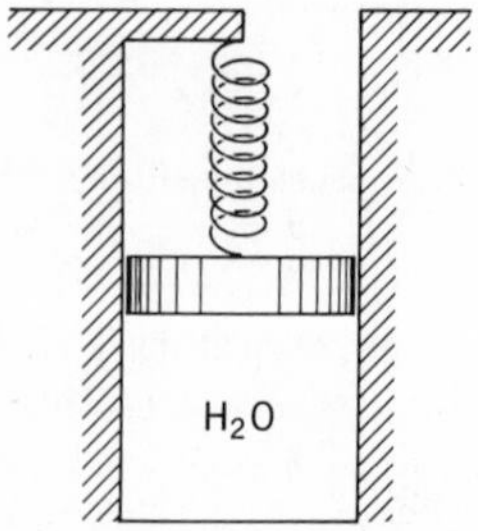

Fig. 5.14 Sketch for Problem 5.22.

on it. Heat is then transferred to the H_2O causing the piston to rise, during which process the spring resisting force is proportional to the distance moved, with a spring constant of 300 lbf/in. The piston area is 0.5 ft^2.

(*a*) What is the temperature in the cylinder when the pressure reaches 40 lbf/in.^{2}.

(*b*) Calculate the heat transfer for this process.

5.23 Consider the insulated vessel indicated in Fig. 5.15. The vessel has an evacuated compartment separated by a membrane from a second compartment which contains 1 lbm of water at 150 F, 100 lbf/in.^{2}. The membrane then ruptures and the water fills the entire volume with a resulting pres-

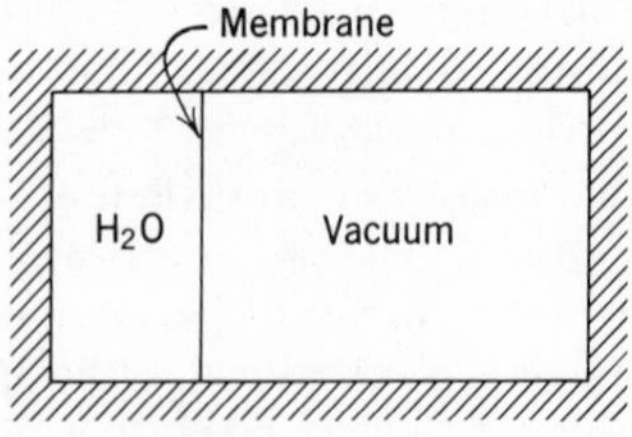

Fig. 5.15 Sketch for Problem 5.23.

sure of 2.0 lbf/in.². Determine the final temperature of the water and the entire volume of the vessel.

5.24 A spherical balloon initially 6 in. in diameter and containing Freon-12 at 15 lbf/in.² is connected to an uninsulated 1 ft³ tank containing Freon-12 at 80 lbf/in.². Both are at 80 F, the temperature of the surroundings.

The valve connecting the two is then opened very slightly and left open until the pressures become equal. During the process, heat is transferred with the surroundings such that the temperature of all the Freon remains at 80 F. It may also be assumed that the balloon diameter is proportional to the pressure inside the balloon at any point during the process. Calculate:

(*a*) The final pressure.
(*b*) Work done by the Freon during the process.
(*c*) Heat transfer to the Freon during the process.

5.25 An insulated and evacuated vessel of 1 ft³ volume contains a capsule of water at 100 lbf/in.², 300 F. The volume of the capsule is 0.1 ft³. The capsule breaks and the contents fill the entire volume. What is the final pressure?

5.26 A frictionless, thermally conducting piston separates the air and water in the cylinder shown in Fig. 5.16. The initial volumes of *A* and *B* are equal,

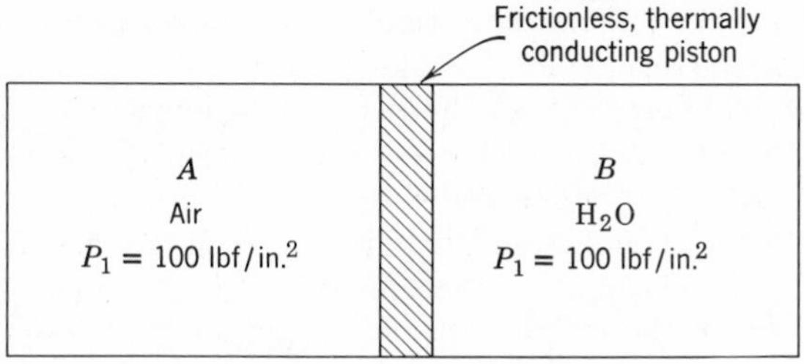

Fig. 5.16 Sketch for Problem 5.26.

the volume of each being 10 ft³. The initial pressure in both *A* and *B* is 100 lbf/in.². The volume of the liquid in *B* is 2% of the total volume of *B*. Heat is transferred to both *A* and *B* until all the liquid in *B* evaporates.

(*a*) Determine the total heat transfer during this process.
(*b*) Determine the work done by the piston on the air and the heat transfer to the air.

5.27 Consider ten lbm of air that is initially at 14.7 lbf/in.², 80 F. Heat is transferred to the air until the temperature reaches 600 F. Determine the change of internal energy, the change in enthalpy, the heat transfer and the work for the following processes:

(*a*) Constant-volume process.
(*b*) Constant-pressure process.

5.28 One lbm of water at 5 lbf/in.², quality of 50%, is heated to 2800 F, 20 lbf/in.².

Calculate the change in internal energy for the process.

5.29 One lbm of nitrogen gas is heated from 100 F to 3000 F. Calculate the change in enthalpy by the following methods:

(*a*) Constant specific heat, value from Table B.6.

(*b*) Constant specific heat, value at the average temperature from the equation, Table B.7.

(*c*) Variable specific heat, integrating the equation, Table B.7.

(*d*) Variable specific heat, values from the ideal gas tables, Table B.9.

5.30 An ideal gas is heated from 500 F to 1500 F. The enthalpy change per lbm is to be calculated assuming constant specific heat, value from Table B.6. Calculate this value, and discuss the accuracy of the result, if the gas is

(*a*) Helium.

(*b*) Nitrogen.

(*c*) Carbon dioxide.

5.31 Air contained in a cylinder fitted with a piston is compressed in a quasi-equilibrium process. During the compression process the relation between pressure and volume is $PV^{1.25}$ = constant. The mass of air in the cylinder is 0.1 lbm. The initial pressure is 15 lbf/in.^{2}, and the initial temperature is 70 F. The final volume is $\frac{1}{8}$ of the initial volume.

Determine the work and the heat transfer.

5.32 Helium which is contained in a cylinder fitted with a piston expands slowly according to the relation $PV^{1.5}$ = constant. The initial volume of the helium is 3 ft^3, the initial pressure is 70 lbf/in.^{2}, and the initial temperature is 500 R. After expansion the pressure is 20 lbf/in.^{2}. Calculate the work done and heat transfer during the expansion.

5.33 Heat is transferred at a given rate to a mixture of liquid and vapor in equilibrium in a closed container. Determine the rate of change of temperature as a function of the thermodynamic properties of the liquid and vapor and the mass of liquid and the mass of vapor.

5.34 Write a computer program to solve the following problem. For one of the substances listed in Table B.7, it is desired to compare the enthalpy change between any two temperatures T_1 and T_2 as calculated by integrating the specific heat equation, by assuming constant specific heat at the average temperature, and by assuming constant specific heat at temperature T_1.

6

First Law Analysis for a Control Volume

In the preceding chapter, we developed the first law analysis for a system undergoing a process. Many applications in thermodynamics do not readily lend themselves to a system approach, but are conveniently handled by the more general control volume techniques, as discussed in Chapter 2. The present chapter is concerned with development of the control volume forms for the conservation of mass and the first law of thermodynamics, and includes a number of special forms of these expressions.

6.1 Conservation of Mass and the Control Volume

A control volume is a volume in space in which one has interest for a particular study or analysis. The surface of this control volume is referred to as a control surface and always consists of a closed surface. The size and shape of the control volume are completely arbitrary, and are so defined as to best suit the analysis which is to be made. The surface may be fixed, or it may move or expand. However, the surface must be defined relative to some coordinate system. In some analyses it may be desirable to consider a rotating or moving coordinate system, and to describe the position of the control surface relative to such a coordinate system.

Mass as well as heat and work can cross the control surface and the mass in the control volume, as well as the properties of this mass, can change with time. Figure 6.1 shows a schematic diagram of a control

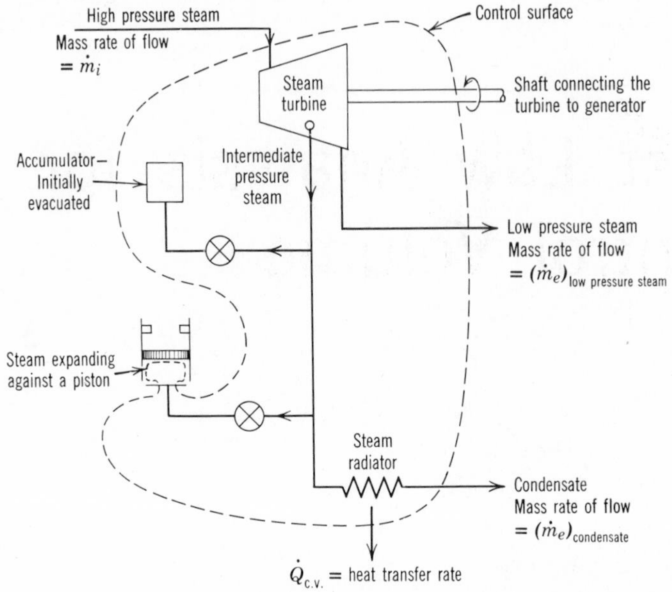

Fig. 6.1 Schematic diagram of a control volume showing mass and energy transfers and accumulation.

volume, with heat transfer, shaft work, accumulation of mass within the control volume and a moving boundary.

Let us first consider the law of the conservation of mass as it relates to the control volume. In doing so we consider the mass flow into and out of the control volume and the net increase of mass within the control volume. During a time interval δt let the mass δm_i, as shown in Fig. 6.2, enter the control volume, and the mass δm_e leave the control volume. Further, let us designate the mass within the control volume at the beginning of this time interval as m_t and the mass within the control volume after this interval as $m_{t+\delta t}$. Then, from the law of the conservation of mass we can write

$$m_t + \delta m_i = m_{t+\delta t} + \delta m_e$$

We can also look at this from the point of view of the net flow across the control surface and the change of mass within the control volume.

Net flow into the control volume during δt = increase of mass within the control volume during δt

$$(\delta m_i - \delta m_e) = m_{t+\delta t} - m_t$$

or

$$(m_{t+\delta t} - m_t) + (\delta m_e - \delta m_i) = 0 \qquad (6.1)$$

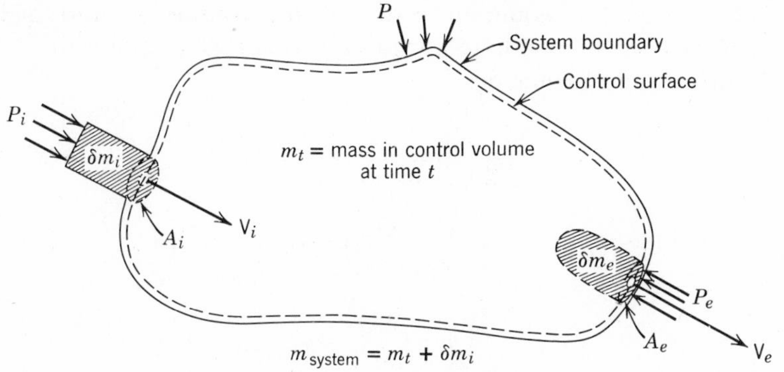

a System and control volume at time t

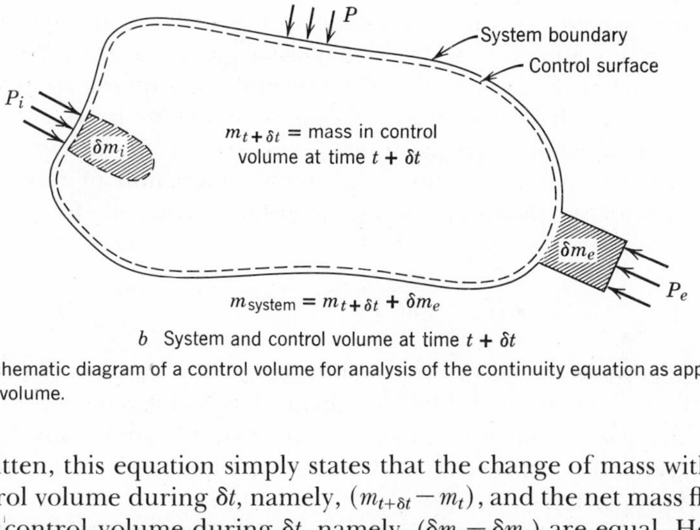

b System and control volume at time $t + \delta t$

Fig. 6.2 Schematic diagram of a control volume for analysis of the continuity equation as applied to a control volume.

As written, this equation simply states that the change of mass within the control volume during δt, namely, $(m_{t+\delta t} - m_t)$, and the net mass flow into the control volume during δt, namely, $(\delta m_i - \delta m_e)$ are equal. However, in many problems requiring a thermodynamic analysis we find it very convenient to have the law of the conservation of mass (as well as the first and second laws of thermodynamics and the momentum equation) expressed as a rate equation for the control volume. This involves the instantaneous rate of mass flow across the control surface and the instantaneous rate of change of mass within the control volume. Let us therefore write an expression for the average rate of change of mass within the control volume during δt and the average mass rates of flow across the control surface during δt by dividing Eq. 6.1 by δt.

$$\left(\frac{m_{t+\delta t} - m_t}{\delta t}\right) + \frac{\delta m_e}{\delta t} - \frac{\delta m_i}{\delta t} = 0 \tag{6.2}$$

To obtain the rate equation for the control volume, we find the limit of each term as δt is made to approach zero, at which point the system and control volume coincide.

$$\lim_{\delta t \to 0} \left(\frac{m_{t+\delta t} - m_t}{\delta t} \right) = \frac{dm_{\text{c.v.}}}{dt}$$

$$\lim_{\delta t \to 0} \left(\frac{\delta m_e}{\delta t} \right) = \dot{m}_e$$

$$\lim_{\delta t \to 0} \left(\frac{\delta m_i}{\delta t} \right) = \dot{m}_i$$

The symbol $m_{\text{c.v.}}$ is used to denote the instantaneous mass inside the control volume; $\dot{m}_i$ is the instantaneous rate of flow entering the control volume across area A_i; and $\dot{m}_e$ is that leaving across area A_e. We recognize that, in practice, there may well be several areas on the control surface across which flow occurs (in mixing processes or processes involving chemical reactions, for example). We account for this possibility by including summation signs, implying a summation of flows over the various discrete flow areas. Thus, the instantaneous rate form becomes

$$\frac{dm_{\text{c.v.}}}{dt} + \Sigma \dot{m}_e - \Sigma \dot{m}_i = 0 \tag{6.3}$$

Equation 6.3 for the conservation of mass is commonly termed the continuity equation. While this form of the equation is sufficient for the majority of our applications in this text, it is often convenient (and instructive) to express each of these mass quantities in terms of local fluid properties.

The state of the matter inside the control volume is not necessarily uniform at any given instant of time. However, we may divide the control volume into arbitrarily small volume elements δV containing mass $\rho \delta V$, where ρ is the local density. Thus, the total mass inside the control volume, $m_{\text{c.v.}}$, at any instant of time is

$$m_{\text{c.v.}} = \int_V \rho \, dV \tag{6.4}$$

The mass flow rate terms are more difficult to express in this manner, since in general neither the thermodynamic state nor the velocity are uniform across a finite area of flow. To further complicate the problem, the flow may not be normal to the control surface, and the control surface may itself be moving.

In order to demonstrate these possibilities, let us consider a fluid stream moving with some nonuniform velocity, as shown in Fig. 6.3. The control surface is moving with velocity V_{cs} and the flow area on the surface is A. Now consider an elemental area δA, sufficiently small that

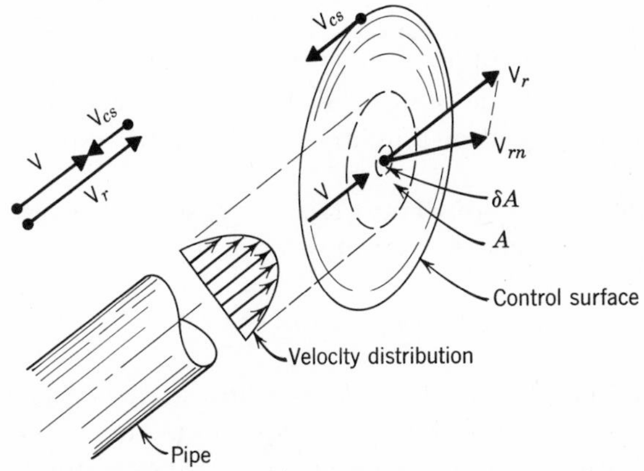

Fig. 6.3 Non-uniform flow across a moving control surface.

the local state and fluid velocity V are uniform. The velocity relative to the control surface is V_r, with a normal component V_{rn}. Thus, the instantaneous rate of flow $\delta\dot{m}$ across this element may be expressed in terms of the local density as

$$\delta\dot{m} = \rho \mathsf{V}_{rn} \delta A$$

The total flow rate across area A, then, is

$$\dot{m} = \int_A \delta\dot{m} = \int_A \rho \mathsf{V}_{rn} dA \tag{6.5}$$

We are now in a position to rewrite the continuity equation in terms of local fluid properties. In regard to the flow terms, if we require that V_{rn} be the outward-directed normal to the control volume, then a single term of the form of Eq. 6.5 integrated over the entire control surface includes all flow terms (V_{rn} is then negative for inward-directed flow and zero at locations on the control surface where no flow occurs.) The result is

$$\frac{d}{dt}\int_V \rho dV + \int_A \rho \mathsf{V}_{rn} dA = 0 \tag{6.6}$$

This form of the continuity equation is basic to subsequent developments in the fields of fluid mechanics and heat transfer, which are concerned with state and velocity distributions. In thermodynamics, however, we are commonly concerned with bulk average values for these quantities, and the continuity equation in the form of Eq. 6.3 is sufficient.

As a result, let us reconsider Eq. 6.5 and make the following assumptions:

1. The control surface is stationary.
2. The flow is normal to the control surface.
3. The thermodynamic state and fluid velocity are uniform (bulk average values) over the flow area at any instant of time.

As a result of assumptions **1** and **2**, $\mathsf{V}_{rn} = \mathsf{V}_n = \mathsf{V}$. As a result of assumption **3**, ρ and V are constant over the area A, such that

$$\dot{m} = \int_A \rho \mathsf{V} dA = \rho A \mathsf{V} \tag{6.7}$$

This result applies to any of the various flow streams entering or leaving the control volume, subject to the conditions assumed. The significance of each of these assumptions should be well understood and borne in mind, as they will be considered again in the development of the first and second laws of thermodynamics for the control volume.

Example 6.1

Air is flowing in a 6-in. diameter pipe at a uniform velocity of 30 ft/min. The temperature and pressure are 80 F and 20 lbf/in.^{2}. Determine the mass flow rate.

From Eq. 6.7,

$$\dot{m} = \rho A \mathsf{V} = \frac{A\mathsf{V}}{v}$$

For air, using R from Table B.6,

$$v = \frac{RT}{P} = \frac{53.34 \times 540}{20 \times 144} = 10.0 \text{ ft}^3/\text{lbm}$$

The cross-section area is

$$A = \frac{\pi}{4}\left(\frac{6}{12}\right)^2 = 0.196 \text{ ft}^2$$

Therefore,

$$\dot{m} = \frac{0.196 \times 30}{10.0} = 0.588 \text{ lbm/min}$$

6.2 The First Law of Thermodynamics for a Control Volume

We have already considered the first law of thermodynamics for a system, which consists of a fixed quantity of mass, and noted, Eq. 5.5, that it may be written

$$ {}_1Q_2 = E_2 - E_1 + {}_1W_2 $$

We have also noted that this may be written as an average rate equation, over the time interval δt by dividing by δt

$$ \frac{\delta Q}{\delta t} = \frac{E_2 - E_1}{\delta t} + \frac{\delta W}{\delta t} \tag{6.8} $$

In order to write the first law as a rate equation for a control volume we proceed in a manner analogous to that used in developing a rate equation for the law of the conservation of mass. A system and control volume are shown in Fig. 6.4. The system consists of all the mass initially in the control volume plus the mass δm_i.

Consider the changes that take place in the system and control volume during the time interval δt. During this time δt the mass δm_i enters the control volume through the discrete area A_i, and the mass δm_e leaves the control volume through the area A_e.

In our analysis we will assume that the increment of mass, δm_i has uniform properties, and similarly that δm_e has uniform properties. The total work done by the system during this process, δW, is that associated with the masses δm_i and δm_e crossing the control surface, usually referred to as flow work, and the work $\delta W_{\text{c.v.}}$ which includes all other forms of work, such as work associated with a rotating shaft that crosses the system boundary, shear forces, electrical, magnetic or surface effects, or expansion or contraction of the control volume. An amount of heat δQ crosses the boundary during δt.

Let us now consider each of the terms of the first law as it is written for the system, and transform each term into an equivalent form that applies to the control volume. Consider first the term $E_2 - E_1$.

Let E_t = the energy in the control volume at time t

$E_{t+\delta t}$ = the energy in the control volume at time $t + \delta t$.

Then

$E_1 = E_t + e_i\,\delta m_i$ = energy of the system at time t

$E_2 = E_{t+\delta t} + e_e \delta m_e$ = energy of the system at time $t + \delta t$.

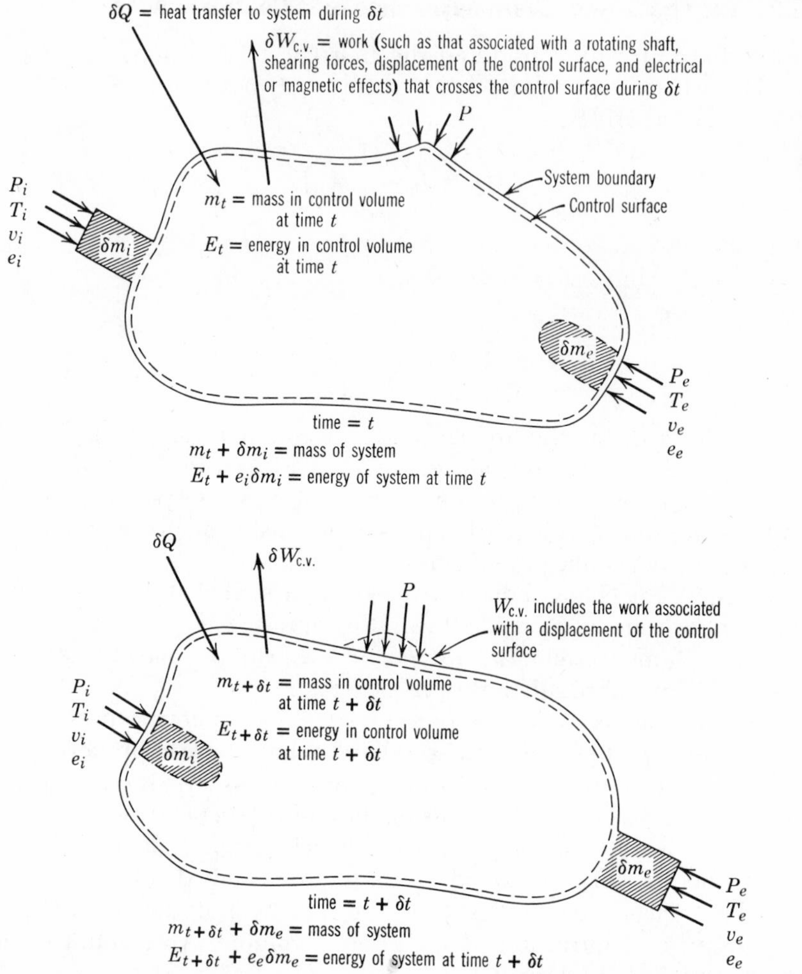

Fig. 6.4 Schematic diagram for a first law analysis of a control volume, showing heat and work as well as mass crossing the control surface.

Therefore

$$E_2 - E_1 = E_{t+\delta t} + e_e\,\delta m_e - E_t - e_i\,\delta m_i$$
$$= (E_{t+\delta t} - E_t) + (e_e\,\delta m_e - e_i\,\delta m_i) \tag{6.9}$$

The term $(e_e\,\delta m_e - e_i\,\delta m_i)$ represents the net flow of energy that crosses the control surface during δt as the result of the masses δm_e and δm_i crossing the control surface.

Let us consider in more detail the work associated with the masses

δm_i and δm_e crossing the control surface. Work is done by the normal force (normal to area A) acting on δm_i and δm_e as these masses cross the control surface. This normal force is equal to the product of the normal tensile stress, $-\sigma_n$, and the area, A. The work done is

$$-\sigma_n A\, dl = -\sigma_n\, \delta V = -\sigma_n v\, \delta m \tag{6.10}$$

A complete analysis of the nature of the normal stress, σ_n, for an actual fluid involves both static pressure and viscous effects, and is beyond the scope of this book. We will assume in this book that the normal stress, σ_n, at a point is equal to the static pressure at this point, P, and simply note that in many applications this is a very reasonable assumption and yields results of good accuracy.

With this assumption the work done on mass δm_i as it enters the control volume is $P_i v_i\, \delta m_i$, and the work done by mass δm_e as it leaves the control volume is $P_e v_e\, \delta m_e$. We shall refer to these terms as the flow work. Various other terms are encountered in the literature such as flow energy and work of introduction and work of expulsion.

Thus the total work done by the system during time δt is

$$\delta W = \delta W_{\text{c.v.}} + [P_e v_e\, \delta m_e - P_i v_i\, \delta m_i] \tag{6.11}$$

Let us now divide Eqs. 6.9 and 6.11 by δt and substitute into the first law, Eq. 6.8. Combining terms and rearranging,

$$\frac{\delta Q}{\delta t} + \frac{\delta m_i}{\delta t}(e_i + P_i v_i) = \left(\frac{E_{t+\delta t} - E_t}{\delta t}\right) + \frac{\delta m_e}{\delta t}(e_e + P_e v_e) + \frac{\delta W_{\text{c.v.}}}{\delta t} \tag{6.12}$$

We recognize that each of the flow terms in this expression can be rewritten in the form

$$e + Pv = u + Pv + \frac{\mathsf{V}^2}{2g_c} + Z\frac{g}{g_c}$$

$$= h + \frac{\mathsf{V}^2}{2g_c} + Z\frac{g}{g_c} \tag{6.13}$$

using the definition of the thermodynamic property enthalpy, Eq. 5.13. It should be apparent that the appearance of the combination $(u + Pv)$ whenever mass flows across a control surface is the principal reason for defining the property enthalpy. Its introduction in Chapter 5 in conjunction with the constant-pressure process was to facilitate use of the tables of thermodynamic properties at that time.

Utilizing Eq. 6.13 for the masses entering and leaving the control volume, Eq. 6.12 becomes

$$\frac{\delta Q}{\delta t}+\frac{\delta m_i}{\delta t}\left(h_i+\frac{\mathsf{V}_i^2}{2g_c}+Z_i\frac{g}{g_c}\right)=\left(\frac{E_{t+\delta t}-E_t}{\delta t}\right)+\frac{\delta m_e}{\delta t}\left(h_e+\frac{\mathsf{V}_e^2}{2g_c}+Z_e\frac{g}{g_c}\right)+\frac{\delta W_{\text{c.v.}}}{\delta t} \tag{6.14}$$

In order to reduce this expression to a rate equation, let us now examine each of the terms and establish the limits as δt is made to approach zero.

As regards the heat transfer, as δt approaches zero the system and the control volume coincide, and the heat transfer rate to the system is also the heat transfer rate to the control volume. Therefore,

$$\lim_{\delta t\to 0}\frac{\delta Q}{\delta t}=\dot{Q}_{\text{c.v.}}$$

the rate of heat transfer to the control volume. Further,

$$\lim_{\delta t\to 0}\left(\frac{E_{t+\delta t}-E_t}{\delta t}\right)=\frac{dE_{\text{c.v.}}}{dt}$$

$$\lim_{\delta t\to 0}\left(\frac{\delta W_{\text{c.v.}}}{\delta t}\right)=\dot{W}_{\text{c.v.}}$$

$$\lim_{\delta t\to 0}\left(\frac{\delta m_i}{\delta t}\left(h_i+\frac{\mathsf{V}_i^2}{2g_c}+Z_i\frac{g}{g_c}\right)\right)=\dot{m}_i\left(h_i+\frac{\mathsf{V}_i^2}{2g_c}+Z_i\frac{g}{g_c}\right)$$

$$\lim_{\delta t\to 0}\left(\frac{\delta m_e}{\delta t}\left(h_e+\frac{\mathsf{V}_e^2}{2g_c}+Z_e\frac{g}{g_c}\right)\right)=\dot{m}_e\left(h_e+\frac{\mathsf{V}_e^2}{2g_c}+Z_e\frac{g}{g_c}\right)$$

We originally assumed uniform properties for the incremental mass δm_i entering the control volume across area A_i, and made a similar assumption for δm_e leaving across area A_e. Consequently, in taking the limits above, this reduces to the restriction of uniform properties across area A_i and also across A_e at an instant of time. These may of course be time dependent.

In utilizing these limiting values to express the rate equation of the first law for a control volume, we again include summation signs on the flow terms to account for the possibility of additional flow streams entering or leaving the control volume. Therefore, the result is

$$\dot{Q}_{\text{c.v.}}+\sum\dot{m}_i\left(h_i+\frac{\mathsf{V}_i^2}{2g_c}+Z_i\frac{g}{g_c}\right)=\frac{dE_{\text{c.v.}}}{dt}+\sum\dot{m}_e\left(h_e+\frac{\mathsf{V}_e^2}{2g_c}+Z_e\frac{g}{g_c}\right)+\dot{W}_{\text{c.v.}} \tag{6.15}$$

which is, for our purposes, the general expression of the first law of thermodynamics. In words, this equation says that the rate of heat transfer into the control volume plus rate of energy flowing in as a result of mass transfer is equal to the rate of change of energy inside the control volume plus rate of energy flowing out as a result of mass transfer plus the power output associated with shaft, shear, and electrical effects and other factors that have already been mentioned.

Equation 6.15 can be integrated over the total time of a process to give the total energy changes that occur during that period. However, to do so requires knowledge of the time dependency of the various mass flow rates and the states at which mass enters and leaves the control volume. An example of this type of process will be considered in Section 6.5.

Another point to be noted is that if there is no mass flow into or out of the control volume, those terms simply drop out of Eq. 6.15, which then reduces to the rate equation form of the first law for a system,

$$\dot{Q} = \frac{dE}{dt} + \dot{W}$$

discussed in Section 5.7.

Since the control volume approach is more general, and reduces to the usual statement of the first law for a system when there is no mass flow across the control surface we will, when making a general statement of the first law, utilize Eq. 6.15 the rate form of the first law for a control volume.

The term flux should also be introduced at this point. Flux is defined as a rate of flow of any quantity per unit area across a control surface. Thus the mass flux is the rate of mass flow per unit area and heat flux is the heat flow rate per unit area across the control surface.

Finally, we should consider the question of uniformity of states and representation of the first law, Eq. 6.15, in terms of local fluid properties, as was done for the continuity equation in Section 6.1. As discussed in that section, the state of the matter inside the control volume is not necessarily uniform at any given instant of time. Thus, the quantity $E_{\text{c.v.}}$, representing the total integrated energy inside the control volume, may be expressed as

$$E_{\text{c.v.}} = \int_V e\rho \, dV \tag{6.16}$$

Analysis of the flow terms for nonuniform states of mass crossing the control surface is the same as that depicted in Fig. 6.2 and leading to Eq. 6.5. For each area of flow A, the resulting first law term is

$$\int_A \left(h + \frac{\mathsf{V}^2}{2g_c} + Z\frac{g}{g_c}\right)\delta m = \int_A \left(h + \frac{\mathsf{V}^2}{2g_c} + Z\frac{g}{g_c}\right)\rho \mathsf{V}_{rn} dA \tag{6.17}$$

in terms of the local values h, V, Z, ρ and V_{rn}. Substitution of this expression for flow terms (with V_{rn} the outward-directed normal, as discussed above, Eq. 6.6) and Eq. 6.16 for the control volume term into Eq. 6.15 results in

$$\dot{Q}_{\text{c.v.}} = \frac{d}{dt}\int_V e\rho\, dV + \int_A \left(h + \frac{\mathsf{V}^2}{2g_c} + Z\frac{g}{g_c}\right)\rho \mathsf{V}_{rn}\, dA + \dot{W}_{\text{c.v.}} \tag{6.18}$$

which is the first law for a control volume stated in terms of local properties. This form of the first law is not required for the majority of applications in the remainder of this text, and in general the simplified form of the first law as given by Eq. 6.15 will be adequate. However, we should always bear in mind the assumptions implicit in Eq. 6.15 and the significance of the various terms in that equation.

6.3 The Steady-State, Steady-Flow Process

Our first application of the control volume equations will be to develop a suitable analytical model for the long-term steady operation of devices such as turbines, compressors, nozzles, boilers, condensers—a very large class of problems of interest in thermodynamic analysis. This model will not include the short-term transient start-up or shutdown of such devices, but only the steady operating period of time.

Let us consider a certain set of assumptions (beyond those leading to Eqs. 6.3 and 6.15) that lead to a reasonable model for this type of process, which we refer to as the steady-state, steady-flow process. For convenience, we shall often refer to this process as the SSSF process.

1. The control volume does not move relative to the coordinate frame.
2. As for the mass in the control volume, the state of the mass at each point in the control volume does not vary with time.
3. As for the mass that flows across the control surface, the mass flux and the state of this mass at each discrete area of flow on the control surface do not vary with time. The rates at which heat and work cross the control surface remain constant.

As an example of a steady-state, steady-flow process consider a centrifugal air compressor that operates with constant mass rate of flow into and out of the compressor, constant properties at each point across the inlet and exit ducts, a constant rate of heat transfer to the surroundings, and a constant power input. At each point in the compressor the properties are constant with time, even though the properties of a given elemental mass of air vary as it flows through the compressor. Usually such a process is referred to simply as a steady flow process, since we are concerned primarily with the properties of the fluid entering and leaving the control volume. On the other hand, in the analysis of certain heat transfer problems in which the same assumptions apply, we are primarily interested in the spatial distribution of properties, particularly temperature, and such a process is often referred to as a steady state process. Since this is an introductory book we will use the term steady-state, steady-flow process in order to emphasize the basic assumptions involved. The student should realize that the terms steady-state process and steady-flow process are both used extensively in the literature.

Let us now consider the significance of each of these assumptions for the SSSF process.

1. The assumption that the control volume does not move relative to the coordinate frame means that all velocities measured relative to the coordinate frame are also velocities relative to the control surface, and there is no work associated with the acceleration of the control volume.
2. The assumption that the state of the mass at each point in the control volume does not vary with time requires that

$$\frac{dm_{\text{c.v.}}}{dt} = \frac{d}{dt}\int_V \rho dV = 0$$

and also

$$\frac{dE_{\text{c.v.}}}{dt} = \frac{d}{dt}\int_V e\rho dV = 0$$

Therefore, we conclude that for the SSSF process, we can write, from Eqs. 6.3 and 6.15

Continuity Equation:

$$\sum \dot{m}_i = \sum \dot{m}_e \tag{6.19}$$

First Law:

$$\dot{Q}_{\text{c.v.}} + \sum \dot{m}_i\left(h_i + \frac{V_i^2}{2g_c} + Z_i\frac{g}{g_c}\right) = \sum \dot{m}_e\left(h_e + \frac{V_e^2}{2g_c} + Z_e\frac{g}{g_c}\right) + \dot{W}_{\text{c.v.}} \tag{6.20}$$

3. The assumption that the various mass flows, states, and rates at which heat and work cross the control surface remain constant requires that every quantity in Eqs. 6.19 and 6.20 is steady with time. This means that application of Eqs. 6.19 and 6.20 to the operation of some device is independent of time.

Many of the applications of the SSSF model are such that there is only one flow stream entering and one leaving the control volume. For this type of process, we can write

Continuity Equation:

$$\dot{m}_i = \dot{m}_e = \dot{m} \tag{6.21}$$

First Law:

$$\dot{Q}_{\text{c.v.}} + \dot{m}\left(h_i + \frac{\mathsf{V}_i^2}{2g_c} + Z_i\frac{g}{g_c}\right) = \dot{m}\left(h_e + \frac{\mathsf{V}_e^2}{2g_c} + Z_e\frac{g}{g_c}\right) + \dot{W}_{\text{c.v.}} \tag{6.22}$$

Rearranging this equation we have:

$$q + h_i + \frac{\mathsf{V}_i^2}{2g_c} + Z_i\frac{g}{g_c} = h_e + \frac{\mathsf{V}_e^2}{2g_c} + Z_e\frac{g}{g_c} + w \tag{6.23}$$

where, by definition,

$$q = \frac{\dot{Q}_{\text{c.v.}}}{\dot{m}} \quad \text{and} \quad w = \frac{\dot{W}_{\text{c.v.}}}{\dot{m}} \tag{6.24}$$

Note that the units for q and w are Btu/lbm. From their definition q and w can be thought of as the heat transfer and work (other than flow work) per unit mass flowing into and out of the control volume for this particular SSSF process.

The symbols q and w are also used for the heat transfer and work per unit mass of a system. However, since it is always evident from the context whether it is a system (fixed mass) or control volume (involving a flow of mass) with which we are concerned, the significance of the symbols q and w will also be readily evident in each situation.

The SSSF process is often used in the analysis of reciprocating machines, such as reciprocating compressors or engines. In this case the rate of flow, which may actually be pulsating, is considered to be the average rate of flow for an integral number of cycles. A similar assumption is made as regards the properties of the fluid flowing across the control surface, and the heat transfer and work crossing the control surface. It is also assumed that for an integral number of cycles the reciprocating device undergoes, the energy and mass within the control volume do not change.

A number of examples are now given to illustrate the analysis of SSSF processes.

Example 6.2

The mass rate of flow into a steam turbine is 10,000 lbm/hr, and the heat transfer from the turbine is 30,000 Btu/hr. The following data are known for the steam entering and leaving the turbine.

	Inlet Conditions	Exit Conditions
Pressure	300 lbf/in.^{2}	15 lbf/in.^{2}
Temperature	700 F	
Quality		100%
Velocity	200 ft/sec	600 ft/sec
Elevation above reference plane	16 ft	10 ft
$g = 32.17$ ft/sec^2		

Determine the power output of the turbine.

Consider a control surface around the turbine, as shown in Fig. 6.5. From the data available it is evident that we can assume an SSSF process. Since flow enters at only one point and leaves at only one point the equation for the first law is given by Eq. 6.22.

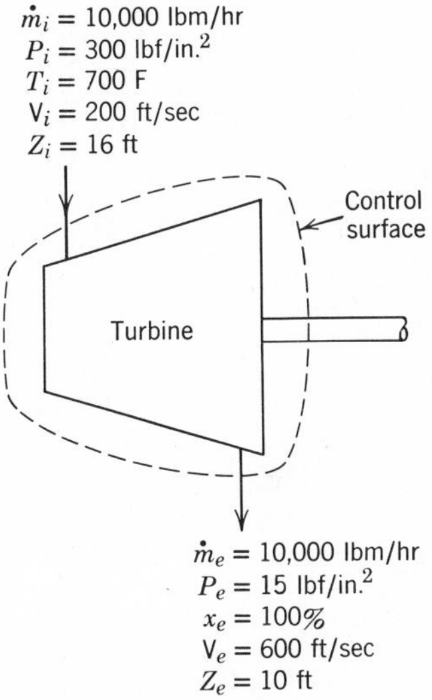

Fig. 6.5 Illustration for Example 6.2.

$$\dot{Q}_{\text{c.v.}} + \dot{m}\left(h_i + \frac{V_i^2}{2g_c} + Z_i\frac{g}{g_c}\right)$$

$$= \dot{m}\left(h_e + \frac{V_e^2}{2g_c} + Z_e\frac{g}{g_c}\right) + \dot{W}_{\text{c.v.}}$$

$$\dot{Q}_{\text{c.v.}} = -30{,}000 \text{ Btu/hr}$$

$h_i = 1368.3$ Btu/lbm (from the steam tables)

$$\frac{V_i^2}{2g_c} = \frac{200 \times 200 \text{ ft}^2/\text{sec}^2}{2 \times 32.17 \text{ lbm-ft/lbf-sec}^2 \times 778 \text{ ft-lbf/Btu}} = 0.799 \text{ Btu/lbm}$$

$$Z_i\frac{g}{g_c} = \frac{16 \text{ ft} \times 32.17 \text{ ft/sec}^2}{32.17 \text{ lbm-ft/lbf-sec}^2 \times 778 \text{ ft-lbf/Btu}} = 0.0206 \text{ Btu/lbm}$$

Similarly

$$h_e = 1150.9 \text{ Btu/lbm}$$

$$\frac{V_e^2}{2g_c} = \frac{(600)^2}{2 \times 32.17 \times 778} = 7.2 \text{ Btu/lbm}$$

$$Z_e \frac{g}{g_c} = \frac{10}{778} = 0.0128 \text{ Btu/lbm}$$

Therefore, substituting into Eq. 6.22,

$$-30{,}000 + 10{,}000(1368.3 + 0.799 + 0.0206)$$

$$= 10{,}000(1150.9 + 7.2 + 0.0128) + \dot{W}_{\text{c.v.}}$$

$$\dot{W}_{\text{c.v.}} = -30{,}000 + 13{,}691{,}000 - 11{,}581{,}000 = 2{,}080{,}000 \text{ Btu/hr}$$

$$= \frac{2{,}080{,}000 \text{ Btu/hr}}{2545 \text{ Btu/hp-hr}} = 816 \text{ hp}$$

If Eq. 6.23 is used, the work per pound mass of fluid flowing is found first.

$$q + h_i + \frac{V_i^2}{2g_c} + Z_i \frac{g}{g_c} = h_e + \frac{V_e^2}{2g_c} + Z_e \frac{g}{g_c} + w$$

$$q = \frac{-30{,}000}{10{,}000} = -3 \text{ Btu/lbm}$$

Therefore, substituting into Eq. 6.23,

$$-3 + 1368.3 + \frac{(200)^2}{2 \times 32.17 \times 778} + \frac{16}{778}$$

$$= 1150.9 + \frac{(600)^2}{2 \times 32.17 \times 778} + \frac{10}{778} + w$$

$$-3 + 1368.3 + 0.799 + 0.0206$$

$$= 1150.9 + 7.2 + 0.0128 + w$$

$$w = 208.0 \text{ Btu/lbm}$$

$$\dot{W}_{\text{c.v.}} = \frac{208.0 \text{ Btu/lbm} \times 10{,}000 \text{ lbm/hr}}{2545 \text{ Btu/hp-hr}}$$

$$= 816 \text{ hp}$$

Two further observations can be made by referring to this example. First, in many engineering problems potential energy changes are insignificant when compared with the other energy quantities. In the above example the potential energy change did not affect any of the significant figures. In most problems where the change in elevation is small the potential energy terms may be neglected.

Second, if velocities are small, say under 100 ft/sec, in many cases the kinetic energy is insignificant compared with other energy quantities. Further, when the velocities entering and leaving the system are essentially the same, the change in kinetic energy is small. Since it is the change in kinetic energy that is important in the SSSF energy equation, the kinetic energy terms can usually be neglected when there is no significant difference between the velocity of the fluid entering and leaving the control volume. Thus in many thermodynamic problems one must make judgments as to which quantities may be negligible for a given analysis.

Example 6.3

Steam at 100 lbf/in.^{2}, 400 F, enters an insulated nozzle with a velocity of 200 ft/sec. It leaves at a pressure of 20 lbf/in.^{2} and a velocity of 2000 ft/sec. Determine the final temperature if the steam is superheated in the final state, and the quality if it is saturated.

Considering a control surface around the nozzle (Fig. 6.6) it is evident that a SSSF process is a reasonable assumption. Further, $\dot{Q}_{\text{c.v.}}$ and $\dot{W}_{\text{c.v.}}$

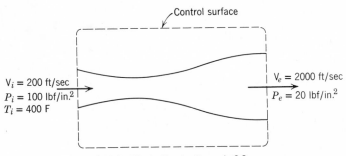

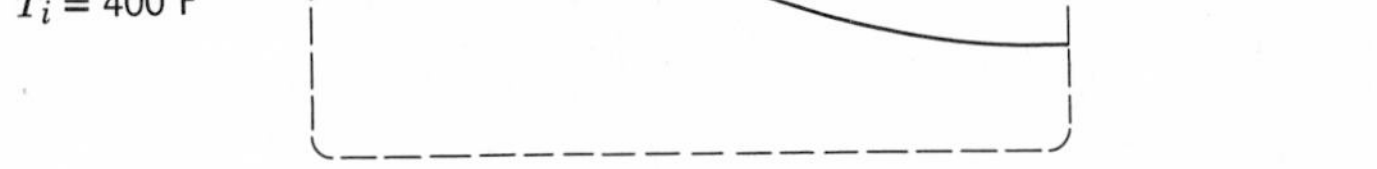

Fig. 6.6 Illustration for Example 6.3.

are zero in this problem. Also, since there is no significant change in elevation between the inlet and exit of the nozzle, the change in potential energy is negligible. Therefore, for this process the first law, Eq. 6.23 reduces to

$$h_i + \frac{V_i^2}{2g_c} = h_e + \frac{V_e^2}{2g_c}$$

$$h_e = 1227.5 + \frac{(200)^2}{2 \times 32.17 \times 778} - \frac{(2000)^2}{2 \times 32.17 \times 778}$$

$$h_e = 1227.5 - 79.2 = 1148.3 \text{ Btu/lbm}$$

The two properties of the fluid leaving which we now know are pressure and enthalpy, and therefore the state of this fluid is determined. Since h_e is less than h_g at 20 lbf/in.^{2}, the quality is calculated.

$$h = h_g - (1 - x)h_{fg}$$

$$1148.3 = 1156.4 - (1 - x)960.1$$

$$(1 - x) = \frac{8.1}{960.1} = 0.8\%$$

$$x = 99.2\%$$

Example 6.4

In a refrigeration system, in which Freon-12 is the refrigerant, the Freon-12 enters the compressor at 30 lbf/in.^{2}, 20 F, and leaves at 160 lbf/in.^{2}, 200 F. The mass rate of flow is 125 lbm/hr, and the power input to the compressor is 1 kw.

On leaving the compressor the Freon-12 enters a water-cooled condenser at 160 lbf/in.^{2}, 180 F, and leaves as a liquid at 150 lbf/in.^{2}, 100 F. Water enters the condenser at 55 F and leaves at 75 F. Determine:

1. The heat transfer from the compressor per hour.
2. The rate at which cooling water flows through the condenser.

Consider first a control volume analysis of the compressor. A steady-state, steady-flow process is a reasonable assumption.

$$w = -\frac{3412}{125} = -27.3 \text{ Btu/lbm}$$

From the Freon-12 tables

$$h_i = 79.76 \text{ Btu/lbm} \qquad h_e = 103.91 \text{ Btu/lbm}$$

Since the velocity of the vapor entering a refrigeration compressor is low and not greatly different from the velocity leaving, kinetic energy changes, as well as potential energy changes, can be neglected. Therefore, the first law for this process, Eq. 6.23 reduces to

$$q + h_i = h_e + w$$

$$q = -27.30 + 103.91 - 79.76 = -3.15 \text{ Btu/lbm}$$

$$Q_{\text{c.v.}} = 125 \text{ lbm/hr} \times (-3.15) \text{ Btu/lbm} = -394 \text{ Btu/hr}$$

Next consider a control volume analysis of the condenser and a steady-state, steady-flow process. The schematic diagram for this condenser is shown in Fig. 6.7.

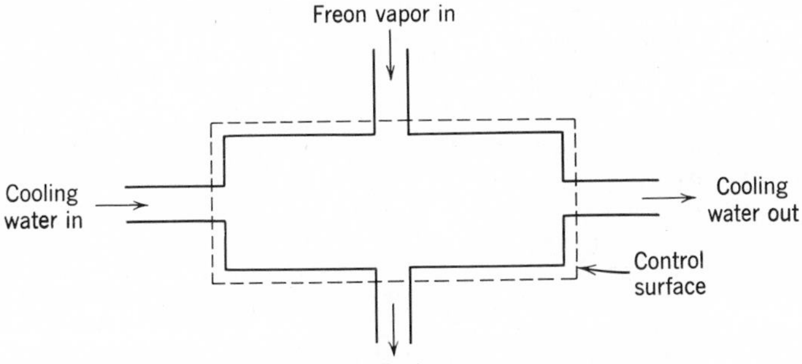

Fig. 6.7 Schematic diagram of a Freon condenser.

With this control volume we have two fluid streams, the Freon and the water, entering and leaving the control volume. It is reasonable to assume that the kinetic and potential energy changes are negligible. We note that the work is zero, and we make the other reasonable assumption that there is no heat transfer across the control surface. Therefore, the first law, Eq. 6.20, reduces to

$$\sum \dot{m}_i h_i = \sum \dot{m}_e h_e$$

In this case we have two streams entering and two streams leaving. Using the subscript r for refrigerant and w for water

$$\dot{m}_r (h_i)_r + \dot{m}_w (h_i)_w = \dot{m}_r (h_e)_r + \dot{m}_w (h_e)_w$$

From the Freon-12 and steam tables we have

$$(h_i)_r = 100.34 \text{ Btu/lbm} \qquad (h_i)_w = 23.07 \text{ Btu/lbm}$$
$$(h_e)_r = 31.10 \qquad (h_e)_w = 43.08$$

Solving the above equation for $\dot{m}_w$, the rate of flow of water, we have

$$\dot{m}_w = \dot{m}_r \frac{(h_i - h_e)_r}{(h_e - h_i)_w} = 125 \text{ lbm/hr} \frac{(100.34 - 31.10) \text{ Btu/lbm}}{(43.08 - 23.07) \text{ Btu/lbm}} = 433 \text{ lbm/hr}$$

This problem can also be solved by considering two separate control volumes, one of which has the flow of Freon-12 across its control surface, and the other having the flow of water across its control surface. Further there is heat transfer from one control volume to the other. This is shown schematically in Fig. 6.8.

The heat transfer for the control volume involving Freon is calculated first. In this case the SSSF energy equation, Eq. 6.22, reduces to

$$\dot{Q}_{\text{c.v.}} = \dot{m}_r (h_e - h_i)_r$$
$$\dot{Q}_{\text{c.v.}} = 125 \text{ lbm/hr } (31.10 - 100.34) \text{ Btu/lbm} = -8655 \text{ Btu/hr}$$

This is also the heat transfer to the other control volume, for which $Q_{\text{c.v.}} = +8655$ Btu/hr.

$$\dot{Q}_{\text{c.v.}} = \dot{m}_w (h_e - h_i)_w$$

$$\dot{m}_w = \frac{8655 \text{ Btu/hr}}{(43.08 - 23.07) \text{ Btu/lbm}} = 433 \text{ lbm/hr}$$

Example 6.5

Consider the simple steam power plant, as shown in Fig. 6.9. The following data are for such a power plant.

Location	Pressure	Temperature or quality
Leaving boiler	300 lbf/in.^{2}	600 F
Entering turbine	280 lbf/in.^{2}	550 F
Leaving turbine, entering condenser	2 lbf/in.^{2}	93%
Leaving condenser, entering pump	1.9 lbf/in.^{2}	110 F
Pump work = 3 Btu/lbm		

Determine the following quantities per pound mass flowing through the unit.

1. Heat transfer in line between boiler and turbine.
2. Turbine work.
3. Heat transfer in condenser.
4. Heat transfer in boiler.

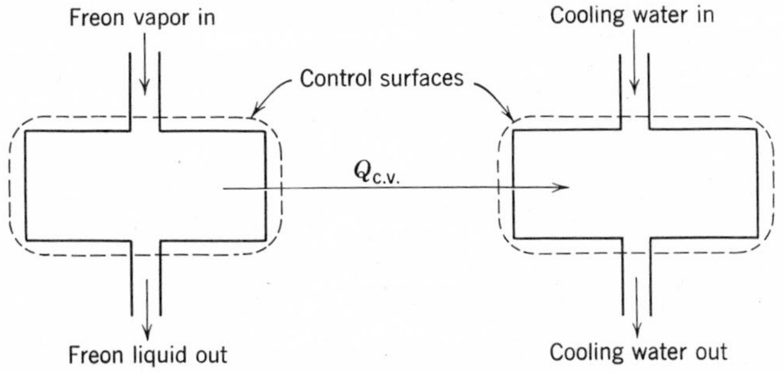

Fig. 6.8 Schematic diagram of Freon condenser considering two control surfaces.

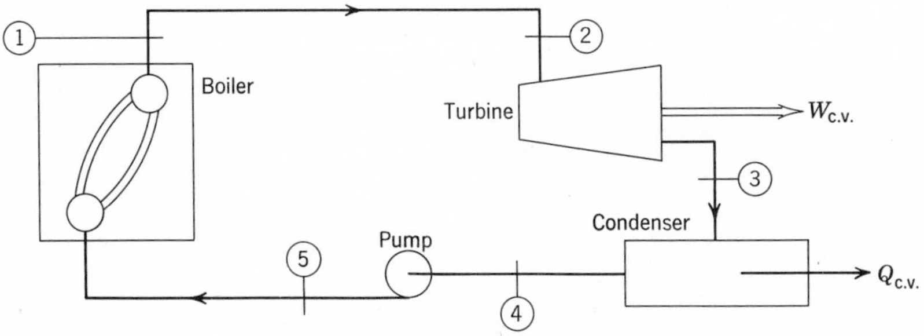

Fig. 6.9 Simple steam power plant.

There is a certain advantage in assigning a number to various points in the cycle. For this reason the subscripts i and e in the steady state, steady-flow energy equation are often replaced by appropriate numbers. Thus for a steady-state, steady-flow process for a control volume around the turbine one can write

$$q_{\text{turb}} + h_2 + \frac{\mathsf{V}_2^2}{2g_c} + Z_2 \frac{g}{g_c} = h_3 + \frac{\mathsf{V}_3^2}{2g_c} + Z_3 \frac{g}{g_c} + w_{\text{turb}}$$

Similarly, for a control volume around the boiler

$$q_{\text{boiler}} + h_5 + \frac{\mathsf{V}_5^2}{2g_c} + Z_5 \frac{g}{g_c} = h_1 + \frac{\mathsf{V}_1^2}{2g_c} + Z_1 \frac{g}{g_c}$$

We might also use the alternate but equally acceptable notation

$$q_{\text{turb}} = {}_2q_3; \qquad q_{\text{boiler}} = {}_5q_1$$

In this problem we shall neglect changes in kinetic energy and potential energy.

The following property values are obtained from the steam tables, where the subscripts refer to Fig. 6.9.

$$
\begin{aligned}
h_1 &= 1314.5 \text{ Btu/lbm} \\
h_2 &= 1288.6 \text{ Btu/lbm} \\
h_3 &= 1116.1 - 0.07(1022.2) = 1044.6 \text{ Btu/lbm} \\
h_4 &= 78.0 \text{ Btu/lbm}
\end{aligned}
$$

1. Considering a control volume that includes the pipe line between boiler and turbine,

$$
{}_1q_2 + h_1 = h_2
$$
$$
{}_1q_2 = h_2 - h_1 = 1288.6 - 1314.5 = -25.9 \text{ Btu/lbm}
$$

2. A turbine is essentially an adiabatic machine, and therefore, considering a control volume around the turbine we can write

$$
h_2 = h_3 + {}_2w_3
$$
$$
{}_2w_3 = 1288.6 - 1044.6 = 244.0 \text{ Btu/lbm}
$$

3. For the condenser the work is zero, and therefore

$$
{}_3q_4 + h_3 = h_4
$$
$$
{}_3q_4 = 78.0 - 1044.6 = -966.6 \text{ Btu/lbm}
$$

4. The enthalpy at point 5 may be found by considering a control volume around the pump

$$
h_4 = h_5 + {}_4w_5
$$
$$
h_5 = 3.0 + 78.0 = 81.0 \text{ Btu/lbm}
$$

For the boiler we can write

$$
{}_5q_1 + h_5 = h_1
$$
$$
{}_5q_1 = 1314.5 - 81.0 = 1233.5 \text{ Btu/lbm}
$$

Example 6.6

The centrifugal air compressor of a gas turbine receives air from the ambient atmosphere where the pressure is 14.5 lbf/in.^{2} and the temperature is 80 F. At the discharge of the compressor the pressure is 54 lbf/in.^{2}, the temperature is 400 F, and the velocity is 300 ft/sec. The mass rate of flow into the compressor is 2000 lbm/min. Determine the power required to drive the compressor.

We consider a control volume around the compressor, and locate the control volume at some distance from the compressor so that the air crossing the control surface has a very low velocity and is essentially at ambient conditions. If we located our control volume directly across the inlet section it would be necessary to know the temperature and velocity at the compressor inlet. We assume an adiabatic steady-state, steady-flow process in which the change in potential energy and the inlet kinetic energy are zero. Therefore, the first law for this process, Eq. 6.23 reduces to

$$h_i = h_e + \frac{V_e^2}{2g_c} + w$$

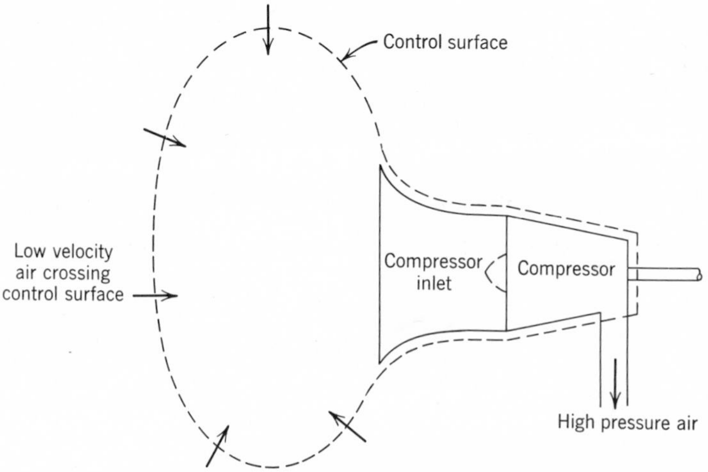

Fig. 6.10 Sketch for Example 6.6.

If we assume constant specific heat, from Table B.6 we have

$$-w = h_e - h_i + \frac{V_e^2}{2g_c} = C_{po}(T_e - T_i) + \frac{V_e^2}{2g_c}$$

$$= 0.24(400 - 80) + \frac{(300)^2}{2 \times 32.17 \times 778}$$

$$= 76.8 + 1.8 = 78.6 \text{ Btu/lbm}$$

$$-\dot{W}_{\text{c.v.}} = \frac{78.6(2000)}{42.4} = 3695 \text{ hp}$$

$$\text{Power input} = 3695 \text{ hp}$$

If we use the values for the enthalpy of air that are given in the Air Tables (Table B.8), we have

$$T_i = 80 + 460 = 540 \text{ R} \qquad T_e = 400 + 460 = 860 \text{ R}$$
$$h_i = 129.06 \text{ Btu/lbm} \qquad h_e = 206.46 \text{ Btu/lbm}$$

$$-w = h_e - h_i + \frac{V_e^2}{2g_c}$$

$$= 206.46 - 129.06 + \frac{(300)^2}{2 \times 32.17 \times 778}$$

$$= 77.40 + 1.8 = 79.2 \text{ Btu/lbm}$$

$$-\dot{W}_{\text{c.v.}} = \frac{79.2(2000)}{42.4} = 3740 \text{ hp}$$

6.4 The Joule-Thomson coefficient and the Throttling Process

The Joule-Thomson coefficient μ_J is defined by the relation

$$\mu_J \equiv \left(\frac{\partial T}{\partial P}\right)_h \tag{6.25}$$

As in the definition of specific heats in Section 5.5, this quantity is defined in terms of thermodynamic properties, and therefore is itself a property of a substance.

The significance of the Joule-Thomson coefficient may be demonstrated by considering a throttling process that is a steady-state, steady-flow process across a restriction, with a resulting drop in pressure. A typical example is the flow through a partially opened valve or a restriction in the line. In most cases this occurs so rapidly and in such a small space, that there is neither sufficient time nor a large enough area for much heat transfer. Therefore, we usually may assume such processes to be adiabatic.

If we consider the control surface shown in Fig. 6.11 we can write the SSSF energy equation for this process. There is no work, no change in potential energy, and we make the reasonable assumption that there is

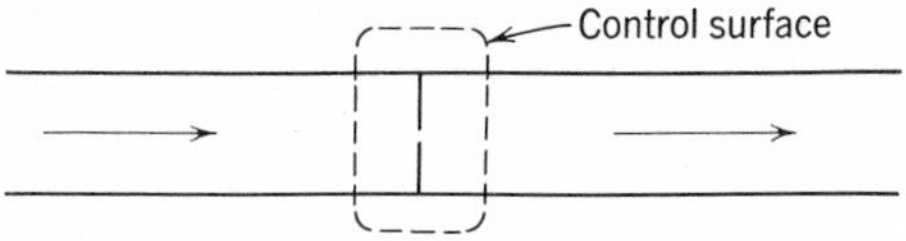

Fig. 6.11 The throttling process.

no heat transfer. The SSSF energy equation, Eq. 6.23, reduces to

$$h_i + \frac{\mathsf{V}_i^2}{2g_c} = h_e + \frac{\mathsf{V}_e^2}{2g_c}$$

If the fluid is a gas, the specific volume always increases in such a process, and, therefore, if the pipe is of constant diameter, the kinetic energy of the fluid increases. In many cases, however, this increase in kinetic energy is small (or perhaps the diameter of the exit pipe is larger than that of the inlet pipe) and we can say with a high degree of accuracy that in this process the final and initial enthalpies are equal.

It is for such a process that the Joule-Thomson coefficient, μ_J, is significant. A positive Joule-Thomson coefficient means that the temperature drops during throttling, and when the Joule-Thomson coefficient is negative the temperature rises during throttling.

Example 6.7

Steam at 100 lbf/in.^{2}, 500 F is throttled to 20 lbf/in.^{2} Changes in kinetic energy are negligible for this process. Determine the final temperature and specific volume of the steam, and the average Joule-Thomson coefficient.

For this process,

$$h_i = h_e = 1279.1 \text{ Btu/lbm}$$

$$P_e = 20 \text{ lbf/in.}^2$$

These two properties determine the final state. From the superheat table for steam

$$T_e = 483.9 \text{ F}$$

$$\mu_{J(\text{av})} = \left(\frac{\Delta T}{\Delta P}\right)_h = \frac{-16.1 \text{ F}}{-80 \text{ lbf/in.}^2} = 0.202 \text{ in.}^2\text{-F/lbf}$$

Frequently a throttling process involves a change in the phase of the fluid. A typical example is the flow through the expansion valve of a vapor compression refrigeration system. The following example deals with this problem.

Example 6.8

Consider the throttling process across the expansion valve or through the capillary tube in a vapor compression refrigeration cycle. In this process the pressure of the refrigerant drops from the high pressure in the condenser to the low pressure in the evaporator, and during this process some of the liquid flashes into vapor. If we consider this process to be adiabatic, the quality of the refrigerant entering the evaporator can be calculated.

Consider the following process, in which ammonia is the refrigerant. The ammonia enters the expansion valve at a pressure of 225 lbf/in.^{2} and a temperature of 90 F. Its pressure on leaving the expansion valve is 38.5 lbf/in.^{2}. Calculate the quality of the ammonia leaving the expansion valve.

Considering a steady-state, steady-flow process for a control surface around the expansion valve or capillary tube, we conclude that $\dot{Q}_{c.v.} = 0$, $\dot{W}_{c.v.} = 0$, $\Delta KE = 0$, $\Delta PE = 0$, and therefore the statement of the first law reduces to

$$h_i = h_e$$

From the ammonia tables

$$h_i = 143.6 \text{ Btu/lbm}$$

(The enthalpy of a slightly compressed liquid is essentially equal to the enthalpy of saturated liquid at the same temperature.)

$$h_e = h_i = 143.6 = 53.8 + x_e(561.1)$$

$$x_e = \frac{89.8}{561.1} = 0.160 = 16.0\%$$

6.5 The Uniform-State, Uniform-Flow Process

In Section 6.3 we considered the steady-state, steady-flow process and several examples of its application. Many processes of interest in thermodynamics involve unsteady flow, and do not fit into this category. A certain group of these—for example, filling closed tanks with a gas or liquid, or discharge from closed vessels—can be reasonably represented to a first approximation by another simplified model. We shall call this process the uniform-state, uniform-flow process, or for convenience, the USUF process. The basic assumptions are as follows:

1. The control volume remains constant relative to the coordinate frame.
2. The state of the mass within the control volume may change with time, but at any instant of time the state is uniform throughout the entire control volume (or over several identifiable regions that make up the entire control volume).
3. The state of the mass crossing each of the areas of flow on the control surface is constant with time although the mass flow rates may be time varying.

Let us examine the consequence of these assumptions and derive an expression for the first law that applies to this process. The assumption that the control volume remains stationary relative to the coordinate frame has already been discussed in Section 6.3. The remaining assumptions lead to the following simplifications for the continuity equations and the first law.

The overall process occurs during time t. At any instant of time during the process, the continuity equation is

$$\frac{dm_{\text{c.v.}}}{dt}+\Sigma\dot{m}_e-\Sigma\dot{m}_i=0$$

where the summation is over all areas on the control surface through which flow occurs. Integrating over time t gives the change of mass in the control volume during the overall process.

$$\int_0^t\left(\frac{dm_{\text{c.v.}}}{dt}\right)dt=(m_2-m_1)_{\text{c.v.}}$$

The total mass leaving the control volume during time t is

$$\int_0^t(\Sigma\dot{m}_e)\,dt=\Sigma m_e$$

and the total mass entering the control volume during time t is

$$\int_0^t(\Sigma\dot{m}_i)\,dt=\Sigma m_i$$

Therefore, for this period of time t we can write the continuity equation for the USUF process as

$$(m_2-m_1)_{\text{c.v.}}+\Sigma m_e-\Sigma m_i=0 \tag{6.26}$$

In writing the first law for the USUF process we consider Eq. 6.15, which applies at any instant of time during the process.

$$\dot{Q}_{\text{c.v.}} + \Sigma \dot{m}_i \left(h_i + \frac{\mathsf{V}_i^2}{2g_c} + Z_i \frac{g}{g_c}\right) = \frac{dE_{\text{c.v.}}}{dt} + \Sigma \dot{m}_e \left(h_e + \frac{\mathsf{V}_e^2}{2g_c} + Z_e \frac{g}{g_c}\right) + \dot{W}_{\text{c.v.}}$$

Since at any instant of time the state within the control volume is uniform, the first law for the USUF process becomes

$$\dot{Q}_{\text{c.v.}} + \Sigma \dot{m}_i \left(h_i + \frac{\mathsf{V}_i^2}{2g_c} + Z_i \frac{g}{g_c}\right) = \Sigma \dot{m}_e \left(h_e + \frac{\mathsf{V}_e^2}{2g_c} + Z_e \frac{g}{g_c}\right)$$

$$+ \frac{d}{dt}\left[(m)\left(u + \frac{\mathsf{V}^2}{2g_c} + Z\frac{g}{g_c}\right)\right]_{\text{c.v.}} + W_{\text{c.v.}}$$

Let us now integrate this equation over time t, during which time we have

$$\int_0^t \dot{Q}_{\text{c.v.}} dt = Q_{\text{c.v.}}$$

$$\int_0^t \left[\Sigma \dot{m}_i \left(h_i + \frac{\mathsf{V}_i^2}{2g_c} + Z_i \frac{g}{g_c}\right)\right] dt = \Sigma m_i \left(h_i + \frac{\mathsf{V}_i^2}{2g_c} + Z_i \frac{g}{g_c}\right)$$

$$\int_0^t \left[\Sigma \dot{m}_e \left(h_e + \frac{\mathsf{V}_e^2}{2g_c} + Z_e \frac{g}{g_c}\right)\right] dt = \Sigma m_e \left(h_e + \frac{\mathsf{V}_e^2}{2g_c} + Z_e \frac{g}{g_c}\right)$$

$$\int_0^t \dot{W}_{\text{c.v.}} dt = W_{\text{c.v.}}$$

$$\int_0^t \frac{d}{dt}\left[(m)\left(u + \frac{\mathsf{V}^2}{2g_c} + Z\frac{g}{g_c}\right)\right]_{\text{c.v.}} dt$$

$$= \left[m_2\left(u_2 + \frac{\mathsf{V}_2^2}{2g_c} + Z_2\frac{g}{g_c}\right) - m_1\left(u_1 + \frac{\mathsf{V}_1^2}{2g_c} + Z_1\frac{g}{g_c}\right)\right]_{\text{c.v.}}$$

Therefore, for this period of time t we can write the first law for the uniform-state, uniform-flow process as

$$Q_{\text{c.v.}} + \Sigma m_i \left(h_i + \frac{\mathsf{V}_i^2}{2g_c} + Z_i \frac{g}{g_c}\right) = \Sigma m_e \left(h_e + \frac{\mathsf{V}_e^2}{2g_c} + Z_e \frac{g}{g_c}\right)$$

$$+ \left[m_2\left(u_2 + \frac{\mathsf{V}_2^2}{2g_c} + Z_2\frac{g}{g_c}\right) - m_1\left(u_1 + \frac{\mathsf{V}_1^2}{2g_c} + Z_1\frac{g}{g_c}\right)\right]_{\text{c.v.}} + W_{\text{c.v.}} \qquad (6.27)$$

As an example of the type of problem for which these assumptions are valid and Eq. 6.27 is appropriate, let us consider the classic problem of flow into an evacuated vessel. This is the subject of Example 6.9.

Example 6.9

Steam at a pressure of 200 lbf/in.², 600 F, is flowing in a pipe, Fig. 6.12. Connected to this pipe through a valve is an evacuated tank. The valve is

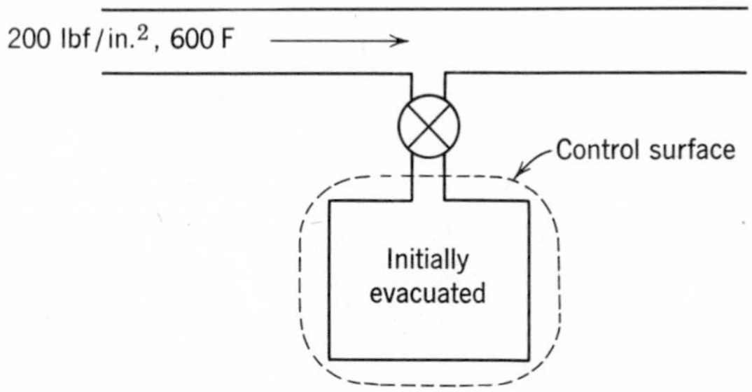

Fig. 6.12 Flow into an evacuated vessel—control volume analysis.

opened and the tank fills with steam until the pressure is 200 lbf/in.², and then the valve is closed. The process takes place adiabatically and kinetic energies and potential energies are negligible. Determine the final temperature of the steam.

Consider first the control volume shown in Fig. 6.12. The assumptions of uniform state in the control volume and across the pipe are both reasonable, and therefore the first law can be written as stated in Eq. 6.27.

$$Q_{\text{c.v.}} + \sum m_i\left(h_i + \frac{\mathsf{V}_i^2}{2g_c} + Z_i\frac{g}{g_c}\right) = \sum m_e\left(h_e + \frac{\mathsf{V}_e^2}{2g_c} + Z_e\frac{g}{g_c}\right)$$

$$+\left[m_2\left(u_2 + \frac{\mathsf{V}_2^2}{2g_c} + Z_2\frac{g}{g_c}\right) - m_1\left(u_1 + \frac{\mathsf{V}_1^2}{2g_c} + Z_1\frac{g}{g_c}\right)\right]_{\text{c.v.}} + W_{\text{c.v.}}$$

We note that $Q_{\text{c.v.}} = 0$, $W_{\text{c.v.}} = 0$, $m_e = 0$, $(m_1)_{\text{c.v.}} = 0$. We further assume that changes in kinetic and potential energy are negligible. Therefore, the statement of the first law for this process reduces to

$$m_i h_i = m_2 u_2$$

From the continuity equation for this process, Eq. 6.26, we conclude that

$$m_2 = m_i$$

Therefore, combining the continuity equation with the first law we have

$$h_i = u_2$$

That is, the final internal energy of the steam in the tank is equal to the enthalpy of the steam entering the tank.

From the steam tables

$$h_i = u_2 = 1322.1 \text{ Btu/lbm}$$

Since the final pressure is given as 200 lbf/in.^{2}, we know two properties of the final state and therefore the final state is determined. The temperature corresponding to a pressure of 200 lbf/in.^{2} and an internal energy of 1322.1 Btu/lbm is found to be 883 F. Had this problem involved a substance for which internal energies are not listed in the thermodynamic tables, it would have been necessary to calculate a few values for u before an interpolation could be made for the final temperature.

This problem can also be solved by considering the steam that enters the tank and the evacuated space as a system, as indicated in Fig. 6.13.

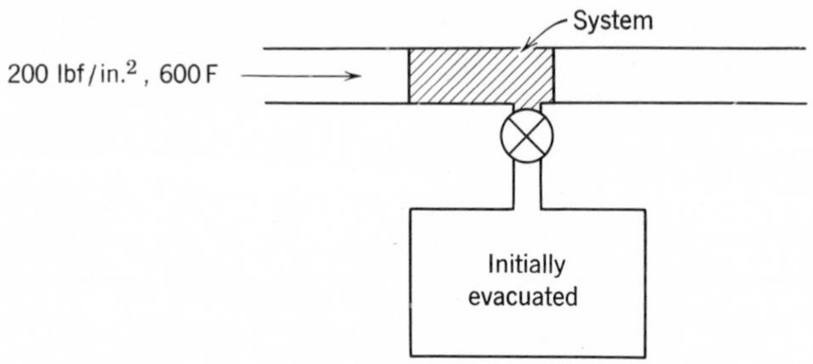

Fig. 6.13 Flow into an evacuated vessel—system analysis.

The process is adiabatic, but we must examine the boundaries for work. If we visualize a piston between the steam that is included in the system and the steam that flows behind, we readily recognize that the boundaries of the system move and that the steam in the pipe does work on the steam that comprises the system. The amount of this work is

$$-W = P_1V_1 = mP_1v_1$$

Writing the first law for the system, Eq. 5.11, and noting that kinetic and potential energies can be neglected, we have

$$
\begin{aligned}
{}_1Q_2 &= U_2 - U_1 + {}_1W_2 \\
0 &= U_2 - U_1 - P_1V_1 \\
0 &= mu_2 - mu_1 - mP_1v_1 = mu_2 - mh_1
\end{aligned}
$$

Therefore,

$$u_2 = h_1$$

which is the same conclusion which was reached using a control volume analysis.

The two other examples that follow illustrate further the uniform-state, uniform-flow process.

Example 6.10

Let the tank of the previous example have a volume of 10 ft^3 and initially contain saturated vapor at 50 lbf/in.^{2}. The valve is then opened and steam from the line at 200 lbf/in.^{2}, 600 F, flows into the tank until the pressure is 200 lbf/in.^{2}

Calculate the mass of steam that flows into the tank.

Let us again consider a control surface around the tank as indicated in Fig. 6.12. In this case, however, the control volume initially contains the mass m_1 and the internal energy m_1u_1.

Again we note that $Q_{\text{c.v.}} = 0$, $W_{\text{c.v.}} = 0$, $m_e = 0$, and we assume that changes in kinetic energy and potential energy are zero. The statement of the first law for this process, Eq. 6.27 reduces to

$$m_ih_i = m_2u_2 - m_1u_1$$

The continuity equation, Eq. 6.26 reduces to

$$m_2 - m_1 = m_i$$

Therefore, combining the continuity equation and the first law we have

$$(m_2 - m_1)h_i = m_2u_2 - m_1u_1$$

$$m_2(h_i - u_2) = m_1(h_i - u_1) \qquad (a)$$

There are two unknowns in this equation; namely, m_2 and u_2. However, we have one additional equation,

$$m_2v_2 = V = 10 \text{ ft}^3 \qquad (b)$$

These two equations, (*a*) and (*b*), can be solved simultaneously by trial and error. The correct solution is shown below:

$$v_1 = 8.515 \text{ ft}^3/\text{lbm} \qquad m_1 = \frac{10}{8.515} = 1.173 \text{ lbm}$$

$$h_i = 1322.1 \text{ Btu/lbm} \qquad u_1 = 1095.3 \text{ Btu/lbm}$$

Assume

$$T_2 = 666 \text{ F}$$

For this temperature and the known value of P_2,

$$v_2 = 3.269, \qquad u_2 = 1235.1$$

Then, from Eq. (*b*)

$$m_2 = \frac{10}{3.269} = 3.06 \text{ lbm}$$

while, from Eq. (*a*)

$$m_2 = m_1 \frac{(h_i - u_1)}{(h_i - u_2)} = \frac{1.173(1322.1 - 1095.6)}{1322.1 - 1235.1} = \frac{266}{87.0} = 3.06 \text{ lbm}$$

and we conclude that the assumed value for $T_2 = 666$ F is correct.

The mass of steam that flows into the tank is

$$m_2 - m_1 = 3.06 - 1.17 = 1.89 \text{ lbm}$$

Example 6.11

A tank of 50 ft^3 volume contains saturated ammonia at a pressure of 200 lbf/in.^{2}. Initially the tank contains 50 per cent liquid and 50 per cent vapor by volume. Vapor is withdrawn from the top of the tank until the pressure is 100 lbf/in.^{2}. Assuming that only vapor (i.e., no liquid) leaves and that the process is adiabatic, calculate the mass of ammonia that is withdrawn.

We consider a control volume around the tank and note that $Q_{c.v.} = 0$, $W_{c.v.} = 0$, $m_i = 0$, and we assume that changes in kinetic and potential energy are negligible. However, the enthalpy of saturated vapor varies with temperature, and therefore we cannot simply assume that the enthalpy of the vapor leaving the tank remains constant. However we note that at 200 lbf/in.^{2}, $h_g = 632.7$ Btu/lbm and at 100 lbf/in.^{2}, $h_g = 626.5$

Btu/lbm. Since the change in h_g during this process is small, we may accurately assume that h_e is the average of the two values given above. Therefore

$$(h_e)_{\text{av}} = 629.6 \text{ Btu/lbm}$$

With this assumption we can assume a uniform-state, uniform-flow process, and write:
First law (from Eq. 6.27):

$$m_e h_e + m_2 u_2 - m_1 u_1 = 0$$

Continuity equation (from Eq. 6.26):

$$(m_2 - m_1)_{\text{c.v.}} + m_e = 0$$

Combining these two equations we have

$$m_2 (h_e - u_2) = m_1 h_e - m_1 u_1$$

The following values are from the ammonia tables:

$$v_{f1} = 0.02732 \text{ ft}^3/\text{lbm} \qquad v_{g1} = 1.502 \text{ ft}^3/\text{lbm}$$

$$v_{f2} = 0.02584 \qquad v_{g2} = 2.952$$

$$u_{f1} = 150.9 - \frac{200 \times 144 \times 0.0273}{778} = 149.9 \text{ Btu/lbm}$$

$$u_{f2} = 104.7 - \frac{100 \times 144 \times 0.0258}{778} = 104.2$$

$$u_{g1} = 632.7 - \frac{200 \times 144 \times 1.502}{778} = 577.1$$

$$u_{g2} = 626.5 - \frac{100 \times 144 \times 2.952}{778} = 571.9$$

$$u_{fg2} = 571.9 - 104.2 = 467.7$$

Calculating first the initial mass, m_1, in the tank: the mass of the liquid initially present, m_{f1}, is

$$\text{m}_{f1} = \frac{25}{0.02732} = 915 \text{ lbm}$$

Similarly, the initial mass of vapor, m_{g1}, is

$$m_{g1} = \frac{25}{1.502} = 16.6 \text{ lbm}$$

$$m_1 = m_{f1} + m_{g1} = 915 + 17 = 932 \text{ lbm}$$
$$m_1 h_e = 932 \times 629.6 = 586{,}800 \text{ Btu}$$
$$m_1 u_1 = (mu)_{f1} + (mu)_{g1} = 915 \times 149.9 + 16.6 \times 577.0 = 146{,}700 \text{ Btu}$$

Substituting these into the first law,

$$m_2(h_e - u_2) = m_1 h_e - m_1 u_1 = 586{,}800 - 146{,}700 = 440{,}100 \text{ Btu}$$

There are two unknowns, m_2 and u_2, in this equation. However,

$$m_2 = \frac{V}{v_2} = \frac{50}{0.0258 + x_2(2.952 - 0.0258)}$$

and

$$u_2 = 104.2 + x_2(467.7)$$

both functions only of x_2, the quality at the final state. Consequently,

$$\frac{50(629.6 - 104.2 - 467.7\, x_2)}{0.0258 + 2.926\, x_2} = 440{,}100$$

Solving, $x_2 = 0.01137$

Therefore,

$$v_2 = 0.0258 + 0.01137\ (2.926) = 0.0590 \text{ ft}^3/\text{lbm}$$

$$m_2 = \frac{50}{0.0590} = 847 \text{ lbm}$$

and the mass of ammonia withdrawn, m_e, is

$$m_e = m_1 - m_2 = 932 - 847 = 85 \text{ lbm}$$

PROBLEMS

6.1 Freon 12 vapor enters a compressor at 30 lbf/in., 60 F, and the mass rate of flow is 5 lbm/min. What is the smallest diameter tubing that can be used if the velocity of refrigerant must not exceed 20 ft/sec?

6.2 Air is heated electrically in a constant-diameter tube in a steady-flow process. At the entrance, the air has a velocity of 10 ft/sec and is at 50 lbf/in.^{2}, 80 F. The air exits at 45 lbf/in.^{2}, 200 F. Calculate the velocity at the exit.

6.3 Water vapor is compressed in a centrifugal compressor. Dry saturated vapor at 100 F enters the compressor and the vapor leaves at 5 lbf/in.^{2}, 400 F. Heat is transferred from the vapor during the compression process at the rate of 2000 Btu/hr. The rate of flow of vapor is 300 lbm/hr. Calculate the horsepower required to drive the compressor.

6.4 Nitrogen gas is heated in a steady-state, steady-flow process. The inlet conditions are 80 lbf/in.^{2}, 100 F, and the exit conditions are 75 lbf/in.^{2}, 2000 F. Kinetic and potential energy changes are negligible. Calculate the required heat transfer per lbm of nitrogen.

6.5 Steam enters the nozzle of a turbine with a low velocity at 400 lbf/in.^{2}, 600 F, and leaves the nozzle at 250 lbf/in.^{2} at a velocity of 1450 ft/sec. The rate of flow of steam is 3000 lbm/hr. Calculate the quality or temperature of the steam leaving the nozzle and the exit area of the nozzle.

6.6 A small, high-speed turbine operating on compressed air produces 1/10 HP. The inlet and exit conditions are 60 lbf/in.^{2}, 80 F and 14.7 lbf/in.^{2}, -60 F, respectively. Assuming the velocities to be low, find the required mass flow rate of air.

6.7 In a commercial refrigerator, Freon-12 enters the compressor as a saturated vapor at -10 F and leaves at a pressure of 200 lbf/in.^{2} and temperature of 160 F. The mass rate of flow is 5 lbm/min and the heat loss from the compressor is 2 Btu/lbm of Freon-12. What horsepower is required to drive the compressor?

6.8 In an air liquefaction plant, air flows through an expansion engine at the rate of 600 lbm/hour. The pressure and temperature of the air entering the expansion engine are 220 lbf/in.^{2}, -80 F, and on leaving the pressure is 25 lbf/in.^{2} and the temperature is -170 F. The heat transfer to the air as it flows through the expansion engine is equal to 10% of the power output of the expansion engine. Determine the power output and the heat transfer per hour from the expansion engine.

6.9 The construction of a one-mile-high skyscraper has been proposed. Suppose that in such a skyscraper heating steam is to be supplied to the top floor via a vertical pipe. Steam enters the pipe at ground level as dry saturated vapor at a pressure of 30 lbf/in.^{2}. At the top of the pipe the pressure is 15 lbf/in.^{2}, and the heat transfer from the steam as it flows up the pipe is 50 Btu/lbm. What is the quality of the steam at the top of the pipe?

6.10 In a nuclear reactor steam generator, 2.94 ft^3/min of water enters a $\frac{3}{4}$ in. diameter tube at a pressure of 1000 lbf/in.^{2} and temperature of 100 F, and leaves the tube as a saturated vapor at 900 lbf/in.^{2}. Find the heat transfer rate to the water in Btu/hr.

6.11 Consider the steady-state, steady-flow heat exchanger shown in Fig. 6.14. Freon-12 enters at point 1 and exits at point 3, while helium enters at point

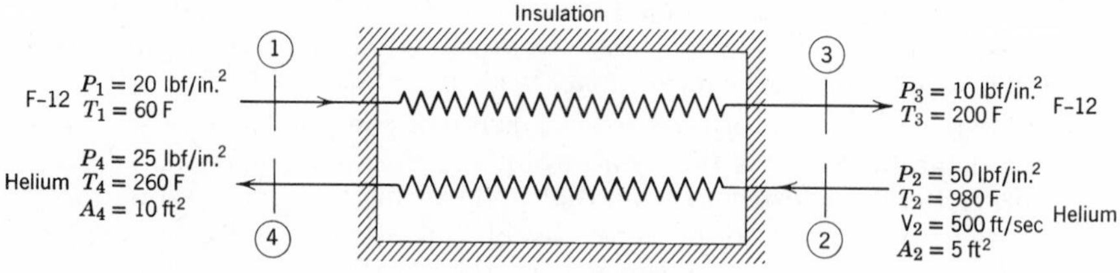

Fig. 6.14 Sketch for Problem 6.11.

2 and exits at point 4. The conditions are as shown in the figure. For this process, determine the following.

(*a*) The exit velocity of the helium.

(*b*) The mass flow rate of the Freon-12.

6.12 Liquid ammonia at a temperature of 60 F and a pressure of 180 lbf/in.^{2} is mixed in a steady-state, steady-flow process with saturated ammonia vapor at a pressure of 180 lbf/in.^{2}. The mass rates of flow of liquid and vapor are equal, and after mixing the pressure is 140 lbf/in.^{2} and the quality is 85%. Determine the heat transfer per lbm of mixture.

6.13 In certain situations, when only superheated steam is available, a need for saturated steam may arise for a specific purpose. This can be accomplished in a desuperheater, in which case water is sprayed into the superheated steam in such amounts that the steam leaving the superheater is dry and saturated. The following data apply to such a desuperheater, which operates as a steady-flow process. Superheated steam at the rate of 2000 lbm/hr, at 400 lbf/in.^{2}, 600 F, enters the desuperheater. Water at 420 lbf/in.^{2}, 100 F, also enters the desuperheater. The dry saturated vapor leaves at 380 lbf/in.^{2}. Calculate the rate of flow of water.

6.14 The following data are for a simple steam power plant as shown in Fig. 6.15.

$P_1 = 900$ lbf/in.^{2}
$P_2 = 890$ lbf/in.^{2}, $T_2 = 115$ F
$P_3 = 860$ lbf/in.^{2}, $T_3 = 350$ F
$P_4 = 830$ lbf/in.^{2}, $T_4 = 920$ F
$P_5 = 800$ lbf/in.^{2}, $T_5 = 900$ F
$P_6 = 1.5$ lbf/in.^{2}, $x_6 = 0.92$, $V_6 = 600$ ft/sec
$P_7 = 1.4$ lbf/in.^{2}, $T_7 = 110$ F
Rate of steam flow = 200,000 lbm/hr
Power to pump = 400 hp
Pipe diameters:
Steam generator to turbine: 8 in.
Condenser to steam generator: 3 in.

Calculate:

(*a*) Power output of the turbine.

(*b*) Heat transfer per hour in condenser, economizer, and steam generator.

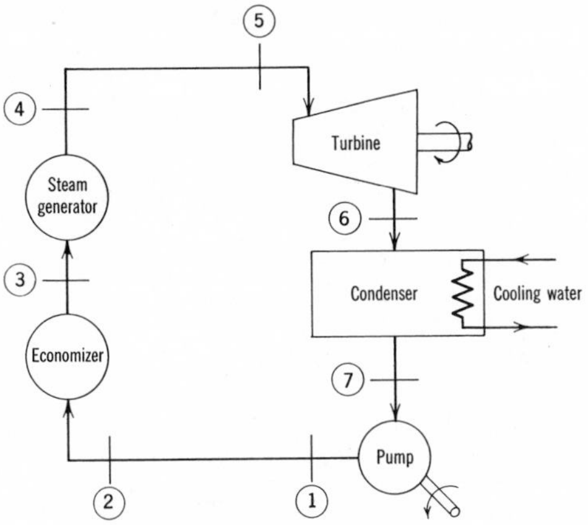

Fig. 6.15 Sketch for Problem 6.14.

(*c*) Diameter of pipe connecting the turbine to the condenser.

(*d*) Gallons of cooling water per minute through the condenser if the temperature of the cooling water increases from 55 F to 75 F in the condenser.

6.15 A somewhat simplified flow diagram for the nuclear power plant shown in Fig. 1.5 is given in Fig. 6.16. The mass flow rates and state of the steam at various points in the cycle are shown in the table below. This cycle involves a number of "heaters." In these units, heat is transferred from steam which leaves the turbine at some intermediate pressure to liquid water which is being pumped from the condenser to the steam drum. The rate of heat transfer to the H_2O in the reactor is 537×10^6 Btu/hr.

(*a*) Assuming no heat transfer from the moisture separator between the high-pressure and low-pressure turbines, determine the enthalpy per pound of steam and the quality of the steam entering the low-pressure turbine.

(*b*) Determine the power output of the high-pressure turbine, assuming no heat transfer from the steam as it flows through the turbine.

(*c*) Determine the power output of the low-pressure turbine, assuming no heat transfer from the steam as it flows through the turbine.

(*d*) Determine the quality of the steam leaving the reactor.

(*e*) Determine the temperature of the water leaving the intermediate-pressure heater, assuming no heat transfer from the heater to the surroundings.

(*f*) Determine the rate of heat transfer (Btu/hr) to the condenser cooling water.

(*g*) What is the ratio of the total power output of the two turbines to the heat transferred to the H_2O in the reactor?

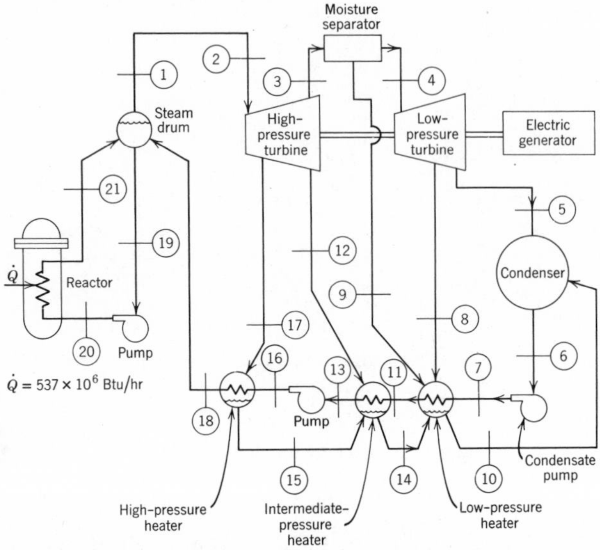

Fig. 6.16 Sketch for Problem 6.15.

Point	$\dot{m}$ – lbm/hr	P – lbf/in.^{2}	T – F	h – Btu/lbm
1	600,000	1050	sat vap	
2	600,000	1000		1189
3	499,000	50		1082
4		45		
5		1		980
6	600,000	1	92	
7		60		60
8	22,000	5		1057
9	37,000	45		240
10		5	130	
11	600,000	55	155	
12	64,000	50		1082
13	600,000	48		
14				150
15	37,000	140	282	251
16	600,000	1150		243
17	37,000	140		1115
18	600,000	1100		296
19	11,000,000	1050	530	
20	11,000,000	1075		525
21	11,000,000	1060		

6.16 A schematic diagram of a hydroelectric plant is shown in Fig. 6.17. Water enters the inlet conduit at the level of the lake, and work is done by the water in the hydraulic turbine. Assume no change in kinetic energy or internal energy u in the inlet conduit, and that the kinetic energy and internal energy of the water leaving the hydraulic turbine is the same as that in the inlet conduit. Calculate the power developed by the hydraulic turbine considering each of the three control volumes shown.

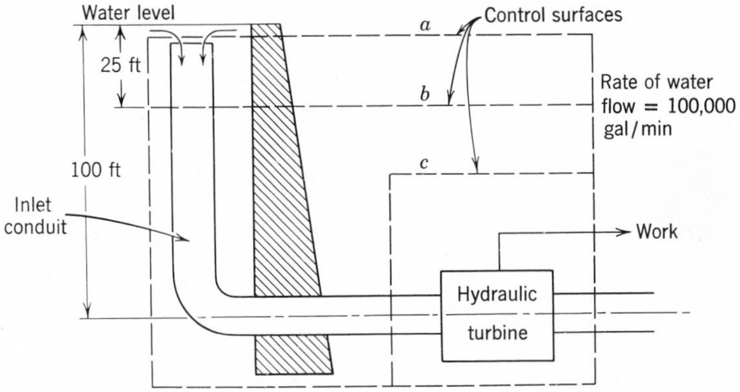

Fig. 6.17 Sketch for Problem 6.16.

6.17 A steam turbine is used to drive a nitrogen compressor, as shown in Fig. 6.18. The states of the steam and nitrogen are given in the diagram, and the mass flow rates are 1000 lbm/hr and 150 lbm/hr through the turbine and compressor, respectively. Heat transfer from the turbine is small and can be neglected. The turbine delivers 16 hp to the compressor and the balance to an electric power generator.

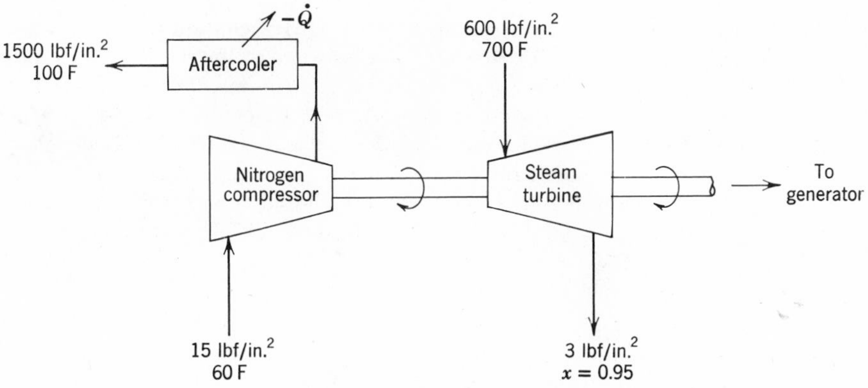

Fig. 6.18 Sketch for Problem 6.17.

(*a*) Determine the power available from the turbine for driving the electric power generator.

(*b*) Determine the rate of heat transfer from the nitrogen as it flows through the compressor and aftercooler.

6.18 A schematic arrangement for a proposed procedure for producing fresh water from salt water which would operate in conjunction with a large steam power plant and utilize a flash evaporator is shown in Fig. 6.19. Cooling water at the rate of 2,500,000 lbm/hr enters the condenser, where its

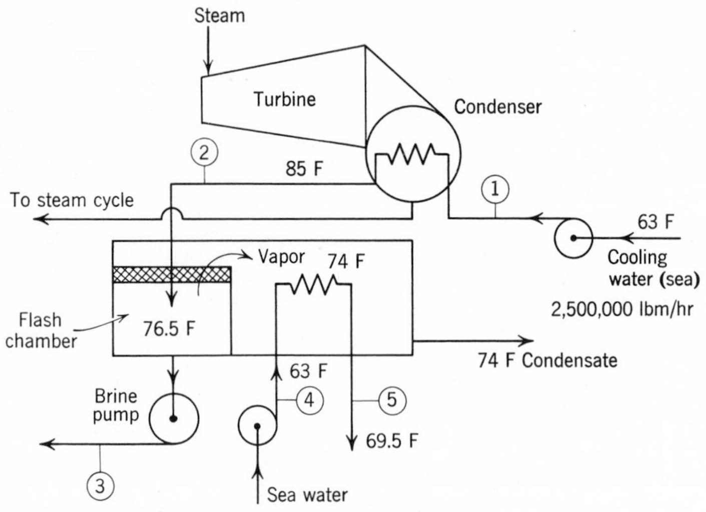

Fig. 6.19 Sketch for Problem 6.18.

temperature is increased from 63 F to 85 F. It then enters a flash evaporator, where the pressure is reduced to that corresponding to a saturation temperature of 76.5 F. During this process some of the liquid flashes into vapor, and the remainder is pumped back to the sea. The vapor is then condensed at 74 F to form the desired fresh water. This takes place by utilizing the sea water which enters at 63 F and leaves at 69.5 F. Calculate:

(*a*) The amount of fresh water which is produced per hour. Assume that the mixture which is formed when the liquid at 85 F is throttled (as it enters the flash evaporator) is in equilibrium at 76.5 F, and that it is perfectly separated.

(*b*) The amount of cooling water which enters at 4.

6.19 (*a*) Air flows in a pipe in which frictional effects are present. At one point in the pipe the pressure is 100 lbf/in.^{2}, 200 F, and the velocity is 300 ft/sec. At a certain point down stream the pressure is 70 lbf/in.^{2}. Assuming no heat transfer during the process, determine the final velocity and temperature of the air. (*b*) What is the temperature change of an incompressible liquid as it flows through a pipe with a pressure drop due to frictional effects?

6.20 Water at a pressure of 1500 lbf/in.^{2}, 300 F, is throttled to a pressure of 30 lbf/in.^{2} in an adiabatic process. What is the quality after throttling?

6.21 Nitrogen at 80 F, 50 lbf/in.^{2}, is throttled to 14.7 lbf/in.^{2}. Calculate the exit temperature, assuming

(*a*) Real gas behavior.

(*b*) Ideal gas behavior.

6.22 A throttling calorimeter is a device which is used to determine the quality of wet steam (i.e., steam which has a small amount of moisture present) flowing through a pipe. This device involves taking a small but continuous flow of steam from the pipe and throttling it in an adiabatic process to approximately atmospheric pressure. After this adiabatic throttling (constant enthalpy) process the pressure and temperature of the steam are measured, and thus the enthalpy of the wet steam in the line is known.

A throttling calorimeter is used to measure the quality of steam in a pipe in which the absolute pressure is 180 lbf/in.^{2}. A mercury manometer is used to measure the pressure in the calorimeter, and shows a pressure in the calorimeter of 2.5 in. Hg above atmospheric pressure. If a minimum of 10 degrees of superheat is required in the calorimeter, what is the minimum quality of steam that can be determined? The barometer reads 29.43 in. Hg.

6.23 A small steam turbine operating at part load produces 100 hp output with a flow rate of 1350 lbm/hr. Steam at 200 lbf/in.^{2}, 450 F, is throttled to 160 lbf/in.^{2} before entering the turbine, and the exhaust pressure is 1 lbf/in.^{2}. Find the quality (or temperature, if superheated) at the turbine outlet.

6.24 The mixing process indicated in Fig. 6.20 is used to obtain saturated liquid nitrogen at 28.12 lbf/in.^{2}. The high pressure nitrogen gas is throttled to

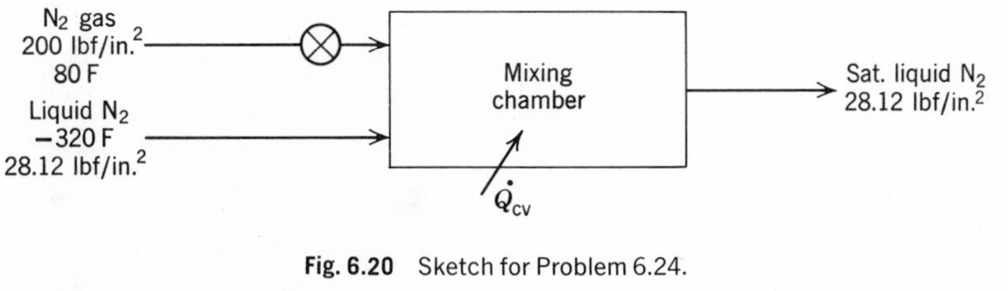

Fig. 6.20 Sketch for Problem 6.24.

the mixing chamber pressure and then mixed with the subcooled liquid, which flows into the chamber at the rate of 6 lbm/min. If the heat transfer to the chamber is 2.7 Btu/min, what is the mass flow rate of nitrogen gas into the chamber?

6.25 The following data are for the Freon-12 refrigeration cycle shown in Fig. 6.21:

$P_1 = 180$ lbf/in.^{2}, $T_1 = 240$ F
$P_2 = 178$ lbf/in.^{2}, $T_2 = 220$ F
$P_3 = 175$ lbf/in.^{2}, $T_3 = 100$ F
$P_4 = 29$ lbf/in.^{2},
$P_5 = 27$ lbf/in.^{2}, $T_5 = 20$ F
$P_6 = 25$ lbf/in.^{2}, $T_6 = 40$ F
Rate of flow of Freon = 200 lbm/hr
Power input to compressor = 2.5 hp

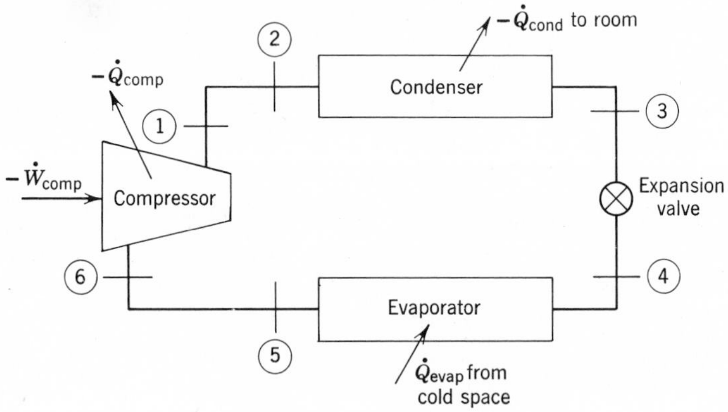

Fig. 6.21 Sketch for Problem 6.25.

Calculate:

(*a*) The heat transfer per hour from the compressor.
(*b*) The heat transfer per hour from the Freon in the condenser.
(*c*) The heat transfer per hour to the Freon in the evaporator.

6.26 The following data are from the test of a large refrigeration unit utilizing ammonia as the refrigerant.

	Pressure	Temperature
Leaving compressor	260 lbf/in.^{2}	
Entering condenser	250 lbf/in.^{2}	200 F
Leaving condenser, entering expansion valve	240 lbf/in.^{2}	105 F
Leaving expansion valve, entering the evaporator	35 lbf/in.^{2}	
Leaving the evaporator, entering compressor	30 lbf/in.^{2}	20 F

Assume no heat transfer from the compressor. The power input to the compressor is 140 hp. The capacity of the plant is 1,000,000 Btu/hr (i.e., the heat transfer to the refrigerant in the evaporator).

(*a*) Determine the rate of flow of the ammonia.

(*b*) What is the temperature of the ammonia leaving the compressor?

(*c*) What is the rate of heat transfer from the ammonia in the condenser?

6.27 A steam turbine driven generator is shown in Fig. 6.22. The turbine consists of a high-pressure turbine and a low-pressure turbine on the same shaft. The steam approaching the turbine in the steam line is at 600 lbf/in.^{2},

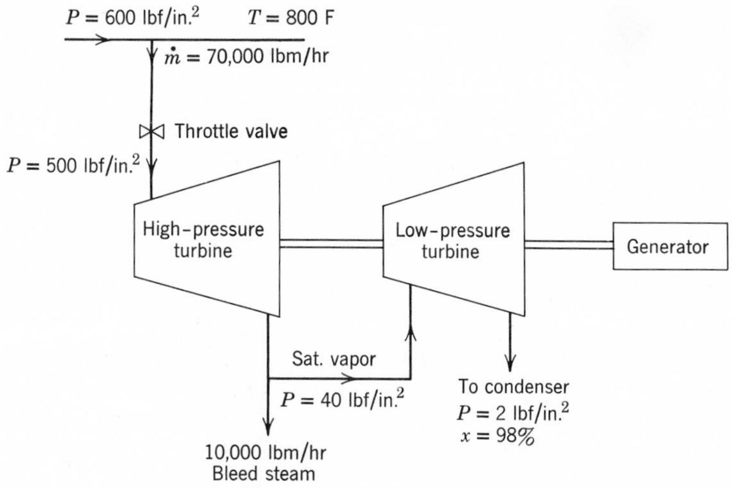

Fig. 6.22 Sketch for Problem 6.27.

800 F. Before entering the turbine the steam flows across a throttle valve causing a reduction in pressure to 500 lbf/in.^{2}. The rate of steam flow to the high-pressure turbine is 70,000 lbm/hr. The steam leaves the high-pressure turbine and enters the low-pressure turbine as saturated vapor at 40 lbf/in.^{2}. At this point 10,000 lbm/hr are bled off to be used in a chemical process and the rest of the steam flows through the low-pressure turbine. The steam leaves the low-pressure turbine at 2 lbf/in.^{2} pressure and a quality of 98% and flows to a condenser.

Determine the power output of each turbine.

6.28 Consider the process indicated in Fig. 6.23 for producing liquid Freon-12. Freon-12 at 400 lbf/in.^{2}, 200 F enters a heat exchanger and is cooled by the saturated vapor being withdrawn from the insulated liquid receiver. The high pressure gas is then throttled across a valve to the liquid receiver pressure. The liquid receiver contains liquid and vapor in equilibrium at −10 F and the saturated vapor leaving the heat exchanger is 180 F. Neglecting all pressure losses except across the expansion valve, determine:

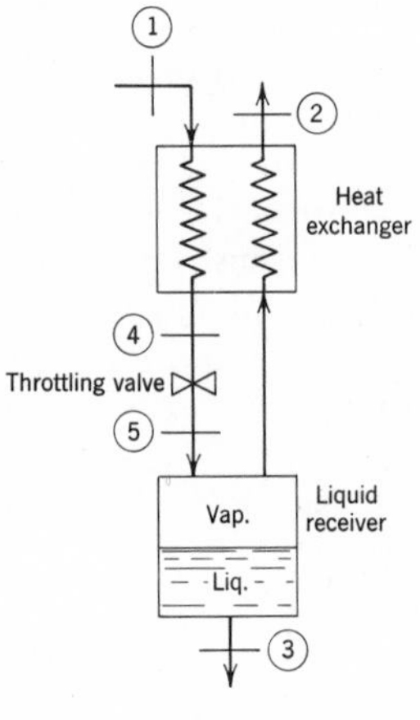

Fig. 6.23 Sketch for Problem 6.28.

(*a*) The fraction of high pressure gas which is liquified.

(*b*) The pressure and temperature (if superheated) or quality (if saturated) at the inlet and outlet of the valve.

6.29 Ammonia vapor flows through a pipe at a pressure of 140 lbf/in.^{2} and a temperature of 160 F. Attached to the pipeline is an evacuated vessel having a volume of 1 ft^3. The valve in the line to this evacuated vessel is opened, and the pressure in the vessel comes to 140 lbf/in.^{2}, at which time the valve is closed. If this process occurs adiabatically, how much ammonia flows into the vessel?

6.30 Air flows in a pipeline at a pressure of 100 lbf/in.^{2} and a temperature of 80 F. Connected to this tank is an evacuated vessel. When the valve on this tank is opened, air flows into the tank until the pressure is 100 lbf/in.^{2}.

(*a*) If this process occurs adiabatically, what is the final temperature of the air?

(*b*) Determine a general expression that gives the relation between the temperature of a gas flowing into an evacuated vessel and the final temperature of the gas in the tank in terms of the thermodynamic properties of the gas.

6.31 A tank having a volume of 200 ft^3 contains saturated vapor steam at a pressure of 20 lbf/in.^{2}. Attached to this tank is a line in which vapor at 100 lbf/in.^{2}, 400 F, flows. Steam from this line enters the vessel until the pressure is 100 lbf/in.^{2}. If there is no heat transfer from the tank and the heat capacity of the tank is neglected, calculate the mass of steam that enters the tank.

6.32 Freon-12 is contained in a 1 ft^3 tank at 80 F, 15 lbf/in.^{2}. It is desired to fill the tank 60% full of liquid (by volume) at this temperature. The tank is connected to a line flowing F-12 at 100 lbf/in.^{2}, 100 F, and the valve opened slightly.

(*a*) Calculate the final mass in the tank at 80 F.

(*b*) Determine the required heat transfer during the filling process if the temperature is to remain at 80 F.

6.33 An evacuated 2-ft³ tank is connected to a line flowing air at room temperature, 80 F, and 1000 lbf/in.². The valve is then opened, allowing air to flow into the tank until the tank pressure reaches 700 lbf/in.², at which time the valve is closed. This filling process occurs rapidly and is essentially adiabatic. The tank is then allowed to sit for a long time with the valve closed, and eventually returns to room temperature. What is the final pressure inside the tank?

6.34 Steam is flowing in a line at 100 lbf/in.², 500 F. An insulated vessel containing a piston and spring is connected to the line, as shown in Fig. 6.24.

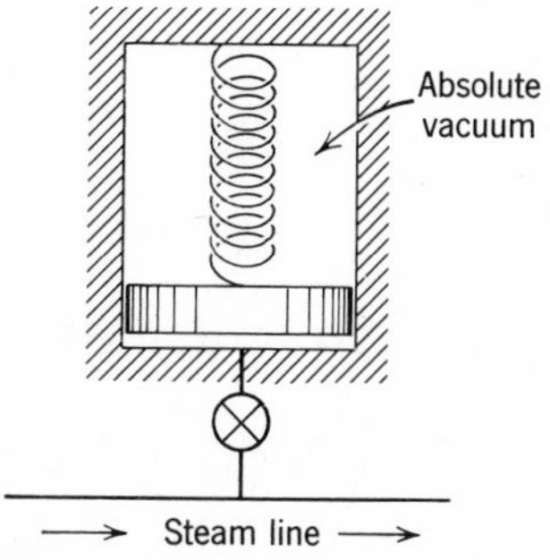

Fig. 6.24 Sketch for Problem 6.34.

Initially, the spring force is zero, but when the valve is opened and steam enters, the resisting force is directly proportional to the distance moved. Find the temperature inside when the pressure reaches 100 lbf/in.².

6.35 Consider the device shown in Fig. 6.25. Steam flows in a steam line at 100

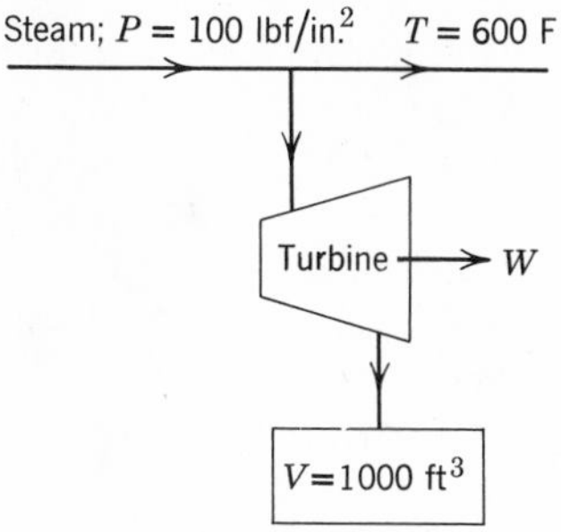

Fig. 6.25 Sketch for Problem 6.35.

lbf/in.², 600 F. From this steam line, steam flows through a steam turbine. The steam exhausts into large chamber having a volume of 1000 ft³. Initially, this chamber is evacuated. The turbine can operate until the pressure in the chamber is 100 lbf/in.². At this point the steam temperature is 550 F. Assume the entire process to be adiabatic.

Determine the work done by the turbine during this process.

6.36 Consider the "do-it-yourself" power plant to be operated on the kitchen stove, shown in Fig. 6.26. A steam kettle having a volume of 1 ft³ contains

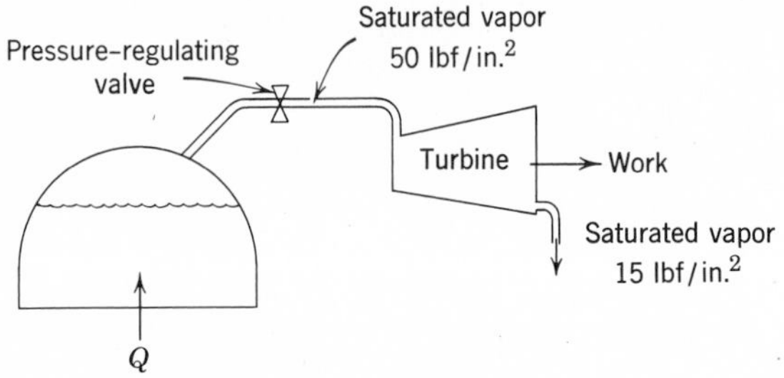

Fig. 6.26 Sketch for Problem 6.36.

90% liquid and 10% vapor (by volume) at 14.7 lbf/in.². The pressure-regulating valve is then adjusted to 50 lbf/in.², and heat is transferred to the water. When the pressure reaches 50 lbf/in.², saturated vapor flows from the kettle to the turbine. The steam leaves the turbine as saturated vapor at 15 lbf/in.². This process is continued until the kettle is filled with 10% liquid and 90% vapor by volume (at 50 lbf/in.²). There is no heat transfer from the turbine.

(*a*) Determine the total work done by the turbine during this process.

(*b*) Determine the total heat transfer during this process.

6.37 An insulated tank having a volume of 20 ft³ contains dry saturated steam at 800 lbf/in.². Steam is withdrawn through a line from the top of the tank until the pressure is 400 lbf/in.². Calculate the mass of steam that is withdrawn, assuming that at any moment the tank contains a homogeneous mixture of liquid and vapor, and that this homogeneous mixture is withdrawn from the tank.

6.38 A 60-ft³ insulated vessel contains saturated vapor steam at 400 lbf/in.². A valve at the top of the vessel is then opened and steam is withdrawn. During the process, liquid collects at the bottom of the vessel, such that only saturated vapor leaves. Calculate the total mass withdrawn when the pressure inside reaches 100 lbf/in.².

6.39 A pressure vessel having a volume of 30 ft³ contains saturated steam at 500 F. The vessel initially contains 50% vapor and 50% liquid by volume.

Liquid is withdrawn slowly from the bottom of the tank, and heat is transferred to the tank in order to maintain constant temperature. Determine the heat transfer to the tank when half of the contents of the tank has been removed.

6.40 Usually a cryogenic fluid is stored in a vacuum insulated tank, Fig. 6.27. One way of discharging the fluid from such a tank is to withdraw some

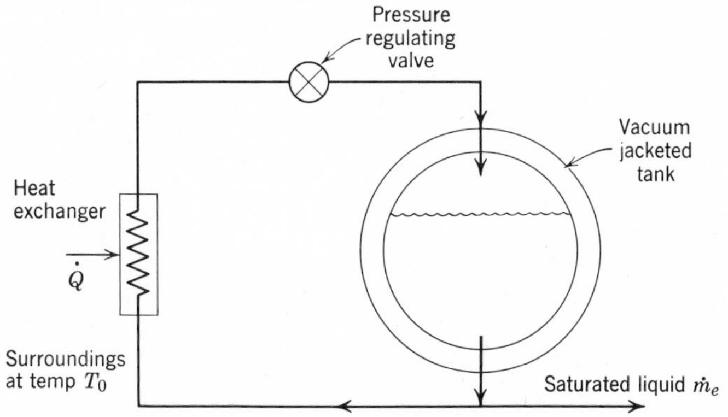

Fig. 6.27 Sketch for Problem 6.40.

of the liquid from the storage tank and vaporize it in a heat exchanger, as shown in the sketch, and then feed the vapor to the top of the tank at such a rate as to maintain constant pressure. The heat transferred to the fluid is from the surroundings which are at temperature T_0. Suppose the fluid in the tank is saturated at a given pressure, and it is desired to withdraw fluid from the tank at this pressure at the rate of $\dot{m}_e$. Assume that the vapor leaving the heat exchanger and entering the storage tank is saturated vapor at the given pressure.

Determine the heat transfer rate to the heat exchanger, $\dot{Q}$, as a function of the mass rate of liquid flow $\dot{m}_e$, and the thermodynamic properties of the fluid.

6.41 A 10-ft^3 low-temperature storage tank initially contains nitrogen at 14.7 lbf/in.^{2}, 80% liquid and 20% vapor by volume. Heat transfer to the tank from the surroundings is constant at the rate of 30 Btu/hr, and causes the tank pressure to rise. The tank is fitted with a relief valve, such that once a pressure of 75 lbf/in.^{2} is reached, saturated vapor will be discharged to the surroundings as necessary to maintain that pressure.

(*a*) How much mass has been discharged from the tank by the time the tank contains 50% liquid, 50% vapor by volume at 75 lbf/in.^{2}?

(*b*) How long does it take to reach this state?

6.42 A popular demonstration involves making ice by pumping a vacuum over liquid water until the pressure is reduced to less than the triple-point pressure. Such an apparatus is shown schematically in Fig. 6.28. In this case the tank has a total volume of 1 ft^3. Initially the tank contains 0.9 ft^3 of saturated vapor (H_2O) and 0.1 ft^3 of saturated liquid. The initial temperature is 80 F. Assume no heat transfer during this process.

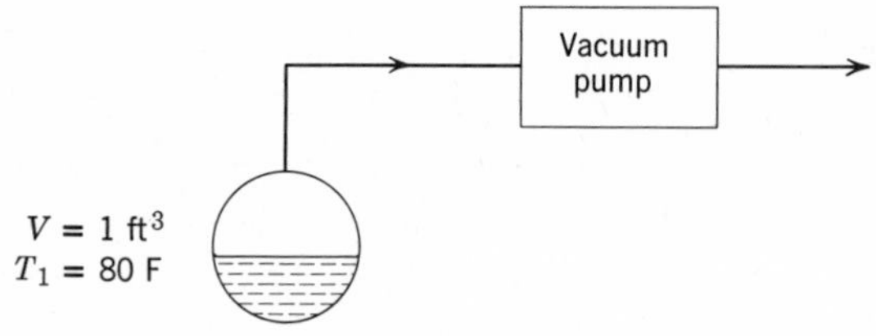

Fig. 6.28 Sketch for Problem 6.42.

(*a*) Determine what fraction of the initial mass will be pumped off when the liquid and vapor first reach 32 F.

(*b*) Determine what fraction of the initial mass can be solidified.

6.43 Liquids are often transferred from a tank by pressurization with a gas. Consider the problem shown in Fig. 6.29. Tanks *A* and *B* both contain

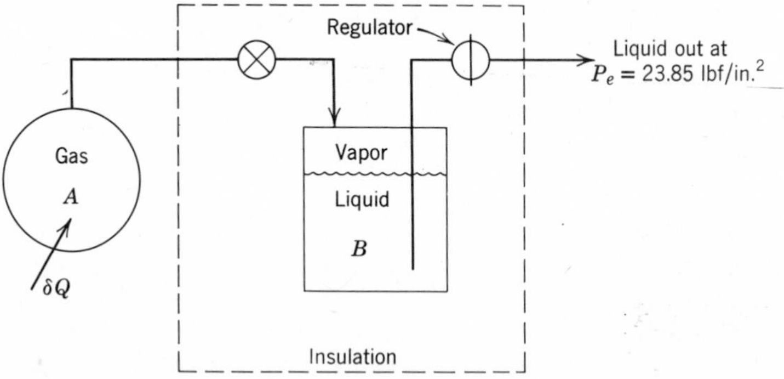

Fig. 6.29 Sketch for Problem 6.43.

Freon-12. Tank *A* has a volume of 3.2 ft^3, and *B* a volume of 2.5 ft^3. The Freon in *A* is initially at 80 F, saturated vapor, and that in *B* is at -40 F with a quality of 0.01. The valve is opened slightly, allowing Freon to flow from *A* to *B*. When the pressure in *B* has built up to 23.85 lb/in.^{2}, liquid begins to flow out. The liquid transfer continues until the pressure in *A* has dropped to 23.85 lbf/in.^{2}. During the process, heat is transferred to the gas inside tank A such that its temperature always remains at 80 F, but tank B is insulated. Determine

(*a*) The quality in tank B at the end of the process.

(*b*) The mass of liquid transferred from tank B.

6.44 Nitrogen is to be stored at low temperature in an insulated, high-pressure spherical tank, 2 ft in diameter. The rate of heat transfer to the tank from the room is 10 Btu/hr, and may be assumed constant. One method proposed for venting the tank is shown in Fig. 6.30. The regulator maintains

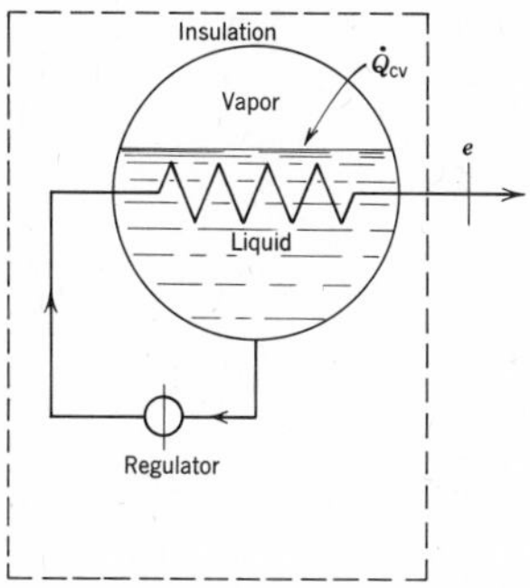

Fig. 6.30 Sketch for Problem 6.44.

tank pressure at 75 lbf/in.^{2} by throttling liquid to low pressure, after which it is passed through the tank and exhausts as a gas at 14.7 lbf/in.^{2}, 165 R. At some particular time, the tank contains 50% liquid, 50% vapor by volume at 75 lbf/in.^{2}. How much nitrogen will be lost during the next 5 days?

6.45 A spherical drop of liquid is suspended in an infinite atmosphere as shown in Fig. 6.31. Owing to differences in temperature it is exchanging heat with its surroundings. It also is losing mass by evaporation to the surroundings. The instantaneous rate of heat transfer $\dot{Q}$ is expressed by

$$\dot{Q} = KA(T_0 - T)$$

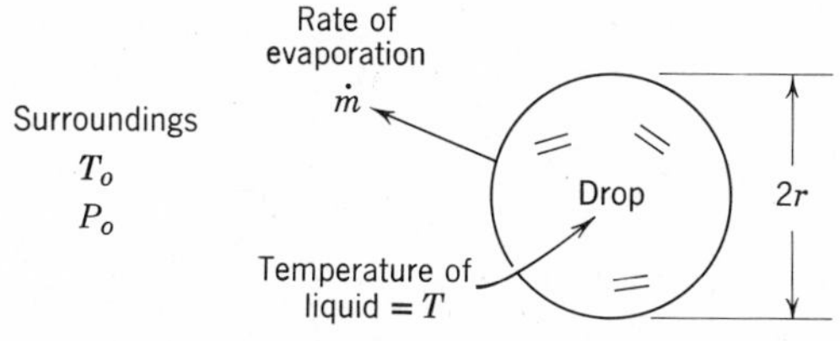

Fig. 6.31 Sketch for Problem 6.45.

where K is a constant and A is the surface area. The instantaneous rate of mass transfer is given by the symbol $\dot{m}$. It may be assumed that at a given instant of time the sphere is uniform in temperature and always at a pressure equal to that of the surroundings, P_0. The vapor of the evaporating liquid in the region adjacent to the drop is saturated.

Derive an expression for the instantaneous time rate of change of the drop temperature, $\partial T/\partial t$ in terms of the significant physical quantities.

6.46 Write a computer program to solve the following problem. An insulated tank of volume V contains a specified ideal gas (with constant specific heat) at P_1, T_1. A valve is opened, allowing the gas to flow out until the pressure inside drops to P_2. Determine T_2 and m_2 using a stepwise solution in increments of pressure between P_1 and P_2, where the number of increments is variable.

7

The Second Law of Thermodynamics

The first law of thermodynamics states that during any cycle that a system undergoes, the cyclic integral of the heat is equal to the cyclic integral of the work. The first law, however, places no restrictions on the direction of flow of heat and work. A cycle in which a given amount of heat is transferred from the system and an equal amount of work is done on the system satisfies the first law just as well as a cycle in which the flows of heat and work are reversed. However, we know from our experience that the fact that a proposed cycle does not violate the first law does not insure that the cycle will actually occur. It is this kind of experimental evidence that has led to the formulation of the second law of thermodynamics. Thus a cycle will occur only if both the first and second laws of thermodynamics are satisfied.

In its broader significance the second law involves the fact that processes proceed in a certain direction but not in the opposite direction. A hot cup of coffee cools by virtue of heat transfer to the surroundings, but heat will not flow from the cooler surroundings to the hotter cup of coffee. Gasoline is used as a car drives up a hill, but on coasting down the hill, the fuel level in the gasoline tank cannot be restored to its original level. Such familiar observations as these, and a host of others, are evidence of the validity of the second law of thermodynamics.

We will first consider this second law for a system undergoing a cycle and in the next two chapters will extend the principles to a system undergoing a change of state and then to a control volume.

7.1 Heat Engines and Refrigerators

Consider the system and the surroundings previously cited in the development of the first law, as shown in Fig. 7.1. Let the gas constitute the system and, as in our discussion of the first law, let this system undergo a cycle in which work is first done on the system by the paddle wheel

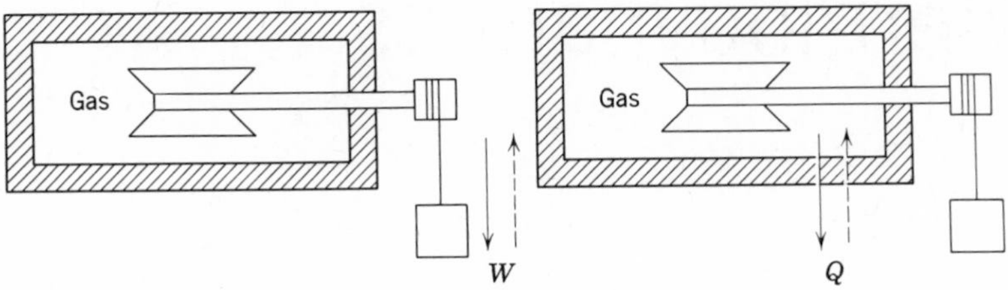

Fig. 7.1 A system that undergoes a cycle involving work and heat.

as the weight is lowered. Then let the cycle be completed by transferring heat to the surroundings.

We know from our experience, however, that we cannot reverse this cycle. That is, if we transfer heat to the gas, as shown by the dotted arrow, the temperature of the gas will increase, but the paddle wheel will not turn and raise the weight. With the given surroundings (the container, the paddle wheel, and the weight) this system can operate in a cycle in which the heat transfer and work are both negative, but it cannot operate in a cycle in which both the heat transfer and work are positive, even though this would not violate the first law.

Consider another cycle, which we know from our experience is impossible to actually accomplish. Let two systems, one at a high temperature and the other at a low temperature, undergo a process in which a quantity of heat is transferred from the high-temperature system to the low-temperature system. We know that this process can take place. We also know that the reverse process, in which heat is transferred from the low-temperature system to the high-temperature system, does not occur, and that it is impossible to complete the cycle by heat transfer only. This is illustrated in Fig. 7.2.

These two illustrations lead us to the consideration of the heat engine and refrigerator, which is also referred to as a heat pump. With the heat engine we can have a system that operates in a cycle and has a net positive work and a net positive heat transfer. With the heat pump we can have a system that operates in a cycle and has heat transferred to it from a low-temperature body and heat transferred from it to a high-temperature

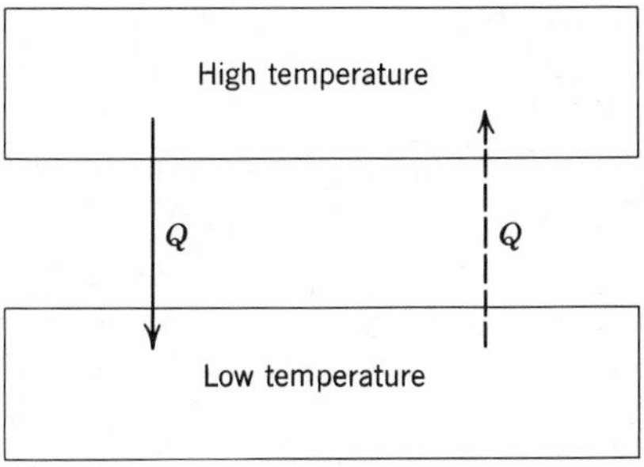

Fig. 7.2 An example showing the impossibility of completing a cycle by transferring heat from a low-temperature body to a high-temperature body.

body, though work is required to do this. Three simple heat engines and two simple refrigerators will be considered.

The first heat engine is shown in Fig. 7.3, and consists of a cylinder fitted with appropriate stops and a piston. Let the gas in the cylinder consistitute the system. Initially the piston rests on the lower stops, with

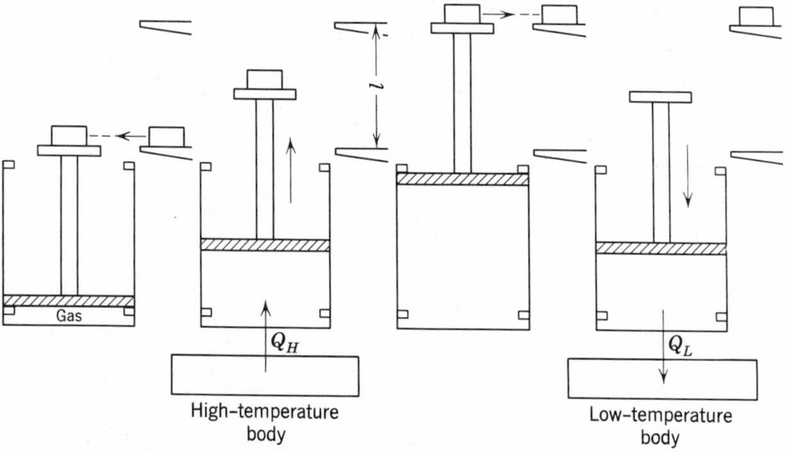

Fig. 7.3 A simple heat engine.

a weight on the platform. Let the system now undergo a process in which heat is transferred from some high-temperature body to the gas, causing it to expand and raise the piston to the upper stops. At this point the weight is removed. Now let the system be restored to its initial state by transferring heat from the gas to a low-temperature body, thus completing the cycle. Since the weight was raised during the cycle, it is evident that work was done by the gas during the cycle. From the first law we conclude that the net heat transfer was positive and equal to the work done during the cycle.

Such a device is called a heat engine, and the substance to which and from which heat is transferred is called the working substance or working fluid. A heat engine may be defined as a device that operates in a thermodynamic cycle and does a certain amount of net positive work as a result of heat transfer from a high-temperature body and to a low-temperature body. Often the term heat engine is used in a broader sense to include all devices that produce work, either through heat transfer or combustion, even though the device does not operate in a thermodynamic cycle. The internal-combustion engine and the gas turbine are examples of such devices, and calling these heat engines is an acceptable use of the term. In this chapter, however, we are concerned with the more restricted form of heat engine, as defined above, which operates on a thermodynamic cycle.

A simple steam power plant is an example of a heat engine in this restricted sense. Each component in this plant may be analyzed by a steady-state, steady-flow process, but considered as a whole it may be considered a heat engine (Fig. 7.4) in which water (steam) is the working

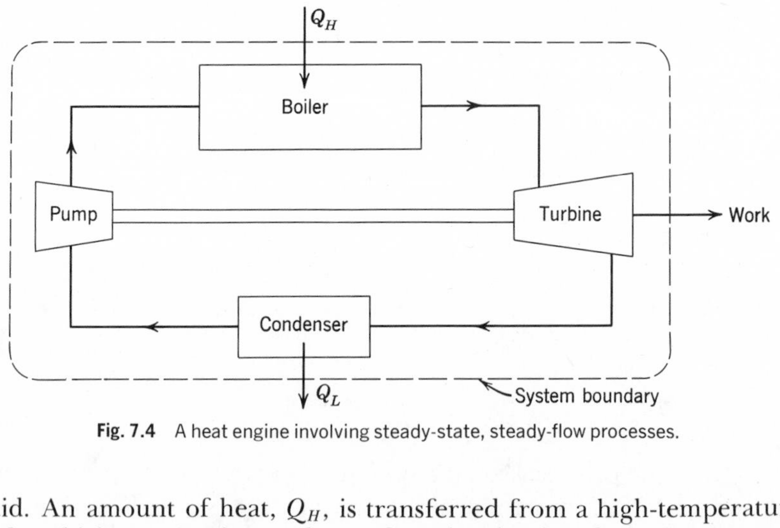

Fig. 7.4 A heat engine involving steady-state, steady-flow processes.

fluid. An amount of heat, Q_H, is transferred from a high-temperature body, which may be the products of combustion in a furnace, a reactor, or a secondary fluid which in turn has been heated in a reactor. In Fig. 7.4 the turbine is shown schematically as driving the pump, indicating that what is significant is the net work that is delivered during the cycle. The quantity of heat Q_L is rejected to a low-temperature body, which is usually the cooling water in a condenser. Thus, the simple steam power plant is a heat engine in the restricted sense, for it has a working fluid, to which and from which heat is transferred, and which does a certain amount of work as it undergoes a cycle.

Another example of a heat engine is the thermoelectric power generation device that was discussed in Chapter 1 and shown schematically in Fig. 1.11. Heat is transferred from a high temperature body to the hot junction (Q_H) and heat is transferred from the cold junction to the surroundings (Q_L). Work is done in the form of electrical energy. Since there is no working fluid we usually do not think of this as a device that operates in a cycle. However, if one adopted a microscopic point of view one could think of a cycle as regards the flow of electrons. Furthermore, as in the case of the steam power plant, the state at each point in the thermoelectric power generator does not change with time under steady state conditions.

Thus, by means of a heat engine we are able to have a system operate in a cycle and have the net work and net heat transfer both positive, which we were not able to do with the system and surroundings of Fig. 7.1.

One should note that in using the symbols Q_H and Q_L we have departed from our sign connotation for heat, because for a heat engine Q_L is negative when the working fluid is considered as the system. In this chapter it will be advantageous to use the symbol Q_H to represent the heat transfer to or from the high-temperature body, and Q_L the heat transfer to or from the low-temperature body. The direction of the heat transfer will be evident in each case from the context.

At this point it is appropriate to introduce the concept of thermal efficiency of a heat engine. In general we say that efficiency is the ratio of output (the energy sought) to input (the energy that costs), but these must be clearly defined. At the risk of oversimplification we may say that in a heat engine the energy sought is the work, and the energy that costs money is the heat from the high-temperature source (indirectly, the cost of the fuel). Thermal efficiency is defined as:

$$\eta_{\text{thermal}} = \frac{W(\text{energy sought})}{Q_H(\text{energy that costs})} = \frac{Q_H - Q_L}{Q_H} = 1 - \frac{Q_L}{Q_H} \tag{7.1}$$

The second cycle we were not able to complete was the one that involved the impossibility of transferring heat directly from a low-temperature body to a high-temperature body. This can of course be done with a refrigerator or heat pump. A vapor-compression refrigerator cycle, which was introduced in Chapter 1 and shown in Fig. 1.8, is also shown in Fig. 7.5. The working fluid is the refrigerant, such as a Freon or ammonia, which goes through a thermodynamic cycle. Heat is transferred to the refrigerant in the evaporator, where its pressure and temperature are low. Work is done on the refrigerant in the compressor and

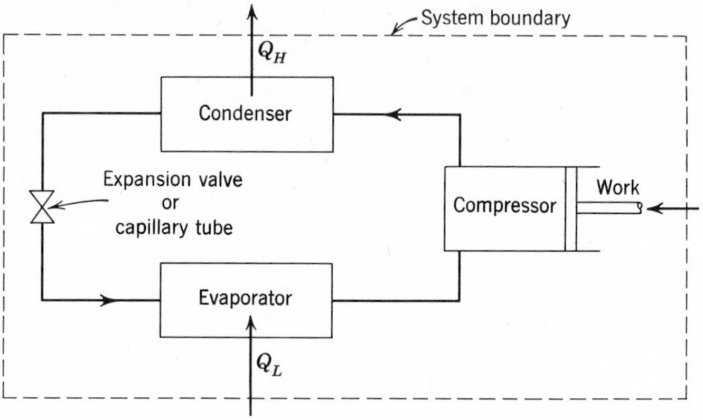

Fig. 7.5 A simple refrigeration cycle.

heat is transferred from it in the condenser, where its pressure and temperature are high. The pressure drop occurs as the refrigerant flows through the throttle valve or capillary tube.

Thus, in a refrigerator or heat pump we have a device that operates in a cycle, that requires work, and that accomplishes the objective of transferring heat from a low-temperature body to a high-temperature body.

The thermoelectric refrigerator, which was discussed in Chapter 1 and is shown schematically in Fig. 1.10, is another example of a device that meets our definition of a refrigerator. The work input to the thermoelectric refrigerator is in the form of electrical energy, and heat is transferred from the refrigerated space to the cold junction (Q_L) and from the hot junction to the surroundings (Q_H).

The "efficiency" of a refrigerator is expressed in terms of the coefficient of performance, which we designate with the symbol β. In the case of a refrigerator the objective (i.e., energy sought) is Q_L, the heat transferred from the refrigerated space, and the energy that costs is the work

W. Thus the coefficient of performance, β,[1] is

$$\beta = \frac{Q_L(\text{energy sought})}{W(\text{energy that costs})} = \frac{Q_L}{Q_H - Q_L} = \frac{1}{Q_H/Q_L - 1} \tag{7.2}$$

Before stating the second law, the concept of a thermal reservoir should be introduced. A thermal reservoir is a body to which and from which heat can be transferred indefinitely without change in the temperature of the reservoir. Thus, a thermal reservoir always remains at constant temperature. The ocean and the atmosphere approach this definition very closely. Frequently it will be useful to designate a high-temperature reservoir and a low-temperature reservoir. Sometimes a reservoir from which heat is transferred is called a source, and a reservoir to which heat is transferred is called a sink.

7.2 Second Law of Thermodynamics

On the basis of the matter considered in the previous section we are now ready to state the second law of thermodynamics. There are two classical statements of the second law, known as the Kelvin-Planck statement and the Clausius statement.

The Kelvin-Planck statement: It is impossible to construct a device that will operate in a cycle and produce no effect other than the raising of a weight and the exchange of heat with a single reservoir.

[1]It should be noted that a refrigeration or heat pump cycle can be used with either of two objectives in mind. It can be used as a refrigerator, in which case the primary objective is Q_L, the heat transferred to the refrigerant from the refrigerated space. It also can be used as a heating system (in which case it usually is referred to as a heat pump) the objective being Q_H, the heat transferred from the refrigerant to the high-temperature body, which is the space to be heated. Q_L is transferred to the refrigerant from the ground, the atmospheric air, or well water. The coefficient of performance is this case, β' is

$$\beta' = \frac{Q_H(\text{energy sought})}{W(\text{energy that costs})} = \frac{Q_H}{Q_H - Q_L} = \frac{1}{1 - Q_L/Q_H}$$

It also follows that for a given cycle,

$$\beta' - \beta = 1$$

Unless otherwise specified, the term coefficient of performance will always refer to a refrigerator as defined by Eq. 7.2.

This statement ties in with our discussion of the heat engine, and, in effect, it states that it is impossible to construct a heat engine that operates in a cycle and receives a given amount of heat from a high-temperature body and does an equal amount of work. The only alternative is that some heat must be transferred from the working fluid at a lower temperature to a low-temperature body. Thus, work can be done by the transfer of heat only if there are two temperature levels involved, and heat is transferred from the high-temperature body to the heat engine and also from the heat engine to the low-temperature body. This implies that it is impossible to build a heat engine that has a thermal efficiency of 100 per cent.

The Clausius statement: It is impossible to construct a device that operates in a cycle and produces no effect other than the transfer of heat from a cooler body to a hotter body.

This statement is related to the refrigerator or heat pump, and in effect states that is is impossible to construct a refrigerator that operates without an input of work. This also implies that the coefficient of performance is always less than infinity.

In regard to these two statements, three observations should be made. The first is that both are negative statements. It is of course impossible to "prove" a negative statement. However, we can say that the second law of thermodynamics (like every other law of nature) rests on experimental evidence. Every relevant experiment that has been conducted has either directly or indirectly verified the second law, and no experiment has ever been conducted that contradicts the second law. The basis of the second law is therefore experimental evidence.

A second observation is that these two statements of the second law are equivalent. Two statements are equivalent if the truth of each statement implies the truth of the other, or if the violation of each statement implies the violation of the other. That a violation of the Clausius statement implies a violation of the Kelvin-Planck statement may be shown as follows. The device at the left in Fig. 7.6 is a refrigerator that requires no work, and thus violates the Clausius statement. Let an amount of heat Q_L be transferred from the low-temperature reservoir to this refrigerator, and let the same amount of heat Q_L be transferred to the high-temperature reservoir. Let an amount of heat Q_H, which is greater than Q_L, be transferred from the high-temperature reservoir to the heat engine, and let the engine reject the amount of heat Q_L as it does an amount of work W (which equals $Q_H - Q_L$). Since there is no net heat transfer to the low-temperature reservoir, the low-temperature reservoir, the heat engine, and the refrigerator can be considered together as a device that operates

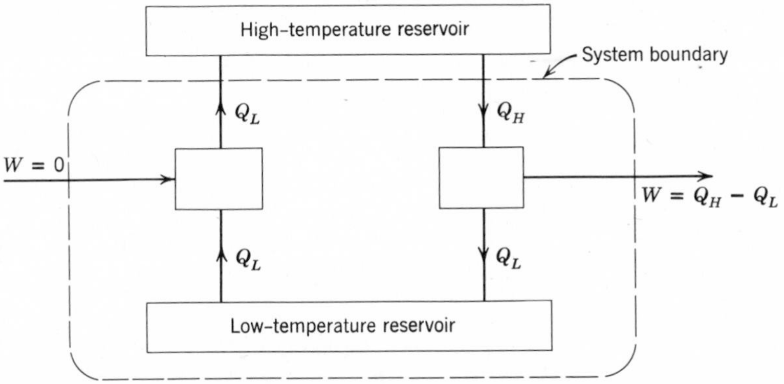

Fig. 7.6 Demonstration of the equivalence of the two statements of the second law.

in a cycle and produces no effect other than the raising of a weight (work) and the exchange of heat with a single reservoir. Thus, a violation of the Clausius statement implies a violation of the Kelvin-Planck statement. The complete equivalence of those two statements is established when it is also shown that a violation of the Kelvin-Planck statement implies a violation of the Clausius statement. This is left as an exercise for the student.

The third observation is that frequently the second law of thermodynamics has been stated as the impossibility of constructing a perpetual motion machine of the second kind. A perpetual motion machine of the first kind would create work from nothing or create mass-energy, thus violating the first law. A perpetual motion machine of the second kind would violate the second law, and a perpetual-motion machine of the third kind would have no friction, and thus run indefinitely but would produce no work.

A heat engine that violated the second law could be made into a perpetual motion machine of the second kind as follows. Consider Fig. 7.7, which might be the power plant of a ship. An amount of heat Q_L is transferred from the ocean to a high-temperature body by means of a heat pump. The work required is W', and the heat transferred to the high-temperature body is Q_H; let the same amount of heat be transferred to a heat engine which violates the Kelvin-Planck statement of the second law, and does an amount of work $W = Q_H$. Of this work an amount of work $Q_H - Q_L$ is required to drive the heat pump, leaving the net work ($W_{\text{net}} = Q_L$) available for driving the ship. Thus, we have a perpetual motion machine in the sense that work is done by utilizing freely available sources of energy such as the ocean or atmosphere.

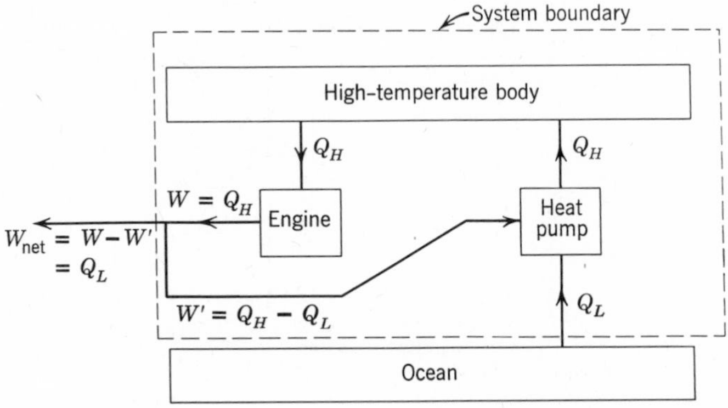

Fig. 7.7 A perpetual-motion machine of the second kind.

7.3 The Reversible Process

The question that now logically arises is this. If it is impossible to have a heat engine of 100 per cent efficiency, what is the maximum efficiency one can have? The first step in the answer to this question is to define an ideal process, which is called a reversible process.

A reversible process for a system is defined as a process, which once having taken place, can be reversed and in so doing leave no change in either the system or surroundings.

Let us illustrate the significance of this definition for a gas contained in a cylinder that is fitted with a piston. Consider first Fig. 7.8, in which a gas

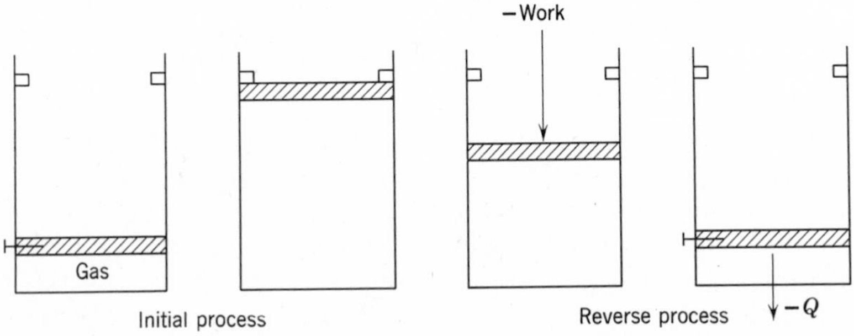

Fig. 7.8 An example of an irreversible process.

(which we define as the system) at high pressure is restrained by a piston that is secured by a pin. When the pin is removed, the piston is raised and forced abruptly against the stops. Some work is done by the system, since the piston has been raised a certain amount. Suppose we wish to restore the system to its initial state. One way of doing this would be to exert a force on the piston, thus compressing the gas until the pin could again be inserted in the piston. Since the pressure on the face of the piston is greater on the return stroke than on the initial stroke, the work done on the gas in this reverse process is greater than the work done by the gas in the initial process. An amount of heat must be transferred from the gas during the reverse stroke in order that the system have the same internal energy it had originally. Thus, the system is restored to its initial state, but the surroundings have changed by virtue of the fact that work was required to force the piston down and heat was transferred to the surroundings. Thus, the initial process is an irreversible one because it could not be reversed without leaving a change in the surroundings.

In Fig. 7.9 let the gas in the cylinder comprise the system and let the piston be loaded with a number of weights. Let the weights be slid off

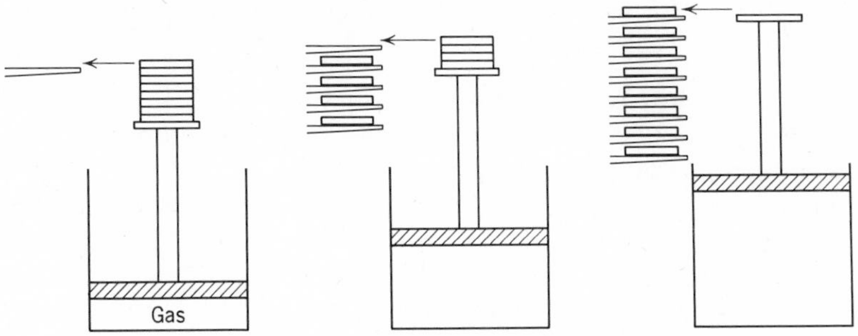

Fig. 7.9 An example of a process that approaches being reversible.

horizontally one at a time, allowing the gas to expand and do work in raising the weights that remain on the piston. As the size of the weights is made smaller and their number is increased, we approach a process that can be reversed, for at each level of the piston during the reverse process there will be a small weight that is exactly at the level of the platform and thus can be placed on the platform without requiring work. In the limit, therefore, as the weights become very small, the reverse process can be accomplished in such a manner that both the system and surroundings are in exactly the same state they were initially. Such a process is a reversible process.

7.4 Factors That Render Processes Irreversible

There are many factors that make processes irreversible, four of which are considered in this section.

Friction

It is readily evident that friction makes a process irreversible, but a brief illustration may amplify the point. Let a block and an inclined plane comprise a system, Fig. 7.10, and let the block be pulled up the inclined plane by weights that are lowered. A certain amount of work is

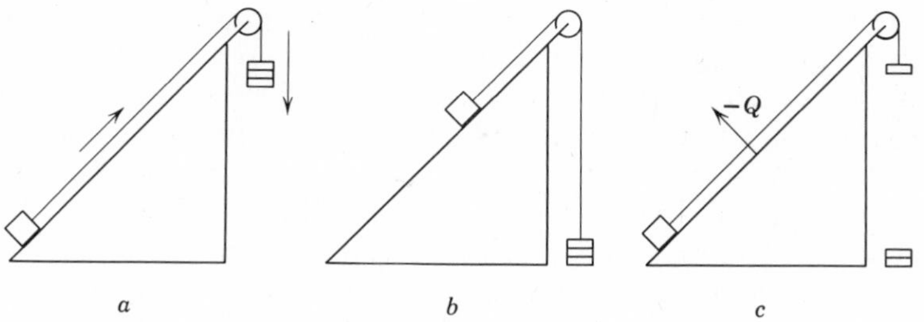

Fig. 7.10 Demonstration of the fact that friction makes processes irreversible.

required to do this. Some of this work is required to overcome the friction between the block and the plane, and some is required to increase the potential energy of the block. The block can be restored to its initial position by removing some of the weights, thus allowing the block to slide down the plane. Some heat transfer from the system to the surroundings will no doubt be required to restore the block to its initial temperature. Since the surroundings are not restored to their initial state at the conclusion of the reverse process, we conclude that friction has rendered the process irreversible. Another type of frictional effect is that associated with the flow of viscous fluids in pipes and passages and in the movement of bodies through viscous fluids.

Unrestrained Expansion

The classic example of an unrestrained expansion is shown in Fig. 7.11, in which a gas is separated from a vacuum by a membrane. Consider the

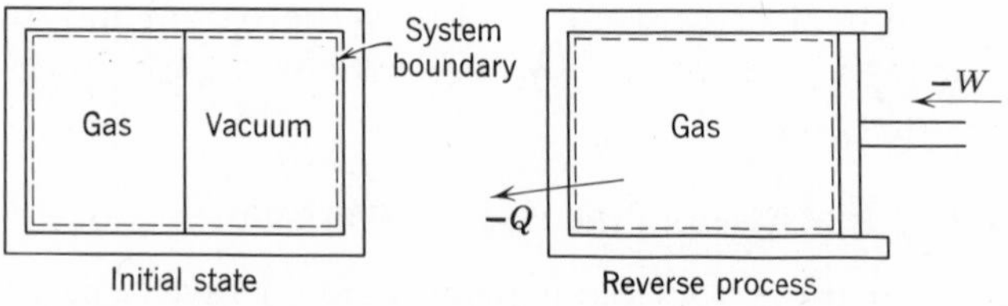

Fig. 7.11 Demonstration of the fact that unrestrained expansion makes processes irreversible.

process that occurs when the membrane breaks and the gas fills the entire vessel. It can be shown that this is an irreversible process by considering the process that would be necessary to restore the system to its original state. This would involve compressing the gas and transferring heat from the gas until its initial state was reached. Since the work and heat transfer involve a change in the surroundings, the surroundings are not restored to their initial state, indicating that the unrestrained expansion was an irreversible process. The process described in Fig. 7.8 is also an example of an unrestrained expansion.

In the reversible expansion of a gas there must be only an infinitesimal difference between the force exerted by the gas and the restraining force, so the rate at which the boundary moves will be infinitesimal. In accordance with our previous definition, this is a quasiequilibrium process. However, actual cases involve a finite difference in forces, which gives rise to a finite rate of movement of the boundary, and thus are irreversible in some degree.

Heat Transfer Through a Finite Temperature Difference

Consider as a system a high-temperature body and a low-temperature body, and let heat be transferred from the high-temperature body to the low-temperature body. The only way in which the system can be restored to its initial state is to provide refrigeration, which requires work from the surroundings, and some heat transfer to the surroundings will also be necessary. Because of the heat transfer and the work, the surroundings are not restored to their original state, indicating that the process was irreversible.

An interesting question now arises. Heat is defined as energy that is transferred due to a temperature difference. We have just shown that heat transfer through a temperature difference is an irreversible process. Therefore, how can we have a reversible heat-transfer process? A heat-transfer process approaches a reversible process as the temperature difference between the two bodies approaches zero. Therefore, we define a reversible heat-transfer process as one in which the heat is transferred through an infinitesimal temperature difference. We realize of course that to transfer a finite amount of heat through an infinitesimal temperature difference would require an infinite amount of time, or infinite area. Therefore, all actual heat-transfer processes are through a finite temperature difference and are therefore irreversible, and the greater the temperature difference the greater the irreversibility. We will find however, that the concept of reversible heat transfer is very useful in describing ideal processes.

Mixing of Two Different Substances

This process is illustrated in Fig. 7.12 in which two different gases are separated by a membrane. Let the membrane break and a homogeneous mixture of oxygen and nitrogen fill the entire volume. This process will be considered in some detail in Chapter 11. We can say here that this

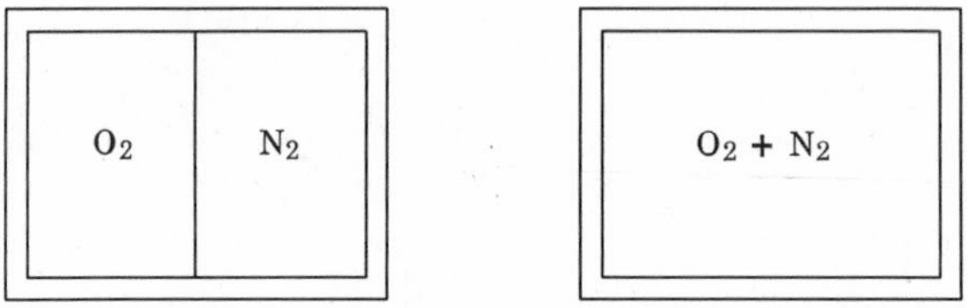

Fig. 7.12 Demonstration of the fact that the mixing of two different substances is an irreversible process.

may be considered as a special case of an unrestrained expansion, for each gas undergoes an unrestrained expansion as it fills the entire volume. A certain amount of work is necessary to separate these gases. Thus an air separation plant such as described in Chapter 1 requires an input of work in order that the separation may be accomplished.

Other Factors

There are a number of other factors that make processes irreversible but they will not be considered in detail here. Hysteresis effects and the i^2R loss encountered in electrical circuits are both factors that make processes irreversible. A combustion process as it ordinarily takes place is also an irreversible process.

It is frequently advantageous to distinguish between internal and external irreversibility. Figure 7.13 shows two identical systems to which heat is transferred. Assuming each system to be a pure substance, the

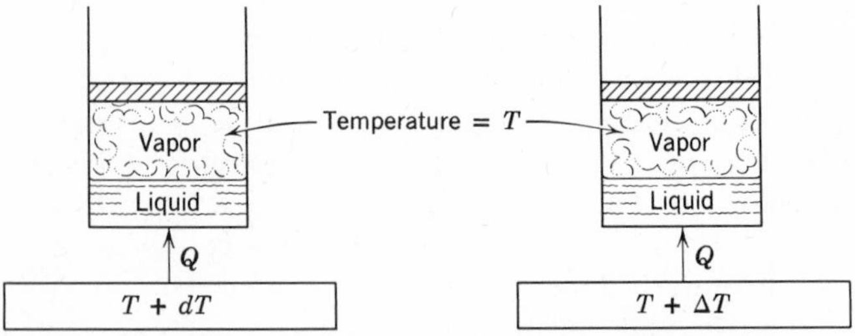

Fig. 7.13 Illustration of the difference between an internally and externally reversible process.

temperature remains constant during the heat-transfer process. In one the heat is transferred from a reservoir at a temperature $T+dT$, and in the other the reservoir is at a much higher temperature, $T+\Delta T$, than the system. The first is a reversible heat-transfer process and the second is an irreversible heat-transfer process. However, as far as the system itself is concerned, it passes through exactly the same states in both processes, which we assume are reversible. Thus, we can say in the second case that the process is internally reversible but externally irreversible because the irreversibility occurs outside the system.

One should also note the general interrelation of reversibility, equilibrium, and time. In a reversible process, the deviation from equilibrium is infinitesimal, and therefore it occurs at an infinitesimal rate. Since it is desirable that actual processes proceed at a finite rate, the deviation from equilibrium must be finite, and therefore the actual process is irreversible in some degree. The greater the deviation from equilibrium, the greater the irreversibility, and the more rapidly the process will occur. It should also be noted that the quasiequilibrium process, which was described in Chapter 2, is a reversible process, and hereafter the term reversible process will be used.

7.5 The Carnot Cycle

Having defined the reversible process and considered some factors that make processes irreversible, let us again pose the question raised in Section 7.3, namely, if the efficiency of all heat engines is less than 100 per cent, what is the most efficient cycle we can have? Let us answer this question for a heat engine that receives heat from a high-temperature reservoir and rejects heat to a low-temperature reservoir. Since we are dealing with reservoirs we recognize that both the high temperature and the low temperature are constant and remain constant regardless of the amount of heat transferred.

Let us assume that this heat engine, which operates between the given high-temperature and low-temperature reservoirs, operates in a cycle in which every process is reversible. If every process is reversible, the cycle is also reversible, and if the cycle is reversed, the heat engine becomes a refrigerator. In the next section we will show that this is the most efficient cycle that can operate between two constant-temperature reservoirs. It is called the Carnot cycle, and is named after a French engineer, Nicolas Leonard Sadi Carnot (1796–1832), who stated the second law of thermodynamics in 1824.

We now turn our attention to a consideration of the Carnot cycle. Figure 7.14 shows a power plant that is similar in many respects to a

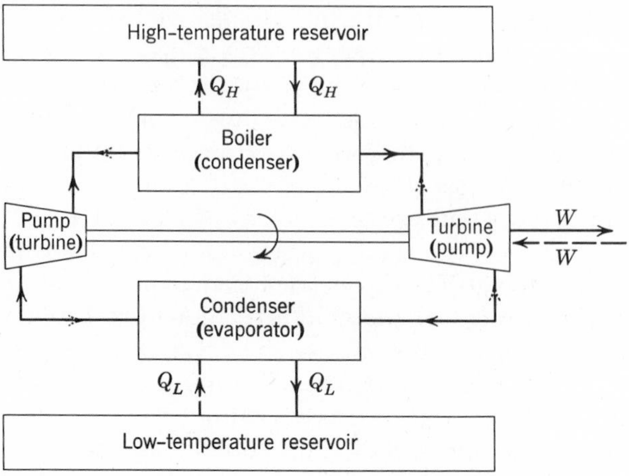

Fig. 7.14 Example of a heat engine that operates on a Carnot cycle.

simple steam power plant and which we assume operates on the Carnot cycle. Assume the working fluid to be a pure substance, such as steam. Heat is transferred from the high-temperature reservoir to the water (steam) in the boiler. For this to be a reversible heat transfer, the temperature of the water (steam) must be only infinitesimally lower than the temperature of the reservoir. This also implies, since the temperature of the reservoir remains constant, that the temperature of the water must remain constant. Therefore, the first process in the Carnot cycle is a reversible isothermal process in which heat is transferred from the high-temperature reservoir to the working fluid. A change of phase from liquid to vapor at constant pressure is of course an isothermal process for a pure substance.

The next process occurs in the turbine. It occurs without heat transfer and is therefore adiabatic. Since all processes in the Carnot cycle are reversible, this must be a reversible adiabatic process, during which the temperature of the working fluid decreases from the temperature of the high-temperature reservoir to the temperature of the low-temperature reservoir.

In the next process heat is rejected from the working fluid to the low-temperature reservoir. This must be a reversible isothermal process in which the temperature of the working fluid is infinitesimally higher than that of low-temperature reservoir. During this isothermal process some of the steam is condensed.

The final process, which completes the cycle, is a reversible adiabatic process in which the temperature of the working fluid increases from the low temperature to the high temperature. If this were to be done with water (steam) as the working fluid, it would involve taking a mixture of liquid and vapor from the condenser and compressing it. (This would be very inconvenient in practice and therefore in all power plants the working fluid is completely condensed in the condenser, and the pump handles only the liquid phase.)

Since the Carnot heat engine cycle is reversible, every process could be reversed, in which case it would become a refrigerator. The refrigerator is shown by the dotted lines and parentheses in Fig. 7.14. The temperature of the working fluid in the evaporator would be infinitesimally less than the temperature of the low-temperature reservoir, and in the condenser it is infinitesimally higher than that of the high-temperature reservoir.

It should be emphasized that the Carnot cycle can be executed in many different ways. Many different working substances can be used, such as a gas, a thermoelectric device, or a paramagnetic substance in a magnetic field such as was described in Chapter 4. There are also various possible arrangements of machinery. For example, a Carnot cycle can be devised that takes place entirely within a cylinder, using a gas as a working substance, as shown in Fig. 7.15.

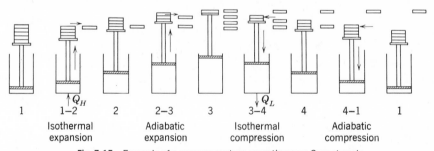

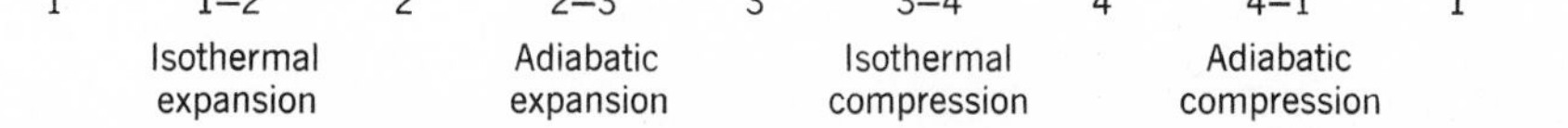

Fig. 7.15 Example of a gaseous system operating on a Carnot cycle.

The important point to be made here is that the Carnot cycle, regardless of what the working substance may be, always has the same four basic processes. These are:

1. A reversible isothermal process in which heat is transferred to or from the high-temperature reservoir.
2. A reversible adiabatic process in which the temperature of the working fluid decreases from the high temperature to the low temperature.
3. A reversible isothermal process in which heat is transferred to or from the low-temperature reservoir.

4. A reversible adiabatic process in which the temperature of the working fluid increases from the low temperature to the high temperature.

7.6 Two Propositions Regarding the Efficiency of a Carnot Cycle

There are two important propositions regarding the efficiency of a Carnot cycle.

First Proposition

It is impossible to construct an engine that operates between two given reservoirs and is more efficient than a reversible engine operating between the same two reservoirs.

The proof of this statement involves a "thought experiment." An initial assumption is made, and it is then shown that this assumption leads to impossible conclusions. The only possible conclusion is that the initial assumption was incorrect.

Let us assume that there is an irreversible engine operating between two given reservoirs that has a greater efficiency than a reversible engine operating between the same two reservoirs. Let the heat transfer to the irreversible engine be Q_H, the heat rejected be Q_L', and the work be W_{IE} (which equals $Q_H - Q_L'$) as shown in Fig. 7.16. Let the reversible engine

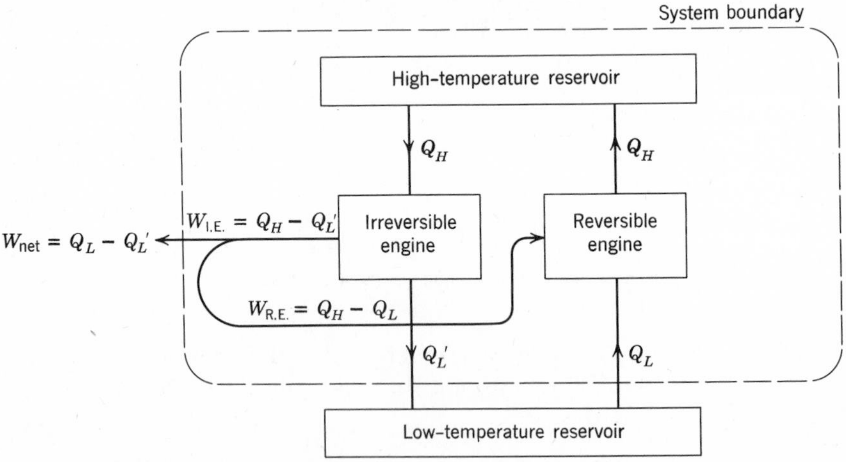

Fig. 7.16 Demonstration of the fact that the Carnot cycle is the most efficient cycle operating between two fixed temperature reservoirs.

operate as a refrigerator (since it is reversible this is possible) and let the heat transfer with the low-temperature reservoir be Q_L, the heat transfer with the high-temperature reservoir be Q_H, and the work required be W_{RE} (which equals $Q_H - Q_L$).

Since the initial assumption was that the irreversible engine is more efficient, it follows (because Q_H is the same for both engines) that $Q'_L < Q_L$ and $W_{IE} > W_{RE}$. Now the irreversible engine can drive the reversible engine and still deliver the net work W_{net} (which equals $W_{IE} - W_{RE} = Q_L - Q'_L$). However, if we consider the two engines and the high-temperature reservoir as a system as indicated in Fig. 7.16, we have a system that operates in a cycle, exchanges heat with a single reservoir, and does a certain amount of work. However, this would constitute a violation of the second law and we conclude that our initial assumption (that the irreversible engine is more efficient than the reversible engine) is incorrect, and therefore we cannot have an irreversible engine that is more efficient than a reversible engine operating between the same two reservoirs.

Second Proposition

All engines that operate on the Carnot cycle between two given constant-temperature reservoirs have the same efficiency. The proof of this proposition is similar to the proof outlined above, and involves the assumption that there is one Carnot cycle that is more efficient than another Carnot cycle operating between the same temperature reservoirs. Let the Carnot cycle with the higher efficiency replace the irreversible cycle of the previous argument, and the Carnot cycle with the lower efficiency operate as the refrigerator. The proof proceeds with the same line of reasoning as in the first proposition. The details are left as an exercise for the student.

7.7 The Thermodynamic Temperature Scale

In discussing the matter of temperature in Chapter 2 it was pointed out that the zeroth law of thermodynamics provides a basis for temperature measurement, but that a temperature scale must be defined in terms of a particular thermometer substance and device. A temperature scale that is independent of any particular substance, which might be called an absolute temperature scale, would be most desirable. In the last paragraph we noted that the efficiency of a Carnot cycle is independent of the working substance and depends only on the temperature. This fact provides the basis for such an absolute temperature scale, which we will call the thermodynamic temperature scale.

The concept of this temperature scale may be developed with the aid of Fig. 7.17, which shows three reservoirs and three engines that operate

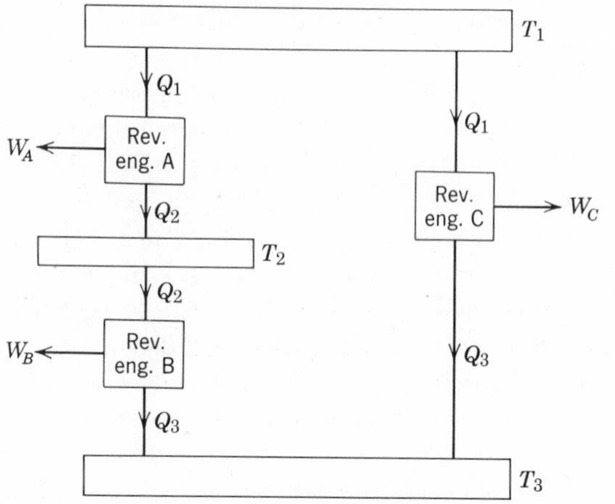

Fig. 7.17 Arrangement of heat engines to demonstrate the thermodynamic temperature scale.

on the Carnot cycle. T_1 is the highest temperature, T_3 is the lowest temperature, and T_2 is an intermediate temperature, and the engines operate between the various reservoirs as indicated. Q_1 is the same for both A and C, and since we are dealing with reversible cycles, Q_3 is the same for B and C.

Since the efficiency of a Carnot cycle is a function of only the temperature we can write

$$\eta_{\text{thermal}} = 1 - \frac{Q_L}{Q_H} = \psi(T_L, T_H) \tag{7.3}$$

where ψ designates a functional relation.

Let us apply this to the three Carnot cycles of Fig. 7.17.

$$\frac{Q_1}{Q_2} = \psi(T_1, T_2)$$

$$\frac{Q_2}{Q_3} = \psi(T_2, T_3)$$

$$\frac{Q_1}{Q_3} = \psi(T_1, T_3)$$

Since,

$$\frac{Q_1}{Q_3} = \frac{Q_1 Q_2}{Q_2 Q_3}$$

it follows that

$$\psi(T_1, T_3) = \psi(T_1, T_2) \times \psi(T_2, T_3) \tag{7.4}$$

Note that the left side is a function of T_1 and T_3 (and not T_2) and therefore the right side of this equation must also be a function of T_1 and T_3, (and not T_2). From this we can conclude that the form of the function ψ must be such that

$$\psi(T_1, T_2) = \frac{f(T_1)}{f(T_2)}$$

$$\psi(T_2, T_3) = \frac{f(T_2)}{f(T_3)}$$

for in this way $f(T_2)$ will cancel from the product of $\psi(T_1, T_2) \times \psi(T_2, T_3)$. Therefore, we conclude that

$$\frac{Q_1}{Q_3} = \psi(T_1, T_3) = \frac{f(T_1)}{f(T_3)} \tag{7.5}$$

In general terms,

$$\frac{Q_H}{Q_L} = \frac{f(T_H)}{f(T_L)} \tag{7.6}$$

Now there are several functional relations which will satisfy this equation. The one that has been selected, which was originally proposed by Lord Kelvin, for the thermodynamic scale of temperature, is the relation

$$\frac{Q_H}{Q_L} = \frac{T_H}{T_L} \tag{7.7}$$

With absolute temperatures so defined the efficiency of a Carnot cycle may be expressed in terms of the absolute temperatures.[2]

[2]Lord Kelvin also proposed a logarithmic scale of the form

$$\frac{Q_H}{Q_L} = \frac{e^{\text{“}T_H\text{”}}}{e^{\text{“}T_L\text{”}}}$$

where "T_H" and "T_L" designate the absolute temperatures on this proposed logarithmic scale. This relation can also be written

$$\ln \frac{Q_H}{Q_L} = \text{“}T_H\text{”} - \text{“}T_L\text{”}$$

(cont'd)

$$\eta_{\text{thermal}} = 1 - \frac{Q_L}{Q_H} = 1 - \frac{T_L}{T_H} \tag{7.8}$$

This means that if the thermal efficiency of a Carnot cycle operating between two given constant-temperature reservoirs is known, the ratio of the two absolute temperatures is also known.

It should be noted that Eq. 7.7 gives us a ratio of absolute temperatures, but it does not give us information about the magnitude of the degree. Let us first consider a qualitative approach to this matter and then a more rigorous statement.

Suppose we had a heat engine operating on the Carnot cycle that received heat at the temperature of the steam point and rejected heat at the temperature of the ice point. (Since a Carnot cycle involves only reversible processes, it is impossible to construct such a heat engine and perform the proposed experiment. However, we can follow the reasoning as a "thought experiment" and gain a further understanding of the thermodynamic temperature scale.) If the efficiency of such an engine could be measured, it would be found to be 26.80 per cent. Therefore, from Eq. 7.8

$$\eta_{\text{th}} = 1 - \frac{T_L}{T_H} = 1 - \frac{T_{\text{ice point}}}{T_{\text{steam point}}} = 0.2680$$

$$\frac{T_{\text{ice point}}}{T_{\text{steam point}}} = 0.7320$$

This gives us one equation involving the two unknowns T_H and T_L. The second equation comes from an arbitrary decision regarding the magnitude of the degree on the thermodynamic temperature scale. If we wish to have the magnitude of the degree on the absolute scale cor-

---

The form Kelvin actually proposed was

$$\log_{10} \frac{Q_H}{Q_L} = \text{"}T_H\text{"} - \text{"}T_L\text{"}$$

Thus, the relation between the scale in use and the proposed logarithmic scale is

$$\text{"}T\text{"} = \log_{10} T + L$$

where L is a constant that determines the level of temperature that corresponds to zero on the logarithmic scale. On this logarithmic scale temperatures range from $-\infty$ to $+\infty$, whereas on the thermodynamic scale in use they vary from 0 to $+\infty$ for ordinary systems.

respond to the magnitude of the degree on the Fahrenheit scale, we can write,

$$T_{\text{steam point}} - T_{\text{ice point}} = 180$$

This scale, the absolute Fahrenheit scale, is referred to as the Rankine scale, and temperatures on this scale are designated by R.

Solving these two equations simultaneously we find

$$T_{\text{steam point}} = 671.67\ \text{R} \quad T_{\text{ice point}} = 491.67\ \text{R}$$

It follows that temperatures on the Fahrenheit and Rankine scales are related as follows:

$$T(°\text{F}) + 459.67 = T(°\text{R})$$

The absolute scale related to the Celsius scale is the Kelvin scale, designated by K. On both these scales there are 100 degrees between the ice point and the steam point. Therefore, if we wished to use our engine operating on the Carnot cycle between the steam point and the ice point we would have the relations:

$$T_{\text{steam point}} - T_{\text{ice point}} = 100$$

$$\frac{T_{\text{ice point}}}{T_{\text{steam point}}} = 0.7320$$

Solving these two equations simultaneously we find

$$T_{\text{steam point}} = 373.15\ \text{K} \quad T_{\text{ice point}} = 273.15\ \text{K}$$

It follows that

$$T(°\text{C}) + 273.15 = T(°\text{K})$$

As already noted, the measurement of efficiencies of Carnot cycles is, however, not a practical way to approach the problem of temperature measurement on the thermodynamic scale of temperature. The actual approach used is based on the ideal gas thermometer and an assigned value for the triple point of water. At the Tenth Conference on Weights and Measures, which was held in 1954, the temperature of the triple point of water was assigned the value 273.16 K. (The triple point of water is approximately 0.01 C above the ice point. The ice point is defined as the temperature of a mixture of ice and water at a pressure of 1 atm of air which is saturated with water vapor.)

Let us now briefly consider the ideal gas scale of temperature. This scale is based upon the fact that as the pressure of a gas approaches zero, its equation of state approaches the ideal gas equation of state, namely,

$$Pv = RT$$

Consider how an ideal gas might be used to measure temperature in a constant-volume gas thermometer, which is shown schematically in Fig. 7.18. Let the gas bulb be placed in the location where the temperature is to be measured, and let the mercury column be so adjusted that

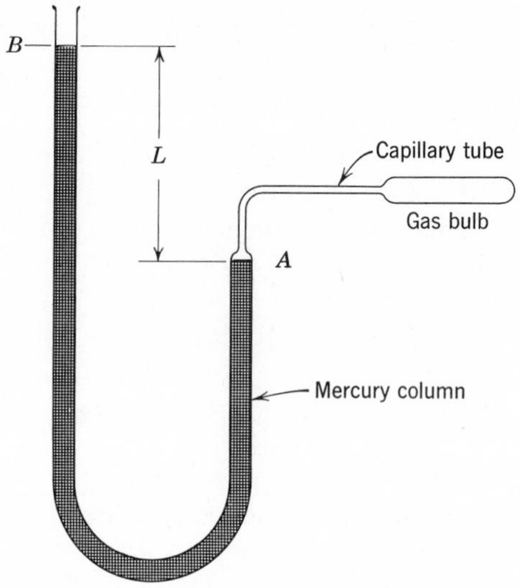

Fig. 7.18 Schematic diagram of a constant-volume gas thermometer.

the level of mercury stands at the reference mark A. Thus the volume of the gas remains constant. Assume that the gas in the capillary tube is at the same temperature as the gas in the bulb. Then the pressure of the gas, which is indicated by the height L of the mercury column, is an indication of the temperature.

Let the pressure that is associated with the temperature of the triple point of water (273.16 K) also be measured and let us designate this pressure $P_{\text{t.p.}}$. Then, from the definition of an ideal gas, any other temperature T could be determined from a pressure measurement P by the relation

$$T = 273.16\left(\frac{P}{P_{\text{t.p.}}}\right)$$

The temperature so measured is referred to as the ideal-gas temperature, and it can be shown that the temperature so measured is exactly equal to the thermodynamic temperature.

From a practical point of view we have the problem that no gas behaves exactly like an ideal gas. However, we do know that as the pressure approaches zero, the behavior of all gases approaches that of an ideal gas. Suppose then, that a series of measurements is made with varying amounts of gas in the gas bulb. This means that the pressure measured at the triple point, and also the pressure at any other temperature, will vary. If the indicated temperature T_i (obtained by assuming that the gas is ideal) is plotted against the pressure of gas with the bulb at the triple point of water, a curve like the one shown in Fig. 7.19 is obtained.

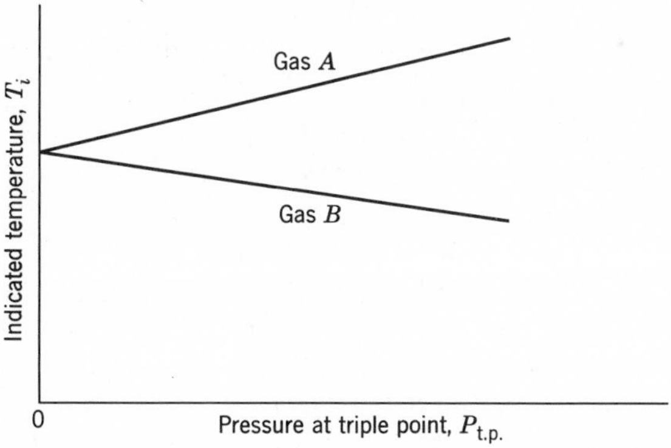

Fig. 7.19 Sketch showing how the ideal-gas temperature is determined.

When this curve is extrapolated to zero pressure, the correct ideal-gas temperature is obtained. Different curves might result from different gases, but they would all indicate the same temperature at zero pressure.

We have outlined only the general features and principles for measuring temperature on the ideal gas scale of temperatures. Precision work in this field is difficult and laborious, and there are only a few laboratories in the world where this precision work is carried on. The International Practical Scale of Temperature, which was presented in Chapter 2, closely approximates the thermodynamic temperature scale and is much easier to work with in actual temperature measurement.

The significance of absolute zero can be indicated by considering a Carnot cycle heat engine that receives a given amount of heat from a given high temperature reservoir. As the temperature at which heat is rejected from the cycle is lowered, the work increases and the amount of heat rejected decreases. In the limit, the heat rejected is zero, and the temperature of the reservoir corresponding to this limit is absolute zero.

Similarly, in the case of a Carnot cycle refrigerator, the amount of work required to produce a given amount of refrigeration increases as the temperature of the refrigerated space decreases. Absolute zero represents the limiting temperature that can be achieved, and the amount of work required to produce a finite amount of refrigeration approaches infinity as the temperature at which refrigeration is provided approaches zero.

PROBLEMS

7.1 Calculate the thermal efficiency of the steam power plant described in Problem 6.14.

7.2 Calculate the coefficient of performance of the Freon-12 refrigeration cycle described in Problem 6.25.

7.3 Prove that a device that violates the Kelvin-Planck statement of the second law also violates the Clausius statement of the second law.

7.4 Discuss the factors that would make the cycle described in Problem 6.14 an irreversible cycle.

7.5 Calculate the thermal efficiency of a Carnot cycle heat engine operating between 920 F and 110 F, and compare the result with that of Problem 7.1.

7.6 Calculate the coefficient of performance of a Carnot cycle refrigerator operating between 20 F and 100 F, and compare the result with that of Problem 7.2.

7.7 One thousand Btu of heat are transferred from a reservoir at 800 F to an engine that operates on the Carnot cycle. The engine rejects heat to a reservoir at 80 F. Determine the thermal efficiency of the cycle and the work done by the engine.

7.8 A refrigerator that operates on a Carnot cycle is required to transfer 10,000 Btu/min from a reservoir at −20 F to the atmosphere at 80 F. What is the power required?

7.9 The maximum allowable temperature of the working fluid is usually determined by metallurgical considerations. In a certain power plant this temperature is 1600 F. Nearby is a river in which the water has a temperature of 48 F. What is the maximum possible efficiency for this power plant?

7.10 An inventor claims to have developed a refrigeration unit which maintains the refrigerated space at 20 F while operating in a room where the temperature is 80 F, and which has a coefficient of performance of 8.5. How do you evaluate his claim? How would you evaluate his claim of a coefficient of performance of 8.0?

7.11 It is proposed to heat a house using a heat pump. The heat transfer from the house is 50,000 Btu/hr. The house is to be maintained at 75 F while the outisde air is at a temperature of 20 F. What is the minimum power required to drive the heat pump?

7.12 A home is to be maintained at a temperature of 70 F by means of a "heat pump" pumping heat from the atmosphere. Heat losses through the walls of the home are estimated at 1200 Btu per hour per F temperature difference between the atmosphere and the inside of the home.

(*a*) If the atmospheric temperature is 40 F, what is the minimum work (Btu/hr) required to drive the pump?

(*b*) It is proposed to use the same heat pump to cool the home in summer. For the same room temperature, the same heat flow rate per degree F through the walls, and the same work rate input to the pump, what is the maximum permissible atmospheric temperature?

7.13 A cyclic machine is used to transfer heat from a higher to a lower temperature reservoir, as shown in Fig. 7.20. Determine whether this machine is reversible, irreversible, or impossible.

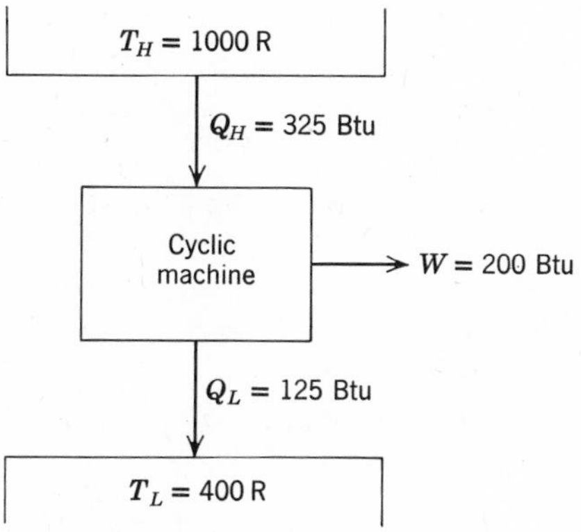

Fig. 7.20 Sketch for Problem 7.13.

7.14 It is desired to produce refrigeration at −20 F. A reservoir is available at a temperature of 400 F and the ambient temperature is 90 F. Thus work can be done by a heat engine operating between the 400 F reservoir and the ambient, and this work can be used to drive the refrigerator. Determine the ratio of the heat transferred from the high temperature reservoir to the heat transferred from the refrigerated space, assuming all processes to be reversible.

7.15 Helium has the lowest normal boiling point of any of the elements, namely 4.2 K. At this temperature it has an enthalpy of evaporation of 19.9 cal/gm mole.

A Carnot refrigeration cycle is to be used in the production of 1 gm mole of liquid helium at 4.2 K from saturated vapor at the same temperature. What is the work input to the refrigerator and the coefficient of performance of this refrigeration cycle? Assume an ambient temperature of 300 K.

7.16 Temperatures of 0.01 K can be achieved by a technique known as magnetic cooling. In this process a strong magnetic field is imposed on a paramagnetic salt which is maintained at 1 K by transferring heat to liquid helium which is boiling at very low pressure. The salt is then thermally isolated from the helium, the magnetic field is removed, and the temperature drops.

Assume that 1×10^{-3} calories are to be removed from the paramagnetic salt at an average temperature of 0.1 K, and that the necessary refrigeration is produced by a Carnot refrigeration cycle. What is the work input to the refrigerator and the coefficient of performance of this refrigeration cycle? Assume an ambient temperature of 300 K.

7.17 The lowest temperature which has been achieved at the present time is about 1×10^{-6} K. Achieving this temperature involved an additional stage to that described in Problem 7.16, namely nuclear cooling. This is similar to magnetic cooling, but involves the magnetic moment associated with the nucleus rather than that associated with certain ions in the paramagnetic salt.

Suppose that 10^{-8} Btu were to be removed from a specimen at an average temperature of 10^{-5} K (10^{-8} Btu is about the amount of energy associated with the dropping of a pin through a distance of $\frac{1}{8}$ in., and is about equal to the energy transferred as heat at this temperature in some experiments). If this amount of refrigeration at an average temperature of 1×10^{-5} K is produced by a Carnot refrigeration cycle, determine the work input and the coefficient of performance of the refrigeration cycle. Assume an ambient temperature of 300 K.

7.18 Consider an engine in outer space which operates on the Carnot cycle. The only way in which heat can be transferred from the engine is by radiation. The rate at which heat is radiated is proportional to the fourth power of the absolute temperature and the area of the radiating surface. Show that for a given power output and a given T_H, the area of the radiator will be a minimum when $T_L/T_H = \frac{3}{4}$.

8

Entropy

Up to this point in our consideration of the second law of thermodynamics we have dealt only with thermodynamic cycles. Although this is a very important and useful approach, we are in many cases concerned with processes rather than cycles. Thus, we might be interested in the second-law analysis of processes we encounter daily, such as the combustion process in an automobile engine, the cooling of a cup of coffee, or the chemical processes that take place in our bodies. It would also be most desirable to be able to deal with the second law quantitatively as well as qualitatively.

In our consideration of the first law, we initially stated the law in terms of a cycle, but then defined a property, the internal energy, which enabled us to use the first law quantitatively for processes. Similarly we have stated the second law for a cycle, and we will now find that the second law leads to another property, entropy, which enables us to treat the second law quantitatively for processes. Energy and entropy are both abstract concepts that man has devised to aid in describing certain observations. As we noted in Chapter 2, thermodynamics can be described as the science of energy and entropy. The significance of this statement will now become increasingly evident.

8.1 Inequality of Clausius

The first step in our consideration of the property we call entropy is to establish the inequality of Clausius, which is

$$\oint \frac{\delta Q}{T} \leq 0$$

The inequality of Clausius is a corollary or consequence of the second law of thermodynamics, and will be demonstrated to be valid for all possible cycles. This includes both reversible and irreversible heat engines and refrigerators. Since any reversible cycle can be represented by a series of Carnot cycles, in this analysis we need only consider a Carnot cycle that leads to the inequality of Clausius.

Consider first a reversible (Carnot) heat engine cycle, operating between reservoirs at temperatures T_H and T_L, as shown in Fig. 8.1.

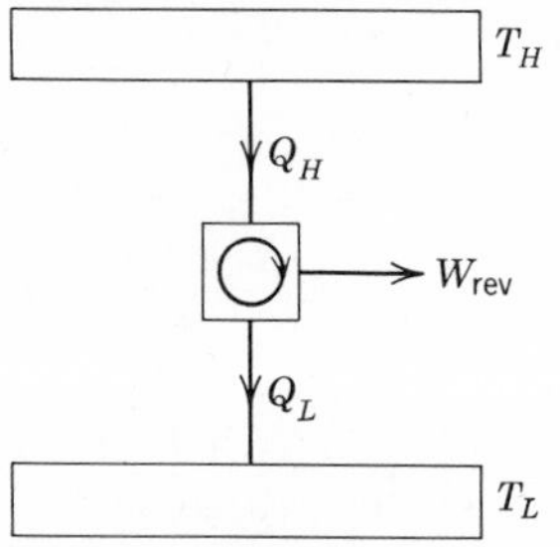

Fig. 8.1 Reversible heat engine cycle for demonstration of the inequality of Clausius.

For this cycle, the cyclic integral of the heat transfer, $\oint \delta Q$, is greater than zero.

$$\oint \delta Q = Q_H - Q_L > 0$$

Since T_H and T_L are constant, it follows from the definition of the absolute temperature scale and the fact that this is a reversible cycle that

$$\oint \frac{\delta Q}{T} = \frac{Q_H}{T_H} - \frac{Q_L}{T_L} = 0$$

If $\oint \delta Q$, the cyclic integral of δQ, is made to approach zero (by making T_H approach T_L), while the cycle remains reversible, the cyclic integral of $\delta Q/T$ remains zero. Thus we conclude that for all reversible heat engine cycles

$$\oint \delta Q \geq 0$$

and

$$\oint \frac{\delta Q}{T} = 0$$

Now consider an irreversible cyclic heat engine operating between the same T_H and T_L as the reversible engine of Fig. 8.1, and receiving the same quantity of heat Q_H. Comparing the irreversible cycle with the reversible one, we conclude from the second law, that

$$W_{\text{irr}} < W_{\text{rev}}$$

Since $Q_H - Q_L = W$ for both the reversible and irreversible cycles, we conclude that

$$Q_H - Q_{L_{\text{irr}}} < Q_H - Q_{L_{\text{rev}}}$$

and therefore

$$Q_{L_{\text{irr}}} > Q_{L_{\text{rev}}}$$

Consequently, for the irreversible cyclic engine,

$$\oint \delta Q = Q_H - Q_{L_{\text{irr}}} > 0$$

$$\oint \frac{\delta Q}{T} = \frac{Q_H}{T_H} - \frac{Q_{L_{\text{irr}}}}{T_L} < 0$$

Suppose that we cause the engine to become more and more irreversible while keeping Q_H, T_H, and T_L fixed. The cyclic integral of δQ then approaches zero, while that for $\delta Q/T$ becomes a progressively larger negative value. In the limit, as the work output goes to zero,

$$\oint \delta Q = 0$$

$$\oint \frac{\delta Q}{T} < 0$$

Thus we conclude that for all irreversible heat engine cycles

$$\oint \delta Q \geqslant 0$$

$$\oint \frac{\delta Q}{T} < 0$$

To complete the demonstration of the inequality of Clausius we must perform similar analyses for both reversible and irreversible refrigeration cycles. For the reversible refrigeration cycle shown in Fig. 8.2,

$$\oint \delta Q = -Q_H + Q_L < 0$$

and

$$\oint \frac{\delta Q}{T} = -\frac{Q_H}{T_H} + \frac{Q_L}{T_L} = 0$$

As the cyclic integral of δQ is made to approach zero reversibly (T_H approaching T_L), the cyclic integral of $\delta Q/T$ remains at zero. In the limit,

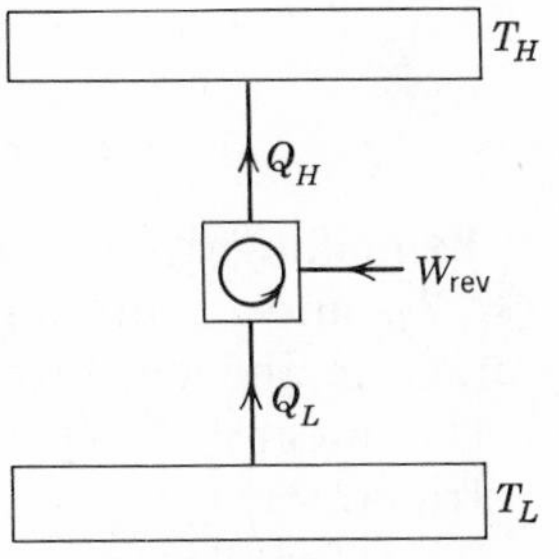

Fig. 8.2 Reversible refrigeration cycle for demonstration of the inequality of Clausius.

$$\oint \delta Q = 0$$

$$\oint \frac{\delta Q}{T} = 0$$

Thus for all reversible refrigeration cycles

$$\oint \delta Q \leqslant 0$$

$$\oint \frac{\delta Q}{T} = 0$$

Finally let an irreversible cyclic refrigerator operate between temperatures T_H and T_L and receive the same amount of heat Q_L as the reversible refrigerator of Fig. 8.2. From the second law, we conclude that the work input required will be greater for the irreversible refrigerator, or

$$W_{\text{irr}} > W_{\text{rev}}$$

Since $Q_H - Q_L = W$ for each cycle, it follows that

$$Q_{H_{\text{irr}}} - Q_L > Q_{H_{\text{rev}}} - Q_L$$

and therefore

$$Q_{H_{\text{irr}}} > Q_{H_{\text{rev}}}$$

That is, the heat rejected by the irreversible refrigerator to the high-temperature reservoir is greater than the heat rejected by the reversible refrigerator. Therefore, for the irreversible refrigerator,

$$\oint \delta Q = -Q_{H_{\text{irr}}} + Q_L < 0$$

$$\oint \frac{\delta Q}{T} = -\frac{Q_{H_{\text{irr}}}}{T_H} + \frac{Q_L}{T_L} < 0$$

By making this machine progressively more irreversible while keeping Q_L, T_H, and T_L constant, the cyclic integrals of δQ and $\delta Q/T$ both become larger in the negative direction. Consequently, a limiting case as the cyclic integral of δQ approaches zero does not exist for the irreversible refrigerator.

Thus for all irreversible refrigeration cycles,

$$\oint \delta Q < 0$$

$$\oint \frac{\delta Q}{T} < 0$$

Summarizing, we note that, as regards the sign of $\oint \delta Q$, we have considered all possible reversible cycles (that is, $\oint \delta Q \gtreqless 0$), and for each of these reversible cycles

$$\oint \frac{\delta Q}{T} = 0$$

We have also considered all possible irreversible cycles for the sign of $\oint \delta Q$ (that is, $\oint \delta Q \gtreqless 0$), and for all these irreversible cycles

$$\oint \frac{\delta Q}{T} < 0$$

Thus for all cycles we can write

$$\oint \frac{\delta Q}{T} \leq 0 \tag{8.1}$$

where the equality holds for reversible cycles and the inequality for irreversible cycles. This relation, Eq. 8.1, is known as the inequality of Clausius.

The significance of the inequality of Clausius may be illustrated by considering the simple steam power plant cycle shown in Fig. 8.3. This

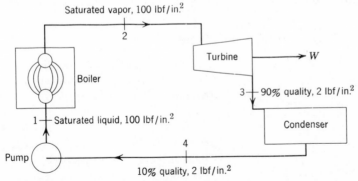

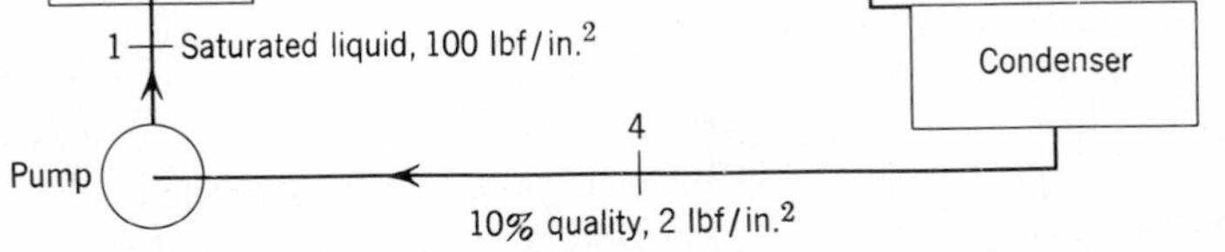

Fig. 8.3 A simple steam power plant that demonstrates the inequality of Clausius.

cycle is slightly different from the usual cycle for steam power plants in that the pump handles a mixture of liquid and vapor in such proportions that saturated liquid leaves the pump and enters the boiler. Suppose that someone reports that the pressure and quality at various points in the cycle are as given in Fig. 8.3. Does this cycle satisfy the inequality of Clausius?

Heat is transferred in two places, the boiler and the condenser. Therefore

$$\oint \frac{\delta Q}{T} = \int \left(\frac{\delta Q}{T}\right)_{\text{boiler}} + \int \left(\frac{\delta Q}{T}\right)_{\text{condenser}}$$

Since the temperature remains constant in both the boiler and condenser this may be integrated as follows:

$$\oint \frac{\delta Q}{T} = \frac{1}{T_1}\int_1^2 \delta Q + \frac{1}{T_3}\int_3^4 \delta Q = \frac{{}_1Q_2}{T_1} + \frac{{}_3Q_4}{T_3}$$

Let us consider a 1-lb mass as the working fluid.

$${}_1q_2 = h_2 - h_1 = 889.2 \text{ Btu/lbm}; \qquad T_1 = 327.9 \text{ F}$$

$${}_3q_4 = h_4 - h_3 = 196.2 - 1013.9 = -817.7 \text{ Btu/lbm}; \quad T_3 = 126.0 \text{ F}$$

Therefore

$$\oint \frac{\delta Q}{T} = \frac{889.2}{327.9 + 459.7} - \frac{817.7}{126.0 + 459.7} = -0.267 \text{ Btu/lbm-R}$$

Thus this cycle satisfies the inequality of Clausius, which is equivalent to saying that it does not violate the second law of thermodynamics.

8.2 Entropy—A Property of a System

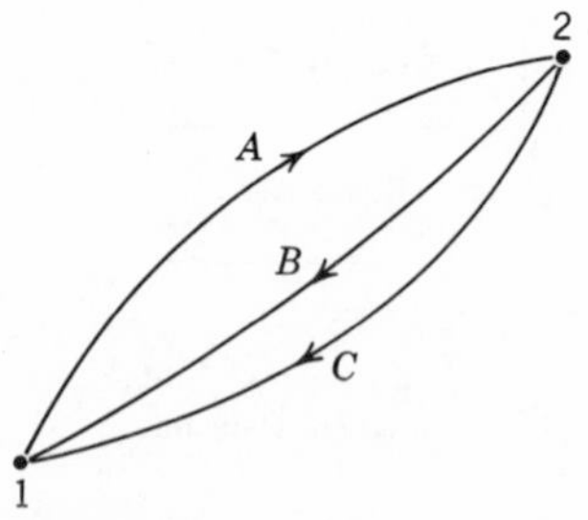

Fig. 8.4 Two reversible cycles demonstrating the fact that entropy is a property of a substance.

By the use of Eq. 8.1 and Fig. 8.4 it may be shown that the second law of thermodynamics leads to a property of a system which we call entropy. Let a system undergo a reversible process from state 1 to state 2 along path A, and let the cycle be completed along path B, which is also reversible.

Since this is a reversible cycle we can write

$$\oint \frac{\delta Q}{T} = 0 = \int_{1A}^{2A} \frac{\delta Q}{T} + \int_{2B}^{1B} \frac{\delta Q}{T}$$

Now consider another reversible cycle, which has the same initial process, but let the cycle be completed along path C. For this cycle we can write

$$\oint \frac{\delta Q}{T} = 0 = \int_{1A}^{2A} \frac{\delta Q}{T} + \int_{2C}^{1C} \frac{\delta Q}{T}$$

Subtracting the second equation from the first we have

$$\int_{2B}^{1B} \frac{\delta Q}{T} = \int_{2C}^{1C} \frac{\delta Q}{T}$$

Since the $\int \delta Q/T$ is the same for all reversible paths between states 2 and 1, we conclude that this quantity is independent of the path and is a function of the end states only, and is therefore a property. This property is called entropy, and is designated S. It follows that entropy may be defined as a property of a substance in accordance with the relation

$$dS \equiv \left(\frac{\delta Q}{T}\right)_{\text{rev}} \tag{8.2}$$

Entropy is an extensive property, and the entropy per unit mass, is designated s. It is important to note that entropy is defined here in terms of a reversible process.

The change in the entropy of a system as it undergoes a change of state may be found by integrating Eq. 8.2. Thus

$$S_2 - S_1 = \int_1^2 \left(\frac{\delta Q}{T}\right)_{\text{rev}} \tag{8.3}$$

In order to perform this integration, the relation between T and Q must be known, and illustrations will be given subsequently. The important point to note here is that since entropy is a property, the change in the entropy of a substance in going from one state to another is the same for all processes, both reversible and irreversible, between these two states. Equation 8.3 enables us to find the change in entropy only along a reversible path. However, once it has been evaluated, this is the magnitude of the entropy change for all processes between these two states.

Equation 8.3 enables us to determine changes of entropy, but tells us nothing about absolute values of entropy. However, from the third law of thermodynamics, which is discussed in Chapter 13, it follows that the entropy of all pure substances can be assigned the value of zero at the absolute zero of temperature. This gives rise to absolute values of entropy and is particularly important when chemical reactions are involved.

However, when no change of composition is involved, it is quite adequate to give values of entropy relative to some arbitrarily selected reference state. This is the procedure followed in most tables of thermodynamic properties, such as the steam tables and ammonia tables. Therefore, until absolute entropy is introduced in Chapter 13, values of entropy will always be given relative to some arbitrary reference state.

A word should be added here regarding the role of T as an integrating factor. We noted in Chapter 4 that Q is a path function, and therefore δQ is an inexact differential. However, since $(\delta Q/T)_{\text{rev}}$ is a thermodynamic property, it is an exact differential. From a mathematical perspective we note that an inexact differential may be converted to an exact differential by the introduction of an integrating factor. Therefore, $1/T$ serves as the integrating factor in converting the inexact differential δQ to the exact differential $\delta Q/T$ for a reversible process.

8.3 The Entropy of a Pure Substance

Entropy is an extensive property of a system. Values of specific entropy (entropy per unit mass) are tabulated in tables of thermodynamic properties in the same manner as specific volume and specific enthalpy. The units of specific entropy in the steam tables, Freon-12 tables, and ammonia tables are Btu/lbm-R, and the values are given relative to an arbitrary reference state. In the steam tables the entropy of saturated liquid at 32.02 F is given the value of zero. For most refrigerants, such as Freon-12 and ammonia, the entropy of saturated liquid at -40 F is assigned the value of zero.

In general, we shall use the term "entropy" to refer to both total entropy and entropy per unit mass, since the context or appropriate symbol will clearly indicate the precise meaning of the term.

In the saturation region the entropy may be calculated using the quality, the relations being similar to those for specific volume and enthalpy.

$$\begin{aligned} s &= (1-x)s_f + xs_g \\ s &= s_f + xs_{fg} \\ s &= s_g - (1-x)s_{fg} \end{aligned} \tag{8.4}$$

The entropy of a compressed liquid is tabulated in the same manner as the other properties. These properties are primarily a function of the temperature, and are not greatly different from those for saturated liquid at the same temperature. Table 4 of the steam tables, which is summarized in Table B. 1.4 of the Appendix, gives the entropy of compressed liquid water in the same manner as for other properties, as has been discussed previously.

The thermodynamic properties of a substance are often shown on a temperature-entropy diagram, and an enthalpy-entropy diagram, which is also called a Mollier diagram, after Richard Mollier (1863–1935) of Germany. Figures 8.5 and 8.6 show the essential elements of

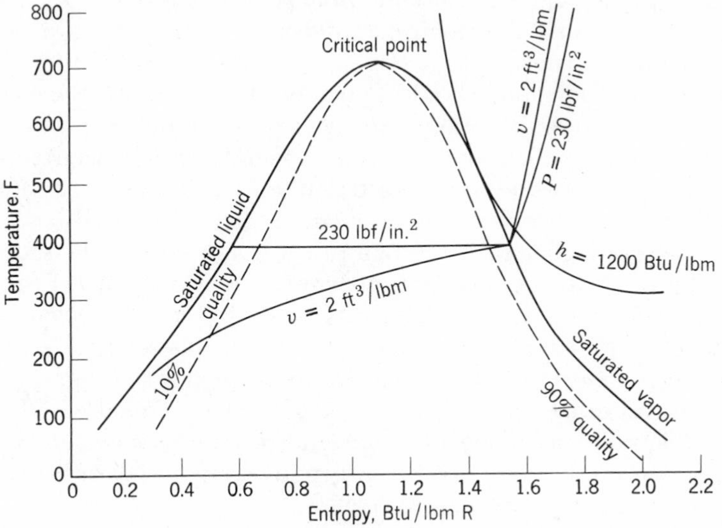

Fig. 8.5 Temperature-entropy diagram for steam.

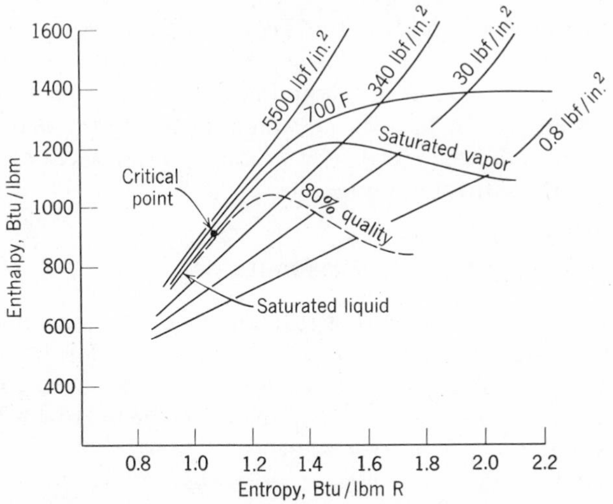

Fig. 8.6 Enthalpy-entropy diagram for steam (not to scale).

temperature-entropy and enthalpy-entropy diagrams for steam. The general features of such diagrams are the same for all pure substances. A more complete temperature-entropy diagram for steam is shown in Fig. B. 1.

These diagrams are valuable both as a means of presenting thermodynamic data and also because they enable one to visualize the changes of state that occur in various processes, and as our study progresses the student should acquire facility in visualizing thermodynamic processes on these diagrams. The temperature-entropy diagram is particularly useful for this purpose.

One more observation should be made here concerning the compressed-liquid lines on the temperature-entropy diagram for water. Reference to Table 4 of the steam tables (Appendix, Table B.1.4) indicates that the entropy of the compressed liquid is less than that of saturated liquid at the same temperature for all the temperatures listed except 32 F. It can be shown that each constant-pressure line crosses the saturated-liquid line at the point of maximum density, which is about 39 F for water (the exact temperature at which maximum density occurs varies with pressure). For temperatures less than the temperature at which the density is a maximum, the entropy of compressed liquid is greater than that of the saturated liquid. Thus, lines of constant pressure in the compressed liquid region appear (on a magnified scale) as shown in Fig. 8.7*a*. It is important to understand the general shape of these lines when showing the pumping process for liquids.

Having made this observation for water, it should be stated that for most substances the difference in the entropy between compressed liquid and saturated liquid at the same temperature is so small that usually a process in which liquid is heated at constant pressure is shown as coinciding with the saturated liquid line until the saturation temperature is reached (Fig. 8.7*b*). Thus, if water at 1500 lbf/in.^{2} is heated from 32 F to the saturation temperature, it would be shown by line *ABD*, which coincides with the saturated-liquid line.

8.4 Entropy Change in Reversible Processes

Having established the fact that entropy is a thermodynamic property of a system, its significance in various processes will now be considered. In this section we will limit ourselves to systems that undergo reversible processes, and consider the Carnot cycle, reversible heat-transfer processes, and reversible adiabatic processes.

Let the working fluid of a heat engine operating on the Carnot cycle comprise the system. The first process is the isothermal transfer of heat

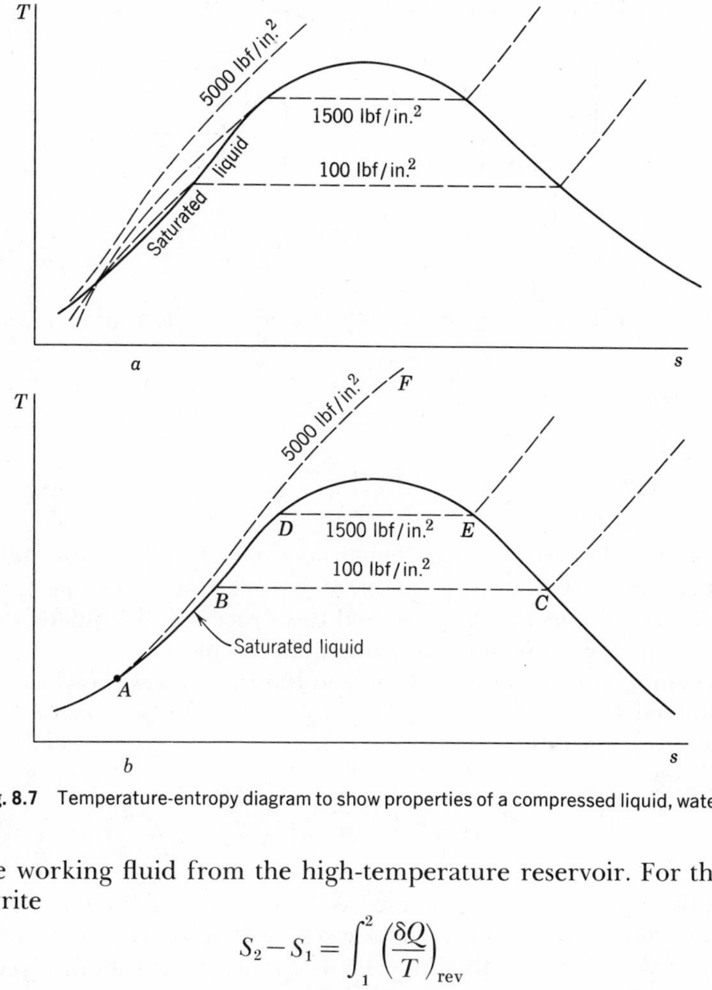

Fig. 8.7 Temperature-entropy diagram to show properties of a compressed liquid, water.

to the working fluid from the high-temperature reservoir. For this we can write

$$S_2 - S_1 = \int_1^2 \left(\frac{\delta Q}{T}\right)_{\text{rev}}$$

Since this is a reversible process in which the temperature of the working fluid remains constant, it can be integrated to give

$$S_2 - S_1 = \frac{1}{T_H}\int_1^2 \delta Q = \frac{{}_1Q_2}{T_H}$$

This process is shown in Fig. 8.8*a*, and the area under line 1-2, area 1-2-*b*-*a*-1, represents the heat transferred to the working fluid during the process.

The second process of a Carnot cycle is a reversible adiabatic one.

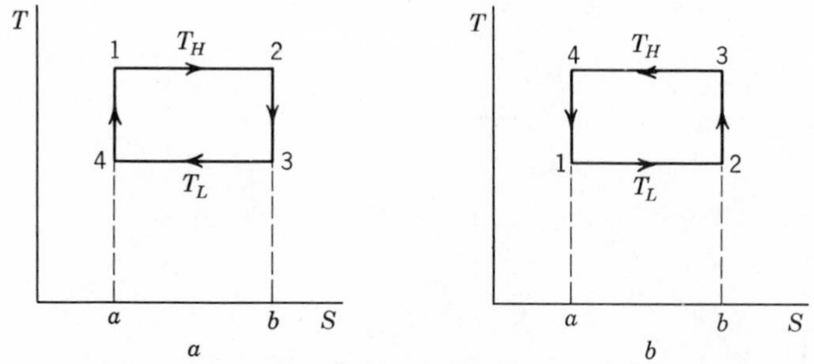

Fig. 8.8 The Carnot cycle on the temperature-entropy diagram.

From the definition of entropy,

$$dS = \left(\frac{\delta Q}{T}\right)_{\text{rev}}$$

it is evident that the entropy remains constant in a reversible adiabatic process. A constant-entropy process is called an isentropic process. Line 2-3 represents this process, and this process is concluded at state 3 when the temperature of the working fluid reaches T_L.

The third process is the reversible isothermal process in which heat is transferred from the working fluid to the low-temperature reservoir. For this we can write

$$S_4 - S_3 = \int_3^4 \left(\frac{\delta Q}{T}\right)_{\text{rev}} = \frac{{}_3Q_4}{T_L}$$

Since during this process the heat transfer is negative (as regards the working fluid) the entropy of the working fluid decreases during this process. Also, since the final process 4-1, which completes the cycle, is a reversible adiabatic process (and therefore isentropic), it is evident that the entropy decrease in process 3-4 must exactly equal the entropy increase in process 1-2. The area under line 3-4, area 3-4-*a*-*b*-3, represents the heat transferred from the working fluid to the low-temperature reservoir.

Since the net work of the cycle is equal to the net heat transfer, it is evident that area 1-2-3-4-1 represents the net work of the cycle. The efficiency of the cycle may also be expressed in terms of areas.

$$\eta_{\text{th}} = \frac{W_{\text{net}}}{Q_H} = \frac{\text{area 1-2-3-4-1}}{\text{area 1-2-}b\text{-}a\text{-1}}$$

Some statements made earlier about efficiencies may now be understood graphically. For example, increasing T_H while T_L remains constant increases the efficiency. Decreasing T_L as T_H remains constant increases the efficiency. It is also evident that the efficiency approaches 100 per cent as the absolute temperature at which heat is rejected approaches zero.

If the cycle is reversed, we have a refrigerator or heat pump, and the Carnot cycle for a refrigerator is shown in Fig. 8.8*b*. Notice in this case that the entropy of the working fluid increases at T_L, since heat is transferred to the working fluid at T_L. The entropy decreases at T_H due to heat transfer from the working fluid.

Let us next consider reversible heat-transfer processes. Actually we are concerned here with processes that are internally reversible, i.e., processes that involve no irreversibilities within the boundary of the system. For such processes the heat transfer to or from a system can be shown as an area on a temperature-entropy diagram. For example, consider the change of state from saturated liquid to saturated vapor at constant pressure. This would correspond to the process 1-2 on the *T*-*s* diagram of Fig. 8.9 (note that absolute temperature is required here), and

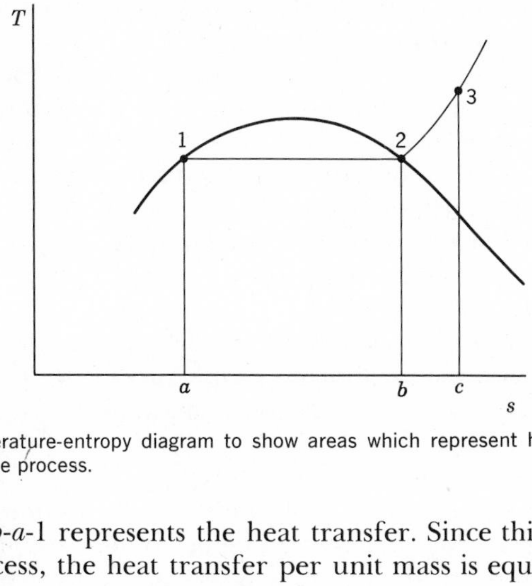

Fig. 8.9 A temperature-entropy diagram to show areas which represent heat transfer for an internally reversible process.

the area 1-2-*b*-*a*-1 represents the heat transfer. Since this is a constant-pressure process, the heat transfer per unit mass is equal to h_{fg}. Thus,

$$s_2 - s_1 = s_{fg} = \frac{1}{m}\int_1^2 \left(\frac{\delta Q}{T}\right)_{\text{rev}} = \frac{1}{mT}\int_1^2 \delta Q = \frac{{}_1q_2}{T} = \frac{h_{fg}}{T}$$

This relation gives a clue as to how s_{fg} is calculated for tabulation in tables of thermodynamic properties. For example, consider steam at 100 lbf/

in.[2] From the steam tables we have

$$h_{fg} = 889.2 \text{ Btu/lbm}$$
$$T = 327.86 + 459.67 = 787.53$$

Therefore,

$$s_{fg} = \frac{h_{fg}}{T} = \frac{889.2}{787.53} = 1.1290 \text{ Btu/lbm-R}$$

This is the value listed for s_{fg} in the steam tables.

If heat is transferred to the saturated vapor at constant pressure, the steam is superheated along line 2-3. For this process we can write

$$_2q_3 = \frac{1}{m}\int_2^3 \delta Q = \int_2^3 T\,ds$$

Since T is not constant this cannot be integrated unless we know a relation between temperature and entropy. However, we do realize that the area under line 2-3, area 2-3-*c*-*b*-2, represents $\int_2^3 T\,ds$, and therefore represents the heat transferred during this reversible process.

The important conclusion to draw here is that for processes that are internally reversible, the area underneath the process line on a temperature-entropy diagram represents the quantity of heat transferred. This is not true for irreversible processes, as will be demonstrated later.

There are many situations in which essentially adiabatic processes take place. We have already noted that in such cases the ideal process, which is a reversible adiabatic process, is isentropic. We shall consider here an example of a reversible adiabatic process for a system and consider in the next chapter the reversible adiabatic process for a control volume. We shall also note in a later section of this chapter that by comparing an actual process with the ideal or isentropic process we have a basis for defining the efficiency of certain classes of machines.

Example 8.1

Consider a cylinder fitted with a piston which contains saturated Freon-12 vapor at 20 F. Let this vapor be compressed in a reversible adiabatic process until the pressure is 150 lbf/in.[2] Determine the work per pound of Freon-12 for this process.

From the first law we conclude that

$$_1q_2 = u_2 - u_1 + {}_1w_2 = 0$$
$$_1w_2 = u_1 - u_2$$

State 1 is specified by the statement of the problem and therefore the initial entropy is known. We also conclude, from the second law that since the process is reversible and adiabatic, $s_1 = s_2$. Therefore we know entropy and pressure in the final state, which is sufficient to specify the final state, since we are dealing with a pure substance.

From the Freon-12 tables

$$u_1 = h_1 - P_1 v_1 = 79.385 - \frac{35.736 \times 144 \times 1.0988}{778} = 72.11 \text{ Btu/lbm}$$

$$s_1 = s_2 = 0.16719 \text{ Btu/lbm-R}$$

$$P_2 = 150 \text{ lbf/in.}^2$$

Therefore, from the superheat tables for Freon-12

$$T_2 = 122.9 \text{ F}; \quad h_2 = 90.330; \quad v_2 = 0.28270$$

$$u_2 = 90.330 - \frac{150 \times 144 \times 0.28270}{778} = 82.48 \text{ Btu/lbm}$$

$${}_1w_2 = u_1 - u_2 = 72.11 - 82.48 = -10.37 \text{ Btu/lbm}$$

8.5 Two Important Thermodynamic Relations

At this point we will derive two important thermodynamic relations for a simple compressible substance. These relations are

$$T\,dS = dU + P\,dV$$
$$T\,dS = dH - V\,dP$$

The first of these relations can be derived by considering a simple compressible substance in the absence of motion or gravitational effects. The first law for a change of state under these conditions can be written

$$\delta Q = dU + \delta W$$

The equations we are deriving here deal first of all with those changes of state in which the state of the substance can be identified at all times. Thus we must consider a quasiequilibrium process, or, to use the term introduced in the last chapter, a reversible process. For a reversible process of a simple compressible substance we can write

$$\delta Q = T\,dS \quad \text{and} \quad \delta W = P\,dV$$

Substituting these relations into the first law equation we have

$$T\,dS = dU + P\,dV \tag{8.5}$$

which is the equation we set out to derive. Note that this equation was derived by assuming a reversible process, and thus this equation can be integrated for any reversible process, for during such a process the state of the substance can be identified at any point during the process. We also note that Eq. 8.5 deals only with properties. Suppose we have an irreversible process taking place between given initial and final states. The properties of a substance depend only on the state, and therefore the change in the properties during a given change of state are the same for an irreversible process as for a reversible process. Therefore Eq. 8.5 is often applied to an irreversible process between two given states, but the integration of Eq. 8.5 is performed along a reversible path between the same two states.

Since enthalpy is defined as

$$H = U + PV$$

it follows that

$$dH = dU + P\,dV + V\,dP$$

Substituting this relation into Eq. 8.5 we have

$$T\,dS = dH - V\,dP \tag{8.6}$$

which is the second equation that we set out to derive.

These equations can also be written for a unit mass,

$$\begin{aligned} T\,ds &= du + P\,dv \\ T\,ds &= dh - v\,dP \end{aligned} \tag{8.7}$$

or on a mole basis,

$$\begin{aligned} T\,d\bar{s} &= d\bar{u} + P\,d\bar{v} \\ T\,d\bar{s} &= d\bar{h} - \bar{v}\,dP \end{aligned} \tag{8.8}$$

These equations will be used extensively in certain subsequent sections of this book.

If we consider substances of fixed composition other than a simple compressible substance we can write other "$T\,dS$" equations than those given above for a simple compressible substance. In Chapter 4, Eq. 4.13, we noted that for a reversible process we can write the following expression for work.

$$\delta W = P\,dV - \mathcal{T}\,dL - \mathcal{S}\,dA - \mu_0\,\mathcal{H}\,d(V\mathcal{M}) - \mathcal{E}\,dZ + \cdots$$

It follows that a more general expression for the "$T\,dS$" equation would be

$$T\,dS = dU + P\,dV - \mathcal{T}\,dL - \mathcal{S}\,dA - \mu_0\,\mathcal{H}\,d(V\mathcal{M}) - \mathcal{E}\,dZ + \cdots \quad (8.9)$$

8.6 Entropy Change of a System During an Irreversible Process

Consider a system which undergoes the cycles shown in Fig. 8.10. The cycle made up of the reversible processes A and B is a reversible cycle. Therefore we can write

$$\oint \frac{\delta Q}{T} = \int_{1A}^{2A} \frac{\delta Q}{T} + \int_{2B}^{1B} \frac{\delta Q}{T} = 0$$

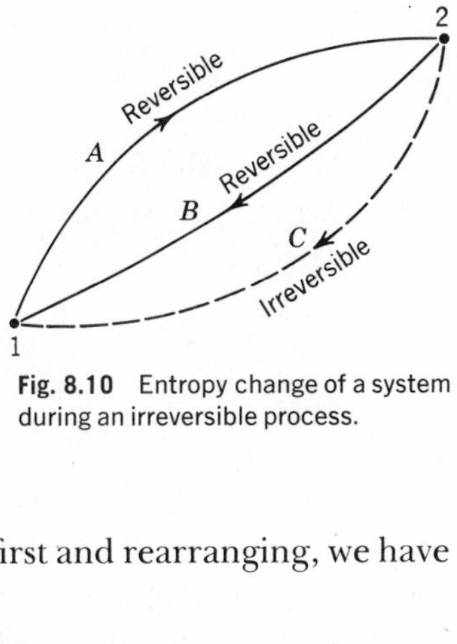

Fig. 8.10 Entropy change of a system during an irreversible process.

The cycle made up of the reversible process A and the irreversible process C is an irreversible cycle. Therefore, for this cycle the inequality of Clausius may be applied, giving the result

$$\oint \frac{\delta Q}{T} = \int_{1A}^{2A} \frac{\delta Q}{T} + \int_{2C}^{1C} \frac{\delta Q}{T} < 0$$

Subtracting the second equation from the first and rearranging, we have

$$\int_{2B}^{1B} \frac{\delta Q}{T} > \int_{2C}^{1C} \frac{\delta Q}{T}$$

Since path B is reversible, and since entropy is a property,

$$\int_{2B}^{1B} \frac{\delta Q}{T} = \int_{2B}^{1B} dS = \int_{2C}^{1C} dS$$

Therefore,

$$\int_{2C}^{1C} dS > \int_{2C}^{1C} \frac{\delta Q}{T}$$

For the general case we can write

$$dS \geqslant \frac{\delta Q}{T}$$

$$S_2 - S_1 \geqslant \int_1^2 \frac{\delta Q}{T} \tag{8.10}$$

In these equations the equality holds for a reversible process and the inequality for an irreversible process.

This is one of the most important equations of thermodynamics and is used to develop a number of concepts and definitions. In essence this equation states the influence of irreversibility on the entropy of a system. Thus, if an amount of heat δQ is transferred to a system at temperature T in a reversible process the change of entropy is given by the relation

$$dS = \frac{\delta Q}{T}$$

However, if while the amount of heat δQ is transferred to the system at temperature T there are any irreversible effects occurring, the change of entropy will be greater than for the reversible process, and we would write

$$dS > \frac{\delta Q}{T}$$

Equation 8.10 holds when $\delta Q = 0$ or when $\delta Q < 0$ as well as when $\delta Q > 0$. If δQ is negative, the entropy will tend to decrease as the result of the heat transfer. However, the influence of irreversibilities is still to increase the entropy of the system, and from the absolute numerical perspective we can still write for $\delta Q < 0$,

$$dS \geqslant \frac{\delta Q}{T}$$

8.7 Lost Work

The significance of the change of entropy for an irreversible process can be amplified by introducing the concept of lost work, to which we give the symbol LW. This concept can be described with the aid of Fig. 8.11, in which a gas is separated from a vacuum by a membrane. Let the membrane have a small rupture so that the gas fills the entire volume,

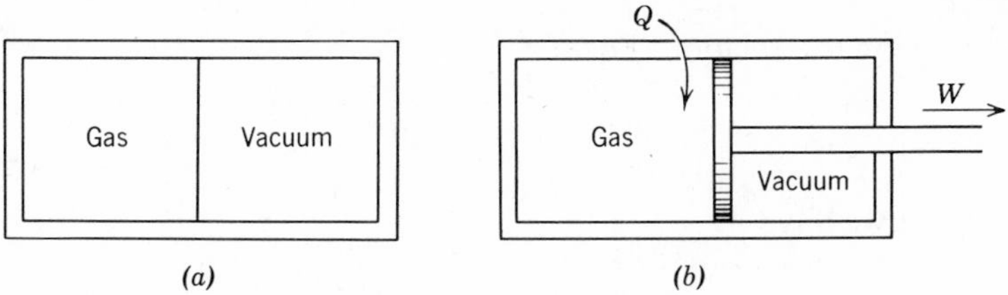

Fig. 8.11 A process that demonstrates lost work.

and let the necessary heat transfer take place so that the final temperature is the same as the initial temperature. (How much heat transfer would be necessary if this were an ideal gas?). Since the work is zero for this process we conclude from the first law that

$$\delta Q = dU$$

We would like to compare this irreversible process with a reversible process between the same two states. This could be achieved by having the gas expand against a frictionless piston while heat is transferred to the gas in the amount necessary to maintain constant temperature. For this process we conclude from the first law that

$$\delta Q = dU + \delta W$$

and, since this is a reversible process we can write

$$\delta Q = T\,dS \quad \text{and} \quad \delta W = P\,dV$$

The major difference between these two processes is that in the irreversible process the work is zero, whereas in the reversible process the greatest possible work is done. Thus, we might speak of the lost work for an irreversible process. Every irreversible process has associated with it a certain amount of lost work. In Fig. 8.11 we have shown the two extreme cases, a reversible process and one in which the work is zero. Between these two extremes are processes that have various degrees of irreversibility, i.e., the lost work may vary from zero to some maximum value. For a simple compressible substance we can therefore write

$$P\,dV = \delta W + \delta LW \tag{8.11}$$

where δLW is the lost work and δW the actual work.

Substituting the relation (Eq. 8.5)

$$T\,dS = dU + P\,dV$$

into Eq. 8.11 we have

$$T\,dS = dU + \delta W + \delta LW$$

But, from the first law

$$\delta Q = dU + \delta W$$

Therefore,

$$T\,dS = \delta Q + \delta LW$$

and

$$dS = \frac{\delta Q}{T} + \frac{\delta LW}{T} \tag{8.12}$$

Thus, we have an expression for the change of entropy for an irreversible process as an equality, whereas in the last section we had the inequality

$$dS \geqslant \frac{\delta Q}{T} \tag{8.13}$$

In a reversible process the lost work is zero and therefore both Eqs. 8.12 and 8.13 reduce to

$$dS = \left(\frac{\delta Q}{T}\right)_{\text{rev}}$$

for a reversible process.

Some important conclusions can now be drawn from Eqs. 8.12 and 8.13. First of all, there are two ways in which the entropy of a system can be increased, namely, by transferring heat to it and by having it undergo an irreversible process. Since the lost work cannot be less than zero, there is only one way in which the entropy of a system can be decreased, and that is to transfer heat from the system.

Secondly, the change in entropy of a system can be separated into the change due to heat transfer and the change due to internal irreversibilities. Frequently the increase in entropy due to irreversibilities is called the irreversible production of entropy.

Finally, as we have already noted, for an adiabatic process $\delta Q = 0$, and in this case the increase in entropy is always associated with the irreversibilities.

One other point, which involves the representation of irreversible processes on *P-V* and *T-S* diagrams, should be made. The work for an irreversible process is not equal to $\int P\, dV$ and the heat transfer is not equal to $\int T\, dS$. Therefore, the area underneath the path does not represent work and heat on the *P-V* and *T-S* diagrams, respectively. In fact, in many cases we are not certain of the exact state through which a system passes when it undergoes an irreversible process. For this reason it is advantageous to show irreversible processes as dotted lines and reversible processes as solid lines. Thus, the area underneath the dotted line will never represent work or heat. For example, the processes of Fig. 8.11 would be shown on *P-V* and *T-S* diagrams as in Fig. 8.12.

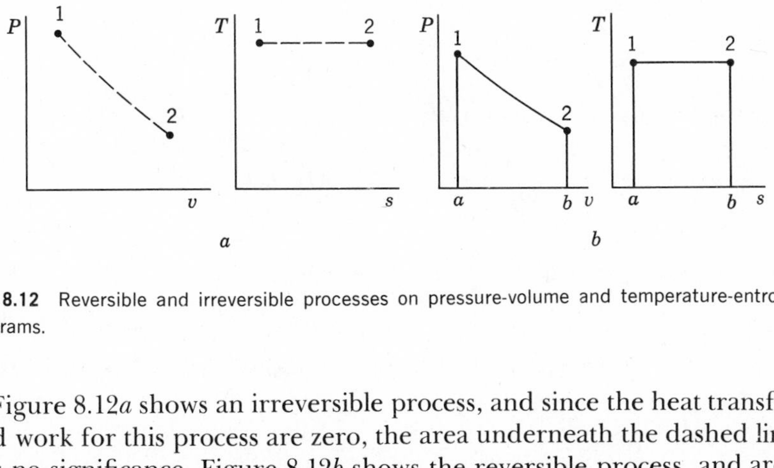

Fig. 8.12 Reversible and irreversible processes on pressure-volume and temperature-entropy diagrams.

Figure 8.12*a* shows an irreversible process, and since the heat transfer and work for this process are zero, the area underneath the dashed line has no significance. Figure 8.12*b* shows the reversible process, and area 1–2–*b*–*a*–1 represents the work on the *P-V* diagram and the heat transfer on the *T-S* diagram.

8.8 Principle of the Increase of Entropy

In this section we consider the total change in the entropy of a system and its surroundings when the system undergoes a change of state. This consideration leads to the principle of the increase of entropy.

Consider the process shown in Fig. 8.13 in which a quantity of heat δQ is transferred from the surroundings at temperature T_0 to the system at temperature T, and let the work done by the system during this

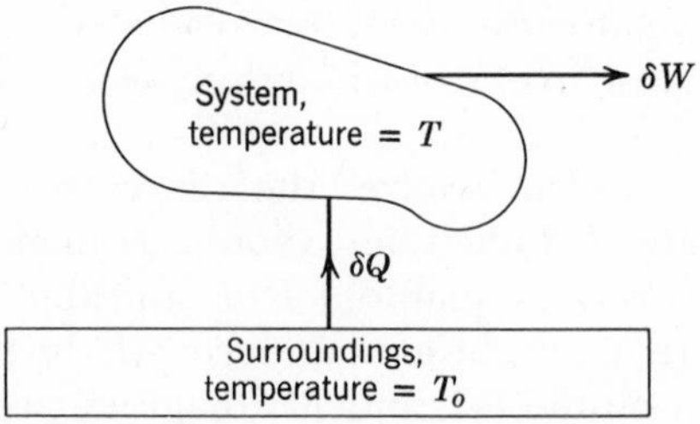

Fig. 8.13 Entropy change for the system plus surroundings.

process be δW. For this process we can apply Eq. 8.10 to the system and write

$$dS_{\text{system}} \geqslant \frac{\delta Q}{T}$$

For the surroundings δQ is negative and we can write

$$dS_{\text{surr}} = \frac{-\delta Q}{T_0}$$

The total change of entropy is therefore

$$dS_{\text{system}} + dS_{\text{surr}} \geqslant \frac{\delta Q}{T} - \frac{\delta Q}{T_0}$$

$$\geqslant \delta Q\left(\frac{1}{T} - \frac{1}{T_0}\right) \tag{8.14}$$

Since $T_0 > T$, the quantity $\left(\frac{1}{T} - \frac{1}{T_0}\right)$ is positive and we conclude that

$$dS_{\text{system}} + dS_{\text{surr}} \geqslant 0$$

If $T > T_0$, the heat transfer is from the system to the surroundings, and both δQ and the quantity $(1/T) - (1/T_0)$ are negative, thus yielding the same result.

Thus we conclude that for all possible processes that a system in a given surroundings can undergo,

$$dS_{\text{system}} + dS_{\text{surr}} \geqslant 0 \tag{8.15}$$

where the equality holds for reversible processes and the inequality for irreversible processes. This is a very important equation, not only for thermodynamics, but also for philosophical thought, and is referred to as the principle of the increase of entropy. The great significance is that the only processes that can take place are those in which the net change in entropy of the system plus its surroundings increases (or in the limit, remains constant). The reverse process, in which both the system and surroundings are returned to their original state, can never be made to occur. In other words, Eq. 8.15 dictates the single direction in which any process can proceed. Thus, the principle of the increase of entropy can be considered as a quantitative general statement of the second law from the macroscopic point of view, and applies to the combustion of fuel in our automobile engines, the cooling of our coffee, and the processes that take place in our body.

Sometimes this principle of the increase of entropy is stated in terms of an isolated system, one in which there is no interaction between the system and its surroundings. In this case, there is no change in entropy of the surroundings, and one then concludes that

$$dS_{\text{isolated system}} \geqslant 0 \tag{8.16}$$

That is, in an isolated system, the only processes that can occur are those that have an increase in entropy associated with them.

Example 8.2

Suppose 1 lbm of saturated water vapor at 212 F is condensed to saturated liquid at 212 F in a constant-pressure process by heat transfer to the surrounding air, which is at 80 F. What is the net increase in entropy of the system plus surroundings?

For the system, from the steam tables

$$\Delta S_{\text{system}} = -s_{fg} = -1.4446 \text{ Btu/lbm-R}$$

Considering the surroundings

$$Q_{\text{to surroundings}} = h_{fg} = 970.3 \text{ Btu/lbm}$$

$$\Delta S_{\text{surr}} = \frac{Q}{T_0} = \frac{970.3}{540} = 1.7980 \text{ Btu/lbm-R}$$

$$\Delta S_{\text{system}} + \Delta S_{\text{surr}} = -1.4446 + 1.7980 = 0.3534 \text{ Btu/lbm-R}$$

This increase in entropy is in accordance with the principle of the increase of entropy, and tells, as does our experience, that this process can take place.

It is interesting to note how this heat transfer from the water to the surroundings might have taken place reversibly. Suppose that an engine operating on the Carnot cycle received heat from the water and rejected heat to the surroundings, as shown in Fig. 8.14. The decrease in the entropy of the water is equal to the increase in the entropy of the surroundings.

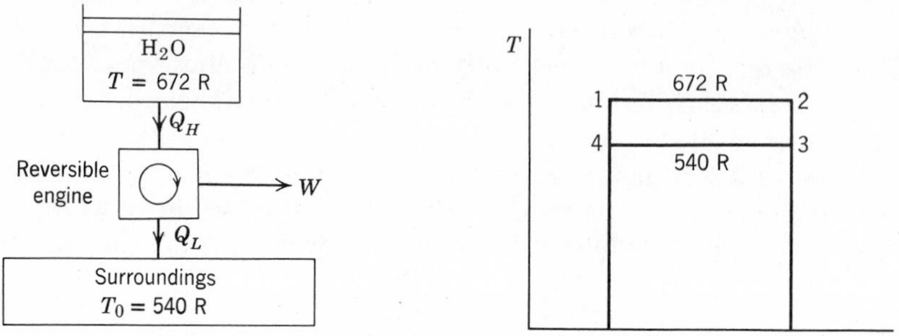

Fig. 8.14 Reversible heat transfer with the surroundings.

$$\Delta S_{\text{system}} = -1.4446 \text{ Btu/lbm-R}$$

$$\Delta S_{\text{surr}} = 1.4446 \text{ Btu/lbm-R}$$

$$Q_{\text{to surroundings}} = T_0 \Delta S = 540(1.4446) = 780.1 \text{ Btu/lbm}$$

$$W = Q_H - Q_L = 970.3 - 780.1 = 190.2 \text{ Btu/lbm}$$

Since this is a reversible cycle, the engine could be reversed and operated as a heat pump. For this cycle the work input to the heat pump would be 190.2 Btu/lbm.

8.9 Entropy Change of an Ideal Gas

Two very useful equations for computing the entropy change of an ideal gas can be developed from Eq. 8.7 by substituting Eqs. 5.17 and 5.21, as follows:

$$T\,ds = du + P\,dv$$

For an ideal gas

$$du = C_{vo}\,dT \quad \text{and} \quad \frac{P}{T} = \frac{R}{v}$$

Therefore,

$$ds = C_{vo}\frac{dT}{T} + \frac{R\,dv}{v} \tag{8.17}$$

$$s_2 - s_1 = \int_1^2 C_{vo}\frac{dT}{T} + R\ln\frac{v_2}{v_1} \tag{8.18}$$

Similarly

$$T\,ds = dh - v\,dP$$

For an ideal gas

$$dh = C_{po}\,dT \quad \text{and} \quad \frac{v}{T} = \frac{R}{P}$$

Therefore,

$$ds = C_{po}\frac{dT}{T} - R\frac{dP}{P} \tag{8.19}$$

$$s_2 - s_1 = \int_1^2 C_{po}\frac{dT}{T} - R\ln\frac{P_2}{P_1} \tag{8.20}$$

In order to integrate Eqs. 8.18 and 8.20, the relation between specific heat and temperature must be known. It follows from Eqs. 8.18 and 8.20 that when C_{po} and C_{vo} are assumed constant that the change in entropy is given by the relations

$$s_2 - s_1 = C_{po}\ln\left(\frac{T_2}{T_1}\right) - R\ln\left(\frac{P_2}{P_1}\right)$$

$$= C_{vo}\ln\left(\frac{T_2}{T_1}\right) + R\ln\left(\frac{v_2}{v_1}\right) \tag{8.21}$$

For those cases in which it is not reasonable to assume constant specific heat, Eqs. 8.18 and 8.20 can be integrated by using the empirical specific heat equations of Table B.7, as illustrated by the following example.

Example 8.3

Consider Example 5.6, in which oxygen is heated from 500 R to 2000 R. Assume that during this process the pressure dropped from 30 lbf/in.^{2} to 20 lbf/in.^{2}. Calculate the change in entropy per lbm.

Using Eq. 8.20 and the appropriate equation from Table B. 7,

$$\bar{s}_2-\bar{s}_1=\int_{T_1}^{T_2}\left(11.515\frac{dT}{T}-\frac{172dT}{T^{3/2}}+\frac{1530dT}{T^2}\right)-\bar{R}\int_{P_1}^{P_2}\frac{dP}{P}$$

$$=11.515\ln\frac{2000}{500}+2\times 172\left[\frac{1}{\sqrt{2000}}-\frac{1}{\sqrt{500}}\right]$$

$$-1530\left[\frac{1}{2000}-\frac{1}{500}\right]-1.987\ln\frac{20}{30}$$

$$=15.98-7.70+2.29+0.80=11.37\text{ Btu/lb mole-R}$$

$$s_2-s_1=\frac{\bar{s}_2-\bar{s}_1}{M}=\frac{11.37}{32.00}=0.356\text{ Btu/lbm-R}$$

Unless one is using a digital computer, it is most inconvenient to integrate these specific heat equations for each entropy calculation. To avoid this difficulty, an approach has been developed that involves integrating Eq. 8.19 along the one-atmosphere pressure line, from a reference temperature T_0 (at which s is taken as zero) to any other temperature T. The entropy value so calculated is called the standard-state entropy, designated s^0,

$$s_T^0=\int_{T_0}^{T}\frac{C_{po}}{T}dT \tag{8.22}$$

which is a function of temperature only, and can be so tabulated. It follows that the entropy change between any two states 1 and 2 is then given by the relation

$$s_2-s_1=(s_{T_2}^0-s_{T_1}^0)-R\ln\left(\frac{P_2}{P_1}\right) \tag{8.23}$$

Values of s^0 for air are given in Table B. 8. Similarly, values of standard-state entropy for other ideal gases are given on a mole basis in Table B. 9. Thus, an ideal gas entropy change that includes the variation of specific heat with temperature can be readily evaluated from these data and Eq. 8.23.

Example 8.4

Calculate the change in entropy per pound as air is heated from 540 R to 1200 R while pressure drops from 50 lbf/in.^{2} to 40 lbf/in.^{2}, assuming

1. constant specific heat
2. variable specific heat

1. From Table B. 6 for air at 80 F,

$$C_{po} = 0.24 \text{ Btu/lbm-R}$$

Therefore, using Eq. 8.21,

$$s_2 - s_1 = 0.24 \ln\left(\frac{1200}{540}\right) - \frac{53.34}{778}\ln\left(\frac{40}{50}\right) = 0.2068 \text{ Btu/lbm-R}$$

2. From Table B. 8,

$$s^0_{T_1} = 0.6008 \text{ Btu/lbm-R} \qquad s^0_{T_2} = 0.7963 \text{ Btu/lbm-R}$$

Using Eq. 8.23,

$$s_2 - s_1 = 0.7963 - 0.6008 - \frac{53.34}{778}\ln\frac{40}{50} = 0.2108 \text{ Btu/lbm-R}$$

At this point it is advantageous to introduce the specific heat ratio k, which is defined as the ratio of the constant-pressure to the constant-volume specific heat at zero pressure.

$$k = \frac{C_{po}}{C_{vo}} \tag{8.24}$$

Since the difference between C_{po} and C_{vo} is a constant (Eq. 5.24) and since C_{po} and C_{vo} are functions of temperature, it follows that k is also a function of temperature. However, when we consider the specific heat to be constant, k is also constant.

From the definition of k and Eq. 5.24 it follows that

$$C_{vo} = \frac{R}{k-1}; \quad C_{po} = \frac{kR}{k-1} \tag{8.25}$$

Some very useful and simple relations for the reversible adiabatic process can be developed when the specific heats are assumed to be constant.

For the reversible adiabatic process, $ds = 0$. Therefore,

$$T\,ds = du + P\,dv = C_{vo}\,dT + P\,dv = 0$$

From the equation of state for an ideal gas,

$$dT = \frac{1}{R}(P\,dv + v\,dP)$$

Therefore,

$$\frac{C_{vo}}{R}(P\,dv + v\,dP) + P\,dv = 0$$

Substituting Eq. 8.25 into the expression and rearranging,

$$\frac{1}{k-1}(P\,dv + v\,dP) + P\,dv = 0$$

$$v\,dP + kP\,dv = 0$$

$$\frac{dP}{P} + k\frac{dv}{v} = 0$$

Since k is constant when the specific heat is constant, this equation can be integrated under these conditions to give

$$Pv^k = \text{constant} \qquad (8.26)$$

Equation 8.26 holds for all reversible adiabatic processes involving an ideal gas with constant specific heat. It is usually advantageous to express this constant in terms of the initial and final states.

$$Pv^k = P_1v_1{}^k = P_2v_2{}^k = \text{constant} \qquad (8.27)$$

From this equation and the ideal gas equation of state the following expressions relating the initial and final states of an isentropic process can be derived.

$$\frac{P_2}{P_1} = \left(\frac{v_1}{v_2}\right)^k = \left(\frac{V_1}{V_2}\right)^k \qquad (8.28)$$

$$\frac{T_2}{T_1} = \left(\frac{P_2}{P_1}\right)^{(k-1)/k} = \left(\frac{v_1}{v_2}\right)^{k-1} \qquad (8.29)$$

With the assumption of constant specific heat some convenient equations can be derived for the work done by an ideal gas during an adiabatic process. Consider first a system consisting of an ideal gas that undergoes a process in which work is done only at the moving boundary.

$${}_1Q_2 = m(u_2 - u_1) + {}_1W_2 = 0$$

$${}_1W_2 = -m(u_2 - u_1) = -mC_{vo}(T_2 - T_1)$$

$$= \frac{mR}{1-k}(T_2 - T_1) = \frac{P_2V_2 - P_1V_1}{1-k} \qquad (8.30)$$

It should be noted that Eq. 8.30 applies to adiabatic processes only. Since no assumption was made regarding reversibility, it applies to both reversible and irreversible processes. Frequently Eq. 8.30 is derived for reversible processes by starting with the relation

$${}_1W_2 = \int_1^2 P\,dV$$

for the system.

8.10 The Reversible Polytropic Process for an Ideal Gas

When a gas undergoes a reversible process in which there is heat transfer, the process frequently takes place in such a manner that a plot of $\log P$ vs. $\log V$ is a straight line as shown in Fig. 8.15. For such a process PV^n = constant.

This is called a polytropic process. An example is the expansion of the combustion gases in the cylinder of a water-cooled reciprocating engine. If the pressure and volume during a polytropic process are measured during the expansion stroke, as might be done with an engine indicator, and the logarithms of the pressure and volume are plotted, the result would be similar to Fig. 8.15. From this figure it follows that

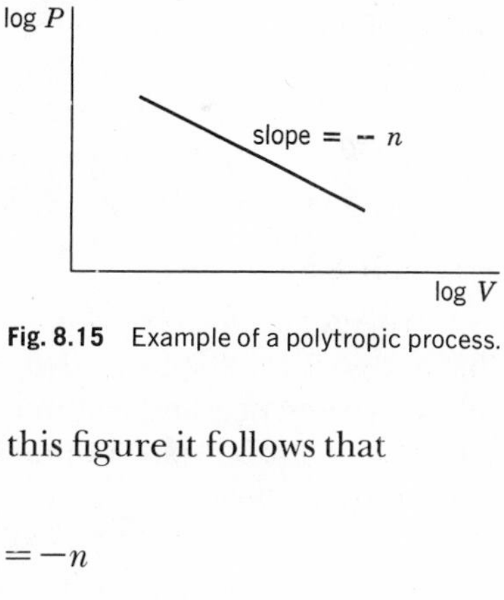

Fig. 8.15 Example of a polytropic process.

$$\frac{d \ln P}{d \ln V} = -n$$

$$d \ln P + n d \ln V = 0$$

If n is a constant (which implies a straight line on the $\log P$ vs. $\log V$ plot), this can be integrated to give the following relation:

$$PV^n = \text{constant} = P_1V_1{}^n = P_2V_2{}^n \qquad (8.31)$$

From this it is evident that the following relations can be written for a polytropic process.

$$\frac{P_2}{P_1} = \left(\frac{V_1}{V_2}\right)^n$$

$$\frac{T_2}{T_1} = \left(\frac{P_2}{P_1}\right)^{(n-1)/n} = \left(\frac{V_1}{V_2}\right)^{n-1} \tag{8.32}$$

For a system consisting of an ideal gas, the work done at the moving boundary during a reversible polytropic process can be derived from the relations

$$_1W_2 = \int_1^2 P\,dV \quad \text{and} \quad PV^n = \text{constant}.$$

$$_1W_2 = \int_1^2 P\,dV = \text{constant} \int_1^2 \frac{dV}{V^n}$$

$$= \frac{P_2V_2 - P_1V_1}{1-n} = \frac{mR(T_2 - T_1)}{1-n} \tag{8.33}$$

for any value of n except $n = 1$.

The polytropic processes for various values of n are shown in Fig. 8.16 on P-v and T-s diagrams. The values of n for some familiar processes are given below.

Isobaric process	$n = 0$
Isothermal process	$n = 1$
Isentropic process	$n = k$
Isometric process	$n = \infty$

Example 8.5

Nitrogen is compressed in a reversible process in a cylinder from 14.7 lbf/in.^{2}, 60 F to 60 lbf/in.^{2}. During the compression process the relation between pressure and volume is $PV^{1.3}$ = constant. Calculate the work and heat transfer per pound, and show this process on P-v and T-s diagrams.

$$P_1 = 14.7 \text{ lbf/in.}^2; \quad P_2 = 60 \text{ lbf/in.}^2$$
$$T_1 = 520 \text{ R}$$

T_2 can be calculated from Eq. 8.32,

$$\frac{T_2}{T_1} = \left(\frac{P_2}{P_1}\right)^{(n-1)/n} = \left(\frac{60}{14.7}\right)^{(1.3-1)/1.3} = 1.383$$

$$T_2 = 520 \times 1.383 = 719 \text{ R}$$

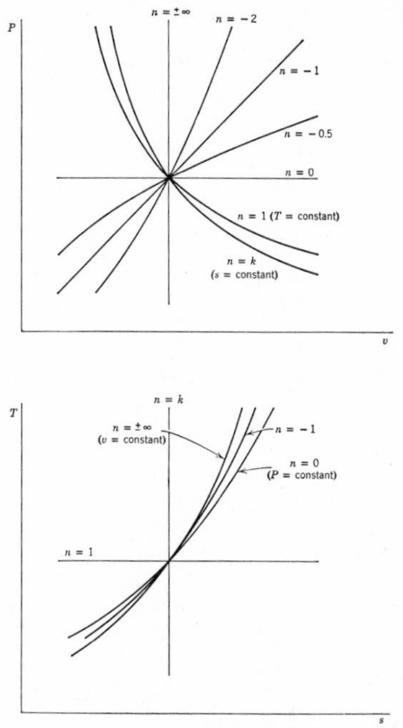

Fig. 8.16 Polytropic processes on *P-v* and *T-s* diagrams.

For this process the work can be found using Eq. 8.33,

$$_1w_2 = \frac{R(T_2 - T_1)}{1-n} = \frac{55.15(719-520)}{(1-1.3)778} = -47.0 \text{ Btu/lbm}$$

The heat transfer can be calculated using the first law. Let us assume C_{vo} to be constant over this range in temperature.

$$_1q_2 = u_2 - u_1 + {_1w_2} = C_{vo}(T_2 - T_1) + {_1w_2}$$

$$= 0.177(719-520) - 47.0 = -11.8 \text{ Btu/lbm}$$

This process is shown on the P-v and T-s diagrams of Fig. 8.17.

The reversible isothermal process for an ideal gas is of particular interest. In this case

$$PV = \text{constant} = P_1V_1 = P_2V_2 \tag{8.34}$$

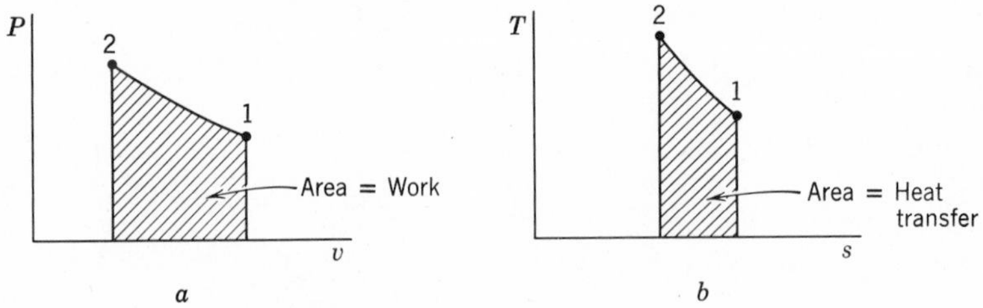

Fig. 8.17 Diagram for Example 8.5.

The work done at the boundary of a simple compressible system during a reversible isothermal process can be found by integrating the equation

$$ {}_1W_2 = \int_1^2 P\,dV $$

The integration is as follows:

$$ {}_1W_2 = \int_1^2 P\,dV = \text{constant} \int_1^2 \frac{dV}{V} = P_1V_1 \ln\frac{V_2}{V_1} = P_1V_1 \ln\frac{P_1}{P_2} \qquad (8.35) $$

$$ {}_1W_2 = mRT \ln\frac{V_2}{V_1} = mRT \ln\frac{P_1}{P_2} $$

Since there is no change in internal energy or enthalpy in an isothermal process, the heat transfer is equal to the work (neglecting changes in kinetic and potential energy). Therefore, we could have derived Eq. 8.35 by calculating the heat transfer.

For example, using Eq. 8.5

$$ \int_1^2 T\,ds = {}_1q_2 = \int_1^2 du + \int_1^2 P\,dv $$

But $du = 0$ and $Pv = \text{constant} = P_1v_1 = P_2v_2$

$$ {}_1q_2 = \int_1^2 P\,dv = P_1v_1 \ln\frac{v_2}{v_1} $$

which yields the same result as Eq. 8.35.

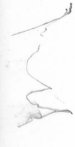

PROBLEMS

8.1 Plot to scale a temperature-entropy diagram for Freon-12, showing the following lines:

(a) Saturated liquid and saturated vapor.

(*b*) 20 lbf/in.^{2} and 200 lbf/in.^{2} constant-pressure lines.
(*c*) 90 Btu/lbm and 100 Btu/lbm constant-enthalpy lines.
(*d*) 1 ft^3/lbm constant-specific volume line.

8.2 A Carnot cycle heat engine receives 500 Btu from a reservoir at 1000 F, and rejects heat at 80 F.

(*a*) Show the cycle on a *T-s* diagram, considering the working fluid as the system.

(*b*) Calculate the work and efficiency of the cycle.

(*c*) Calculate the change in entropy of the high-temperature and low-temperature reservoirs.

(*d*) Suppose the engine operates on the cycle indicated above, but the temperature of the high-temperature reservoir is increased to 1500 F. (Heat is then transferred from the reservoir at 1500 F to the working fluid at 1000 F.) The low-temperature reservoir remains at 80 F. Determine the entropy change for each reservoir.

8.3 A Carnot cycle utilizes steam as the working fluid, and has an efficiency of 20%. Heat is transferred to the working fluid at 400 F and during this process the working fluid changes from saturated liquid to saturated vapor.

(*a*) Show this cycle on a *T-s* diagram that includes the saturated-liquid and saturated-vapor lines.

(*b*) Calculate the quality at the beginning and end of the heat-rejection process.

(*c*) Calculate the work per lbm of steam.

8.4 A Carnot cycle refrigerator utilizes Freon-12 as the working fluid. Heat is rejected from the Freon at 100 F, and during this process the Freon changes from saturated vapor to saturated liquid. Heat transfer to the Freon is at 0 F.

(*a*) Show this cycle on a *T-s* diagram.

(*b*) What is the quality at the beginning and end of the isothermal process at 0 F?

(*c*) What is the coefficient of performance of this cycle?

8.5 A cylinder fitted with a piston contains steam initially at 140 lbf/in.^{2}, 600 F, at which point the volume is 1 ft^3. The steam then expands in a reversible adiabatic process to a final pressure of 20 lbf/in.^{2}. Determine the work done during the process.

8.6 Saturated vapor ammonia at 30 lbf/in.^{2} is compressed in a cylinder by a piston to a pressure of 220 lbf/in.^{2} in a reversible adiabatic process. Determine the work done per lbm of ammonia.

8.7 A pressure vessel contains steam at 225 lbf/in.^{2}, 800 F. A valve at the top of the pressure vessel is opened, allowing steam to escape. Assume that at any instant the steam that remains in the pressure vessel has undergone a reversible adiabatic process. Determine the fraction of steam that has escaped when the steam remaining in the pressure vessel is saturated vapor.

8.8 Freon-12 at 80 F, 80 lbf/in.^{2}, is contained in a cylinder fitted with a piston. The volume at this condition is 0.5 ft^3. The Freon then undergoes a reversible, adiabatic expansion to −20 F. At the completion of this process, the Freon is allowed to warm back to 80 F at constant pressure. Show the two processes on a *T-s* diagram, and calculate the work and heat transfer for each process.

8.9 A mass of 100 lbm is held in position in a length of 5 in. diameter tube by a pin. Below this mass is a volume of 0.5 ft³ of steam at 800 lbf/in.², 800 F. The pin is released and the mass is accelerated upwards, leaving the top

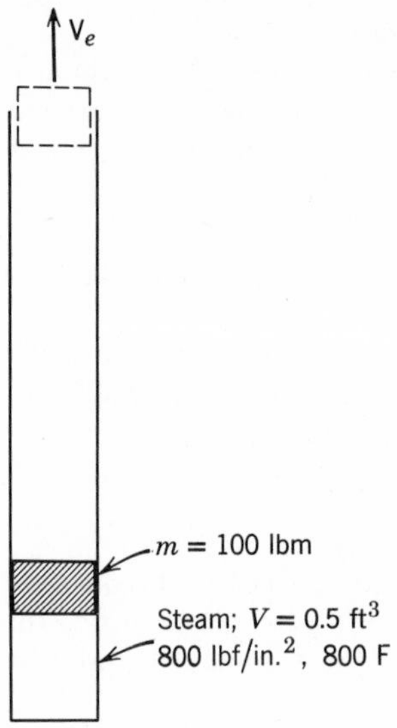

Fig. 8.18 Sketch for Problem 8.9.

of the tube with a certain velocity. If the steam undergoes a reversible adiabatic expansion, until a pressure of 100 lbf/in.² is reached, what is the exit velocity of the mass?

8.10 A cylinder fitted with a piston contains one pound of Freon-12 at 15 lbf/in.², 200 F. The piston is slowly moved and the necessary heat transfer takes place so that the Freon-12 is compressed in a reversible isothermal process until the Freon-12 exists as saturated vapor.

(*a*) Show the process on a temperature-entropy diagram that is approximately to scale.

(*b*) Calculate the final pressure and specific volume of the Freon-12.

(*c*) Determine the work and the heat transfer for this process.

8.11 Solve Problem 4.12 without using the relation ${}_1W_2 = \int_1^2 P\,dV$.

8.12 Tank *A* initially contains 10 lbm of steam at 100 lbf/in.², 600 F, and is connected through a valve to a cylinder fitted with a frictionless piston, as shown in Fig. 8.19. A pressure of 20 lbf/in.^{2} is required to balance the weight of the piston.

The connecting valve is opened until the pressure in A equals 20 lbf/in.^{2}. Assume the entire process to be adiabatic, and that the steam that finally remains in A has undergone a reversible adiabatic process.

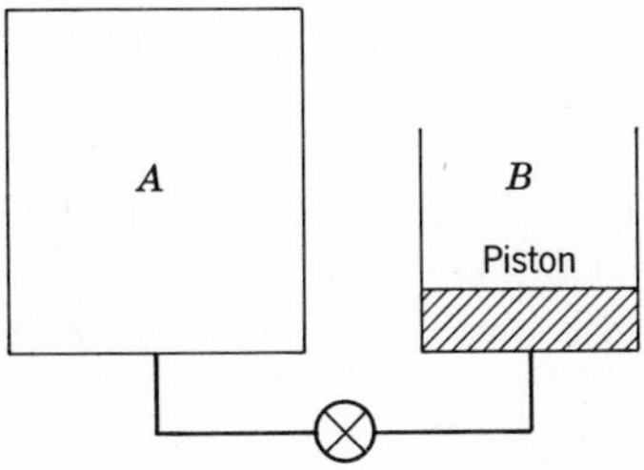

Fig. 8.19 Sketch for Problem 8.12.

Determine the work done against the piston and the final temperature of the steam in cylinder B.

8.13 Repeat Problem 8.12, but let the piston initially rest on stops so that the initial volume of the cylinder B is 10 ft^3, as shown in Fig. 8.20. Assume that this volume is initially evacuated.

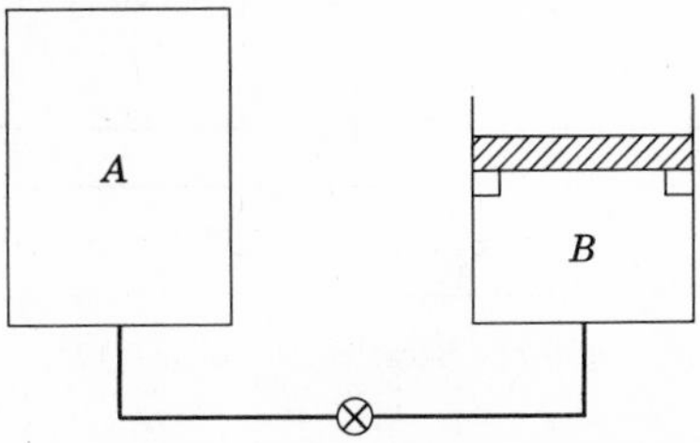

Fig. 8.20 Sketch for Problem 8.13.

8.14 A neophyte engineer has a problem to solve which involves the entropy of superheated steam at pressures below 1 lbf/in.^{2}, but he finds that the steam table which he has does not give the properties of superheated steam below this pressure.

(*a*) How would you advise him to proceed?

(*b*) What assumption does this procedure involve?

(*c*) Set up, in accordance with the assumptions made, a line similar to the 1 lbf/in.^{2} line in Table B.1.3 which gives v, u, h, and s at 0.5 lbf/in.^{2} at temperatures of 200 F, 500 F, 1000 F.

8.15 It is desired to quickly cool a given quantity of material to 40 F. The required heat transfer is 2000 Btu. One possibility is to immerse it in a mixture of ice and water, in which case the heat transfer from the material results in the melting of ice. Another possibility is to cool the material by evaporating Freon-12 at 0 F, in which case the heat transfer results in changing Freon-12 from saturated liquid to saturated vapor. A third possibility is to do the same with liquid nitrogen at one atmosphere pressure.

(*a*) Calculate the change of entropy of the cooling medium in each of the three cases.

(*b*) What is the significance of these results?

8.16 An insulated cylinder fitted with a piston has an initial volume of 5 ft^3 and contains steam at 60 lbf/in.^{2}, 400 F. The steam is now expanded adiabatically, and during this process the work output is carefully measured and found to be 30 Btu. It is claimed that at the final state the steam is in the two-phase (liquid plus vapor) region. What is your evaluation of this claim?

8.17 Consider the scheme shown in Fig. 8.21 for raising a weight by heat transfer from the reservoir at 300 F to the Freon-12. The pressure on the Freon-12 due to the weight plus the atmosphere is 200 lbf/in.^{2}. Initially the

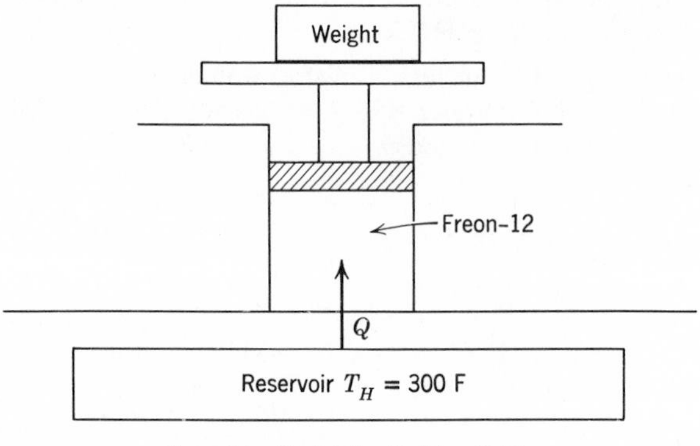

Fig. 8.21 Sketch for Problem 8.17.

temperature of the Freon-12 is 160 F and its volume is 0.4 ft^3. Heat is transferred until the temperature of the Freon-12 is 300 F.

(*a*) Determine the heat transfer and work for this process.

(*b*) Determine the net change of entropy.

(*c*) Devise a scheme for accomplishing this with no net change of entropy.

8.18 A cylinder contains 2 lbm of ammonia at 15 lbf/in.^{2} and 80 F, the temperature of the surroundings. The ammonia is then compressed isothermally to saturated vapor, with a work input of 320 Btu. During this process, any heat transfer takes place with the surroundings. Is this process reversible, irreversible, or impossible?

8.19 A Carnot engine has 1 lbm of air as the working fluid. Heat is received at 1400 R and rejected at 550 R. At the beginning of the heat addition process the pressure is 120 lbf/in.^{2}, and during this process the volume triples. Calculate the net cycle work per lbm of air.

8.20 Twenty ft^3 of air at 600 lbf/in.^{2}, 600 F expand in a reversible isothermal process in a cylinder to 14.7 lbf/in.^{2}. Calculate

(*a*) The heat transfer during the process.

(*b*) The change of entropy of the air.

8.21 A large tank having a volume of 30 ft^3 is connected to a small tank having a volume of 10 ft^3. The large tank contains air initially at 100 lbf/in.^{2}, 70 F and the small tank is initially evacuated. A valve in the connecting pipe between the two tanks is suddenly opened and is closed when the tanks come to pressure equilibrium. The air in the large tank may be assumed to have undergone a reversible process and the entire process is adiabatic. What is the final mass of air in the small tank? What is the final temperature of the air in the small tank?

8.22 One lbm of air contained in a cylinder at 75 lbf/in.^{2}, 2000 R, expands in a reversible adiabatic process to 15 lbf/in.^{2}. Calculate the final temperature and the work done during the process, using

(*a*) The air tables, Table B.8.

(*b*) Constant specific heat, value from Table B.6.

(*c*) Discuss why the method of part *b* results in a poor value for final temperature and yet a relatively good value for the work.

8.23 It is desired to obtain a supply of cold helium gas by the following technique: helium contained in a cylinder at ambient conditions, 14.7 lbf/in.^{2}, 80 F, is compressed in an isothermal process to 100 lbf/in.^{2}, after which the gas is expanded back to 14.7 lbf/in.^{2} in an adiabatic process. Both processes are reversible.

(*a*) Show the process on T-s coordinates.

(*b*) Calculate the final temperature and the net work per lbm.

(*c*) If a diatomic gas had been used instead of helium, would the final temperature be higher or lower?

8.24 An insulated tank having a volume of 30 ft^3 contains air at 120 lbf/in.^{2}, 80 F. A valve on the tank is then opened and the pressure inside drops quickly to 20 lbf/in.^{2}.

(*a*) Assuming that the air remaining inside the tank has undergone a reversible adiabatic expansion, calculate the mass withdrawn.

(*b*) Calculate the mass withdrawn by a first law analysis, and compare the result with part *a*.

8.25 An insulated cylinder having an initial volume of 2 ft^3 contains carbon dioxide at 15 lbf/in.^{2}, 440 F. The gas is then compressed to 200 lbf/in.^{2} in a reversible adiabatic process. Calculate the final temperature and the work, assuming

(*a*) Variable specific heat, Table B.9.

(*b*) Constant specific heat, value from Table B.6.

(*c*) Constant specific heat, value at an intermediate temperature from Table B.7.

8.26 In 1819 Desormes and Clement measured the value of the specific heat ratio k for air in the following manner. The air in a large pressure vessel was maintained at a pressure P_1, slightly above atmospheric pressure, and both the air and the vessel were at atmospheric temperature. The valve in a line connecting the vessel to the atmosphere was quickly opened until the pressure in the tank equaled the atmospheric pressure P_0, at which moment the valve was closed. After the vessel and the air had again come to thermal equilibrium with the atmosphere (the atmospheric temperature remains constant), the final pressure in the vessel, P_2, was noted. Find an expression for k in terms of P_1, P_2, and P_0. It may be assumed that there is no heat transfer to the air during the time when air is escaping from the vessel.

8.27 Water at 70 F is pumped from a lake to an elevated storage tank. The average elevation of the water in the tank is 100 ft above the surface of the lake, and the volume of the tank is 10,000 gal. Initially the tank contains air at 14.7 lbf/in.^{2}, 70 F, and the tank is closed so that the air is compressed as the water enters the bottom of the tank. The pump is operated until the tank is three-quarters full. The temperature of the air and water remain constant at 70 F. Determine the work input to the pump.

8.28 The power-stroke expansion of exhaust gases in an internal combustion engine can reasonably be represented as a polytropic expansion. Consider air contained in a cylinder volume of 5 in.^{3} at 1000 lbf/in.^{2}, 3000 R. The air then expands in a reversible polytropic process, with exponent $n = 1.30$, through a volume ratio of 8 : 1. Show the process on P-v and T-s coordinates, and calculate the work and heat transfer.

8.29 Argon gas at 15 lbf/in.^{2}, 70 F, is compressed reversibly in a cylinder to 60 lbf/in.^{2}. Calculate the required work per lbm if the process is

(*a*) Adiabatic.
(*b*) Isothermal.
(*c*) Polytropic with $n = 1.30$.

8.30 An ideal gas having a specific heat of

$$\bar{C}_{po} = 10.0 \text{ Btu/lb mole-R}$$

undergoes a reversible polytropic compression in which the polytropic exponent $n = 1.40$. Will the heat transfer for the process be positive, negative or zero?

8.31 At one time a certain scientist felt that the equivalent of the following process violated the second law of thermodynamics. (See Fig. 8.22.) Initially compartments A and B, which are separated by a frictionless nonconducting piston of negligible mass, each contain 1 mole of nitrogen at 1 atm pressure. The initial temperature in compartment A is 200 F and in compartment B it is 100 F. Heat is transferred from a constant-temperature reservoir at 200 F to the nitrogen in compartment B until it reaches a temperature of 200 F. During the heat transfer process the piston will move

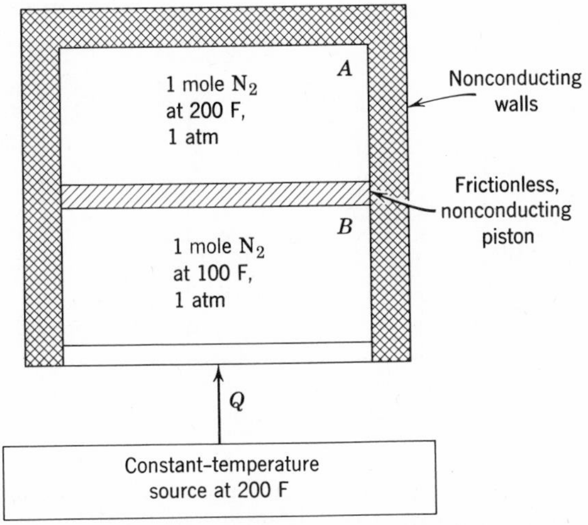

Fig. 8.22 Sketch for Problem 8.31.

as necessary to maintain the same pressure in both compartments.

(*a*) Determine the final temperature in compartment *A*.

(*b*) Prove that even though the final temperature in compartment *A* is greater than 200 F this process does not violate the second law.

8.32 A cylinder divided by a frictionless, thermally insulated piston contains air on one side and water on the other, as shown in Fig. 8.23. The cylinder is well insulated, except for the end of the part containing the water. Each volume is initially 5 ft^3, at which point the air is at 100 F, and the water is at 200 F with a quality of 10%. Heat is now slowly transferred to the water, until at the final state it exists as saturated vapor. Calculate the final pressure and the amount of heat transferred.

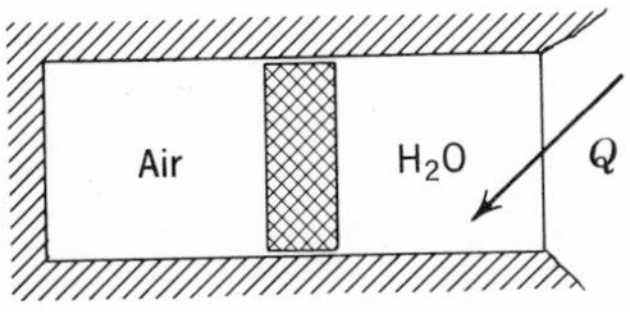

Fig. 8.23 Sketch for Problem 8.32.

8.33 Consider the system shown in Fig. 8.24. Tank *A* has a volume of 10 ft^3 and initially contains air at 100 $lbf/in.^2$, 100 F. Cylinder *B* is fitted with a frictionless piston resting on the bottom, at which position the spring is fully extended. The piston has a cross-sectional area of 100 $in.^2$ and mass of 100 lbm, and the spring constant K_s is 100 lbf/in. Atmospheric pressure is 14.7 $lbf/in.^2$.

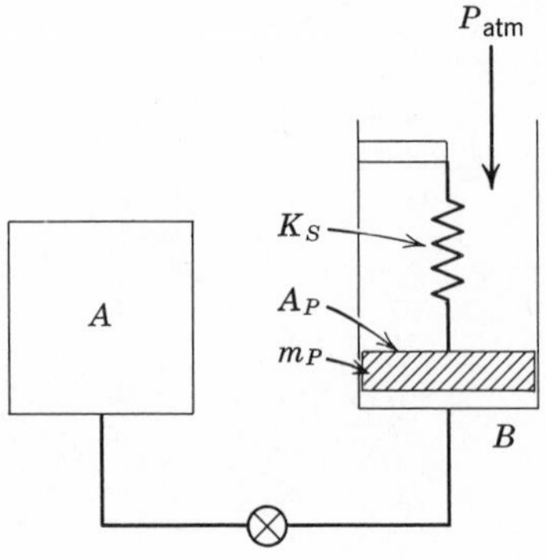

Fig. 8.24 Sketch for Problem 8.33.

The valve is opened and air flows into the cylinder until the pressures in A and B become equal, after which the valve is closed. During this process, the air finally remaining in A may be considered to have undergone a reversible adiabatic process, and the entire process is adiabatic. The spring force is proportional to the displacement.

Determine the final pressure in the system and the temperature in cylinder B.

8.34 Write a computer program to solve the following problem. One of the gases listed in Table B.7 undergoes a reversible adiabatic process in a cylinder from P_1, T_1, to P_2. It is desired to calculate the final temperature and the work for the process by integrating the specific heat equation, by assuming constant specific heat at temperature T_1, and finally by assuming constant specific heat at an average temperature (by iteration).

9

Second Law Analysis for a Control Volume

In the preceding two chapters we have discussed the second law of thermodynamics and the thermodynamic property entropy. As was done with the first law analysis, we now consider the more general application of these concepts, the control volume analysis, and a number of special cases of interest. We shall also discuss the topic of thermodynamic availability and two related thermodynamic properties, the Gibbs and Helmholtz functions.

9.1 The Second Law of Thermodynamics for a Control Volume

The second law of thermodynamics can be applied to a control volume by a similar procedure to that used in writing the first law for a control volume in Section 6.2. The second law for a system analysis has been stated (Eq. 8.12) in the form

$$dS = \frac{\delta Q}{T} + \frac{\delta LW}{T}$$

For a change in entropy, $S_2 - S_1$, that occurs during a time interval δt we can write

$$\frac{S_2 - S_1}{\delta t} = \frac{1}{\delta t}\left(\frac{\delta Q}{T}\right) + \frac{1}{\delta t}\left(\frac{\delta LW}{T}\right) \tag{9.1}$$

Consider the system and control volume shown in Fig. 9.1. During time δt the mass δm_i enters the control volume across the discrete area

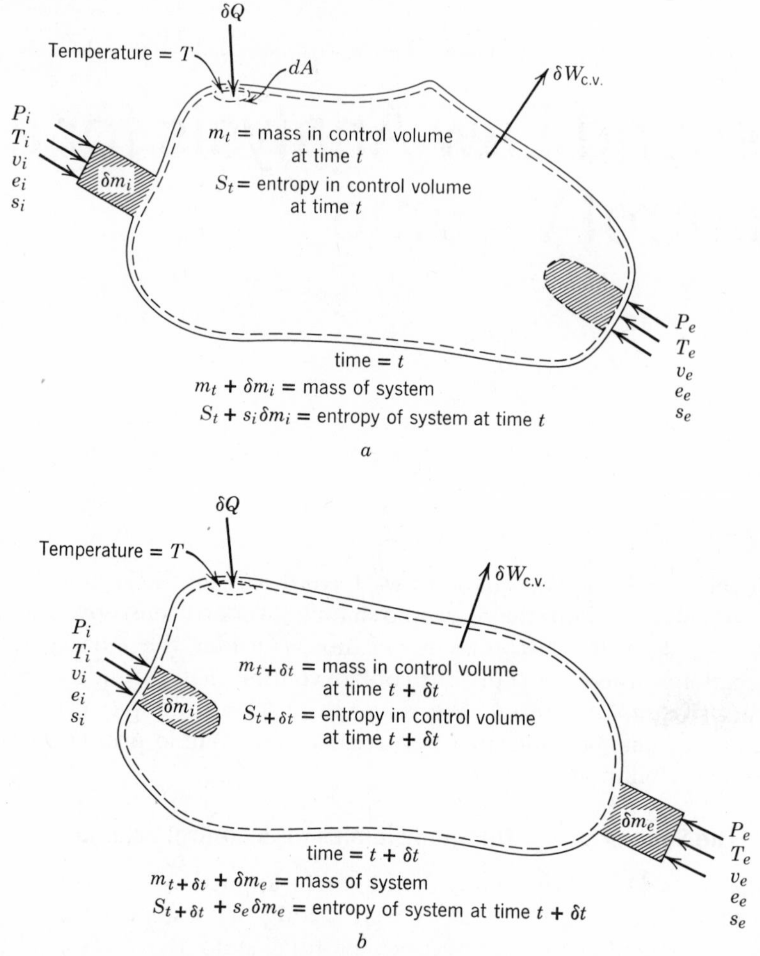

Fig. 9.1 Schematic diagram for a second law analysis of a control volume.

A_i, the mass δm_e leaves across the discrete area A_e, an amount of heat δQ is transferred to the system across an element of area where the surface temperature is T, and the work δW is done by the system. As in the analysis of Section 6.2, we assume that the increment of mass δm_i has uniform properties, and similarly that δm_e has uniform properties.

Let

$$S_t = \text{the entropy in the control volume at time } t$$

$$S_{t+\delta t} = \text{the entropy in the control volume at time } t + \delta t$$

Then

$$S_1 = S_t + s_i \delta m_i = \text{entropy of the system at time } t$$

$$S_2 = S_{t+\delta t} + s_e \delta m_e = \text{entropy of the system at time } t + \delta t$$

Therefore

$$S_2 - S_1 = (S_{t+\delta t} - S_t) + (s_e \delta m_e - s_i \delta m_i) \tag{9.2}$$

The term $(s_e \delta m_e - s_i \delta m_i)$ represents the net flow of entropy out of the control volume during δt as the result of the masses δm_e and δm_i crossing the control surface.

The significance of the remaining terms of Eq. 9.1 must be carefully considered when applied to a control volume. When we write the terms $(\delta Q/T)$ and $(\delta LW/T)$ for a system we normally consider a quantity of mass at uniform temperature at any instant of time. As noted before, in the case of a control volume analysis we deviate from a strictly classical concept of thermodynamics, and are prepared to consider a variation in temperature throughout the control volume, the assumption of "local" equilibrium.

As a consequence, for the heat transfer term we must consider each element of area on the surface across which heat flows and the surface temperature of this local area. Therefore this term in Eq. 9.1 must be written as the summation of all local $(\delta Q/T)$ terms over the control surface, or

$$\frac{1}{\delta t}\left(\frac{\delta Q}{T}\right) = \frac{1}{\delta t}\int_A \left(\frac{\delta Q/A}{T}\right) dA \tag{9.3}$$

Similarly, for the lost work term, which represents irreversibilities occurring inside the control volume, the result must be written as the summation of all local $(\delta LW/T)$ terms throughout the interior of the control volume, or

$$\frac{1}{\delta t}\left(\frac{\delta LW}{T}\right) = \frac{1}{\delta t}\int_V \left(\frac{\delta LW/V}{T}\right) dV \tag{9.4}$$

Substituting Eqs. 9.2, 9.3, and 9.4 into Eq. 9.1, we have

$$\left(\frac{S_{t+\delta t} - S_t}{\delta t}\right) + \frac{s_e \delta m_e}{\delta t} - \frac{s_i \delta m_i}{\delta t} = \frac{1}{\delta t}\int_A \left(\frac{\delta Q/A}{T}\right) dA + \frac{1}{\delta t}\int_V \left(\frac{\delta LW/V}{T}\right) dV \tag{9.5}$$

In order to reduce this expression to a rate equation, we establish the limit for each term as δt is made to approach zero, at which time the system and control volume coincide.

$$\lim_{\delta t\to 0}\left(\frac{S_{t+\delta t}-S_t}{\delta t}\right)=\frac{dS_{\text{c.v.}}}{dt}$$

$$\lim_{\delta t\to 0}\left(\frac{s_e\delta m_e}{\delta t}\right)=\dot{m}_e s_e$$

$$\lim_{\delta t\to 0}\left(\frac{s_i\delta m_i}{\delta t}\right)=\dot{m}_i s_i$$

$$\lim_{\delta t\to 0}\left[\frac{1}{\delta t}\int_A\left(\frac{\delta Q/A}{T}\right)dA\right]=\int_A\left(\frac{\dot{Q}_{\text{c.v.}}/A}{T}\right)dA$$

$$\lim_{\delta t\to 0}\left[\frac{1}{\delta t}\int_V\left(\frac{\delta LW/V}{T}\right)dV\right]=\int_V\left(\frac{L\dot{W}_{\text{c.v.}}/V}{T}\right)dV$$

The assumptions of uniform properties for the incremental mass δm_i entering the control volume across area A_i and also for δm_e leaving across area A_e reduce in the limit to the restriction of uniform properties across area A_i and also across A_e at an instant of time. The heat transfer term in the limit becomes the integral over the control surface of the local rate of heat transfer divided by the local surface temperature. Similarly, the internal irreversibility term becomes the integral throughout the control volume of the local lost work rate divided by the local interior temperature.

In utilizing these limiting values to express the rate form of the entropy equation for a control volume, we again include summation signs on the flow terms, to account for the possibility of additional flow streams entering or leaving the control volume. The resulting expression is

$$\frac{dS_{\text{c.v.}}}{dt}+\sum \dot{m}_e s_e-\sum \dot{m}_i s_i=\int_A\left(\frac{\dot{Q}_{\text{c.v.}}/A}{T}\right)dA+\int_V\left(\frac{L\dot{W}_{\text{c.v.}}/V}{T}\right)dV \tag{9.6}$$

which, for our purposes, is the general expression of the second law entropy equation. In words, this expression states that the rate of change of entropy inside the control volume plus the net rate of entropy flow out is equal to the sum of two terms, the integrated heat transfer surface term and the positive internal irreversibility production term. Since this latter term is necessarily positive (or zero) and is difficult if not impossible to evaluate quantitatively, Eq. 9.6 is frequently written in the form

$$\frac{dS_{\text{c.v.}}}{dt}+\sum \dot{m}_e s_e-\sum \dot{m}_i s_i\geqslant\int_A\left(\frac{\dot{Q}_{\text{c.v.}}/A}{T}\right)dA \tag{9.7}$$

where the equality applies to internally reversible processes and the inequality to internally irreversible processes.

Note that if there is no mass flow into or out of the control volume and if temperature is considered uniform at any instant of time, then Eq. 9.6 reduces to a rate form of the system entropy equation, Eq. 8.12. In this sense, Eq. 9.6 (or 9.7) can be considered a general expression, as was also pointed out for the control volume form of the first law.

Representation of Eq. 9.6 in terms of local fluid properties follows readily from the corresponding representation of the first law. As in Eq. 6.16 for energy, the total entropy $S_{\text{c.v.}}$ inside the control volume at any time may be expressed as the following volume integral:

$$S_{\text{c.v.}} = \int_V s\rho \, dV \tag{9.8}$$

As in Eq. 6.17 for energy, the entropy flow across each area of flow A is

$$\int_A s\delta\dot{m} = \int_A s\rho \mathsf{V}_{rn} \, dA \tag{9.9}$$

Considering each V_{rn} as the outward-directed normal, Eqs. 9.8 and 9.9 may be substituted into Eq. 9.6, with the result being

$$\frac{d}{dt}\int_V s\rho \, dV + \int_A s\rho \mathsf{V}_{rn} \, dA = \int_A \left(\frac{\dot{Q}_{\text{c.v.}}/A}{T}\right) dA + \int_V \left(\frac{L\dot{W}_{\text{c.v.}}/V}{T}\right) dV \tag{9.10}$$

the second law entropy equation for a control volume stated in terms of local fluid properties. This form of the equation should prove beneficial in pointing out the assumptions implicit in Eq. 9.6 and the significance of the notation and terms in that expression.

9.2 The Steady-State, Steady-Flow Process and the Uniform-State, Uniform-Flow Process

We now consider the application of the second law control volume equation, Eq. 9.6 or 9.7, to the two control volume model processes developed in Chapter 6.

For the steady-state, steady-flow process, which has been defined in Section 6.3, we conclude that there is no change with time of the entropy per unit mass at any point within the control volume, and therefore the first term of Eq. 9.7 equals zero. That is

$$\frac{dS_{\text{c.v.}}}{dt} = \frac{d}{dt}\int_V s\rho \, dV = 0 \tag{9.11}$$

so that, for the SSSF process

$$\sum \dot{m}_e s_e - \sum \dot{m}_i s_i \geq \int_A \left(\frac{\dot{Q}_{\text{c.v.}}/A}{T}\right) dA \tag{9.12}$$

in which the various mass flows, heat transfer rate, and states are all constant with time.

If in a steady-state, steady-flow process there is only one area over which mass enters the control volume at a uniform rate and only one area over which mass leaves the control volume at a uniform rate we can write

$$\dot{m}(s_e - s_i) \geq \int_A \left(\frac{\dot{Q}_{\text{c.v.}}/A}{T}\right) dA \tag{9.13}$$

For an adiabatic process with these assumptions it follows that

$$s_e \geq s_i \tag{9.14}$$

where the equality holds for a reversible adiabatic process.

Example 9.1

Steam enters a steam turbine at a pressure of 100 lbf/in.², a temperature of 500 F, and a velocity of 200 ft/sec. The steam leaves the turbine at a pressure of 20 lbf/in.² and a velocity of 600 ft/sec. Determine the work per pound of steam flowing through the turbine, assuming the process to be reversible and adiabatic.

In solving a problem such as this it is usually advisable to draw a schematic diagram of the apparatus, and to show the process or, in the case of irreversible process, the various states involved on a T-s diagram. The next step in the solution is to write the continuity equation, the first and second laws of thermodynamics, and the property relation as they apply to the particular problem.

The schematic diagram and a T-s diagram are shown in Fig. 9.2. The various equations that apply to this process are:

Continuity eq.: $\dot{m}_e = \dot{m}_i = \dot{m}$

First law: $h_i + \frac{\mathsf{V}_i^2}{2g_c} = h_e + \frac{\mathsf{V}_e^2}{2g_c} + w$

Second law: $s_e = s_i$

$P_i = 100 \ \text{lbf/in.}^2$
$T_i = 500 \ \text{F}$
$V_i = 200 \ \text{ft/sec}$

W

$P_e = 20 \ \text{lbf/in.}^2$
$V_e = 600 \ \text{ft/sec}$

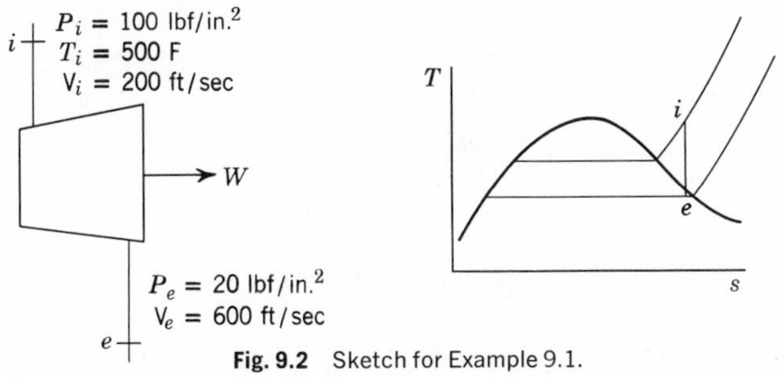

Fig. 9.2   Sketch for Example 9.1.

Property relation: Steam tables
From the steam tables.

$$h_i = 1279.1 \ \text{Btu/lbm} \quad s_i = 1.7085 \ \text{Btu/lbm-R}$$

The two properties known in the final state are pressure and entropy.

$$P_e = 20 \ \text{lbf/in.}^2 \quad s_e = s_i = 1.7085 \ \text{Btu/lbm-R}$$

Therefore, the quality and enthalpy of the steam leaving the turbine can be determined.

$$s_e = 1.7085 = s_g - (1-x)_e s_{fg} = 1.7320 - (1-x)_e 1.3962$$

$$(1-x)_e = \frac{0.0235}{1.3962} = 0.01685$$

$$h_e = h_g - (1-x)_e h_{fg} = 1156.4 - 0.01685(960.1) = 1140.2 \ \text{Btu/lbm}$$

Therefore, the work per pound of steam for this isentropic process may be found using the equation for the first law as given above

$$w = 1279.1 - 1140.2 + \frac{(200)^2 - (600)^2}{2 \times 32.17 \times 778} = 132.5 \ \text{Btu/lbm}$$

Example 9.2

Consider the reversible adiabatic flow of steam through a nozzle. Steam enters the nozzle at 100 lbf/in.², 500 F, with a velocity of 100 ft/sec. The pressure of the steam at the nozzle exit is 40 lbf/in.². Determine the exit velocity of the steam from the nozzle, assuming a reversible, adiabatic, steady-state, steady-flow process.

This process is shown schematically and on a T-s diagram in Fig. 9.3.

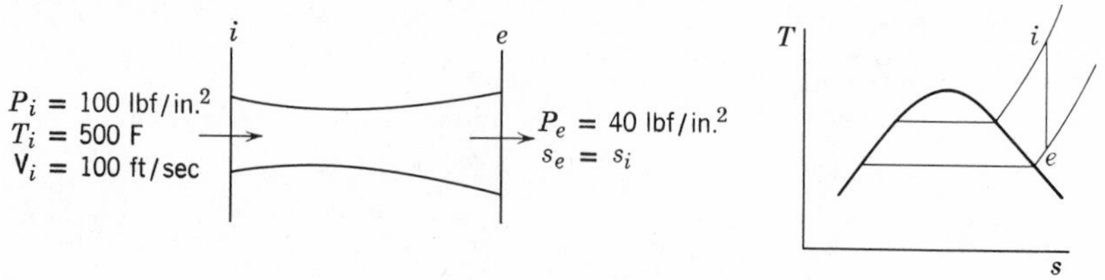

Fig. 9.3 Sketch for Example 9.2.

Since this is a steady-state, steady-flow process in which the work, the heat transfer, and the changes in potential energy are zero, we can write,

Continuity eq.: $\dot{m}_e = \dot{m}_i = \dot{m}$

First law:
$$h_i + \frac{\mathsf{V}_i^2}{2g_c} = h_e + \frac{\mathsf{V}_e^2}{2g_c}$$

Second law: $s_e = s_i$

Property relation: Steam tables

$$h_i = 1279.1 \text{ Btu/lbm}; \quad s_i = 1.7085 \text{ Btu/lbm-R}$$

The two properties that we know in the final state are

$$s_e = s_i = 1.7085 \text{ Btu/lbm-R}; \quad P_e = 40 \text{ lbf/in.}^2$$

Therefore,

$$T_e = 354.2 \text{ F}; \quad h_e = 1193.9 \text{ Btu/lbm}$$

Substituting into the equation for the first law as given above we have,

$$\frac{\mathsf{V}_e^2}{2g_c} = h_i - h_e + \frac{\mathsf{V}_i^2}{2g_c}$$

$$= 1279.1 - 1193.9 + \frac{100 \times 100}{2 \times 32.17 \times 778} = 85.4 \text{ Btu/lbm}$$

$$\mathsf{V}_e = \sqrt{2 \times 32.17 \times 778 \times 85.4} = 2070 \text{ ft/sec}$$

Example 9.3

An inventor reports that he has a refrigeration compressor that receives saturated Freon-12 vapor at 0 F and delivers the vapor at 150 lbf/in.^{2}, 120 F. The compression process is adiabatic. Does the process described violate the second law?

Since this is a steady-state, steady-flow, adiabatic process we can write:

Second law: $s_e \geqslant s_i$

Property relation: Freon-12 tables
From the Freon-12 tables,

$$s_e = 0.16629 \text{ Btu/lbm R}; \; s_i = 0.16888 \text{ Btu/lbm R}$$

Therefore, $s_e < s_i$, whereas, for this process, the second law requires that $s_e \geqslant s_i$. The process described would involve a violation of the second law, and would not be possible.

Example 9.4

Air expands in an air turbine from a pressure of 50 lbf/in.^{2} and a temperature of 600 F to an exhaust pressure of 20 lbf/in.^{2}. Assume the process to be reversible and adiabatic, with negligible changes in kinetic and potential energy. Calculate the work per pound of air flowing through the turbine.

Since this is a steady-state, steady-flow reversible adiabatic process we can write

Continuity eq.: $\dot{m}_e = \dot{m}_i = \dot{m}$

First law: $h_i = h_e + w$

Second law: $s_e = s_i$

Property relation: Gas Tables
From Table B.8,

$$T_i = 1060 \text{ R}; \quad h_i = 255.96 \text{ Btu/lbm}; \quad s_i^\circ = 0.76496$$

From Eq. 8.23

$$s_e - s_i = 0 = s_e^\circ - 0.76496 - \frac{53.34}{778} \ln \frac{20}{50}$$

$$s_e^\circ = 0.7022$$

From Table B.8,

$$T_e = 822 \text{ R}; \quad h_e = 197.18$$

Therefore

$$w = h_i - h_e = 255.96 - 197.18 = 58.8 \text{ Btu/lbm}$$

For the uniform-state, uniform-flow process, which was described in Section 6.5, the second law for a control volume, Eq. 9.7, can be written in the following form.

$$\frac{d}{dt}[ms]_{\text{c.v.}} + \sum \dot{m}_e s_e - \sum \dot{m}_i s_i \geq \int_A \left(\frac{\dot{Q}_{\text{c.v.}}/A}{T}\right) dA \tag{9.15}$$

If this is integrated over the time interval t we have

$$\int_0^t \frac{d}{dt}[ms]_{\text{c.v.}}\, dt = [m_2 s_2 - m_1 s_1]_{\text{c.v.}}$$

$$\int_0^t \left(\sum \dot{m}_e s_e\right) dt = \sum m_e s_e; \qquad \int_0^t \left(\sum \dot{m}_i s_i\right) dt = \sum m_i s_i \tag{9.16}$$

Therefore, for this period of time t we can write the second law for the uniform-state, uniform-flow process as

$$[m_2 s_2 - m_1 s_1]_{\text{c.v.}} + \sum m_e s_e - \sum m_i s_i \geq \int_0^t \left[\int_A \left(\frac{\dot{Q}_{\text{c.v.}}/A}{T}\right) dA\right] dt \tag{9.17}$$

However, since in this process the temperature is uniform throughout the control volume at any instant of time, the integral on the right reduces to

$$\int_0^t \left[\int_A \left(\frac{\dot{Q}_{\text{c.v.}}/A}{T}\right) dA\right] dt = \int_0^t \left[\frac{1}{T}\int_A \left(\frac{\dot{Q}_{\text{c.v.}}}{A}\right) dA\right] dt = \int_0^t \left(\frac{\dot{Q}_{\text{c.v.}}}{T}\right) dt$$

and therefore the second law for the uniform-state, uniform-flow process can be written

$$[m_2 s_2 - m_1 s_1]_{\text{c.v.}} + \sum m_e s_e - \sum m_i s_i \geq \int_0^t \left(\frac{\dot{Q}_{\text{c.v.}}}{T}\right) dt \tag{9.18}$$

By introducing the lost work we can write this as an equality. In doing so, we note that since the temperature is uniform throughout the control volume at any instant of time,

$$\int_0^t \left[\int_V \left(\frac{L\dot{W}_{\text{c.v.}}/V}{T} \right) dV \right] dt = \int_0^t \left[\frac{1}{T} \int_V \left(\frac{L\dot{W}_{\text{c.v.}}}{V} \right) dV \right] dt = \int_0^t \left(\frac{L\dot{W}_{\text{c.v.}}}{T} \right) dt$$

Therefore,

$$[m_2 s_2 - m_1 s_1]_{\text{c.v.}} + \sum m_e s_e - \sum m_i s_i = \int_0^t \left(\frac{\dot{Q}_{\text{c.v.}} + L\dot{W}_{\text{c.v.}}}{T} \right) dt \tag{9.19}$$

9.3 The Reversible Steady-State, Steady-Flow Process

An expression can be derived for the work in a reversible adiabatic, steady-state, steady-flow process which is of great help in understanding the significant variables in such a process. We have noted that when a steady-state, steady-flow process involves a single flow of fluid into and out of the control volume, the first law can be written, Eq. 6.23,

$$q + h_i + \frac{\mathsf{V}_i^2}{2g_c} + Z_i \frac{g}{g_c} = h_e + \frac{\mathsf{V}_e^2}{2g_c} + Z_e \frac{g}{g_c} + w$$

and the second law, Eq. 9.13 is

$$\dot{m}(s_e - s_i) \geqslant \int_A \left(\frac{\dot{Q}_{\text{c.v.}}/A}{T} \right) dA$$

Let us now consider two types of flow, a reversible adiabatic process and a reversible isothermal process.

If the process is reversible and adiabatic, the second law equation above reduces to

$$s_e = s_i$$

and it follows from the property relation

$$T\,ds = dh - v\,dP$$

that

$$h_e - h_i = \int_i^e v\,dP \tag{9.20}$$

Substituting these relations into Eq. 6.23 and noting that $q = 0$ we have

$$w = (h_i - h_e) + \frac{(\mathsf{V}_i^2 - \mathsf{V}_e^2)}{2g_c} + (Z_i - Z_e)\frac{g}{g_c}$$

$$= -\int_i^e v\,dP + \frac{(\mathsf{V}_i^2 - \mathsf{V}_e^2)}{2g_c} + (Z_i - Z_e)\frac{g}{g_c} \tag{9.21}$$

If, instead, the process is reversible and isothermal, the second law reduces to

$$\dot{m}(s_e - s_i) = \frac{1}{T}\int_A (\dot{Q}_{\text{c.v.}}/A)\,dA = \frac{\dot{Q}_{\text{c.v.}}}{T}$$

or

$$T(s_e - s_i) = \frac{\dot{Q}_{\text{c.v.}}}{\dot{m}} = q \tag{9.22}$$

and the property relation can be integrated to give

$$T(s_e - s_i) = (h_e - h_i) - \int_i^e v\,dP \tag{9.23}$$

Substituting Eqs. 9.22 and 9.23 into the first law, Eq. 6.23, then results in the same expression as for the reversible adiabatic process, Eq. 9.21. We further note that any other reversible process can be constructed, in the limit, from a series of alternate adiabatic and isothermal processes. Thus we may conclude that Eq. 9.21 is valid for any reversible, steady-state, steady-flow process, without the restriction that it be either adiabatic or isothermal.

This expression has a wide range of application. If we consider a reversible steady-state, steady-flow process in which the work is zero (such as flow through a nozzle) and the fluid is incompressible ($v =$ constant), Eq. 9.21 can be integrated to give

$$v(P_e - P_i) + \frac{(\mathsf{V}_e^2 - \mathsf{V}_i^2)}{2g_c} + (Z_e - Z_i)\frac{g}{g_c} = 0 \tag{9.24}$$

This is known as the Bernoulli equation, (after Daniel Bernoulli) and is a very important equation in fluid mechanics.

Equation 9.21 is also frequently applied to the large class of flow processes involving work (such as turbines and compressors) in which changes in kinetic and potential energies of the working fluid are small. The model process for these machines is then a reversible, SSSF process with no change in kinetic or potential energy (and commonly, although not necessarily, adiabatic as well). For this process, Eq. 9.21 reduces to the form

$$w = -\int_i^e v\,dP \tag{9.25}$$

From this result, we conclude that the shaft work associated with this type of process is closely related to the specific volume of the fluid during the process. To further amplify this point, consider the simple steam power plant shown in Fig. 9.4. Suppose this is an ideal power plant, with no pressure drop in the piping, the boiler, or the condenser. Thus the pressure increase in the pump is equal to the pressure decrease in the turbine. Neglecting kinetic and potential energy changes, the work done in each of these processes is given by Eq. 9.25. Since the pump handles liquid, which has a very small specific volume as compared to the vapor

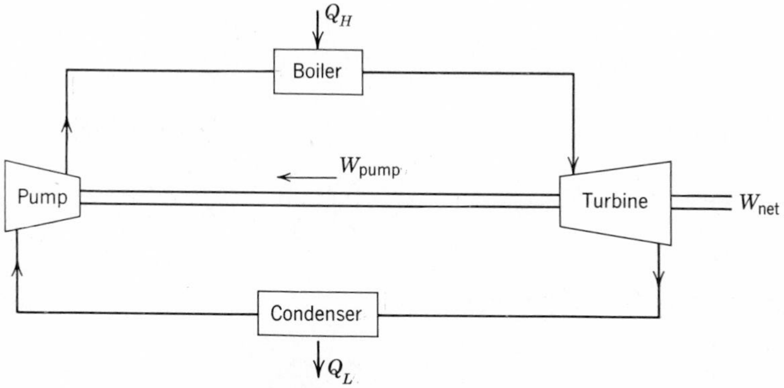

Fig. 9.4 Simple steam power plant.

that flows through the turbine, the power input to the pump is much less than the power output of the turbine, the difference being the net power output of the power plant.

This same line of reasoning can be qualitatively applied to actual devices that involve steady-state, steady-flow processes, even though the processes are not exactly reversible and adiabatic.

Example 9.5

Calculate the work per pound to pump water isentropically from 100 lbf/in.^{2}, 80 F to 1000 lbf/in.^{2}.

From the steam tables, $v_1 = 0.01607$ ft^3/lbm. Assuming the specific volume to remain constant and using Eq. 9.25 we have,

$$-w = \int_1^2 v\,dP = v(P_2 - P_1) = 0.01607(1000 - 100) \times \frac{144}{778}$$

$$= 2.68 \text{ Btu/lbm}$$

As a final application of Eq. 9.21, we recall the reversible polytropic process for an ideal gas, discussed in Section 8.10 for a system process.

For the SSSF process with no change in kinetic and potential energies, from the relations

$$w = -\int_i^e v\,dP \quad \text{and} \quad Pv^n = \text{constant} = C^n.$$

$$w = -\int_i^e v\,dP = -C\int_i^e \frac{dP}{P^{1/n}}$$

$$= -\frac{n}{n-1}(P_e v_e - P_i v_i) = -\frac{nR}{n-1}(T_e - T_i) \tag{9.26}$$

If the process is isothermal, then $n = 1$ and the integral becomes

$$w = -\int_i^e v\,dP = -\text{constant}\int_i^e \frac{dP}{P} = -P_i v_i \ln\frac{P_e}{P_i} \tag{9.27}$$

These evaluations of the integral

$$\int_i^e v\,dP$$

may also be used in conjunction with Eq. 9.21 in cases for which kinetic and potential energy changes are not negligibly small.

9.4 Principle of the Increase of Entropy

The principle of the increase of entropy for a system analysis was discussed in Section 8.8. The same general conclusion is reached in the case of a control volume analysis. To demonstrate this, consider a control volume, Fig. 9.5, which exchanges both mass and heat with its surround-

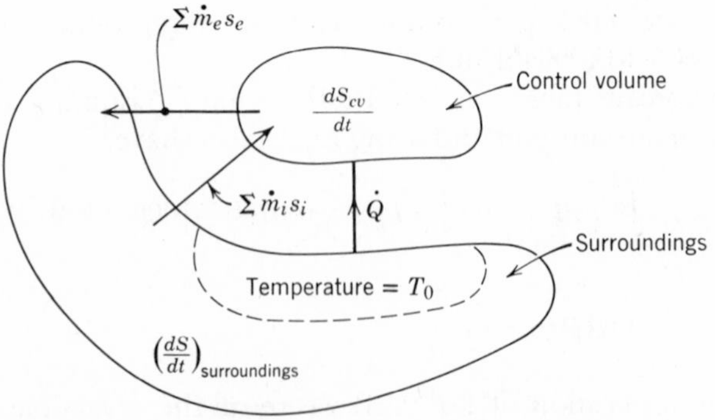

Fig. 9.5 Entropy change for a control volume plus surroundings.

ings. At the point in the surroundings where the heat transfer occurs the temperature is T_0. From Eq. 9.7 the second law for this process is

$$\frac{dS_{c.v.}}{dt} + \sum \dot{m}_e s_e - \sum \dot{m}_i s_i \geq \int_A \left(\frac{\dot{Q}_{c.v.}/A}{T}\right) dA$$

We recall that the first term represents the rate of change of entropy within the control volume, and the next terms the net entropy flow out of the control volume as the result of the mass flow. Therefore, for the surroundings we can write

$$\frac{dS_{surr}}{dt} = \sum \dot{m}_e s_e - \sum \dot{m}_i s_i - \frac{\dot{Q}_{c.v.}}{T_0} \tag{9.28}$$

Adding Eqs. 9.7 and 9.28 we have

$$\frac{dS_{c.v.}}{dt} + \frac{dS_{surr}}{dt} \geq \int_A \left(\frac{\dot{Q}_{c.v.}/A}{T}\right) dA - \frac{\dot{Q}_{c.v.}}{T_0} \tag{9.29}$$

Since $\dot{Q}_{c.v.} > 0$ when $T_0 > T$, and $\dot{Q}_{c.v.} < 0$ when $T_0 < T$, it follows that

$$\frac{dS_{c.v.}}{dt} + \frac{dS_{surr}}{dt} \geq 0 \tag{9.30}$$

which can be termed the general statement of the principle of the increase of entropy.

9.5 Efficiency

In Chapter 7 we noted that the second law of thermodynamics led to the concept of thermal efficiency for a heat engine cycle, namely

$$\eta_{th} = \frac{W_{net}}{Q_H}$$

where W_{net} is the net work of the cycle and Q_H is the heat transfer from the high-temperature body.

In this chapter we have extended our consideration of the second law to processes, and this leads us now to consider the efficiency of a process. For example, we might be interested in the efficiency of a turbine in a steam power plant, or the compressor in a gas turbine engine.

In general we can say that the efficiency of a machine in which a process takes place involves a comparison between the actual performance of the machine under given conditions and the performance that would have been achieved in an ideal process. It is in the definition of this ideal process that the second law becomes a major consideration. For example, a steam turbine is intended to be an adiabatic machine. The only heat transfer is the unavoidable heat transfer that takes place between the given turbine and the surroundings. Also, we note that for a given steam turbine operating in a steady-state, steady-flow manner, the state of the steam entering the turbine and the exhaust pressure are fixed. Therefore the ideal process would be a reversible adiabatic process, which is an isentropic process, between the inlet state and the turbine exhaust pressure. If we denote the actual work done per unit mass of steam flow through the turbine as w_a and the work that would have been done in a reversible adiabatic process between the inlet state and the turbine exhaust pressure as w_s the efficiency of the turbine is defined as

$$\eta_{\text{turbine}} = \frac{w_a}{w_s} \tag{9.31}$$

The same relation would hold for a gas turbine.

A few other examples may help to clarify this point. In a nozzle the objective is to have the maximum kinetic energy leaving the nozzle for the given inlet conditions and exhaust pressure. The nozzle is also an adiabatic device and therefore the ideal process is a reversible adiabatic or isentropic process. The efficiency of a nozzle is the ratio of the actual kinetic energy leaving the nozzle, $V_a^2/2g_c$, to the kinetic energy for an isentropic process between the same inlet conditions and exhaust pressure, namely, $V_s^2/2g_c$.

$$\eta_{\text{nozzle}} = \frac{V_a^2/2g_c}{V_s^2/2g_c} \tag{9.32}$$

In compressors for air or other gases, there are two ideal processes to which the actual performance can be compared. If no effort is made to cool the gas during compression (that is, when the process is adiabatic), the ideal process is a reversible adiabatic or isentropic process between the given inlet state and exhaust pressure. If we denote the work per unit mass of gas flow through the compressor for this isentropic process as w_s, and the actual work as w_a (the actual work input will be greater than the work input for an isentropic process), the efficiency is defined by the relation

$$\eta_{\text{adiabatic compressor}} = \frac{w_s}{w_a} \tag{9.33}$$

If an attempt is made to cool the air during compression by use of a water jacket or fins, the ideal process is considered a reversible isothermal process. If w_t is the work for the reversible isothermal process between the given inlet state and exhaust pressure, and w_a the actual work, the efficiency is defined by the relation

$$\eta_{\text{cooled compressor}} = \frac{w_t}{w_a} \tag{9.34}$$

Thus we note that the efficiency of a device that involves a process (rather than a cycle) involves a comparison of the actual performance to that which would be achieved in a related, but well defined ideal process.

Example 9.6

A steam turbine receives steam at a pressure of 100 lbf/in.^{2}, 500 F. The steam leaves the turbine at a pressure of 2 lbf/in.^{2}. The work output of the turbine is measured and is found to be 172 Btu per pound of steam flowing through the turbine. Determine the efficiency of the turbine.

The efficiency of the turbine is given by Eq. 9.31.

$$\eta_{\text{turbine}} = \frac{w_a}{w_s}$$

Thus a determination of the turbine efficiency involves a calculation of the work which would be done in an isentropic process between the given inlet state and final pressure. For this isentropic process:

Continuity eq.: $\dot{m}_1 = \dot{m}_2 = \dot{m}$

First law: $h_1 = h_{2s} + w_s$

Second law: $s_1 = s_{2s}$

Property relation: Steam tables

$$h_1 = 1279.1; \quad s_1 = 1.7085$$

$$s_{2s} = s_1 = 1.7085 = 1.9198 - (1-x)_{2s}1.7448$$

$$(1-x)_{2s} = \frac{0.2113}{1.7448} = 0.1210$$

$$h_{2s} = 1116.1 - 0.1210(1022.1) = 992.3$$

$$w_s = h_1 - h_{2s} = 1279.1 - 992.3 = 286.8 \text{ Btu/lbm}$$

$$w_a = 172 \text{ Btu/lbm}$$

$$\eta_{\text{turbine}} = \frac{w_a}{w_s} = \frac{172}{286.8} = 0.60 = 60\%$$

9.6 Reversible Work and Irreversibility

From this concept of efficiency of a device such as a turbine or compressor we proceed to develop the concept of irreversibility, which enables one to evaluate the "efficiency" of any process, including those that involve no work, such as a heat transfer process or a chemical reaction. We proceed by referring to Fig. 9.6. In order to focus on the

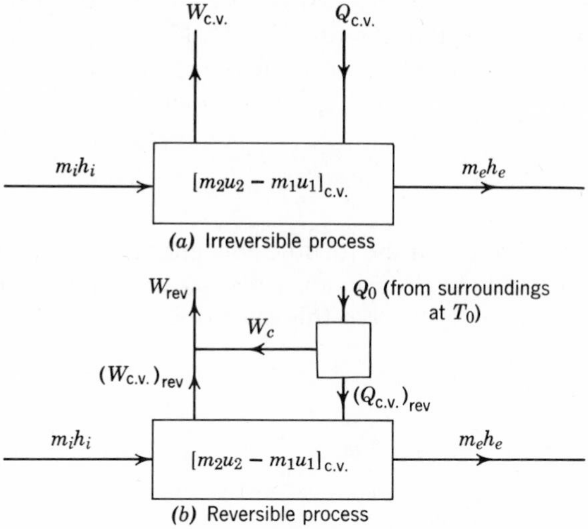

Fig. 9.6 Two uniform state–uniform flow processes to demonstrate the concept of reversible work.

concepts and simplify the presentation, the two uniform-state, uniform-flow processes in Fig. 9.6 have been simplified by ignoring kinetic and potential energies and by having a single flow entering and leaving the control volume.[1] It is essential to note that the two processes have identical changes of state within the control volume, and identical conditions of the fluid entering and leaving the control volume. The only difference is that in Fig. 9.6(*a*) the process is irreversible and in Fig. 9.6(*b*) all processes are reversible. In order that the heat transfer

[1]A similar development, but without these simplifying assumptions, has been given by the authors in their book, "Fundamentals of Classical Thermodynamics," p. 239 ff.

between the control volume and the surroundings may occur reversibly when there is a difference between the temperature in the control volume and the temperature of the surroundings, it is necessary that this heat transfer take place through a reversible heat engine. The work output of this reversible heat engine is designated W_c. The sum of the work crossing the control surface for the reversible case and the work output of the reversible engine is called the reversible work and is designated W_{rev}. That is,

$$W_{rev} = (W_{c.v.})_{rev} + W_c \tag{9.35}$$

As one would expect, the work for the reversible process, W_{rev}, is greater than the work for the actual process $W_{c.v.}$. The irreversibility, I, for the given process is defined as the difference between the reversible work and the actual work.

$$I = W_{rev} - W_{c.v.} \tag{9.36}$$

It is evident that irreversibility is a measure of the "inefficiency" of an actual process, since the less the actual work for a given change of state, the greater the irreversibility. The irreversibility would be zero for a reversible process.

In order to develop an expression for the irreversibility, let us first find an expression for the reversible work. Considering the first law for the control volume of Fig. 9.6(*b*), we have

$$(Q_{c.v.})_{rev} + m_i h_i = (W_{c.v.})_{rev} + m_e h_e + [m_2 u_2 - m_1 u_1]_{c.v.} \tag{9.37}$$

From the second law we have

$$m_2 s_2 - m_1 s_1 + m_e s_e - m_i s_i = \int_0^t \left(\frac{\dot{Q}_{c.v.}}{T}\right)_{rev} dt \tag{9.38}$$

Note that the equality is used because the process is reversible.

We also note that for the Carnot engine,

$$W_c = Q_0 - (Q_{c.v.})_{rev}$$

and

$$\frac{Q_0}{T_0} = \int_0^t \left(\frac{\dot{Q}_{c.v.}}{T}\right)_{rev} dt$$

It follows that

$$W_c = T_0 \int_0^t \left(\frac{\dot{Q}_{c.v.}}{T}\right)_{rev} dt - (Q_{c.v.})_{rev}$$

Substituting this into Eq. 9.38 we have

$$W_c = T_0[m_2s_2 - m_1s_1 + m_es_e - m_is_i] - (Q_{c.v.})_{rev}$$

Substituting this into Eq. 9.37 and rearranging, yields an expression for W_{rev}.

$$W_{rev} = m_i(h_i - T_0s_i) - m_e(h_e - T_0s_e) - [m_2(u_2 - T_0s_2) - m_1(u_1 - T_0s_1)]_{c.v.} \tag{9.39}$$

This is the expression for reversible work that we set out to develop. Note that it involves T_0, the temperature of the surroundings, as well as the thermodynamic properties enthalpy, entropy, and internal energy.

For a system there is no mass flow across the boundary, and this equation reduces to:

$$w_{rev} = {}_1\left(\frac{W_{rev}}{m}\right)_2 = (u_1 - T_0s_1) - (u_2 - T_0s_2) \tag{9.40}$$

Had we included kinetic and potential energies in our original development, this equation would have been:

$$w_{rev} = {}_1\left(\frac{W_{rev}}{m}\right)_2 = \left(u_1 - T_0s_1 + \frac{V_1^2}{2g_c} + Z_1\frac{g}{g_c}\right) - \left(u_2 - T_0s_2 + \frac{V_2^2}{2g_c} + Z_2\frac{g}{g_c}\right) \tag{9.41}$$

For a steady-state, steady-flow process, Eq. 9.39 reduces to:

$$w_{rev} = \frac{\dot{W}_{rev}}{\dot{m}} = (h_i - T_0s_i) - (h_e - T_0s_e) \tag{9.42}$$

Had we included in our consideration, kinetic and potential energies, this equation for the steady-state, steady-flow process would have been:

$$w_{rev} = \frac{\dot{W}_{rev}}{\dot{m}} = \left(h_i - T_0s_i + \frac{V_i^2}{2g_c} + Z_i\frac{g}{g_c}\right) - \left(h_e - T_0s_e + \frac{V_e^2}{2g_c} + Z_e\frac{g}{g_c}\right) \tag{9.43}$$

If we had more than one flow into and out of the control volume, the corresponding equation would be:

$$\dot{W}_{rev} = \sum \dot{m}_i\left(h_i - T_0s_i + \frac{V_i^2}{2g_c} + Z_i\frac{g}{g_c}\right) - \sum \dot{m}_e\left(h_e - T_0s_e + \frac{V_e^2}{2g_c} + Z_e\frac{g}{g_c}\right)$$

We now proceed to find an expression for the irreversibility, I, which we have defined as:

$$I = W_{rev.} - W_{c.v.} \tag{9.44}$$

The work for the actual process of Fig. 9.6(*a*) is found by writing the first law for the process.

$$W_{c.v.} = m_i h_i - m_e h_e - [m_2 u_2 - m_1 u_1] + Q_{c.v.} \tag{9.45}$$

Substituting Eqs. 9.39 and 9.45 in Eq. 9.44 we have

$$\begin{aligned} I &= m_e T_0 s_e - m_i T_0 s_i + [m_2 T_0 s_2 - m_1 T_0 s_1]_{c.v.} - Q_{c.v.} \\ &= T_0[(m_e s_e - m_i s_i) + (m_2 s_2 - m_1 s_1)_{c.v.}] - Q_{c.v.} \end{aligned} \tag{9.46}$$

This is a general expression for the irreversibility in a uniform-state, uniform-flow process. The same expression results when the kinetic and potential energies are considered for the processes of Fig. 9.6. For the system this reduces to

$${}_1I_2 = mT_0(s_2 - s_1) - {}_1Q_2 \tag{9.47}$$

For the steady-state, steady-flow process with a simple flow of fluid across the boundary this reduces to

$$\frac{I}{m} = T_0(s_e - s_i) - \frac{Q_{c.v.}}{m} \tag{9.48}$$

The concepts of reversible work and irreversibility are illustrated in the following two examples. Unless otherwise stated, we shall use 77F(25C) for the temperature of the surroundings. This temperature has been selected because thermochemical data are frequently given relative to this base and it is also a reasonable temperature to assume for the surroundings.

Example 9.7

Referring to Example 9.6, determine the reversible work and the irreversibility for the actual change of state that takes place as the steam flows through the turbine (using a temperature of the surroundings of 80F). In this solution it will be helpful to refer to the temperature entropy diagram of Fig. 9.7.

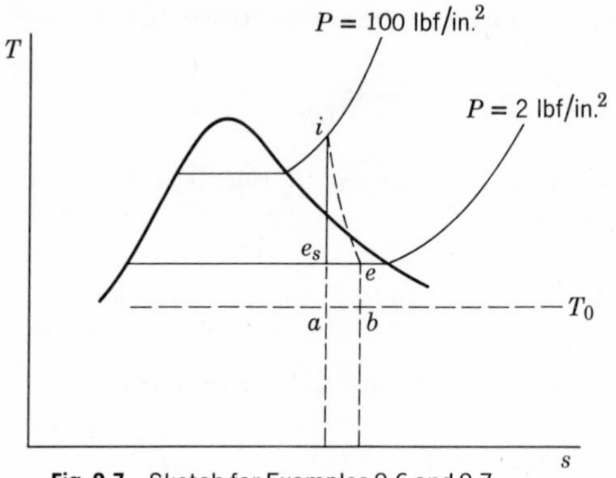

Fig. 9.7 Sketch for Examples 9.6 and 9.7.

Let us first determine the actual state of the steam as it leaves the turbine.

$$P_e = 2 \text{ lbf/in.}^2$$

$$h_e = h_i - w$$

$$= 1279.1 - 172 = 1107.1 \text{ Btu/lbm}$$

$$h_e = 1107.1 = 1116.1 - (1-x)_e 1022.1$$

$$(1-x)_e = 0.0088$$

$$s_e = 1.9198 - 0.0088(1.7448) = 1.9044 \text{ Btu/lbm R}$$

We first calculate the reversible work using Eq. 9.42

$$\frac{W_{\text{rev}}}{m} = (h_i - T_0 s_i) - (h_e - T_0 s_e)$$

$$= (h_i - h_e) - T_0(s_i - s_e)$$

$$= (1279.1 - 1107.1) - 540(1.7085 - 1.9044)$$

$$= 172.0 + 105.8 = 277.8 \text{ Btu/lbm}$$

The irreversibility can be found using Equation 9.48

$$\frac{I}{m} = T_0(s_e - s_i) - \frac{Q_{\text{c.v.}}}{m}$$

$$= 540(1.7085 - 1.9044) = 105.8 \text{ Btu/lbm}$$

Note also that we could use our basic definition of the irreversibility.

$$\frac{I}{m} = \frac{W_{\text{rev}}}{m} - \frac{W_{\text{c.v.}}}{m}$$

$$= 277.8 - 172 = 105.8 \text{ Btu/lbm}$$

It may be of help in understanding these concepts to conceive of a device by which we could achieve this amount of reversible work. Consider the device shown in Fig. 9.8, in which the steam expands in a

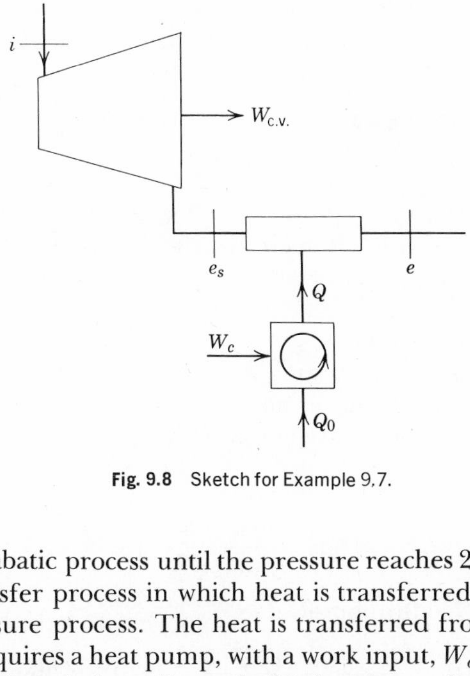

Fig. 9.8 Sketch for Example 9.7.

reversible adiabatic process until the pressure reaches 2 lbf/in.^{2}, followed by a heat transfer process in which heat is transferred to the steam in a constant pressure process. The heat is transferred from the surroundings, which requires a heat pump, with a work input, W_c.

We already found, from Example 9.6 that the work for the reversible adiabatic process is 286.7 Btu/lbm. The work required for the heat pump is area e_sabee_s on Fig. 9.7.

$$W_c = (T_e - T_0)(s_e - s_i)$$

$$= (586 - 540)(1.9044 - 1.7085) = 9.0 \text{ Btu/lbm}$$

$$W_{\text{rev}} = 286.8 - 9.0 = 277.8 \text{ Btu/lbm}$$

Example 9.8

1. Considering Fig. 9.9, tank A has a volume of 100 ft^3 and initially contains Freon-12 at a pressure of 10 lbf/in.^{2} and a temperature of 80 F. The compressor evacuates tank A and charges tank B. Tank B is initially evacuated and is of such a volume that the final pressure of the Freon-12 in tank B is 80 lbf/in.^{2} when the temperature reaches its final value of 80 F. The temperature of the surroundings is 80 F. Determine the minimum work input to the compressor.

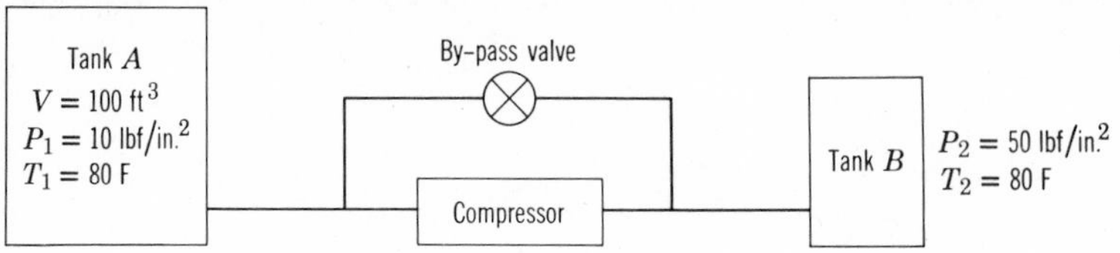

Fig. 9.9 Sketch for Example 9.8.

2. After tank B is charged and tank A is evacuated, a by-pass valve around the compressor is left open, and the two tanks come to a uniform pressure at a temperature of 80 F. Determine the irreversibility for this process.

The solution is as follows:

1. Consider a system consisting of the two tanks and the connecting piping. For this system there is no change in volume. Since changes in kinetic and potential energy are not significant, we have, from Eq. 9.40,

$$_1(w_{\text{rev}})_2 = (u_1 - u_2) - T_0(s_1 - s_2)$$

$$u_1 = h_1 - P_1v_1 = 89.596 - \frac{10 \times 144 \times 4.7248}{778} = 80.85 \text{ Btu/lbm}$$

$$s_1 = 0.20746 \text{ Btu/lbm-R}$$

$$u_2 = 86.316 - \frac{80 \times 144 \times 0.52795}{778} = 78.51 \text{ Btu/lbm}$$

$$s_2 = 0.16885 \text{ Btu/lbm-R}$$

$$_1(w_{\text{rev}})_2 = (80.85 - 78.51) - 540(0.20746 - 0.16885)$$

$$= 2.34 - 20.85 = -18.51 \text{ Btu/lbm}$$

$$m = \frac{V_1}{v_1} = \frac{100}{4.7248} = 21.15 \text{ lbm}$$

$$_1(W_{\text{rev}})_2 = 21.15(-18.51) = -392 \text{ Btu}$$

Note that the reversible work is a negative number. This means that the minimum work input to the compressor to accomplish this change of state is 392 Btu. If there are any irreversibilities this work input will increase, say to 500 Btu. Thus the reversible work can still be thought of as the maximum work, inasmuch as −392 is a larger number than −500.

2. When the Freon-12 flows from tank B to tank A through the by-pass valve, we recognize that this is an irreversible process. We could calculate this irreversibility either by calculating the reversible work from Eq. 9.40 and the irreversibility from Eq. 9.36 or we could calculate the irreversibility directly from Eq. 9.47. Let us do both calculations in order to gain further insight into these concepts. The state of the Freon-12 in tank B is designated state 2 as above, and the final state is designated state 3.

We must first find the final pressure in the system. The two properties we know are specific volume and temperature. The volume of tank B is found from the results of part **1**.

$$V_B = mv_{2B} = 21.15(0.52795) = 11.17\ \text{ft}^3$$

$$V_3 = V_A + V_B = 100 + 11.17 = 111.17\ \text{ft}^3$$

$$v_3 = \frac{V_3}{m} = \frac{111.17}{21.15} = 5.26\ \text{ft}^3/\text{lbm}$$

By interpolation from the Freon-12 table, $P_3 = 9.44\ \text{lbf/in.}^2$

$$h_3 = 89.620\ \text{Btu/lbm} \qquad s_3 = 0.20871\ \text{Btu/lbm-R.}$$

$$u_3 = 89.620 - \frac{9.44 \times 144 \times 5.26}{778} = 80.44\ \text{Btu/lbm}$$

From Eq. 9.40,

$$\begin{aligned} {}_2(w_{\text{rev}})_3 &= u_2 - u_3 - T_0(s_2 - s_3) \\ &= 78.51 - 80.44 - 540(0.16885 - 0.20871) \\ &= 19.60\ \text{Btu/lbm} \end{aligned}$$

$${}_2(W_{\text{rev}})_3 = 21.15(19.60) = 415\ \text{Btu}$$

The irreversibility can now be calculated from Eq. 9.36.

$${}_2I_3 = {}_2(W_{\text{rev}})_3 - {}_2W_3 = 415 - 0 = 415\ \text{Btu}$$

The irreversibility can also be calculated from Eq. 9.47.

$$_2I_3 = mT_0(s_3 - s_2) - {_2Q_3}$$

In order to determine $_2Q_3$ we apply the equation for the first law to this process.

$$_2Q_3 = m(u_3 - u_2) + {_2W_3}$$

$$= 21.15(80.44 - 78.51) + 0 = +40.9 \text{ Btu.}$$

$$_2I_3 = 21.15 \times 540(0.20871 - 0.16885) - 40.9 = 415 \text{ Btu.}$$

9.7 Availability

What is the maximum reversible work that can be done by a system in a given state? In Section 9.6 we developed an expression for the reversible work for a given change of state of a system. But the question that arises is, what final state will make this reversible work the maximum?

The answer to this question is that when a system is in equilibrium with the environment, no spontaneous change of state will occur, and the system will not be capable of doing any work. Therefore, if a system in a given state undergoes a completely reversible process until it reaches a state in which it is in equilibrium with the environment, the maximum reversible work will have been done by the system.

If a system is in equilibrium with the surroundings, it must certainly be in pressure and temperature equilibrium with the surroundings, that is, at pressure P_0 and temperature T_0. It must also be in chemical equilibrium with the surroundings, which implies that no further chemical reaction will take place. Equilibrium with the surroundings also requires that the system have zero velocity and minimum potential energy. Similar requirements could be set forth regarding magnetic, electrical, and surface effects if these are relevant to a given problem.

The same general remarks can be made in regard to a quantity of mass that undergoes a steady-state, steady-flow process. With a given state for the mass entering the control volume, the reversible work will be a maximum when this mass leaves the control volume in equilibrium with the surroundings. This means that as the mass leaves the control volume it must be at the pressure and temperature of the surroundings, in chemical equilibrium with the surroundings, and have minimum potential energy, and zero velocity. (The mass leaving the control volume must of necessity have some velocity but it can be made to approach zero.)

Let us first consider the availability associated with a steady-state, steady-flow process. In Eq. 9.43 we noted that when we consider a single flow,

$$w_{rev} = \left(h_i - T_0 s_i + \frac{V_i^2}{2g_c} + Z_i \frac{g}{g_c}\right) - \left(h_e - T_0 s_e + \frac{V_e^2}{2g_c} + Z_e \frac{g}{g_c}\right)$$

This reversible work will be a maximum when the mass leaving the control volume is in equilibrium with the surroundings. If we designate this state in which the fluid is in equilibrium with the surroundings with subscript 0, the reversible work will be a maximum when $h_e = h_0$, $s_e = s_0$, $V_e = 0$, and $Z_e = Z_0$. Designating this maximum reversible work per unit mass flow as the availability per unit mass flow, and assigning this the symbol ψ, we have

$$\psi = \left(h - T_0 s + \frac{V^2}{2g_c} + Z \frac{g}{g_c}\right) - \left(h_0 - T_0 s_0 + Z_0 \frac{g}{g_c}\right) \tag{9.49}$$

The initial state is designated without a subscript to indicate that this is the availability associated with a substance in any state as it enters a control volume in a steady-state, steady-flow process. It also follows that the reversible work per unit mass flow between any two states is equal to the decrease in availability between these two states.

$$w_{rev} = \psi_i - \psi_e \tag{9.50}$$

If we have more than one flow into and out of the control volume in a steady-state, steady-flow process we can write

$$W_{rev} = \sum m_i \psi_i - \sum m_e \psi_e \tag{9.51}$$

The availability associated with a system is developed in a similar way, except for one factor. When the volume of a system increases, some work is done by the system against the surroundings, and this is not available for doing useful work.

To simplify this analysis let us also assume that the change in kinetic and potential energy of the system is negligible. In this case w_{rev}, as given by Eq. 9.40, is

$${}_1(w_{rev})_2 = (u_1 - T_0 s_1) - (u_2 - T_0 s_2)$$

If the final state is in equilibrium with the surroundings $u_2 = u_0$ and $s_2 = s_0$. In this case the reversible work is a maximum. If we designate this maximum reversible work as $(w_{rev})_{max}$, and write it without subscript to indicate a general state we have

$$(w_{rev})_{max} = (u - T_0 s) - (u_0 - T_0 s_0) \tag{9.52}$$

The availability per unit mass for a system is equal to this maximum reversible work minus the work done against the surroundings. This work done against the surroundings, W_{surr}, is

$$W_{surr} = P_0(V_0 - V) = -mP_0(v - v_0) \tag{9.53}$$

The availability per unit mass for a system in the absence of kinetic and potential energy changes is designated ϕ

$$\begin{aligned} \phi &= (w_{rev})_{max} - w_{surr} \\ \phi &= (u - T_0 s) - (u_0 - T_0 s_0) + P_0(v - v_0) \\ \phi &= (u + P_0 v - T_0 s) - (u_0 + P_0 v_0 - T_0 s_0) \\ \phi &= (u - u_0) + P_0(v - v_0) - T_0(s - s_0) \end{aligned} \tag{9.54}$$

It follows that

$$_1(w_{rev})_2 = \phi_1 - \phi_2 - P_0(v_1 - v_2) + \frac{\mathsf{V}_1^2 - \mathsf{V}_2^2}{2g_c} + (Z_1 - Z_2)\frac{g}{g_c} \tag{9.55}$$

The use of availability and irreversibility in an actual thermodynamic problem is shown in Fig. 9.10. A theoretical analysis was made of a reciprocating internal-combustion automotive engine to see what happened to the availability of the air-fuel mixture that entered the engine and where the irreversibility occurred during the process. The abscissa on Fig. 9.10 is crank angle, the left side representing bottom dead center when the cylinder is assumed to be filled with an air-fuel mixture having the availability indicated on the ordinate. As the compression process takes place, the availability of this mixture increases as the result of the work done in compressing the mixture. When the piston passes top dead center the expansion process takes place. The beginning and end of combustion are also indicated on the diagram. During the combustion and expansion process irreversibilities occur. Those that are associated with the combustion process itself and those associated with heat transfer to the cooling water or surroundings are both indicated. The work done during the expansion process and availability at the end of the expansion are indicated on the right ordinate. Note that this availability that remains in the cylinder at the end of the expansion stroke is exhausted to the atmosphere.

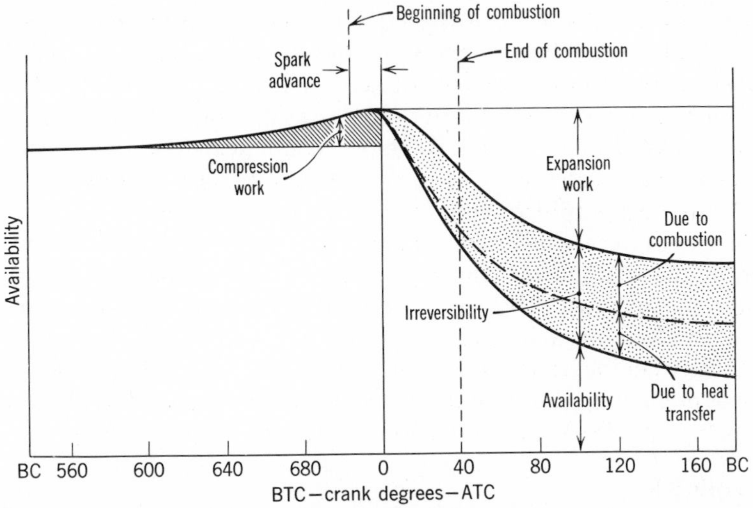

Fig. 9.10 Availability vs. crank angle of the charge in a spark-ignited internal combustion engine. From D. J. Patterson and G. J. Van Wylen, "A Digital Computer Simulation for Spark Ignited Engine Cycles," *SAE Progress in Technology Series*, 7, p. 88. Published by SAE Inc., New York, 1964.

The less the irreversibility associated with a given change of state, the greater the work that will be done (or the less work that will be required). This is significant in at least two regards. The first is that availability is one of our natural resources. This availability is found in such forms as oil reserves, coal reserves, and uranium reserves. Suppose we wish to accomplish a given objective that requires a certain amount of work. If this work is produced reversibly while drawing on one of the availability reserves, the decrease in availability would be exactly equal to the reversible work. However, since there are irreversibilities involved in producing this required amount of work, the actual work will be less than the reversible work, and the decrease in availability will be greater (by the amount of the irreversibility) than if this work had been produced reversibly. Thus the more irreversibilities we have in all of our processes, the greater will be the decrease in our availability reserves.[2] The conservation and effective use of these availability reserves is an important responsibility for all of us.

The second reason that it is desirable to accomplish a given objective with the smallest irreversibility is an economic one. Work costs money,

[2]In many popular talks reference is made to our energy reserves. From a thermodynamic point of view availability reserves would be a much more acceptable term. There is much energy in the atmosphere and the ocean, but relatively little availability.

and in many cases a given objective can be accomplished at less cost when the irreversibility is less. It should be noted, however, that many factors enter into the total cost of accomplishing a given objective, and an optimization process that involves consideration of many factors is often necessary in arriving at the most economical design. For example, in a heat transfer process, the smaller the temperature difference across which the heat is transferred, the less the irreversibility. However, for a given rate of heat transfer, a smaller temperature difference will require a larger (and therefore more expensive) heat exchanger, and these various factors must all be considered in the development of the optimum and most economical design.

In many engineering decisions other factors, such as the impact on the environment (e.g., air pollution and water pollution) and the impact on society must be considered in developing the optimum design.

Example 9.9

A steam turbine, Fig. 9.11, receives 200,000 lbm of steam per hour at 400 lbf/in.², 600 F. At the point in the turbine where the pressure is 60

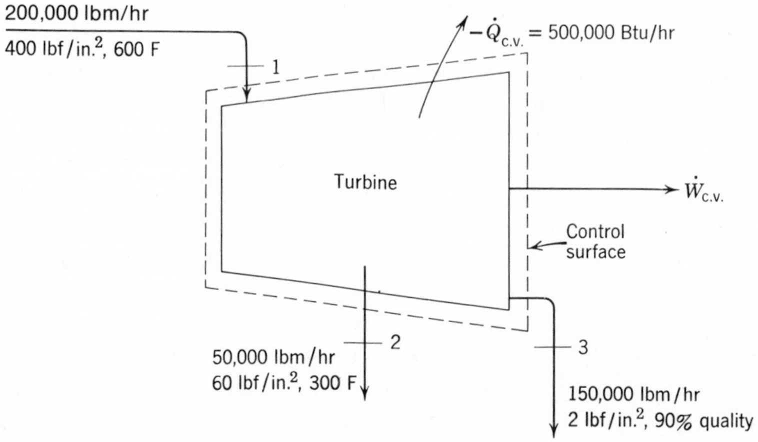

Fig. 9.11 Sketch for Example 9.9.

lbf/in.², steam is bled off for use in processing equipment at the rate of 50,000 lbm/hr. The temperature of this bled steam is 300 F. The balance of the steam leaves the turbine at 2 lbf/in.², 90 per cent quality. The heat transfer to the surroundings is 500,000 Btu/hr. Determine the availability per pound of the steam entering and at both points at which steam leaves the turbine and the reversible work per pound of steam for the given change of state.

Let us designate the states as shown in Fig. 9.11. The availability at any point for the steam entering or leaving the turbine is given by Eq. 9.49.

$$\psi = (h - h_0) - T_0(s - s_0) + \frac{V^2}{2g_c} + (Z - Z_0)\frac{g}{g_c}$$

Since changes in kinetic and potential energy are not involved in this problem this equation reduces to

$$\psi = (h - h_0) - T_0(s - s_0)$$

At the pressure and temperature of the surroundings, namely, 14.7 lbf/in.², 77 F, the water is a slightly compressed liquid, and the properties of the water are essentially equal to those for saturated liquid at 77 F.

$$h_0 = 45.1 \text{ Btu/lbm}; \quad s_0 = 0.0877 \text{ Btu/lbm-R}$$

$$\psi_1 = (1306.6 - 45.1) - 537(1.5892 - 0.0877) = 1261.5 - 806.0$$
$$= 455.5 \text{ Btu/lbm}$$

$$\psi_2 = (1181.9 - 45.1) - 537(1.6495 - 0.0877) = 1136.8 - 840.0$$
$$= 296.8 \text{ Btu/lbm}$$

$$\psi_3 = (1013.9 - 45.1) - 537(1.7453 - 0.0877) = 968.8 - 889.0$$
$$= 79.8 \text{ Btu/lbm}$$

The reversible work can be found from Eq. 9.51.

$$\frac{W_{\text{rev}}}{m_1} = \psi_1 - \frac{m_2}{m_1}\psi_2 - \frac{m_3}{m_1}\psi_3$$

$$\frac{W_{\text{rev}}}{m_1} = 455.5 - 0.25(296.8) - 0.75(79.8) = 321.4 \text{ Btu/lbm}$$

Example 9.10

A lead storage battery of the type used in an automobile is able to deliver 1440 watt-hours of electrical energy. This energy is available for starting the car.

Suppose we wish to use compressed air for doing an equivalent amount of work in starting the car. The compressed air is to be stored at 1000 lbf/in.², 77 F. What volume of tank would be required to have the compressed air have an availability of 1440 watt-hours?

$$1440 \text{ watt hours} = 1.440 \times 3412 = 4910 \text{ Btu.}$$

Let us first find the availability of air at 1000 lbf/in.², 77 F. From Eq. 9.54,

$$\phi = (u - u_0) - T_0(s - s_0) + P_0(v - v_0)$$

$$v = \frac{RT}{P} = \frac{53.34 \times 537}{1000 \times 144} = 0.1975 \text{ ft}^3/\text{lbm}$$

$$v_0 = \frac{RT_0}{P_0} = \frac{53.34 \times 537}{14.7 \times 144} = 13.38 \text{ ft}^3/\text{lbm}$$

$$\phi = 0 - 537\left(\frac{-53.34}{778} \ln \frac{1000}{14.7}\right) + \frac{14.7 \times 144}{778}(0.1975 - 13.38)$$

$$= 0 + 155.3 - 35.9 = 119.4 \text{ Btu/lbm}$$

In order to have an availability of 4910 Btu, the mass of air is

$$m = \frac{4910}{119.4} = 41.1 \text{ lbm}$$

$$V = \frac{mRT}{P} = \frac{41.1 \times 53.34 \times 537}{1000 \times 144} = 8.21 \text{ ft}^3$$

9.8 Processes Involving Chemical Reactions

Although chemical reactions will not be considered in detail until Chapter 13 some preliminary remarks regarding availability in such processes can be made here. In the first place, we observe that the reactants are often in pressure and temperature equilibrium with the surroundings before the reaction takes place, and the same is true of the products after the reaction. An automobile engine would be an example of such a process if we visualized the products being cooled to atmospheric temperature before being discharged from the engine.

Let us first consider for a system the implications of temperature equilibrium with the surroundings during a chemical reaction in a system. The temperature of the system T is equal to T_0, the temperature of the surroundings. Therefore, from Eq. 9.41 we can write (noting that in this case $T_0 = T$),

$${}_1(w_{\text{rev}})_2 = \left(u_1 - T_1 s_1 + \frac{\mathsf{V}_1^2}{2g_c} + Z_1 \frac{g}{g_c}\right) - \left(u_2 - T_2 s_2 + \frac{\mathsf{V}_2^2}{2g_c} + Z_2 \frac{g}{g_c}\right)$$

The quantity $(U - TS)$ is a thermodynamic property of a substance and is called the Helmholtz function. It is an extensive property and we designate it by the symbol A.

Thus

$$A = U - TS$$
$$a = u - Ts \tag{9.56}$$

Therefore, when a system undergoes a change of state while in temperature equilibrium with the surroundings, the reversible work is given by the relation

$${}_1(w_{\text{rev}})_2 = \left(a_1 + \frac{V_1^2}{2g_c} + Z_1\frac{g}{g_c}\right) - \left(a_2 + \frac{V_2^2}{2g_c} + Z_2\frac{g}{g_c}\right)$$

In those cases where the kinetic and potential energy changes are not significant, this reduces to

$${}_1(W_{\text{rev}})_2 = A_1 - A_2 = m(a_1 - a_2) \tag{9.57}$$

Let us now consider a system that undergoes a chemical reaction while in both pressure and temperature equilibrium with the surroundings. The availability function ϕ for a system has been defined as

$$\phi = (u + P_0 v - T_0 s) - (u_0 + P_0 v_0 - T_0 s_0)$$

If $P = P_0$ and $T = T_0$, then

$$\phi = (u + Pv - Ts) - (u_0 + P_0 v_0 - T_0 s_0)$$
$$\phi = (h - Ts) - (h_0 - T_0 s_0) \tag{9.58}$$

The quantity $h - Ts$ is a thermodynamic property and is termed the Gibbs function, designated by the symbol g.

$$G = H - TS$$
$$g = h - Ts \tag{9.59}$$

Introducing the Gibbs function into Eq. 9.58 we have, when a system is in pressure and temperature equilibrium with the surroundings,

$$\phi = g - g_0$$

From Eq. 9.55 it follows that under these same conditions

$${}_1(w_{\text{rev}})_2 = (g_1 - g_2) - P_0(v_1 - v_2) + \frac{V_1^2 - V_2^2}{2g_c} + (Z_1 - Z_2)\frac{g}{g_c} \tag{9.60}$$

The Gibbs function is also of significance in an SSSF process that takes place in temperature equilibrium with the surroundings. Since in this case $T_i = T_e = T_0$, Eq. 9.42 reduces to

$$W_{\text{rev}} = m_i\left(h_i - T_i s_i + \frac{V_i^2}{2g_c} + Z_i\frac{g}{g_c}\right) - m_e\left(h_e - T_e s_e + \frac{V_e^2}{2g_c} + Z_e\frac{g}{g_c}\right)$$

Introducing the Gibbs function we have

$$W_{\text{rev}} = m_i\left(g_i + \frac{V_i^2}{2g_c} + Z_i\frac{g}{g_c}\right) - m_e\left(g_e + \frac{V_e^2}{2g_c} + Z_e\frac{g}{g_c}\right) \tag{9.61}$$

Thus we have introduced two new properties, the Helmholtz function, A, and the Gibbs function, G. Both these functions are very important in the thermodynamics of chemical reactions and will be used extensively in later chapters of this book.

9.9 Some General Comments Regarding Entropy

It is quite possible at this point that a student may have a good grasp of the material that has been covered, and yet he may have only a vague understanding of the significance of entropy. In fact, the question "What is entropy?" is frequently raised by students with the implication that no one really knows! This section has been included in an attempt to give insight into the qualitative and philosophical aspects of the concept of entropy, and to illustrate the broad application of entropy to many different disciplines.

First of all, we recall that the concept of energy rises from the first law of thermodynamics and the concept of entropy from the second law of thermodynamics. Actually it is just as difficult to answer the question "What is energy?" as it is to answer the question "What is entropy?" However, since we regularly use the term energy and are able to relate this term to phenomena that we observe every day, the word energy has a definite meaning to us and thus serves as an effective vehicle for thought and communication. The word entropy could serve in the same capacity. If, when we observed a highly irreversible process (such as cooling coffee by placing an ice cube in it), we said, "That surely increases the entropy," we would soon be as familiar with the word *entropy* as we are with the word *energy*. In many cases when we speak about a higher efficiency we are actually speaking about accomplishing a given objective with a smaller total increase in entropy.

A second point to be made regarding entropy is that in statistical thermodynamics, the property entropy is defined in terms of probability. Although this topic will be examined in detail in Chapter 17, a few brief remarks regarding entropy and probability may prove helpful here.

From this point of view the net increase in entropy that occurs during an irreversible process can be associated with a change of state from a less probable state to a more probable state. For example, to use a previous example, one is more likely to find gas on both sides of the ruptured membrane of Fig. 7.11 than to find a gas on one side and a vacuum on the other. Thus, when the membrane ruptures, the direction of the process is from a less probable state to a more probable state and associated with this process is an increase in entropy. Similarly, the more probable state is that a cup of coffee will be at the same temperature as its surroundings than at a higher (or lower) temperature. Therefore, as the coffee cools as the result of a transferring of heat to the surroundings, there is a change from a less probable to a more probable state, and associated with this is an increase in entropy.

The final point to be made is that the second law of thermodynamics and the principle of the increase of entropy have philosophical implications. Does the second law of thermodynamics apply to the universe as a whole? Are there processes unknown to us that occur somewhere in the universe, such as "continual creation," that have a decrease in entropy associated with them, and thus offset the continual increase in entropy that is associated with the natural processes that are known to us? If the second law is valid for the universe (we of course do not know if the universe can be considered as an isolated system) how did it get in the state of low entropy? On the other end of the scale, if all processes known to us have an increase in entropy associated with them, what is the future of the natural world as we know it?

Quite obviously it is impossible to give conclusive answers to these questions on the basis of the second law of thermodynamics alone. However, the authors see the second law of thermodynamics as man's description of the prior and continuing work of a creator, who also holds the answer to the future destiny of man and the universe.

PROBLEMS

9.1 The flow rate of Freon-12 in a refrigeration cycle is 150 lbm/hr. The compressor inlet conditions are 30 lbf/in.^{2}, 20 F and the exit pressure is 175 lbf/in.^{2}. Assuming the compression process to be reversible and adiabatic, what horsepower motor is required to drive the compressor?

9.2 Nitrogen flows through a nozzle at the rate of 1 lbm/sec. The inlet conditions are 60 lbf/in.^{2}, 300 F, the velocity at this point is 100 ft/sec, and the exit pressure is 15 lbf/in.^{2}. Assuming the process to be reversible and adiabatic, find the exit velocity and the exit area of the nozzle.

9.3 A steam turbine receives steam at 100 lbf/in.^{2}, 500 F. The steam expands in a reversible adiabatic process and leaves the turbine at 14.7 lbf/in.^{2}. The power output of the turbine is 20,000 hp. What is the rate of steam flow in lbm/hr to the turbine?

9.4 To aid in the reduction of pollutants from automotive exhaust, it has been proposed to utilize a small centrifugal compressor to take in ambient air at 14.7 lbf/in.^{2}, 70 F, compress it by 5 lbf/in.^{2}, and discharge the compressed air into the engine exhaust gases. It is estimated that the compressor must handle 50 ft^3/min of air at the inlet conditions. Determine the horsepower required to drive the compressor.

9.5 Consider the turbine process in a gas turbine power plant, assuming the gas to have the same properties as air. The turbine inlet conditions are 1600 F, 80 lbf/in.^{2}, and the exit pressure is 14.7 lbf/in.^{2}. Calculate the turbine exit temperature and the work output per lbm, assuming

(*a*) Variable specific heat, Table B.8.

(*b*) Constant specific heat, value from Table B.6.

9.6 A diffuser is a device in which a fluid flowing at high velocity is decelerated in such a way that the pressure increases during the process. Steam at 20 lbf/in.^{2}, 300 F enters the diffuser with a velocity of 2000 ft/sec and leaves with a velocity of 200 ft/sec. If the process is reversible and occurs without heat transfer, what is the final pressure and temperature of the steam? A Mollier diagram may be helpful in solving this problem.

9.7 An ideal gas with constant specific heat enters a nozzle with a velocity V_i and leaves with a velocity V_e after undergoing a reversible adiabatic expansion. Show that the velocity leaving is given by the relation

$$V_e = \sqrt{V_i^2 + \frac{2g_c kRT_i}{k-1}\left[1 - \left(\frac{P_e}{P_i}\right)^{(k-1)/k}\right]}$$

9.8 A steam turbine can be operated at part-load conditions by throttling the steam to a lower pressure before it enters the turbine, as shown in Fig. 9.12. The line conditions are 200 lbf/in.^{2}, 600 F, and the turbine exhaust pressure

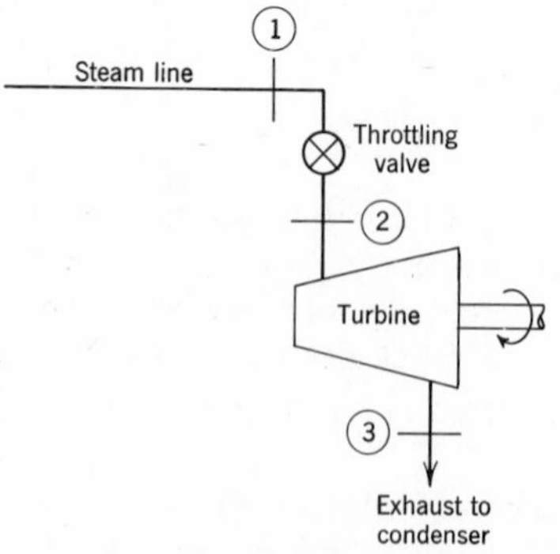

Fig. 9.12 Sketch for Problem 9.8.

is fixed at 1 lbf/in.^{2}. Assuming the expansion inside the turbine to be reversible and adiabatic, calculate

(*a*) The full-load work output of the turbine per lbm of steam.

(*b*) The pressure to which the steam must be throttled to produce 75 per cent of the full-load output. Show both processes on a *T-s* diagram.

9.9 An insulated cylinder is divided into two compartments *A* and *B* by a frictionless nonconducting piston as shown in Fig. 9.13. Compartment *A*

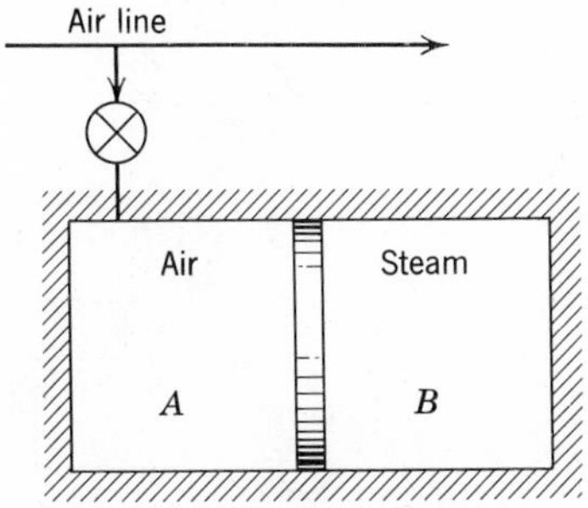

Fig. 9.13 Sketch for Problem 9.9.

contains air at 14.7 lbf/in.^{2}, 80 F and *B* contains saturated water vapor at 14.7 lbf/in.^{2}. Each side has an initial volume of 1 ft^3. Side *A* is connected by a valve to a line in which air flows at 100 lbf/in.^{2}, 80 F.

The valve is opened slightly and air from the line flows into *A* until the pressure reaches 100 lbf/in.^{2}. Calculate the final temperature in each of the compartments and the mass entering *A*.

9.10 A turbo-supercharger is to be utilized to boost the inlet air pressure to an automobile engine. This device consists of an exhaust gas-driven turbine directly connected to an air compressor, as shown in Fig. 9.14. At a certain

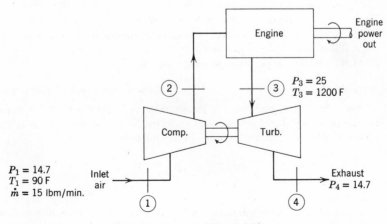

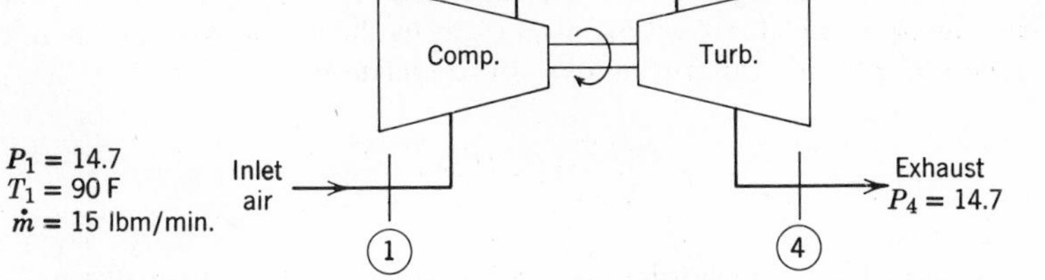

Fig. 9.14 Sketch for Problem 9.10.

engine load, the conditions are as shown in the figure. It may be assumed that the exhaust gas properties are the same as those of air. If it is further assumed that the turbine and compressor are both reversible and adiabatic, calculate

(*a*) The turbine exit temperature and horsepower output.
(*b*) The compressor exit pressure and temperature.

9.11 An insulated closed cylinder fitted with a frictionless piston contains nitrogen at 14.7 lbf/in.^{2}, 60 F, as shown in Fig. 9.15. The piston is thermally

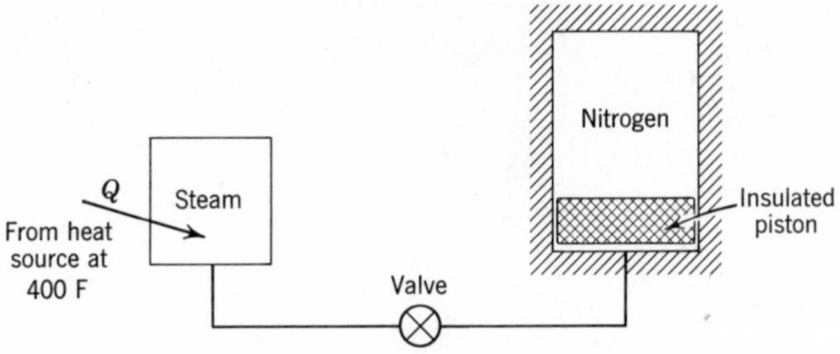

Fig. 9.15 Sketch for Problem 9.11.

insulated and rests at the bottom of the cylinder, and the total volume of the cylinder is 2 ft^3. The valve is now opened to a 2-ft^3 tank containing steam at 60 lbf/in.^{2}, 400 F, and the valve is closed when the pressures equalize. During this process, heat is transferred to the steam inside the tank such that its temperature remains at 400 F. It may be assumed that the nitrogen is compressed reversibly in the cylinder. Determine

(*a*) The final pressure.
(*b*) The final temperatures of the nitrogen and the steam in the cylinder.

9.12 Consider the pump in a steam power plant cycle. Water leaves the condenser and enters the pump at 2 lbf/in.^{2}, 100 F, and leaves the pump at 500 lbf/in.^{2}. Assuming the process to be reversible and adiabatic, calculate

(*a*) The work required per lbm.
(*b*) The water temperature leaving the pump.

9.13 Starting with the relation $w = -\int_i^e v\,dP$, show that the work done per lbm of fluid flow in a steady-state, steady-flow, reversible adiabatic process, involving an ideal gas with constant specific heat, and with no changes in kinetic or potential energy, is given by the relation

$$w = \frac{kRT_i}{k-1}\left[1 - \left(\frac{P_e}{P_i}\right)^{(k-1)/k}\right]$$

9.14 A steam turbine powerplant operating at supercritical pressure is shown in Fig. 9.16. The turbine and pump processes are both reversible and

$P_2 = P_3 = 1 \text{ lbf/in.}^2$, $T_3 = 90 \text{ F}$

$P_4 = P_1 = 3500 \text{ lbf/in.}^2$, $T_1 = 1200 \text{ F}$

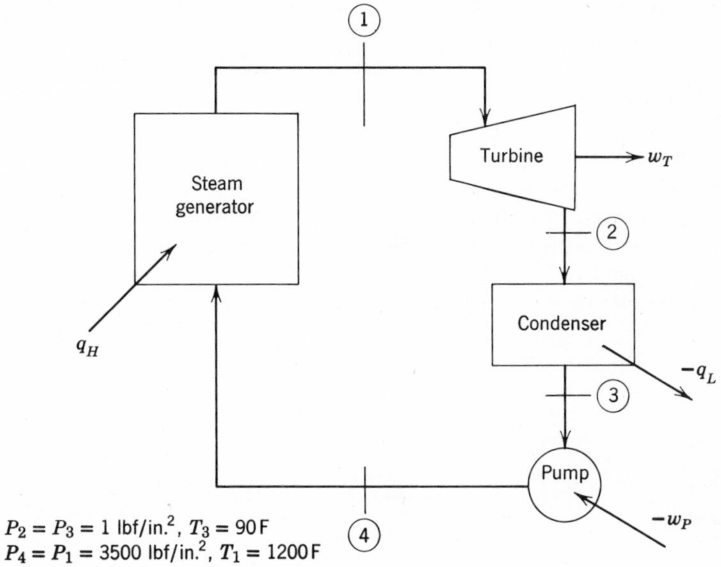

Fig. 9.16   Sketch for Problem 9.14.

adiabatic. Neglecting any changes in kinetic and potential energy, calculate

(a) The work output of the turbine, and the state at the turbine exit.

(b) The work input to the pump and the enthalpy of the liquid at the pump exit.

(c) The thermal efficiency of the cycle.

9.15 A centrifugal pump delivers liquid oxygen to a rocket engine at the rate of 100 lbm/sec. The oxygen enters the pump as saturated liquid at one atmosphere pressure, and the discharge pressure is 500 lbf/in.². Determine the power required to drive the pump if the process is reversible and adiabatic. The specific volume of the liquid is 0.01406 ft³/lbm.

9.16 Air is compressed in a reversible steady-state, steady-flow process from 15 lbf/in.², 80 F to 120 lbf/in.². Calculate the work of compression per pound, the change of entropy, and the heat transfer per pound of air compressed, assuming the following processes.

(a) Isothermal.

(b) Polytropic, $n = 1.25$.

(c) Adiabatic.

(d) Show all these processes on a P-v and a T-s diagram.

9.17 In a refrigeration plant liquid ammonia enters the expansion valve at 250 lbf/in.², 100 F. The pressure leaving the expansion valve is 30 lbf/in.², and changes in kinetic energy are negligible. What is the increase in entropy per lbm? Show this process on a temperature-entropy diagram.

9.18 A salesman reports that he has a steam turbine available that delivers 3800 hp. The steam enters the turbine at 100 lbf/in.^{2}, 500 F and leaves the turbine at a pressure of 2 lbf/in.^{2}, and the required rate of steam flow is 30,000 lbm/hr.

(*a*) How do you evaluate his claim?

(*b*) Suppose he changed his claim and said the required steam flow was 34,000 lbm/hr.

9.19 One type of feedwater heater for preheating the water before entering a boiler operates on the principle of mixing the water with steam. For the states as shown in Fig. 9.17, calculate the rate of increase in entropy per hour, assuming the process to be steady-flow and adiabatic.

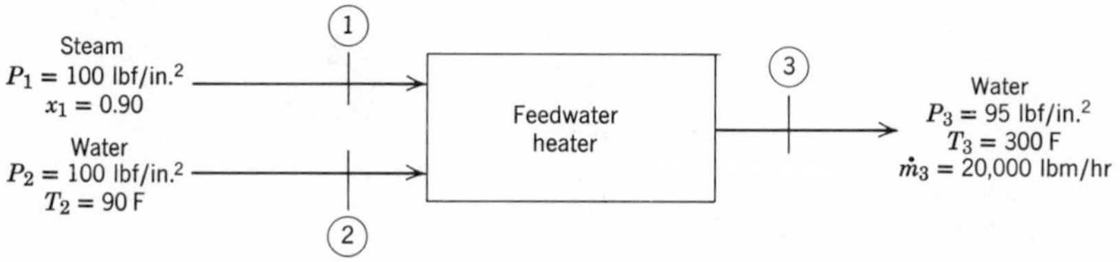

Fig. 9.17 Sketch for Problem 9.19.

9.20 Steam enters a turbine at 100 lbf/in.^{2}, 800 F, and exhausts at 20 lbf/in.^{2}, 400 F. Any heat transfer is with the surroundings at 80 F, and changes in kinetic and potential energy are negligible. It is claimed that this turbine produces 100 hp with a mass flow of 1800 lbm/hr. Does this process violate the second law of thermodynamics?

9.21 Steam at 100 lbf/in.^{2}, 500 F is flowing in a line. Connected to the line is a 50 ft^3 tank containing steam at 14.7 lbf/in.^{2}, 360 F. The valve is opened, allowing steam to flow into the tank until the final pressure in the tank is 100 lbf/in.^{2}. During this process, heat is transferred from the tank at such a rate that the temperature of its contents remains constant at 360 F. The surroundings are at a constant temperature of 80 F.

(*a*) Find the mass of steam that flows into the tank, and the heat transferred during the process.

(*b*) Find the change in entropy of the control volume (tank) during the process.

(*c*) Show that this process does not violate the second law of thermodynamics.

9.22 The equipment shown in Fig. 9.18 is used to fill bottles of Freon-12 for shipping. The shipping bottles are initially evacuated. Each bottle contains 100 lbm of Freon-12 when it is filled, at which time the volume of liquid

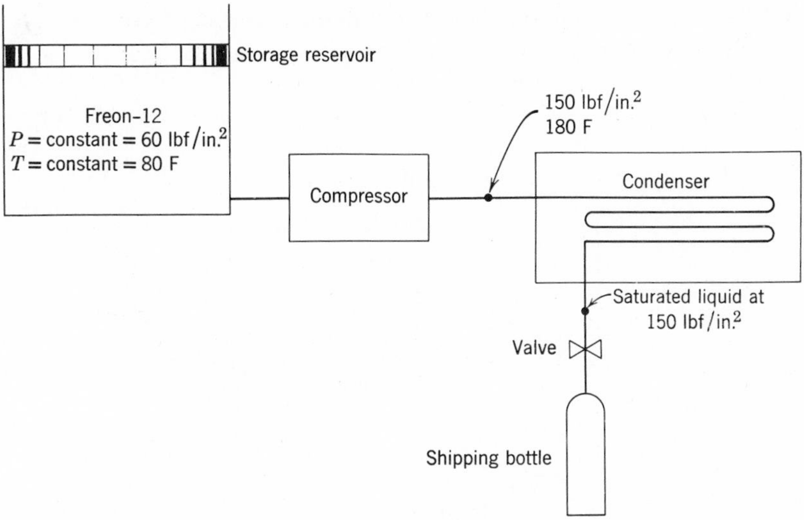

Fig. 9.18 Sketch for Problem 9.22.

Freon-12 is 85% of the total, and the volume of vapor is 15% of the total. Heat is transferred from the bottle during the filling process so that the temperature of the Freon-12 when the filling process has been completed is 80 F. The compression process is adiabatic.

(*a*) Determine the volume of a shipping bottle.

(*b*) Determine the work of compression for filling one bottle.

(*c*) Determine the heat transfer in the condenser for filling one bottle.

(*d*) Determine the heat transfer from the bottle for filling one bottle.

(*e*) If all the heat transfer from the condenser and bottle is to the surroundings at a temperature of 80 F, determine the net increase in entropy of the system and the surroundings for the filling of one bottle.

(*f*) Prove that the compression process indicated does not violate the second law of thermodynamics.

9.23 A 5 ft^3 tank containing steam at 1 atm pressure, 1% quality, is fitted with a relief valve. Heat is transferred to the tank from a large source at 500 F. When the pressure in the tank reaches 300 lbf/in.^{2}, the relief valve opens, saturated vapor at 300 lbf/in.^{2} is throttled across the valve and discharged at atmospheric pressure. The process continues until the quality in the tank is 90%.

(*a*) Calculate the mass discharged from the tank.

(*b*) Determine the heat transfer to the tank during the process.

(*c*) Considering a control volume that contains the tank and valve, calculate the entropy change within the control volume and that of the surroundings. Show that the process does not violate the second law.

9.24 Consider the scheme shown in Fig. 9.19 for producing fresh water from saltwater. The conditions are as shown in the figure. Assume that the properties of saltwater are the same as for pure water, and also that the pump is reversible and adiabatic.

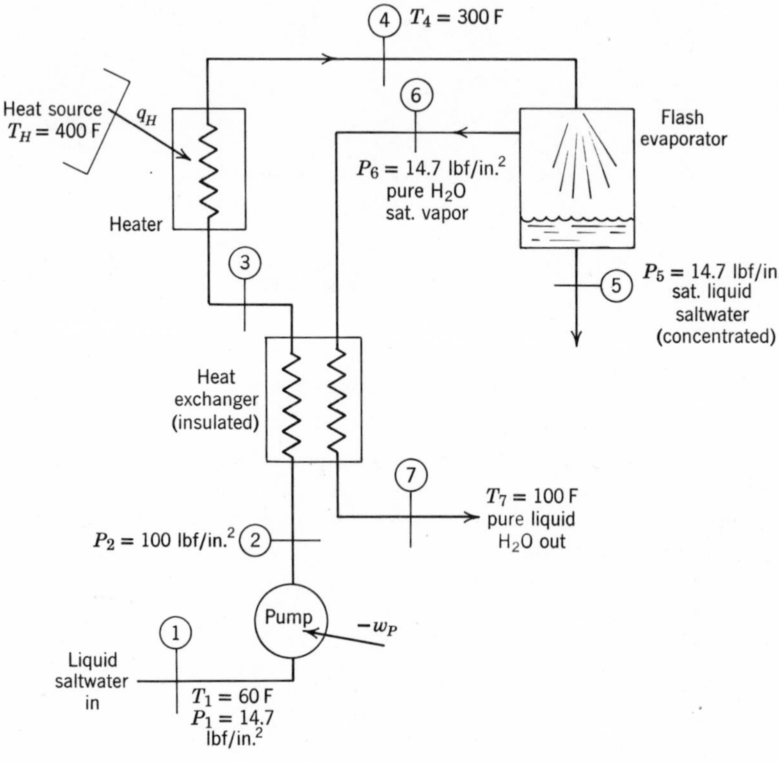

Fig. 9.19 Sketch for Problem 9.24.

(*a*) Determine the ratio $\dot{m}_7/\dot{m}_1$, that is, the fraction of saltwater purified by the process.

(*b*) Determine the input quantities, w_P and q_H.

(*c*) Make a second law analysis of the overall system.

9.25 Steam enters a turbine at 80 lbf/in.^{2}, 500 F and exhausts at a pressure of 1 lbf/in.^{2}. The efficiency of the turbine is 75 per cent. Determine:

(*a*) The work output per lbm of steam.

(*b*) The state of the steam (quality if saturated, temperature if superheated) leaving the turbine.

9.26 Steam enters an insulated nozzle at 100 lbf/in.^{2}, 440 F, at low velocity and exits at 30 lbf/in.^{2}. Calculate the exit velocity and temperature (quality if saturated) assuming a nozzle efficiency of 95 per cent.

9.27 Air enters the compressor of a gas turbine at 14.0 lbf/in.^{2}, 60 F at the rate of 4000 ft^3/min, and leaves at 60 lbf/in.^{2}. The process is adiabatic, and changes in kinetic and potential energy are negligible. Calculate the power required to drive this compressor and the exit temperature, assuming:
   (*a*) That the process is reversible.
   (*b*) That the compressor has an efficiency of 82%.

9.28 Water enters a pump at 80 F, 14.7 lbf/in.^{2} and exits at 500 lbf/in.^{2}. If the efficiency (adiabatic) of the pump is 70 per cent, determine the enthalpy of the water at the exit.

9.29 Repeat Problem 9.10 assuming that the turbine has an efficiency of 85% and the compressor has an efficiency of 80%.

9.30 In one type of heat-powered refrigeration system, part of the working fluid is expanded through a turbine to drive the compressor of the refrigeration cycle. Consider the turbine-compressor unit of such a device, as shown in Fig. 9.20. The turbine produces just enough power to drive

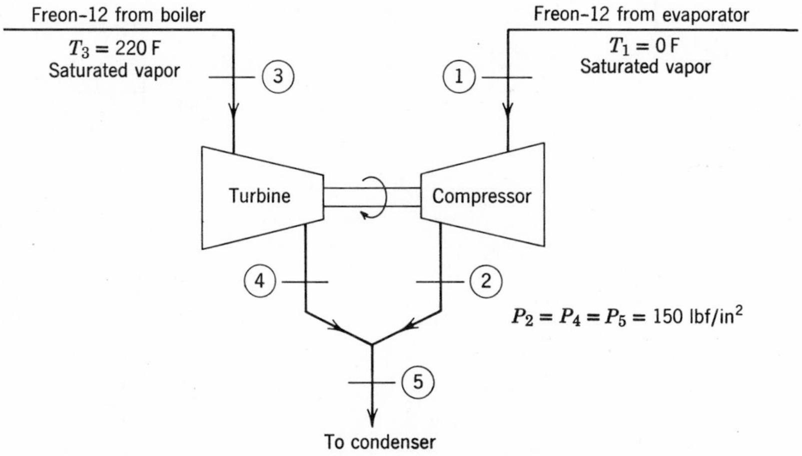

Fig. 9.20 Sketch for Problem 9.30.

the compressor, and both exit streams then mix together. Specifying any assumptions, find the ratio $\dot{m}_3/\dot{m}_1$ and T_5 (x_5 if in the two-phase region) if
   (*a*) The turbine and compressor are reversible and adiabatic.
   (*b*) The turbine and compressor each have an efficiency of 70%.

9.31 Repeat Problem 9.14, assuming that the turbine and pump each have an efficiency of 80 per cent.

9.32 The cycle shown in Fig. 9.21 has been proposed for an auxiliary power supply in a spacecraft. The working fluid is argon (ideal gas throughout the cycle). The compressor and turbine processes are both adiabatic and the heater and cooler processes are both constant-pressure. The following data are known:

$P_1 = P_4 = 5$ lbf/in.^{2}
$P_2 = P_3 = 20$ lbf/in.^{2}
$T_1 = 500$ R
$T_3 = 2000$ R
$\eta_{\text{turbine}} = 80\%$
$\eta_{\text{compressor}} = 80\%$

(*a*) Show this cycle on a *T-s* diagram.
(*b*) Calculate the net work output of the cycle and the thermal efficiency of the cycle.

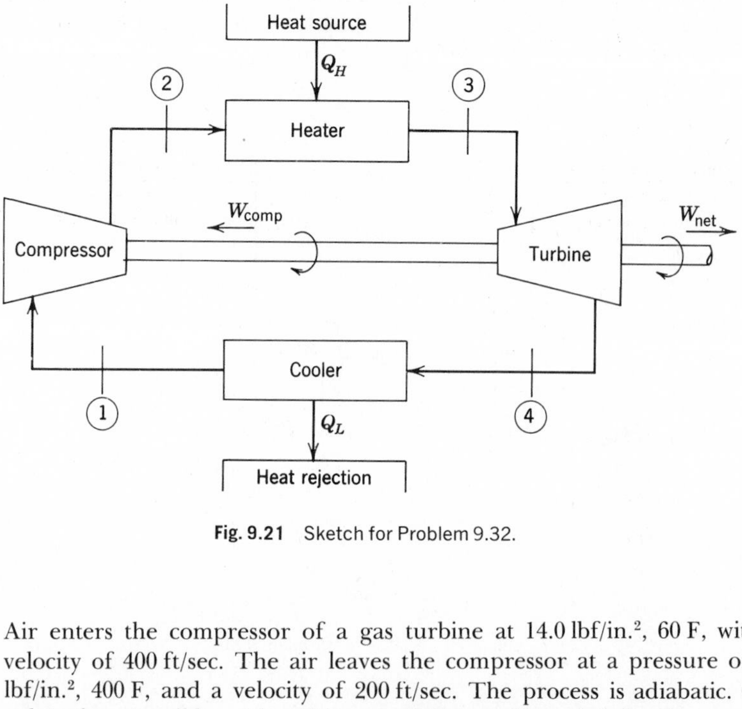

Fig. 9.21 Sketch for Problem 9.32.

9.33 Air enters the compressor of a gas turbine at 14.0 lbf/in.^{2}, 60 F, with a velocity of 400 ft/sec. The air leaves the compressor at a pressure of 60 lbf/in.^{2}, 400 F, and a velocity of 200 ft/sec. The process is adiabatic. Calculate the reversible work and irreversibility per lbm of air for this process.

9.34 Consider a steam turbine that has a throttling governor. (That is, the power output of the turbine is controlled by throttling the inlet steam.) The steam in the pipeline flowing to the turbine has a pressure of 800 lbf/in.^{2} and a temperature of 1000 F. At a certain load the steam is throttled in an adiabatic process to 600 lbf/in.^{2}. Calculate the reversible work and irreversibility per lbm of steam for this process. Show the initial and final states of the steam on a T-s diagram. Room temperature is 77 F.

9.35 Determine the reversible work and the irreversibility for the process described in Problem 5.19. The temperature of the surroundings is 80 F.

9.36 Air enters an adiabatic compressor at ambient conditions, 14.7 lbf/in.², 70 F, and leaves at 65 lbf/in.². The mass flow rate is 50 lbm/hr and the efficiency of the compressor is 75%. After leaving the compressor, the air is cooled to 100 F in an aftercooler. Calculate:

(*a*) The horsepower required to drive the compressor.

(*b*) The rate of irreversibility for the overall process (compressor plus cooler).

9.37 A 10 ft³ tank containing saturated vapor Freon-12 at 20 F is connected to a line flowing liquid Freon-12 at 80 F, 140 lbf/in.². The valve is then opened and Freon flows into the tank. During the process, heat is transferred with the surroundings at 80 F such that when the valve is closed the tank contains 50% liquid, 50% vapor, by volume, at 80 F. Calculate the heat transfer and the irreversibility for this process.

9.38 Two blocks of metal, each having a mass of 10 lbm and a specific heat of 0.1 Btu/lbm-R, are at a temperature of 100 F. A reversible refrigerator receives heat from one block and rejects heat to the other. Calculate the work required to cause a temperature difference of 200 F between the two blocks.

9.39 Consider two identical blocks of copper, A and B, each having a mass of 10 lbm and specific heat of 0.1 Btu/lbm-R. The temperature of block A is 2000 R and that of block B is 500 R. The blocks are then brought to the same temperature.

(*a*) If the process is reversible, find the final temperature and the work done during the process.

(*b*) If the process is accomplished by simply bringing the blocks into thermal communication, find the final temperature and the irreversibility of the process.

9.40 At a certain location the temperature of the water supply is 60 F. Ice is to be made from this water supply by the process shown in Fig. 9.22. The

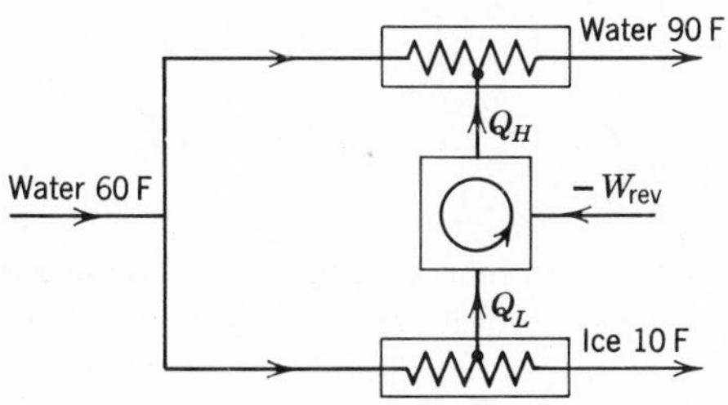

Fig. 9.22 Sketch for Problem 9.40.

final temperature of the ice is 10 F, and the final temperature of the water that is used as cooling water in the condenser is 90 F.

What is the minimum work required to produce one ton of ice?

9.41 A pressure vessel has a volume of 30 ft³ and contains air at 200 lbf/in.², 350 F. The air is cooled to 77 F by heat transfer to the surroundings at 77 F. Calculate the availability in the initial and final states and the irreversibility of this process.

9.42 Liquid nitrogen at 1 atm pressure is to be vaporized and delivered to a pipeline at 500 lbf/in.², 0 F. Three possible schemes for doing this are shown in Fig. 9.23. The first scheme involves pumping the liquid and then

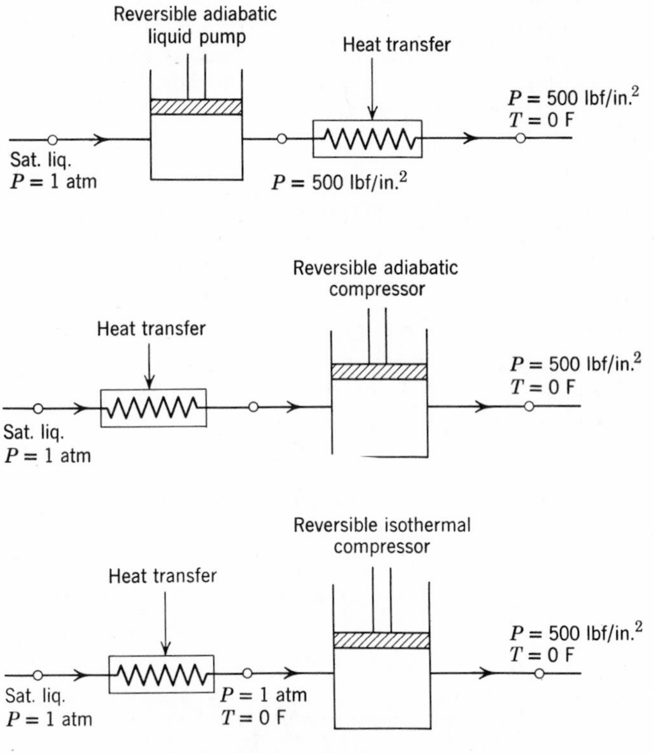

Fig. 9.23 Sketch for Problem 9.42.

vaporizing it. The second involves vaporizing the liquid and superheating it just enough so that it would leave a reversible adiabatic compressor at 500 lbf/in.², 0 F. The third involves vaporizing it and superheating it to 0 F, followed by a reversible isothermal compressor. Assume an ambient temperature of 0 F.

(*a*) What is the availability of liquid nitrogen at 1 atm pressure?

(*b*) What is the minimum work necessary to deliver 1 lbm of nitrogen at 500 lbf/in.^{2}, 0 F?

Show a schematic arrangement of how you would do this with this minimum work input.

(*c*) Determine the work and the irreversibility per pound of nitrogen delivered for each of the three suggested ways of accomplishing this.

9.43 An air preheater is used to cool the products of combustion from a furnace while heating the air to be used for combustion. The rate of flow of products is 100,000 lbm/hr, and the products are cooled from 600 F to 400 F, and for the products at this temperature $C_p = 0.26$ Btu/lbm-R. The rate of air flow is 93,000 lbm/hr, the initial air temperature is 100 F, and for the air $C_p = 0.24$ Btu/lbm-R.

(*a*) What is the initial and final availability of the products (Btu/hr)?

(*b*) What is the irreversibility for this process?

(*c*) Suppose this heat transfer from the products took place reversibly through heat engines. What would be the final temperature of the air? What power would be developed by the heat engines?

9.44 One mole of carbon dust is burned with 1 mole of oxygen to form 1 mole of carbon dioxide in a steady-flow process. The temperature of the carbon and oxygen before combustion is 77 F and the temperature of the carbon dioxide after combustion is also 77 F. The heat transferred during the process is −169,297 Btu, and the entropy of the carbon dioxide after combustion is 0.709 Btu/lb mole-R higher than the entropy of carbon and oxygen before combustion. Calculate the reversible work and the irreversibility for this process.

10

Some Power and Refrigeration Cycles

Some power plants, such as the simple steam power plant, which we have considered several times, operate in a cycle. That is, the working fluid undergoes a series of processes and finally returns to the initial state. In other power plants, such as the internal combustion engine and gas turbine the working fluid does not go through a thermodynamic cycle, even though the engine itself may operate in a mechanical cycle. In this case the working fluid has a different composition or is in a different state at the conclusion of the process than at the beginning. Such equipment is sometimes said to operate on the open cycle (the word cycle is really a misnomer), whereas the steam power plant operates on a closed cycle. The same distinction between open and closed cycles can be made regarding refrigeration devices. For both the open- and closed-cycle type of apparatus, however, it is advantageous to analyze the performance of an idealized closed cycle similar to the actual cycle. Such a procedure is particularly advantageous in determining the influence of certain variables on performance. For example, the spark-ignition internal combustion engine is usually approximated by the Otto cycle. From an analysis of the Otto cycle one concludes that increasing the compression ratio increases the efficiency. This is also true for the actual engine, even though the Otto cycle efficiencies may deviate significantly from the actual efficiencies.

This chapter is concerned with these idealized cycles, both for power and refrigeration apparatus. The working fluids considered are both vapors and ideal gases. An attempt will be made to point out how the processes in actual apparatus deviate from the ideal. Consideration is also given to certain modifications of the basic cycles which are intended to improve performance. These involve the use of such devices as regenerators, multistage compressors and expanders, and intercoolers. The order in which the cycles will be considered is: (1) vapor power cycles, (2) vapor refrigeration cycles, (3) air-standard power cycles, and (4) air-standard refrigeration cycles.

VAPOR POWER CYCLES

10.1 The Rankine Cycle

The ideal cycle for a simple steam power plant is the Rankine cycle, shown in Fig. 10.1. The processes that comprise the cycle are:

1–2: Reversible adiabatic pumping process in the pump.

2–3: Constant-pressure transfer of heat in the boiler.

3–4: Reversible adiabatic expansion in the turbine (or other prime mover such as a steam engine).

4–1: Constant-pressure transfer of heat in the condenser.

The Rankine cycle also includes the possibility of superheating the vapor, as cycle 1–2–3′–4′–1.

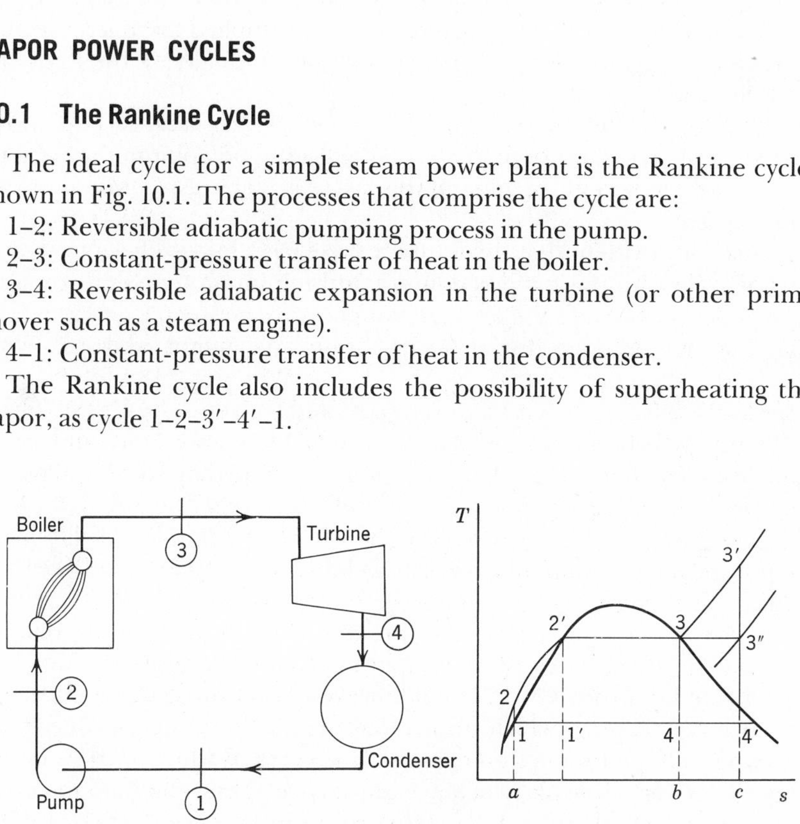

Fig. 10.1 Simple steam power plant which operates on the Rankine cycle.

If changes of kinetic and potential energy are neglected, heat transfer and work may be represented by various areas on the *T*-*s* diagram. The heat transferred to the working fluid is represented by area *a*–2–2′–3–*b*–*a*, and the heat transferred from the working fluid by area *a*–1–4–*b*–*a*. From the first law we conclude that the area representing the work is the difference between these two areas, namely, area 1–2–2′–3–4–1. The thermal efficiency is defined by the relation

$$\eta_{\text{th}} = \frac{w_{\text{net}}}{q_H} = \frac{\text{area } 1\text{–}2\text{–}2'\text{–}3\text{–}4\text{–}1}{\text{area } a\text{–}2\text{–}2'\text{–}3\text{–}b\text{–}a} \tag{10.1}$$

In analyzing the Rankine cycle it is helpful to think of efficiency as depending on the average temperature at which heat is supplied and the average temperature at which heat is rejected. Any changes that increase the average temperature at which heat is supplied or decrease the average temperature at which heat is rejected will increase the Rankine cycle efficiency.

It should be stated that in analyzing the ideal cycles in this chapter the changes in kinetic and potential energies from one point in the cycle to another are neglected. In general this is a reasonable assumption for the actual cycles.

It is readily evident that the Rankine cycle has a lower efficiency than a Carnot cycle with the same maximum and minimum temperatures as a Rankine cycle, because the average temperature between 2 and 2′ is less than the temperature during evaporation. We might well ask, why choose the Rankine cycle as the ideal cycle? Why not select the Carnot cycle 1′–2′–3–4–1′? At least two reasons can be given. The first involves the pumping process. State 1′ is a mixture of liquid and vapor, and great difficulties are encountered in building a pump that will handle the mixture of liquid and vapor at 1′ and deliver saturated liquid at 2′. It is much easier to completely condense the vapor and handle only liquid in the pump, and the Rankine cycle is based on this fact. The second reason involves superheating the vapor. In the Rankine cycle the vapor is superheated at constant pressure, process 3–3′. In the Carnot cycle all the heat transfer is at constant temperature, and therefore the vapor is superheated in process 3–3″. Note, however, that during this process the pressure is dropping, which means that the heat must be transferred to the vapor as it undergoes an expansion process in which work is done. This also is very difficult to achieve in practice. Thus, the Rankine cycle is the ideal cycle that can be approximated in practice. In the sections that follow we will consider some variations on the Rankine cycle that enable one to more closely approach the Carnot-cycle efficiency.

Before discussing the influence of certain variables on the performance of the Rankine cycle, an example is given.

Example 10.1

Determine the efficiency of a Rankine cycle utilizing steam as the working fluid in which the condenser pressure is 1 lbf/in.^{2}. The boiler pressure is 300 lbf/in.^{2}. The steam leaves the boiler as saturated vapor.

In solving Rankine-cycle problems we will let w_p denote the work into the pump per pound of fluid flowing and q_L the heat rejected from the working fluid per pound of fluid flowing.

In solving this problem we consider, in succession, a control surface around the pump, the boiler, the turbine, and the condenser. In each case the property relation used is the steam table.

Consider a control surface around the pump.

First law: $w_p = h_2 - h_1$

Second law: $s_2 = s_1$

Since

$$s_2 = s_1, \quad h_2 - h_1 = \int_1^2 v\,dP$$

Therefore, assuming the fluid to be incompressible,

$$w_p = v(P_2 - P_1) = 0.01614(300-1)\tfrac{144}{778} = 0.893 \text{ Btu/lbm}$$

$$h_2 = h_1 + w_p = 69.7 + 0.9 = 70.6 \text{ Btu/lbm}$$

Next consider a control surface around the boiler.

First law: $q_H = h_3 - h_2 = 1203.9 - 70.6 = 1133.3$ Btu/lbm

Consider a control surface around the turbine.

First law: $w_t = h_3 - h_4$

Second law: $s_3 = s_4$

We can determine the quality at state 4 as follows:

$$\begin{aligned}
s_3 = s_4 = 1.5115 &= [s_g - (1-x)s_{fg}]_4 \\
&= 1.9779 - (1-x)_4 1.8453 \\
(1-x)_4 &= \frac{0.4664}{1.8453} = 0.253 \\
h_4 &= [h_g - (1-x)h_{fg}]_4 \\
&= 1105.8 - 0.253(1036.0) \\
&= 843.8 \text{ Btu/lbm} \\
w_t &= 1203.9 - 843.8 = 360.1 \text{ Btu/lbm}
\end{aligned}$$

Finally, consider a control surface around the condenser.

First law: $q_L = h_4 - h_1 = 843.8 - 69.7 = 774.1$ Btu/lbm

We can now calculate the thermal efficiency

$$\eta_{\text{th}} = \frac{w_{\text{net}}}{q_H} = \frac{q_H - q_L}{q_H} = \frac{w_t - w_p}{q_H} = \frac{360.1 - 0.9}{1133.3} = 31.6\%$$

We could also write an expression for thermal efficiency in terms of properties at various points in the cycle

$$\eta_{\text{th}} = \frac{(h_3 - h_2) - (h_4 - h_1)}{h_3 - h_2} = \frac{(h_3 - h_4) - (h_2 - h_1)}{h_3 - h_2}$$

$$= \frac{1133.3 - 774.1}{1133.3} = \frac{360.1 - 0.9}{1133.3} = 31.6\%$$

10.2 Effect of Pressure and Temperature on the Rankine Cycle

Let us first consider the effect of exhaust pressure and temperature on the Rankine cycle. This effect is shown on the *T-s* diagram of Fig. 10.2.

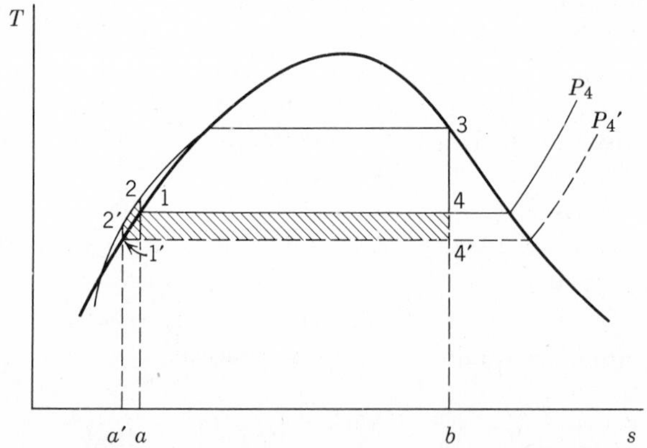

Fig. 10.2 Effect of exhaust pressure on Rankine-cycle efficiency.

Let the exhaust pressure drop from P_4 to P_4', with the corresponding decrease in temperature at which heat is rejected. The net work is increased by area 1–4–4′–1′–2′–2–1 (shown by the crosshatching). The heat transferred to the steam is increased by area a'–2′–2–a–a'. Since these two areas are approximately equal, the net result is an increase in cycle efficiency. This is also evident from the fact that the average temperature at which heat is rejected is decreased. Note however, that lowering the back pressure causes an increase in the moisture content in the steam leaving the turbine. This is a significant factor because if the moisture in the low-pressure stages of the turbine exceeds about 10 per cent, not only is there a decrease in turbine efficiency, but also erosion of the turbine blades may be a very serious problem.

Next consider the effect of superheating the steam in the boiler, as shown in Fig. 10.3. It is readily evident that the work is increased by area

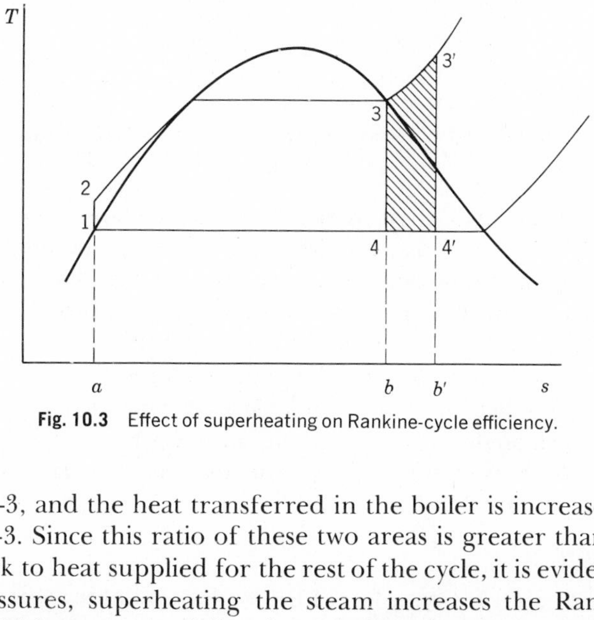

Fig. 10.3 Effect of superheating on Rankine-cycle efficiency.

3–3′–4′–4–3, and the heat transferred in the boiler is increased by area 3–3′–b'–b–3. Since this ratio of these two areas is greater than the ratio of net work to heat supplied for the rest of the cycle, it is evident that for given pressures, superheating the steam increases the Rankine-cycle efficiency. This would also follow from the fact that the average temperature at which heat is transferred to the steam is increased. Note also that when the steam is superheated the quality of the steam leaving the turbine increases.

Finally, the influence of the maximum pressure of the steam must be considered, and this is shown in Fig. 10.4. In this analysis the maximum temperature of the steam, as well as the exhaust pressure, is held constant. The heat rejected decreases by area b'–4′–4–b–b'. The net work increases by the amount of the single crosshatching and decreases by the amount of the double crosshatching. Therefore the net work tends to remain the same, but the heat rejected decreases, and therefore

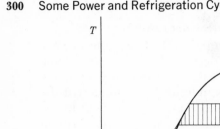

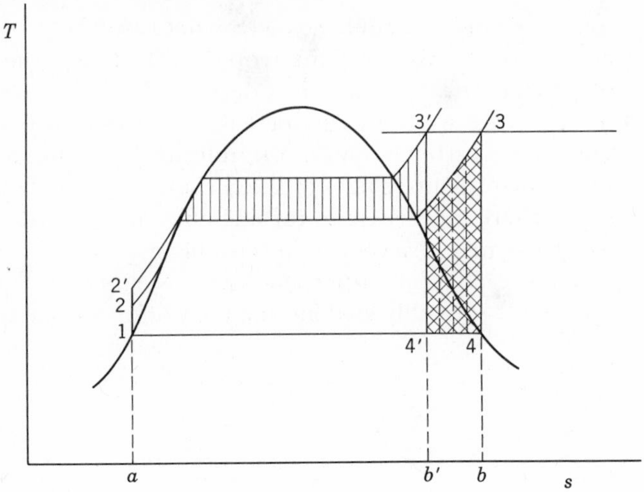

Fig. 10.4 Effect of boiler pressure on Rankine-cycle efficiency.

the Rankine-cycle efficiency increases with an increase in maximum pressure. Note that in this case also the average temperature at which heat is supplied increases with an increase in pressure. The quality of the steam leaving the turbine decreases as the maximum pressure increases.

To summarize this section we can say that the Rankine-cycle efficiency can be increased by lowering the exhaust pressure, increasing the pressure during heat addition, and by superheating the steam. The quality of the steam leaving the turbine is increased by superheating the steam, and decreased by lowering the exhaust pressure and by increasing the pressure during heat addition.

Example 10.2

In a Rankine cycle steam leaves the boiler and enters the turbine at 600 lbf/in.^{2}, 800 F. The condenser pressure is 1 lbf/in.^{2}. Determine the cycle efficiency.

To determine the cycle efficiency we must calculate the turbine work, the pump work, and the heat transfer to the steam in the boiler. We do this by considering a control surface around each of these components and assuming steady-state, steady-flow processes.

Consider a control surface around the pump.

First law: $w_p = h_2 - h_1$

Second law: $s_2 = s_1$

Since

$$s_2 = s_1, \quad h_2 - h_1 = \int_1^2 v\,dP = v(P_2 - P_1)$$

Therefore,

$$w_p = v(P_2 - P_1) = 0.01614(600 - 1) \times \tfrac{144}{778} = 1.8 \text{ Btu/lbm}$$
$$h_1 = 69.70$$
$$h_2 = 69.7 + 1.8 = 71.5 \text{ Btu/lbm.}$$

Consider a control surface around the turbine.

First law: $w_t = h_3 - h_4$

Second law: $s_4 = s_3$

$$h_3 = 1407.6 \quad s_3 = 1.6343$$
$$s_3 = s_4 = 1.6343 = 1.9779 - (1 - x)_4 1.8453$$
$$(1 - x)_4 = 0.1861$$
$$h_4 = 1105.8 - 0.1861(1036.0) = 913.0$$
$$w_t = h_3 - h_4 = 1407.6 - 913.0 = 494.6 \text{ Btu/lbm}$$
$$w_{\text{net}} = w_t - w_p = 494.6 - 1.8 = 492.8 \text{ Btu/lbm}$$

Consider a control surface around the boiler.

$$q_H = h_3 - h_2 = 1407.6 - 71.5 = 1336.1 \text{ Btu/lbm}$$

$$\eta_{\text{th}} = \frac{w_{\text{net}}}{q_H} = \frac{492.8}{1336.1} = 36.9\%$$

The net work could also be determined by calculating the heat rejected in the condenser, q_L, and noting, from the first law, that the net work for the cycle is equal to the net heat transfer. Considering a control surface around the condenser.

$$q_L = h_4 - h_1 = 913.0 - 69.7 = 843.3 \text{ Btu/lbm}$$

Therefore,

$$w_{\text{net}} = q_H - q_L = 1336.1 - 843.3 = 492.8 \text{ Btu/lbm}$$

10.3 The Reheat Cycle

In the last paragraph we noted that the efficiency of the Rankine cycle could be increased by increasing the pressure during the addition of heat. However, this also increases the moisture content of the steam in the low-pressure end of the turbine. The reheat cycle has been developed to take advantage of the increased efficiency with higher pressures, and yet avoid excessive moisture in the low-pressure stages of the turbine. This cycle is shown schematically and on a *T-s* diagram in Fig. 10.5. The unique feature of this cycle is that the steam is expanded to some intermediate pressure in the turbine, and is then reheated in the boiler, after

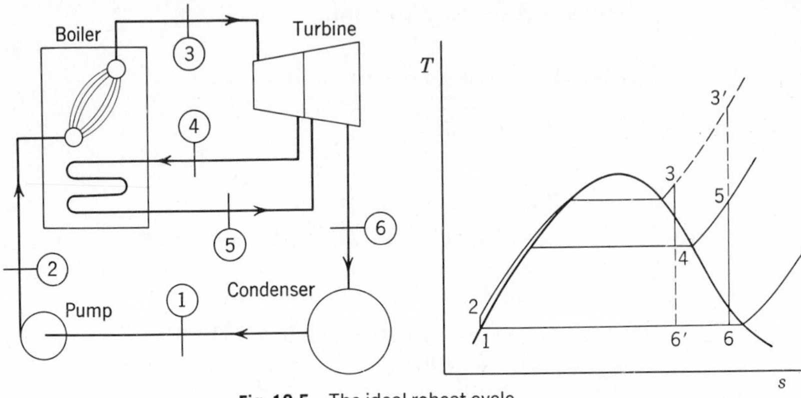

Fig. 10.5 The ideal reheat cycle.

which it expands in the turbine to the exhaust pressure. It is evident from the *T-s* diagram that there is very little gain in efficiency from reheating the steam, because the average temperature at which heat is supplied is not greatly changed. The chief advantage is in decreasing the moisture content in the low-pressure stages of the turbine to a safe value. Note also, that if metals could be found that would enable one to superheat the steam to 3′, the simple Rankine cycle would be more efficient than the reheat cycle, and there would be no need for the reheat cycle.

Example 10.3

Consider a reheat cycle utilizing steam. Steam leaves the boiler and enters the turbine at 600 lbf/in.², 800 F. After expansion in the turbine to 60 lbf/in.², the steam is reheated to 800 F and then expanded in the low-pressure turbine to 1 lbf/in.². Determine the cycle efficiency.

Designating the states in accordance with Fig. 10.5 we proceed as follows:

Consider a control surface around the turbine.

First law: $w_t = (h_3 - h_4) + (h_5 - h_6)$

Second law: $s_4 = s_3$

$s_6 = s_5$

To find w_t we proceed as follows:

$$h_3 = 1407.6 \quad s_3 = 1.6343$$

$$s_4 = s_3 = 1.6343 = 1.6444 - (1-x)_4 1.2170$$

$$(1-x)_4 = 0.0083$$

$$h_4 = 1178.0 - 0.0083(915.8) = 1170.4$$

$$h_5 = 1431.2 \quad s_5 = 1.9022$$

$$s_5 = s_6 = 1.9022 = 1.9779 - (1-x)_6 1.8453$$

$$(1-x)_6 = 0.0411$$

$$h_6 = 1105.8 - 0.0411(1036.0) = 1063.3$$

$$w_t = (h_3 - h_4) + (h_5 - h_6)$$

$$w_t = (1407.6 - 1170.4) + (1431.2 - 1063.3) = 605.1 \text{ Btu/lbm}$$

Consider a control surface around the pump.

First law: $w_p = h_2 - h_1$
Second law: $s_2 = s_1$

Since

$$s_2 = s_1, \quad h_2 - h_1 = \int_1^2 v\, dP = v(P_2 - P_1)$$

Therefore,

$$w_p = v(P_2 - P_1) = 0.01614(600 - 1)\tfrac{144}{778} = 1.8 \text{ Btu/lbm.}$$

$$h_2 = 69.7 + 1.8 = 71.5 \text{ Btu/lbm}$$

Consider a control surface around the boiler.

$$q_H = (h_3 - h_2) + (h_5 - h_4)$$

$$q_H = (1407.6 - 71.5) + (1431.2 - 1170.4) = 1596.9 \text{ Btu/lbm}$$

$$w_{\text{net}} = w_t - w_p = 605.1 - 1.8 = 603.3 \text{ Btu/lbm}$$

$$\eta_{\text{th}} = \frac{w_{\text{net}}}{q_H} = \frac{603.3}{1596.9} = 37.8\%$$

Note by comparison with Example 10.2 that the gain in efficiency from reheating is relatively small, but that the moisture content leaving the turbine is decreased as a result of reheating from 18.6 per cent to 4.1 per cent.

10.4 The Regenerative Cycle

Another important variation from the Rankine cycle is the regenerative cycle, which involves the use of feedwater heaters. The basic concepts of this cycle can be demonstrated by considering the Rankine cycle without superheat as shown in Fig. 10.6. During the process between states 2 and 2′ the working fluid is heated while in the liquid phase, and

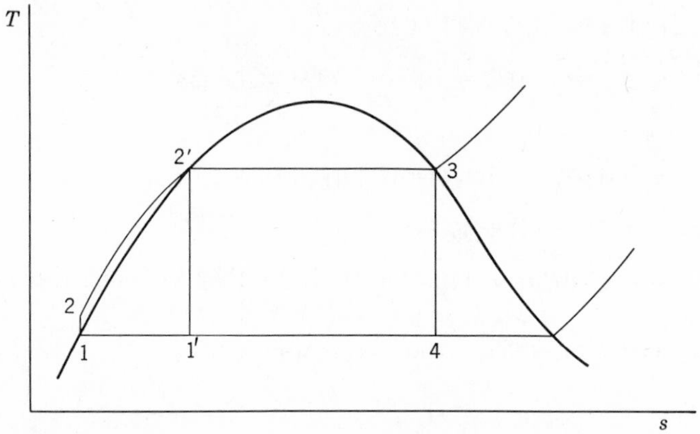

Fig. 10.6 Temperature-entropy diagram showing the relationship between Carnot-cycle efficiency and Rankine-cycle efficiency.

the average temperature of the working fluid is much lower during this process than during the vaporization process 2′–3. This causes the average temperature at which heat is supplied in the Rankine cycle to be lower than in the Carnot cycle 1′–2′–3–4–1′, and consequently the efficiency of the Rankine cycle is less than that of the corresponding Carnot cycle. In the regenerative cycle the working fluid enters the boiler at some state between 2 and 2′, and consequently the average temperature at which heat is supplied is increased.

Consider first an idealized regenerative cycle, shown in Fig. 10.7. The unique feature of this cycle compared to the Rankine cycle is that after leaving the pump, the liquid circulates around the turbine casing, counterflow to the direction of vapor flow in the turbine. Thus, it is possible to transfer heat from the vapor as it flows through the turbine to the liquid flowing around the turbine. Let us assume for the moment that this is a reversible heat transfer; that is, at each point the temperature of the vapor is only infinitesimally higher than the temperature of the liquid. In this case line 4–5 on the T-s diagram of Fig. 10.7, which represents the states of the vapor flowing through the turbine, is exactly

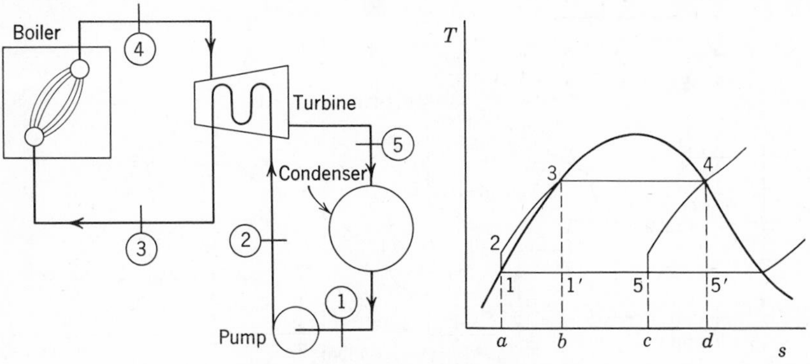

Fig. 10.7 The ideal regenerative cycle.

parallel to line 1–2–3, which represents the pumping process (1–2) and the states of the liquid flowing around the turbine. Consequently areas 2–3–*b*–*a*–2 and 5–4–*d*–*c*–5 are not only equal but congruous, and these areas respectively represent the heat transferred to the liquid and from the vapor. Note also that heat is transferred to the working fluid at constant temperature in process 3–4, and area 3–4–*d*–*b*–3 represents this heat transfer. Heat is transferred from the working fluid in process 5–1, and area 1–5–*c*–*a*–1 represents this heat transfer. Note that this area is exactly equal to area 1′–5′–*d*–*b*–1′, which is the heat rejected in the related Carnot cycle 1′–3–4–5′–1′. Thus, this idealized regenerative cycle has an efficiency exactly equal to the efficiency of the Carnot cycle with the same heat-supply and heat-rejection temperatures.

Quite obviously this idealized regenerative cycle is not practical. First of all, it would not be possible to effect the necessary heat transfer from the vapor in the turbine to the liquid feedwater. Furthermore, the moisture content of the vapor leaving the turbine is considerably increased as a result of the heat transfer, and the disadvantage of this has been noted previously. The practical regenerative cycle involves the extraction of some of the vapor after it has partially expanded in the turbine and the use of feedwater heaters, as shown in Fig. 10.8.

Steam enters the turbine at state 5. After expansion to state 6, some of the steam is extracted and enters the feedwater heater. The steam that is not extracted is expanded in the turbine to state 7 and is then condensed in the condenser. This condensate is pumped into the feedwater heater where it mixes with the steam extracted from the turbine. The proportion of steam extracted is just sufficient to cause the liquid leaving the feedwater heater to be saturated at state 3. Note that the liquid has not been pumped to the boiler pressure, but only to the intermediate

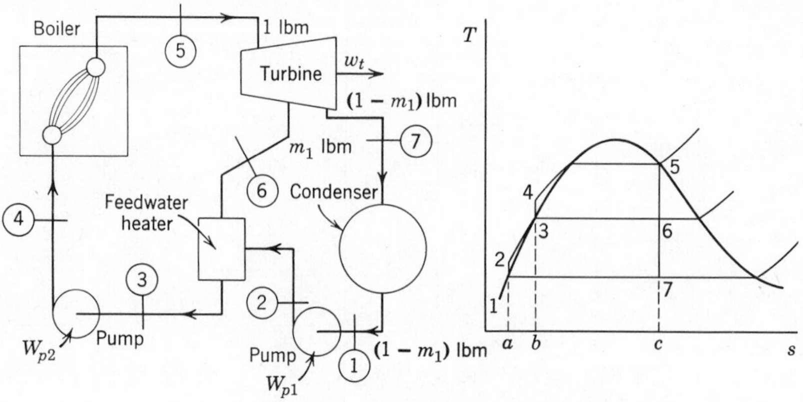

Fig. 10.8 Regenerative cycle with open feedwater heater.

pressure corresponding to state 6. Another pump is required to pump the liquid leaving the feedwater heater to boiler pressure. The significant point is that the average temperature at which heat is supplied has been increased.

This cycle is somewhat difficult to show on a *T-s* diagram because the mass of steam flowing through the various components is not the same. The *T-s* diagram of Fig. 10.8 simply shows the state of the fluid at the various points

Area 4–5–*c*–*b*–4 in Fig. 10.8 represents the heat transferred per pound mass of working fluid. Process 7–1 is the heat-rejection process, but since not all the steam passes through the condenser, area 1–7–*c*–*a*–1 represents the heat transfer per pound mass flowing through the condenser, which does not represent the heat transfer per pound mass of working fluid entering the turbine. Note also that between states 6 and 7 only part of the steam is flowing through the turbine. The example that follows illustrates the calculations involved in the regenerative cycle.

Example 10.4

Consider a regenerative cycle utilizing steam as the working fluid. Steam leaves the boiler and enters the turbine at 600 lbf/in.^{2}, 800 F. After expansion to 60 lbf/in.^{2} some of the steam is extracted from the turbine for the purpose of heating the feedwater in an open feedwater heater. The pressure in the feedwater heater is 60 lbf/in.^{2} and the water leaving it is saturated liquid at 60 lbf/in.^{2}. The steam not extracted expands to 1 lbf/in.^{2}. Determine the cycle efficiency.

The line diagram and *T-s* diagram for this cycle are shown in Fig. 10.8.

From Examples 10.2 and 10.3 we have the following properties:

$$h_5 = 1407.6 \quad h_6 = 1170.4 \quad h_7 = 913.0 \quad h_1 = 69.70$$

Consider a control surface around the low pressure pump.

First law: $w_{p1} = (h_2 - h_1)$
Second law: $s_2 = s_1$
Therefore,

$$h_2 - h_1 = \int_1^2 v\,dP = v(P_2 - P_1)$$

$$w_{p1} = v(P_2 - P_1) = 0.01614(60 - 1)\tfrac{144}{778} = 0.2 \text{ Btu/lbm}$$

$$h_2 = h_1 + w_p = 69.7 + 0.2 = 69.9$$

$$h_3 = 262.2$$

Consider a control surface around the turbine.

First law: $w_t = (h_5 - h_6) + (1 - m_1)(h_6 - h_7)$
Second law: $s_5 = s_6 = s_7$

It follows that, as has been calculated in Examples 10.2 and 10.3,

$$h_6 = 1170.4; \quad h_7 = 913.0$$

Next consider a control surface around the feedwater heater.

First law: $m_1(h_6) + (1 - m_1)h_2 = h_3$

$$m_1(1170.4) + (1 - m_1)69.9 = 262.2$$

$$m_1 = \frac{262.2 - 69.9}{1170.4 - 69.9} = 0.1747$$

We can now calculate the turbine work.

$$w_t = (h_5 - h_6) + (1 - m_1)(h_6 - h_7)$$
$$= (1407.6 - 1170.4) + (1 - 0.1747)(1170.4 - 913.0) = 449.7$$

Consider a control surface around the high pressure pump.

First law: $w_{p2} = (h_4 - h_3)$
Second law: $s_4 = s_3$

$$w_{p2} = v(P_4 - P_3) = 0.01738(600 - 60)\tfrac{144}{778} = 1.7 \text{ Btu/lbm}$$

$$w_{\text{net}} = w_t - (1 - m_1)w_{p1} - w_{p2} = 449.7 - 0.825(0.2) - 1.7 = 447.8$$

Consider a control surface around the boiler.

$$q_H = h_5 - h_4 = 1407.6 - 263.9 = 1143.7$$

$$\eta_{\text{th}} = \frac{w_{\text{net}}}{q_H} = \frac{447.8}{1143.7} = 39.1\%$$

Note the increase in efficiency over the Rankine cycle of Example 10.2.

Up to this point the discussion and example have tacitly assumed that the extraction steam and feedwater are mixed in the feedwater heater. Another much-used type of feedwater heater, known as a closed heater, is one in which the steam and feedwater do not mix, but rather heat is transferred from the extracted steam as it condenses on the outside of tubes as the feedwater flows through the tubes. In a closed heater, a schematic sketch of which is shown in Fig. 10.9, the steam and feed-

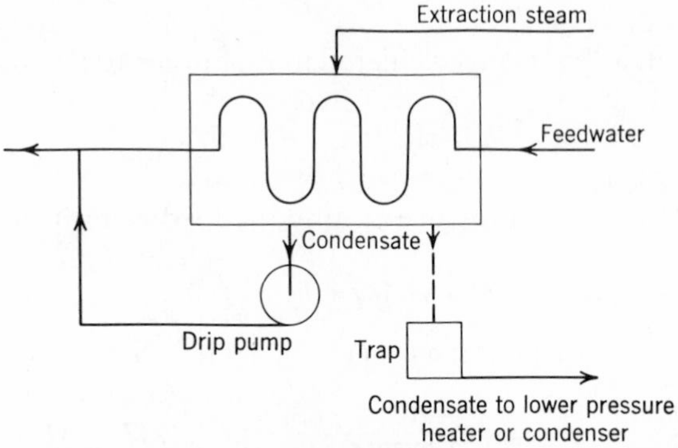

Fig. 10.9 Schematic arrangement for a closed feedwater heater.

water may be at considerably different pressures. The condensate may be pumped into the feedwater line, or it may be removed through a trap (a device that permits liquid but no vapor to flow to a region of lower pressure) to a lower pressure heater or to the main condenser.

Open feedwater heaters have the advantage of being less expensive and having better heat-transfer characteristics compared to closed feedwater heaters. They have the disadvantage of requiring a pump to handle the feedwater between each heater.

In many power plants a number of stages of extraction are used, though only rarely more than five. The number is, of course, determined by economic considerations. It is evident that by using a very large number of extraction stages and feedwater heaters, the cycle efficiency would approach that of the idealized regenerative cycle of Fig. 10.7, where the feedwater enters the boiler as saturated liquid at the maximum pressure. However, in practice this could not be economically justified because the savings effected by the increase in efficiency would be more than offset by the cost of additional equipment (feedwater heaters, piping, etc.).

A typical arrangement of the main components in an actual power plant is shown in Fig. 10.10. Note that one open feedwater heater is a

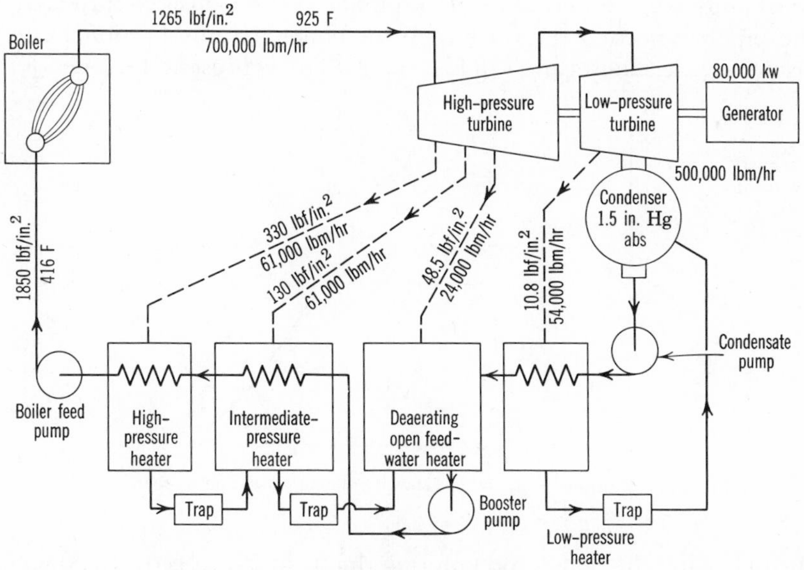

Fig. 10.10 Arrangement of heaters in an actual power plant utilizing regenerative feedwater heaters.

deaerating feedwater heater, and this has the dual purpose of heating and removing the air from the feedwater. Unless the air is removed excessive corrosion occurs in the boiler. Note also that the condensate from the high-pressure heater drains (through a trap) to the intermediate heater, and the intermediate heater drains to the deaerating feedwater heater. The low-pressure heater drains to the condenser.

In many cases an actual power plant combines one reheat stage with a number of extraction stages. The principles already considered are readily applied to such a cycle.

10.5 Deviation of Actual Cycles from Ideal Cycles

Before leaving the matter of vapor power cycles, a few comments are in order regarding the ways in which an actual cycle deviates from an ideal cycle. (The losses associated with the combustion process are considered in a later chapter.) The most important of these are as follows.

Piping Losses

Pressure drop due to frictional effects and heat transfer to the surroundings are the most important piping losses. Consider for example the pipe connecting the turbine to the boiler. If only frictional effects occurred, the states *a* and *b* in Fig. 10.11 would represent the states of the

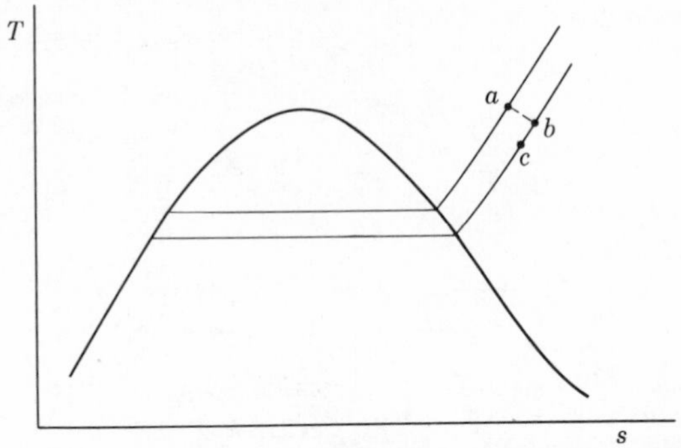

Fig. 10.11 Temperature-entropy diagram showing effect of losses between boiler and turbine.

steam leaving the boiler and entering the turbine respectively. Note that this causes an increase in entropy. Heat transferred to the surroundings at constant pressure can be represented by process *bc*. This effect causes a decrease in entropy. Both the pressure drop and heat transfer cause a decrease in the availability of the steam entering the turbine, and the irreversibility of this process can be calculated by the methods outlined in Chapter 9.

A similar loss is the pressure drop in the boiler. Because of this pressure drop, the water entering the boiler must be pumped to a much higher pressure than the desired steam pressure leaving the boiler, and this requires additional pump work.

Turbine Losses

The losses in the turbine are primarily those associated with the flow of the working fluid through the turbine. Heat transfer to the surroundings also represents a loss, but this is usually of secondary importance. The effects of these two losses are the same as those outlined for piping losses, and the process might be as represented in Fig. 10.12, where 4_s represents the state after an isentropic expansion and state 4 represents the actual state leaving the turbine. The governing procedures may also cause a loss in the turbine, particularly if a throttling process is used to govern the turbine.

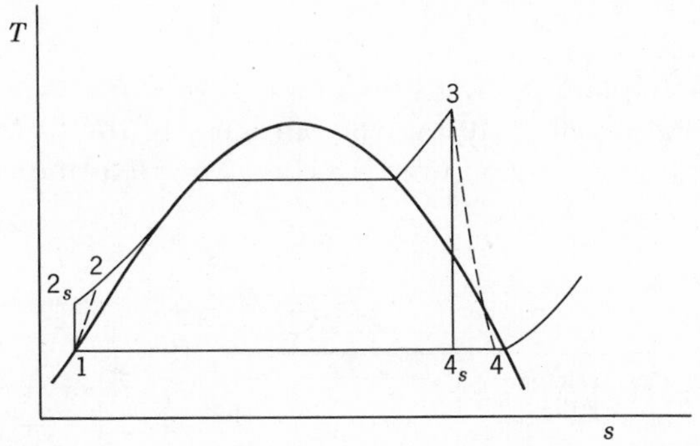

Fig. 10.12 Temperature-entropy diagram showing effect of turbine and pump inefficiencies on cycle performance.

The efficiency of the turbine has been defined (in Chapter 9) as

$$\eta_t = \frac{w_t}{h_3 - h_{4s}}$$

where the states are as designated in Fig. 10.12.

Pump Losses

The losses in the pump are similar to those of the turbine, and are primarily due to the irreversibilities associated with the fluid flow. Heat transfer is usually a minor loss.

The pump efficiency is defined as

$$\eta_p = \frac{h_{2s} - h_1}{w_p}$$

where the states are as shown in Fig. 10.12, and w_p is the actual work input per pound of fluid.

Condenser Losses

The losses in the condenser are relatively small. One of these minor losses is the cooling below the saturation temperature of the liquid leaving the condenser. This represents a loss because additional heat transfer is necessary to bring the water to its saturation temperature.

The influence of these losses on the cycle is illustrated in the following example, which should be compared to Example 10.2.

Example 10.5

A steam power plant operates on a cycle with pressure and temperatures as designated in Fig. 10.13. The efficiency of the turbine is 86 per cent and the efficiency of the pump is 80 per cent. Determine the thermal efficiency of this cycle.

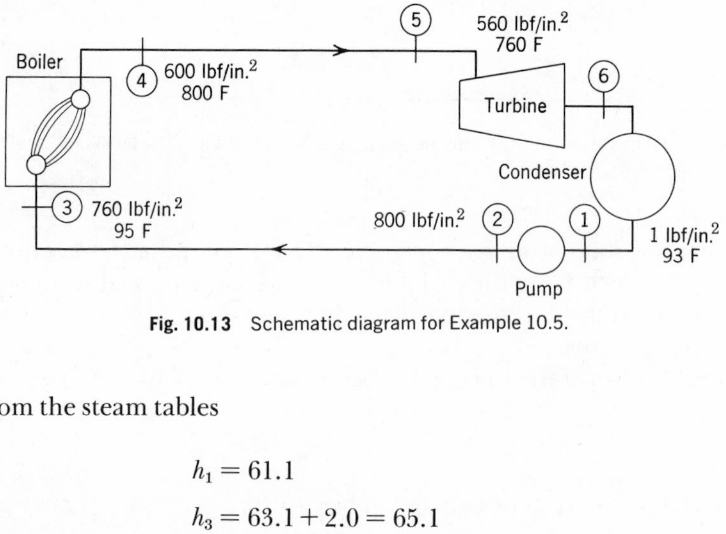

Fig. 10.13 Schematic diagram for Example 10.5.

From the steam tables

$$h_1 = 61.1$$

$$h_3 = 63.1 + 2.0 = 65.1$$

$$h_4 = 1407.6 \quad s_4 = 1.6343$$
$$h_5 = 1386.8 \quad s_5 = 1.6248$$

This cycle is shown on the T-s diagram of Fig. 10.14. Consider a control surface around the turbine.

First law: $w_t = h_6 - h_5$

Second law: $s_{6s} = s_5$

$$\eta_t = \frac{w_t}{h_5 - h_{6s}} = \frac{h_5 - h_6}{h_5 - h_{6s}}$$

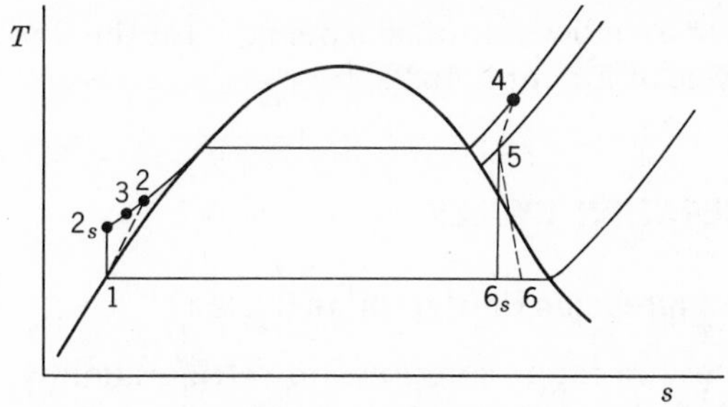

Fig. 10.14 Temperature-entropy diagram for Example 10.5.

$$s_{6s} = s_5 = 1.6248 = 1.9779 - (1-x)_{6s}1.8453$$

$$(1-x)_{6s} = \frac{0.3531}{1.8453} = 0.1912$$

$$h_{6s} = 1105.8 - 0.1912(1036.0) = 907.6$$

$$w_t = \eta_t(h_5 - h_{6s}) = 0.86(1386.8 - 907.6)$$

$$= 0.86(479.2) = 412.1 \text{ Btu/lbm}$$

Consider a control surface around the pump.

First law: $w_p = h_2 - h_1$
Second law: $s_{2s} = s_1$

$$\eta_p = \frac{h_{2s} - h_1}{w_p} = \frac{h_{2s} - h_1}{h_2 - h_1}$$

Since

$$s_{2s} = s_1, \quad h_{2s} - h_1 = v(P_2 - P_1)$$

Therefore,

$$w_p = \frac{h_{2s} - h_1}{\eta_p} = \frac{v(P_2 - P_1)}{\eta_p} = \frac{0.01615(800-1)144}{0.8 \times 778}$$

$$= \frac{2.4}{0.8} = 3.0 \text{ Btu/lbm}$$

$$w_{\text{net}} = w_t - w_p = 412.1 - 3.0 = 409.1 \text{ Btu/lbm}$$

If we apply the first law to a control surface around the boiler, we have

$$q_H = h_4 - h_3 = 1407.6 - 65.1 = 1342.5 \text{ Btu/lbm}$$

$$\eta_{\text{th}} = \frac{409.1}{1342.5} = 30.4\%$$

This compares to an efficiency of 36.9 per cent for the Rankine efficiency of the similar cycle of Example 10.2.

VAPOR REFRIGERATION CYCLES

10.6 Vapor Compression Refrigeration Cycles

The ideal cycle for vapor compression refrigeration is shown in Fig. 10.15 as cycle 1–2–3–4–1. Saturated vapor at low pressure enters the

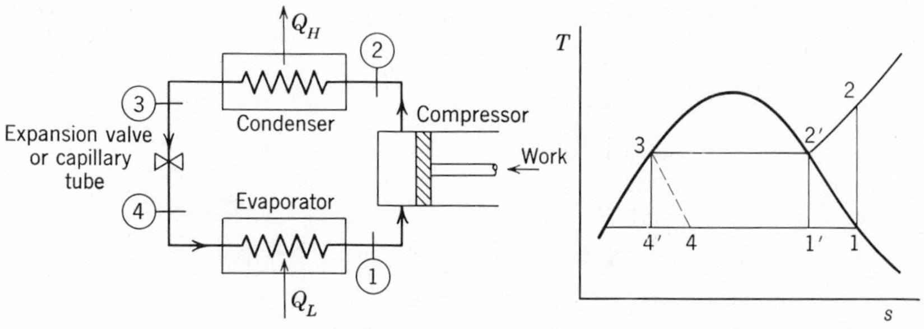

Fig. 10.15 The ideal vapor-compression refrigeration cycle.

compressor and undergoes a reversible adiabatic compression, 1–2. Heat is then rejected at constant pressure in process 2–3, and the working fluid leaves the condenser as saturated liquid. An adiabatic throttling process follows, process 3–4, and the working fluid is then evaporated at constant pressure, process 4–1, to complete the cycle.

The similarity between this cycle and the Rankine cycle is evident, for it is essentially the same cycle in reverse, except that an expansion valve replaces the pump. This throttling process is irreversible, whereas the pumping process of the Rankine cycle is reversible. The deviation of this ideal cycle from the Carnot cycle 1′–2′–3–4′–1′ is evident from the *T-s* diagram. The reason for the deviation is that it is much more expedient to have a compressor handle only vapor than a mixture of liquid and vapor as would be required in process 1′–2′ of the Carnot cycle. It is virtually impossible to compress (at a reasonable rate) a mixture such as that represented by state 1′ and maintain equilibrium between the liquid and vapor, because there must be a heat and mass transfer across the phase boundary. It is also much simpler to have the expansion process take place irreversibly through an expansion valve than to have an expansion device that receives saturated liquid and discharges a mixture of liquid and vapor, as would be required in process 3–4′. For these reasons the ideal cycle for vapor-compression refrigeration is as shown in Fig. 10.15 by cycle 1–2–3–4–1.

As we saw in Chapter 7, the performance of a refrigeration cycle is given in terms of the coefficient of performance, β, which is defined for a refrigeration cycle as

$$\beta = \frac{q_L}{w_c} \tag{10.2}$$

The capacity of a refrigeration plant is usually given in tons of refrigeration. This term had its origin in the ice-making industry, where cooling capacity was given in terms of tons of ice melting per day. The unit is now defined as

$$\begin{aligned} 1 \text{ ton refrigeration} &= 288{,}000 \text{ Btu of refrigeration/day} \\ &= 12{,}000 \text{ Btu of refrigeration/hr} \end{aligned}$$

Example 10.6

Consider an ideal refrigeration cycle that utilizes Freon-12 as the working fluid. The temperature of the refrigerant in the evaporator is 0 F and in the condenser it is 100 F. The refrigerant is circulated at the rate of 200 lbm/hr. Determine the coefficient of performance and the capacity of the plant in tons of refrigeration.

From the thermodynamic tables for Freon-12 we find the following properties for the states as designated in Fig. 10.15.

$$h_1 = 77.271 \quad s_1 = 0.16888 \quad P_2 = 131.86$$

Consider a control surface around the compressor.

First law: $w_c = h_2 - h_1$
Second law: $s_2 = s_1$

Therefore,

$$s_2 = s_1 = 0.16888, \text{ and } h_2 = 90.306$$

$$w_c = h_2 - h_1 = 90.306 - 77.271 = 13.035$$

Considering a control surface around the expansion valve, we conclude from the first law that

$$h_3 = h_4 = 31.100$$

Finally, consider a control surface around the evaporator.

First law: $q_L = h_1 - h_4 = 77.271 - 31.100 = 46.171$

Therefore,

$$\beta = \frac{q_L}{w_c} = \frac{46.171}{13.035} = 3.55$$

$$\text{Capacity} = \frac{46.171 \times 200}{12{,}000} = 0.768 \text{ tons.}$$

10.7 Working Fluids for Vapor-Compression Refrigeration Systems

A much larger number of different working fluids (refrigerants) are utilized in vapor-compression refrigeration systems than in vapor power cycles. Ammonia and sulfur dioxide were important in the early days of vapor-compression refrigeration. Today, however, the main refrigerants are the halogenated hydrocarbons, which are marketed under the trade names of Freon and Genatron. For example, dichlorodifluoromethane (CCl_2F_2) is known as Freon-12 and Genatron-12. Two important considerations in selecting a refrigerant are the temperature at which refrigeration is desired and the type of equipment to be used.

Since the refrigerant undergoes a change of phase during the heat-transfer process, the pressure of the refrigerant will be the saturation pressure during the heat-supply and heat-rejection processes. Low pressures mean large specific volumes and correspondingly large equipment. High pressures mean smaller equipment but it must be designed to withstand higher pressure. In particular, the pressures should be well below the critical pressure. For extremely low temperature applications a binary fluid system may be used by cascading two separate systems.

The type of compressor used has a particular bearing on the refrigerant. Reciprocating compressors are best adapted to low specific volumes, which means higher pressures, whereas centrifugal compressors are most suitable for low pressures and high specific volumes.

It is also important that the refrigerants used in domestic appliances be nontoxic. Other important characteristics are tendency to cause corrosion, miscibility with compressor oil, dielectric strength, stability, and cost. Also, for given temperatures during evaporation and condensation, not all refrigerants have the same coefficient of performance for the ideal cycle. It is, of course, desirable to utilize the refrigerant with the highest coefficient of performance, other factors permitting.

10.8 Deviation of the Actual Vapor-Compression Refrigeration Cycle from the Ideal Cycle

The actual refrigeration cycle deviates from the ideal cycle primarily because of pressure drops associated with fluid flow and heat transfer to or from the surroundings. The actual cycle might approach the one shown in Fig. 10.16.

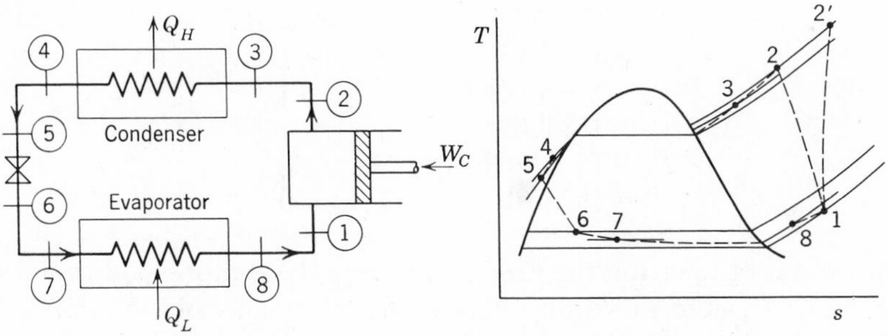

Fig. 10.16 The actual vapor-compression refrigeration cycle.

The vapor entering the compressor will probably be superheated. During the compression process there are irreversibilities and heat transfer either to or from the surroundings, depending on the temperature of the refrigerant and the surroundings. Therefore, the entropy might increase or decrease during this process, for the irreversibility and heat transfer to the refrigerant cause an increase in entropy, and heat transfer from the refrigerant causes a decrease in entropy. These possibilities are represented by the two dotted lines 1–2 and 1–2′. The pressure of the liquid leaving the condenser will be less than the pressure of the vapor entering, and the temperature of the refrigerant in the condenser will be somewhat above that of the surroundings to which heat is being transferred. Usually the temperature of the liquid leaving the condenser is lower than the saturation temperature, and it might drop somewhat more in the piping between the condenser and expansion valve. This represents a gain, however, because as a result of this heat transfer the refrigerant enters the evaporator with a lower enthalpy, thus permitting more heat transfer to the refrigerant in the evaporator.

There is some drop in pressure as the refrigerant flows through the evaporator. It may be slightly superheated as it leaves the evaporator, and due to heat transfer from the surroundings the temperature will increase in the piping between the evaporator and compressor. This heat transfer represents a loss, because it increases the work of the compressor as a result of the increased specific volume of the fluid entering it.

Example 10.7

A refrigeration cycle utilizes Freon-12 as the working fluid. Following are the properties at various points of the cycle designated in Fig. 10.16.

$$
\begin{array}{ll}
P_1 = 18\ \text{lbf/in.}^2 & T_1 = 20\ \text{F} \\
P_2 = 180\ \text{lbf/in.}^2 & T_2 = 220\ \text{F} \\
P_3 = 175\ \text{lbf/in.}^2 & T_3 = 180\ \text{F} \\
P_4 = 170\ \text{lbf/in.}^2 & T_4 = 110\ \text{F} \\
P_5 = 168\ \text{lbf/in.}^2 & T_5 = 104\ \text{F} \\
P_6 = P_7 = 20\ \text{lbf/in.}^2 & x_6 = x_7 \\
P_8 = 19\ \text{lbf/in.}^2 & T_8 = 0\ \text{F}
\end{array}
$$

The heat transfer from the Freon-12 during the compression process is 2.0 Btu/lbm. Determine the coefficient of performance of this cycle.

From the Freon-12 tables the following properties are found.

$$
\begin{array}{ll}
h_1 = 80.527 & h_2 = 106.896 \\
h_5 = h_6 = h_7 = 32.067 & h_8 = 77.621
\end{array}
$$

Consider a control surface around the compressor.

First law: $q + h_1 = h_2 + w$

$$
\begin{aligned}
w_c &= -w = h_2 - h_1 - q \\
&= 106.896 - 80.527 - (-2) = 28.4\ \text{Btu/lbm}
\end{aligned}
$$

To find q_L, consider a control surface around the evaporator.

$$q_L = h_8 - h_7 = 77.621 - 32.067 = 45.6$$

Therefore,

$$\beta = \frac{q_L}{w_c} = \frac{45.6}{28.4} = 1.61$$

10.9 The Ammonia-Absorption Refrigeration Cycle

The ammonia-absorption refrigeration cycle differs from the vapor compression cycle in the manner in which compression is achieved. In the absorption cycle the low-pressure ammonia vapor is absorbed in water and the liquid solution is pumped to a high pressure by a liquid pump. Figure 10.17 shows a schematic arrangement of the essential elements of such a system.

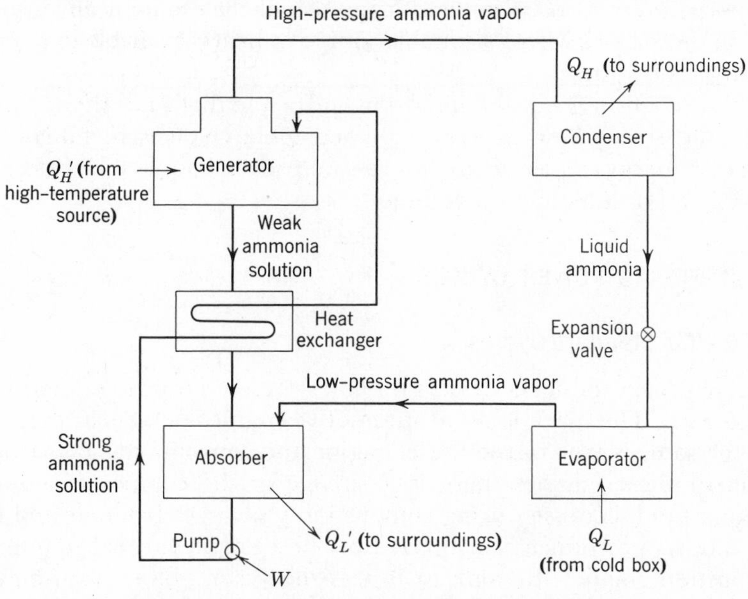

Fig. 10.17 The ammonia-absorption refrigeration cycle.

The low-pressure ammonia vapor leaving the evaporator enters the absorber where it is absorbed in the weak ammonia solution. This process takes place at a temperature slightly above that of the surroundings and heat must be transferred to the surroundings during this process. The strong ammonia solution is then pumped through a heat exchanger to the generator where a higher pressure and temperature are maintained. Under these conditions ammonia vapor is driven from the solution as a result of heat transfer from a high-temperature source. The ammonia vapor goes to the condenser where it is condensed, as in a vapor-compression system, and then to the expansion valve and evaporator. The weak ammonia solution is returned to the absorber through the heat exchanger.

The distinctive feature of the absorption system is that very little work input is required because the pumping process involves a liquid. This follows from the fact that for a reversible steady-flow process with negligible changes in kinetic and potential energy, the work is equal to $-\int v\,dP$, and the specific volume of the liquid is much less than the specific volume of the vapor. On the other hand, a relatively high-temperature source of heat must be available (200 F to 400 F). The equipment involved in an absorption system is somewhat greater than in a vapor-compression system and it can usually be economically justified only in those cases where a suitable source of heat is available that would otherwise be wasted.

This cycle brings out the important principle that since the work in a reversible steady-flow process with negligible changes in kinetic and potential energy is $-\int v\,dP$, a compression process should take place with the smallest possible specific volume.

AIR-STANDARD POWER CYCLES

10.10 Air-Standard Cycles

Many work-producing devices (engines) utilize a working fluid that is always a gas. The spark-ignition automotive engine is a familiar example, and the same is true of the Diesel engine and conventional gas turbine. In all of these engines there is a change in the composition of the working fluid, because during combustion it changes from air and fuel to combustion products. For this reason these engines are called internal combustion engines. In contrast to this the steam power plant may be called an external-combustion engine, because heat is transferred from the products of combustion to the working fluid. External-combustion engines using a gaseous working fluid (usually air) have been built. To date they have had very limited application, but the use of the gas turbine cycle in conjunction with a nuclear reactor has been investigated extensively. Other external-combustion engines are currently receiving serious attention in an effort to combat the air pollution problem.

Because the working fluid does not go through a complete thermodynamic cycle in the engine (even though the engine operates in a mechanical cycle) the internal-combustion engine operates on the so-called open cycle. However, in order to analyze internal-combustion engines it is advantageous to devise closed cycles that closely approximate the open cycles. One such approach is the air-standard cycle, which is based on the following assumptions:

1. A fixed mass of air is the working fluid throughout the entire cycle, and the air is always an ideal gas. Thus there is no inlet process or exhaust process.
2. The combustion process is replaced by a heat-transfer process from an external source.
3. The cycle is completed by heat transfer to the surroundings (in contrast to the exhaust and intake process of an actual engine).
4. All processes are internally reversible.

5. The additional assumption is usually made that air has a constant specific heat.

The main value of the air-standard cycle is to enable us to examine qualitatively the influence of a number of variables on performance. The results obtained from the air-standard cycle, such as efficiency and mean effective pressure, will differ a great deal from those of the actual engine. The emphasis, therefore, in our consideration of the air-standard cycle will be primarily on the qualitative aspects.

The term "mean effective pressure," which is used in conjunction with reciprocating engines, is defined as the pressure which, if it acted on the piston during the entire power stroke, would do an amount of work equal to that actually done on the piston. The work for one cycle is found by multiplying this mean effective pressure by the area of the piston (minus the area of the rod on the crank end of a double-acting engine) and by the stroke.

10.11 The Air-Standard Carnot Cycle

The air-standard Carnot cycle is shown on the P-v and T-s diagrams of Fig. 10.18. Such a cycle could be achieved in either a reciprocating or steady-flow device, as shown in the same figure.

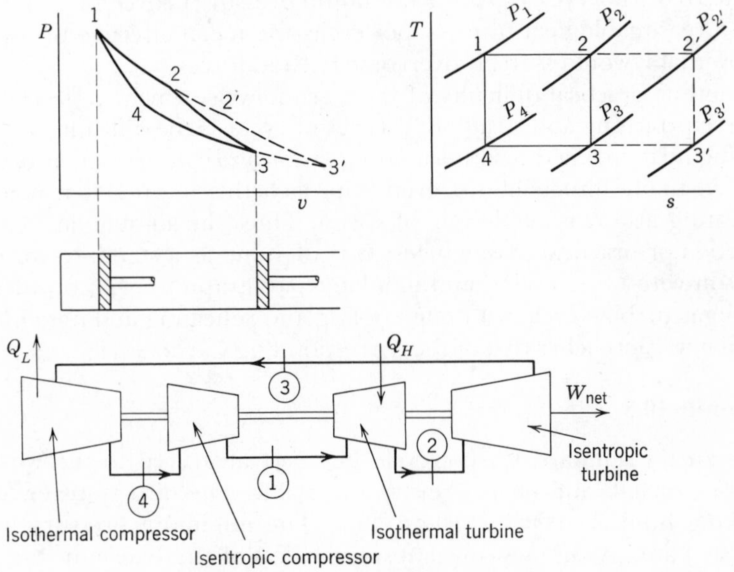

Fig. 10.18 The air-standard Carnot cycle.

In Chapter 7 it was noted that the efficiency of a Carnot cycle depends only on the temperatures at which heat is supplied and rejected, and is given by the relation

$$\eta_{th} = 1 - \frac{T_L}{T_H} = 1 - \frac{T_4}{T_1} = 1 - \frac{T_3}{T_2}$$

where the subscripts refer to Fig. 10.18. The efficiency may also be expressed by the pressure ratio or compression ratio during the isentropic processes. This follows from the fact that

$$\text{Isentropic pressure ratio} = r_{ps} = \frac{P_1}{P_4} = \frac{P_2}{P_3} = \left(\frac{T_3}{T_2}\right)^{k/(1-k)}$$

$$\text{Isentropic compression ratio} = r_{vs} = \frac{V_4}{V_1} = \frac{V_3}{V_2} = \left(\frac{T_3}{T_2}\right)^{1/(1-k)}$$

Therefore

$$\eta_{th} = 1 - r_{ps}^{(1-k)/k} = 1 - r_{vs}^{1-k} \tag{10.3}$$

One other important variable in the air-standard Carnot cycle is the amount of heat transferred to the working fluid per cycle. It is evident from Fig. 10.18 that increasing the heat transfer per cycle at a given T_H causes a larger change in volume during the cycle, and this causes a lower mean effective pressure in a reciprocating engine. In fact, for air-standard Carnot cycles having a minimum pressure between 1 and 10 atm and a reasonable heat transfer per cycle, the mean effective pressure is so low that it would scarcely overcome friction forces.

Another practical difficulty of the Carnot cycle, which applies to both the reciprocating and steady-flow types of cycle, is the difficulty of transferring heat during the isothermal expansion and compression processes. It is virtually impossible to even approach this in an actual machine operating at a reasonable rate of speed. Thus, the air-standard Carnot cycle is not practical. Nonetheless it is of value as a standard for comparison with other cycles, and in a later paragraph we will consider the ideal gas turbine cycle with intercooling and reheating and note how its efficiency approaches that of the corresponding Carnot cycle.

Example 10.8

In an air-standard Carnot cycle heat is transferred to the working fluid at 2000 R, and heat is rejected at 500 R. The heat transfer to the working fluid at 2000 R is 50 Btu/lbm. The minimum pressure in the cycle is 1 atm. Assuming constant specific heat of air, determine the cycle efficiency and the mean effective pressure.

Designating the states as in Fig. 10.18 we have

$$P_3 = 14.7 \text{ lbf/in.}^2 \qquad T_3 = T_4 = 500 \text{ R}$$

$$\frac{T_2}{T_3} = \left(\frac{P_2}{P_3}\right)^{(k-1)/k} = 4 \qquad T_1 = T_2 = 2000 \text{ R}$$

$$\therefore \frac{P_2}{P_3} = 128$$

$$P_2 = 14.7(128) = 1882 \text{ lbf/in.}^2$$

Consider the process that occurs between states 1 and 2.

$${}_1q_2 = RT \ln \frac{V_2}{V_1} = RT \ln \frac{P_1}{P_2} = \frac{53.34 \times 2000}{778} \ln \frac{P_1}{P_2} = 50 \text{ Btu/lbm}$$

Therefore,

$$\frac{P_1}{P_2} = 1.44 \qquad P_1 = 1882(1.44) = 2710 \text{ lbf/in.}^2$$

$$\frac{T_1}{T_4} = \left(\frac{P_1}{P_4}\right)^{(k-1)/k} = 4 \quad \therefore \frac{P_1}{P_4} = 128$$

$$P_4 = \frac{2710}{128} = 21.2 \text{ lbf/in.}^2$$

$$\eta_{\text{th}} = 1 - \frac{T_L}{T_H} = 1 - \frac{500}{2000} = 0.75$$

The product of mean effective pressure and the piston displacement is equal to the net work.

$$\text{mep } (v_3 - v_1) = w_{\text{net}} = \eta_{\text{th}} \times {}_1q_2 = 37.5 \text{ Btu/lbm}$$

$$v_3 = \frac{RT_3}{P_3} = \frac{53.34 \times 500}{14.7 \times 144} = 12.6 \text{ ft}^3\text{/lbm}$$

$$v_1 = \frac{RT_1}{P_1} = \frac{53.34 \times 2000}{2710 \times 144} = 0.273 \text{ ft}^3\text{/lbm}$$

$$\text{mep} = \frac{37.5 \times 778}{(12.6 - 0.273)} = 2360 \text{ lbf/ft}^2 = 16.4 \text{ lbf/in.}^2$$

This is a much lower mean effective pressure than can be effectively utilized in a reciprocating engine.

10.12 The Air-Standard Otto Cycle

The air-standard Otto cycle is an ideal cycle that approximates a spark-ignition internal-combustion engine. This cycle is shown on the *P-v* and *T-s* diagrams of Fig. 10.19. Process 1–2 is an isentropic compression of the air as the piston moves from crank-end dead center to head-end dead center. Heat is then added at constant volume while the piston is momentarily at rest at head-end dead center. (This process corresponds to the ignition of the fuel-air mixture by the spark and the subsequent burning in the actual engine.) Process 3–4 is an isentropic expansion, and process 4–1 is the rejection of heat from the air while the piston is at crank-end dead center.

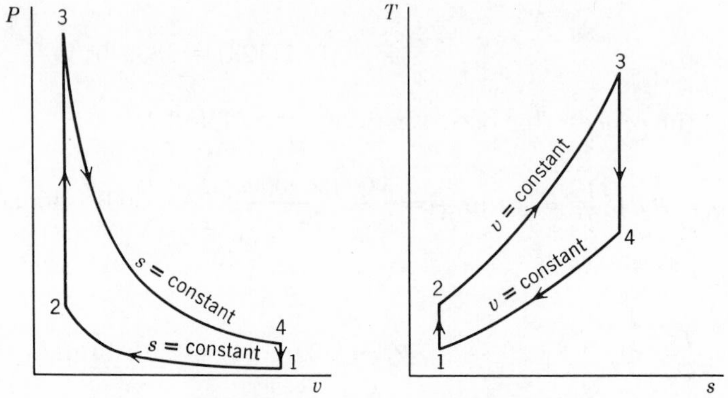

Fig. 10.19 The air-standard Otto cycle.

The thermal efficiency of this cycle is found as follows, assuming constant specific heat of air.

$$\eta_{\text{th}} = \frac{Q_H - Q_L}{Q_H} = 1 - \frac{Q_L}{Q_H} = 1 - \frac{mC_v(T_4 - T_1)}{mC_v(T_3 - T_2)}$$

$$= 1 - \frac{T_1\,(T_4/T_1 - 1)}{T_2\,(T_3/T_2 - 1)}$$

We note further that

$$\frac{T_2}{T_1} = \left(\frac{V_1}{V_2}\right)^{k-1} = \left(\frac{V_4}{V_3}\right)^{k-1} = \frac{T_3}{T_4}$$

Therefore,

$$\frac{T_3}{T_2} = \frac{T_4}{T_1}$$

and

$$\eta_{th} = 1 - \frac{T_1}{T_2} = 1 - (r_v)^{1-k} = 1 - \frac{1}{r_v^{k-1}} \qquad (10.4)$$

where r_v = compression ratio $= \dfrac{V_1}{V_2} = \dfrac{V_4}{V_3}$

The important thing to note is that the efficiency of the air-standard Otto cycle is a function only of the compression ratio, and that the efficiency is increased by increasing the compression ratio. Figure 10.20

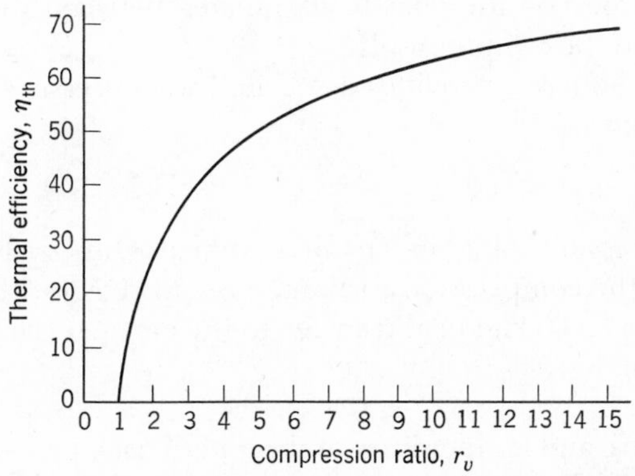

Fig. 10.20 Thermal efficiency of the Otto cycle as a function of compression ratio.

is a plot of the air-standard cycle thermal efficiency vs. compression ratio. It is also true of an actual spark-ignition engine that the efficiency can be increased by increasing the compression ratio. The trend toward higher compression ratios is prompted by the effort to obtain higher thermal efficiency. In the actual engine there is an increased tendency towards detonation of the fuel as compression ratio is increased. Detonation is characterized by an extremely rapid burning of the fuel and strong pressure waves present in the engine cylinder which give rise to the so-called spark knock. Therefore, the maximum compression ratio that can be used is fixed by the fact that detonation must be avoided. The advance in compression ratios over the years in the actual engine has been made possible by developing fuels with better antiknock characteristics, primarily through the addition of tetraethyl lead. More recently, however, non-leaded gasolines with fairly good antiknock characteristics have been developed in an effort to reduce atmospheric contamination.

Some of the most important ways in which the actual open-cycle spark-ignition engine deviates from the air-standard cycle are as follows:

1. The specific heats of the actual gases increase with an increase in temperature.
2. The combustion process replaces the heat-transfer process at high temperature, and combustion may be incomplete.
3. Each mechanical cycle of the engine involves an inlet and an exhaust process, and due to the pressure drop through the valves a certain amount of work is required to charge the cylinder with air and exhaust the products of combustion.
4. There will be considerable heat transfer between the gases in the cylinder and the cylinder walls.
5. There will be irreversibilities associated with pressure and temperature gradients.

Example 10.9

The compression ratio in an air-standard Otto cycle is 8. At the beginning of the compression stroke the pressure is 14.7 lbf/in.^{2} and the temperature is 60 F. The heat transfer to the air per cycle is 800 Btu/lbm air. Determine:

1. The pressure and temperature at the end of each process of the cycle.
2. The thermal efficiency.
3. The mean effective pressure.

Designating the states as in Fig. 10.19 we have

$$P_1 = 14.7\ \text{lbf/in.}^2 \qquad T_1 = 520\ \text{R}$$

$$v_1 = \frac{53.34 \times 520}{14.7 \times 144} = 13.08\ \text{ft}^3/\text{lbm}$$

$$\frac{T_2}{T_1} = \left(\frac{V_1}{V_2}\right)^{k-1} = 8^{0.4} = 2.3 \qquad T_2 = 2.3(520) = 1197\ \text{R}$$

$$\frac{P_2}{P_1} = \left(\frac{V_1}{V_2}\right)^{k} = 8^{1.4} = 18.4 \qquad P_2 = 18.4(14.7) = 270.3\ \text{lbf/in.}^2$$

$$v_2 = \frac{13.08}{8} = 1.637\ \text{ft}^3/\text{lbm}$$

$$ {}_2q_3 = C_v(T_3 - T_2) = 800 \text{ Btu/lbm} $$

$$ T_3 - T_2 = \frac{800}{0.171} = 4690 \text{ R} \qquad T_3 = 1197 + 4690 = 5887 \text{ R} $$

$$ \frac{T_3}{T_2} = \frac{P_3}{P_2} = \frac{5887}{1197} = 4.92 \qquad P_3 = 4.92(270.3) = 1331 \text{ lbf/in.}^2 $$

$$ \frac{T_3}{T_4} = \left(\frac{V_4}{V_3}\right)^{k-1} = 8^{0.4} = 2.3 \qquad T_4 = \frac{5887}{2.3} = 2558 \text{ R} $$

$$ \frac{P_3}{P_4} = \left(\frac{V_4}{V_3}\right)^{k} = 8^{1.4} = 18.4 \qquad P_4 = \frac{1331}{18.4} = 73.0 \text{ lbf/in.}^2 $$

$$ \eta_{\text{th}} = 1 - \frac{1}{r_v^{k-1}} = 1 - \frac{1}{8^{0.4}} = 1 - \frac{1}{2.3} = 1 - 0.435 = 0.565 $$

This can be checked by finding the heat rejected.

$$ {}_4q_1 = C_v(T_1 - T_4) = 0.171(520 - 2558) = -348 \text{ Btu/lbm} $$

$$ \eta_{\text{th}} = 1 - \frac{348}{800} = 1 - 0.435 = 0.565 $$

$$ w_{\text{net}} = 800 - 348 = 452 \text{ Btu/lbm} = (v_1 - v_2) \text{ mep} $$

$$ \text{mep} = \frac{452 \times 778}{(13.08 - 1.637)144} = 213.5 \text{ lbf/in.}^2 $$

Note how much higher this mean effective pressure is than for the Carnot cycle of Example 10.8. A low mean effective pressure means a large piston displacement for a given power output, and a large piston displacement means high frictional losses in an actual engine. In fact, the mean effective pressure of the Carnot cycle of Example 10.8 would scarcely overcome the friction in an actual engine.

10.13 The Air-Standard Diesel Cycle

The air-standard Diesel cycle is shown in Fig. 10.21. This is the ideal cycle for the Diesel engine, which is also called the compression-ignition engine.

In this cycle the heat is transferred to the working fluid at constant pressure. This process corresponds to the injection and burning of the

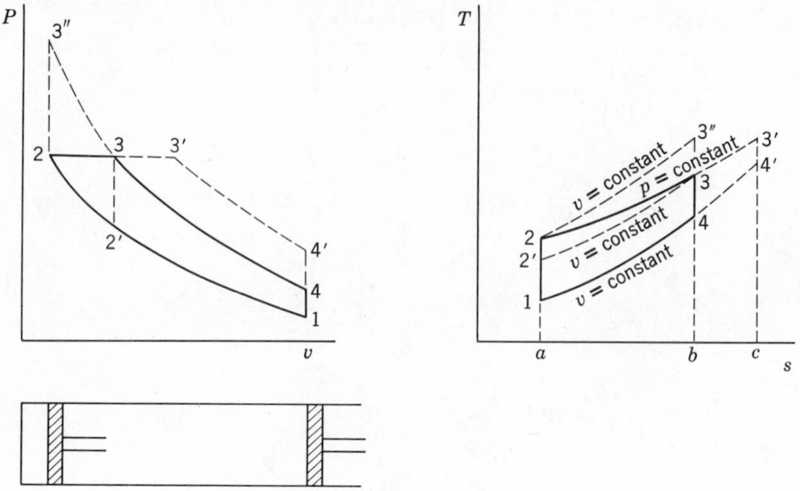

Fig. 10.21 The air-standard Diesel cycle.

fuel in the actual engine. Since the gas is expanding during the heat addition in the air-standard cycle, the heat transfer must be just sufficient to maintain constant pressure. When state 3 is reached the heat addition ceases and the gas undergoes an isentropic expansion, process 3–4, until the piston reaches crank-end dead center. As in the air-standard Otto cycle, a constant-volume rejection of heat at crank-end dead center replaces the exhaust and intake processes of the actual engine.

The efficiency of the Diesel cycle is given by the relation

$$\eta_{\text{th}} = 1 - \frac{Q_L}{Q_H} = 1 - \frac{C_v(T_4 - T_1)}{C_p(T_3 - T_2)} = 1 - \frac{T_1(T_4/T_1 - 1)}{kT_2(T_3/T_2 - 1)} \tag{10.5}$$

It is important to note that the isentropic compression ratio is greater than the isentropic expansion ratio in the Diesel cycle. Also, for a given state before compression and a given compression ratio (i.e., given states 1 and 2) the cycle efficiency decreases as the maximum temperature increases. This is evident from the *T-s* diagram, because the constant-pressure and constant-volume lines converge, and increasing the temperature from 3 to 3′ requires a large addition of heat (area 3–3′–*c*–*b*–3) and results in a relatively small increase in work (area 3–3′–4′–4–3).

There are a number of comparisons between the Otto cycle and the Diesel cycle, but here we will note only two. Consider Otto cycle 1–2–3″–4–1 and Diesel cycle 1–2–3–4–1, which have the same state at the beginning of the compression stroke and the same piston displacement and compression ratio. It is evident from the *T-s* diagram that the Otto cycle has the higher efficiency. In practice, however, the Diesel engine can operate on a higher compression ratio than the spark-ignition engine. The reason is that in the spark-ignition engine an air-fuel mixture is compressed, and detonation (spark knock) becomes a serious problem if too high a compression ratio is used. This problem does not exist in the Diesel engine because only air is compressed during the compression stroke.

Therefore, we might compare an Otto cycle with a Diesel cycle and in each case select a compression ratio that might be achieved in practice. Such a comparison can be made by considering Otto cycle 1–2′–3–4–1 and Diesel cycle 1–2–3–4–1. The maximum pressure and temperature is the same for both cycles, which means that the Otto cycle has a lower compression ratio than the Diesel cycle. It is evident from the *T-s* diagram that in this case the Diesel cycle has the higher efficiency. Thus, the conclusions drawn from a comparison of these two cycles must always be related to the basis on which the comparison has been made.

The actual compression-ignition open cycle differs from the air-standard Diesel cycle in much the same way that the spark-ignition open cycle differs from the air-standard Otto cycle.

Example 10.10

An air-standard Diesel cycle has a compression ratio of 15, and the heat transferred to the working fluid per cycle is 800 Btu/lbm. At the beginning of the compression process the pressure is 14.7 lbf/in.^{2} and the temperature is 60 F. Determine:

1. The pressure and temperature at each point in the cycle.
2. The thermal efficiency.
3. The mean effective pressure.

Designating the cycle as in Fig. 10.21 we have

$$P_1 = 14.7 \text{ lbf/in.}^2 \quad T_1 = 520 \text{ R}$$

$$v_1 = \frac{53.34 \times 520}{14.7 \times 144} = 13.08 \text{ ft}^3\text{/lbm}$$

$$v_2 = \frac{v_1}{15} = \frac{13.08}{15} = 0.872 \text{ ft}^3\text{/lbm}$$

$$\frac{T_2}{T_1} = \left(\frac{V_1}{V_2}\right)^{k-1} = 15^{0.4} = 2.955$$

$$T_2 = 2.955(520) = 1535 \text{ R}$$

$$\frac{P_2}{P_1} = \left(\frac{V_1}{V_2}\right)^{k} = 15^{1.4} = 44.2$$

$$P_2 = 44.2(14.7) = 650 \text{ lbf/in.}^2$$

$$q_H = {}_2q_3 = C_p(T_3 - T_2) = 800 \text{ Btu/lbm}$$

$$T_3 - T_2 = \frac{800}{0.24} = 3333 \text{ R}$$

$$T_3 = 3333 + 1535 = 4868 \text{ R}$$

$$\frac{V_3}{V_2} = \frac{T_3}{T_2} = \frac{4868}{1535} = 3.17$$

$$v_3 = 3.17(0.872) = 2.77 \text{ ft}^3\text{/lbm}$$

$$\frac{T_3}{T_4} = \left(\frac{V_4}{V_3}\right)^{k-1} = \left(\frac{13.08}{2.83}\right)^{0.4} = 1.860 \quad T_4 = \frac{4868}{1.860} = 2620 \text{ R}$$

$$q_L = {}_4q_1 = C_v(T_1 - T_4) = 0.171(520 - 2620) = -359 \text{ Btu/lbm}$$

$$w_{\text{net}} = 800 - 359 = 441 \text{ Btu/lbm}$$

$$\eta_{\text{th}} = \frac{w_{\text{net}}}{q_H} = \frac{441}{800} = 0.551$$

$$\text{mep} = \frac{w_{\text{net}}}{v_1 - v_2} = \frac{441 \times 778}{(13.08 - 0.87)144} = 195 \text{ lbf/in.}^2$$

10.14 The Ericsson and Stirling Cycles

We shall briefly consider these two cycles, not so much because of their extensive use, but rather because they serve to demonstrate how a regenerator can often be incorporated in a cycle to give a significant increase in efficiency. This principle finds extensive application in turbines as well as in certain reciprocating devices. The Stirling cycle is shown on the *P-v* and *T-s* diagrams of Fig. 10.22. Heat is transferred to the working fluid during the constant-volume process 2–3 and during the isothermal expansion process 3–4. Heat is rejected during the constant-volume process 4–1 and during the isothermal compression

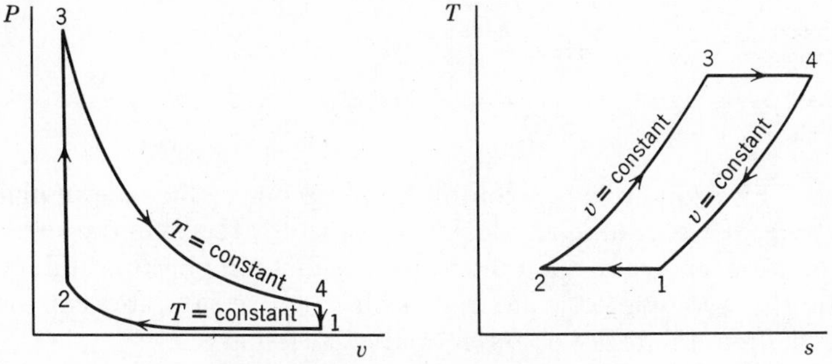

Fig. 10.22 The air-standard Stirling cycle.

process 1–2. The significance of this cycle in conjunction with a regenerator is discussed in the next paragraph.

The Ericsson cycle is shown on the *P-v* and *T-s* diagrams of Fig. 10.23. This cycle differs from the Stirling cycle in that the constant-volume processes of the Stirling cycle are replaced by constant-pressure processes.

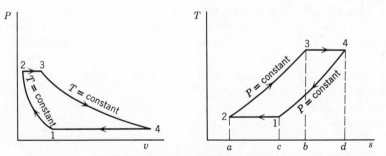

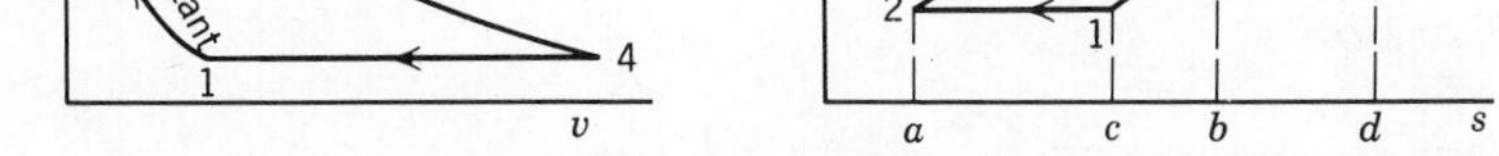

Fig. 10.23 The air-standard Ericsson cycle.

cesses. In both cycles there is an isothermal compression and expansion.

The importance of both cycles is the possibility of including a regenerator; by so doing the air-standard Stirling and Ericsson cycles may have an efficiency equal to that of a Carnot cycle operating between the same temperatures. This may be demonstrated by considering Fig. 10.24, in

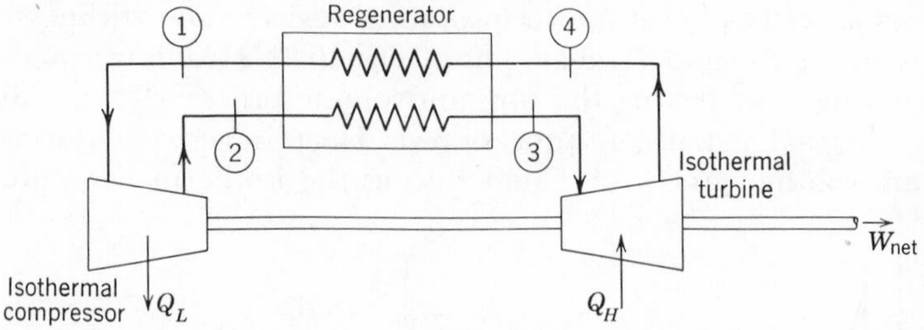

Fig. 10.24 Schematic arrangement of an engine operating on the Ericsson cycle and utilizing a regenerator.

which the Ericsson cycle is accomplished in a device that is essentially a gas turbine. If we assume an ideal heat-transfer process in the regenerator, i.e., no pressure drop and an infinitesimal temperature difference between the two streams, and reversible compression and expansion processes, then this device operates on the Ericsson cycle.

Note that the heat transfer to the gas between states 2 and 3, area $23ba2$, is exactly equal to the heat transfer from the gas between states 4 and 1, area $14dc1$. Thus Q_H all takes place in the isothermal turbine between states 3 and 4, and Q_L takes place in the isothermal compressor between states 1 and 2. Since all the heat is supplied and rejected isothermally, the efficiency of this cycle will equal the efficiency of a Carnot cycle operating between the same temperatures. A similar cycle could be developed which would approximate the Stirling cycle.

The difficulties in achieving such a cycle are primarily those associated with heat transfer. It is difficult to achieve an isothermal compression or expansion in a machine operating at a reasonable speed, and there will be pressure drops in the regenerator and a temperature difference between the two streams flowing through the regenerator. However, the gas turbine with intercooling and regenerators, which is described in Section 10.17, is a practical attempt to approach the Ericsson cycle. There have also been attempts to approach the Stirling cycle with the use of regenerators.

10.15 The Brayton Cycle

The air-standard Brayton cycle is the ideal cycle for the simple gas turbine. The simple open-cycle gas turbine utilizing an internal-combustion process and the simple closed-cycle gas turbine, which utilizes heat-transfer processes, are both shown schematically in Fig. 10.25. The air-standard Brayton cycle is shown on the P-v and T-s diagrams of Fig. 10.26.

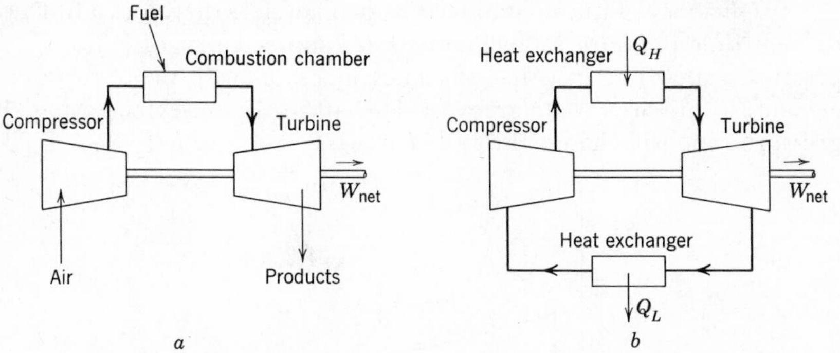

Fig. 10.25 A gas turbine operating on the Brayton cycle. (*a*) Open cycle. (*b*) Closed cycle.

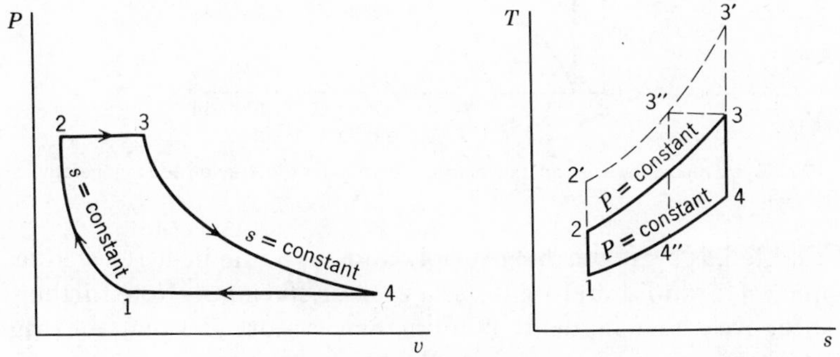

Fig. 10.26 The air-standard Brayton cycle.

The efficiency of the air-standard Brayton cycle is found as follows:

$$\eta_{\text{th}} = 1 - \frac{Q_L}{Q_H} = 1 - \frac{C_p(T_4 - T_1)}{C_p(T_3 - T_2)} = 1 - \frac{T_1(T_4/T_1 - 1)}{T_2(T_3/T_2 - 1)}$$

We note, however, that

$$\frac{P_3}{P_4} = \frac{P_2}{P_1}$$

$$\frac{P_2}{P_1} = \left(\frac{T_2}{T_1}\right)^{k/(k-1)} = \frac{P_3}{P_4} = \left(\frac{T_3}{T_4}\right)^{k/(k-1)}$$

$$\frac{T_3}{T_4} = \frac{T_2}{T_1} \quad \therefore \frac{T_3}{T_2} = \frac{T_4}{T_1} \quad \text{and} \quad \frac{T_3}{T_2} - 1 = \frac{T_4}{T_1} - 1$$

$$\eta_{\text{th}} = 1 - \frac{T_1}{T_2} = 1 - \frac{1}{(P_2/P_1)^{(k-1)/k}} \qquad (10.6)$$

The efficiency of the air-standard Brayton cycle is therefore a function of isentropic pressure ratio; Figure 10.27 shows a plot of efficiency vs. pressure ratio. The fact that efficiency increases with pressure ratio is evident from the *T-s* diagram of Fig. 10.26, because increasing the pressure ratio will change the cycle from 1–2–3–4–1 to 1–2′–3′–4–1. The

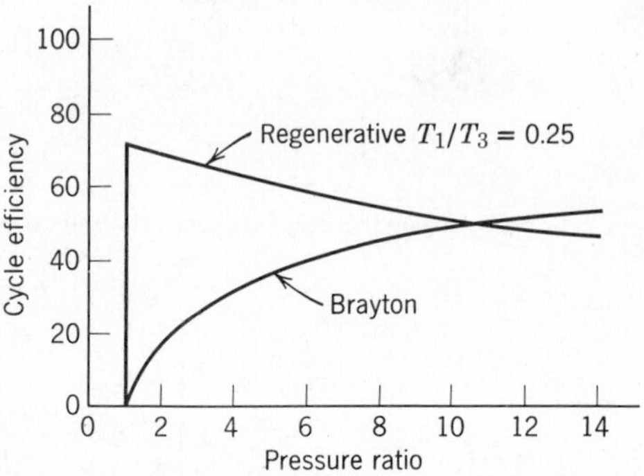

Fig. 10.27 Cycle efficiency as a function of pressure ratio for the Brayton and regenerative cycles.

latter cycle has a greater heat supply and the same heat rejected as the original cycle, and therefore it has a greater efficiency. Note further that the latter cycle has a higher maximum temperature (T_3') than the original cycle (T_3). In an actual gas turbine the maximum temperature of the gas entering the turbine is fixed by metallurgical considerations. Therefore, if we fix the temperature T_3 and increase the pressure ratio, the resulting cycle is 1–2′–3″–4″–1. This cycle would have a higher efficiency than the original cycle, but the work per pound of working fluid is thereby changed.

With the advent of nuclear reactors the closed-cycle gas turbine has become more important. Heat is transferred, either directly or via a second fluid, from the fuel in the nuclear reactor to the working fluid in the gas turbine. Heat is rejected from the working fluid to the surroundings.

The actual gas-turbine engine differs from the ideal cycle primarily because of irreversibilities in the compressor and turbine, and because of pressure drop in the flow passages and combustion chamber (or in the heat exchanger of a closed-cycle turbine). Thus, the state points in a simple open-cycle gas turbine might be as shown in Fig. 10.28.

The efficiencies of the compressor and turbine are defined in relation to isentropic processes. Designating the states as in Fig. 10.28, the definitions of compressor and turbine efficiencies are as follows:

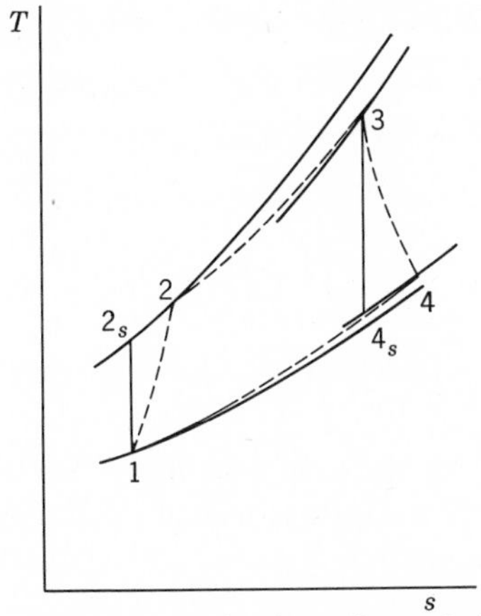

Fig. 10.28 Effect of inefficiencies on the gas-turbine cycle.

$$\eta_{\text{comp}} = \frac{h_{2s} - h_1}{h_2 - h_1} \tag{10.7}$$

$$\eta_{\text{turb}} = \frac{h_3 - h_4}{h_3 - h_{4s}} \tag{10.8}$$

One other important feature of the Brayton cycle is the large amount of compressor work (also called back work) compared to the turbine work. Thus, the compressor might require from 40 per cent to 80 per cent of the output of the turbine. This is particularly important when the actual cycle is considered, because the effect of the losses is to require a larger amount of compression work from a smaller amount of turbine work, and thus the over-all efficiency drops very rapidly with a decrease in the efficiencies of the compressor and turbine. In fact, if these efficiencies drop below about 60 per cent, all the work of the turbine will be required to drive the compressor, and the over-all efficiency will be zero. This is in sharp contrast to the Rankine cycle, where only 1 or 2 per cent of the turbine work is required to drive the pump. The reason for this is that for a reversible steady-state, steady-flow process with negligible change in kinetic and potential energy, the work is equal to $-\int v\,dP$. Because we are pumping a liquid in the Rankine cycle, the specific volume is very low compared to the specific volume of the gas in a gas turbine. This matter is illustrated in the following examples.

Example 10.11

In an air-standard Brayton cycle the air enters the compressor at 14.7 lbf/in.^{2}, 60 F. The pressure leaving the compressor is 70 lbf/in.^{2} and the maximum temperature in the cycle is 1600 F. Determine:

1. The pressure and temperature at each point in the cycle.
2. The compressor work, turbine work, and cycle efficiency.

Designating the points as in Fig. 10.26 we have

$$P_1 = P_4 = 14.7 \text{ lbf/in.}^2 \quad T_1 = 520 \text{ R}$$

$$P_2 = P_3 = 70 \text{ lbf/in.}^2 \qquad T_3 = 2060 \text{ R}$$

In the solution we consider successively a control surface around the compressor, the turbine, and the two heat exchangers. For the property relation we assume that air is an ideal gas with constant specific heat.

Control surface: Compressor

First law: $w_c = h_2 - h_1$. (Note that the compressor work w_c is here defined as work input to the compressor.)

Second law: $s_2 = s_1$

Therefore

$$\left(\frac{P_2}{P_1}\right)^{(k-1)/k} = \left(\frac{70}{14.7}\right)^{0.286} = \frac{T_2}{T_1} = 1.563, \quad T_2 = 1.563(520) = 814 \text{ R}$$

$$w_c = h_2 - h_1 = C_p(T_2 - T_1) = 0.24(814 - 520)$$

$$= 70.6 \text{ Btu/lbm}$$

Control surface: Turbine

First law: $w_t = h_3 - h_4$

Second law: $s_3 = s_4$

Therefore,

$$\left(\frac{P_3}{P_4}\right)^{(k-1)/k} = \left(\frac{70}{14.7}\right)^{0.286} = \frac{T_3}{T_4} = 1.563 \quad T_4 = \frac{2060}{1.563} = 1318 \text{ R}$$

$$w_t = h_3 - h_4 = C_p(T_3 - T_4) = 0.24(2060 - 1318)$$

$$= 178.1 \text{ Btu/lbm}$$

$$w_{\text{net}} = w_t - w_c = 178.1 - 70.6 = 107.5 \text{ Btu/lbm}$$

From a first law analysis of the two heat exchangers we have

$$q_H = h_3 - h_2 = C_p(T_3 - T_2) = 0.24(2060 - 814) = 299.0 \text{ Btu/lbm}$$

$$q_L = h_4 - h_1 = C_p(T_4 - T_1) = 0.24(1318 - 520) = 191.5 \text{ Btu/lbm}$$

$$\eta_{\text{th}} = \frac{w_{\text{net}}}{q_H} = \frac{107.5}{299.0} = 36.0\%$$

This may be checked by using Eq. 10.6

$$\eta_{\text{th}} = 1 - \frac{1}{(P_2/P_1)^{(k-1)/k}} = 1 - \frac{1}{(70/14.7)^{0.286}} = 1 - \frac{1}{1.563}$$

$$= 1 - 0.64 = 36.0\%$$

Example 10.12

Consider a gas turbine with air entering the compressor under the same conditions as in Example 10.11, and leaving at a pressure of 70 lbf/in.^{2}. The maximum temperature is 1600 F. Assume a compressor efficiency of 80 per cent, a turbine efficiency of 85 per cent, and a pressure drop between the compressor and turbine of 2 lbf/in.^{2}. Determine the compressor work, turbine work, and cycle efficiency.

Designating the states as in Fig. 10.28 and proceeding as we did in Example 10.11 we have

Control surface: Compressor

First law: $w_c = h_2 - h_1$

Second law: $s_{2s} = s_1$

$$\eta_c = \frac{h_{2s} - h_1}{h_2 - h_1}$$

$$\left(\frac{P_2}{P_1}\right)^{(k-1)/k} = \frac{T_{2s}}{T_1} = \left(\frac{70}{14.7}\right)^{0.286} = 1.563 \quad T_{2s} = 1.563(520) = 814 \text{ R}$$

$$\eta_c = \frac{h_{2s} - h_1}{h_2 - h_1} = \frac{T_{2s} - T_1}{T_2 - T_1} = \frac{814 - 520}{T_2 - T_1} = 0.80$$

$$T_2 - T_1 = \frac{294}{0.80} = 367 \text{ R} \quad T_2 = 520 + 367 = 887 \text{ R}$$

$$w_c = h_2 - h_1 = C_p(T_2 - T_1) = 0.24(887 - 520)$$

$$= 87.9 \text{ Btu/lbm}$$

Control surface: Turbine

First law: $w_t = h_3 - h_4$

Second law: $s_3 = s_{4s}$

$$\eta_t = \frac{h_3 - h_4}{h_3 - h_{4s}}$$

$$P_3 = P_2 - \text{pressure drop} = 70 - 2 = 68 \text{ lbf/in.}^2$$

$$\left(\frac{P_3}{P_4}\right)^{(k-1)/k} = \frac{T_3}{T_{4s}} = \left(\frac{68}{14.7}\right)^{0.286} = 1.550 \quad T_{4s} = \frac{2060}{1.550} = 1330 \text{ R}$$

$$\eta_t = \frac{h_3 - h_4}{h_3 - h_{4s}} = \frac{T_3 - T_4}{T_3 - T_{4s}} = 0.85$$

$$T_3 - T_4 = 0.85(2060 - 1330) = 620 \text{ R}$$

$$T_4 = 2060 - 620 = 1440 \text{ R}$$

$$w_t = h_3 - h_4 = C_p(T_3 - T_4) = 0.24(2060 - 1440) = 148.8$$

$$w_{\text{net}} = w_t - w_c = 148.8 - 87.9 = 60.9 \text{ Btu/lbm}$$

Control surface: Combustion Chamber

First law: $q_H = h_3 - h_2 = C_p(T_3 - T_2)$

$$= 0.24(2060 - 887) = 281.6 \text{ Btu/lbm}$$

$$\eta_{\text{th}} = \frac{w_{\text{net}}}{q_H} = \frac{60.9}{281.6} = 21.6\%$$

The following comparisons can be made between Examples 10.11 and 10.12.

	w_c	w_t	w_{net}	q_H	η_{th}
Example 10.11 (Ideal)	70.6	178.1	107.5	299.0	36.0
Example 10.12 (Actual)	87.9	148.8	60.9	281.6	21.6

As stated previously the result of the irreversibilities is to decrease the turbine work and increase the compressor work. Since the net work is the difference between these two it decreases very rapidly as compressor and turbine efficiencies decrease. The development of compressors and turbines of high efficiency is therefore an important aspect of the development of gas turbines.

Note also that in the ideal cycle (Example 10.11) about 40 per cent of the turbine work is required to drive the compressor and 60 per cent is delivered as net work. In the actual turbine (Example 10.12) 59 per cent of the turbine work is required to drive the compressor and 41 per cent is delivered as net work. Thus, if the net power of this unit is to be 10,000 hp, a 25,000-hp turbine and a 15,000-hp compressor are required. This demonstrates the statement that a gas turbine has a high back-work ratio.

10.16 The Simple Gas-Turbine Cycle with Regenerator

The efficiency of the gas-turbine cycle may be improved by introducing a regenerator. The simple open-cycle gas-turbine cycle with regenerator is shown in Fig. 10.29, and the corresponding ideal air-

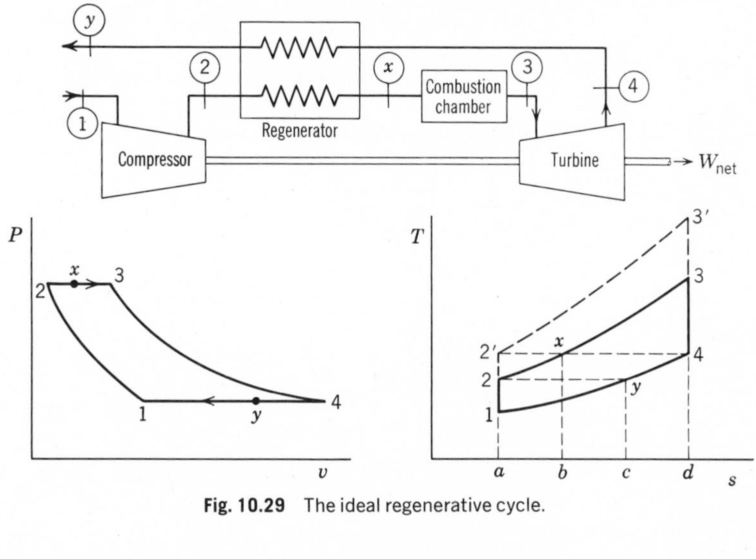

Fig. 10.29 The ideal regenerative cycle.

standard cycle with regenerator is shown on the *P-v* and *T-s* diagrams. Note that in cycle 1–2–*x*–3–4–*y*–1, the temperature of the exhaust gas leaving the turbine in state 4 is higher than the temperature of the gas leaving the compressor. Therefore, heat can be transferred from the exhaust gases to the high-pressure gases leaving the compressor. If this is done in a counterflow heat exchanger, which is known as a regenerator, the temperature of the high-pressure gas leaving the regenerator, T_x, may, in the ideal case, have a temperature equal to T_4, the temperature of the gas leaving the turbine. In this case heat transfer from the external source is necessary only to increase the temperature from T_x to T_3, and this heat transfer is represented by area *x*–3–*d*–*b*–*x*. Area *y*–1–*a*–*c*–*y* represents the heat rejected.

The influence of pressure ratio on the simple gas-turbine cycle with regenerator is shown by considering cycle 1–2′–3′–4–1. In this cycle the temperature of the exhaust gas leaving the turbine is just equal to the temperature of the gas leaving the compressor; therefore there is no possibility of utilizing a regenerator. This may be shown more exactly by determining the efficiency of the ideal gas-turbine cycle with regenerator.

The efficiency of this cycle with regeneration is found as follows, where the states are as given in Fig. 10.29.

$$\eta_{th} = \frac{w_{net}}{q_H} = \frac{w_t - w_c}{q_H}$$

$$q_H = C_p(T_3 - T_x)$$

$$w_t = C_p(T_3 - T_4)$$

But for ideal regenerator $T_4 = T_x$, and therefore $q_H = w_t$. Therefore,

$$\eta_{th} = 1 - \frac{w_c}{w_t} = 1 - \frac{C_p(T_2 - T_1)}{C_p(T_3 - T_4)}$$

$$= 1 - \frac{T_1(T_2/T_1 - 1)}{T_3(1 - T_4/T_3)} = 1 - \frac{T_1\,[(P_2/P_1)^{(k-1)/k} - 1]}{T_3\,[1 - (P_1/P_2)^{(k-1)/k}]}$$

$$\eta_{th} = 1 - \frac{T_1}{T_3}\left(\frac{P_2}{P_1}\right)^{(k-1)/k}$$

Thus, we see that for the ideal cycle with regeneration the thermal efficiency depends not only upon the pressure ratio, but also upon the ratio of the minimum to maximum temperature. We also note, in contrast to the Brayton cycle, that the efficiency decreases with an increase in pressure ratio. The thermal efficiency vs. pressure ratio for this cycle is plotted in Fig. 10.27 for a value of

$$\frac{T_1}{T_3} = 0.25$$

The effectiveness or efficiency of a regenerator is given by the term regenerator efficiency. This may best be defined by reference to Fig. 10.30. State x represents the high-pressure gas leaving the regenerator. In the ideal regenerator there would be only an infinitesimal temperature difference between the two streams, and the high-pressure gas would

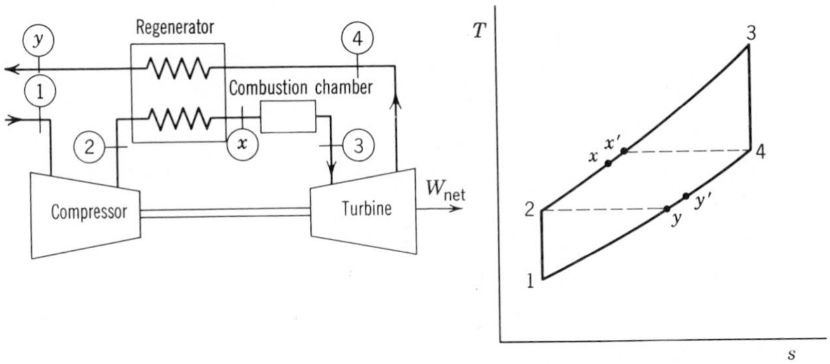

Fig. 10.30 Temperature-entropy diagram to illustrate the definition of regenerator efficiency.

leave the regenerator at temperature T'_x, and $T'_x = T_4$. In an actual regenerator, which must operate with a finite temperature difference, T_x, the actual temperature leaving the regenerator is therefore less than T'_x. The regenerator efficiency is defined by

$$\eta_{\text{reg}} = \frac{h_x - h_2}{h'_x - h_2}$$

If the specific heat is assumed to be constant, the regenerator efficiency is also given by the relation

$$\eta_{\text{reg}} = \frac{T_x - T_2}{T'_x - T_2}$$

It should also be pointed out that a higher efficiency can be achieved by using a regenerator with greater heat-transfer area. However, this also increases the pressure drop, which represents a loss, and both the pressure drop and the regenerator efficiency must be considered in determining which regenerator gives maximum thermal efficiency for the cycle. From an economic point of view, the cost of the regenerator must be weighed against the saving that can be effected by its use.

Example 10.13

If an ideal regenerator is incorporated into the cycle of Example 10.11, determine the thermal efficiency of the cycle.

Designating the states as in Fig. 10.30,

$$T_x = T_4 = 1318 \text{ R}$$
$$q_H = h_3 - h_x = C_p(T_3 - T_x) = 0.24(2060 - 1318) = 178.1 \text{ Btu/lbm}$$
$$w_{\text{net}} = 107.5 \text{ Btu/lbm (from Example 10.11)}$$
$$\eta_{\text{th}} = \frac{107.5}{178.1} = 60.4\%$$

10.17 The Ideal Gas-Turbine Cycle Using Multistage Compression with Intercooling, Multistage Expansion with Reheating, and Regenerator

In Section 10.14 it was pointed out that when an ideal regenerator was incorporated into the Ericsson cycle, an efficiency equal to the efficiency of the corresponding Carnot cycle could be attained. It was also pointed out that it is impossible to even closely approach in practice the reversible isothermal compression and expansion required in the Ericsson cycle.

The practical approach to this cycle is to use multistage compression with intercooling between stages, multistage expansion with reheat between stages, and a regenerator. Figure 10.31 shows a cycle with two stages of compression and two stages of expansion. The air-standard cycle is shown on the corresponding *T-s* diagram. It may be shown that for this cycle the maximum efficiency is obtained if equal pressure ratios are maintained across the two compressors and the two turbines. In this ideal cycle it is assumed that the temperature of the air leaving the intercooler, T_3, is equal to the temperature of the air entering the first stage of compression, T_1, and that the temperature after reheating, T_8, is equal to the temperature entering the first turbine, T_6. Further, in the ideal cycle it is assumed that the temperature of the high-pressure air leaving the regenerator, T_5, is equal to the temperature of the low-pressure air leaving the turbine, T_9.

If a large number of stages of compression and expansion are used, it is evident that the Ericsson cycle is approached. This is shown in Fig. 10.32. In practice the economical limit to the number of stages is usually two or three. The turbine and compressor losses and pressure drops that have already been discussed would be involved in any actual unit employing this cycle.

There are a variety of ways in which the turbines and compressors using this cycle can be utilized. Two possible arrangements for closed cycles are shown in Fig. 10.33. One advantage frequently sought in a given arrangement is ease of control of the unit under various loads. Detailed discussion of this point, however, is beyond the scope of this text.

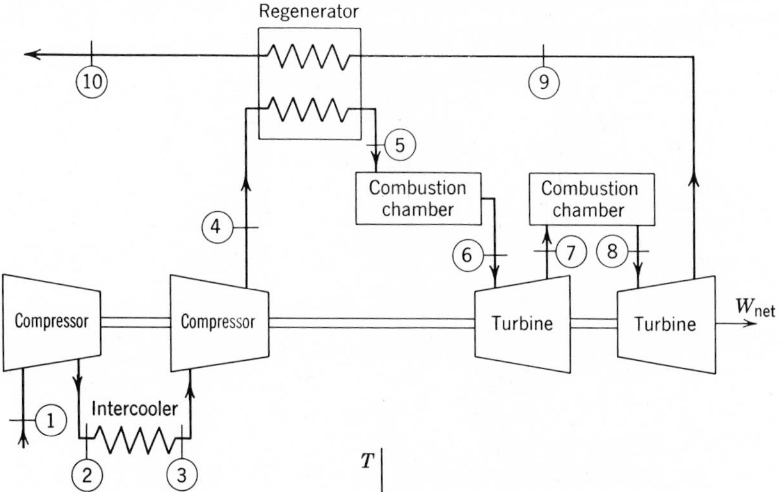

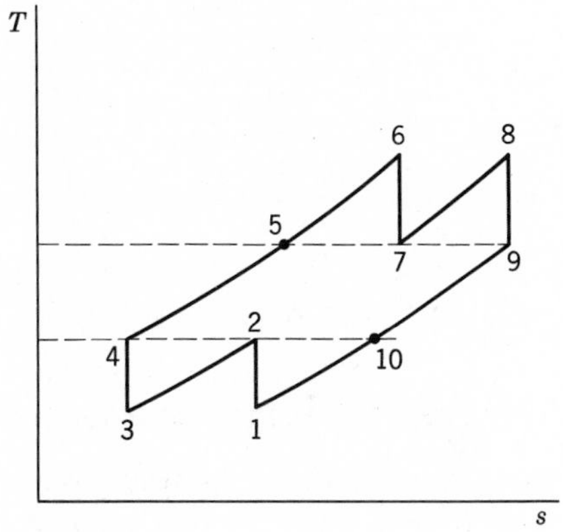

Fig. 10.31 The ideal gas-turbine cycle utilizing intercooling, reheat, and a regenerator.

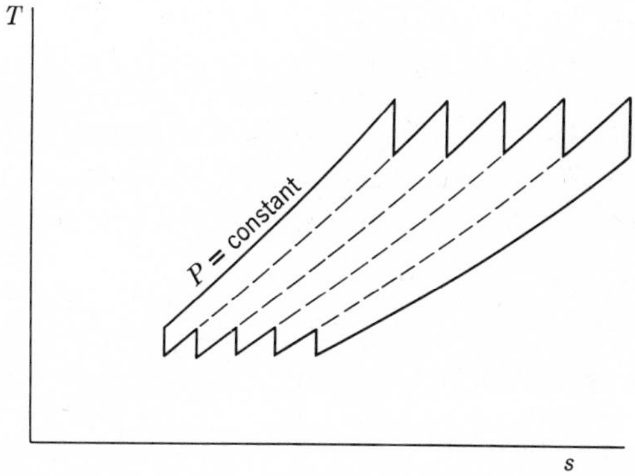

Fig. 10.32 Temperature-entropy diagram which shows how the gas-turbine cycle with many stages approaches the Ericsson cycle.

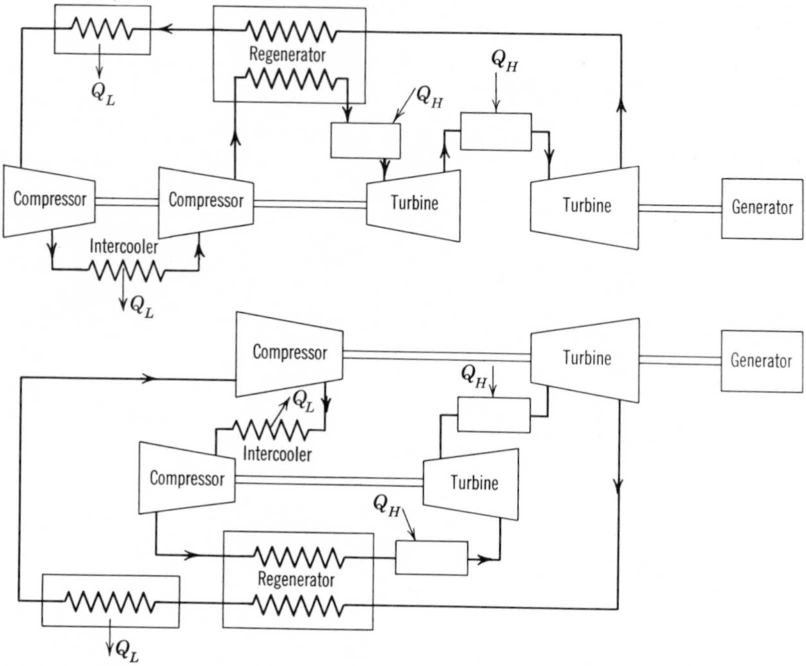

Fig. 10.33 Some arrangements of components that may be utilized in stationary gas turbine power plants.

10.18 The Air-Standard Cycle for Jet Propulsion

The last air-standard power cycle we will consider is utilized in jet propulsion. In this cycle the work done by the turbine is just sufficient to drive the compressor. The gases are expanded in the turbine to a pressure such that the turbine work is just equal to the compressor work. The exhaust pressure of the turbine will then be above that of the surroundings, and the gas can be expanded in a nozzle to the pressure of the surroundings. Since the gases leave at a high velocity, the change in momentum the gases undergo results in a thrust upon the aircraft in which the engine is installed. The air-standard cycle for this is shown in Fig. 10.34.

Since all the principles involved in this cycle have already been covered, the example that follows will conclude our consideration of air-standard power cycles.

Example 10.14

Consider an ideal cycle in which air enters the compressor at 14.7 lbf/in.^{2}, 60 F. The pressure leaving the compressor is 70 lbf/in.^{2} and the

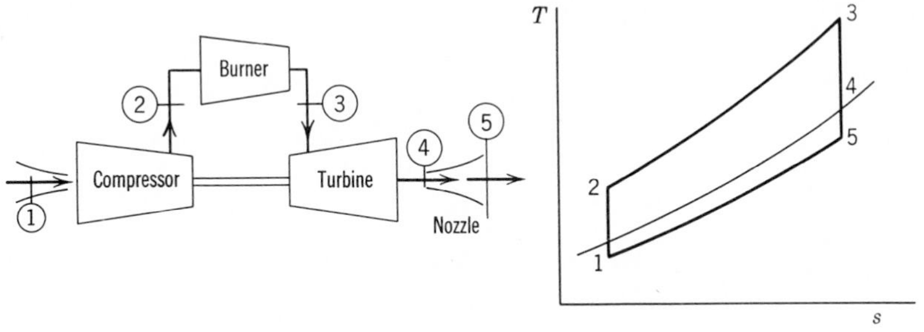

Fig. 10.34 The ideal gas-turbine cycle for a jet engine.

maximum temperature is 1600 F. The air expands in the turbine to such a pressure that the turbine work is just equal to the compressor work. On leaving the turbine the air expands in a reversible adiabatic process in a nozzle to 14.7 lbf/in.^{2} Determine the velocity of the air leaving the nozzle.

From Example 10.11 we have the following, where the states are as designated in Fig. 10.34.

$$P_1 = 14.7 \text{ lbf/in.}^2 \quad T_1 = 520 \text{ R}$$
$$P_2 = 70 \text{ lbf/in.}^2 \quad T_2 = 814 \text{ R}$$
$$w_c = 70.6 \text{ Btu/lbm}$$
$$P_3 = 70 \text{ lbf/in.}^2 \qquad T_3 = 2060 \text{ R}$$
$$w_c = w_t = C_p(T_3 - T_4)$$
$$= 70.6 \text{ Btu/lbm}$$

$$T_3 - T_4 = \frac{70.6}{0.24} = 294 \text{ R}$$

$$T_4 = 2060 - 294 = 1766 \text{ R}$$

$$\frac{T_3}{T_4} = \left(\frac{P_3}{P_4}\right)^{(k-1)/k} = \frac{2060}{1766} = 1.167$$

$$\frac{P_3}{P_4} = 1.715 \quad P_4 = \frac{70}{1.715} = 40.8 \text{ lbf/in.}^2$$

Next consider a control surface around the nozzle. Let us assume that the velocity of the air entering the nozzle is low.

First law: $$h_4 = h_5 + \frac{V_5^2}{2g_c}$$

Second law: $$s_4 = s_5$$

Therefore (from Example 10.11), $T_5 = 1318$ R

$$V_5^2 = 2g_cC_p(T_4 - T_5)$$
$$V_5^2 = 2 \times 32.17 \times 778 \times 0.24(1766 - 1318)$$
$$V_5 = 2318 \text{ ft/sec}$$

AIR-STANDARD REFRIGERATION CYCLE

10.19 The Air-Standard Refrigeration Cycle

The final section of this chapter concerns the air-standard refrigeration cycle. Its main use in practice is in the liquefaction of air and other gases and in certain special situations that require refrigeration.

The simplest form of the air-standard refrigeration cycle, which is essentially the reverse of the Brayton cycle, is shown in Fig. 10.35. The

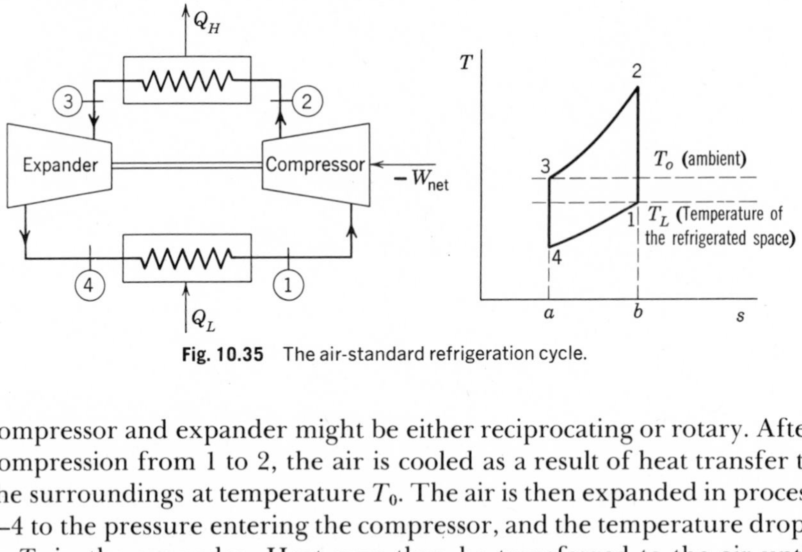

Fig. 10.35 The air-standard refrigeration cycle.

compressor and expander might be either reciprocating or rotary. After compression from 1 to 2, the air is cooled as a result of heat transfer to the surroundings at temperature T_0. The air is then expanded in process 3–4 to the pressure entering the compressor, and the temperature drops to T_4 in the expander. Heat may then be transferred to the air until temperature T_L is reached. The work for this cycle is represented by area 1–2–3–4–1, and the refrigeration effect is represented by area 4–1–b–a–4. The coefficient of performance is the ratio of these two areas.

In practice this cycle has been utilized for the cooling of aircraft in an open cycle, a simplified form of which is shown in Fig. 10.36. Upon leaving the expander the cool air is blown directly into the cabin thus providing the cooling effect where needed.

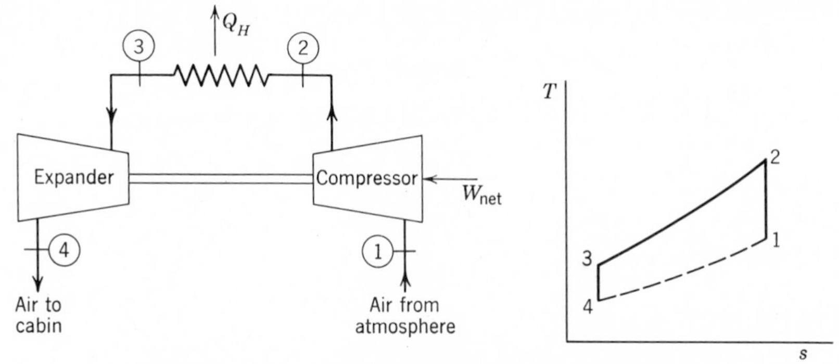

Fig. 10.36 An air refrigeration cycle that might be utilized for aircraft cooling.

When counterflow heat exchangers are incorporated, very low temperatures can be obtained. This is essentially the cycle used in low-pressure air liquefaction plants and in other liquefaction devices such as the Collins helium liquefier. The ideal cycle in this case is as shown in Fig. 10.37. It is evident that the expander operates at very low temperature, which presents unique problems to the designer in regard to lubrication and materials.

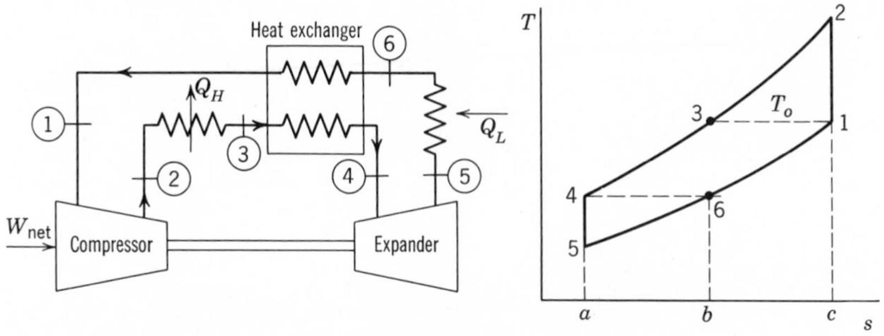

Fig. 10.37 The air refrigeration cycle utilizing a heat exchanger.

Example 10.15

Consider the simple air-standard refrigeration cycle of Fig. 10.35. Air enters the compressor at 14.7 lbf/in.^{2}, 0 F, and leaves at 80 lbf/in.^{2} Air enters the expander at 60 F. Determine:

1. The coefficient of performance for this cycle.
2. The rate at which air must enter the compressor in order to provide one ton of refrigeration.

$$P_1 = P_4 = 14.7 \text{ lbf/in.}^2; \quad T_1 = 460 \text{ R}$$
$$P_2 = P_3 = 80.0 \text{ lbf/in.}^2; \quad T_3 = 520 \text{ R}$$

Control surface: Compressor

First law: $w_c = h_2 - h_1$

Second law: $s_1 = s_2$

Therefore,

$$\frac{T_2}{T_1} = \left(\frac{P_2}{P_1}\right)^{(k-1)/k} = \left(\frac{80}{14.7}\right)^{0.286} = 1.624$$

$$T_2 = 460(1.624) = 747 \text{ R}$$

$$w_c = h_2 - h_1 = C_p(T_2 - T_1) = 0.24(747 - 460) = 68.9 \text{ Btu/lbm}$$

(w_c designates work into the compressor)

Control surface: Expander

First law: $w_t = h_3 - h_4$

Second law: $s_3 = s_4$

$$T_3 = 520 \text{ R}$$

$$\frac{T_3}{T_4} = \left(\frac{P_3}{P_4}\right)^{(k-1)/k} = \left(\frac{80}{14.7}\right)^{0.286} = 1.624$$

$$T_4 = \frac{520}{1.624} = 320 \text{ R}$$

$$w_t = h_3 - h_4 = 0.24(520 - 320) = 48.0 \text{ Btu/lbm}$$

(w_t designates work done by expander)

From the first law as applied to a control surface around each of the heat exchangers we conclude that

$$q_H = h_2 - h_3 = C_p(T_2 - T_3) = 0.24(747 - 520) = 54.5 \text{ Btu/lbm}$$

$$q_L = h_1 - h_4 = C_p(T_1 - T_4) = 0.24(460 - 320) = 33.6 \text{ Btu/lbm}$$

$$w_{\text{net}} = w_c - w_t = 68.9 - 48.0 = 20.9 \text{ Btu/lbm}$$

$$\beta = \frac{q_L}{w_{\text{net}}} = \frac{33.6}{20.9} = 1.61$$

$$1 \text{ ton of refrigeration} = 12{,}000 \text{ Btu/hr} = 200 \text{ Btu/min}$$

$$\frac{\text{lbm air/min}}{\text{ton refrigeration}} = \frac{200}{33.6} = 5.95$$

PROBLEMS

10.1 It is desired to determine the effect of turbine exhaust pressure on the performance of a steam Rankine cycle. Steam enters the turbine at 500 lbf/in.^{2}, 700 F. Calculate the cycle thermal efficiency and the moisture content of the steam leaving the turbine for turbine exhaust pressures of 1 lbf/in.^{2}, 2 lbf/in.^{2}, 5 lbf/in.^{2} and 14.7 lbf/in.^{2}. Plot thermal efficiency versus turbine exhaust pressure for the specified turbine inlet pressure and temperature.

10.2 It is desired to determine the effect of turbine inlet pressure on the performance of a steam Rankine cycle. Steam enters the turbine at 700 F and exhausts at 2 lbf/in.^{2}. Calculate the cycle thermal efficiency and the moisture content of the steam leaving the turbine for turbine inlet pressures of 100 lbf/in.^{2}, 500 lbf/in.^{2}, 1000 lbf/in.^{2}, 2000 lbf/in.^{2}, and saturated vapor. Plot thermal efficiency versus turbine inlet pressure for the specified turbine inlet temperature and exhaust pressure.

10.3 It is desired to determine the effect of turbine inlet temperature on the performance of a steam Rankine cycle. Steam enters the turbine at 500 lbf/in.^{2} and exhausts at 2 lbf/in.^{2}. Calculate the cycle thermal efficiency and the moisture content leaving the turbine for turbine inlet temperatures of saturated vapor, 700 F, 1000 F, and 1400 F. Plot thermal efficiency versus turbine inlet temperature for the specified turbine inlet pressure and exhaust pressure.

10.4 Consider a reheat cycle utilizing steam as the working fluid. Steam enters the high-pressure turbine at 500 lbf/in.^{2}, 700 F, and expands to 120 lbf/in.^{2}. It is then reheated to 700 F and then expanded to 2 lbf/in.^{2} in the low-pressure turbine. Calculate the cycle thermal efficiency and the moisture content leaving the low-pressure turbine.

10.5 Consider a regenerative cycle utilizing steam as the working fluid. Steam enters the turbine at 500 lbf/in.^{2}, 700 F and exhausts to the condenser at 2 lbf/in.^{2}. Steam is extracted at 120 lbf/in.^{2} and also at 20 lbf/in.^{2} for purposes of heating the feedwater in two open feedwater heaters. The feedwater leaves each heater at the temperature of the condensing steam. The appropriate pumps are used for the water leaving the condenser and the two feedwater heaters. Calculate the cycle thermal efficiency and the net work per lbm of steam.

10.6 Repeat Problem 10.5 assuming closed instead of open feedwater heaters. A single pump is used to pump the water leaving the condenser to 500 lbf/in.^{2}. Condensate from the high-pressure heater is drained through a trap to the low-pressure heater, and that from the low-pressure heater is drained through a trap to the condenser.

10.7 Consider a combined reheat and regenerative cycle utilizing steam as the working fluid. Steam enters the high-pressure turbine at 500 lbf/in.^{2}, 700 F, and is extracted for purposes of feedwater heating at 120 lbf/in.^{2}. The remainder of the steam is reheated at this pressure to 700 F and fed to the low-pressure turbine. Steam is extracted from the low-pressure turbine at 20 lbf/in.^{2} for feedwater heating. The condenser pressure is 2 lbf/in.^{2} and both feedwater heaters are open heaters. Calculate the cycle thermal efficiency and the net work per lbm of steam.

10.8 It is desired to study the influence of the number of feedwater heaters on thermal efficiency for a cycle in which the steam leaves the steam generator at 2500 lbf/in.^{2}, 1000 F, and which has a condenser pressure of 1 lbf/in.^{2}. Assume that open feedwater heaters are used. Determine the thermal efficiency for each of the following cycles.

(*a*) No feedwater heater.

(*b*) One feedwater heater which operates at 140 lbf/in.^{2}.

(*c*) Two feedwater heaters, one operating at 450 lbf/in.^{2} and the other at 35 lbf/in.^{2}.

10.9 A nuclear reactor to produce power is so designed that the maximum temperature in the steam cycle is 800 F. The minimum temperature of the steam cycle is 100 F. It is planned to build a steam power plant which has one open feedwater heater.

Select what you consider to be a reasonable cycle within these specifications, determine the ideal cycle efficiency, and explain why the particular pressures involved were selected.

10.10 Consider an ideal steam cycle that combines the reheat cycle and the regenerative cycle. The net power output of the turbine is 100,000 kw. Steam enters the high-pressure turbine at 1200 lbf/in.^{2}, 1000 F. After expansion to 90 lbf/in.^{2}, some of the steam goes to an open feedwater heater and the balance is reheated to 700 F, after which it expands to 1 lbf/in.^{2}.

(*a*) Draw a line diagram of the unit and show the state points on a *T-s* diagram.

(*b*) What is the steam flow rate to the high-pressure turbine?

(*c*) What hp motor is required to drive each of the pumps?

(*d*) If there is a 20 F rise in the temperature of the cooling water, what is the rate of flow of cooling water through the condenser?

(*e*) The velocity of the steam flowing from the turbine to the condenser is limited to a maximum of 400 ft/sec. What is the diameter of this connecting pipe?

10.11 Steam leaves the boiler of a power plant at 500 lbf/in.^{2}, 700 F, and when it enters the turbine it is at 475 lbf/in.^{2}, 650 F. The turbine efficiency is 85% and the exhaust pressure is 2 lbf/in.^{2}. Condensate leaves the condenser and enters the pump at 100 F, 2 lbf/in.^{2}. The pump efficiency is 75% and the discharge pressure is 550 lbf/in.^{2}. The feedwater enters the boiler at 530 lbf/in.^{2}, 95 F. Calculate:

(*a*) The thermal efficiency of the cycle.

(*b*) The irreversibility of the process between the boiler exit and the turbine inlet, assuming an ambient temperature of 77 F.

10.12 Consider the following reheat-cycle power plant. Steam enters the high-pressure turbine at 500 lbf/in.^{2}, 700 F, and expands to 70 lbf/in.^{2}, after which it is reheated to 700 F. The steam is then expanded through the low-pressure turbines to 1 lbf/in.^{2}. Liquid leaves the condenser at 90 F, is pumped to 500 lbf/in.^{2}, and returned to the boiler. Each turbine is adiabatic, with an efficiency of 85%, and the pump efficiency is 80%. If the total power output of the turbines is 1000 hp, determine:

(*a*) Mass flow rate of steam.

(*b*) Pump horsepower.

(*c*) Thermal efficiency of the unit.

10.13 Repeat Problem 10.8 assuming a turbine efficiency of 85%.

10.14 In a nuclear power plant heat is transferred in the reactor to liquid sodium. The liquid sodium is then pumped to a heat exchanger where heat is transferred to steam. The steam leaves this heat exchanger as saturated vapor at 800 lbf/in.^{2}, and is then superheated in an external gas fired superheater to 1200 F. The steam then enters the turbine, which has one extraction point at 60 lbf/in.^{2}, where steam flows to an open feedwater heater. The turbine efficiency is 75% and the condenser pressure is 1 lbf/in.^{2}.

Determine the heat transfer in the reactor and in the superheater to produce a power output of 80,000 kw.

10.15 Consider the preliminary design for a supercritical steam power plant cycle. The maximum pressure will be 4000 lbf/in.^{2}, and the maximum temperature will be 1100 F. The cooling water temperature is such that 1 lbf/in.^{2} pressure can be maintained in the condensers. Turbine efficiencies of at least 85% can be expected.

(*a*) Do you recommend any reheat for this cycle? If so, how many stages of reheat and at what pressures? Give reasons for your decisions.

(*b*) Do you recommend feedwater heaters? If so, how many and at what pressures would they operate? Would they be open or closed feedwater heaters?

(*c*) Estimate the thermal efficiency of the cycle which you recommend.

10.16 In an ideal refrigeration cycle the temperature of the condensing vapor is 110 F and the temperature during evaporation is 10 F. Determine the coefficient of performance of this cycle for the working fluids Freon-12 and ammonia.

10.17 A study is to be made of the influence on power requirements of the difference between the temperature of the refrigerant in the condenser and the temperature of the surroundings. For this purpose consider the ideal cycle of Fig. 10.15 with Freon-12 as the refrigerant. The temperature of the surroundings is 100 F and the temperature of the refrigerant in the evaporator is 10 F. Plot a curve of hp per ton of refrigeration for temperature differences of 0 to 100 F between the refrigerant in the condenser and the surroundings.

10.18 A study is to be made of the influence on power requirements of the temperature difference between the cold box and the refrigerant in the evaporator. For this purpose consider the ideal cycle of Fig. 10.15 with Freon-12 as the refrigerant. The temperature of the cold box is 10 F and the temperature of the refrigerant in the condenser is 160 F. Plot a curve of hp per ton of refrigeration for temperature differences of 0 to 50 F between the cold box and the refrigerant in the evaporator.

10.19 In a conventional refrigeration cycle which uses Freon-12 as the refrigerant, the temperature of the evaporating fluid is −10 F. It leaves the evaporator as saturated vapor at −10 F and enters the compressor. The pressure in the condenser is 200 lbf/in.². The liquid leaves the condenser and enters the expansion valve at a temperature of 100 F.

It is proposed to modify this cycle as shown in Fig. 10.38. In this case an additional exchanger is supplied in which the cold vapor leaving the evaporator cools the liquid which enters the expansion valve.

Compare the coefficient of performance of these two cycles.

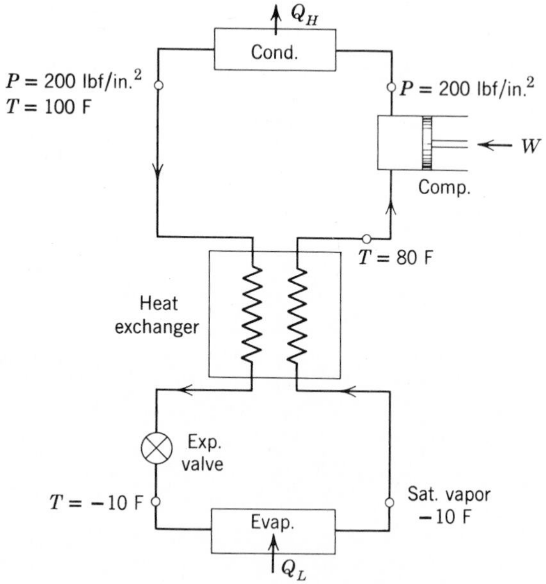

Fig. 10.38 Sketch for Problem 10.19.

10.20 Consider an ideal dual-loop heat-powered refrigeration cycle utilizing Freon-12, as is shown schematically in Fig. 10.39. Saturated vapor at 220 F leaves the boiler and expands in the turbine to the condenser pressure. Saturated vapor at 0 F leaves the evaporator and is compressed to the

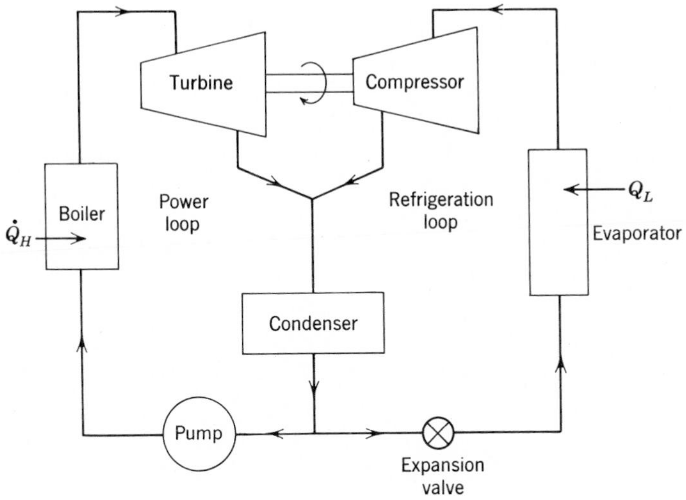

Fig. 10.39 Sketch for Problem 10.20.

condenser pressure. The ratio of flows through the two loops is such that the turbine produces just enough power to drive the compressor. The two exiting streams mix together and enter the condenser. Saturated liquid at 110 F leaving the condenser is then separated into two streams in the necessary proportions. Calculate:

(*a*) The ratio of mass flow rate through the power loop to that through the refrigeration loop.

(*b*) The performance of the cycle, in terms of the ratio $\dot{Q}_L/\dot{Q}_H$.

10.21 In an actual refrigeration cycle using Freon-12 as a working fluid the rate of flow of refrigerant is 300 lbm/hr. The refrigerant enters the compressor at 25 lbf/in.^{2}, 20 F and leaves at 175 lbf/in.^{2}, 170 F. The power input to the compressor is 2.5 hp. The refrigerant enters the expansion valve at 165 lbf/in.^{2}, 100 F and leaves the evaporator at 27 lbf/in.^{2}, 10 F. Determine:

(*a*) The irreversibility during the compression process.

(*b*) The refrigeration capacity in tons.

(*c*) The coefficient of performance for this cycle.

10.22 In both power and refrigeration cycles that operate over a wide temperature range it is frequently advantageous to utilize more than one working fluid. In refrigeration cycles this is commonly referred to as a cascade system. Consider such a system comprised of two ideal refrigeration cycles, as shown in Fig. 10.40. The high-temperature cycle uses Freon-12 and the low-temperature cycle uses Freon-13, with conditions as shown in the diagram. Calculate:

(*a*) The ratio of mass flow rates through the two cycles.

(*b*) The coefficient of performance of the system.

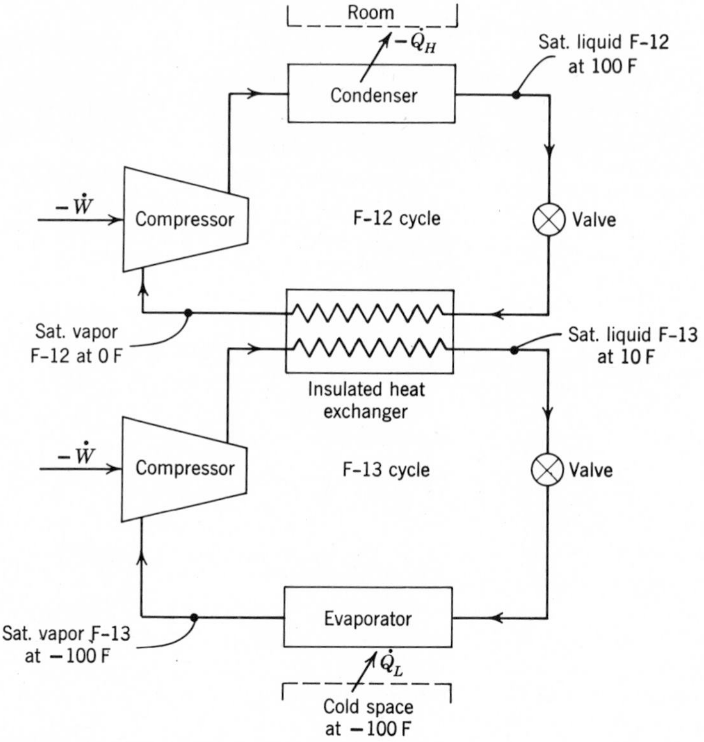

Fig. 10.40 Sketch for Problem 10.22.

10.23 An ammonia-absorption system has an evaporator temperature of 10 F and a condenser temperature of 120 F. The generator temperature is 302 F. In this cycle, 0.42 Btu is transferred to the ammonia in the evaporator for each Btu transferred to the ammonia solution in the generator from the high-temperature source.

We wish to compare the performance of this cycle with the performance of a similar vapor-compression cycle. To do this, assume that a reservoir is available at 302 F, and that heat is transferred from this reservoir to a reversible engine which rejects heat to the surroundings at 77 F. This work is then used to drive an ideal vapor-compression system with ammonia as the refrigerant. Compare the amount of refrigeration that can be achieved per Btu from the high-temperature source in this case with the 0.42 Btu that can be achieved in the adsorption system.

10.24 A stoichiometric mixture of fuel and air has an enthalpy of combustion of approximately -1200 Btu/lbm of mixture. In order to approximate an actual spark-ignition engine using such a mixture, consider an air-standard Otto cycle that has a heat addition of 1200 Btu per lbm of air, a compression ratio of 8, and a pressure and temperature at the beginning of the compression process of 14.7 lbf/in.^{2}, 60 F. Determine the following:

The maximum pressure and temperature for this cycle.
The thermal efficiency.
The mean effective pressure.

Assume:

(*a*) Constant specific heat as given in Table B.6.

(*b*) Variable specific heat. The Air Tables B.8 are recommended for this calculation.

10.25 It has been found that the power stroke in an internal combustion engine can be approximated reasonably as a polytropic process with a value of the polytropic exponent n somewhat smaller than the specific heat ratio k. Repeat Problem 10.24, assuming that the expansion process is reversible and polytropic (instead of the isentropic expansion of the Otto Cycle) with a value of n equal to 1.25.

10.26 An air-standard Diesel cycle has a compression ratio of 14. The pressure at the beginning of the compression stroke is 14.7 lbf/in.^{2}, and the temperature is 60 F. The maximum temperature is 4500 R. Determine the thermal efficiency and the mean effective pressure for this cycle.

10.27 Consider an air-standard Ericsson cycle with an ideal regenerator incorporated. The temperature at which heat is supplied is 2200 R. Heat is rejected at 540 R. The pressure at the beginning of the isothermal compression process is 14.7 lbf/in.^{2}. Determine the compressor and turbine work per lbm of air and the thermal efficiency of the cycle. $Q_H = 400$ Btu/lbm.

10.28 The pressure ratio across the compressor of an air-standard Brayton cycle is 4 to 1. The pressure of the air entering the compressor is 14.7 lbf/in.^{2}, and the temperature is 60 F. The maximum temperature in the cycle is 1500 F. The rate of air flow is 20 lbm/sec. Determine the following, assuming constant specific heat:

(*a*) The compressor work, turbine work, and thermal efficiency of the cycle.

(*b*) If this cycle were utilized for a reciprocating machine, what would be the mean effective pressure? Would you recommend this cycle for a reciprocating machine?

10.29 Repeat Problem 10.28, assuming variable specific heat for the air.

10.30 An ideal regenerator is incorporated into the air-standard Brayton cycle of Problem 10.28. Determine the thermal efficiency of the cycle with this modification.

10.31 A stationary gas-turbine power plant operates on the Brayton cycle and delivers 20,000 hp to an electric generator. The maximum temperature is 1540 F and the minimum temperature is 60 F. The minimum pressure is 14.0 lbf/in.^{2} and the maximum pressure is 60 lbf/in.^{2}.

(*a*) What is the power output of the turbine?

(*b*) What fraction of the output of the turbine is used to drive the compressor?

(*c*) What is the mass rate of air flow to the compressor per min? What is the volume rate of air flow to the compressor per min?

10.32 Repeat Problem 10.31 assuming that a regenerator of 75% efficiency is added to the cycle, and the efficiency of the compressor is 80% and the efficiency of the turbine is 85%.

10.33 A gas turbine is to be used for pumping natural gas through a cross-country pipeline. The required power input to the natural gas compressor is 1000 hp. It has been decided to use a gas turbine to provide the necessary power input. A simple open cycle will be used with a regenerator of 50% efficiency. Since a supply of fuel is readily available, and since some of these units may be located in relatively isolated places, low maintenance costs are more important than a high efficiency.

Select a cycle which you would recommend, making appropriate assumptions for compressor and turbine efficiencies, and determine (for the gas turbine unit) the power output of the turbine, the power input to the compressor, and the efficiency.

10.34 Consider a gas-turbine cycle with two stages of compression and two stages of expansion. The pressure ratio across each turbine and each compressor is 2. The temperature entering each compressor is 60 F and the temperature entering each turbine is 1500 F. An ideal regenerator is incorporated into the cycle. Determine the compressor work, turbine work, and thermal efficiency. $P_1 = 14.7$ lbf/in.^{2}.

10.35 Repeat Problem 10.34 assuming that each compressor has an efficiency of 80%, each turbine has an efficiency of 85%, and the regenerator has an efficiency of 60%.

10.36 A gas turbine cycle for use as an automotive engine is shown in Fig. 10.41. In the first turbine, the gas expands to just a low enough pressure P_5 for that turbine to drive the compressor. The gas is then expanded through a second turbine connected to the drive wheels. The data for this engine are as shown in the figure. Consider the working fluid to be air throughout the entire cycle, and assume that all processes are ideal. Determine the following:

(*a*) Pressure P_5.

(*b*) Net work per lbm and mass flow rate.

(*c*) Temperature T_3 and cycle thermal efficiency.
(*d*) The *T-s* diagram for the cycle.

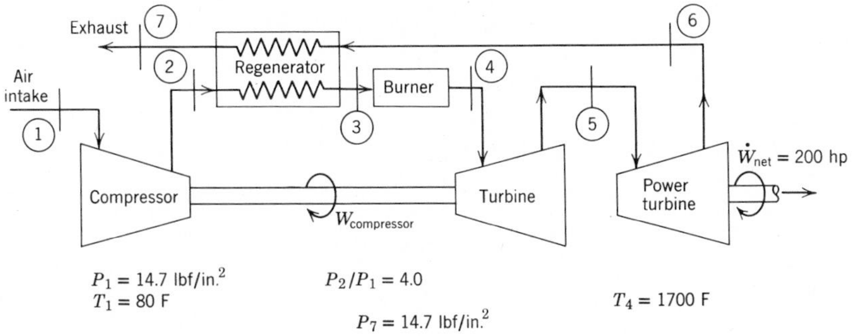

Fig. 10.41 Sketch for Problem 10.36.

10.37 Repeat Problem 10.36 assuming that the compressor has an efficiency of 82%, that both turbines have efficiencies of 87%, and that the regenerator efficiency is 70%. Also include the following pressure drops:

$$\Delta P_{23} = 0.02\,P_2$$
$$\Delta P_{34} = 0.02\,P_3$$
$$\Delta P_{67} = 0.02\,P_6$$

10.38 Consider the air-standard cycle for a gas turbine-jet propulsion unit shown in Fig. 10.34. The pressure and temperature entering the compressor are 14.7 lbf/in.^{2}, 60 F respectively. The pressure ratio across the compressor is 6 to 1 and the temperature at the turbine inlet is 2000 F. On leaving the turbine the air enters the nozzle and expands to 14.7 lbf/in.^{2}. Determine the pressure at the nozzle inlet and the velocity of the air leaving the nozzle.

10.39 Repeat Problem 10.38, assuming that the efficiency of the compressor and turbine are both 85%, and that the nozzle efficiency is 95%.

10.40 A jet aircraft is flying at an altitude of 16,000 ft, where the ambient pressure is approximately 8 lbf/in.^{2} and the ambient temperature is 0 F. The velocity of the aircraft is 700 ft/sec, and the pressure ratio across the compressor is 5.

Devise an air-standard cycle that approximates this cycle, and determine the velocity (relative to the aircraft) of the air leaving the engine, assuming that it has been expanded to the ambient pressure. Assume a maximum gas temperature of 1600 F.

10.41 A heat exchanger is incorporated into an air-standard refrigeration cycle as shown in Fig. 10.42. Assume that both the compression and expansion are reversible adiabatic processes. Determine the coefficient of performance for this cycle.

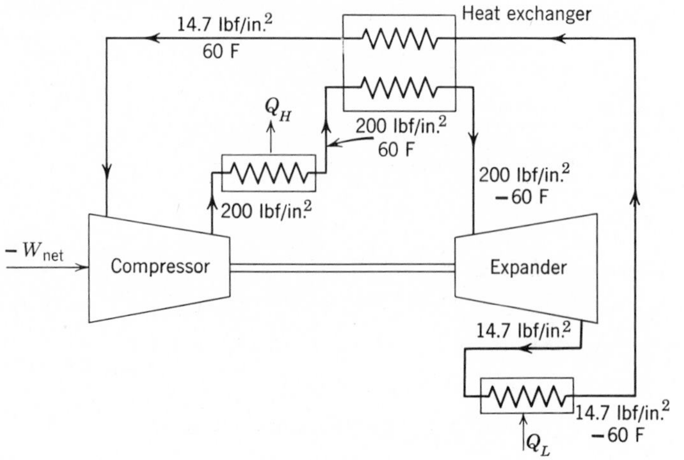

Fig. 10.42 Schematic arrangement of Problem 10.41.

10.42 Repeat Problem 10.41 assuming an isentropic efficiency for the compressor and expander of 75%.

10.43 Repeat Problems 10.41 and 42 assuming that helium is used as the working fluid instead of air. Discuss the results.

10.44 Write a computer program to solve the following problem. It is desired to determine the effects of varying parameters on the performance of an air-standard Brayton cycle. Consider a compressor inlet condition of 14.7 lbf/in.², 70 F, and assume constant specific heat. The cycle thermal efficiency and net work output per lbm are to be determined for all combinations of the following variables:

(*a*) Compressor pressure ratio of 4, 6, 8, and 10.
(*b*) Maximum cycle temperature of 1200 F, 1600 F, 2000 F, and 2400 F.
(*c*) Compressor and turbine efficiencies each 100%, 90%, 80%, and 70%.

10.45 Write a computer program to solve the following problem. Consider an ideal Otto cycle operating with nitrogen gas as the working fluid. The condition at the beginning of the compression process is 14.7 lbf/in.², 70 F. Assume variable specific heat according to the equation in Table B.7. Determine the net work output per lbm and the cycle thermal efficiency for various combinations of compression ratio and maximum cycle temperature, and compare the results with those found assuming constant specific heat.

10.46 Write a computer program to solve the following problem. It is desired to investigate the effect of the number of stages on the performance of an ideal gas turbine cycle having intercooling, reheat, and a regenerator, as shown in Fig. 10.31 (for two stages). Assume that the inlet condition, point 1, is 14.7 lbf/in.², 70 F, and that the total pressure ratio, the maximum cycle temperature, and the number of stages (equal number of compressor and turbine stages) are all variable. Each stage is to have an equal pressure ratio. Determine the net work output per lbm and the cycle thermal efficiency for various combinations of these variables.

11

An Introduction to the Thermodynamics of Mixtures

Up to this point in our development of thermodynamics we have limited our consideration primarily to pure substances. However, a large number of processes involve mixtures of different pure substances. Sometimes these mixtures are referred to as solutions—particularly in the liquid and solid phases.

This chapter involves an introduction to the thermodynamics of mixtures and solutions. Perhaps the word introduction should be emphasized, because the thermodynamics of mixtures and solutions is a rather complex subject, and a detailed presentation is beyond the scope of this book. The focus will be on mixtures of ideal gases and a simplified, but very useful model of certain mixtures, such as air and water vapor, which may involve a condensed (solid or liquid) phase of one of the components. However, an attempt has been made to present the material so that this introduction can lead directly to a more comprehensive study of mixtures.

An understanding of the material in this chapter is a necessary foundation for the consideration of chemical reactions and chemical and phase equilibrium. These topics are covered in the chapters that follow.

11.1 General Considerations and Mixtures of Ideal Gases.

In any mixture, the mole fraction of component i is defined as

$$y_i = \frac{n_i}{n} \tag{11.1}$$

where n_i is the number of moles of component i, and n is the total number of moles in the mixture.

Similarly, the mass fraction mf_i is

$$mf_i = \frac{m_i}{m} \tag{11.2}$$

where m_i is the mass of component i and m is the total mass of the mixture.

Consider a mixture of two gases (not necessarily ideal gases) such as shown in Fig. 11.1.

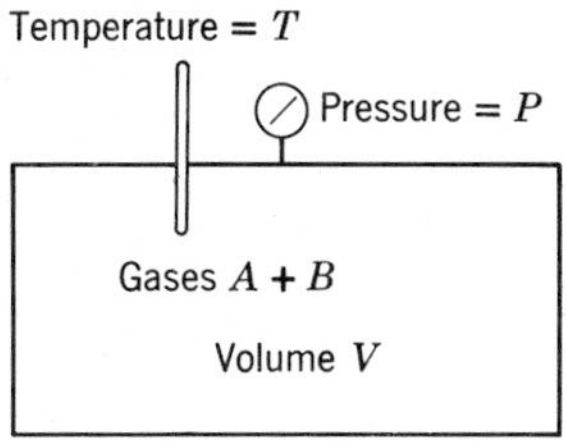

Fig. 11.1 A mixture of two gases.

What properties can we experimentally measure for such a mixture? Certainly we can measure the pressure, temperature, volume, and mass of the mixture. We can also experimentally measure the composition of the mixture, and thus determine the mole and mass fractions.

Suppose that this mixture undergoes a process or a chemical reaction and we wish to perform a thermodynamic analysis of this process or reaction. What type of thermodynamic data would we use in performing such an analysis? One possibility would be to have tables of thermodynamic properties of mixtures. However, the number of different mixtures that is possible, both as regards the substances involved and the relative amounts of each is such that one would need a library full of tables of thermodynamic properties to handle all possible situations. It would be much simpler if we could determine the thermodynamic properties of a mixture from the properties of the pure components. This is in essence the approach that is used in dealing with ideal gases and certain other simplified models of mixtures.

One exception to this procedure is the case where a particular mixture is encountered very frequently, the most familiar being air. Tables and charts of the thermodynamics properties of air are available. However, even in this case it is necessary to define the composition of the "air" for which the tables are given, since the composition of the atmosphere varies with altitude, with the number of pollutants, and with other variables at a given location. The composition of air on which air tables are usually based is as follows:

Component	**% on Mole Basis**
Nitrogen	78.09
Oxygen	20.95
Argon	0.93
CO_2 & trace elements	0.03

Consider again the mixture of Fig. 11.1. For the general case the properties of the mixture are defined in terms of the partial molal properties of the individual components. A partial molal property is defined as the value of a property, such as internal energy, for a given component as it exists in the mixture. With this definition, the internal energy of the mixture of Fig. 11.1 would be

$$U_{\text{mix}} = n_A \bar{U}_A + n_B \bar{U}_B \tag{11.3}$$

where $\bar{U}$ designates the partial molal internal energy. Similar equations can be written for other properties, and this matter will be further developed in Chapters 12 and 14.

In this chapter we focus on mixtures of ideal gases. We assume that each component is uninfluenced by the presence of the other components, and that each component can be treated as an ideal gas. In an actual case of a gaseous mixture at high pressure this assumption would probably not be true because of the nature of the interaction between the molecules of the different components.

There are two models used in conjunction with the mixtures of gases, namely the Dalton model and the Amagat model.

Dalton Model

In the case of the Dalton model, the properties of each component are considered as though each component existed separately at the volume and temperature of the mixture, as shown in Fig. 11.2.

Consider this model for the special case in which both the mixture and the separated components can be considered an ideal gas.

For the mixture:

$$PV = n\bar{R}T$$
$$n = n_A + n_B \tag{11.4}$$

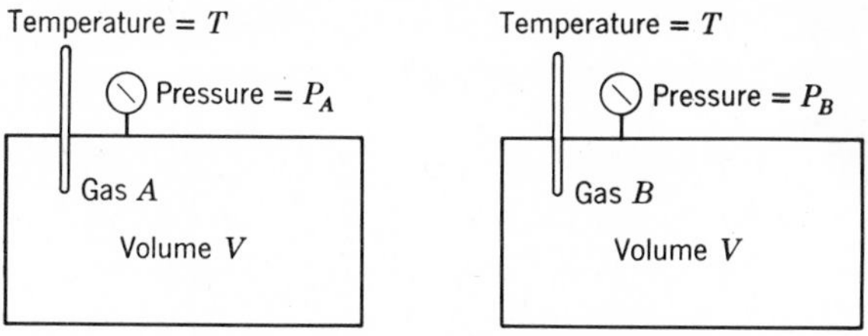

Fig. 11.2 The Dalton model.

For the components:

$$P_A V = n_A \bar{R} T$$
$$P_B V = n_B R T \tag{11.5}$$

On substituting we have

$$n = n_A + n_B$$

$$\frac{PV}{\bar{R}T} = \frac{P_A V}{\bar{R}T} + \frac{P_B V}{\bar{R}T} \tag{11.6}$$

or

$$P = P_A + P_B$$

where P_A and P_B are referred to as partial pressures.

Thus for a mixture of ideal gases, the pressure is the sum of the partial pressures of the individual components.

It should be stressed that the term partial pressure has relevance only for ideal gases. In effect this concept assumes that the molecules of each component are uninfluenced by the other components, and that the total pressure is the sum of partial pressures of the individual components. It should also be noted that partial pressure is not a partial molal property in the sense as defined by Eq. 11.3, since partial molal properties relate only to extensive properties.

Amagat Model

In the Amagat model the properties of each component are considered as though each component existed separately at the pressure and temperature of the mixture, as shown in Fig. 11.3. The volumes of A and B under these conditions are V_A and V_B respectively.

In the general case the sum of the volumes when separated, namely, $V_A + V_B$, need not be equal to the volume of the mixture. However, let us consider the special case in which both the separated components and the mixture are considered to be ideal gases. In this case we can write:

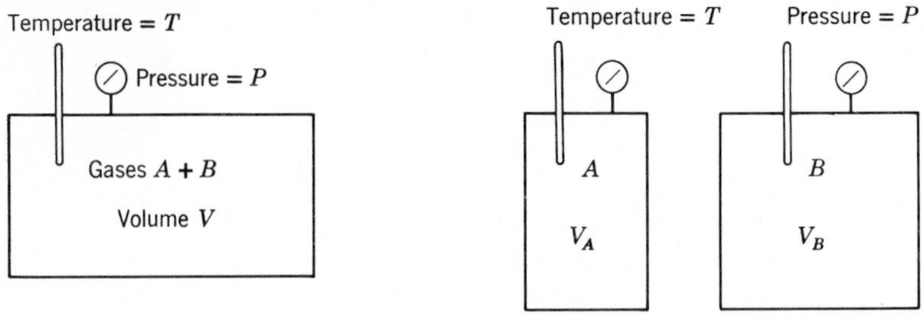

Fig. 11.3 The Amagat model.

For the mixture:
$$PV = n\bar{R}T \qquad n = n_A + n_B \tag{11.7}$$

For the components:
$$PV_A = n_A\bar{R}T \qquad PV_B = n_B\bar{R}T \tag{11.8}$$

On substituting we have,

$$n = n_A + n_B$$

$$\frac{PV}{\bar{R}T} = \frac{PV_A}{\bar{R}T} + \frac{PV_B}{\bar{R}T}$$

Therefore

$$V_A + V_B = V$$

or

$$\frac{V_A}{V} + \frac{V_B}{V} = 1 \tag{11.9}$$

V_A/V and V_B/V are referred to as the volume fractions.

Thus in the case of ideal gases, Amagat's model leads to the conclusion that the sum of the volume fractions is unity, and that there would be no volume change if the components were mixed while holding the temperature and pressure constant.

From Eqs. 11.4, 11.5, 11.7 and 11.8 it is evident that

$$\frac{V_A}{V} = \frac{n_A}{n} = \frac{P_A}{P}$$

$$\frac{V_A}{V} = y_A = \frac{P_A}{P} \tag{11.10}$$

That is, for each component of a mixture of ideal gases, the volume fraction, the mole fraction, and the ratio of the partial pressure to the total pressure are equal.

In determining the internal energy, enthalpy, and entropy of a mixture of ideal gases, the Dalton model proves useful because the assumption is made that each constituent behaves as though it occupied the entire volume by itself. Thus the internal energy, enthalpy, and entropy can be evaluated as the sum of the respective properties of the constituent gases at the condition at which the component exists in the mixture. Since for ideal gases the internal energy and enthalpy are functions only of temperature, it follows that

$$U = n\bar{u} = n_A\bar{u}_A + n_B\bar{u}_B \tag{11.11}$$

$$H = n\bar{h} = n_A\bar{h}_A + n_B\bar{h}_B \tag{11.12}$$

where $\bar{u}_A$ and $\bar{h}_A$ are the internal energy and enthalpy per mole for pure A and $\bar{u}_B$ and $\bar{h}_B$ are the same quantities for pure B, all at the temperature of the mixture.

The entropy of an ideal gas is a function of pressure as well as temperature. Since each component exists in the mixture at its partial pressure,

$$S = n\bar{s} = n_A\bar{s}_A + n_B\bar{s}_B \tag{11.13}$$

where $\bar{s}_A$ is the entropy per mole for pure A at T, P_A (the partial pressure of A), and $\bar{s}_B$ that for pure B at T and P_B.

Example 11.1

A volumetric analysis of a gaseous mixture yields the following results:

CO_2	12.0%
O_2	4.0
N_2	82.0
CO_2	2.0

Determine the analysis on a mass basis, and the molecular weight and the gas constant on a mass basis for the mixture. Assume ideal gas behavior.

The table shown below is a very convenient way to solve this problem.

Table 11.1

Constituent	Per Cent by Volume	Mole Fraction		Molecular Weight		Mass (lbm) per Mole of Mixture	Analysis on Mass Basis, Per Cent
CO_2	12	0.12	×	44.0	=	5.28	$\frac{5.28}{30.08} = 17.55$
O_2	4	0.04	×	32.0	=	1.28	$\frac{1.28}{30.08} = 4.26$
N_2	82	0.82	×	28.0	=	22.96	$\frac{22.96}{30.08} = 76.33$
CO	2	0.02	×	28.0	=	0.56	$\frac{0.56}{30.08} = 1.86$
						30.08	100.00

Molecular weight of mixture = 30.08

$$R \text{ for mixture} = \frac{\bar{R}}{M} = \frac{1545}{30.08} = 51.4 \text{ ft-lbf/lbm R}$$

If the analysis had been given on a mass basis, and the mole faction or volumetric analysis was desired, the following procedure could be used.

Table 11.2

Constituent	Mass Fraction		Molecular Weight		Lb Moles per Pound of Mixture	Mole Fraction	Volumetric Analysis, Per Cent
CO_2	0.1755	÷	44.0	=	0.00399	0.120	12.0
O_2	0.0426	÷	32.0	=	0.00133	0.040	4.0
N_2	0.7633	÷	28.0	=	0.02726	0.820	82.0
CO	0.0186	÷	28.0	=	0.00066	0.020	2.0
					0.03324	1.000	100.0

$$M = \frac{1}{\text{moles/lbm mixture}} = \frac{1}{0.03324} = 30.08$$

$$R = \frac{\bar{R}}{M} = \frac{1545}{30.08} = 51.4 \text{ ft-lbf/lbm R}$$

Example 11.2

Let n_A moles of gas A at a given pressure and temperature be mixed with n_B moles of gas B at the same pressure and temperature in an adiabatic constant-volume process, shown in Fig. 11.4. Determine the increase in entropy for this process.

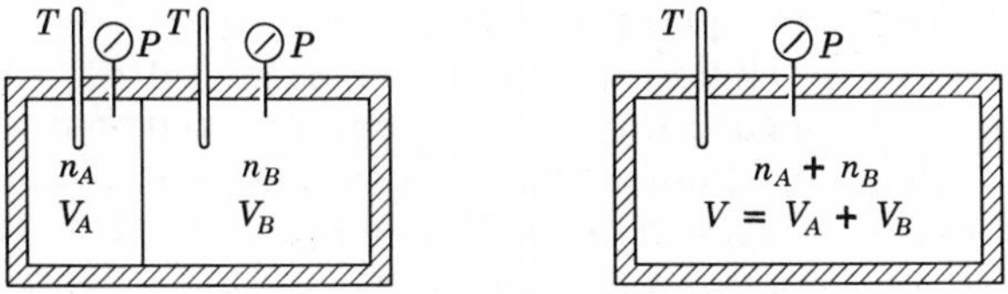

Fig. 11.4 Sketch for Example 11.2.

The final partial pressure of gas A is P_A and for gas B it is P_B. Since there is no change in temperature, Eq. 8.21 reduces to

$$(S_2 - S_1)_A = -n_A \bar{R} \ln \frac{P_A}{P} = -n_A \bar{R} \ln y_A$$

$$(S_2 - S_1)_B = -n_B \bar{R} \ln \frac{P_B}{P} = -n_B \bar{R} \ln y_B$$

The total change in entropy is the sum of the entropy changes for gases A and B.

$$S_2 - S_1 = -\bar{R}(n_A \ln y_A + n_B \ln y_B) \tag{11.14}$$

This equation can be written for the general case of mixing any number of components at the same pressure and temperature as

$$S_2 - S_1 = -\bar{R} \sum_k n_k \ln y_k \tag{11.15}$$

The interesting thing about this equation is that the increase in entropy depends only on the number of moles of component gases, and is independent of the composition of the gas. For example, when 1 mole of oxygen and 1 mole of nitrogen are mixed, the increase in entropy is the same as when 1 mole of hydrogen and 1 mole of nitrogen are mixed. But we also know that if 1 mole of nitrogen is "mixed" with another mole of nitrogen there is no increase in entropy. The question that arises is how dissimilar must the gases be in order to have an increase in entropy? The answer lies in our ability to distinguish between the two gases. The entropy increases whenever we can distinguish between the gases being mixed. When we cannot distinguish between the gases, there is no increase in entropy.

11.2 A Simplified Model of a Mixture Involving Gases and a Vapor

Let us now consider a simplification, which in many cases is a reasonable one, of the problem involving a mixture of ideal gases that is in contact with a solid or liquid phase of one of the components. The most familiar example is a mixture of air and water vapor in contact with liquid water or ice, such as the problems encountered in air conditioning or drying. We are all familiar with the condensation of water from the atmosphere when it is cooled on a summer day.

This problem, and a number of similar problems, can be analyzed quite simply and with considerable accuracy if the following assumptions are made:

1. The solid or liquid phase contains no dissolved gases.
2. The gaseous phase can be treated as a mixture of ideal gases.
3. When the mixture and the condensed phase are at a given pressure and temperature, the equilibrium between the condensed phase and its vapor is not influenced by the presence of the other component. This means that when equilibrium is achieved the partial pressure of the vapor will be equal to the saturation pressure corresponding to the temperature of the mixture.

Since this approach is used extensively and with considerable accuracy, let us give some attention to the terms that have been defined and the type of problems for which this approach is valid and relevant. In our discussion we will refer to this as a gas-vapor mixture.

The dew point of a gas-vapor mixture is the temperature at which the vapor condenses or solidifies when it is cooled at constant pressure. This is shown on the *T-s* diagram for the vapor shown in Fig. 11.5. Suppose that the temperature of the gaseous mixture and the partial pressure of the vapor in the mixture are such that the vapor is initially superheated at state 1. If the mixture is cooled at constant pressure the partial pressure of the vapor remains constant until point 2 is reached, and then condensation begins. The temperature at state 2 is the dew-point temperature. Line 1–3 on the diagram indicates that if the mixture is cooled at constant volume the condensation begins at point 3, which is slightly lower than the dew-point temperature.

If the vapor is at the saturation pressure and temperature, the mixture is referred to as a saturated mixture, and for an air-water vapor mixture, the term "saturated air" is used.

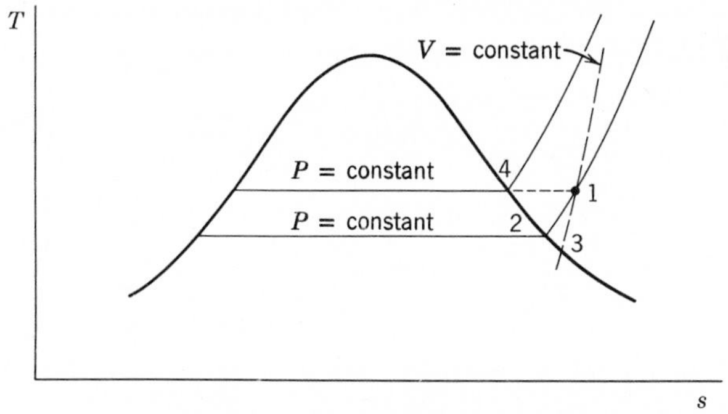

Fig. 11.5 Temperature-entropy diagram to show definition of the dew point.

The relative humidity ϕ is defined as the ratio of the mole fraction of the vapor in the mixture to the mole fraction of vapor in a saturated mixture at the same temperature and total pressure. Since the vapor is considered an ideal gas, the definition reduces to the ratio of the partial pressure of the vapor as it exists in the mixture P_v, to the saturation pressure of the vapor at the same temperature P_g.

$$\phi = \frac{P_v}{P_g}$$

In terms of the numbers on the T-s diagram of Fig. 11.5, the relative humidity ϕ would be

$$\phi = \frac{P_1}{P_4}$$

Since we are considering the vapor to be an ideal gas, the relative humidity can also be defined in terms of specific volume or density.

$$\phi = \frac{P_v}{P_g} = \frac{\rho_v}{\rho_g} = \frac{v_g}{v_v} \tag{11.16}$$

The humidity ratio ω of an air-water vapor mixture is defined as the ratio of the mass of water vapor m_v to the mass of dry air m_a. The term "dry air" is used to emphasize that this refers only to air and not to the water vapor. The term "specific humidity" is used synonymously with humidity ratio.

$$\omega = \frac{m_v}{m_a} \tag{11.17}$$

This definition is identical for any other gas-vapor mixture, and the subscript a refers to the gas, exclusive of the vapor. Since we are considering both the vapor and the mixture to be ideal gases, a very useful expression for humidity ratio in terms of partial pressures and molecular weights can be developed.

$$m_v = \frac{P_v V}{R_v T} = \frac{P_v V M_v}{\bar{R} T} \qquad m_a = \frac{P_a V}{R_a T} = \frac{P_a V M_a}{\bar{R} T}$$

Then

$$\omega = \frac{P_v V / R_v T}{P_a V / R_a T} = \frac{R_a P_v}{R_v P_a} = \frac{M_v P_v}{M_a P_a} \tag{11.18}$$

For an air-water vapor mixture this reduces to

$$\omega = 0.622 \frac{P_v}{P_a} \tag{11.19}$$

The degree of saturation is defined as the ratio of the actual humidity ratio to the humidity ratio of a saturated mixture at the same temperature and total pressure.

An expression for the relation between the relative humidity ϕ and the humidity ratio ω can be found by solving Eqs. 11.16 and 11.19 for P_v and equating them. The resulting relation for an air-water vapor mixture is

$$\phi = \frac{\omega P_a}{0.622 P_g} \tag{11.20}$$

A few words should also be said about the nature of the process that occurs when a gas-vapor mixture is cooled at constant pressure. Suppose that the vapor is initially superheated at state 1 in Fig. 11.6. As the mix-

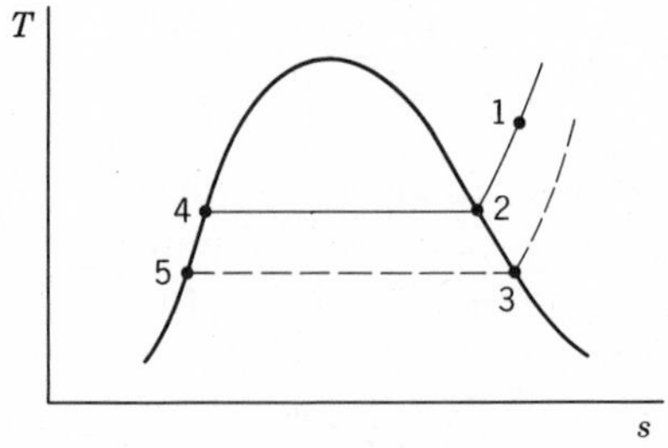

Fig. 11.6 Temperature-entropy diagram to show the cooling of a gas-vapor mixture at a constant pressure.

ture is cooled at constant pressure the partial pressure of the vapor remains constant until the dew point is reached at point 2, at which point the vapor in the mixture is saturated. The initial condensate is at state 4, and is in equilibrium with the vapor at state 2. As the temperature is lowered further, more of the vapor condenses, which lowers the partial pressure of the vapor in the mixture. The vapor that remains in the mixture is always saturated, and the liquid or solid is in equilibrium with it. For example, when the temperature is reduced to T_3, the vapor in the mixture is at state 3, and its partial pressure is the saturation pressure corresponding to T_3. The liquid in equilibrium with it is at state 5.

Example 11.3

Consider 2000 ft^3 of an air-water vapor mixture at 14.70 lbf/in.^{2}, 90 F, 70 per cent relative humidity. Calculate the humidity ratio, dew point, mass of air, and mass of vapor.

From Eq. 11.16 and the steam tables,

$$\phi = 0.70 = \frac{P_v}{P_g}$$

$$P_v = 0.70(0.6988) = 0.4892 \text{ lbf/in.}^2$$

The dew point is the saturation temperature corresponding to this pressure, which is 78.9 F.

The partial pressure of the air is

$$P_a = P - P_v = 14.70 - 0.49 = 14.21 \text{ lbf/in.}^2$$

The humidity ratio can be calculated from Eq. 11.19.

$$\omega = 0.622 \times \frac{P_v}{P_a} = 0.622 \times \frac{0.4892}{14.21} = 0.02135$$

The mass of air is

$$m_a = \frac{P_a V}{R_a T} = \frac{14.21 \times 144 \times 2000}{53.34 \times 550} = 139.6 \text{ lbm}$$

The mass of the vapor can be calculated by using the humidity ratio or by using the ideal gas equation of state.

$$m_v = \omega m_a = 0.02135(139.6) = 2.98 \text{ lbm}$$

$$m_v = \frac{0.4892 \times 144 \times 2000}{85.7 \times 550} = 2.98 \text{ lbm}$$

Example 11.4

Calculate the amount of water vapor condensed if the mixture of Example 11.3 is cooled to 40 F in a constant-pressure process.

At 40 F the mixture is saturated, since this is below the dew-point temperature. Therefore

$$P_{v2} = P_{g2} = 0.1217 \text{ lbf/in.}^2$$

$$P_{a2} = 14.7 - 0.12 = 14.58 \text{ lbf/in.}^2$$

$$\omega_2 = 0.622 \times \frac{0.1217}{14.58} = 0.00520$$

The amount of water vapor condensed is equal to the difference between the initial and final mass of water vapor.

$$\text{mass of vapor condensed} = m_a(\omega_1 - \omega_2) = 139.6(0.02135 - 0.0052)$$
$$= 2.25 \text{ lbm}$$

11.3 The First Law Applied to Gas-Vapor Mixtures

In applying the first law of thermodynamics to gas-vapor mixtures it is helpful to realize that because of our assumption that ideal gases are involved, the various components can be treated separately when calculating changes of internal energy and enthalpy. Therefore, in dealing with air-water vapor mixtures the changes in enthalpy of the water vapor can be found from the steam tables and the ideal gas relations can be applied to the air. This is illustrated by the examples that follow.

Example 11.5

An air-conditioning unit is shown in Fig. 11.7, with pressure, temperature, and relative humidity data. Calculate the heat transfer per pound

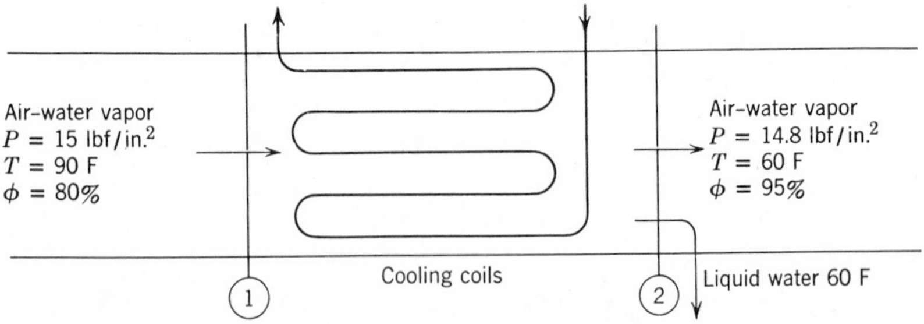

Fig. 11.7 Sketch for Example 11.5.

of dry air, assuming that changes in kinetic energy are negligible.

Let us consider a steady-state, steady-flow process for a control volume that excludes the cooling coils.

Since this is a steady-flow process, the continuity equation for the air and water (vapor and liquid) are

$$m_{a1} = m_{a2}$$

$$m_{v1} = m_{v2} + m_{l2}$$

The first law for this steady-state, steady-flow process is

$$Q_{\text{c.v.}} + \sum m_i h_i = \sum m_e h_e$$

$$Q_{\text{c.v.}} + m_a h_{a1} + m_{v1} h_{v1} = m_a h_{a2} + m_{v2} h_{v2} + m_{l2} h_{l2}$$

If we divide this equation by m_a, introduce the continuity equation for the H_2O, and note that $m_v = \omega m_a$, we can write the first law in the form

$$\frac{Q_{\text{c.v.}}}{m_a} + h_{a1} + \omega_1 h_{v1} = h_{a2} + \omega_2 h_{v2} + (\omega_1 - \omega_2) h_{l2}$$

The air can be considered an ideal gas with constant specific heat, and the values of enthalpy for the H_2O can be taken from the steam tables. (Since the water vapor at these low pressures is being considered an ideal gas, the enthalpy of the water vapor is a function of the temperature only. Therefore, the enthalpy of slightly superheated water vapor is equal to the enthalpy of saturated vapor at the same temperature.)

$$P_{v1} = \phi P_{g1} = 0.80(0.6988) = 0.5590 \text{ lbf/in.}^2$$

$$\omega_1 = \frac{R_a}{R_v}\frac{P_{v1}}{P_{a1}} = 0.622 \times \left(\frac{0.5590}{15.00 - 0.56}\right) = 0.0241$$

$$P_{v2} = \phi_2 P_{g2} = 0.95(0.2563) = 0.2435$$

$$\omega_2 = \frac{R_a}{R_v} \times \frac{P_{v2}}{P_{a2}} = 0.622 \times \left(\frac{0.2435}{14.80 - 0.24}\right) = 0.0104$$

$$Q_{\text{c.v.}}/m_a + h_{a1} + \omega_1 h_{v1} = h_{a2} + \omega_2 h_{v2} + (\omega_1 - \omega_2) h_{l2}$$

$$\begin{aligned} Q_{\text{c.v.}}/m_a &= 0.240(60 - 90) + 0.0104(1087.7) - 0.0241(1100.7) \\ &\quad + (0.0241 - 0.0104)28.08 \\ &= -7.20 + 11.32 - 26.53 + 0.38 = -22.03 \text{ Btu/lbm dry air} \end{aligned}$$

Example 11.6

A tank has a volume of 10 ft^3 and contains nitrogen and water vapor. The temperature of the mixture is 120 F and the total pressure is 30 lbf/in.^{2}. The partial pressure of the water vapor is 0.8 lbf/in.^{2}. Calculate the heat transfer when the contents of the tank are cooled to 50 F.

This is a constant-volume process, and since the work is zero the first law reduces to

$$Q = U_2 - U_1 = m_{N_2} C_{v(N_2)}(T_2 - T_1) + (m_2 u_2)_v + (m_2 u_2)_l - (m_1 u_1)_v$$

This equation assumes that some of the vapor condensed. This must be checked, however, as shown below.

The mass of nitrogen and water vapor can be calculated using the ideal-gas equation of state.

$$m_{N_2} = \frac{P_{N_2} V}{R_{N_2} T} = \frac{29.2 \times 144 \times 10}{55.15 \times 580} = 1.314 \text{ lbm}$$

$$m_{v1} = \frac{P_{v1} V}{R_v T} = \frac{0.8 \times 144 \times 10}{85.76 \times 580} = 0.0232 \text{ lbm}$$

If condensation takes place the final state of the vapor will be saturated vapor at 50 F. In this case

$$m_{v2} = \frac{P_{v2} V}{R_v T} = \frac{0.1780 \times 144 \times 10}{85.76 \times 510} = 0.00586 \text{ lbm}$$

Since this is less than the original mass of vapor there must have been condensation.

The mass of liquid that is condensed, m_{l_2}, is

$$m_{l_2} = m_{v_1} - m_{v_2} = 0.0232 - 0.0059 = 0.0173 \text{ lbm}$$

The internal energy of the water vapor is equal to the internal energy of saturated water vapor at the same temperature. Therefore,

$$u_{v_1} = 1049.9 \text{ Btu/lbm}$$

$$u_{v_2} = 1027.2 \text{ Btu/lbm}$$

$$\begin{aligned} Q_{\text{c.v.}} &= 1.314 \times 0.177(50 - 120) + 0.00586(1027.2) \\ &\quad + 0.0173(18.06) - 0.0232(1049.9) \\ &= -16.31 + 6.02 + 0.31 - 24.26 = -34.24 \text{ Btu} \end{aligned}$$

11.4 The Adiabatic Saturation Process

An important process involving an air-water vapor mixture is the adiabatic saturation process, in which an air-vapor mixture comes in contact with a body of water in a well-insulated duct (Fig. 11.8). If the initial relative humidity is less than 100 per cent some of the water will

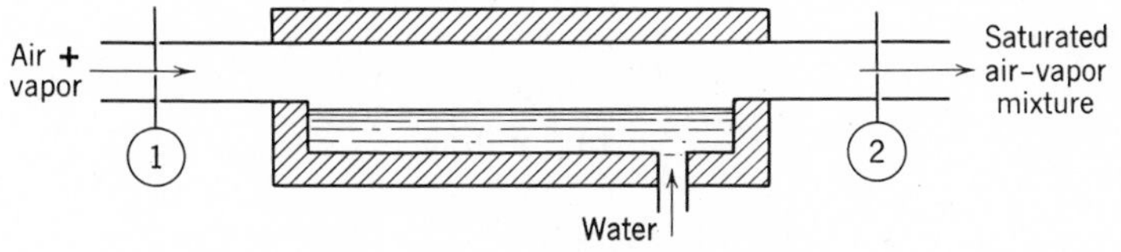

Fig. 11.8 The adiabatic saturation process.

evaporate and the temperature of the air-vapor mixture will decrease. If the mixture leaving the duct is saturated and if the process is adiabatic, the temperature of the mixture on leaving is known as the adiabatic saturation temperature. In order for this to take place as a steady-flow process, make-up water at the adiabatic saturation temperature is added at the same rate at which water is evaporated. The pressure is assumed to be constant.

Considering the adiabatic saturation process to be a steady-state, steady-flow process, and neglecting changes in kinetic and potential energy, the first law reduces to

$$h_{a1} + \omega_1 h_{v1} + (\omega_2 - \omega_1) h_{l2} = h_{a2} + \omega_2 h_{v2}$$

$$\omega_1 (h_{v1} - h_{l2}) = C_{pa}(T_2 - T_1) + \omega_2 (h_{v2} - h_{l2})$$

$$\omega_1 (h_{v1} - h_{l2}) = C_{pa}(T_2 - T_1) + \omega_2 h_{fg2} \qquad (11.21)$$

The most significant point to be made about the adiabatic saturation process is that the adiabatic saturation temperature, the temperature of the mixture when it leaves the duct, is a function of the pressure, temperature, and relative humidity of the entering air-vapor mixture and of the exit pressure. Thus, the relative humidity and the humidity ratio of the entering air-vapor mixture can be determined from the measurements of the pressure and temperature of the air-vapor mixture entering and leaving the adiabatic saturator. Since these measurements are relatively easy to make, this is one means of determining the humidity of an air-vapor mixture.

Example 11.7

The pressure of the mixture entering and leaving the adiabatic saturator is 14.7 lbf/in.^{2}, the entering temperature is 84 F, and the temperature leaving is 70 F, which is the adiabatic saturation temperature. Calculate the humidity ratio and relative humidity of the air-water vapor mixture entering.

Since the water vapor leaving is saturated, $P_{v2} = P_{g2}$ and ω_2 can be calculated.

$$\omega_2 = 0.622 \times \frac{0.3632}{14.7 - 0.36} = 0.01573$$

ω_1 can be calculated using Eq. 11.21.

$$\omega_1 = \frac{C_{pa}(T_2 - T_1) + \omega_2 h_{fg2}}{(h_{v1} - h_{l2})}$$

$$\omega_1 = \frac{0.24(70 - 84) + 0.01573 \times 1054.0}{1098.1 - 38.1} = \frac{-3.36 + 16.60}{1060.0} = 0.0125$$

$$\omega_1 = 0.622 \times \left(\frac{P_{v1}}{14.7 - P_{v1}}\right) = 0.0125$$

$$P_{v1} = 0.289$$

$$\phi_1 = \frac{P_{v1}}{P_{g1}} = \frac{0.289}{0.584} = 0.495$$

11.5 Wet-Bulb and Dry-Bulb Temperatures

The humidity of an air-water vapor mixture is usually found from dry-bulb and wet-bulb data. These data are obtained by use of a psychrometer, which involves the flow of air past wet-bulb and dry-bulb thermometers. The bulb of the wet-bulb thermometer is covered with a cotton wick that is saturated with water. The dry-bulb thermometer is used simply to measure the temperature of the air. The flow of air may be maintained by a fan, as in the continuous-flow psychrometer shown in Fig. 11.9, or by moving the thermometer through the air, as in the sling psychrometer, which consists of wet-bulb and dry-bulb thermometers mounted so that they can be whirled.

The processes that take place at the wet-bulb thermometer are somewhat involved. First of all, if the air-water vapor mixture is not saturated, some of the water in the wick evaporates and diffuses into the surrounding air. A drop in the temperature of the water in the wick will

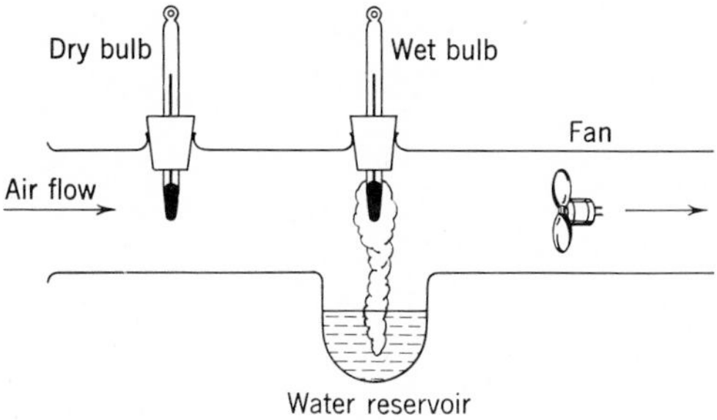

Fig. 11.9 Steady-flow apparatus for measuring wet- and dry-bulb temperatures.

be associated with this evaporation. However, as soon as the temperature of the water drops, heat is transferred to the water from both the air and the thermometer. Finally, a steady state, determined by heat and mass transfer rates, will be reached. In general, air velocities exceeding 700 ft per min are required in order that the convective heat transfer be large in comparison to the radiant heat transfer.

The psychrometric chart is the most convenient method of determining relative humidity and humidity ratio from wet-bulb and dry-bulb data, although equations that have been developed can also be used. A psychrometric chart is included in the Appendix, Fig. B.7.

The difference between the wet-bulb temperature and adiabatic saturation temperature should be carefully noted. The wet-bulb temperature is influenced by heat and mass transfer rates, whereas the adiabatic saturation temperature simply involves equilibrium between the entering air-vapor mixture and water at the adiabatic saturation temperature. However, it happens that the wet-bulb temperature and the adiabatic saturation temperature are approximately equal for air-water vapor mixtures at atmospheric temperature and pressure. This is not necessarily true at temperatures and pressures that deviate significantly from ordinary atmospheric conditions, or for other gas-vapor mixtures.

11.6 The Psychrometric Chart

Properties of air-water vapor mixtures are given in graphical form on psychrometric charts. These are available in a number of different forms, and only the main features are considered here.

The basic psychrometric chart consists of a plot of dry-bulb temperature (abscissa) and humidity ratio (ordinate). If we fix the total pressure for which the chart is to be constructed (which is usually one standard atmosphere), lines of constant relative humidity and wet-bulb temperature can be drawn on the chart, because for a given dry-bulb temperature, total pressure, and humidity ratio, the relative humidity and wet-bulb temperature are fixed. The partial pressure of the water vapor is fixed by the humidity ratio and total pressure, and therefore a second ordinate scale that indicates the partial pressure of the water vapor can be constructed.

Most psychrometric charts give the enthalpy of an air-vapor mixture per pound of dry air. The values given assume that the enthalpy of the dry air is zero at 0 F, and the enthalpy of the vapor is taken from the steam tables (which are based on the assumption that the enthalpy of saturated liquid is zero at 32 F). This procedure is satisfactory because we are usually concerned only with differences in enthalpy. The fact that lines of constant enthalpy are essentially parallel to lines of constant wet-bulb temperature is evident from the fact that the wet-bulb temperature is essentially equal to the adiabatic saturation temperature. Thus, in Fig. 11.8, if we neglect the enthalpy of the liquid entering the adiabatic saturator, the enthalpy of the air-vapor mixture entering is equal to the enthalpy of the saturated air-vapor mixture leaving, and a given adiabatic saturation temperature fixes the enthalpy of the mixture entering.

Some charts are available that give corrections for variation from standard atmospheric pressures. Before using a given chart one should fully understand the assumptions made in constructing it, and that it is applicable to the particular problem at hand.

PROBLEMS

11.1 An analysis of the products of a certain combustion process yields the following composition on a volumetric basis:

Constituent	Percent by Volume
N_2	70
CO_2	15
O_2	11
CO	4

Determine:

(*a*) The composition of this mixture on a mass basis.

(*b*) The mass of 10 ft^3 of this mixture at a condition of 14.7 lbf/in.^{2}, 70 F.

(*c*) The heat transfer required to heat the mixture at constant volume from the initial state to 300 F.

11.2 The mixture of Problem 11.1 is compressed in a reversible adiabatic process in a cylinder from the initial state to a volume of 5 ft^3. Calculate the final temperature and the work for the process.

11.3 A vessel contains one lb mole of nitrogen and one lb mole of helium, each at 1 atm pressure and 77 F and separated by a membrane.

Determine the change of entropy which occurs when the membrane ruptures and a homogeneous mixture fills the vessel. Assume no heat transfer during the process.

11.4 The vacuum insulating space of the vessel shown in Fig. 11.10 contained carbon dioxide with a trace of helium at ambient conditions when the vessel was empty. When the vessel was then filled with liquid hydrogen, essentially all the CO_2 froze out on the cold inner wall, leaving only the helium gas in the vacuum space. If the gas initially contained 0.01% helium by volume, what is the pressure in the vacuum space, if the average temperature in this space is assumed to be 80 K?

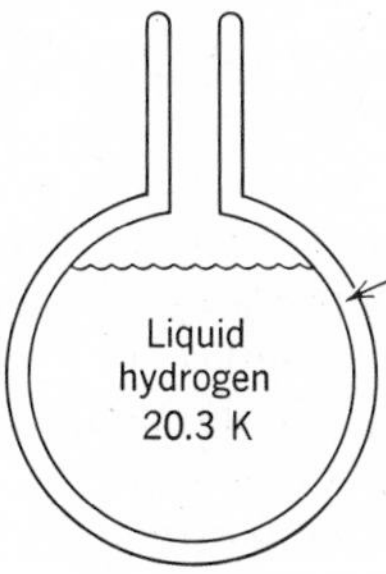

Fig. 11.10 Sketch for Problem 11.4.

11.5 A gas mixture containing 50% argon, 50% carbon dioxide by volume is contained in a cylinder at 40 lbf/in^2, 300 F. The gas is then expanded to 20 lbf/in.^{2} in a reversible adiabatic process. Determine the final temperature, the work done per mole of mixture, and the entropy changes of the Ar and the CO_2 during the process.

11.6 Nitrogen and hydrogen are mixed in a steady-flow adiabatic process in the ratio of 3 lbm of hydrogen per lbm of nitrogen. The hydrogen enters at 20 lbf/in.^{2}, 100 F, and the nitrogen at 20 lbf/in.^{2}, 500 F. The pressure after mixing is 18 lbf/in.^{2}. Determine the final temperature of the mixture and the net entropy change per lbm of mixture.

11.7 Consider the compression of a fuel-air mixture which takes place in an internal combustion engine. Assume that the fuel is ethane and that the air-fuel ratio on a mass basis is 15 to 1. The engine has a compression ratio of 9 to 1 and before the compression stroke begins the pressure in the cylinder is 13.0 lbf/in.^{2} and the temperature is 100 F. Determine the pressure and temperature after compression and the work of compression per pound of mixture.

11.8 An insulated 5-ft^3 tank contains nitrogen gas at 30 lbf/in.^{2}, 80 F. Carbon dioxide at 80 lbf/in.^{2}, 200 F flows from a pipe into the tank until the pressure inside reaches 60 lbf/in.^{2}, at which time the valve is closed. Calculate the final temperature inside the tank and the change in entropy for the process.

11.9 (*a*) 0.79 mole of nitrogen at 14.7 lbf/in.^{2}, 77 F are separated from 0.21 mole of oxygen at 14.7 lbf/in.^{2}, 77 F by a membrane. The membrane ruptures and the gases mix in an adiabatic process to form a uniform mixture. Determine the reversible work and irreversibility for this process.

(*b*) Determine the minimum work required to separate 1 mole of air (assume composition to be 79% nitrogen and 21% oxygen by volume) at 14.7 lbf/in.^{2}, 77 F into nitrogen and oxygen at 14.7 lbf/in.^{2}, 77 F.

(*c*) How would you evaluate the performance of an air separation plant regarding work input?

11.10 A room of dimensions 12 ft × 20 ft × 8 ft contains an air-water vapor mixture at a total pressure of 14.7 lbf/in.^{2} and a temperature of 80 F. The partial pressure of the water vapor is 0.2 lbf/in.^{2}. Calculate:

(*a*) The humidity ratio.

(*b*) The dew point.

(*c*) The total mass of water vapor in the room.

11.11 A certain apparatus involves a precise knowledge of the amount of water vapor in an electrical conductivity test cell. This is obtained by charging the bomb with a mixture of nitrogen and water vapor at 1000 lbf/in.^{2} pressure at a temperature of 80 F. The water vapor in the mixture is saturated. What is the humidity ratio of the mixture in the cell?

11.12 One method of removing moisture from atmospheric air is to cool the air so that the moisture condenses or freezes out. Suppose an experiment requires a humidity ratio of 0.0001. To what temperature must the air be cooled at a pressure of 1 atm in order to achieve this humidity? To what temperature must it be cooled if the pressure is 100 atm?

11.13 An air-water vapor mixture at 80 F, 14.7 lbf/in.^{2}, 50% relative humidity is compressed to 120 F, 40 lbf/in.^{2}, and is then cooled at constant pressure. At what temperature will water begin to condense?

11.14 An air-water vapor mixture enters an air-conditioning unit at a pressure of 20 lbf/in.^{2}, a temperature of 90 F, and a relative humidity of 80%. The mass of dry air entering per min is 100 lbm. The air-vapor mixture leaves the air-conditioning unit at 18 lbf/in.^{2}, 55 F, 100% relative humidity. The moisture condensed leaves at 55 F. Determine the heat transfer per min from the air.

11.15 An air-water vapor mixture enters a heater-humidifier unit at 35 F, 14.7 lbf/in.^{2}, 50% relative humidity. The flow rate of dry air is 10 lbm/min. Liquid water at 50 F is sprayed into the mixture at the rate of 0.25 lbm/min. The mixture leaves the unit at 90 F, 14.7 lbf/in.^{2}. Calculate:

(*a*) The relative humidity at the outlet.

(*b*) The rate of heat transfer to the unit.

11.16 (*a*) Determine (by a first-law analysis) the humidity ratio and relative humidity of an air-water vapor mixture that has a dry-bulb temperature of 90 F, an adiabatic saturation temperature of 78 F, and a pressure of 14.7 lbf/in.^{2}.

(*b*) By use of the psychrometric chart determine the humidity ratio and relative humidity of an air-water vapor mixture that has a dry-bulb temperature of 90 F, a wet-bulb temperature of 78 F, and a pressure of 14.7 lbf/in.^{2}.

11.17 In areas where the temperature is high and the humidity is low, some measure of air conditioning can be achieved by evaporative cooling. This involves spraying water into the air, which subsequently evaporates with a resulting decrease in the temperature of the mixture. Such a scheme is shown in Fig. 11.11.

Consider the case of atmospheric air at 100 F, 10% relative humidity, and 14.7 lbf/in.^{2}. Cooling water at 50 F is sprayed into the air. If the air-water vapor mixture is to be cooled to 80 F, what will be the relative humidity? What are the disadvantages of this approach to air conditioning?

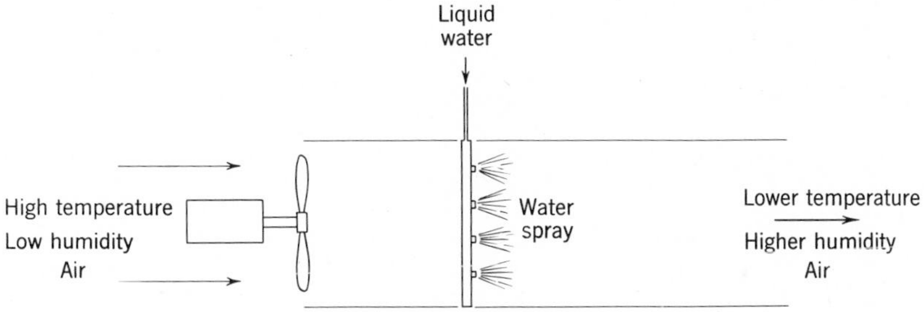

Fig. 11.11 Sketch for Problem 11.17.

11.18 An air-water vapor mixture at 14.7 lbf/in.^{2}, 80 F, and 60% relative humidity is contained in a 10 ft^3 closed tank. If the tank is cooled, find the temperature at which H_2O will begin to condense out of the mixture. If the tank is then cooled 10 degrees further, how much H_2O will be condensed?

11.19 An uninsulated tank having a volume of 10 ft^3 contains an air-water vapor mixture with a relative humidity of 90% at 100 F, 15 lbf/in.^{2}. Dry air at 100 F, 40 lbf/in.^{2} flows from a pipe into the tank until the pressure reaches 40 lbf/in.^{2}. Heat is transferred during this process such that the temperature of the material within the tank remains constant at 100 F. Determine the amount of heat transferred, in Btu, as well as the relative humidity and humidity ratio existing at the completion of the process.

11.20 Atmospheric air enters a two-stage compressor at 14.2 lbf/in.^{2}, 90 F, and 70% relative humidity. The volume rate of flow into the compressor is 500 ft^3/min. On leaving the first stage the air enters the intercooler. The pressure at the exit of the intercooler is 60 lbf/in.^{2}. The air leaves the second stage at 255 lbf/in.², 320 F and then enters the aftercooler. The air leaves the aftercooler at 250 lbf/in.², 100 F. This is shown schematically in Fig. 11.12.

(*a*) What will be the temperature of the air leaving the intercooler if the relative humidity on leaving is 100%, but no moisture is condensed in the intercooler?

(*b*) How much moisture is condensed per hr in the aftercooler, assuming no condensation in the intercooler?

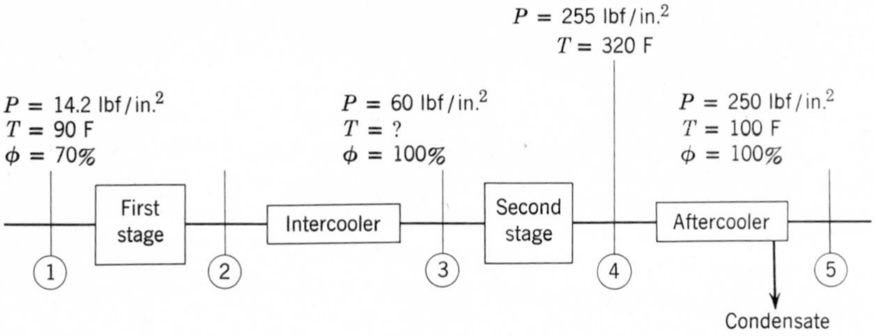

Fig. 11.12 Sketch for Problem 11.20.

11.21 A combination air cooler and dehumidifier unit receives outside air at 100 F, 14.7 lbf/in.², relative humidity of 90%. The air-water vapor mixture is first cooled to a low temperature to condense the proper amount of water, after which the air-vapor mixture is heated, leaving the unit at 70 F, 14.7 lbf/in.², relative humidity of 30%. The volume flow rate of the air-vapor mixture at the outlet is 10 ft³/min.

(*a*) Find the temperature to which the mixture is initially cooled, and the mass of water condensed per lbm of dry air. Show the process the H_2O undergoes on a *T-s* diagram.

(*b*) If all the liquid condensed leaves the unit at the minimum temperature, calculate the heat transfer rate in Btu/hr.

11.22 The initial process in an air liquefaction and separation plant involves compressing ambient air (14.7 lbf/in.², 80 F, 50% relative humidity) at the rate of 5000 ft³/min to a pressure of 3000 lbf/in.². At this point, the air is cooled to 90 F in an aftercooler and then fed to a heat exchanger in which the air will be cooled to −150 F. It may be assumed that any water entering the heat exchanger will freeze out on the tubes of the heat exchanger at such a low temperature. The designer of this plant is trying to decide whether to put a liquid trap between the aftercooler and the heat exchanger. What do you recommend? How much water will freeze on the heat exchanger tubes per hour with and without the liquid trap?

11.23 Air at 100 F, 40 lbf/in.^{2}, with a relative humidity of 35% is to be expanded in a reversible adiabatic nozzle. To how low a pressure can the gas be expanded if no condensation is to take place? What is the exit velocity at this condition?

11.24 The process of injection cooling may be demonstrated by the following problem: Consider a column of water 10 ft high which is insulated from the surroundings. Let dry air be bubbled through the water. During this process some of the water will evaporate and be carried away by the air and the temperature of the remaining liquid will decrease.

Assume that the water is initially at a temperature of 90 F, and that the dry air enters at 90 F and a pressure slightly above 1 atm. Assume that the air vapor mixture leaves the surface of the water at 1 atm pressure and the same temperature as the water and at a relative humidity of 100%.

(*a*) What is the heat transfer from the water for each pound of air entering under the initial conditions?

(*b*) What will be the ratio of the mass of air bubbled through the liquid to the mass of water cooled if the temperature of the water is decreased 10 F?

11.25 Consider the compressor and aftercooler shown in Fig. 11.13. Atmospheric air at a pressure of 14.7 lbf/in.^{2}, a temperature of 90 F, and a relative humidity of 75% enters the compressor. Under these conditions the volume flow into the compressor is 1000 ft^3/min. The rate of heat transfer from the compressor is 800 Btu/min, and the air-vapor mixture leaves the compressor at 100 lbf/in.^{2}, 400 F. It then enters the aftercooler, where it is cooled to 100 F in a constant pressure process. The air-vapor mixture leaving the aftercooler has a relative humidity of 100%, and the condensate leaves at 100 F.

(*a*) What is the power input to the compressor?

(*b*) Determine the amount of condensate per hour.

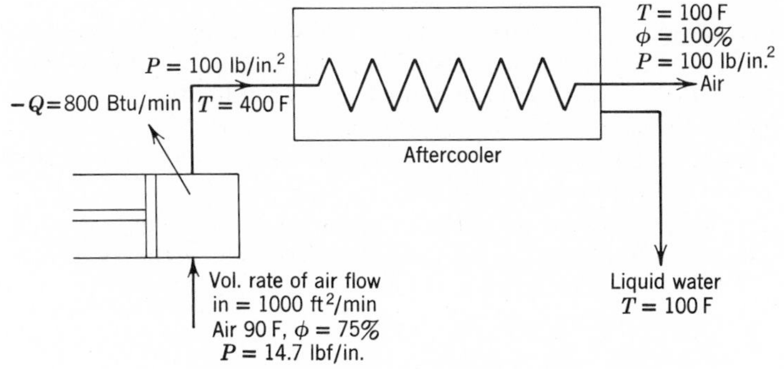

Fig. 11.13 Sketch for Problem 11.25.

11.26 A problem of current interest is that of thermal pollution of riverwater by using the water for power plant condenser cooling water and then returning it to the river at a higher temperature. The use of a cooling tower presents a solution to this particular problem. Consider the case shown in Fig. 11.14, in which 20,000 lbm/hr of water at 100 F (from the condenser) enters the top of the cooling tower, and the cool water leaves the bottom at 60 F. The air-water vapor mixture enters the bottom of the cooling tower at 14.7 lbf/in.^{2}, and has a dry-bulb temperature of 72 F and a wet-bulb temperature of 60 F. The air-water vapor mixture leaving the tower has a pressure of 14.3 lbf/in.^{2}, a temperature of 90 F, and a relative humidity of 80%. Determine the lbm of dry air per min that must be used and the fraction of the incoming water that evaporates. Assume the process to be adiabatic.

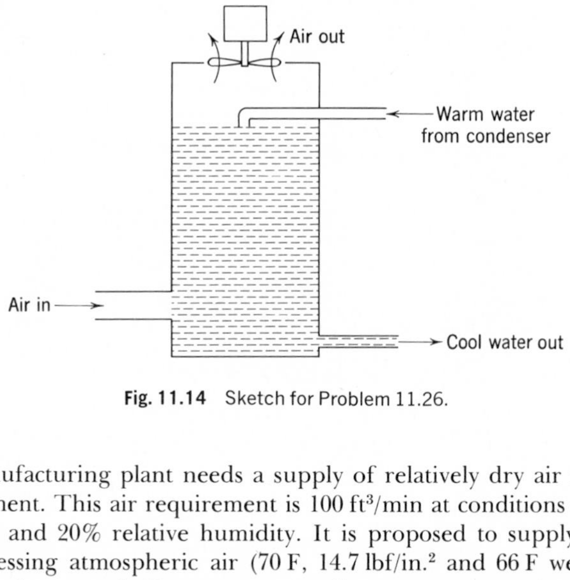

Fig. 11.14 Sketch for Problem 11.26.

11.27 A manufacturing plant needs a supply of relatively dry air for purging equipment. This air requirement is 100 ft^3/min at conditions of 70 F, 14.7 lbf/in.^{2} and 20% relative humidity. It is proposed to supply this air by compressing atmospheric air (70 F, 14.7 lbf/in.^{2} and 66 F wet-bulb temperature) to a suitable pressure, cooling at constant pressure to 70 F, removing the water that condenses, and throttling the air down to atmospheric pressure, as shown in Fig. 11.15.

(*a*) What pressure is required at point 2 to achieve the desired condition at the exit, point 5?

(*b*) How many gallons per hour of condensate are removed by the trap?

(*c*) What is the required rate of heat rejection from the cooler?

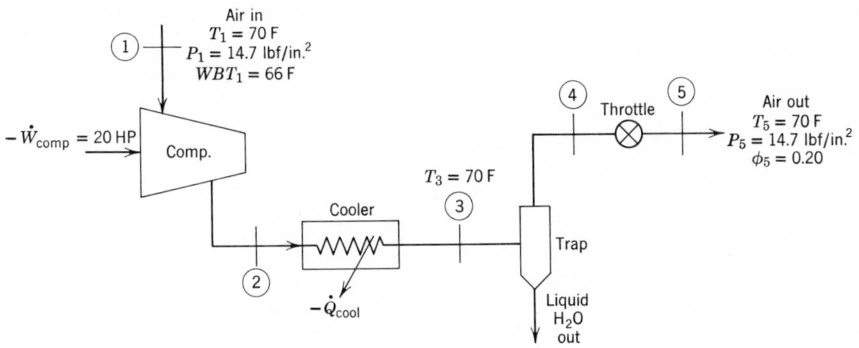

Fig. 11.15 Sketch for Problem 11.27.

11.28 One means for condensing steam during an emergency blowdown of a nuclear reactor is by use of a containment vessel as shown in Fig. 11.16. The vessel is insulated and has a total volume of 1000 ft³, with liquid H_2O initially occupying 30 ft³. The initial state in the vessel is 100 F, 14.7 lbf/in.².

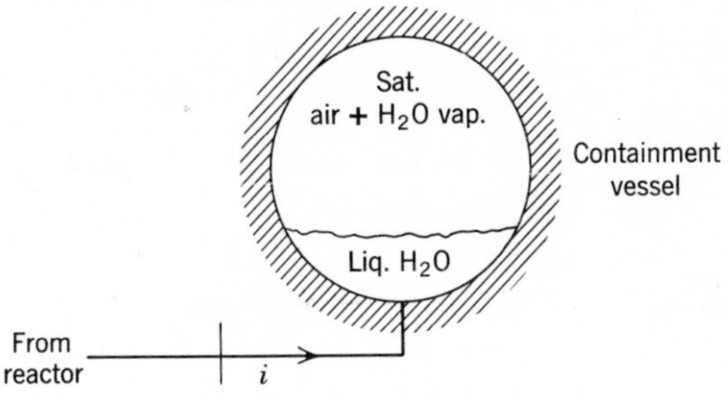

Fig. 11.16 Sketch for Problem 11.28.

Now, during a short period of time, 120 lbm of H_2O enters the vessel at an average state of $P_i = 100$ lbf/in.², $x_i = 0.50$. Find the temperature and pressure in the vessel at the end of this time.

12

Thermodynamic Relations

We have already defined and used several thermodynamic properties. Among these are pressure, specific volume, density, temperature, mass, internal energy, enthalpy, entropy, constant-pressure and constant-volume specific heats, and Joule-Thomson coefficient. Two other properties, the Helmholtz function and the Gibbs function have been introduced and will be used more extensively in the following chapters. We have also had occasion to use tables of thermodynamic properties for a number of different substances.

One important question is now raised, namely, which of the thermodynamic properties can be experimentally measured? We can answer this question by considering the measurements we can make in the laboratory. Some of the properties such as internal energy and entropy, cannot be measured directly, and must be calculated from other experimental data. If we carefully consider all these thermodynamic properties, we conclude that there are only four that can be directly measured; pressure, temperature, volume, and mass.

This leads to a second question, namely, how can values of the thermodynamic properties that cannot be measured be determined from experimental data on those properties which can be measured? In answering this question we will develop certain general thermodynamic relations. In view of the fact that there are millions of such equations that can be written, our study will be limited to certain basic considerations, with particular reference to the determination of thermodynamic properties from experimental data. We shall also consider such related matters as generalized charts and equations of state.

12.1 Two Important Relations

This chapter involves partial derivatives, and two important relations are reviewed here. Consider a variable z which is a continuous function of x and y.

$$z = f(x, y)$$

$$dz = \left(\frac{\partial z}{\partial x}\right)_y dx + \left(\frac{\partial z}{\partial y}\right)_x dy$$

It is convenient to write this in the form

$$dz = M\,dx + N\,dy \tag{12.1}$$

$M = \left(\dfrac{\partial z}{\partial x}\right)_y$ = partial derivative of z with respect to x (the variable y being held constant)

$N = \left(\dfrac{\partial z}{\partial y}\right)_x$ = partial derivative of z with respect to y (the variable x being held constant)

The physical significance of partial derivatives as they relate to the properties of a pure substance can be explained by referring to Fig. 12.1, which shows a *P-v-T* surface of the superheated vapor region of a pure substance. It shows a constant-temperature, constant-pressure, and a constant-specific volume plane which intersect at point b on the surface. Thus, the partial derivative $(\partial P/\partial v)_T$ is the slope of curve *abc* at point b. Line *de* represents the tangent to curve *abc* at point b. A similar interpretation can be made of the partial derivatives $(\partial P/\partial T)_v$ and $(\partial v/\partial T)_P$.

It should also be noted that if we wish to evaluate the partial derivative along a constant-temperature line the rules for ordinary derivatives can be applied. Thus, we can write for a constant-temperature process:

$$\left(\frac{\partial P}{\partial v}\right)_T = \frac{dP_T}{dv_T}$$

and the integration can be performed as usual. This point will be demonstrated later in a number of examples.

Let us return to the consideration of the relation

$$dz = M\,dx + N\,dy$$

If x, y, and z are all point functions (that is, quantities that depend only on the state and are independent of the path), the differentials are exact

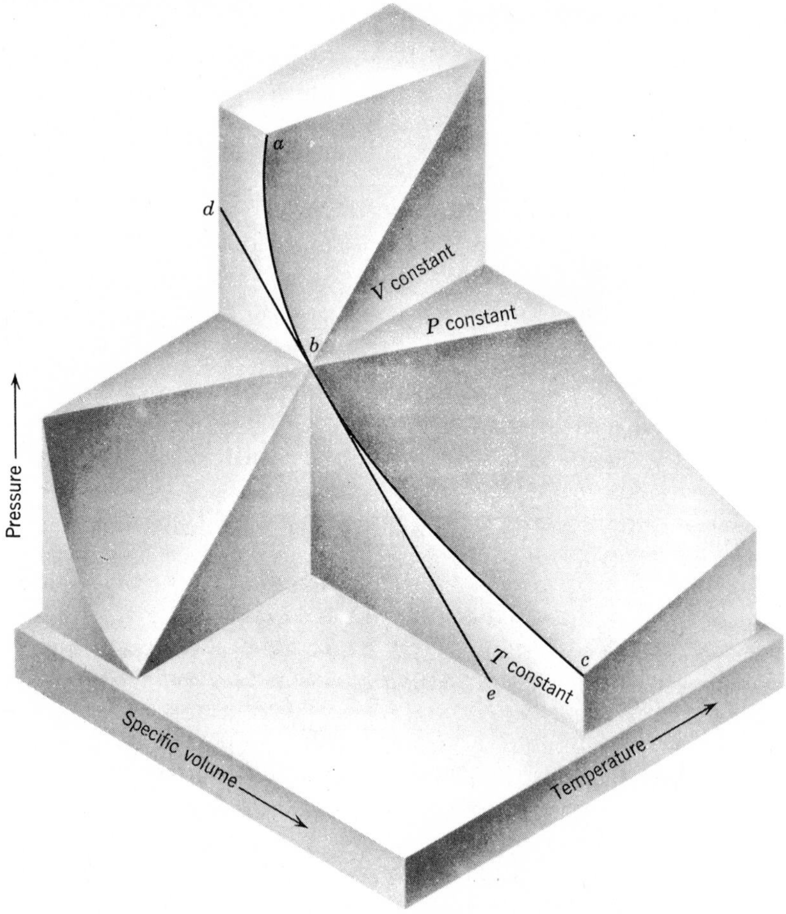

Fig. 12.1 Schematic representation of partial derivatives.

differentials. If this is the case, the following important relation holds:

$$\left(\frac{\partial M}{\partial y}\right)_x = \left(\frac{\partial N}{\partial x}\right)_y$$

The proof of this is as follows:

$$\left(\frac{\partial M}{\partial y}\right)_x = \frac{\partial^2 z}{\partial x\, \partial y}$$

$$\left(\frac{\partial N}{\partial x}\right)_y = \frac{\partial^2 z}{\partial y\, \partial x}$$

Since the order of differentiation makes no difference when point functions are involved, it follows that

$$\frac{\partial^2 z}{\partial x\,\partial y} = \frac{\partial^2 z}{\partial y\,\partial x}$$

$$\left(\frac{\partial M}{\partial y}\right)_x = \left(\frac{\partial N}{\partial x}\right)_y$$

The second important mathematical relation is

$$\left(\frac{\partial x}{\partial y}\right)_z \left(\frac{\partial y}{\partial z}\right)_x \left(\frac{\partial z}{\partial x}\right)_y = -1 \tag{12.2}$$

The proof of this relation is as follows. Consider three variables x, y, and z. Suppose there exists a relation between the variables of the form

$$x = f(y, z)$$

then

$$dx = \left(\frac{\partial x}{\partial y}\right)_z dy + \left(\frac{\partial x}{\partial z}\right)_y dz \tag{12.3}$$

If this relationship between the three variables is written in the form

$$y = f(x, z)$$

it follows that

$$dy = \left(\frac{\partial y}{\partial x}\right)_z dx + \left(\frac{\partial y}{\partial z}\right)_x dz \tag{12.4}$$

Substituting Eq. 12.4 into Eq. 12.3 we have

$$dx = \left(\frac{\partial x}{\partial y}\right)_z \left[\left(\frac{\partial y}{\partial x}\right)_z dx + \left(\frac{\partial y}{\partial z}\right)_x dz\right] + \left(\frac{\partial x}{\partial z}\right)_y dz$$

$$= \left(\frac{\partial x}{\partial y}\right)_z \left(\frac{\partial y}{\partial x}\right)_z dx + \left[\left(\frac{\partial x}{\partial y}\right)_z \left(\frac{\partial y}{\partial z}\right)_x + \left(\frac{\partial x}{\partial z}\right)_y\right] dz$$

There are two independent variables, and we select x and z as these variables. Suppose that $dz = 0$ and $dx \neq 0$. It then follows that

$$\left(\frac{\partial x}{\partial y}\right)_z \left(\frac{\partial y}{\partial x}\right)_z = 1 \tag{12.5}$$

Similarly, suppose that $dx = 0$ and $dz \neq 0$. It then follows that

$$\left(\frac{\partial x}{\partial y}\right)_z \left(\frac{\partial y}{\partial z}\right)_x + \left(\frac{\partial x}{\partial z}\right)_y = 0$$

$$\left(\frac{\partial x}{\partial y}\right)_z \left(\frac{\partial y}{\partial z}\right)_x = -\left(\frac{\partial x}{\partial z}\right)_y$$

$$\left(\frac{\partial x}{\partial y}\right)_z \left(\frac{\partial y}{\partial z}\right)_x \left(\frac{\partial z}{\partial x}\right)_y = -1$$

This is Eq. 12.2 which we set out to derive.

12.2 The Maxwell Relations

Consider a simple compressible system of fixed chemical composition. The Maxwell relations, which can be written for such a system, are four equations relating the properties P, v, T, and s.

The Maxwell relations are most easily derived by considering four relations involving thermodynamic properties. Two of these relations have already been derived and are

$$du = T\,ds - P\,dv \tag{12.6}$$

$$dh = T\,ds + v\,dP \tag{12.7}$$

The other two are derived from the definition of the Helmholtz function, a, and the Gibbs function, g.

$$a = u - Ts$$

$$da = du - T\,ds - s\,dT$$

Substituting Eq. 12.6 into this relation gives the third relation.

$$da = -P\,dv - s\,dT \tag{12.8}$$

Similarly,

$$g = h - Ts$$

$$dg = dh - T\,ds - s\,dT$$

Substituting Eq. 12.7 yields the fourth relation.

$$dg = v\,dP - s\,dT \tag{12.9}$$

Since Eqs. 12.6, 12.7, 12.8, and 12.9 are relations involving properties, we conclude that these are exact differentials, and, therefore, are of the general form

$$dz = M\,dx + N\,dy$$

Since

$$\left(\frac{\partial M}{\partial y}\right)_x = \left(\frac{\partial N}{\partial x}\right)_y \tag{12.10}$$

it follows from Eq. 12.6 that

$$\left(\frac{\partial T}{\partial v}\right)_s = -\left(\frac{\partial P}{\partial s}\right)_v \tag{12.11}$$

Similarly, from Eqs. 12.7, 12.8, and 12.9 we can write

$$\left(\frac{\partial T}{\partial P}\right)_s = \left(\frac{\partial v}{\partial s}\right)_P \tag{12.12}$$

$$\left(\frac{\partial P}{\partial T}\right)_v = \left(\frac{\partial s}{\partial v}\right)_T \tag{12.13}$$

$$\left(\frac{\partial v}{\partial T}\right)_P = -\left(\frac{\partial s}{\partial P}\right)_T \tag{12.14}$$

These four equations are known as the Maxwell relations for a simple compressible system, and the great utility of these equations will be demonstrated in later sections of this chapter. In particular, it should be noted that pressure, temperature, and specific volume can be measured by experimental methods, whereas entropy cannot be determined experimentally. By using the Maxwell relations changes in entropy can be determined from quantities that can be measured, i.e., pressure, temperature, and specific volume.

There are a number of other very useful relations that can be derived from Eqs. 12.6 through 12.9. For example, from Eq. 12.6 we can write the relations

$$\left(\frac{\partial u}{\partial s}\right)_v = T, \quad \left(\frac{\partial u}{\partial v}\right)_s = -P \tag{12.15}$$

Similarly, from the other equations we have the following:

$$\left(\frac{\partial h}{\partial s}\right)_P = T, \quad \left(\frac{\partial h}{\partial P}\right)_s = v$$

$$\left(\frac{\partial a}{\partial v}\right)_T = -P, \quad \left(\frac{\partial a}{\partial T}\right)_v = -s$$

$$\left(\frac{\partial g}{\partial P}\right)_T = v, \quad \left(\frac{\partial g}{\partial T}\right)_P = -s \tag{12.16}$$

As already noted, the Maxwell relations just presented are written for a simple compressible substance. It is readily evident, however, that similar Maxwell relations can be written for substances involving other effects, such as electrical and magnetic effects. For example, Eq. 8.9 can be written in the form

$$dU = T\,dS - P\,dV + \mathscr{T}\,dL + \mathscr{S}\,dA + \mu_0\mathscr{H}\,d(V\mathscr{M}) + \mathscr{E}\,dZ + \cdots \tag{12.17}$$

Thus, at constant volume for a substance involving only magnetic effects we can write

$$dU = T\,dS + \mu_0 V\,\mathscr{H}\,d\mathscr{M}$$

and it follows that for such a substance

$$\left(\frac{\partial T}{\partial \mathscr{M}}\right)_S = \mu_0 V\left(\frac{\partial \mathscr{H}}{\partial S}\right)_{\mathscr{M}}$$

Other Maxwell relations similar to Eqs. 12.12 through 12.14 could be written for such a substance. The extension of this approach to other systems, as well as to the interrelation between the various effects that may occur in a given system, is readily evident. For example, suppose a system involved both magnetic and surface effects. For such a system we could consider a constant entropy process and write

$$\left(\frac{\partial \mathscr{S}}{\partial \mathscr{M}}\right)_{S,A} = \mu_0 V\left(\frac{\partial \mathscr{H}}{\partial A}\right)_{S,\mathscr{M}}$$

Example 12.1

From an examination of the properties of compressed liquid water, as given in Table B.1.4 of the Appendix, we find that the entropy of compressed liquid is greater than the entropy of saturated liquid for a temperature of 32 F, and is less than that of saturated liquid for all the other temperatures listed. Explain why this follows from other thermodynamic data.

Suppose we increase the pressure of liquid water that is initially saturated, while keeping the temperature constant. The change of entropy for the water during this process can be found by integrating the following Maxwell relation, Eq. 12.14,

$$\left(\frac{\partial s}{\partial P}\right)_T = -\left(\frac{\partial v}{\partial T}\right)_P$$

Therefore, the sign of the entropy change depends on the sign of the term $(\partial v/\partial T)_P$. The physical significance of this term is that it involves the change in specific volume of water as the temperature changes while the pressure remains constant. Now, as water at moderate pressures and 32 F is heated in a constant-pressure process, the specific volume decreases until the point of maximum density is reached at approximately 39 F, after which it increases. This is shown on a v-T diagram in Fig. 12.2. Thus, the quantity $(\partial v/\partial T)_P$ is the slope of the curve in Fig.

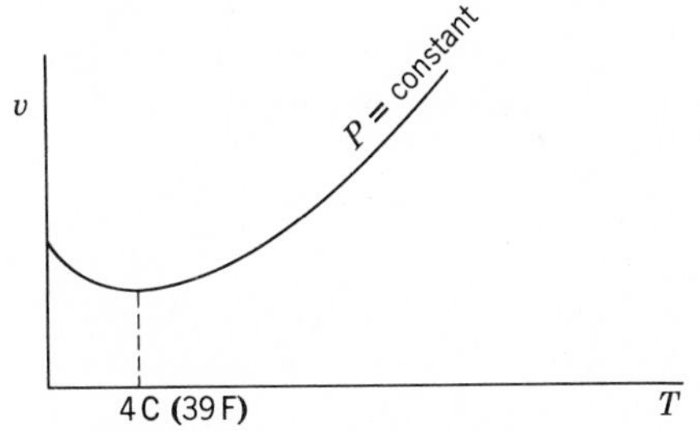

Fig. 12.2 Sketch for Example 12.1.

12.2. Since this slope is negative at 32 F, the quantity $(\partial s/\partial P)_T$ is positive at 32 F. At the point of maximum density the slope is zero, and therefore, the constant-pressure line shown in Fig. 8.7 crosses the saturated-liquid line at the point of maximum density.

12.3 The Property Relation for Mixtures

In Chapter 11 the partial molal property was briefly introduced. However, the consideration of mixtures in this chapter was limited to ideal gases, and there was no need at that point for further expansion of the subject. We now continue this subject with a view toward developing the property relations for mixtures. This will be particularly relevant to our consideration of chemical equilibrium in Chapter 14.

For a mixture, any extensive property X is a function of the temperature and pressure of the mixture and the number of moles of each component. Thus, for a mixture of two components,

$$X = f(T, P, n_A, n_B)$$

Therefore,

$$dX_{T,P} = \left(\frac{\partial X}{\partial n_A}\right)_{T,P,n_B} dn_A + \left(\frac{\partial X}{\partial n_B}\right)_{T,P,n_A} dn_B \qquad (12.18)$$

Since at constant temperature and pressure an extensive property is directly proportional to the mass, Eq. 12.18 can be integrated to give

$$X_{T,P} = \bar{X}_A n_A + \bar{X}_B n_B \tag{12.19}$$

where

$$\bar{X}_A = \left(\frac{\partial X}{\partial n_A}\right)_{T,P,n_B}; \quad \bar{X}_B = \left(\frac{\partial X}{\partial n_B}\right)_{T,P,n_A}$$

$\bar{X}$ is defined as the partial molal property for a component in a mixture. It is particularly important to note that the partial molal property is defined under conditions of constant temperature and pressure. Note that Eq. 12.19 is of the same form as Eq. 11.3.

The partial molal property is particularly significant when dealing with a mixture undergoing a chemical reaction. Suppose a mixture consists of components A and B, and a chemical reaction takes place so that the number of moles of A is changed by dn_A and the number of moles of B by dn_B, while the temperature and pressure remain constant. What would the change of internal energy of the mixture be during this process?

From Eq. 12.19 we conclude that

$$dU_{T,P} = \bar{U}_A dn_A + \bar{U}_B dn_B \tag{12.20}$$

where $\bar{U}_A$ and $\bar{U}_B$ are the partial molal internal energy of A and B respectively. Equation 12.20 suggests that the partial molal internal energy of each component can also be defined as the internal energy of the component as it exists in the mixture.

In Section 12.2 we considered a number of property relations for systems of fixed mass such as

$$dU = T\,dS - P\,dV$$

We note that in this equation temperature is the intensive property or potential function associated with entropy and pressure is the intensive property associated with volume. Suppose we have a chemical reaction such as described in the last paragraph. How would we modify this property relation for this situation? Intuitively we might write the equation

$$dU = T\,dS - P\,dV + \mu_A dn_A + \mu_B dn_B \tag{12.21}$$

where μ_A is the intensive property or potential function associated with n_A, and similarly μ_B for n_B. This potential function is called the chemical potential.

To derive an expression for this chemical potential, we examine Eq. 12.21 and conclude that it might be reasonable to write an expression for U in the form

$$U = f(S, V, n_A, n_B)$$

Therefore,

$$dU = \left(\frac{\partial U}{\partial S}\right)_{V,n_A,n_B} dS + \left(\frac{\partial U}{\partial V}\right)_{S,n_A,n_B} dV + \left(\frac{\partial U}{\partial n_A}\right)_{S,V,n_B} dn_A + \left(\frac{\partial U}{\partial n_B}\right)_{S,V,n_A} dn_B$$

Since the expressions

$$\left(\frac{\partial U}{\partial S}\right)_{V,n_A,n_B} \quad \text{and} \quad \left(\frac{\partial U}{\partial V}\right)_{S,n_A,n_B}$$

imply constant composition, it follows from Eq. 12.15 that

$$\left(\frac{\partial U}{\partial S}\right)_{V,n_A,n_B} = T \quad \text{and} \quad \left(\frac{\partial U}{\partial V}\right)_{S,n_A,n_B} = -P$$

Thus

$$dU = T\,dS - P\,dV + \left(\frac{\partial U}{\partial n_A}\right)_{S,V,n_B} dn_A + \left(\frac{\partial U}{\partial n_B}\right)_{S,V,n_A} dn_B \qquad (12.22)$$

On comparing this equation with Eq. 12.21 it follows that the chemical potential can be defined by the relation

$$\mu_A = \left(\frac{\partial U}{\partial n_A}\right)_{S,V,n_B} \quad ; \quad \mu_B = \left(\frac{\partial U}{\partial n_B}\right)_{S,V,n_A} \qquad (12.23)$$

We can also relate the chemical potential to the partial molal Gibbs function. To do so we proceed as follows.

$$G = U + PV - TS$$

$$dG = dU + P\,dV + V\,dP - T\,dS - S\,dT$$

Substituting Eq. 12.11 into this relation we have

$$dG = -S\,dT + V\,dP + \mu_A dn_A + \mu_B dn_B \qquad (12.24)$$

This equation suggests that we write an expression for G in the following form.

$$G = f(T, P, n_A, n_B)$$

Proceeding as we did in the case of a similar expression for internal energy, Eq. 12.22, we have

$$dG = \left(\frac{\partial G}{\partial T}\right)_{P,n_A,n_B} dT + \left(\frac{\partial G}{\partial P}\right)_{T,n_A,n_B} dP + \left(\frac{\partial G}{\partial n_A}\right)_{T,P,n_B} dn_A + \left(\frac{\partial G}{\partial n_B}\right)_{T,P,n_A} dn_B$$

$$= -S\,dT + V\,dP + \left(\frac{\partial G}{\partial n_A}\right)_{T,P,n_B} dn_A + \left(\frac{\partial G}{\partial n_B}\right)_{T,P,n_A} dn_B$$

Comparing this with Eq. 12.24 it follows that

$$\mu_A = \left(\frac{\partial G}{\partial n_A}\right)_{T,P,n_B} \quad ; \quad \mu_B = \left(\frac{\partial G}{\partial n_B}\right)_{T,P,n_A}$$

Note that, since partial molal properties are defined at constant temperature and pressure, the quantities $(\partial G/\partial n_A)_{T,P,n_B}$ and $(\partial G/\partial n_B)_{T,P,n_A}$ are the partial molal Gibbs functions for the two components. That is, the chemical potential is equal to the partial molal Gibbs function.

$$\mu_A = \overline{G}_A = \left(\frac{\partial G}{\partial n_A}\right)_{T,P,n_B} ; \quad \mu_B = \overline{G}_B = \left(\frac{\partial G}{\partial n_B}\right)_{T,P,n_A} \tag{12.25}$$

Although μ can also be defined in terms of other properties, such as in Eq. 12.23, this expression is not the partial molal internal energy, since the pressure and temperature are not constant in this partial derivative. The partial molal Gibbs function is an extremely important property in the thermodynamic analysis of chemical reactions because at constant temperature and pressure (the conditions under which many chemical reactions occur) it is a measure of the chemical potential or the driving force tending to cause a chemical reaction to take place.

12.4 Clapeyron Equation

The Clapeyron equation is an important relation involving the saturation pressure and temperature, the change of enthalpy associated with a change of phase, and the specific volumes of the two phases. In particular it is an example of how a change in a property that cannot be measured directly, the enthalpy in this case, can be determined from measurements of pressure, temperature, and specific volume. It can be derived in a number of ways. Here we proceed by considering one of the Maxwell relations, Eq. 12.13.

$$\left(\frac{\partial P}{\partial T}\right)_v = \left(\frac{\partial s}{\partial v}\right)_T$$

Consider, for example, the change of state from saturated liquid to saturated vapor of a pure substance. This is a constant-temperature process, and therefore Eq. 12.13 can be integrated between the saturated-liquid and saturated-vapor state. We also note that when saturated states are involved, pressure and temperature are independent of volume. Therefore,

$$\left(\frac{dP}{dT}\right)_{\text{sat}} = \frac{s_g - s_f}{v_g - v_f} = \frac{s_{fg}}{v_{fg}} = \frac{h_{fg}}{Tv_{fg}} \tag{12.26}$$

The significance of this equation is that $(dP/dT)_{\text{sat}}$ is the slope of the vapor-pressure curve. Thus, h_{fg} at a given temperature can be determined from the slope of the vapor-pressure curve and the specific volume of saturated liquid and saturated vapor at the given temperature.

There are several different changes of phase that can occur at constant temperature and constant pressure. If we designate the two phases with superscripts $''$ and $'$, we can write the Clapeyron equation for the general case.

$$\left(\frac{dP}{dT}\right)_{\text{sat}} = \frac{s'' - s'}{v'' - v'}$$

We also note that $T(s'' - s') = h'' - h'$. Therefore

$$\left(\frac{dP}{dT}\right)_{\text{sat}} = \frac{h'' - h'}{T(v'' - v')} \tag{12.27}$$

If the phase designated $''$ is vapor, then at low pressure the equation is usually simplified by assuming that $v'' \gg v'$ and assuming also that $v'' = RT/P$. The relation then becomes

$$\left(\frac{dP}{dT}\right)_{\text{sat}} = \frac{h_g - h'}{T(RT/P)}$$

$$\left(\frac{dP}{P}\right)_{\text{sat}} = \frac{(h_g - h')}{R}\left(\frac{dT}{T^2}\right)_{\text{sat}} \tag{12.28}$$

Example 12.2

Determine the saturation pressure of water vapor at -100 F using data available in the steam tables. Table 6 of the steam tables (Appendix Table B.1.5) does not give saturation pressures for temperatures less than -40 F. However, we do notice that h_{ig} is relatively constant in this range, and, therefore, we proceed to use Eq. 12.28 and integrate between the limits -40 F and -100 F.

$$\int_1^2 \frac{dP}{P} = \int_1^2 \frac{h_{ig}}{R} \frac{dT}{T^2} = \frac{h_{ig}}{R} \int_1^2 \frac{dT}{T^2}$$

$$\ln \frac{P_2}{P_1} = \frac{h_{ig}}{R}\left(\frac{T_2 - T_1}{T_1 T_2}\right)$$

Let

$$P_2 = 0.0019 \text{ lbf/in.}^2 \quad T_2 = 420 \text{ R}$$

$$P_1 = ? \quad T_1 = 360 \text{ R}$$

Then

$$\ln \frac{P_2}{P_1} = \frac{1220.6 \times 778}{85.76}\left(\frac{420 - 360}{420 \times 360}\right)$$

$$\ln \frac{P_2}{P_1} = 4.40; \quad \frac{P_2}{P_1} = 81.5; \quad P_1 = \frac{0.0019}{81.5} = 2.33 \times 10^{-5} \text{ lbf/in.}^2$$

12.5 Some Thermodynamic Relations Involving Enthalpy, Internal Energy, and Entropy

Let us first derive two equations, one involving C_p and the other involving C_v.

We have defined C_p as

$$C_p \equiv \left(\frac{\partial h}{\partial T}\right)_P$$

We have also noted that for a pure substance

$$T\,ds = dh - v\,dP$$

Therefore,

$$C_p = \left(\frac{\partial h}{\partial T}\right)_P = T\left(\frac{\partial s}{\partial T}\right)_P \tag{12.29}$$

Similarly, from the definition of C_v,

$$C_v \equiv \left(\frac{\partial u}{\partial T}\right)_v$$

and the relation

$$T ds = du + P\,dv,$$

it follows that

$$C_v = \left(\frac{\partial u}{\partial T}\right)_v = T\left(\frac{\partial s}{\partial T}\right)_v \tag{12.30}$$

We will now derive a general relation for the change of enthalpy of a pure substance. We first note that for a pure substance

$$h = h(T, P)$$

Therefore,

$$dh = \left(\frac{\partial h}{\partial T}\right)_P dT + \left(\frac{\partial h}{\partial P}\right)_T dP$$

From the relation

$$T\,ds = dh - v\,dP$$

it follows that

$$\left(\frac{\partial h}{\partial P}\right)_T = v + T\left(\frac{\partial s}{\partial P}\right)_T$$

Substituting the Maxwell relation, Eq. 12.14, we have

$$\left(\frac{\partial h}{\partial P}\right)_T = v - T\left(\frac{\partial v}{\partial T}\right)_P \tag{12.31}$$

On substituting this equation and Eq. 12.29 we have

$$dh = C_p\,dT + \left[v - T\left(\frac{\partial v}{\partial T}\right)_P\right] dP \tag{12.32}$$

Along an isobar we have

$$dh_P = C_p\,dT_P$$

and along an isotherm,

$$dh_T = \left[v - T\left(\frac{\partial v}{\partial T}\right)_P\right] dP_T \tag{12.33}$$

The significance of Eq. 12.32 is that this equation can be integrated to give the change in enthalpy associated with a change of state

$$h_2 - h_1 = \int_1^2 C_p\,dT + \int_1^2 \left[v - T\left(\frac{\partial v}{\partial T}\right)_P\right] dP \tag{12.34}$$

The information needed to integrate the first term is a constant pressure specific heat along one (and only one) isobar. The integration of the second integral requires that an equation of state giving the relation between P, v, and T be known. Furthermore, it is advantageous to have this equation of state explicit in v, for then the derivative $(\partial v/\partial T)_P$ is readily evaluated.

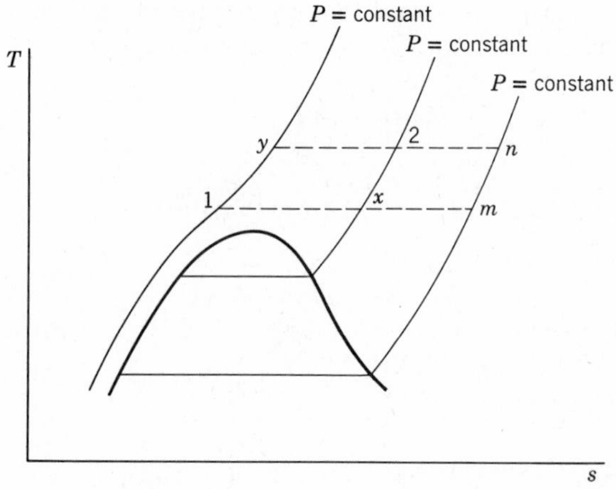

Fig. 12.3 Sketch showing various paths by which a given change of state can take place.

This can be further illustrated by reference to Fig. 12.3. Suppose we wished to know the change of enthalpy between states 1 and 2. We might do so along path 1–x–2, which consists of one isotherm (1–x) and one isobar (x–2). Thus we could integrate Eq. 12.34,

$$h_2 - h_1 = \int_{T_1}^{T_2} C_p \, dT + \int_{P_1}^{P_2} \left[v - T\left(\frac{\partial v}{\partial T}\right)_P \right] dP$$

Since $T_1 = T_x$ and $P_2 = P_x$ this can be written

$$h_2 - h_1 = \int_{T_x}^{T_2} C_p \, dT + \int_{P_1}^{P_x} \left[v - T\left(\frac{\partial v}{\partial T}\right)_P \right] dP$$

The second term in this equation gives the change in enthalpy along the isotherm 1–x and the first term the change in enthalpy along the isobar x–2. When these are added together, the result is the net change in enthalpy between 1 and 2. It should be noted that in this case the constant-pressure specific heat must be known along the isobar passing through 2 and x. One could also find the change in enthalpy by following path 1–y–2, in which case the constant-pressure specific heat must be known along the 1–y isobar. If the constant-pressure specific heat is known at another pressure, say the isobar passing through m–n, the change in enthalpy could be found by following path 1–m–n–2. This would involve calculating the change of enthalpy along two isotherms, namely, 1–m and n–2.

Let us now derive a similar relation for the change of internal energy. All the steps in this derivation are given, but without detailed comment. Note that the starting point is to write $u = u(T, v)$ whereas in the case of enthalpy the starting point was $h = h(T, P)$.

$$u = f(T, v)$$

$$du = \left(\frac{\partial u}{\partial T}\right)_v dT + \left(\frac{\partial u}{\partial v}\right)_T dv$$

$$T\,ds = du + P\,dv$$

Therefore,

$$\left(\frac{\partial u}{\partial v}\right)_T = T\left(\frac{\partial s}{\partial v}\right)_T - P \tag{12.35}$$

Substituting the Maxwell relation, Eq. 12.13,

$$\left(\frac{\partial u}{\partial v}\right)_T = T\left(\frac{\partial P}{\partial T}\right)_v - P$$

Therefore,

$$du = C_v\,dT + \left[T\left(\frac{\partial P}{\partial T}\right)_v - P\right] dv \tag{12.36}$$

Along an isometric this reduces to

$$du_v = C_v\,dT_v$$

and along an isotherm we have

$$du_T = \left[T\left(\frac{\partial P}{\partial T}\right)_v - P\right] dv_T \tag{12.37}$$

In a manner similar to that outlined above for changes in enthalpy, the change of internal energy for a given change of state for a pure substance can be determined from Eq. 12.36 if the constant-volume specific heat is known along one isometric and an equation of state explicit in P (to obtain the derivative $(\partial P/\partial T)_v$) is available in the region involved. A diagram similar to Fig. 12.3 could be drawn, with the isobars replaced with isometrics, and the same general conclusions would be reached.

To summarize, we have derived Eqs. 12.32 and 12.36,

$$dh = C_p\,dT + \left[v - T\left(\frac{\partial v}{\partial T}\right)_P\right] dP$$

$$du = C_v\,dT + \left[T\left(\frac{\partial P}{\partial T}\right)_v - P\right] dv$$

The first of these involves the change of enthalpy, the constant-pressure specific heat, and is particularly suited to an equation of state explicit in v. The second involves the change of internal energy, the constant-volume specific heat, and is particularly suited to an equation of state explicit in P. If the first of these equations is used to determine the change of enthalpy, the internal energy is readily found by noting that

$$u_2 - u_1 = h_2 - h_1 - (P_2 v_2 - P_1 v_1)$$

If the second equation is used to find changes of internal energy, the change of enthalpy is readily found from this same relation. Which of these two equations is used to determine changes in internal energy and enthalpy will depend on the information available for specific heat and an equation of state (or other P-v-T data).

Two parallel expressions can be found for the change of entropy.

$$s = s(T, P)$$

$$ds = \left(\frac{\partial s}{\partial T}\right)_P dT + \left(\frac{\partial s}{\partial P}\right)_T dP$$

Substituting Eqs. 12.29 and 12.14 we have

$$ds = C_p \frac{dT}{T} - \left(\frac{\partial v}{\partial T}\right)_P dP \tag{12.38}$$

$$s_2 - s_1 = \int_1^2 C_p \frac{dT}{T} - \int_1^2 \left(\frac{\partial v}{\partial T}\right)_P dP \tag{12.39}$$

Along an isobar we have

$$(s_2 - s_1)_P = \int_1^2 C_p \frac{dT_P}{T}$$

and along an isotherm,

$$(s_2 - s_1)_T = -\int_1^2 \left(\frac{\partial v}{\partial T}\right)_P dP$$

Note from Eq. 12.39 that if a constant-pressure specific heat is known along one isobar and an equation of state explicit in v is available, the change of entropy can be evaluated. This is analogous to the expression for the change of enthalpy given in Eq. 12.32.

The second equation for the change of entropy can be found as follows.

$$s = s(T, v)$$

$$ds = \left(\frac{\partial s}{\partial T}\right)_v dT + \left(\frac{\partial s}{\partial v}\right)_T dv$$

Substituting Eqs. 12.30 and 12.13,

$$ds = C_v \frac{dT}{T} + \left(\frac{\partial P}{\partial T}\right)_v dv \qquad (12.40)$$

$$s_2 - s_1 = \int_1^2 C_v \frac{dT}{T} + \int_1^2 \left(\frac{\partial P}{\partial T}\right)_v dv \qquad (12.41)$$

This expression for change of entropy involves the change of entropy along an isometric where the constant-volume specific heat is known and along an isotherm where an equation of state explicit in P is known, and thus it is analogous to the expression for change of internal energy given in Eq. 12.36.

It should also be noted that it follows directly from Eq. 12.32, that the enthalpy of an ideal gas is a function of only the temperature. From Eq. 12.32,

$$dh = C_p\, dT + \left[v - T\left(\frac{\partial v}{\partial T}\right)_P\right] dP$$

For an ideal gas, for which the equation of state is $Pv = RT$, it follows that

$$T\left(\frac{\partial v}{\partial T}\right)_P = T\left(\frac{R}{P}\right) = v$$

Therefore

$$\left[v - T\left(\frac{\partial v}{\partial T}\right)_P\right] = 0$$

and for an ideal gas

$$dh = C_p dT$$

Let us also apply Eq. 12.38 to an ideal gas.

$$ds = C_p \frac{dT}{T} - \left(\frac{\partial v}{\partial T}\right)_P dP$$

For an ideal gas

$$\left(\frac{\partial v}{\partial T}\right)_P = \frac{R}{P}$$

and

$$ds = C_p \frac{dT}{T} - R \frac{dP}{P}$$

This equation was derived by an alternate procedure when discussing ideal gas behavior in Chapter 8, Eq. 8.19.

Example 12.3

Over a certain range of pressures and temperatures the equation of state of a certain substance is given with considerable accuracy by the relation

$$\frac{Pv}{RT} = 1 - C' \frac{P}{T^4}$$

or

$$v = \frac{RT}{P} - \frac{C}{T^3}$$

where C and C' are constants.

Derive an expression for the change of enthalpy and entropy of this substance in an isothermal process.

Since the equation of state is explicit in v, Eq. 12.33 would be particularly relevant to this problem. On integrating this equation we have,

$$(h_2 - h_1)_T = \int_1^2 \left[v - T \left(\frac{\partial v}{\partial T} \right)_P \right] dP_T$$

From the equation of state,

$$\left(\frac{\partial v}{\partial T} \right)_P = \frac{R}{P} + \frac{3C}{T^4}$$

Therefore,

$$(h_2 - h_1)_T = \int_1^2 \left[v - T \left(\frac{R}{P} + \frac{3C}{T^4} \right) \right] dP_T$$

$$= \int_1^2 \left[\frac{RT}{P} - \frac{C}{T^3} - \frac{RT}{P} - \frac{3C}{T^3} \right] dP_T$$

$$(h_2 - h_1)_T = \int_1^2 -\frac{4C}{T^3} dP_T = -\frac{4C}{T^3} (P_2 - P_1)_T$$

For entropy we use Eq. 12.39, which is particularly relevant for an equation of state explicit in v.

$$(s_2 - s_1)_T = -\int_1^2 \left(\frac{\partial v}{\partial T} \right)_P dP_T = -\int_1^2 \left(\frac{R}{P} + \frac{3C}{T^4} \right) dP_T$$

$$(s_2 - s_1)_T = -R \ln \left(\frac{P_2}{P_1} \right)_T - \frac{3C}{T^4} (P_2 - P_1)_T$$

12.6 Some Thermodynamic Relations Involving Specific Heat

Some important relations involving specific heats can also be developed. We have noted that the specific heat of an ideal gas is a function of the temperature only. For real gases the specific heat varies with pressure as well as temperature, and frequently we are interested in the variation of specific heat with pressure or volume. These relations can be derived as follows. Consider Eq. 12.38,

$$ds = \left(\frac{C_p}{T}\right) dT - \left(\frac{\partial v}{\partial T}\right)_P dP$$

Since this equation is of the general form $dz = M\,dx + N\,dy$, we can proceed as follows to find a relation that gives the variation of the constant-pressure specific heat with pressure at constant temperature.

$$\left(\frac{\partial (C_p/T)}{\partial P}\right)_T = -\left[\frac{\partial}{\partial T}\left(\frac{\partial v}{\partial T}\right)_P\right]_P$$

$$\left(\frac{\partial C_p}{\partial P}\right)_T = -T\left(\frac{\partial^2 v}{\partial T^2}\right)_P \qquad (12.42)$$

The variation of the constant-volume specific heat with volume as the temperature remains constant can be found in a similar manner. Consider Eq. 12.40,

$$ds = \left(\frac{C_v}{T}\right) dT + \left(\frac{\partial P}{\partial T}\right)_v dv$$

Since this is of the form $dz = M\,dx + N\,dy$,

$$\left(\frac{\partial (C_v/T)}{\partial v}\right)_T = \left[\frac{\partial}{\partial T}\left(\frac{\partial P}{\partial T}\right)_v\right]_v$$

$$\left(\frac{\partial C_v}{\partial v}\right)_T = T\left(\frac{\partial^2 P}{\partial T^2}\right)_v \qquad (12.43)$$

The important thing to note about Eqs. 12.42 and 12.43 is that the variation of the constant-volume and constant-pressure specific heats at constant-temperature can be found from the equation of state.

Example 12.4

Determine the variation of C_p with pressure at constant temperature for a substance such as the one in Example 12.3 over the range where the equation of state is given by the relation

$$v=\frac{RT}{P}-\frac{C}{T^3}$$

Using Eq. 12.42

$$\left(\frac{\partial C_p}{\partial P}\right)_T=-T\left(\frac{\partial^2 v}{\partial T^2}\right)_P$$

$$\left(\frac{\partial v}{\partial T}\right)_P=\frac{R}{P}+\frac{3C}{T^4}$$

$$\left(\frac{\partial^2 v}{\partial T^2}\right)_P=-\frac{12C}{T^5}$$

$$\left(\frac{\partial C_p}{\partial P}\right)_T=-T\left(-\frac{12C}{T^5}\right)=\frac{12C}{T^4}$$

A final interesting and useful relation involving the difference between C_p and C_v can be derived by equating Eqs. 12.38 and 12.40.

$$C_p\frac{dT}{T}-\left(\frac{\partial v}{\partial T}\right)_P dP=C_v\frac{dT}{T}+\left(\frac{\partial P}{\partial T}\right)_v dv$$

$$dT=\frac{T(\partial P/\partial T)_v}{C_p-C_v}dv+\frac{T(\partial v/\partial T)_P}{C_p-C_v}dP$$

But

$$T=f(v,P)$$

$$dT=\left(\frac{\partial T}{\partial v}\right)_P dv+\left(\frac{\partial T}{\partial P}\right)_v dP$$

Therefore,

$$\left(\frac{\partial T}{\partial v}\right)_P=\frac{T(\partial P/\partial T)_v}{C_p-C_v}$$

and

$$\left(\frac{\partial T}{\partial P}\right)_v=\frac{T(\partial v/\partial T)_P}{C_p-C_v}$$

When these equations are solved for C_p-C_v, they yield the same result.

$$C_p-C_v=T\left(\frac{\partial v}{\partial T}\right)_P\left(\frac{\partial P}{\partial T}\right)_v \qquad (12.44)$$

But from Eq. 12.2,

$$\left(\frac{\partial P}{\partial T}\right)_v=-\left(\frac{\partial v}{\partial T}\right)_P\left(\frac{\partial P}{\partial v}\right)_T$$

Therefore,

$$C_p - C_v = -T\left(\frac{\partial v}{\partial T}\right)_P^2\left(\frac{\partial P}{\partial v}\right)_T \tag{12.45}$$

From this equation we draw several conclusions:

1. For liquids and solids $(\partial v/\partial T)_P$ is usually relatively small, and, therefore, for these phases the difference between the constant-pressure and constant-volume specific heats is small. For this reason many tables simply give the specific heat of a solid or a liquid without designating that it is at constant volume or pressure. Further, $C_p = C_v$ exactly when $(\partial v/\partial T)_P = 0$, as is true at the point of maximum density of water.
2. $C_p \to C_v$ as $T \to 0$, and, therefore, we conclude that the constant-pressure and constant-volume specific heats are equal at absolute zero.
3. The difference between C_p and C_v is always positive because $(\partial v/\partial T)_P{}^2$ is always positive and $(\partial P/\partial v)_T$ is negative for all known substances.
4. For an ideal gas

$$\left(\frac{\partial v}{\partial T}\right)_P^2 = \frac{R^2}{P^2} \quad \text{and} \quad \left(\frac{\partial P}{\partial v}\right)_T = -\frac{R}{Tv^2}$$

Therefore,

$$C_p - C_v = -T\left(\frac{R}{P}\right)^2\left(-\frac{RT}{v^2}\right) = R$$

This relationship was derived by another procedure in arriving at Eq. 5.24.

12.7 Volume Expansivity and Isothermal and Adiabatic Compressibility

The student has most likely encountered the coefficient of linear expansion in his studies of strength of materials. This coefficient indicates how the length of a solid body is influenced by a change in temperature while the pressure remains constant. In terms of the notation of partial derivatives the *coefficient of linear expansion*, δ_T, is defined as follows:

$$\delta_T = \frac{1}{L}\left(\frac{\partial L}{\partial T}\right)_P \tag{12.46}$$

A similar coefficient can be defined for changes in volume, and such a coefficient is applicable to liquids and gases as well as to solids. This coefficient of volume expansion α, also called the volume expansivity, is an indication of the change in volume that results from a change in temperature while the pressure remains constant. The definition of *volume expansivity* is

$$\alpha \equiv \frac{1}{V}\left(\frac{\partial V}{\partial T}\right)_P = \frac{1}{v}\left(\frac{\partial v}{\partial T}\right)_P \tag{12.47}$$

The isothermal compressibility β_T is an indication of the change in volume that results from a change in pressure while the temperature remains constant. The definition of the *isothermal compressibility* is

$$\beta_T \equiv -\frac{1}{V}\left(\frac{\partial V}{\partial P}\right)_T = -\frac{1}{v}\left(\frac{\partial v}{\partial P}\right)_T \tag{12.48}$$

The reciprocal of the isothermal compressibility is called the *isothermal bulk modulus* B_T.

$$B_T \equiv -v\left(\frac{\partial P}{\partial v}\right)_T \tag{12.49}$$

The *adiabatic compressibility* β_s is an indication of the change in volume that results from a change in pressure while the entropy remains constant, and is defined as follows:

$$\beta_s \equiv -\frac{1}{v}\left(\frac{\partial v}{\partial P}\right)_s \tag{12.50}$$

The *adiabatic bulk modulus* B_s is the reciprocal of the adiabatic compressibility.

$$B_s \equiv -v\left(\frac{\partial P}{\partial v}\right)_s \tag{12.51}$$

Both the volume expansivity and isothermal compressibility are thermodynamic properties of a substance, and for a simple compressible substance are functions of two independent properties. Values of these properties are found in the standard handbooks of physical properties. The following example gives an indication of the use and significance of the volume expansivity and isothermal compressibility.

Example 12.5

Show that $C_p - C_v$ can be expressed in terms of the volume expansivity α, the specific volume v, the temperature T, and the isothermal compressibility β_T, by the relation

$$C_p - C_v = \frac{\alpha^2 v T}{\beta_T}$$

From Eq. 12.45

$$C_p - C_v = -T\left(\frac{\partial v}{\partial T}\right)_P^2\left(\frac{\partial P}{\partial v}\right)_T$$

$$\alpha \equiv \frac{1}{v}\left(\frac{\partial v}{\partial T}\right)_P; \quad \left(\frac{\partial v}{\partial T}\right)^2 = \alpha^2 v^2$$

$$\beta_T \equiv -\frac{1}{v}\left(\frac{\partial v}{\partial P}\right)_T; \quad \left(\frac{\partial P}{\partial v}\right)_T = -\frac{1}{\beta_T v}$$

Therefore

$$C_p - C_v = -T(\alpha^2 v^2)\left(-\frac{1}{\beta_T v}\right) = \frac{\alpha^2 v T}{\beta_T}$$

Example 12.6

The pressure on a block of copper having a mass of 1 lbm is increased in a reversible process from 1 atm to 1000 atm while the temperature is held constant at 60 F. Determine the work done on the copper during this process, the change in entropy per pound of copper, the heat transfer, the change of internal energy per pound, and the value of $C_p - C_v$ at this temperature.

Over the range of pressures and the temperature involved in this problem the following data can be used:

Volume expansivity $= \alpha = 2.8 \times 10^{-5} R^{-1}$
Isothermal compressibility $= \beta_T = 5.9 \times 10^{-8}$ in.^{2}/lbf
Specific volume $= 1.82 \times 10^{-3}$ ft^3/lbm

The work done during the isothermal expansion is

$$w = \int P\, dv_T$$

The isothermal compressibility has been defined:

$$\beta_T = -\frac{1}{v}\left(\frac{\partial v}{\partial P}\right)_T$$

$$v\beta_T\, dP_T = -dv_T$$

Therefore, for this isothermal process

$$w = -\int_1^2 v\beta_T P\, dP_T$$

Since v and β_T remain essentially constant, this is readily integrated:

$$w = -\frac{v\beta_T}{2}(P_2^2 - P_1^2)$$

$$= -1.82 \times 10^{-3}\ \text{ft}^3/\text{lbm} \times 5.9 \times 10^{-8}\ \text{in.}^2/\text{lbf} \times \frac{1000^2 - 1^2}{2}$$

$$\times (14.7)^2\ \text{lbf}^2/\text{in.}^4 \times 144\ \text{in.}^2/\text{ft}^2 = -1.68\ \text{ft-lbf}$$

The change of entropy can be found by considering the Maxwell relation, Eq. 12.14, and the definition of volume expansivity.

$$\left(\frac{\partial s}{\partial P}\right)_T = -\left(\frac{\partial v}{\partial T}\right)_P = -\frac{v}{v}\left(\frac{\partial v}{\partial T}\right)_P = -v\alpha$$

$$ds_T = -v\alpha dP_T$$

This can be readily integrated as follows if we assume that v and α remain constant:

$$(s_2 - s_1)_T = -v\alpha(P_2 - P_1)_T$$

$$= -1.82 \times 10^{-3}\ \text{ft}^3/\text{lbm} \times \frac{2.8 \times 10^{-5}}{R} \times (1000 - 1)$$

$$\times 14.7 \times 144\ \text{lbf/ft}^2 = -0.108\ \text{ft-lbf/lbm-R}$$

The heat transfer for this reversible isothermal process is $T(s_2 - s_1)$.

$$q = T(s_2 - s_1) = 520(-0.108) = -56.1\ \text{ft-lbf}$$

The change in internal energy follows directly from the first law.

$$(u_2 - u_1) = q - w = -56.1 - (-1.7) = -54.4\ \text{ft-lbf}$$

From Example 12.5

$$C_p - C_v = \frac{\alpha^2 v T}{\beta_T}$$

Substituting the data for copper as given above,

$$C_p - C_v = \left(\frac{2.8}{10^5}\right)^2 \times \frac{1.82}{10^3} \times \frac{10^8}{5.9} \times 520 \times \frac{144}{778} = 0.0023 \frac{\text{Btu}}{\text{lbm-R}}$$

This is consistent with the earlier observation that $C_p - C_v$ is relatively small for solids.

12.8 Developing Tables of Thermodynamic Properties from Experimental Data

There are many ways in which tables of thermodynamic properties can be developed from experimental data. The purpose of this section is to convey some general principles and concepts by considering only the liquid and vapor phases.

Let us assume that the following data for a pure substance have been obtained in the laboratory.

1. Vapor-pressure data. That is, saturation pressures and temperatures have been measured over a wide range.
2. Pressure, specific volume, temperature data in the vapor region. These data are usually obtained by determining the mass of the substance in a closed vessel (which means a fixed specific volume) and then measuring the pressure as the temperature is varied. This is done for a large number of specific volumes.
3. Density of the saturated liquid and the critical pressure and temperature.
4. Zero-pressure specific heat for the vapor. This might be obtained either calorimetrically or from spectroscopic data.

From these data a complete set of thermodynamic tables for the saturated liquid, saturated vapor, and superheated vapor can be calculated. The first step is to determine an equation for the vapor-pressure curve that accurately fits the data. It may be necessary to use one equation for one portion of the vapor-pressure curve and a different equation for another portion of the curve.

One form of equation that has been used is

$$\ln P_{\text{sat}} = A + \frac{B}{T} + C \ln T + DT$$

Once an equation has been found that accurately represents the data, the saturation pressure for any given temperature can be found by solving this equation. Thus, the saturation pressures in Table 1 of the Steam Tables would be determined for the given temperatures. The second step is to determine an equation of state for the vapor region that accurately represents the P-v-T data. There are many possible forms of the equation of state which may be selected. The important considerations are that the equation of state accurately represents the data, and that it be of such a form that the differentiations required can be performed (i.e., in some cases it may be desirable to have an equation of state explicit in v, whereas on other occasions an equation of state that is explicit in P may be more desirable).

Once an equation of state has been determined, the specific volume of superheated vapor at given pressures and temperatures can be determined by solving the equation and tabulating the results as in the superheat tables for steam, ammonia, and Freon-12. The specific volume of saturated vapor at a given temperature may be found by finding the saturation pressure from the vapor-pressure curve and substituting this saturation pressure and temperature into the equation of state.

The procedure followed in determining enthalpy and entropy is best explained with the aid of Fig. 12.4. Let us assume that the enthalpy and

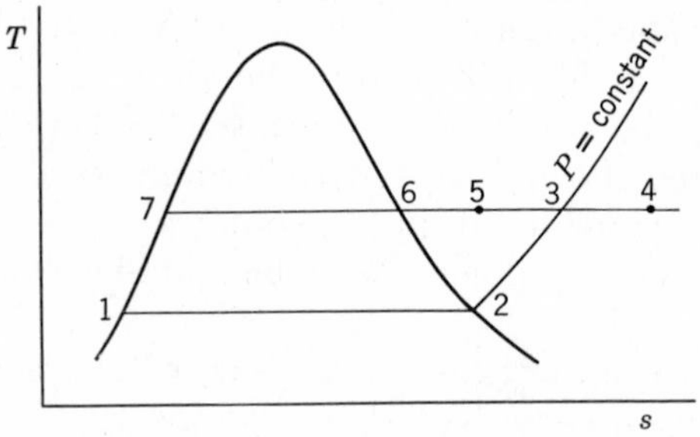

Fig. 12.4 Sketch showing procedure for developing a table of thermodynamic properties from experimental data.

entropy of saturated liquid in state 1 are zero. The enthalpy of saturated vapor in state 2 can be found from the Clapeyron equation.

$$\left(\frac{dP}{dT}\right)_{\text{sat}} = \frac{h_{fg}}{T(v_g - v_f)}$$

The left side of this equation is found by differentiating the vapor-pressure curve. The specific volume of the saturated vapor is found by the procedure outlined in the last paragraph, and it is assumed the specific volume of the saturated liquid has been measured. Thus, the enthalpy of evaporation, h_{fg}, can be found for this particular temperature, and the enthalpy at state 2 is equal to the enthalpy of evaporation (since the enthalpy in state 1 is assumed to be zero). The entropy at state 2 is readily found, since

$$s_{fg} = \frac{h_{fg}}{T}$$

The change in enthalpy between states 2 and 3 is readily found if the specific heat at this pressure is known. We have assumed that the zero-pressure specific heat, C_{po}, is known. Therefore, at a given pressure the specific heat, C_p, can be found by using Eq. 12.42.

$$\left(\frac{\partial C_p}{\partial P}\right)_T = -T\left(\frac{\partial^2 v}{\partial T^2}\right)_P$$

$$(dC_p)_T = -T\left(\frac{\partial^2 v}{\partial T^2}\right)_P dP_T$$

$$C_p - C_{po} = -\int_{P=0}^{P} T\left(\frac{\partial^2 v}{\partial T^2}\right)_P dP_T$$

The quantity $(\partial^2 v/\partial T^2)_P$ is found from the equation of state. If this equation of state is of the form $v = f(P,T)$ this differentiation can be readily obtained, and this is the reason for the previous statement regarding the form of the equation of state.

With the specific-heat equation for the given pressure, the change of enthalpy and entropy along the constant-pressure line is readily found.

$$(h_3 - h_2)_P = \int_2^3 C_p dT_P$$

$$(s_3 - s_2)_P = \int_2^3 C_p \frac{dT_P}{T}$$

From state 3 changes in enthalpy and entropy at constant temperature are readily found by using Eqs. 12.33 and 12.39 which were developed in the last section. For example:

$$h_3 - h_5 = \int_5^3 \left[v - T\left(\frac{\partial v}{\partial T}\right)_P\right] dP_T$$

$$s_3 - s_5 = \int_5^3 -\left(\frac{\partial v}{\partial T}\right)_P dP_T$$

The enthalpy and entropy of the saturated vapor in state 6 is found by this same procedure. Finally, the enthalpy and entropy of the saturated liquid in state 7 is found by applying the Clapeyron equation between states 6 and 7.

Thus, values for the pressure, temperature, specific volume, enthalpy, entropy, and internal energy of saturated liquid, saturated vapor, and superheated vapor can be tabulated for the entire region for which experimental data were obtained. It is evident that the accuracy of such a

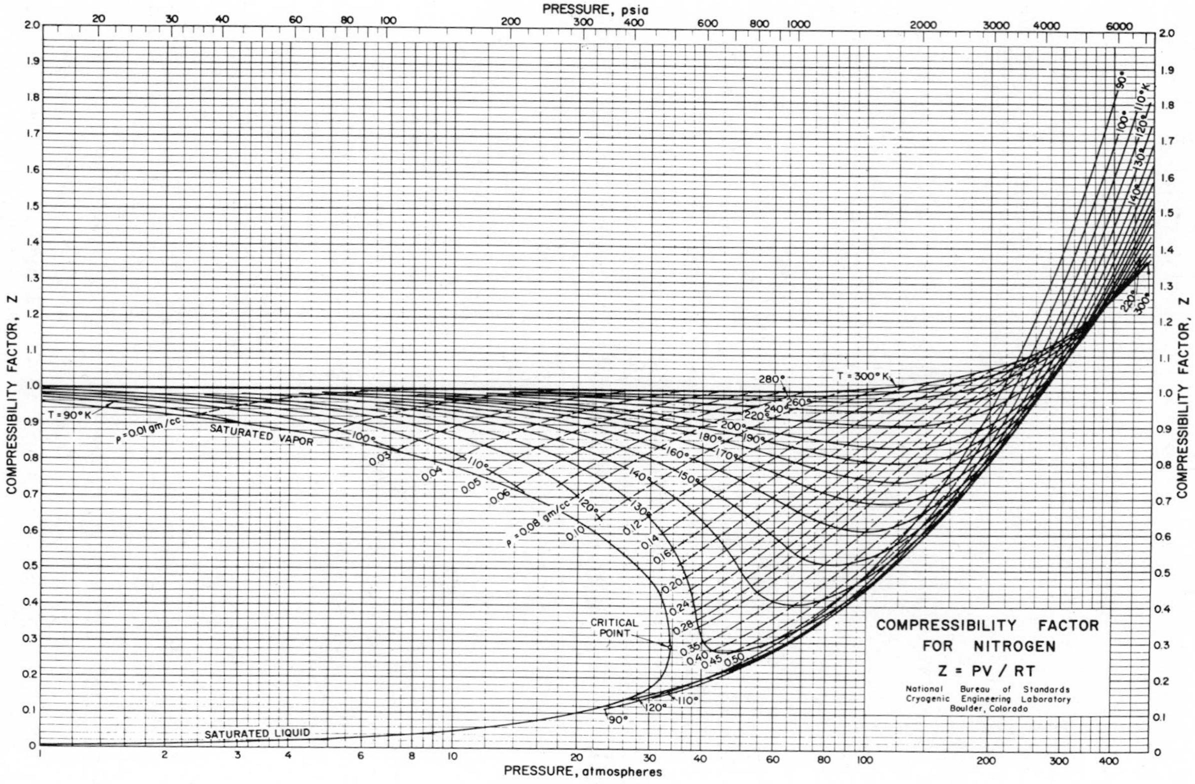

Fig. 12.5 Compressibility diagram for nitrogen.

table depends both on the accuracy of the experimental data and the degree to which the equation for the vapor pressure and the equation of state represent the experimental data.

12.9 *P-v-T* Behavior of Real Gases

In dealing with real gases we utilize the compressibility factor, Z, which is defined, in Eq. 3.6, as

$$Z = \frac{Pv}{RT}$$

The compressibility factor has the value of unity for an ideal gas. That is, as $P \to 0, Z \to 1$.

A plot of lines of constant temperature on a *Z-P* diagram is one effective way of showing *P-v-T* behavior for a real gas. Figure 12.5 shows such a diagram for nitrogen. Note that as the pressure approaches zero, Z approaches unity for all isotherms, which is consistent with our observations made regarding ideal gases. However, the slope of the isotherm as the pressure approaches zero is zero for only one isotherm, which is referred to as the Boyle temperature. At pressures less than the critical pressure both saturated liquid and saturated vapor states can be shown on a *Z-P* diagram.

The behavior of the critical isotherm on a *P-v* diagram in the vicinity of the critical point is particularly important in a study of real gases. Consider the *P-v* plot for a pure substance shown in Fig. 12.6.

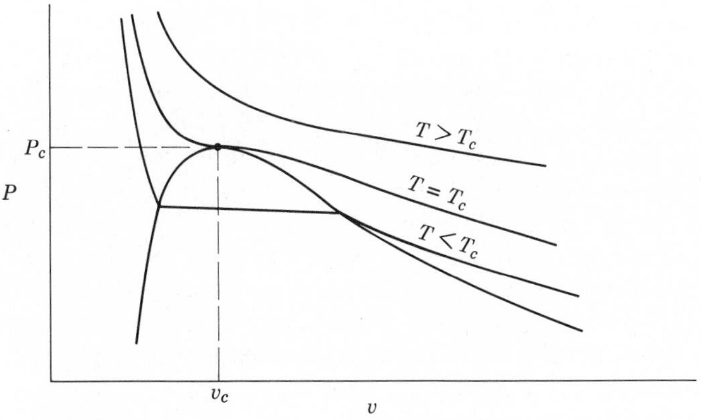

Fig. 12.6 Plot of isotherms in the region of the critical point on pressure-volume coordinates for a typical pure substance.

From a careful study of experimental data it is evident that at least the first two derivatives are zero at the critical point. This means that the critical isotherm goes through a point of inflection at the critical point.

$$\left(\frac{\partial P}{\partial v}\right)_T = 0$$

$$\left(\frac{\partial^2 P}{\partial v^2}\right)_T = 0$$

One of the most important characteristics of the *P-v-T* behavior of real gases involves the concept of reduced pressure, reduced temperature, and reduced specific volume. These are defined as follows:

Reduced pressure $= P_r = \dfrac{P}{P_c}$ $\qquad P_c =$ Critical pressure

Reduced temperature $= T_r = \dfrac{T}{T_c}$ $\qquad T_c =$ Critical temperature

Reduced specific volume $= v_r = \dfrac{v}{v_c}$ $\qquad v_c =$ Critical specific volume

These equations state that the reduced property for a given state is the value of that property in this state divided by the value of that same property at the critical point.

If lines of constant T_r are plotted on a Z vs. P_r diagram, a plot such as Fig. B.8 is obtained. The striking fact is that when such Z vs. P_r diagrams are prepared for a large number of different substances, these diagrams very nearly coincide. This leads to the development of a generalized compressibility chart. Figure B.8 is actually a generalized diagram, which means that it represents the average diagram for a number of different substances. It should be noted that when such a diagram is used for a particular substance, the results might be slightly in error. On the other hand, if *P-v-T* information is required for a substance in a region where no experimental measurements have been made, the generalized compressibility diagram will in all probability give relatively accurate results. It is only necessary to know the critical pressure and the critical temperature in order to use the generalized chart. It should be emphasized that whenever one has accurate thermodynamic data for a given substance, these should be used rather than the generalized charts.

This behavior of pure substances that leads to the generalized compressibility chart is sometimes referred to as the rule of corresponding states, which may be expressed as

$$v_r = f(P_r, T_r)$$

That is, if the rule of corresponding states held exactly, there would be a single functional relation between v_r, P_r, and T_r which would apply to all substances. The generalized chart is one way of expressing this relationship, and the fact that this is only an approximate relation indicates the approximate nature of the rule of corresponding states.

The compressibility factor at the critical point varies considerably from one substance to another, generally in the range from 0.23–0.33, and therefore the generalized chart is usually not accurate in the immediate vicinity of the critical point. Also, since liquid compressibilities do not generalize well as functions of only T_r and P_r, the liquid region is often not included on a generalized chart.

Example 12.7

1. Volume unknown. Using the generalized compressibility chart, determine the specific volume of propane at a pressure of 1000 lbf/in.^{2} and a temperature of 300 F, and compare this with the specific volume given by the ideal-gas equation of state.

For propane, from Table B.5,

$$T_c = 666 \text{ R}$$

$$P_c = 617 \text{ lbf/in.}^2$$

$$R = 35.1 \text{ ft-lbf/lbm-R}$$

$$T_r = \tfrac{760}{666} = 1.141 \quad P_r = \tfrac{1000}{617} = 1.62$$

From the compressibility chart

$$Z = 0.54$$

$$Pv = ZRT$$

$$v = \frac{0.54 \times 35.1 \times 760}{1000 \times 144} = 0.100 \text{ ft}^3/\text{lbm}$$

The ideal-gas equation would give the value

$$v = \frac{35.1 \times 760}{1000 \times 144} = 0.185 \text{ ft}^3/\text{lbm}$$

2. Pressure unknown. What pressure is required in order that propane have a specific volume of 0.100 ft^3/lbm at a temperature of 300 F?

$$Pv = ZRT \quad T_r = \tfrac{760}{666} = 1.141$$

$$P = P_c P_r$$

Therefore,

$$P_r = \frac{ZRT}{vP_c} = \frac{Z \times 35.1 \times 760}{0.100 \times 617 \times 144} = 3.00Z$$

$$Z = \frac{P_r}{3.00} = 0.33P_r$$

By a trial and error procedure, or by plotting a few points and drawing the curve representing this equation, P_r is found to be 1.62 at the point where $T_r = 1.141$. Therefore,

$$P = P_r P_c = 1.62(617) = 1000 \text{ lbf/in.}^2$$

3. Temperature unknown. What will be the temperature of propane when it has a specific volume of 0.100 ft³/lbm and a pressure of 1000 lbf/in.²?

$$Pv = ZRT \quad T = T_c T_r$$

$$Pv = ZRT_c T_r$$

$$T_r = \frac{Pv}{ZRT_c} = \frac{1000 \times 144 \times 0.100}{Z \times 35.1 \times 666} = \frac{0.615}{Z}$$

When this line is plotted on the compressibility chart, the state point is given by the intersection of this line with the known reduced-pressure line ($P_r = 1.62$ in this case).

The reduced temperature is thus found to be

$$T_r = 1.141$$

$$T = T_r \times T_c = 1.141 \times 666 = 760 \text{ R}$$

12.10 Equations of State

An accurate equation of state, which is an analytical representation of P-v-T behavior, is often desirable from a computational standpoint. Many different equations of state have been developed. Most of these are accurate only to some density less than the critical density, though a few are reasonably accurate to approximately 2.5 times the critical density. All equations of state fail badly when the density exceeds the maximum density for which the equation was developed.

Three broad classifications of equations of state can be identified, namely, generalized, empirical, and theoretical.

The best known of the generalized equations of state is also the oldest, namely, the van der Waals equation, which was presented in 1873 as a semi-theoretical improvement over the ideal gas equation. The van der Waals equation of state is:

$$P=\frac{RT}{v-b}-\frac{a}{v^2} \tag{12.52}$$

The constant b is intended to correct for the volume occupied by the molecules, and the term a/v^2 is a correction that accounts for the intermolecular forces of attraction. As might be expected in the case of a generalized equation, the constants a and b are evaluated from the general behavior of gases. In particular these constants are evaluated by noting that the critical isotherm passes through a point of inflection at the critical point, and that the slope is zero at this point. Thus, for the van der Waals equation of state we have

$$\left(\frac{\partial P}{\partial v}\right)_T=-\frac{RT}{(v-b)^2}+\frac{2a}{v^3} \tag{12.53}$$

$$\left(\frac{\partial^2 P}{\partial v^2}\right)_T=\frac{2RT}{(v-b)^3}-\frac{6a}{v^4} \tag{12.54}$$

Since both of these derivatives are equal to zero at the critical point we can write

$$-\frac{RT_c}{(v_c-b)^2}+\frac{2a}{v_c^{\,3}}=0$$

$$\frac{2RT_c}{(v_c-b)^3}-\frac{6a}{v_c^{\,4}}=0 \tag{12.55}$$

$$P_c=\frac{RT_c}{(v_c-b)}-\frac{a}{v_c^{\,2}}$$

Solving these three equations we find

$$v_c=3b$$

$$a=\frac{27}{64}\frac{R^2T_c^{\,2}}{P_c} \tag{12.56}$$

$$b=\frac{RT_c}{8P_c}$$

The compressibility factor at the critical point for the van der Waals equation is

$$Z_c = \frac{P_c v_c}{RT_c} = \frac{3}{8}$$

which is higher than the actual value for essentially all substances.

Van der Waals' equation can be written in terms of the compressibility factor and the reduced pressure and reduced temperature as follows:

$$Z^3 - \left(\frac{P_r}{8T_r} + 1\right)Z^2 + \left(\frac{27P_r}{64T_r^2}\right)Z - \frac{27P_r^2}{512T_r^3} = 0 \qquad (12.57)$$

It is significant to note that this is of the same form as the generalized compressibility chart, namely, $Z = f(P_r, T_r)$, though the functional relation may be quite different from the generalized chart. This concept that different substances will have the same compressibility factor at the same reduced pressure and reduced temperature is another way of expressing the rule of corresponding states.

One of the best known empirical equations of state is the Beattie-Bridgeman equation, which is a pressure-explicit equation with five constants that are determined by a graphical procedure from experimental data for each substance. This equation was introduced in Chapter 3 in the form

$$P = \frac{RT(1-\epsilon)}{v^2}(v+B) - \frac{A}{v^2} \qquad (12.58)$$

where

$$A = A_0(1 - a/v)$$

$$B = B_0(1 - b/v)$$

$$\epsilon = c/vT^3$$

This equation may also be written in the form

$$P = \frac{RT}{v} + \frac{\beta}{v^2} + \frac{\gamma}{v^3} + \frac{\delta}{v^4}$$

where

$$\beta = B_0RT - A_0 - cR/T^2$$

$$\gamma = -B_0bRT + A_0a - B_0cR/T^2$$

$$\delta = B_0bcR/T^2$$

Table 3.3 gives the values of the five constants for a number of different substances. The Beattie-Bridgeman equation of state is quite accurate for densities less than about 0.8 times the critical density.

Many other empirical equations of state have been developed and used to represent the P-v-T behavior of various substances.

A different approach to this problem is from the theoretical point of view. The theoretical equation of state, which is derived from kinetic theory or statistical thermodynamics, is written here in the form of a power series in reciprocal volume:

$$Z = \frac{Pv}{RT} = 1 + \frac{B}{v} + \frac{C}{v^2} + \frac{D}{v^3} + \cdots \tag{12.59}$$

where B, C, D, . . . are temperature dependent, and are called virial coefficients. B is termed the second virial coefficient and is due to binary interactions on the molecular level. The virial coefficients can be expressed in terms of intermolecular forces from statistical mechanics, and evaluated upon selection of an empirical potential function model. For all but the simplest functions the mathematical expressions become so complex that very little has been accomplished beyond the third virial coefficient. When the virial equation is used with the second and third virial coefficients, P-v-T behavior can be accurately represented at densities as high as 0.7 of the critical density.

12.11 The Generalized Chart for Changes of Enthalpy at Constant Temperature

In Section 12.5, Eq. 12.33 was derived for the change of enthalpy at constant temperature.

$$(h_2 - h_1)_T = \int_1^2 \left[v - T\left(\frac{\partial v}{\partial T}\right)_P \right] dP_T$$

A similar equation for the change of enthalpy in an isothermal process can be developed as follows:

$$dh = T\,ds + v\,dP$$

$$\left(\frac{\partial h}{\partial v}\right)_T = T\left(\frac{\partial s}{\partial v}\right)_T + v\left(\frac{\partial P}{\partial v}\right)_T$$

Substituting one of the Maxwell relations, Eq. 12.13, we have

$$\left(\frac{\partial h}{\partial v}\right)_T = T\left(\frac{\partial P}{\partial T}\right)_v + v\left(\frac{\partial P}{\partial v}\right)_T$$

Therefore,

$$(h_2-h_1)_T = \int_1^2 \left[T\left(\frac{\partial P}{\partial T}\right)_v + v\left(\frac{\partial P}{\partial v}\right)_T\right] dv_T$$

One essential point about these two equations for $(h_2-h_1)_T$ is that the change of enthalpy at constant temperature is a function of the P-v-T behavior of the gas. This means that if the equation of state is known for a substance, the change in enthalpy at constant temperature can be determined by the appropriate differentiation and integration. It also follows that since the generalized chart is a representation of the P-v-T behavior of the gas, this chart can be used to prepare a generalized chart for the change of enthalpy in an isothermal process.

The procedure for the development of such a chart is as follows: It has been shown that

$$\left(\frac{\partial h}{\partial P}\right)_T = v - T\left(\frac{\partial v}{\partial T}\right)_P$$

But

$$v = \frac{ZRT}{P}$$

and

$$\left(\frac{\partial v}{\partial T}\right)_P = \frac{ZR}{P} + \frac{RT}{P}\left(\frac{\partial Z}{\partial T}\right)_P$$

Therefore

$$\left(\frac{\partial h}{\partial P}\right)_T = \frac{ZRT}{P} - \frac{ZRT}{P} - \frac{RT^2}{P}\left(\frac{\partial Z}{\partial T}\right)_P = -\frac{RT^2}{P}\left(\frac{\partial Z}{\partial T}\right)_P$$

$$dh_T = -\frac{RT^2}{P}\left(\frac{\partial Z}{\partial T}\right)_P dP_T$$

But $T = T_c T_r$ and $P = P_c P_r$. Therefore,

$$\frac{dT}{T} = \frac{dT_r}{T_r}; \qquad \frac{dP}{P} = \frac{dP_r}{P_r}$$

$$dT = T_c\, dT_r; \quad dP = P_c\, dP_r$$

Substituting these values we have

$$dh_T = -\frac{RT_c^2 T_r^2}{P_c P_r}\left(\frac{\partial Z}{T_c \partial T_r}\right)_{P_r} P_c (dP_r)_{T_r}$$

$$= -RT_c T_r^2 \left(\frac{\partial Z}{\partial T_r}\right)_{P_r} (d \ln P_r)_{T_r}$$

Integrating at constant temperature we have

$$\frac{\Delta \bar{h}_T}{T_c} = -\bar{R} \int_1^2 T_r^2 \left(\frac{\partial Z}{\partial T_r}\right)_{P_r} d \ln P_r \tag{12.60}$$

Let us designate by an asterisk, $\bar{h}^*$, the enthalpy per mole at a given temperature and a very low pressure, so that $P_r \to 0$, and let us designate without superscript the enthalpy at any state at the same temperature and a given pressure P_r. Then the integration is as follows:

$$\frac{\bar{h}^* - \bar{h}}{T_c} = \bar{R} \int_{P_r=0}^{P_r} T_r^2 \left(\frac{\partial Z}{\partial T_r}\right)_{P_r} d \ln P_r \tag{12.61}$$

The right side of this equation can be obtained by graphical integration of the compressibility chart, and is a function of only the reduced pressure and temperature. The left side represents the change in enthalpy as the pressure is increased from zero to the given pressure while the temperature is held constant. Sometimes this quantity is called the enthalpy departure. Figure B.9 shows a plot of this quantity, $(\bar{h}^* - \bar{h})/T_c$, as a function of P_r for various lines of constant T_r.

In many cases the specific heat of a substance is known at low pressure. This information, along with the generalized residual enthalpy chart, enables one to find the change of enthalpy for any given change of state.

Example 12.8

Nitrogen is throttled from 3000 lbf/in.², −100 F, to 200 lbf/in.² in an adiabatic, SSSF process. Determine the final temperature of the nitrogen. The zero-pressure specific heat of nitrogen in this temperature range is 0.248 Btu/lbm-R. This is a constant-enthalpy process, and in solving such a problem it is usually helpful to use a temperature-entropy diagram such as the one shown in Fig. 12.7.

$$P_1 = 3000 \text{ lbf/in.}^2 \quad P_2 = 200 \text{ lbf/in.}^2$$

$$P_{r1} = \tfrac{3000}{492} = 6.1 \quad P_{r2} = \tfrac{200}{492} = 0.406$$

$$T_1 = 360 \text{ R}$$

$$T_{r1} = \tfrac{360}{227} = 1.586$$

From the generalized chart for enthalpy change at constant temperature,

$$\frac{\bar{h}_1^* - \bar{h}_1}{T_c} = 3.6 \text{ Btu/lb mole-R}, \qquad h_1^* - h_1 = \frac{3.6 \times 227}{28} = 29.2 \text{ Btu/lbm}$$

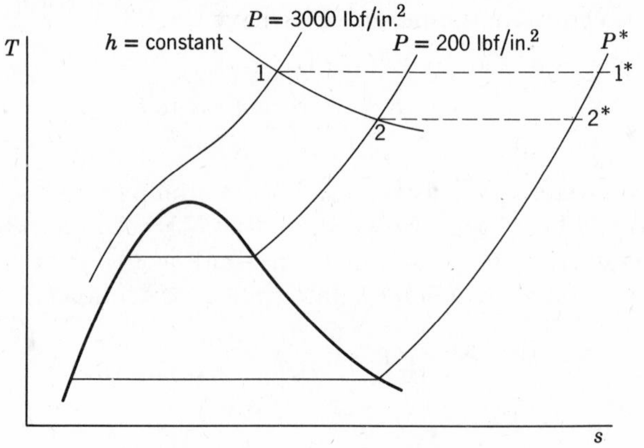

Fig. 12.7 Sketch for Example 12.8.

It is now necessary to assume a final temperature and check to see if the net change in enthalpy for the process is zero. Let us assume that $T_2 = 265$ R. Then the change in enthalpy between 1* and 2* can be found from the zero-pressure specific-heat data.

$$h_1^* - h_2^* = C_{po}(T_1^* - T_2^*) = 0.248(360 - 265) = 23.6 \text{ Btu/lbm}$$

(The variation in C_{po} with temperature can be taken into account when necessary.)

We now find the enthalpy change between 2* and 2.

$$T_{r2} = \tfrac{265}{227} = 1.17 \quad P_{r2} = 0.406$$

Therefore, from the enthalpy departure chart, Fig. B.9,

$$\frac{\bar{h}_2^* - \bar{h}_2}{T_c} = 0.7 \text{ Btu/lb mole-R}$$

$$h_2^* - h_2 = \frac{0.7(227)}{28} = 5.7 \text{ Btu/lbm}$$

We now check to see if the net change in enthalpy for the process is zero.

$$h_1 - h_2 = 0 = -(h_1^* - h_1) + (h_1^* - h_2^*) + (h_2^* - h_2)$$

$$= -29.2 + 23.6 + 5.7 = 0.1 \text{ Btu/lbm}$$

This essentially checks, and we conclude that the final temperature is approximately 265 R.

12.12 Fugacity and the Generalized Fugacity Chart

At this point a new thermodynamic property, the fugacity, f, is introduced. The fugacity is particularly important when considering mixtures and equilibrium, which is discussed in Chapter 14. However, in view of the fact that a generalized fugacity chart can be developed, the fugacity is introduced here.

The fugacity is essentially a pseudo-pressure. When the fugacity is substituted for pressure, one can, in effect, use the same equations for real gases that one normally uses for ideal gases. The concept of fugacity can be introduced as follows:

Consider the relation
$$dg = -s\,dT + v\,dP$$

At constant temperature
$$dg_T = v\,dP_T \tag{12.62}$$

For an ideal gas this last equation may be written
$$dg_T = \frac{RT}{P}\,dP_T = RT\,d(\ln P)_T \tag{12.63}$$

and for a real gas, with the equation of state $Pv = ZRT$, this equation becomes
$$dg_T = ZRT\frac{dP_T}{P} = ZRT\,d(\ln P)_T \tag{12.64}$$

Fugacity, f, is defined as
$$dg_T \equiv RT\,d(\ln f)_T \tag{12.65}$$

It follows from these last two equations that
$$\left(\frac{\partial \ln f}{\partial \ln P}\right)_T = Z \tag{12.66}$$

Since $Z \to 1$ as $P \to 0$, it follows that
$$\lim_{P\to 0}(f/P) = 1$$

Therefore, as $P \to 0, f \to 0$.

Let us consider the change in the Gibbs function of a real gas during an isothermal process at temperature T that involves a change in pressure from a very low pressure, P^* (where ideal gas behavior can be assumed) to a higher pressure P. Let the Gibbs function at this low pressure be designated g^*. The value of the Gibbs function at this temperature T and at the pressure P can be found in terms of the fugacity at P and T and g^*. This is evident if one integrates Eq. 12.65 from the very low pressure P^* to the pressure P.
$$\int_{g^*}^{g} dg_T = \int_{f^*=P^*}^{f} RT(d\ln f)_T \tag{12.67}$$
$$g = g^* + RT\ln(f/P^*)$$

A generalized fugacity coefficient chart based on the generalized compressibility chart, Fig. B.10, can be developed by the following procedure. From Eqs. 12.62 and 12.65,

$$dg_T = v\, dP_T = \frac{ZRT\, dP_T}{P} = RT\, d(\ln f)_T$$

Therefore,

$$Z\, d(\ln P)_T = d(\ln f)_T$$

But

$$\frac{dP}{P} = \frac{dP_r}{P_r}; \quad d \ln P = d \ln P_r$$

Therefore,

$$Z(d \ln P_r)_T = d(\ln f)_T$$

$$Z(d \ln P_r)_T - d(\ln P_r)_T = d(\ln f)_T - d(\ln P_r)_T \tag{12.68}$$

$$(Z-1)(d \ln P_r)_T = d(\ln f/P)_T$$

Integrating at constant temperature from $P = 0$ to a finite pressure we have

$$\int_{f/P=1}^{f/P} d(\ln f/P)_T = \int_0^{P_r} (Z-1)\, d \ln P_{rT} \tag{12.69}$$

$$\ln (f/P) = \int_0^{P_r} (Z-1)\, d \ln P_{rT}$$

At any temperature the right hand side of this equation can be integrated graphically using the generalized compressibility chart to find Z at each P_r. The result is the generalized chart shown in Fig. B.10.

The reversible isothermal process is one type of problem in which the fugacity can be used to advantage.

Example 12.9

Calculate the work of compression and the heat transfer per pound when ethane is compressed reversibly and isothermally from 14.7 lbf/in.^{2} to 1000 lbf/in.^{2} at a temperature of 110 F in an SSSF process.

The first law for this process is,

$$q + h_1 + \text{KE}_1 + \text{PE}_1 = h_2 + \text{KE}_2 + \text{PE}_2 + w$$

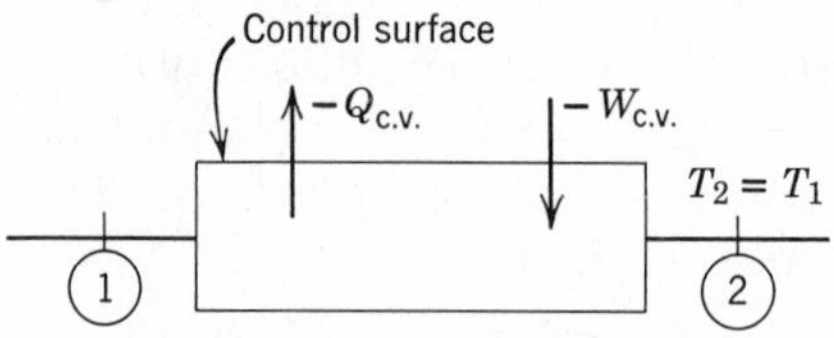

Fig. 12.8 Sketch for Example 12.9.

Since the process is reversible and isothermal,

$$q = T(s_2 - s_1) = T_2 s_2 - T_1 s_1$$

Therefore,

$$-w = h_2 - h_1 - (T_2 s_2 - T_1 s_1) + \Delta\text{KE} + \Delta\text{PE}$$
$$-w = g_2 - g_1 + \Delta\text{KE} + \Delta\text{PE}$$

We assume that the changes in kinetic and potential energy are negligible.

$$-w = g_2 - g_1 = RT \ln (f_2/f_1)$$

The fugacity in state 1 and state 2 can be found from the generalized fugacity chart in the Appendix, Fig. B.10.

For ethane,

$$P_c = 708 \text{ lbf/in.}^2 \quad T_c = 549.8 \text{ R}$$

Therefore,

$$P_{r1} = \frac{14.7}{708} = 0.021 \quad T_{r1} = T_{r2} = \frac{570}{549.8} = 1.037$$

$$P_{r2} = \frac{1000}{708} = 1.412$$

From the generalized enthalpy departure chart, Fig. B.9,

$$\frac{\bar{h}_2^* - \bar{h}_2}{T_c} = 6.8 \qquad \frac{\bar{h}_1^* - \bar{h}_1}{T_c} < 0.1$$

Since $h_1^* = h_2^* = h_1$

$$h_2^* - h_2 = h_1 - h_2 = \frac{6.8(549.8)}{30} = 124.8 \text{ Btu/lbm}$$

From the generalized fugacity chart, Fig. B. 10,

$$f_2/P_2 = 0.58 \quad f_2 = 1000(0.58) = 580 \text{ lbf/in.}^2$$

$$f_1/P_1 = 1.0 \quad f_1 = 14.7(1.0) = 14.7 \text{ lbf/in.}^2$$

Therefore,

$$-w = g_2 - g_1 = RT \ln f_2/f_1 = \frac{1.987}{30} \times 570 \ln \frac{580}{14.7}$$
$$= 138.8 \text{ Btu/lbm}$$

$$g_2 - g_1 = (h_2 - h_1) - T(s_2 - s_1) = (h_2 - h_1) - q$$

$$138.8 = -124.8 - T(s_2 - s_1) = -124.8 - q$$

$$q = T_2(s_2 - s_1) = -124.8 - 138.8 = -263.6 \text{ Btu/lbm}$$

$$s_2 - s_1 = \frac{-263.6}{570} = -0.462 \text{ Btu/lbm-R}$$

12.13 The Generalized Chart for Changes of Entropy at Constant Temperature

A generalized chart for entropy can be developed that gives the difference between the entropy of an ideal gas and an actual gas as the pressure is increased from zero pressure along an isotherm. The thermodynamic analysis for the development of such a chart is as follows.

From Eq. 12.38

$$ds_T = -\left(\frac{\partial v}{\partial T}\right)_P dP_T$$

Integrating at constant temperature, T, from $P = 0$ to P we have

$$(s_P - s_0^*)_T = -\left[\int_{P=0}^{P}\left(\frac{\partial v}{\partial T}\right)_P dP_T\right] \tag{12.70}$$

The entropy at zero pressure can be designated with an * (which we use to refer to ideal gas behavior) since ideal gas behavior is approached as $P \to 0$. However, $s \to \infty$ as $P \to 0$, and therefore this equation is not particularly useful in this form.

If we repeat this integration for an ideal gas we have

$$(s_P^* - s_0^*)_T = -\left[\int_{P=0}^{P}\left(\frac{\partial v}{\partial T}\right)_P dP_T\right] = -\left[\int_{P=0}^{P} R\frac{dP_T}{P}\right] \tag{12.71}$$

Subtracting Eq. 12.70 from Eq. 12.71,

$$(s_P^* - s_P) = -\int_{P=0}^{P}\left[\frac{R}{P} - \left(\frac{\partial v}{\partial T}\right)_P\right] dP_T$$

Therefore, with the equation of state $Pv = ZRT$,

$$(s_P^* - s_P) = -\int_{P=0}^{P}\left[\frac{R}{P} - \frac{ZR}{P} - \frac{RT}{P}\left(\frac{\partial Z}{\partial T}\right)_P\right] dP_T$$

or,

$$(\bar{s}_P^* - \bar{s}_P) = \bar{R}\int_0^{P_r}(Z-1)\frac{dP_r}{P_r} + \bar{R}T_r\int_0^{P_r}\left(\frac{\partial Z}{\partial T_r}\right)_{P_r}\frac{dP_r}{P_r}$$

But, from Eq. 12.69

$$\bar{R}\int_0^{P_r}(Z-1)\frac{dP_r}{P_r} = \bar{R}\ln(f/P)$$

and from Eq. 12.61,

$$\bar{R}T_r\int_0^{P_r}\left(\frac{\partial Z}{\partial T_r}\right)_{P_r}\frac{dP_r}{P_r} = \frac{\bar{h}^* - \bar{h}}{T_c T_r}$$

Therefore,

$$(\bar{s}_P^* - \bar{s}_P) = \bar{R}\ln f/P + \frac{\bar{h}^* - \bar{h}}{T_c T_r}$$

Thus, for the equation of state $Pv = ZRT$, the quantity $(\bar{s}^*_P - \bar{s}_P)$ can be obtained from the generalized enthalpy departure and the generalized fugacity coefficient charts. Figure B. 11 shows such a chart.

Example 12.10

Nitrogen at 1000 lbf/in.², −200 F is throttled to 100 lbf/in.². After passing through a short length of pipe the temperature is measured and found to be −260 F. Determine the heat transfer and the change of entropy using the generalized charts and compare these results with those obtained by using the nitrogen tables.

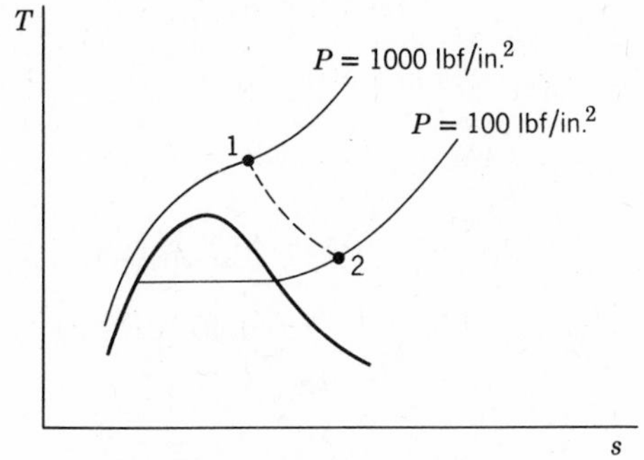

Fig. 12.9 Sketch for Example 12.10.

$$P_{r1} = \frac{1000}{492} = 2.03 \quad T_{r1} = \frac{260}{227.1} = 1.146$$

$$P_{r2} = \frac{100}{492} = 0.203 \quad T_{r2} = \frac{200}{227.1} = 0.882$$

From the first law

$$q = h_2 - h_1 = -(h_2^* - h_2) + (h_2^* - h_1^*) + (h_1^* - h_1)$$

From Fig. B. 9 in the Appendix,

$$\frac{\bar{h}_1^* - \bar{h}_1}{T_c} = 4.85 \text{ Btu/lb mole-R}; \; h_1^* - h_1 = \frac{227.1(4.85)}{28.0} = 41.0 \text{ Btu/lbm}$$

$$\frac{\bar{h}_2^* - \bar{h}_2}{T_c} = 0.7 \text{ Btu/lb mole-R}; \quad h_2^* - h_2 = \frac{227.1(0.7)}{28.0} = 5.7 \text{ Btu/lbm}$$

Assuming a constant specific heat for the ideal gas,

$$h_2^* - h_1^* = C_{p0}(T_2 - T_1) = 0.248(200 - 260) = -14.9 \text{ Btu/lbm}$$

$$q = -5.7 - 14.9 + 41.0 = 20.4 \text{ Btu/lbm}$$

From the nitrogen tables, Table B. 4, we can find the change of enthalpy directly.

$$q = h_2 - h_1 = 109.93 - 85.54 = 24.39 \text{ Btu/lbm}$$

To calculate the change of entropy using the generalized charts we proceed as follows:

$$\bar{s}_2 - \bar{s}_1 = -(\bar{s}^*_{P_2,T_2} - \bar{s}_2) + (\bar{s}^*_{P_2,T_2} - \bar{s}^*_{P_1,T_1}) + (\bar{s}^*_{P_1,T_1} - \bar{s}_1)$$

From Fig. B. 11 in the Appendix,

$$\bar{s}^*_{P_1,T_1} - \bar{s}_1 = 3.35 \text{ Btu/lb mole-R}$$

$$\bar{s}^*_{P_2,T_2} - \bar{s}_2 = 0.65 \text{ Btu/lb mole-R}$$

Assuming a constant specific heat for the ideal gas,

$$\bar{s}^*_{P_2,T_2} - \bar{s}^*_{P_1,T_1} = \bar{C}_{p0} \ln \frac{T_2}{T_1} - \bar{R} \ln \frac{P_2}{P_1}$$

$$= 6.95 \times \ln \tfrac{200}{260} - 1.987 \times \ln \tfrac{100}{1000} = 2.76 \text{ Btu/lb mole-R}$$

$$\bar{s}_2 - \bar{s}_1 = -0.65 + 2.76 + 3.35 = 5.46 \text{ Btu/lb mole-R}$$

$$s_2 - s_1 = \frac{5.46}{28.0} = 0.195 \text{ Btu/lbm-R}$$

From Table B. 4

$$s_2 - s_1 = 0.6585 - 0.4419 = 0.2166 \text{ Btu/lbm-R.}$$

12.14 Equations of State and Pseudocritical State for Mixtures

Frequently it is desirable to have an equation of state for a mixture of gases. The question arises as to how an equation of state for a mixture can be developed from the equations of state for the pure components. For example, suppose the equations of state of the components are available in a given form, with given constants for each component. How should these constants be combined to provide an equation of state that accurately represents the P-v-T behavior of the mixture?

An extensive discussion of this question is beyond the scope of this text. Instead, we shall simply note that various empirical combining rules have been proposed for the different equations of state and compared with experimental data for mixtures. As an example, for van der Waals' equation, the pure substance constants are commonly combined according to the relations,

$$a_m = \left(\sum_i y_i a_i^{1/2} \right)^2, \qquad b_m = \sum_i y_i b_i \qquad (12.72)$$

When limited P-v-T data are available, the generalized compressibility chart can often be used with considerable accuracy. The mixture may be treated as a pseudo-pure substance, for which a set of critical constants are determined by some empirical combinations of the pure substance constants. W. B. Kay in 1936 first suggested a simple linear combination.

$$(P_c)_{\text{mix}} = \sum_i y_i P_{ci} \tag{12.73}$$

$$(T_c)_{\text{mix}} = \sum_i y_i T_{ci}$$

These relations give results of reasonable accuracy for mixtures. A number of other procedures for defining pseudocritical constants for mixtures have also been proposed. These procedures are in general more complicated to use than Kay's rule, but may yield somewhat more accurate results.

Example 12.11

A mixture of 59.39% CO_2 and 40.61% CH_4 (mole basis) is maintained at 310.94 K, 85.06 atm., at which condition the specific volume has been measured as 0.2205 lit/mole. Calculate the per cent deviation if the specific volume had been calculated by (1) Kay's Rule, and (2) van der Waals' equation of state.

1. For convenience, let

$$CO_2 = A, \quad CH_4 = B$$

Then

$$T_{c_A} = 304.2 \text{ K}, \quad P_{c_A} = 72.9 \text{ atm}$$

$$T_{c_B} = 191.1 \text{ K}, \quad P_{c_B} = 45.8 \text{ atm}$$

For Kay's Rule, Eq. 12.73,

$$\begin{aligned} T_{c_m} &= \sum_i y_i T_{c_i} = y_A T_{c_A} + y_B T_{c_B} \\ &= 0.5939(304.2) + 0.4061(191.1) \\ &= 258.0 \text{ K} \end{aligned}$$

$$\begin{aligned} P_{c_m} &= \sum_i y_i P_{c_i} = y_A P_{c_A} + y_B P_{c_B} \\ &= 0.5939(72.9) + 0.4061(45.8) \\ &= 62.0 \text{ atm.} \end{aligned}$$

Therefore, the pseudo-reduced properties of the mixture are

$$T_{r_m} = \frac{T}{T_{c_m}} = \frac{310.94}{258.0} = 1.205$$

$$P_{r_m} = \frac{P}{P_{c_m}} = \frac{85.06}{62.0} = 1.372$$

From the generalized chart,

$$Z_m = 0.705$$

and

$$\bar{v} = \frac{Z_m \bar{R} T}{P} = \frac{0.705(0.08206)(310.94)}{85.06}$$

$$= 0.2115 \text{ lit/mole}$$

The per cent deviation from the experimental value is

$$\% \text{ Dev.} = \left(\frac{0.2205 - 0.2115}{0.2205}\right) \times 100 = 4.08\%$$

2. For van der Waals' equation, the pure substance constants are

$$a_A = \frac{27\,\bar{R}^2 T_{cA}{}^2}{64 P_{cA}} = 3.606 \frac{\text{atm-lit}^2}{\text{mole}^2}$$

$$b_A = \frac{\bar{R} T_{cA}}{8 P_{cA}} = 0.0428 \text{ lit/mole}$$

and

$$a_B = \frac{27\,\bar{R}^2 T_{cB}{}^2}{64 P_{cB}} = 2.256 \frac{\text{atm-lit}^2}{\text{mole}^2}$$

$$b_B = \frac{\bar{R} T_{cB}}{8 P_{cB}} = 0.04271 \text{ lit/mole}$$

Therefore, for the mixture, from Eq. 12.72,

$$a_m = (y_A \sqrt{a_A} + y_B \sqrt{a_B})^2$$

$$= (0.5939\sqrt{3.606} + 0.4061\sqrt{2.256}\,)^2$$

$$= 3.030 \frac{\text{atm-lit}^2}{\text{mole}^2}$$

$$\begin{aligned} b_m &= y_A b_A + y_B b_B \\ &= 0.5939(0.0428) + 0.4061(0.04271) \\ &= 0.04276 \text{ lit/mole} \end{aligned}$$

The equation of state for the mixture of this composition is

$$P = \frac{\bar{R}T}{\bar{v} - b_m} - \frac{a_m^{\ 2}}{\bar{v}^2}$$

$$85.06 = \frac{0.08206(310.94)}{\bar{v} - 0.04276} - \frac{3.03}{\bar{v}^2}$$

Solving for $\bar{v}$ by trial and error,

$$\bar{v} = 0.2050 \text{ lit/mole}$$

$$\% \text{ Dev.} = \left(\frac{0.2205 - 0.2050}{0.2205}\right) \times 100 = 7.03\%$$

As a point of interest from the ideal gas law, $\bar{v} = 0.30$ lit/mole, which is a deviation of 36 per cent from the measured value.

We must be careful not to draw too general a conclusion from the results of this example. We have calculated per cent deviation in $\bar{v}$ at only a single point for only one mixture. We do note, however, that the methods used give quite different results. From a more general study of these models for a number of mixtures, we find that the results found here are fairly typical, at least qualitatively. Kay's rule is very useful because it is fairly accurate and yet relatively simple. The van der Waals' equation is too simplified an expression to accurately represent P-v-T behavior except at moderate densities, but it is useful to demonstrate the procedures followed in utilizing more complex analytical equations of state. Both methods are much better than assuming an ideal gas mixture.

The more sophisticated generalized behavior models and empirical equations of state will represent mixture P-v-T behavior to within about one per cent over a wide range of density, but they are of course more difficult to use than the methods considered in Example 12.11. The generalized models have the advantage of being easier to use, and they are suitable for hand computations. Calculations with the complex empirical equations of state become very involved, but have the advantage of expressing the P-v-T-composition relations in analytical form, which is of great value when using a digital computer for such calculations.

PROBLEMS

12.1 Using the first Maxwell relation derived, Eq. 12.11, and the mathematical relation Eq. 12.2, derive the other three Maxwell relations, Eqs. 12.11–12.14.

12.2 Show that on a Mollier diagram (*h-s* diagram) the slope of a constant-pressure line increases with temperature in the superheat region and that the constant-pressure line is straight in the two-phase region.

12.3 Show that the chemical potential μ_A for component A in a mixture can be expressed as

$$\mu_A = \left(\frac{\partial H}{\partial n_A}\right)_{S,P,n_B} = \left(\frac{\partial A}{\partial n_A}\right)_{T,V,n_B}$$

in addition to the relations of Eqs. 12.23 and 12.25.

12.4 A transfer line for liquid nitrogen is so designed that it has a double walled pipe as indicated in Fig. 12.10. The annular space is so constructed

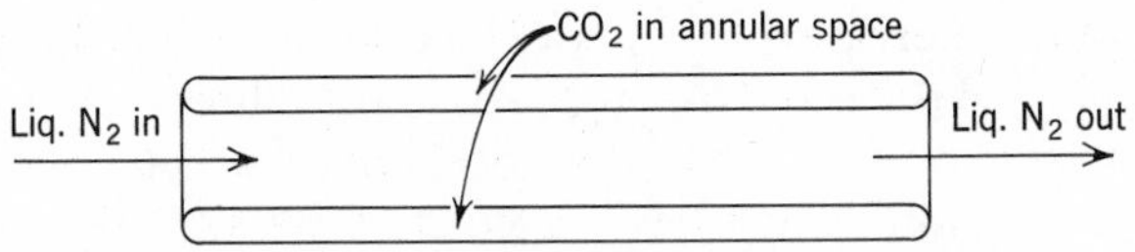

Fig. 12.10 Sketch for Problem 12.4.

that when the pipeline is at ambient temperature it contains carbon dioxide at 1 atm pressure. When liquid nitrogen at 1 atm pressure flows through the pipe the carbon dioxide will freeze on the inner walls. The pressure of the carbon dioxide in the annular space will essentially be equal to the vapor pressure of carbon dioxide at the saturation temperature of nitrogen at 1 atm pressure. Suppose that the tables of thermodynamic properties which are available do not go down to this temperature. However, at 5 lbf/in.^{2} pressure the saturation temperature is −132 F and the enthalpy of sublimation is 247 Btu/lbm. Estimate the pressure of the carbon dioxide in the annular space when liquid nitrogen is flowing through the tube.

12.5 At 1 atm pressure helium boils at 4.22 K, and the latent heat of vaporization is 19.9 cal/gm mole. Suppose one wishes to determine how low a temperature could be obtained by producing a vacuum over liquid helium, thus causing it to boil at a lower pressure and temperature. Determine what pressure would be necessary to produce a temperature of 1 K and 0.1 K.

What do you think is the lowest temperature that could be obtained by pumping a vacuum over liquid helium?

12.6 The following data are available at the triple point of water:

Pressure:	0.0887 lbf/in.^{2}	
Temperature:	32.02 F	
Enthalpy:	Liquid:	0.01 Btu/lbm
	Solid:	−143.3 Btu/lbm
Specific Volume:	Liquid:	0.01602 ft^3/lbm
	Solid:	0.01747 ft^3/lbm

A man weighing 180 lbm is ice skating on "hollow ground" blades which have a total area in contact with the ice of 0.032 in.^{2} The temperature of the ice is 28 F. Will the ice melt under the blades?

12.7 Consider Fig. 3.4, and prove that for water the slope of the sublimation line is greater than the slope of the vaporization line at the triple point.

12.8 An experiment is to be conducted at −160 F using Freon-12 as the bath liquid in a closed tank (with some vapor at the top of the tank).

(*a*) What is the pressure in the tank?

(*b*) When the experiment is completed, the tank will be allowed to warm to room temperature, 80 F. If the pressure in the tank is never to exceed 100 lbf/in.^{2}, what is the maximum per cent liquid (by volume) that can be used in the tank at −160 F?

12.9 Derive expressions for $(\partial u/\partial P)_T$ and $(\partial u/\partial T)_P$ that do not contain the properties h, u, or s.

12.10 Derive expressions for $(\partial T/\partial v)_u$ and $(\partial h/\partial s)_v$ that do not contain the properties h, u, or s.

12.11 Show that the Joule-Thomson coefficient μ_J is given by the relation

$$\mu_J = \frac{1}{C_p}\left[T\left(\frac{\partial v}{\partial T}\right)_P - v\right]$$

12.12 Show that if the volume expansivity α is constant, the following relation is valid:

$$\left(\frac{\partial C_p}{\partial P}\right)_T = -vT\alpha^2$$

12.13 From the data given in Table B.1.4 of the Appendix determine the average value of the isothermal compressibility at 32 F, 200 F, and 500 F. Does the isothermal compressibility vary significantly with pressure at a given temperature?

12.14 It can be shown that the velocity of sound, V_s in a fluid is given by the relation

$$\frac{\mathsf{V}_s^2}{g_c} = \left(\frac{\partial P}{\partial \rho}\right)_s$$

(*a*) Using this relation, show that the velocity of sound can be written as

$$V_s = \sqrt{\frac{g_c}{\beta_s \rho}}$$

where β_s is the adiabatic compressibility.

(*b*) At a temperature of 100 F the velocity of sound in water is 5040 ft/sec. Verify this by determining the adiabatic compressibility of water at 100 F by using Tables B.1.1 and B.1.4. (Note that the state of water after an isentropic process can be determined by using Tables B.1.1 and B.1.4.)

12.15 The following data are available for ethyl alcohol at 20 C.

$$\rho = 49.2 \text{ lbm/ft}^3$$
$$\beta_s = 6.5 \times 10^{-6} \text{ in.}^2/\text{lbf}$$
$$\beta_T = 7.7 \times 10^{-6} \text{ in.}^2/\text{lbf}$$

(*a*) What is the velocity of sound (see Problem 12.14) in ethyl alcohol?

(*b*) The pressure on a cylinder having a volume of 1 ft³ containing ethyl alcohol at 20 C is increased from 1 atm to 200 atm while the temperature remains constant. Determine the work done on the ethyl alcohol during this process.

12.16 It is desired to develop a set of tables of thermodynamic properties for a substance without using any elevated pressure values of specific heat, as found from C_{p0} and Eq. 12.42. Assuming that the same data as specified by items $a-d$ of Section 12.8 are known, discuss the procedure to be followed. Why would it be desirable to use this procedure instead of that discussed in Section 12.8?

12.17 Suppose that the following data are available for a given pure substance:
A vapor-pressure curve.
An equation of state of the form $P = f(v, T)$.
Specific volume of saturated liquid.
Critical pressure and temperature.
Constant-volume specific heat of the vapor at one specific volume.
Outline the steps you would follow in order to develop a table of thermodynamic properties comparable to Tables 1, 2, and 3 of the steam tables.

12.18 Five ft³ of ethylene at 330 lbf/in.², 80 F, is cooled at constant volume to saturated vapor, after which the C_2H_4 is condensed at constant pressure to saturated liquid. Determine the final pressure, temperature, and volume using the generalized compressibility chart.

12.19 A 2-ft³ rigid tank contains propane at 1350 lbf/in.², 1000 R. The propane is then allowed to cool to 600 R due to heat transfer with the surroundings. Determine the quality at the final state and the mass of liquid in the tank.

12.20 Saturated liquid carbon dioxide at the temperature of the surroundings, 60 F, is contained in a cylinder as shown in Fig. 12.11. The volume at this state is 0.01 ft³. The piston has a mass of 10 lbm, area of 20 in.², and ambient

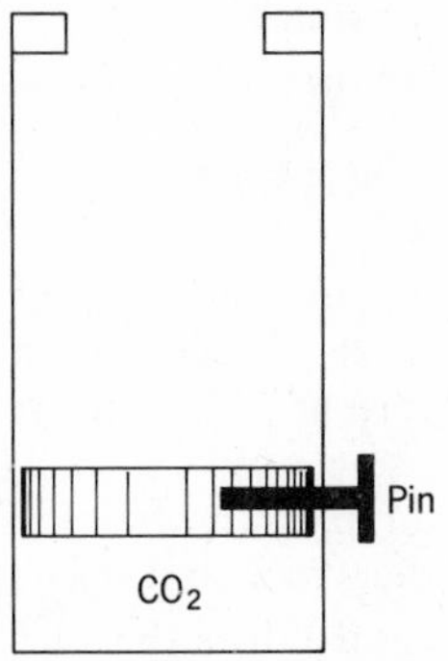

Fig. 12.11 Sketch for Problem 12.20.

pressure is 1 atmosphere. The pin is then released from the piston, allowing it to rise and hit the stops, at which point the volume is 0.15 ft^3. Eventually, the temperature of the CO_2 returns to that of the surroundings.

(*a*) Calculate the pressure (if superheated) or quality (if saturated at the final state.

(*b*) Calculate the work done during the process.

12.21 Show that for van der Waals' equation the following relations apply.

$$(a)\ (h_2-h_1)_T=(P_2v_2-P_1v_1)+a\left(\frac{1}{v_1}-\frac{1}{v_2}\right)$$

$$(b)\ (s_2-s_1)_T=R\ln\left(\frac{v_2-b}{v_1-b}\right)$$

$$(c)\ \left(\frac{\partial C_v}{\partial v}\right)_T=0$$

$$(d)\ C_p-C_v=\frac{R}{1-2a(v-b)^2/RTv^3}$$

12.22 Derive expressions for the changes in enthalpy and entropy at constant temperature for a substance that follows the Beattie-Bridgeman equation of state.

12.23 The residual volume α, defined as

$$\alpha=\frac{RT}{P}-v$$

is a parameter used in analyzing thermodynamic behavior. Find the limiting value of α and also that of compressibility factor Z as the pressure goes to zero (at constant temperature) for:

(*a*) An ideal gas.

(*b*) The van der Waals equation of state.

(*c*) The virial equation of state.

12.24 The single temperature for which

$$\lim_{P\to 0}\left(\frac{\partial Z}{\partial P}\right)_T = 0$$

is termed the Boyle temperature.

(*a*) Using this relation, show that residual volume α as defined in Problem 12.23 tends to zero at zero pressure only at the Boyle temperature.

(*b*) Using the results of part (*a*) and Problem 12.23, find the Boyle temperature for the van der Waals and virial equations of state.

12.25 At what reduced temperature does the van der Waals equation of state predict a limiting value of zero (ideal gas value) for the Joule-Thomson coefficient as the pressure approaches zero?

12.26 Methane at a pressure of 1200 lbf/in.^{2}, 100 F is throttled in a steady-flow process to 300 lbf/in.^{2}. Determine the final temperature of the methane.

12.27 Ethane at 800 lbf/in.^{2}, 300 F is cooled in a heat exchanger to 120 F. The volume rate of flow entering the heat exchanger is 100 ft^3/min, and the pressure drop through the heat exchanger is small. Determine the heat transfer per min.

12.28 Saturated liquid nitrogen at −260 F is throttled to 100 lbf/in.^{2}. Calculate the quality at the outlet, using

(*a*) The generalized charts.

(*b*) The nitrogen tables.

12.29 A number of tanks are to be filled with ethylene for use in another location. The procedure is as follows: a 2-ft^3 evacuated tank is connected to a line carrying C_2H_4 at 2100 lbf/in.^{2} and room temperature, 80 F. The valve is then opened, and ethylene flows in rapidly, such that the process is essentially adiabatic. The valve is to be closed when the pressure reaches a certain value P_2, and the tank is then disconnected. After a period of time, the temperature inside the tank will return to room temperature because of heat transfer with the surroundings. At this time, the pressure inside the tank must be 1500 lbf/in.^{2}. What is the pressure P_2 at which the valve should be closed while filling the tank?

12.30 Carbon dioxide is to be liquefied in a steady-flow insulated condenser. CO_2 enters the condenser at 170 F, 900 lbf/in.^{2} at the rate of 1 ft^3/min, and leaves at 900 lbf/in.^{2}. Cooling water enters at 40 F and leaves at 70 F. The flow rate of the cooling water is 1800 lbm/hr.

Determine the temperature of the CO_2 leaving the condenser and show the process for the CO_2 on a *T-s* diagram.

12.31 Propane flows through a pipe of constant cross-sectional area. At the inlet section of this pipe the velocity is 80 ft/sec, the pressure is 400 lbf/in.^{2}, and the temperature is 220 F. At the exit section the pressure is 100 lbf/in.^{2}. There is no heat transfer to or from the gas as it flows in the pipe. Determine the temperature and velocity of the propane leaving the pipe.

12.32 High-pressure liquid methane is to be produced at the rate of 2000 lbm/hr. The available supply of CH_4 gas at 80 F, 100 lbf/in.^{2} is cooled and condensed in a heat exchanger at constant pressure, after which the liquid is pumped to 500 lbf/in.^{2}.

(*a*) Calculate the required heat transfer from the methane in flowing through the heat exchanger.

(*b*) Assuming a pump efficiency of 80%, calculate the horsepower required to drive the pump.

12.33 An uninsulated cylinder with a volume of 2 ft^3 contains ethylene at 1000 lbf/in.^{2}, 80 F, the temperature of the surroundings. The cylinder valve leaks very slightly so that after a considerable length of time the cylinder pressure drops to 500 lbf/in.^{2}. Determine:

(*a*) The mass of ethylene which escaped from the cylinder.

(*b*) The heat transferred during the process.

12.34 Consider a 10-ft^3 tank in which methane is stored at low temperature. The pressure is 100 lbf/in.^{2} and the tank contains 25% liquid, 75% vapor on a volume basis.

(*a*) The tank now slowly warms because of heat transfer from the ambient. What is the temperature inside when the pressure reaches 1500 lbf/in.^{2}?

(*b*) Calculate the heat transferred during this process.

12.35 Ethane is contained in an insulated cylinder, as shown in Fig. 12.12. The

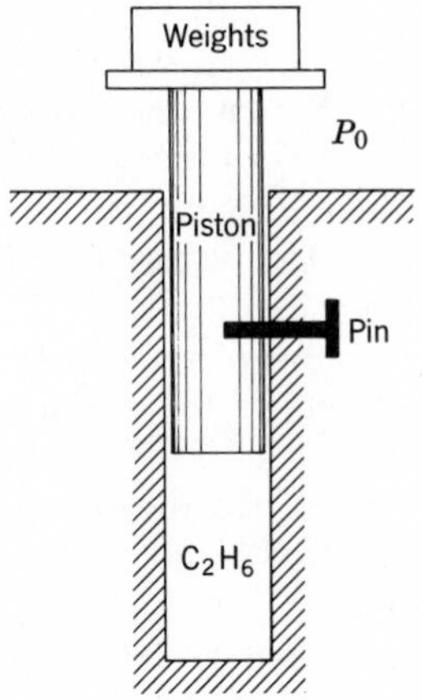

Fig. 12.12 Sketch for Problem 12.35.

initial conditions are 35 F, quality of 90%, and volume of 0.01 ft^3. The piston area is 0.5 in.^{2} and the total mass of piston and weights is 500 lbm. Ambient pressure is 1 atm. The pin holding the piston is now removed. Determine the final pressure, volume, and temperature.

12.36 Freon-12, which has a molecular weight of 121, expands in a cylinder from 300 lbf/in.^{2}, 180 F to 35 lbf/in.^{2} is an isothermal process. During this process the heat transfer to the Freon-12 is 20 Btu/lbm. The process is not reversible.

Determine, for this process, the work per pound of Freon-12 using the following sources for thermodynamic data:

(*a*) The Freon-12 tables.
(*b*) The generalized charts.
(*c*) The relation

$$\left(\frac{\partial u}{\partial v}\right)_T = T\left(\frac{\partial P}{\partial T}\right)_v - P$$

and van der Waals' equation of state.

12.37 A 1 ft^3 evacuated tank is to be filled from a supply line containing ethane at 110 F, 2000 lbf/in.^{2}. Gas flows into the tank until the pressure reaches 1800 lbf/in.^{2}, at which time the valve is closed. The tank is well insulated so that heat transfer is negligible. Using the generalized charts, determine the final temperature in the tank and the mass of ethane entering.

12.38 An insulated tank having a volume of 2 ft^3 contains carbon dioxide at 200 lbf/in.^{2}, 100 F. It is pressurized from a line in which carbon dioxide is flowing at 1200 lbf/in.^{2}, 100 F. The valve is closed when the pressure in the tank reaches 1200 lbf/in.^{2}.

Using the generalized charts determine the amount of carbon dioxide which flows into the tank.

12.39 Propane is compressed in a reversible isothermal steady-flow process from 10 atm pressure, 250 F to 50 atm pressure. Determine the work of compression and heat transfer per lbm of propane.

12.40 Propane is compressed in a reversible adiabatic steady-flow process from 10 atm pressure, 250 F to 50 atm pressure. Determine the work of compression per lbm of propane.

12.41 Determine the minimum work of refrigeration required to perform the process described in Problem 12.18, assuming the ambient temperature to be 80 F.

12.42 Using the generalized charts and the nitrogen tables, estimate the per cent correction (from saturated liquid values) in v, h, s for liquid nitrogen at -260 F, 1000 lbf/in.^{2}.

12.43 Consider the process described in Problem 12.19. Calculate:

(*a*) The heat transferred during the process.
(*b*) The irreversibility of the process, assuming that heat is rejected from the tank to the surroundings at 77 F.

12.44 Propane gas expands in a turbine from 400 lbf/in.^{2}, 300 F to 50 lbf/in.^{2}, 210 F. Assume the process to be adiabatic. Using the generalized charts determine:

(*a*) The work done per pound of propane entering the turbine.

(*b*) The increase of entropy per pound of propane as it flows through the turbine.

(*c*) The isentropic turbine efficiency.

12.45 An uninsulated 5-ft^3 tank contains CO_2 at 60 F with an initial quality of 10%. Saturated vapor is withdrawn and expanded through a turbine to 200 lbf/in.^{2} in a reversible adiabatic process as shown in Fig. 12.13. If the process ends when the tank quality reaches 100%, calculate the total work output of the turbine during the process.

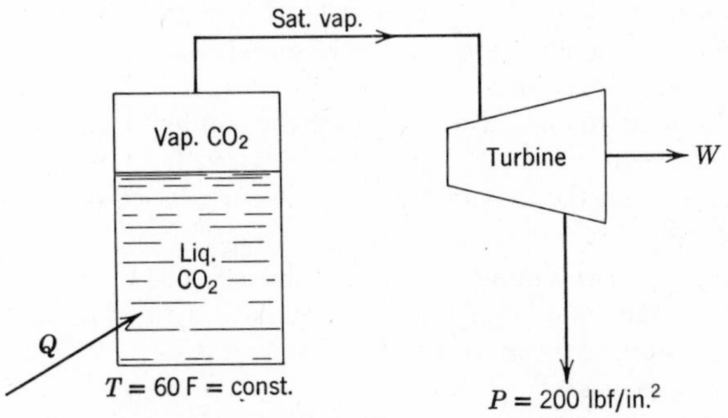

Fig. 12.13 Sketch for Problem 12.45.

12.46 It is desired to evaluate a recently developed compound for possible use as a working fluid in a portable, closed-cycle power plant, shown schematically in Fig. 12.14.

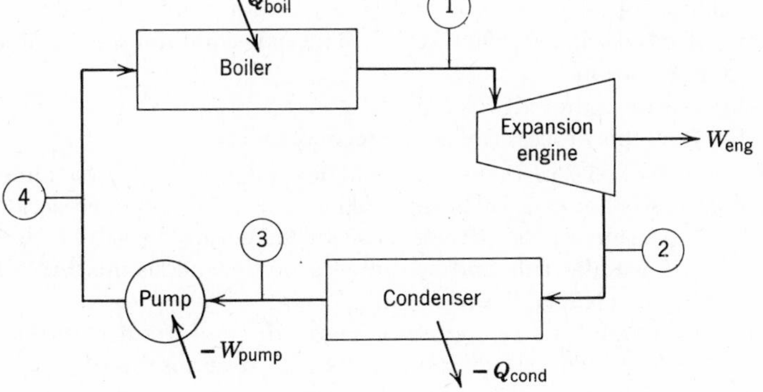

Fig. 12.14 Sketch for Problem 12.46.

The only information available on this compound is the following:

$$T_c = 320 \text{ F}$$
$$P_c = 300 \text{ lbf/in.}^2$$
$$M = 100$$
$$k = 1.10 \text{ (low-pressure value)}$$

The condenser temperature is fixed at 125 F, and the maximum cycle temperature at 350 F. The isentropic efficiency of the expansion engine is estimated to be 80% and the minimum allowable quality of the exiting fluid (actual state) is 90%.

(*a*) Calculate the heat transfer from the condenser, assuming saturated liquid at the exit.

(*b*) Determine the maximum cycle pressure, based on the specifications listed above.

12.47 A 1.4 ft^3 tank containing air at 2200 lbf/in.^{2}, 537 R, is cooled to 350 R. Determine the final pressure in the tank and the heat transfer for the process, assuming the air to be

(*a*) An ideal gas.

(*b*) A pseudo-pure substance using Kay's rule.

(*c*) A pseudo-pure substance using the Beattie-Bridgeman equation of state.

12.48 One lb mole of mixture containing 50% CO_2, 50% C_2H_6 (mole basis) is compressed in a cylinder in a reversible isothermal process. The initial state is 100 lbf/in.^{2}, 100 F and the final pressure is 800 lbf/in.^{2}. Calculate the heat transfer and work for the process, using the van der Waals equation of state, and compare the results with ideal gas behavior.

12.49 A mixture of 50% N_2, 50% O_2 on a mole basis flows through a long pipe. At the pipe inlet, the mixture is at 500 lbf/in.^{2}, −150 F, with a velocity of 100 ft/sec. If the pressure drop in the pipe is estimated to be 100 lbf/in.^{2}, determine the exit temperature and velocity. Assume Kay's rule.

12.50 A mixture of 50% methane and 50% nitrogen (mole basis) enters an insulated nozzle at 0 F, 2900 lbf/in.^{2}. The exit conditions are stated to be −120 F, 850 lbf/in.^{2}.

(*a*) Calculate the exit velocity.

(*b*) Does this process violate the second law?

12.51 A mixture of 85% methane, 15% ethane on a mole basis is contained in an insulated 5 ft^3 tank at 500 lbf/in.^{2}, 100 F. The valve is opened accidentally, and the pressure quickly drops to 300 lbf/in.^{2} before the valve is closed.

(*a*) Using Kay's rule and assuming a homogeneous mixture, calculate the mass that escaped from the tank.

(*b*) Eventually the temperature inside the tank returns to that of the surroundings, 100 F. What is the tank pressure at this time?

12.52 An insulated vertical cylinder fitted with a piston is divided into two parts by a partition as indicated in Fig. 12.15. Side *A* contains argon at 560 R,

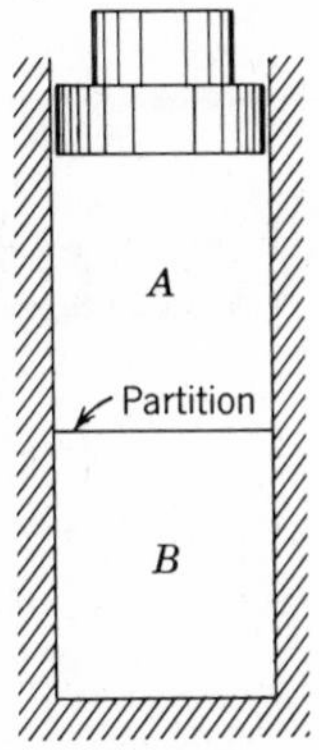

Fig. 12.15 Sketch for Problem 12.52.

2000 lbf/in.^{2} and side B contains carbon dioxide at 560 R, 1000 lbf/in.^{2}. Each has a volume of 1 ft^3. A leak develops in the partition, and the two gases mix together.

(*a*) Assuming that the resulting mixture is represented by Kay's rule, find the pseudocritical pressure and temperature.

(*b*) Determine the final temperature in the system.

12.53 Write a computer program to solve the following problem. It is desired to obtain a plot of pressure versus volume at various temperatures (all on a generalized reduced basis) as predicted by the van der Waals equation of state. (Temperatures less than the critical should be included in the results.)

12.54 Write a computer program to solve the following problem. For one of the substances listed in Table 3.3, calculate enthalpy changes along several isotherms at various integral pressures, using the Beattie–Bridgeman equation of state.

13

Chemical Reactions

Many thermodynamic problems involve chemical reactions. Among the most familiar of these is the combustion of hydrocarbon fuels, for this process is utilized in most of our power generating devices. However, we can all think of a host of other processes involving chemical reactions, including those that occur in the human body.

It is our purpose in this chapter to consider a first and second law analysis of systems undergoing a chemical reaction. In many respects, this chapter is simply an extension of our previous consideration of the first and second laws. However, a number of new terms are introduced, and it will also be necessary to introduce the third law of thermodynamics.

In this chapter the combustion process is considered in detail. There are two reasons for this emphasis. The first is that the combustion process is of great significance in many problems and devices with which the engineer is concerned. The second is that the combustion process provides an excellent vehicle for teaching the basic principles of the thermodynamics of chemical reactions. The student should keep both of these objectives in mind as the study of this chapter progresses.

Chemical equilibrium will be considered in Chapter 14, and, therefore, the matter of dissociation will be deferred until then.

13.1 Fuels

A thermodynamics textbook is not the place for a detailed treatment of fuels. However, some knowledge of them is a prerequisite to a consideration of combustion, and this section is therefore devoted to a brief discussion of some of the hydrocarbon fuels. Most fuels fall into one of the three categories—coal, liquid hydrocarbons, or gaseous hydrocarbons.

Most liquid and gaseous hydrocarbon fuels are a mixture of many different hydrocarbons. For example, gasoline consists primarily of a mixture of about forty hydrocarbons, with many others present in very small quantities. In discussing hydrocarbon fuels, therefore, brief consideration should be given to the most important families of hydrocarbons, which are summarized in Table 13.1.

Table 13.1
Characteristics of Some of the Hydrocarbon Families

Family	Formula	Structure	Saturated
Paraffin	C_nH_{2n+2}	Chain	Yes
Olefin	C_nH_{2n}	Chain	No
Diolefin	C_nH_{2n-2}	Chain	No
Naphthene	C_nH_{2n}	Ring	Yes
Aromatic			
Benzene	C_nH_{2n-6}	Ring	No
Naphthalene	C_nH_{2n-12}	Ring	No

Three terms should be defined. The first pertains to the structure of the molecule. The important types are the ring and chain structures; the difference between the two is illustrated in Fig. 13.1. The same figure

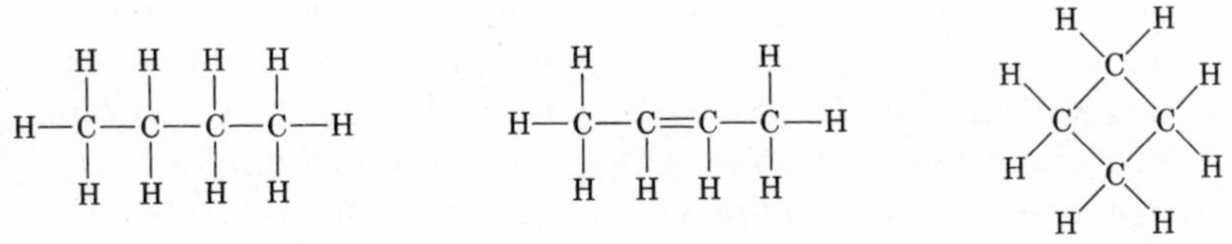

Chain structure, saturated

Chain structure, unsaturated

Ring structure, saturated

Fig. 13.1 Molecular structure of some hydrocarbon fuels.

illustrates the definition of saturated and unsaturated hydrocarbons. An unsaturated hydrocarbon has two or more adjacent carbon atoms joined by a double or triple bond, whereas in a saturated hydrocarbon all the carbon atoms are joined by a single bond. The third term to be defined is an isomer. Two hydrocarbons with the same number of carbon and hydrogen atoms and different structures are called isomers. Thus, there are several different octanes (C_8H_{18}) each having 8 carbon atoms and 18 hydrogen atoms, but each has a different structure.

The various hydrocarbon families are identified by a common suffix. The compounds comprising the paraffin family all end in "-ane" (as propane and octane). Similarly, the compounds comprising the olefin family end in "-ylene" or "-ene" (as propene and octene), and the diolefin family ends in "-diene" (as butadiene). The naphthene family has the same chemical formula as the olefin family, but has a ring rather than chain structure. The hydrocarbons in the naphthene family are named by adding the prefix "cyclo-" (as cyclopentane).

The aromatic family includes the benzene series (C_nH_{2n-6}) and the naphthalene series (C_nH_{2n-12}). The benzene series has a ring structure and is unsaturated.

Alcohols are sometimes used as a fuel in internal combustion engines. The characteristic feature of the alcohol family is that one of the hydrogen atoms is replaced by an OH radical. Thus methyl alcohol, also called methanol, is CH_3OH.

Most liquid hydrocarbon fuels are mixtures of hydrocarbons that are derived from crude oil through distillation and cracking processes. Thus, from a given crude oil a variety of different fuels can be produced, some of the common ones being gasoline, kerosene, diesel fuel, and fuel oil. Within each of these classifications there is a wide variety of grades, and each is made up of a large number of different hydrocarbons. The important distinction between these fuels is the distillation curve, Fig. 13.2. The distillation curve is obtained by slowly heating a sample of fuel so that it vaporizes. The vapor is then condensed and the amount measured. The more volatile hydrocarbons are vaporized first, and thus the temperature of the nonvaporized fraction increases during the process. The distillation curve, which is a plot of the temperature of the nonvaporized fraction vs. the amount of vapor condensed, is an indication of the volatility of the fuel.

In dealing with combustion of liquid fuels it is convenient to express the composition in terms of a single hydrocarbon, even though it is a mixture of many hydrocarbons. Thus gasoline is usually considered to be octane, C_8H_{18}, and diesel fuel is considered to be dodecane, $C_{12}H_{26}$. The composition of a hydrocarbon fuel may also be given in terms of percentage of carbon and hydrogen.

The two primary sources of gaseous hydrocarbon fuels are natural gas wells and certain chemical manufacturing processes. Table 13.2

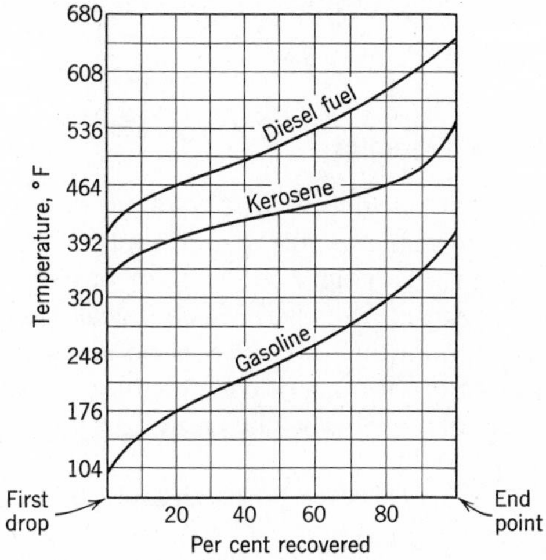

Fig. 13.2 Typical distillation curves of some hydrocarbon fuels.

Table 13.2
Volumetric Analyses of Some Typical Gaseous Fuels

Constituent	Various Natural Gases A	B	C	D	Producer Gas from Bituminous Coal	Carbureted Water Gas	Coke-Oven Gas
Methane	93.9	60.1	67.4	54.3	3.0	10.2	32.1
Ethane	3.6	14.8	16.8	16.3			
Propane	1.2	13.4	15.8	16.2			
Butanes plus[a]	1.3	4.2		7.4			
Ethene						6.1	3.5
Benzene						2.8	0.5
Hydrogen					14.0	40.5	46.5
Nitrogen		7.5		5.8	50.9	2.9	8.1
Oxygen					0.6	0.5	0.8
Carbon monoxide					27.0	34.0	6.3
Carbon dioxide					4.5	3.0	2.2

[a]This includes butane and all heavier hydrocarbons.

gives the composition of a number of gaseous fuels. The major constituent of natural gas is methane, which distinguishes it from manufactured gas.

13.2 The Combustion Process

The combustion process involves the oxidation of constituents in the fuel that are capable of being oxidized, and can therefore be represented by a chemical equation. During a combustion process the mass of each element remains the same. Thus, writing chemical equations and solving problems involving quantities of the various constituents basically involves the conservation of mass of each element. A brief review of this subject, particularly as it applies to the combustion process, is presented in this chapter.

Consider first the reaction of carbon with oxygen.

$$\begin{array}{ccc} \text{Reactants} & & \text{Products} \\ C+O_2 & \rightarrow & CO_2 \end{array}$$

This equation states that one mole of carbon reacts with one mole of oxygen to form one mole of carbon dioxide. This also means that 12 lbm of carbon react with 32 lbm of oxygen to form 44 lbm of carbon dioxide. All the initial substances that undergo the combustion process are called the reactants, and the substances that result from the combustion process are called the products.

When a hydrocarbon fuel is burned both the carbon and the hydrogen are oxidized. Consider the combustion of methane as an example.

$$CH_4+2O_2 \rightarrow CO_2+2H_2O \tag{13.1}$$

In this case the products of combustion include both carbon dioxide and water. The water may be in the vapor, liquid, or solid phases, depending on the temperature and pressure of the products of combustion.

It should be pointed out that in the combustion process there are many intermediate products formed during the chemical reaction. In this book we are concerned with the initial and final products and not with the intermediate products, but this aspect is very important in a detailed consideration of combustion.

In most combustion processes the oxygen is supplied as air rather than as pure oxygen. The composition of air on a molal basis is approximately 21 per cent oxygen, 78 per cent nitrogen, and 1 per cent argon. We will assume that the nitrogen and the argon do not undergo chemical reaction (except for dissociation which will be considered in Chapter 14).

They do leave at the same temperature as the other products, however, and therefore undergo a change of state if the products are at a temperature other than the original air temperature. It should be pointed out that at the high temperatures achieved in internal combustion engines, there is actually some reaction between the nitrogen and oxygen, and this gives rise to the air pollution problem associated with the oxides of nitrogen in the engine exhaust.

In combustion calculations involving air, the argon is usually neglected, and the air is considered to be composed of 21 per cent oxygen and 79 per cent nitrogen by volume. When this assumption is made, the nitrogen is sometimes referred to as "atmospheric nitrogen." Atmospheric nitrogen has a molecular weight of 28.16 (which takes the argon into account) as compared to 28.016 for pure nitrogen. This distinction will not be made in this text, and we will consider the 79 per cent nitrogen to be pure nitrogen.

The assumption that air is 21.0 per cent oxygen and 79.0 per cent nitrogen by volume leads to the conclusion that for each mole of oxygen, $79.0/21.0 = 3.76$ moles of nitrogen are involved. Therefore, when the oxygen for the combustion of methane is supplied as air, the reaction can be written

$$CH_4 + 2O_2 + 2(3.76)N_2 \rightarrow CO_2 + 2H_2O + 7.52N_2 \tag{13.2}$$

The minimum amount of air that supplies sufficient oxygen for the complete combustion of all the carbon, hydrogen, and any other elements in the fuel that may oxidize is called the "theoretical air." When complete combustion is achieved with theoretical air, the products contain no oxygen. In practice, it is found that complete combustion is not likely to be achieved unless the amount of air supplied is somewhat greater than the theoretical amount. The amount of air actually supplied is expressed in terms of per cent theoretical air. Thus 150 per cent theoretical air means that the air actually supplied is 1.5 times the theoretical air. The complete combustion of methane with 150 per cent theoretical air is written

$$CH_4 + 2(1.5)O_2 + 2(3.76)(1.5)N_2 \rightarrow CO_2 + 2H_2O + O_2 + 11.28N_2 \tag{13.3}$$

The amount of air actually supplied may also be expressed in terms of per cent excess air. The excess air is the amount of air supplied over and above the theoretical air. Thus 150 per cent theoretical air is equivalent to 50 per cent excess air. The terms theoretical air and excess air are both in current usage.

Two important parameters applied to combustion processes are the air-fuel ratio (designated AF) and its reciprocal, the fuel-air ratio (designated FA). The air-fuel ratio is usually expressed on a mass basis, but a mole basis is also used at times. The theoretical air-fuel ratio is the ratio of the mass (or moles) of theoretical air to the mass (or moles) of fuel.

When the amount of air supplied is less than the theoretical air required, the combustion is incomplete. If there is only a slight deficiency of air, the usual result is that some of the carbon unites with the oxygen to form carbon monoxide (CO) instead of carbon dioxide (CO_2). If the air supplied is considerably less than the theoretical air, there may also be some hydrocarbons in the products of combustion.

Even when some excess air is supplied there may be small amounts of carbon monoxide present, the exact amount depending on a number of factors including the mixing and turbulence during combustion. Thus, the combustion of methane with 110 per cent theoretical air might be as follows:

$$CH_4 + 2(1.1)O_2 + 2(1.1)3.76N_2 \rightarrow$$
$$+0.95CO_2 + 0.05CO + 2H_2O + 0.225O_2 + 8.27N_2 \qquad (13.4)$$

The material covered so far in this section is illustrated by the following examples.

Example 13.1

Calculate the theoretical air-fuel ratio for the combusuon of octane, C_8H_{18}.

The combustion equation is

$$C_8H_{18} + 12.5O_2 + 12.5(3.76)N_2 \rightarrow 8CO_2 + 9H_2O + 47.0N_2$$

The air-fuel ratio on a mole basis is

$$AF = \frac{12.5 + 47.0}{1} = 59.5 \text{ moles air/mole fuel}$$

The theoretical air-fuel ratio on a mass basis is found by introducing the molecular weight of the air and fuel.

$$AF = \frac{59.5(28.97)}{114.2} = 15.0 \text{ lbm air/lbm fuel}$$

Example 13.2

Determine the molal analysis of the products of combustion when octane, C_8H_{18}, is burned with 200 per cent theoretical air, and determine the dew point of the products if the pressure is 14.7 lbf/in.^{2}.

The equation for the combustion of octane with 200 per cent theoretical air is

$$C_8H_{18}+12.5(2)O_2+12.5(2)(3.76)N_2 \rightarrow 8CO_2+9H_2O+12.5O_2+94.0N_2$$

Total moles of product $= 8+9+12.5+94.0 = 123.5$
Molal analysis of products:

$$\begin{aligned} CO_2 &= 8/123.5 &&= 6.47\% \\ H_2O &= 9/123.5 &&= 7.29 \\ O_2 &= 12.5/123.5 &&= 10.12 \\ N_2 &= 94/123.5 &&= \underline{76.12} \\ & && 100.00\% \end{aligned}$$

The partial pressure of the H_2O is $14.7(0.0729) = 1.072$ lbf/in.^{2}.

The saturation temperature corresponding to this pressure is 104 F, which is also the dew-point temperature.

The water condensed from the products of combustion usually contains some dissolved gases and therefore may be quite corrosive. For this reason the products of combustion are often kept above the dew point until discharged to the atmosphere.

Example 13.3

Producer gas from bituminous coal (see Table 13.2) is burned with 20 per cent excess air. Calculate the air-fuel ratio on a volumetric basis and on a mass basis.

To calculate the theoretical air requirement, let us write the combustion equation for the combustible substances in 1 mole of fuel.

$$\begin{aligned} 0.14H_2+0.070O_2 &\rightarrow 0.14H_2O \\ 0.27CO+0.135O_2 &\rightarrow 0.27CO_2 \\ 0.03CH_4+\underline{0.06O_2} &\rightarrow 0.03CO_2+0.06H_2O \\ 0.265 &= \text{moles oxygen required/mole fuel} \\ \underline{-0.006} &= \text{oxygen in fuel/mole fuel} \\ 0.259 &= \text{moles oxygen required from air/mole fuel} \end{aligned}$$

Therefore, the complete combustion equation for 1 mole of fuel is

$$\overbrace{0.14H_2+0.27CO+0.03CH_4+0.006O_2+0.509N_2+0.045CO_2}^{\text{fuel}}$$

$$+\overbrace{0.259O_2+0.259(3.76)N_2}^{\text{air}} \rightarrow 0.20H_2O+0.345CO_2+1.482N_2$$

$$\left(\frac{\text{moles air}}{\text{mole fuel}}\right)_{\text{theo}} = 0.259 \times \frac{1}{0.21} = 1.233$$

If the air and fuel are at the same pressure and temperature, this also represents the ratio of the volume of air to the volume of fuel.

$$\text{For 20 per cent excess air, } \frac{\text{moles air}}{\text{mole fuel}} = 1.233 \times 1.200 = 1.48$$

The air-fuel ratio on a mass basis is

$$AF = \frac{1.48(28.97)}{0.14(2) + 0.27(28) + 0.03(16) + 0.006(32) + 0.509(28) + 0.045(44)}$$

$$= \frac{1.48(28.97)}{24.74} = 1.73 \text{ lbm air/lbm fuel}$$

An analysis of the products of combustion affords a very simple method for calculating the actual amount of air supplied in a combustion process. There are various experimental methods by which such an analysis can be made. Some yield results on a "dry" basis; i.e., the fractional analysis of all the components, except for water vapor. Other experimental procedures give results that include the water vapor. In this presentation we are not concerned with the experimental devices and procedures, but rather with the use of such information in a thermodynamic analysis of the chemical reaction. The examples that follow illustrate how an analysis of the products can be used to determine the chemical reaction and the composition of the fuel.

In using the analysis of the products of combustion to obtain the actual fuel-air ratio, the basic principle is the conservation of the mass of each of the elements. Thus, in changing from reactants to products we can make a carbon balance, hydrogen balance, oxygen balance, and nitrogen balance (plus any other elements that may be involved). Further, we recognize that there is a definite ratio between the amounts of some of these elements. Thus, the ratio between the nitrogen and oxygen supplied in the air is fixed, as well as the ratio between carbon and hydrogen if the composition of a hydrocarbon fuel is known.

Example 13.4

Methane (CH_4) is burned with atmospheric air. The analysis of the products on a "dry" basis is as follows:

CO_2	10.00%
O_2	2.37
CO	0.53
N_2	87.10
	100.00%

Calculate the air-fuel ratio, the per cent theoretical air, and determine the combustion equation.

The solution involves writing the combustion equation for 100 moles of dry products, introducing letter coefficients for the unknown quantities, and then solving for them.

From the analysis of the products, the following equation can be written, keeping in mind that this analysis is on a dry basis.

$$a\mathrm{CH_4} + b\mathrm{O_2} + c\mathrm{N_2} \rightarrow 10.0\mathrm{CO_2} + 0.53\mathrm{CO} + 2.37\mathrm{O_2} + d\mathrm{H_2O} + 87.1\mathrm{N_2}$$

A balance for each of the elements involved will enable us to solve for all the unknown coefficients:

Nitrogen balance: $c = 87.1$

Since all the nitrogen comes from the air,

$$\frac{c}{b} = 3.76; \quad b = \frac{87.1}{3.76} = 23.16$$

Carbon balance: $a = 10.00 + 0.53 = 10.53$

Hydrogen balance: $d = 2a = 21.06$

Oxygen balance: All the unknown coefficients have been solved for, and in this case the oxygen balance provides a check on the accuracy. Thus, b can also be determined by an oxygen balance.

$$b = 10.00 + \frac{0.53}{2} + 2.37 + \frac{21.06}{2} = 23.16$$

Substituting these values for a, b, c, d, and e we have,

$$10.53\mathrm{CH_4} + 23.16\mathrm{O_2} + 87.1\mathrm{N_2} \rightarrow$$
$$10.0\mathrm{CO_2} + 0.53\mathrm{CO} + 2.37\mathrm{O_2} + 21.06\mathrm{H_2O} + 87.1\mathrm{N_2}$$

Dividing through by 10.53 yields the combustion equation per mole of fuel.

$$\mathrm{CH_4} + 2.2\mathrm{O_2} + 8.27\mathrm{N_2} \rightarrow 0.95\mathrm{CO_2} + 0.05\mathrm{CO} + 2\mathrm{H_2O} + 0.225\mathrm{O_2} + 8.27\mathrm{N_2}$$

The air-fuel ratio on a mole basis is

$$2.2 + 8.27 = 10.47 \text{ moles air/mole fuel}$$

The air-fuel ratio on a mass basis is found by introducing the molecular weights.

$$AF = \frac{10.47 \times 28.97}{16.0} = 18.97 \text{ lbm air/lbm fuel}$$

The theoretical air-fuel ratio is found by writing the combustion equation for theoretical air.

$$CH_4 + 2O_2 + 2(3.76)N_2 \rightarrow CO_2 + 2H_2O + 7.52N_2$$

$$AF_{\text{theo}} = \frac{(2 + 7.52)28.97}{16.0} = 17.23 \text{ lbm air/lbm fuel}$$

$$\text{The per cent theoretical air is } \frac{18.97}{17.23} = 110\%$$

Example 13.5

The products of combustion of a hydrocarbon fuel of unknown composition have the following composition as measured on a dry basis:

CO_2	8.0%
CO	0.9
O_2	8.8
N_2	82.3
	100.0%

Calculate (1) The air-fuel ratio, (2) The composition of the fuel on a mass basis, and (3) The per cent theoretical air on a mass basis.

Let us first write the combustion equation for 100 moles of dry product.

$$C_aH_b + dO_2 + cN_2 \rightarrow 8.0CO_2 + 0.9CO + 8.8O_2 + eH_2O + 82.3N_2$$

Nitrogen balance: $c = 82.3$

From composition of air, $\frac{c}{d} = 3.76;\quad d = \frac{82.3}{3.76} = 21.9$

Oxygen balance: $21.9 = 8.0 + \frac{0.9}{2} + 8.8 + \frac{e}{2};\quad e = 9.3$

Carbon balance: $a = 8.0 + 0.9 = 8.9$

Hydrogen balance: $b = 2e = 2 \times 9.3 = 18.6$

Thus the composition of the fuel could be written $C_{8.9}H_{18.6}$

The air-fuel ratio can now be found, using the molecular weights.

$$AF = \frac{(21.9 + 82.3)28.97}{8.9(12) + 18.6(1)} = 24.1 \text{ lbm air/lbm fuel}$$

On a mass basis the composition of fuel is:

$$\text{Carbon} = \frac{8.9(12)}{8.9(12) + 18.6(1)} = 85.2\%$$

$$\text{Hydrogen} = \frac{18.6(1)}{8.9(12) + 18.6(1)} = 14.8\%$$

The theoretical air requirement can be found by writing the theoretical combustion equation.

$$C_{8.9}H_{18.6} + 13.5O_2 + 13.5(3.76)N_2 \rightarrow 8.9CO_2 + 9.3H_2O + 50.8N_2$$

$$AF_{\text{theo}} = \frac{(13.5 + 50.8)28.97}{8.9(12) + 18.6(1)} = 14.9 \text{ lbm air/lbm fuel}$$

$$\% \text{ theoretical air} = \frac{24.1}{14.9} = 162\%$$

13.3 Enthalpy of Formation

In the first twelve chapters of this book the problems considered always involved a fixed chemical composition, and never involved a change of composition due to chemical reaction. Therefore, in dealing with a thermodynamic property we made use of tables of thermodynamic properties for the given substance, and in each of these tables the thermodynamic properties were given relative to some arbitrary base. In the steam tables, for example, the internal energy of saturated liquid at 32.02 F is assumed to be zero. This procedure is quite adequate when no change in composition is involved, because we are concerned with the changes in the properties of a given substance. However, it is evident that this procedure is quite inadequate when dealing with a chemical reaction, because the composition changes during the process.

To understand the procedure used in the case of chemical reactions, consider the simple SSSF combustion process shown in Fig. 13.3. This reaction involves the combustion of carbon and oxygen to form CO_2. The carbon and oxygen each enter the control volume at 25C and 1 atm pressure, and the heat transfer is such that the CO_2 leaves at 25C, 1 atm

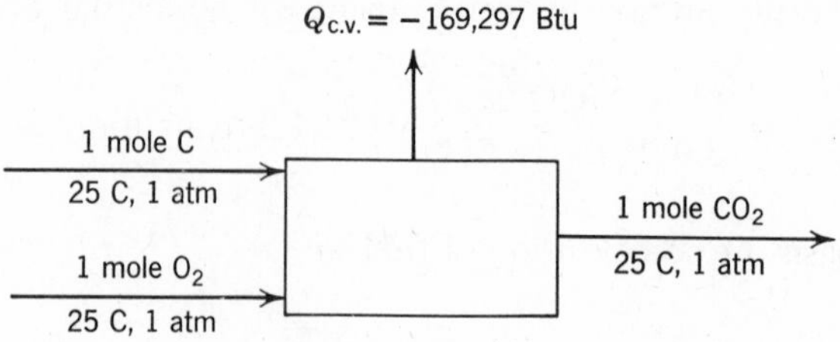

Fig. 13.3 Example of combustion process.

pressure. If the heat transfer were accurately measured, it would be $-169,297$ Btu per mole of CO_2 formed. This reaction can be written

$$C + O_2 \rightarrow CO_2$$

Applying the first law to this process we have

$$Q_{c.v.} + H_R = H_P \tag{13.5}$$

where the subscripts R and P refer to the reactants and products respectively. We will find it convenient to also write the first law for such a process in the form

$$Q_{c.v.} + \sum_R n_i \bar{h}_i = \sum_P n_e \bar{h}_e \tag{13.6}$$

where the summations refer respectively to all the reactants or all the products.

Thus a measurement of the heat transfer really gives us the difference between the enthalpy of the products and the reactants.

However, suppose we assign the value of zero to the enthalpy of all the elements at 25C and 1 atm pressure. In this case the enthalpy of the reactants is zero, and

$$Q_{c.v.} = H_P = -169,297 \text{ Btu/mole.}$$

The enthalpy of CO_2 at 25C and 1 atm pressure (with reference to this base in which the enthalpy of the elements is assumed to be zero) is called the enthalpy of formation. We designate this with the symbol $\bar{h}_f^\circ$ Thus, for CO_2,

$$\bar{h}_f^\circ = -169,297 \text{ Btu/mole}$$

The enthalpy of CO_2 in any other state, relative to this base in which the enthalpy of the elements is zero, would be found by adding the change of enthalpy between 25C, 1 atm and the given state to the enthalpy of formation. That is, the enthalpy at any temperature and pressure, $\bar{h}_{T,P}$ is

$$\bar{h}_{T,P} = (\bar{h}_f^\circ)_{298,1\,\text{atm}} + (\Delta\bar{h})_{298,1\,\text{atm}\rightarrow T,P} \tag{13.7}$$

where the term $(\Delta\bar{h})_{298,1\,\text{atm}\rightarrow T,P}$ represents the difference in enthalpy between any given state and the enthalpy at 298 K, 1 atm. For convenience we usually drop the subscripts in the examples that follow.

The procedure which we have demonstrated for CO_2 can be applied to any compound.

Table 13.3 gives values of the enthalpy of formation for a number of substances in both the units cal/gm mole and Btu/lb mole.

Three further observations should be made in regard to enthalpy of formation.

1. We have demonstrated the concept of enthalpy of formation in terms of the measurement of the heat transfer in a chemical reaction in which a compound is formed from the elements. Actually, the enthalpy of formation is usually found by the application of statistical thermodynamics. This matter is covered in Chapter 19.
2. The justification of this procedure of arbitrarily assigning the value of zero to the enthalpy of the elements at 25C, 1 atm rests on the fact that in the absence of nuclear reactions the mass of each element is conserved in a chemical reaction. No conflicts or ambiguities arise with this choice of reference state, and it proves to be very convenient in studying chemical reactions from a thermodynamic point of view.
3. In certain cases an element or compound can exist in more than one state at 25 C, 1 atm. Carbon, for example, can be in the form of graphite or diamond. It is essential that the state to which a given value is related be clearly identified. Thus, in Table 13.3 the enthalpy of formation of graphite is given the value of zero, and the enthalpy of each substance that contains carbon is given relative to this base.

It will be noted from Table 13.3 that two values for H_2O at 14.7 lbf/in.^{2}, 77 F, are given; one is for liquid H_2O and the other for gaseous H_2O. We know from the thermodynamic properties of water that under equilibrium conditions water exists as a slightly compressed liquid at 14.7 lbf/in.^{2}, 77 F. However, in many cases, we find it convenient to consider a pseudo-equilibrium state, in which water is a vapor at 14.7 lbf/in.^{2}, 77 F. In doing so we assume that the substance behaves as an ideal gas between saturated vapor at 77 F and this pseudo-equilibrium state at 14.7 lbf/in.^{2}, 77 F. Thus, the enthalpy in the pseudo-equilibrium state is the same as that of saturated vapor at this same temperature, and the change in

Table 13.3
Enthalpy of Formation, Gibbs Function of Formation and Absolute Entropy of Various Substances at 77 F (25 C) and 1 Atmosphere Pressure

				$\bar{h}_f^\circ$		$\bar{g}_f^\circ$		$\bar{s}^\circ$
Substance	Formula	M	State	Cal/gm mole	Btu/lb mole	Cal/gm mole	Btu/lb mole	Cal/gm mole-K Btu/lb mole-R
Carbon monoxide[a]	CO	28.011	gas	−26,417	−47,551	−32,783	−59,009	47.214
Carbon dioxide[a]	CO_2	44.011	gas	−94,054	−169,297	−94,265	−169,677	51.072
Water[a,b]	H_2O	18.016	gas	−57,798	−104,036	−54,636	−98,345	45.106
Water[b]	H_2O	18.016	liq.	−68,317	−122,971	−56,690	−102,042	16.716
Methane[a]	CH_4	16.043	gas	−17,895	−32,211	−12,145	−21,861	44.490
Acetylene[a]	C_2H_2	26.038	gas	54,190	97,542	49,993	89,987	48.004
Ethene[a]	C_2H_4	28.054	gas	12,496	22,493	16,281	29,306	52.447
Ethane[c]	C_2H_6	30.070	gas	−20,236	−36,425	−7,860	−14,148	54.85
Propane[c]	C_3H_8	44.097	gas	−24,820	−44,676	−5,614	−10,105	64.51
Butane[c]	C_4H_{10}	58.124	gas	−30,150	−54,270	−4,100	−7,380	74.12
Octane[c]	C_8H_{18}	114.23	gas	−49,820	−89,680	3,950	7,110	111.55
Octane[c]	C_8H_{18}	114.23	liq.	−59,740	−107,532	1,580	2,844	86.23
Carbon[a] (graphite)	C	12.011	solid	0	0	0	0	1.359

[a]From JANAF *Thermochemical Data*, The Dow Chemical Company, Thermal Laboratory, Midland, Mich.
[b]From *Circular 500*, National Bureau of Standards.
[c]From F. D. Rossini et al., *API Research Project 44*.

entropy between these two states would be found from the appropriate ideal gas relations. It also follows that the difference between the enthalpy of formation of $H_2O(l)$ and $H_2O(g)$ at 14.7 lbf/in.^{2}, 77 F is h_{fg}, the enthalpy of evaporation at 77 F. (The correction due to the fact that H_2O is a compressed liquid at 14.7 lbf/in.^{2}, 77 F, is very small and may be neglected.) This same procedure is used in the case of other substances which have a vapor pressure of less than one atmosphere at 77 F.

Frequently students are bothered by the minus sign when the enthalpy of formation is negative. For example, the enthalpy of formation of the CO_2 is negative. This is quite evident because the heat transfer was negative during the steady-flow chemical reaction, and the enthalpy of the CO_2 must be less than the sum of enthalpy of the carbon and oxygen initially, both of which are assigned the value of zero. This is quite analogous to the situation we would have in the steam tables if we let the enthalpy of saturated vapor be zero at 1 atm pressure, for in this case the enthalpy of the liquid would be negative, and we would simple use the negative value for the enthalpy of the liquid when solving problems.

13.4 First Law Analysis of Reacting Systems

The significance of the enthalpy of formation is that it is most convenient in performing a first law analysis of a reacting system, for the enthalpies of different substances can be added or subtracted, since they are all given relative to the same base.

In such problems we will write the first law for a steady-state, steady-flow process in the form

$$Q_{\text{c.v.}} + H_R = W_{\text{c.v.}} + H_P$$

or

$$Q_{\text{c.v.}} + \sum_R n_i \bar{h}_i = W_{\text{c.v.}} + \sum_P n_e \bar{h}_e$$

where R and P refer to the reactants and products respectively. In each problem it is necessary to choose one parameter as the basis of the solution. Usually this is taken as one mole of fuel.

Example 13.6

Consider the following reaction, which occurs in a steady-state, steady-flow process.

$$CH_4 + 2O_2 \rightarrow CO_2 + 2H_2O(l)$$

The reactants and products are each at a total pressure of 1 atm and 25 C. Determine the heat transfer per lb mole of fuel entering the combustion chamber.

Applying the first law,

$$Q_{\text{c.v.}} + \sum_R n_i \bar{h}_i = \sum_P n_e \bar{h}_e$$

From Table 13.3

$$\sum_R n_i \bar{h}_i = (\bar{h}_f^\circ)_{CH_4} = -32{,}211 \text{ Btu/lb mole fuel}$$

$$\sum_P n_e \bar{h}_e = (\bar{h}_f^\circ)_{CO_2} + 2(\bar{h}_f^\circ)_{H_2O(l)}$$

$$= -169{,}297 + 2(-122{,}971) = -415{,}239 \text{ Btu/lb mole fuel}$$

$$Q_{\text{c.v.}} = -415{,}239 - (-32{,}211) = -383{,}028 \text{ Btu/lb mole fuel}$$

In most cases however, the substances which comprise the reactants and products in a chemical reaction are not at a temperature of 77 F and a pressure of 1 atm (the state at which the enthalpy of formation is given). Therefore the change of enthalpy between 77 F and 1 atm and the given state must be known. In the case of a solid or liquid, this change of enthalpy can usually be found from a table of thermodynamic properties or from specific heat data. In the case of gases, this change of enthalpy can usually be found by one of the following procedures.

1. Assume ideal gas behavior between 77 F, 1 atm, and the given state. In this case, the enthalpy is a function of the temperature only, and can be found by use of an equation for C_{po} or from tabulated values of enthalpy as a function of temperature (which assumes ideal gas behavior). Table B.7 gives an equation for $\bar{C}_{po}$ for a number of substances and Table B.9 gives values of $(\bar{h}^\circ - \bar{h}^\circ_{298})$, (that is, the $\Delta\bar{h}$ of Eq. 13.7) in cal/gm mole and $(\bar{h}^\circ - \bar{h}^\circ_{537})$ in Btu/lb mole ($\bar{h}^\circ_{298}$ refers to 25 C or 298.15K, and $\bar{h}^\circ_{537}$ refers to 77 F or 536.7 R. For simplicity these are designated $\bar{h}^\circ_{298}$ and $\bar{h}^\circ_{537}$ respectively). The superscript ° is used to designate that this is the enthalpy at one atmosphere pressure, based on ideal gas behavior, that is, the standard-state enthalpy.
2. If a table of thermodynamic properties is available, Δh can be found directly from these tables. This is necessary when the deviation from ideal gas behavior is significant.
3. If the deviation from ideal gas behavior is significant, but no tables of thermodynamic properties are available, the value for $\Delta\bar{h}$ can be found from the generalized charts and the values for $\bar{C}_{po}$ or $\Delta\bar{h}$ at one atmosphere pressure as indicated above.

Thus, in general, for applying the first law to a steady-state, steady-flow process involving a chemical reaction and negligible changes in kinetic and potential energy we can write

$$Q_{\text{c.v.}} + \sum_R n_i[\bar{h}_f^\circ + \Delta\bar{h}]_i = W_{\text{c.v.}} + \sum_P n_e[\bar{h}_f^\circ + \Delta\bar{h}]_e \tag{13.8}$$

Example 13.7

Calculate the enthalpy of H_2O (on a lb mole basis) at 500 lbf/in.^{2}, 600 F, relative to the 77 F and 1 atm base, using the following procedures.

1. Assume the steam to be an ideal gas with the value of $\bar{C}_{po}$ given in the Appendix, Table B.7.
2. Assume the steam to be an ideal gas with the value for $\Delta\bar{h}$ as given in the Appendix, Table B.9.
3. The steam tables.
4. The specific heat equations given in (1) above and the generalized charts.

For each of these procedures we can write,

$$\bar{h}_{T,P} = [\bar{h}_f^\circ + \Delta\bar{h}]$$

The only difference is in the procedure by which we calculate $\Delta\bar{h}$. From Table 13.3 we note that

$$(\bar{h}_f^\circ)_{H_2O(g)} = -104{,}036 \text{ Btu/lb mole.}$$

1. Using the specific heat equation for $H_2O(g)$ from Table B.7,

$$\bar{C}_{po} = 19.86 - \frac{597}{\sqrt{T}} + \frac{7500}{T}$$

Therefore,

$$\Delta\bar{h} = \int_{537}^{1060} \left(19.86 + \frac{597}{\sqrt{T}} + \frac{7500}{T}\right) dT$$

$$= 4260 \text{ Btu/lb mole}$$

$$\bar{h}_{T,P} = -104{,}036 + 4260 = -99{,}776 \text{ Btu/lb mole.}$$

2. Using Table B.9 for $H_2O(g)$,

$$\Delta\bar{h} = 4345 \text{ Btu/lb mole}$$

$$\bar{h}_{T,P} = -104{,}036 + 4345 = -99{,}691 \text{ Btu/lb mole}$$

3. Using the steam tables

$$\Delta \bar{h} = 18.016(1298.3 - 1095.1)$$
$$= 3661 \text{ Btu/lb mole}$$
$$\bar{h}_{T,P} = -104{,}036 + 3661 = -100{,}375 \text{ Btu/lb mole}$$

4. When using the generalized charts we use the notation introduced in Chapter 12.

$$\bar{h}_{T,P} = \bar{h}_f^\circ - (\bar{h}_2^* - \bar{h}_2) + (\bar{h}_2^* - \bar{h}_1^*) + (\bar{h}_1^* - \bar{h}_1)$$

where subscript 2 refers to the state at 500 lbf/in.^{2}, 600 F and state 1 refers to the state at 1 atm, 77 F.

From part (1) $\bar{h}_2^* - \bar{h}_1^* = 4260$ Btu/lb mole.;

$$\bar{h}_1^* - \bar{h}_1 = 0 \text{ (ideal gas reference)}$$

$$P_{r2} = \tfrac{500}{3204} = 0.156; \quad T_{r2} = \tfrac{1060}{1165} = 0.91$$

From the generalized enthalpy chart,

$$\frac{\bar{h}_2^* - \bar{h}_2}{T_c} = 0.53; \quad \bar{h}_2^* - \bar{h}_2 = 0.53(1165) = 619 \text{ Btu/lb mole}$$

$$\bar{h}_{T,P} = -104{,}036 - 619 + 4260 = -100{,}395 \text{ Btu/lb mole}$$

The particular approach that is used in a given problem will depend on the data available for the given substance.

Example 13.8

A small gas turbine uses $C_8H_{18}(l)$ for fuel, and 400 per cent theoretical air. The air and fuel enter at 77 F, and the products of combustion leave at 1100 F. The output of the engine and the fuel consumption are measured and it is found that the specific fuel consumption is one pound of fuel per horsepower-hour. Determine the heat transfer from the engine per pound mole of fuel. Assume complete combustion.
The combustion equation is

$$C_8H_{18}(l) + 4(12.5)O_2 + 4(12.5)(3.76)N_2$$
$$\rightarrow 8CO_2 + 9H_2O + 37.5O_2 + 188.0N_2$$

We consider that we have a steady-state, steady-flow process involving ideal gases (except for the $C_8H_{18}(l)$). Therefore we can write:

First law: $$Q_{\text{c.v.}} + \sum_R n_i[(\bar{h}_f^\circ + \Delta\bar{h})]_i = W_{\text{c.v.}} + \sum_P n_e[\bar{h}_f^\circ + \Delta\bar{h}]_e$$

Property relation: Table B.9 for ideal gases and Table 13.3 for $C_8H_{18}(l)$. Since the air is composed of elements and enters at 77 F the enthalpy of the reactants is equal to that of the fuel,

$$\sum_R n_i[\bar{h}_f^\circ + \Delta\bar{h}]_i = (\bar{h}_f^\circ)_{C_8H_{18}(l)} = -107{,}532 \text{ Btu/lb mole.}$$

Considering the products

$$\begin{aligned}\sum_P n_e[\bar{h}_f^\circ + \Delta\bar{h}]_e &= n_{CO_2}[\bar{h}_f^\circ + \Delta\bar{h}]_{CO_2} + n_{H_2O}[\bar{h}_f^\circ + \Delta\bar{h}]_{H_2O} + n_{O_2}[\Delta\bar{h}]_{O_2} \\ &\qquad + n_{N_2}[\Delta\bar{h}]_{N_2} \\ &= 8(-169{,}297 + 11{,}315) + 9(-104{,}036 + 8869) \\ &\qquad + 37.5(7792) + 188(7384) \\ &= -439{,}969 \text{ Btu/lb mole fuel.}\end{aligned}$$

$$W_{\text{c.v.}} = 2545 \times 114.23 = 290{,}715 \text{ Btu/lb mole fuel}$$

Therefore, from the first law,

$$\begin{aligned}Q_{\text{c.v.}} &= -439{,}969 + 290{,}715 - (-107{,}532) \\ &= -41{,}722 \text{ Btu/lb mole fuel.}\end{aligned}$$

Example 13.9

A mixture of one mole of gaseous ethene and three moles of oxygen at 298 K react in a constant volume bomb. Heat is transferred until the products are cooled to 600 K. Determine the amount of heat transfer in cal/gm mole from the reactants.

The chemical reaction is

$$C_2H_4 + 3O_2 \rightarrow 2CO_2 + 2H_2O(g)$$

First law: $$Q + U_R = U_P$$

$$Q + \sum_R n[\bar{h}_f^\circ + \Delta\bar{h} - \bar{R}T] = \sum_P n[\bar{h}_f^\circ + \Delta\bar{h} - \bar{R}T]$$

$$\begin{aligned}\sum_R n[\bar{h}_f^\circ + \Delta\bar{h} - \bar{R}T] &= [\bar{h}_f^\circ - \bar{R}T]_{C_2H_4} - n_{O_2}[\bar{R}T]_{O_2} = (\bar{h}_f^\circ)_{C_2H_4} - 4\bar{R}T \\ &= 12{,}496 - 4 \times 1.987 \times 298 = 10{,}130 \text{ cal/gm mole fuel.}\end{aligned}$$

$$\sum_P n[\bar{h}_f^\circ + \Delta\bar{h} - \bar{R}T] = 2[(\bar{h}_f^\circ)_{CO_2} + \Delta\bar{h}_{CO_2}] + 2[(\bar{h}_f^\circ)_{H_2O(g)} + \Delta\bar{h}_{H_2O(g)}] - 4\bar{R}T$$

$$= 2[-94{,}054 + 3087] + 2[-57{,}798 + 2509] - 4 \times 1.987 \times 600$$

$$= -181{,}934 - 110{,}578 - 4766 = -297{,}278 \text{ cal/gm mole fuel.}$$

Therefore,

$$Q = -297{,}278 - 10{,}130 = -307{,}408 \text{ cal/gm mole fuel}$$

13.5 Adiabatic Flame Temperature

Consider a given combustion process that takes place adiabatically and with no work or changes in kinetic or potential energy involved. For such a process the temperature of the products is referred to as the adiabatic flame temperature. With the assumptions of no work and no changes in kinetic or potential energy, this is the maximum temperature that can be achieved for the given reactants because any heat transfer from the reacting substances and any incomplete combustion would tend to lower the temperature of the products.

For a given fuel and given pressure and temperature of the reactants, the maximum adiabatic flame temperature that can be achieved is with a stoichiometric mixture. The adiabatic flame temperature can be controlled by the amount of excess air that is used. This is important, for example, in gas turbines, where the maximum permissible temperature is determined by metallurgical considerations in the turbine, and close control of the temperature of the products is essential.

Example 13.10 shows how the adiabatic flame temperature may be found. The dissociation that takes place in the combustion products, which has a significant effect on the adiabatic flame temperature, will be considered in the next chapter.

Example 13.10

Liquid octane at 77 F is burned with 400 per cent theoretical air at 77 F in a steady-flow process. Determine the adiabatic flame temperature.

The reaction

$$C_8H_{18}(l) + 4(12.5)O_2 + 4(12.5)(3.76)N_2 \rightarrow$$
$$8CO_2 + 9H_2O(g) + 37.5O_2 + 188.0N_2$$

We assume ideal gas behavior for all the constituents except the liquid octane.

In this case the first law reduces to

$$H_R = H_P$$

$$\sum_R n_i[\bar{h}_f^\circ + \Delta\bar{h}]_i = \sum_P n_e[\bar{h}_f^\circ + \Delta\bar{h}]_e$$

where $\Delta\bar{h}_e$ refers to each constituent in the products at the adiabatic flame temperature.

$$H_R = \sum_R n_i[\bar{h}_f^\circ + \Delta\bar{h}]_i = (\bar{h}_f^\circ)_{C_8H_{18}(l)} = -107{,}532 \text{ Btu/mole fuel}$$

$$H_P = \sum_P n_e[\bar{h}_f^\circ + \Delta\bar{h}]_e = 8[-169{,}297 + \Delta\bar{h}_{CO_2}] + 9[-104{,}036 + \Delta\bar{h}_{H_2O}] + 37.5\Delta\bar{h}_{O_2} + 188.0\Delta\bar{h}_{N_2}$$

By trial-and-error solution, a temperature of the products is found that satisfies this equation. Assume

$$T_P = 1620 \text{ R}$$

$$\begin{aligned} H_P &= \sum_P n_e[\bar{h}_f^\circ + \Delta\bar{h}]_e \\ &= 8(-169{,}297 + 12{,}063) + 9(-104{,}036 + 9432) \\ &\quad + 37.5(8281) + 188.0(7839) \\ &= -1{,}257{,}872 - 851{,}436 + 310{,}537 + 1{,}473{,}732 \\ &= -325{,}039 \end{aligned}$$

Assume

$$T_P = 1800 \text{ R}$$

$$\begin{aligned} H_P &= \sum_P n_e[\bar{h}_f^\circ + \Delta\bar{h}]_e \\ &= 8(-169{,}297 + 14{,}371) + 9(-104{,}036 + 11{,}176) \\ &\quad + 37.5(9769) + 188(9232) \\ &= -1{,}239{,}408 - 835{,}740 + 366{,}337 + 1{,}735{,}616 \\ &= +26{,}807 \end{aligned}$$

Since $H_P = H_R = -107{,}532$, we find, by interpolation that the adiabatic flame temperature is 1731 R.

13.6 Enthalpy and Internal Energy of Combustion; Heat of Reaction

The enthalpy of combustion, h_{RP}, is defined as the difference between the enthalpy of the products and the enthalpy of the reactants when complete combustion occurs at a given temperature and pressure. That is,

$$\bar{h}_{RP} = H_P - H_R$$

$$\bar{h}_{RP} = \sum_P n_e[\bar{h}_f^\circ + \Delta\bar{h}]_e - \sum_R n_i[\bar{h}_f^\circ + \Delta\bar{h}]_i \tag{13.9}$$

The usual parameter for expressing the enthalpy of combustion is a unit mass of fuel, such as a lbm (h_{RP}) or a lb mole ($\bar{h}_{RP}$) of fuel.

The tabulated values of the enthalpy of combustion of fuels are usually given for a temperature of 25 C and a pressure of one atmosphere. The enthalpy of combustion for a number of hydrocarbon fuels at this temperature and pressure, which we designate h_{RP_0}, is given in Table 13.4.

The internal energy of combustion is defined in a similar manner.

$$\bar{u}_{RP} = U_P - U_R$$

$$= \sum_P n_e[\bar{h}_f^\circ + \Delta\bar{h} - P\bar{v}]_e - \sum_R n_i[\bar{h}_f^\circ + \Delta\bar{h} - P\bar{v}]_i \tag{13.10}$$

When all the gaseous constituents can be considered as ideal gases, and the volume of the liquid and solid constituents is negligible compared to the value of the gaseous constituents, this relation for $\bar{u}_{RP}$ reduces to

$$\bar{u}_{RP} = \bar{h}_{RP} - \bar{R}T\,(n_{\text{gaseous products}} - n_{\text{gaseous reactants}}) \tag{13.11}$$

Frequently the term "heating value" or "heat of reaction" is used. This represents the heat transferred from the chamber during combustion or reaction at constant temperature. In the case of a constant pressure or steady-flow process, we conclude from the first law of thermodynamics that this is equal to the negative of the enthalpy of combustion. For this reason this heat transfer is sometimes designated the constant-pressure heating value for combustion processes.

In the case of a constant volume process the heat transfer is equal to the negative of the internal energy of combustion. This is sometimes designated the constant-volume heating value in the case of combustion.

When the term heating value is used the terms "higher" and "lower" heating value are used. The higher heating value is the heat transfer with liquid H_2O in the products, and the lower heating value is the heat transfer with vapor H_2O in the products.

Example 13.11

Calculate the enthalpy of combustion of propane at 77 F on both a lb mole and lbm basis under the following conditions.

1. Liquid propane with liquid H_2O in the products.
2. Liquid propane with gaseous H_2O in the products.
3. Gaseous propane with liquid H_2O in the products.
4. Gaseous propane with gaseous H_2O in the products.

Table 13.4
Enthalpy of Combustion of Some Hydrocarbons at 25 C (77 F)

		Liquid H_2O in Products (Negative of Higher Heating Value)		Vapor H_2O in Products (Negative of Lower Heating Value)	
Hydrocarbon	Formula	Liquid Hydro-carbon, Btu/lbm fuel	Gaseous Hydro-carbon, Btu/lbm fuel	Liquid Hydro-carbon, Btu/lbm fuel	Gaseous Hydro-carbon, Btu/lbm fuel
Paraffin Family					
Methane	CH_4		−23,861		−21,502
Ethane	C_2H_6		−22,304		−20,416
Propane	C_3H_8	−21,490	−21,649	−19,773	−19,929
Butane	C_4H_{10}	−21,134	−21,293	−19,506	−19,665
Pentane	C_5H_{12}	−20,914	−21,072	−19,340	−19,499
Hexane	C_6H_{14}	−20,772	−20,930	−19,233	−19,391
Heptane	C_7H_{16}	−20,668	−20,825	−19,157	−19,314
Octane	C_8H_{18}	−20,591	−20,747	−19,100	−19,256
Decane	$C_{10}H_{22}$	−20,484	−20,638	−19,020	−19,175
Dodecane	$C_{12}H_{26}$	−20,410	−20,564	−18,964	−19,118
Olefin Family					
Ethene	C_2H_4		−21,626		−20,276
Propene	C_3H_6		−21,033		−19,683
Butene	C_4H_8		−20,833		−19,483
Pentene	C_5H_{10}		−20,696		−19,346
Hexene	C_6H_{12}		−20,612		−19,262
Heptene	C_7H_{14}		−20,552		−19,202
Octene	C_8H_{16}		−20,507		−19,157
Nonene	C_9H_{18}		−20,472		−19,122
Decene	$C_{10}H_{20}$		−20,444		−19,094
Alkylbenzene Family					
Benzene	C_6H_6	−17,985	−18,172	−17,259	−17,446
Methylbenzene	C_7H_8	−18,247	−18,423	−17,424	−17,601
Ethylbenzene	C_8H_{10}	−18,488	−18,659	−17,596	−17,767
Propylbenzene	C_9H_{12}	−18,667	−18,832	−17,722	−17,887
Butylbenzene	$C_{10}H_{14}$	−18,809	−18,970	−17,823	−17,984

This example is designed to show how enthalpy of combustion can be determined from enthalpies of formation.

The enthalpy of evaporation of propane in 159 Btu/lbm.

The basic combustion equation is:

$$C_3H_8 + 5O_2 \rightarrow 3CO_2 + 4H_2O$$

From Table 13.3, $(\bar{h}_f^\circ)_{C_3H_8(g)} = -44{,}676$ Btu/lb mole. Therefore,

$$(\bar{h}_f^\circ)_{C_3H_8(l)} = -44{,}676 - 44.10(159) = -51{,}688 \text{ Btu/lb mole.}$$

1. Liquid propane – liquid H_2O

$$\bar{h}_{RP_0} = 3(\bar{h}_f^\circ)_{CO_2} + 4(\bar{h}_f^\circ)_{H_2O(l)} - (\bar{h}_f^\circ)_{C_3H_8(l)}$$

$$= 3(-169{,}297) + 4(-122{,}971) - (-51{,}688) = -948{,}087 \text{ Btu/lb mole}$$

$$= \frac{-948{,}087}{44.10} = -21{,}490 \text{ Btu/lbm}$$

The higher heating value of liquid propane is therefore 21,490 Btu/lbm.

2. Liquid propane – gaseous H_2O

$$\bar{h}_{RP_0} = 3(\bar{h}_f^\circ)_{CO_2} + 4(\bar{h}_f^\circ)_{H_2O(g)} - (\bar{h}_f^\circ)_{C_3H_8(l)}$$

$$= 3(-169{,}297) + 4(-104{,}036) - (-51{,}688)$$

$$= -872{,}347 \text{ Btu/lb mole} = \frac{-872{,}347}{44.10} = -19{,}770 \text{ Btu/lbm.}$$

The lower heating value of liquid propane is 19,770 Btu/lbm.

3. Gaseous propane – liquid H_2O

$$\bar{h}_{RP_0} = 3(\bar{h}_f^\circ)_{CO_2} + 4(\bar{h}_f^\circ)_{H_2O(l)} - (\bar{h}_f^\circ)_{C_3H_8(g)}$$

$$= 3(-169{,}297) + 4(-122{,}971) - (-44{,}676)$$

$$= -955{,}099 \text{ Btu/lb mole} = \frac{-955{,}099}{44.10} = -21{,}650 \text{ Btu/lbm}$$

The higher heating value of gaseous propane is 21,650 Btu/lbm

4. Gaseous propane – gaseous H_2O

$$\bar{h}_{RP_0} = 3(\bar{h}_f^\circ)_{CO_2} + 4(\bar{h}_f^\circ)_{H_2O(g)} - (\bar{h}_f^\circ)_{C_3H_8(g)}$$

$$= 3(-169{,}297) + 4(-104{,}036) - (-44{,}676) = -879{,}359 \text{ Btu/mole}$$

$$= -19{,}930 \text{ Btu/lbm.}$$

The lower heating value of gaseous propane is 19,930 Btu/lbm.

Each of the four values calculated in this example corresponds to the appropriate value given in Table 13.4.

Example 13.12

Calculate the enthalpy of combustion of gaseous propane at 400 F. (At this temperature all the water formed during combustion will be in the vapor phase.) This example will demonstrate how the enthalpy of combustion of propane varies with temperature. The average constant-pressure specific heat of propane between 77 F and 400 F is 0.5 Btu/lbm-R.

$$(\bar{h}_{RP})_T = \sum_P n_e[\bar{h}_f^\circ + \Delta\bar{h}]_e - \sum_R n_i[\bar{h}_f^\circ + \Delta\bar{h}]_i$$

$$C_3H_8(g) + 5O_2 \rightarrow 3CO_2 + 4H_2O(g)$$

$$\bar{h}_{R_{400}} = [\bar{h}_f^\circ + \bar{C}_{p\text{av}}(400-77)]_{C_3H_8(g)} + n_{O_2}(\Delta\bar{h})_{O_2}$$

$$= -44{,}676 + 0.5 \times 44.10(400-77) + 5(2325)$$

$$= -44{,}676 + 7122 + 11{,}605 = -25{,}929 \text{ Btu/lb mole.}$$

$$\bar{h}_{P_{400}} = n_{CO_2}[\bar{h}_f^\circ + \Delta\bar{h}]_{CO_2} + n_{H_2O}[\bar{h}_f^\circ + \Delta\bar{h}]_{H_2O}$$

$$= 3(-169{,}297 + 3166) + 4[-104{,}036 + 2646]$$

$$= -498{,}393 - 405{,}560 = -903{,}953 \text{ Btu/lb mole.}$$

$$\bar{h}_{RP_{400}} = -903{,}953 - (-25{,}929) = -878{,}024 \text{ Btu/lb mole}$$

$$h_{RP_{400}} = \frac{-878{,}024}{44.10} = -19{,}901 \text{ Btu/lbm}$$

This compares with a value of −19,929 at 77 F.

This problem could also have been solved using the given value of the enthalpy of combustion at 77 F by noting that

$$(\bar{h}_{RP})_{400} = (H_P)_{400} - (H_R)_{400}$$

$$= n_{CO_2}[\bar{h}_f^\circ + \Delta\bar{h}]_{CO_2} + n_{H_2O}[\bar{h}_f^\circ + \Delta\bar{h}]_{H_2O}$$

$$- [\bar{h}_f^\circ + \bar{C}_{p\text{av}}(400-77)]_{C_3H_8(g)} - n_{O_2}(\Delta\bar{h})_{O_2}$$

$$= \bar{h}_{RP_0} + n_{CO_2}\Delta\bar{h}_{CO_2} + n_{H_2O}\Delta\bar{h}_{H_2O}$$

$$- \bar{C}_{p\text{av}}(400-77)_{C_3H_8(g)} - n_{O_2}\Delta\bar{h}_{O_2}$$

$$(\bar{h}_{RP_0})_{400} = -19{,}929 \times 44.10 + 3(3166) + 4(2646)$$

$$- 0.5 \times 44.1(400-77) - 5(2325)$$

$$(h_{RP_0})_{400} = -19{,}929 + \frac{1235}{44.1} = -19{,}929 + 28 = -19{,}901 \text{ Btu/lbm.}$$

13.7 The Third Law of Thermodynamics and Absolute Entropy

As we consider a second law analysis of chemical reactions, we face the same problem we did in regard to the first law, namely, what base shall we use for the entropy of the various substances. This question leads directly to a consideration of the third law of thermodynamics.

The third law of thermodynamics was formulated during the early part of the twentieth century. The initial work was done primarily by W. H. Nernst (1864–1941) and Max Planck (1858–1947). The third law deals with the entropy of substances at the absolute zero of temperature, and in essence states that the entropy of a perfect crystal is zero at the absolute zero of temperature. From a statistical point of view this means that the crystal structure is such that it has the maximum degree of order. Further, since the temperature is absolute zero, the thermal energy is minimum. It also follows that a substance that does not have a perfect crystalline structure at absolute zero, but instead has a degree of randomness, such as a solid solution or a glassy solid, has a finite value of entropy at absolute zero. The experimental evidence on which the third law rests is primarily data on chemical reactions at low temperatures and measurements of heat capacity at temperatures approaching absolute zero. It should be noted that in contrast to the first and second laws, which lead respectively to the properties of internal energy and entropy, the third law deals only with the question of entropy at absolute zero. However, the implications of the third law are quite profound, particularly in regard to chemical equilibrium.

For our consideration at this point, the particular relevance of the third law is that it provides an absolute base from which the entropy of each substance can be measured. The entropy relative to this base is termed the absolute entropy. The increase in entropy between absolute zero and any given state can be found either from calorimetric data or by procedures based on statistical thermodynamics. The calorimetric method involves precise measurements of specific heat data over the temperature range, as well as the energy associated with phase transformations. The statistical approach, along with general considerations of this subject from a statistical point of view, is considered in Chapters 17 and 19.

Table 13.3 gives the absolute entropy at 77 F (25) and 1 atm pressure for a number of substances. Table B.9 gives the absolute entropy for a number of gases at 1 atm pressure and various temperatures. In the case of gases, ideal gas behavior is assumed at 1 atm pressure in all these tables. If the saturation pressure at a given temperature is less that 1 atm pressure, the ideal gas data at 1 atm pressure and this given temperature refers to a state that does not exist under equilibrium conditions. However, by applying ideal gas relations to the data as given, the absolute entropy at pressures less than 1 atm is readily found. From one of these states, the absolute entropy in any other is readily found by procedures that are outlined in a subsequent paragraph. The absolute entropy at 1 atm pressure as given in these tables, is designated $\bar{s}^\circ$. The temperature is designated with a subscript such as $\bar{s}^\circ_{1000}$.

With the value of the absolute entropy known as a function of temperature at 1 atm pressure, as given in Table B.9, how does one find the absolute entropy at any other state? Let us first consider the case in which the assumption of ideal gas behavior at 1 atm pressure yields no significant error. Actually, this assumption is valid for most of the examples and problems given in this book, as well as most problems encountered in professional practice. In this case, we can write

$$\bar{s}_{T,P} = \bar{s}^\circ_T + (\Delta\bar{s})_{T,1\,\text{atm}\rightarrow T,P} \tag{13.12}$$

where $\bar{s}^\circ_{T,P}$ refers to the absolute entropy at 1 atm and temperature T, and $(\Delta s)_{T,\,1\,\text{atm}\rightarrow T,P}$ refers to the change of entropy along the isotherm T as the pressures changes from 1 atm to pressure P. This is shown schematically in Fig. 13.4.

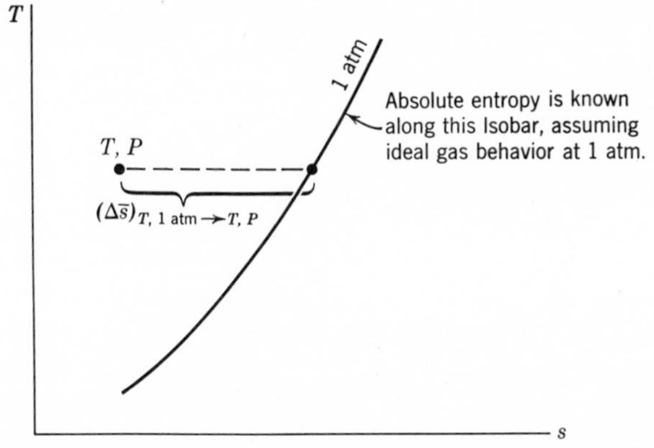

Fig. 13.4 Sketch showing the procedure for the calculation of absolute entropy when ideal gas assumption at 1 atm pressure yields no significant error.

The quantity $(\Delta s)_{T,\,1\,\text{atm}\rightarrow T,P}$ can be found in various ways. If the assumption of ideal gas behavior along this isotherm yields no significant error, this change of entropy can be found from the ideal gas relation.

$$\bar{s}_2 - \bar{s}_1 = -\bar{R}\ln\frac{P_2}{P_1} \tag{13.13}$$

If P_2 refers to the pressure in the given state T,P, and if this pressure is expressed in atmospheres, this relation reduces to

$$(\Delta\bar{s})_{T,\ 1\ \text{atm} \rightarrow T,P} = -\bar{R}\ln P \tag{13.14}$$

When the assumption of ideal gas behavior along the isotherm yields significant error, there are several procedures that can be used. If we have a table of thermodynamic properties that covers the change of state along the isotherm, this change of entropy can be found directly from the tables. If we have an equation of state that is valid in this range, we can follow the procedures outlined in Section 12.5 to find the change of entropy along the isotherm. One can also use the generalized charts, as presented in Section 12.13 to find this change of entropy.

If the assumption of ideal gas behavior is not valid at 1 atm pressure, one must first apply ideal gas relations to find the absolute entropy at a pressure that is low enough to make the assumption of ideal gas behavior valid. From this state, one can use the same procedures outlined above to find the absolute entropy in any other state. This procedure is shown schematically in Fig. 13.5.

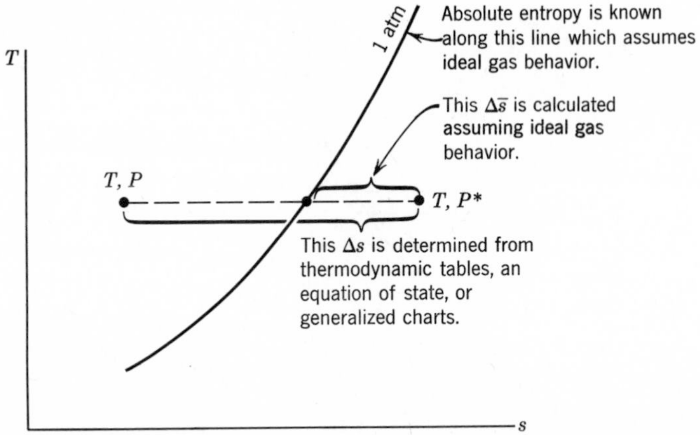

Fig. 13.5 Sketch showing procedure for calculation of absolute entropy when the assumption of ideal gas behavior at 1 atm pressure is not valid.

If the absolute entropy is known at only one temperature, such as 25 C and 1 atm, we would need appropriate specific heat data, as well as an equation of state to find the absolute entropy in other states.

In the case of mixtures of ideal gases, the absolute entropy of each component in the mixture can be found at the given temperature and pressure at which it exists in the mixture. Suppose, for example, that the mole fraction of oxygen in a mixture of ideal gases is 0.1, that the temperature of the mixture is 1000 K, and that the total pressure is 4 atm. What is the absolute entropy of the oxygen as it exists in the mixture?

Since the mole fraction is 0.1 and the total pressure is 4 atm, the partial pressure of the oxygen is 0.4 atm. Therefore, from Eqs. 13.12 and 13.14,

$$\begin{aligned}\bar{s} &= \bar{s}_T^\circ + (\Delta \bar{s})_{T,1\,\mathrm{atm} \to T,P}\\ &= \bar{s}_T^\circ - \bar{R} \ln P\\ &= 58.192 - 1.987 \ln 0.4\\ &= 60.012 \text{ Btu/lb mole R}\end{aligned}$$

13.8 Second Law Analysis of Reacting Systems

The concepts of reversible work, irreversibility, and availability were introduced in Chapter 9. These concepts involved both the first and second laws of thermodynamics. At the conclusion of Chapter 9 the Gibbs function was introduced and the statement was made that this property would be particularly relevant in dealing with chemical reactions. We proceed now to develop this matter further, and we shall be particularly concerned with determining the maximum work (availability) that can be done through a combustion process and with examining the irreversibilities associated with such processes.

The reversible work, Eq. 9.43, for a steady-state, steady-flow process, in the absence of changes in kinetic and potential energy is,

$$W_{\mathrm{rev}} = \sum m_i(h_i - T_0 s_i) - \sum m_e(h_e - T_0 s_e)$$

Applying this to an SSSF process involving a chemical reaction, and introducing the symbols we have been using in this chapter we have

$$W_{\mathrm{rev}} = \sum_R n_i[\bar{h}_f^\circ + \Delta\bar{h} - T_0\bar{s}]_i - \sum_P n_e[\bar{h}_f^\circ + \Delta\bar{h} - T_0\bar{s}]_e \qquad (13.15)$$

Similarly, the irreversibility for such a process becomes

$$I = \sum_P n_e T_0 \bar{s}_e - \sum_R n_i T_0 \bar{s}_i - Q_{\mathrm{c.v.}} \qquad (13.16)$$

The availability, Ψ, in the absence of kinetic and potential energy changes, for an SSSF process was defined, Eq. 9.49, as

$$\Psi = (h - T_0 s) - (h_0 - T_0 s_0)$$

It was also indicated in Chapter 9 that when an SSSF chemical reaction takes place in such a manner that both the reactants and products are in temperature equilibrium with the surroundings, the Gibbs function ($g = h - Ts$) becomes a significant variable, Eq. 9.61. For such a process, in the absence of changes in kinetic and potential energy, the reversible work is given by the relation

$$W_{\text{rev}} = \sum_R n_i \bar{g}_i - \sum_P n_e \bar{g}_e \tag{13.17}$$

Since the Gibbs function is so relevant for processes involving chemical reactants the Gibbs function of formation, $\bar{g}_f^\circ$, has been defined by a procedure similar to that used in defining the enthalpy of formation. That is, the Gibbs function of each of the elements at 25 C and 1 atm pressure is assumed to be zero, and the Gibbs function of each element is found relative to this base. Table 13.3 lists the Gibbs function of formation of a number of substances at 25 C (77 F) and 1 atm pressure. It is also evident that the Gibbs function can be found directly from data for $\bar{h}_f^\circ$ and $\bar{s}^\circ$ at the given temperature, as indicated in the following example.

Example 13.13

Determine the Gibbs function of formation of CO_2.

Consider the reaction

$$C + O_2 \rightarrow CO_2$$

Assume that the carbon and oxygen are each initially at 77 F and 1 atm pressure, and that the CO_2 is finally at 77 F and 1 atm pressure.

The change in Gibbs function for this reaction is found first.

$$G_P - G_R = (H_P - H_R) - T_0(S_P - S_R)$$

$$\sum_P n_e(\bar{g}_f^\circ)_e - \sum_R n_i(\bar{g}_f^\circ)_i$$

$$= \sum_P n_e(\bar{h}_f^\circ)_e - \sum_R n_i(\bar{h}_f^\circ)_i - T_0\Big[\sum_P n_e(\bar{s}_{537}^\circ)_e - \sum_R n_i(\bar{s}_{537}^\circ)_i\Big]$$

Therefore,

$$\begin{aligned} G_P - G_R &= (\bar{h}_f^\circ)_{CO_2} - 536.7[(\bar{s}_{537}^\circ)_{CO_2} - (\bar{s}_{537}^\circ)_C - (\bar{s}_{537}^\circ)_{O_2}] \\ &= -169{,}297 - 536.7(51.072 - 1.359 - 49.004) \\ &= -169{,}297 - 380 = -169{,}677 \text{ Btu/lb mole.} \end{aligned}$$

Since the Gibbs function of the reactants, G_R, is zero (in accordance with the assumption that the Gibbs function of the elements is zero at

25 C and 1 atm pressure), it follows that

$$G_P = (\bar{g}_f^\circ)_{CO_2} = -169{,}667 \text{ Btu/lb mole}$$

This is the value given in Table 13.3.

Let us now consider the question of the maximum work that can be done during a chemical reaction. For example, consider one lb mole of hydrocarbon fuel and the necessary air for complete combustion, each at one atmosphere pressure and 77 F, the pressure and temperature of the surroundings. What is the maximum work that can be done as this fuel reacts with the air? From our considerations in Chapter 9 we conclude that the maximum work would be done if this chemical reaction took place reversibly and the products were finally in pressure and temperature equilibrium with the surroundings. From Eq. 13.17 we conclude this reversible work could be calculated from the relation

$$W_{rev} = \sum_R n_i \bar{g}_i - \sum_P n_e \bar{g}_e$$

However, since the final state is in equilibrium with the surroundings we could consider this amount of work to be the availability of the fuel and air.

Example 13.14

Ethene (g) at 77 F and 1 atm pressure is burned with 400 per cent theoretical air at 77 F and 1 atm pressure. Assume that this reaction takes place reversibly at 77 F and that the products leave at 77 F and 1 atm pressure. To further simplify this problem assume that the oxygen and nitrogen are separated before the reaction takes place (each at 1 atm, 77 F), that the constituents in the products are separated, and that each is at 77 F and 1 atm pressure. Thus, the reaction takes place as shown in Fig. 13.6. For purposes of comparison between this and the two subsequent examples, we consider all of the H_2O in the products to be a gas (a hypothetical situation in this example and Example 13.16).

Determine the reversible work for this process (i.e., the work that would be done if this chemical action took place reversibly and isothermally).

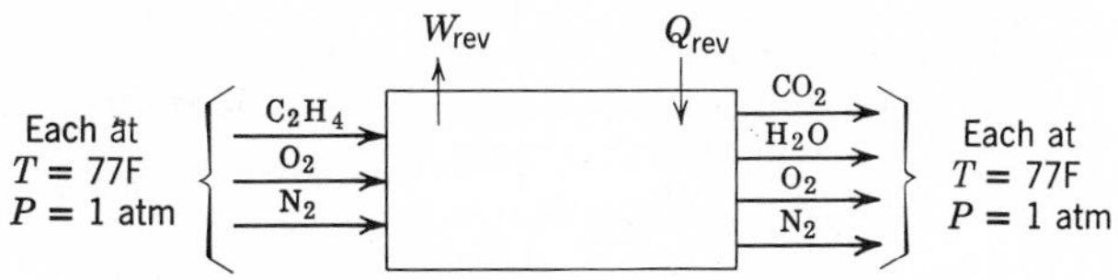

Fig. 13.6 Sketch for Example 13.14.

The equation for this chemical reaction is

$$C_2H_4(g) + 3(4)O_2 + 3(4)(3.76)N_2 \rightarrow 2CO_2 + 2H_2O(g) + 9O_2 + 45.1N_2$$

The reversible work for this process is equal to the decrease in Gibbs function during this reaction, Eq. 13.17. The values for the Gibbs function can be taken directly from Table 13.3, since each is at 77 F and 1 atm pressure. From Eq. 13.17

$$W_{\text{rev}} = \sum_R n_i \bar{g}_i - \sum_P n_e \bar{g}_e$$

Since each of the reactants and each of the products are at one atmosphere pressure and 77 F this reduces to

$$\begin{aligned} W_{\text{rev}} &= (\bar{g}_f^\circ)_{C_2H_4} - 2(\bar{g}_f^\circ)_{CO_2} - 2(\bar{g}_f^\circ)_{H_2O(g)} \\ &= 29{,}306 - 2(-169{,}677) - 2(-98{,}345) \\ &= 565{,}350 \text{ Btu/lb mole} = 20{,}140 \text{ Btu/lbm} \end{aligned}$$

Therefore we might say that when the one pound of ethene is at 77 F, one atm pressure, the temperature and pressure of the surroundings, it has an availability of 20,140 Btu.

Thus it would seem logical to rate the efficiency of a device designed to do work by utilization of a combustion process, such as an internal-combustion engine or a steam power plant, as the ratio of actual work to the decrease in Gibbs function for the given reaction, rather than to the heating value, as is current practice. However, as is evident from the preceding example, the difference between the decrease in Gibbs function and the heating value is small for hydrocarbon fuels and the efficiency defined in terms of heating value is essentially equal to that defined in terms of decrease in Gibbs function.

It is of particular interest to study the irreversibility that takes place during a combustion process. The following examples are presented to illustrate this matter, where we consider the same hydrocarbon fuel that was used in Example 13.14, namely Ethene (g) at 77 F and 1 atm. We determined the availability and found that it was 20,140 Btu/lbm. Now let us burn this fuel with 400 per cent theoretical air in a steady-state, steady-flow adiabatic process. We can determine the irreversibility of this process in two ways. The first is to calculate the increase in entropy for the process. Since the process is adiabatic, the increase in entropy is due entirely to the irreversibilities for the process and we can find the irreversibility from Eq. 13.16. We can also calculate the availability of the products of combustion at the adiabatic flame temperature, and note that they are less than the availability of the fuel and air before the combustion process, the difference being the irreversibility that occurs during the combustion process.

Example 13.15

Consider the same combustion process as in Example 13.14, but let it take place adiabatically, and assume that each constituent in the products is at 1 atm pressure and at the adiabatic flame temperature. This combustion process is shown schematically in Fig. 13.7. The temperature of the surroundings is 77 F.

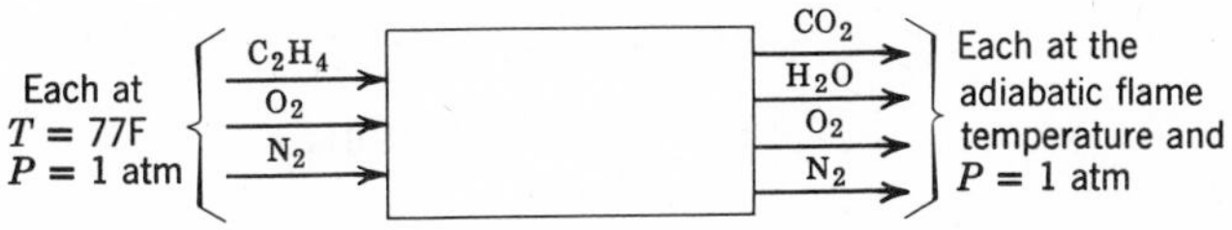

Fig. 13.7 Sketch for Example 13.15.

For this combustion process determine: (1) The increase in entropy during combustion. (2) The availability of the products of combustion.

The combustion equation is

$$C_2H_4 + 12O_2 + 12(3.76)N_2 \rightarrow 2CO_2 + 2H_2O(g) + 9O_2 + 45.1N_2$$

The adiabatic flame temperature is determined first.

$$H_R = H_P$$

$$\sum_R n_i(\bar{h}_f^\circ)_i = \sum_P n_e[\bar{h}_f^\circ + \Delta\bar{h}]_e$$

$$22{,}493 = 2[-169{,}297 + \Delta\bar{h}_{CO_2}] + 2[-104{,}036 + \Delta\bar{h}_{H_2O(g)}]$$

$$+ 9\Delta\bar{h}_{O_2} + 45.1\Delta\bar{h}_{N_2}$$

By a trial and error solution we find the adiabatic flame temperature to be 1829 R.

We now proceed to find the change in entropy during this adiabatic combustion process.

$$S_R = \sum_R (n_i\bar{s}_i^\circ)_{537} = (\bar{s}_{C_2H_4}^\circ + 12\bar{s}_{O_2}^\circ + 45.1\bar{s}_{N_2}^\circ)_{537}$$

$$= 52.447 + 12(49.004) + 45.1(45.770)$$

$$= 2704.72 \text{ Btu/lb mole fuel-R}$$

$$S_P = \sum_P (n_e \bar{s}_e^\circ)_{1829} = (2\bar{s}_{CO_2}^\circ + 2\bar{s}_{H_2O(g)}^\circ + 9\bar{s}_{O_2}^\circ + 45.1\bar{s}_{N_2}^\circ)_{1829}$$

$$= 2(64.546) + 2(55.752) + 9(53.321) + 45.1(54.628)$$

$$= 3229.21 \text{ Btu/lb mole fuel-R}$$

$$S_P - S_R = 524.49 \text{ Btu/lb mole fuel-R.}$$

Since this is an adiabatic process, the increase in entropy indicates the irreversibility of the adiabatic combustion process. This irreversibility can be found from Eq. 13.16.

$$I = T_0\left[\sum_P n_e\bar{s}_e - \sum_R n_i\bar{s}_i\right]$$

$$= 536.7(524.49) = 281{,}494 \text{ Btu/lb mole fuel}$$

$$= 10{,}020 \text{ Btu/lbm fuel}$$

Therefore, the availability after the combustion process is

$$\Psi_P = 20{,}140 - 10{,}020 = 10{,}120 \text{ Btu/lbm fuel.}$$

The availability of the products can also be found from the relation

$$\Psi_P = \sum_P \left[(\bar{h}_e - T_0\bar{s}_e) - (\bar{h}_0 - T_0\bar{s}_0)\right]$$

Since in this problem the products are separated and each is at one atmosphere pressure and the adiabatic flame temperature of 1829 R, this reduces to

$$\Psi_P = \sum_P n_e\left[(\bar{h}_e^\circ - \bar{h}_0^\circ) - T_0(\bar{s}_e^\circ - \bar{s}_0^\circ)\right]$$

$$= 2(14{,}751) + 2(11{,}466) + 9(10{,}012) + 45.1(9460)$$

$$- 536.7[2(64.546 - 51.072) + 2(55.752 - 45.106)$$

$$+ 9(58.321 - 49.004) + 45.1(54.628 - 45.770)]$$

$$= 569{,}182 - 285{,}283 = 283{,}899 \text{ Btu/lb mole fuel}$$

$$= 10{,}120 \text{ Btu/lbm fuel.}$$

In other words, if every process after the adiabatic combustion process were reversible, the maximum amount of work that could be done is

10,120 Btu/lbm fuel. This compares to a value of 20,140 Btu/lbm for the reversible isothermal reaction. This means that if we had an engine that had the indicated adiabatic combustion process, and if all other processes were completely reversible, the efficiency would be about 50 per cent.

In the two prior examples we made the assumption, for purposes of simplifying the calculation, that the constituents in the reactants and products were separated, and each was at one atmosphere pressure. This of course is not a realistic problem, and in the following example, Example 13.14 is repeated with the assumption that the reactants and products each consist of a mixture at one atmosphere pressure.

Example 13.16

Consider the same combustion process of Example 13.14, but assume that the reactants consist of a mixture at 1 atm pressure and 77 F and that the products also consist of a mixture at 1 atm and 77 F. Thus, the combustion process is as shown in Fig. 13.8. Assume each constituent to be an ideal gas.

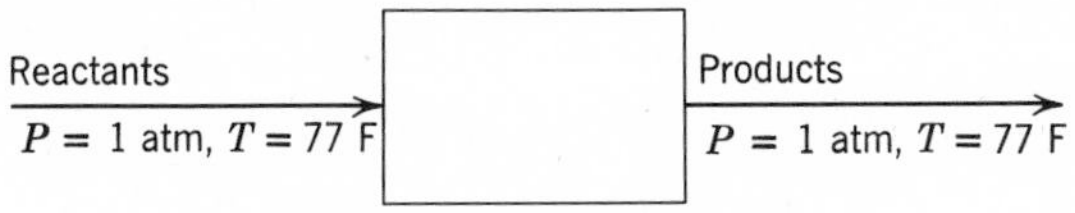

Fig. 13.8 Sketch for Example 13.16.

Determine the work that would be done if this combustion process took place reversibly and in pressure and temperature equilibrium with the surroundings.

The combustion equation, as noted previously, is

$$C_2H_4(g) + 3(4)O_2 + 3(4)(3.76)N_2 \rightarrow 2CO_2 + 2H_2O(g) + 9O_2 + 45.1N_2$$

In this case we must find the entropy of each substance as it exists in the mixture; i.e., at its partial pressure and the given temperature of 77 F. Since the absolute entropies given in Tables 13.3 and B.9 are at 1 atm pressure and 77 F, the entropy of each constituent in the mixture can be found, using Eq. 13.12, from the relation

$$\bar{s} - \bar{s}^\circ = -\bar{R} \ln y \frac{P}{P^\circ}$$

where $\bar{s}$ = entropy of the constituent in the mixture
$\bar{s}^\circ$ = absolute entropy at the same temperature and 1 atm pressure
P = pressure of the mixture
P° = 1 atm pressure
y = mole fraction of the constituent

Since P° and the pressure of the mixture are both 1 atm pressure, the entropy of each constituent can be found by the relation

$$\bar{s} = \bar{s}^\circ - \bar{R} \ln y = \bar{s}^\circ + \bar{R} \ln \frac{1}{y}$$

For the reactants:

	n	$1/y$	$\bar{R} \ln 1/y$	$\bar{s}^\circ$	$\bar{s}$
C_2H_4	1	58.100	8.060	52.447	60.507
O_2	12	4.850	3.135	49.004	52.139
N_2	45.1	1.285	0.498	45.770	46.268
	58.1				

For the products:

	n	$1/y$	$\bar{R} \ln 1/y$	$\bar{s}^\circ$	$\bar{s}$
CO_2	2	29.050	6.690	51.072	57.762
H_2O	2	29.050	6.690	45.106	51.796
O_2	9	6.460	3.700	49.004	52.704
N_2	45.1	1.285	0.498	45.770	46.268
	58.1				

With the assumption of ideal gas behavior, the enthalpy of each constituent is equal to the enthalpy of formation at 77 F. The values of entropy are as calculated above. Therefore, from Eq. 13.15

$$W_{\text{rev}} = \sum_R n_i(\bar{h}_f^\circ)_i - \sum_P n_e(\bar{h}_f^\circ)_e - T_0\Big[\sum_R n_i\bar{s}_i - \sum_P n_e\bar{s}_e\Big]$$

$$= (\bar{h}_f^\circ)_{C_2H_4} - 2(\bar{h}_f^\circ)_{CO_2} - 2(\bar{h}_f^\circ)_{H_2O(g)}$$

$$- 536.7[\bar{s}_{C_2H_4} + 12\bar{s}_{O_2} + 45.1\bar{s}_{N_2} - 2\bar{s}_{CO_2} - 2\bar{s}_{H_2O(g)} - 9\bar{s}_{O_2} - 45.1\bar{s}_{N_2}]$$

$$= 22{,}493 - 2(-169{,}297) - 2(-104{,}036)$$

$$- 536.7[(60.507) + 12(52.139) + 45.1(46.268)$$

$$- 2(57.762) - 2(51.796) - 9(52.704) - 45.1(46.268)]$$

$$= 569{,}159 - 536.7[2772.862 - 2779.139]$$

$$569{,}159 + 3369 = 572{,}528 \text{ Btu/lb mole fuel}$$

$$= \frac{572{,}528}{28.054} = 20{,}200 \text{ Btu/lbm fuel}$$

Note that this is essentially the same value that was obtained in Example 13.14, when the reactants and products were each separated and at one atmosphere pressure.

These examples raise the question of the possibility of a reversible chemical reaction. Some reactions can be made to approach reversibility by having them take place in an electrolytic cell, as described in Chapter 1. When a potential exactly equal to the electromotive force of the cell is applied, no reaction takes place. When the applied potential is increased slightly, the reaction proceeds in one direction, and if the applied potential is decreased slightly, the reaction proceeds in the opposite direction. The work involved is the electrical energy supplied or delivered.

Much effort is being directed toward the development of fuel cells in which carbon, hydrogen, or hydrocarbons will react with oxygen and produce electricity directly. If a fuel cell can be developed that utilizes a hydrocarbon fuel and has a sufficiently high efficiency and capacity (for a given volume or weight) drastic changes will be possible in our techniques for generating electricity on a commercial scale. At the present time, however, fuel cells are not competitive with conventional power plants for the production of electricity on a large scale.

13.9 Evaluation of Actual Combustion Processes

In evaluating the performance of an actual combustion process a number of different parameters can be defined, depending on the nature of the process and the system considered. In the combustion chamber of a gas turbine, for example, the objective is to raise the temperature of the products to a given temperature (usually the maximum temperature the metals in the turbine can withstand). If we had a combustion process in which complete combustion was achieved and which was adiabatic, the temperature of the products would be the adiabatic flame temperature. Let us designate the fuel-air ratio needed to reach a given temperature under these conditions as the ideal fuel-air ratio. In the actual combustion chamber the combustion will be incomplete to some extent and there will be some heat transfer to the surroundings. Therefore more fuel will be required to reach the given temperature, and this we designate as the actual fuel-air ratio. In this case, the combustion efficiency, η_{comb}, is defined as

$$\eta_{\text{comb}} = \frac{FA_{\text{ideal}}}{FA_{\text{actual}}} \tag{13.18}$$

On the other hand, in the furnace of a steam generator (boiler) the purpose is to transfer the maximum possible amount of heat to the steam (water). In practice, the efficiency of a steam generator is defined as the ratio of the heat transferred to the steam to the higher heating value of the fuel. For a coal this is the heating value as measured in a bomb calorimeter, which is the constant-volume heating value, and it corresponds to the internal energy of combustion. One observes a minor inconsistency, since the boiler involves a flow process, and the change in enthalpy is the significant factor. However, in most cases the error thus introduced is less than the experimental error involved in measuring the heating value, and the efficiency of a steam generator is defined by the relation,

$$\eta_{\text{steam generator}} = \frac{\text{heat transferred to steam/lbm fuel}}{\text{higher heating value of the fuel}} \tag{13.19}$$

In an internal combustion engine the purpose is to do work. The logical way to evaluate the performance of an internal-combustion engine would be to compare the actual work done to the maximum work which would be done by a reversible change of state from the reactants to the products. This, as we noted previously is equal to the decrease in Gibbs function $(h-Ts)$.

However, in practice the efficiency of an internal combustion engine is defined as the ratio of the actual work to the negative of the enthalpy of combustion of the fuel (i.e., the constant-pressure heating value). This is usually called the thermal efficiency, η_{th}

$$\eta_{\text{th}} = \frac{w}{(-h_{RP_0})} = \frac{w}{\text{heating value}} \tag{13.20}$$

The over-all efficiency of a gas turbine or steam power plant is defined in the same way. It should be pointed out that in an internal combustion engine or fuel-burning steam power plant the fact that the combustion is itself irreversible is a significant factor in the relatively low thermal efficiency of these devices.

One other factor should be pointed out regarding efficiency. We have noted that the enthalpy of combustion of a hydrocarbon fuel varies considerably with the phase of the water in the products (which leads to the concept of higher and lower heating values). Therefore, in considering the thermal efficiency of an engine, the heating value used to determine this efficiency must be borne in mind. Two engines made by different manufacturers may have identical performance, but if one manufacturer bases his efficiency on the higher heating value and the other on the lower heating value, the latter will be able to claim a higher thermal efficiency. This claim is not significant, of course, as the performance is the same, and a consideration of the way in which the efficiency was defined would reveal this.

The whole matter of efficiencies of devices involving combustion processes is treated in detail in textbooks dealing with particular applications, and our discussion is intended only as an introduction to the subject. However, a few examples will be cited to illustrate these remarks.

Example 13.17

The combustion chamber of a gas turbine uses a liquid hydrocarbon fuel which has an approximate composition of C_8H_{18}. On test the following data are obtained.

$$T_{\text{air}} = 260 \text{ F}(720 \text{ R}) \qquad T_{\text{products}} = 1520 \text{ F}(1980 \text{ R})$$

$$\mathsf{V}_{\text{air}} = 300 \text{ ft/sec} \qquad \mathsf{V}_{\text{products}} = 450 \text{ ft/sec}$$

$$T_{\text{fuel}} = 120 \text{ F} \qquad FA_{\text{actual}} = 0.0211 \text{ lbm fuel/lbm air}$$

Calculate the combustion efficiency for this process.

For the ideal chemical reaction the heat transfer is zero. Therefore, writing the first law for a control volume that includes the combustion chamber we have,

$$H_R + \text{KE}_R = H_P + \text{KE}_P$$

$$H_R + \text{KE}_R = \sum_R n_i \left[\bar{h}_f^\circ + \Delta\bar{h} + \frac{M\mathsf{V}^2}{2g_c} \right]_i$$

$$= [\bar{h}_f^\circ + \bar{C}_p(120-77)]_{C_8H_{18}(l)} + n_{O_2}\left[\Delta\bar{h} + \frac{M\mathsf{V}^2}{2g_c}\right]_{O_2}$$

$$+ 3.76 n_{O_2}\left[\Delta\bar{h} + \frac{M\mathsf{V}^2}{2g_c}\right]_{N_2}$$

$$= -107{,}532 + 0.5(114.23)(43) + n_{O_2}\left[1303 + \frac{32.0 \times (300)^2}{2 \times 32.17 \times 778}\right]$$

$$+ 3.76 n_{O_2}\left[1278 + \frac{28.02 \times (300)^2}{2 \times 32.17 \times 778}\right]$$

$$= (-105{,}076 + 6355 n_{O_2}) \text{ Btu/lb mole fuel}$$

$$H_P + KE_P = \sum_P n_e \left[\bar{h}_f^\circ + \Delta\bar{h} + \frac{MV^2}{2g_c} \right]_e$$

$$= 8\left[\bar{h}_f^\circ + \Delta\bar{h} + \frac{MV^2}{2g_c} \right]_{CO_2}$$

$$+ 9\left[\bar{h}_f^\circ + \Delta\bar{h} + \frac{MV^2}{2g_c} \right]_{H_2O}$$

$$+ (n_{O_2} - 12.5)\left[\Delta\bar{h} + \frac{MV^2}{2g_c} \right]_{O_2}$$

$$+ 3.76 n_{O_2}\left[\Delta\bar{h} + \frac{MV^2}{2g_c} \right]_{N_2}$$

$$= 8\left[-169{,}297 + 16{,}733 + \frac{44.01 \times (450)^2}{2 \times 32.17 \times 778} \right]$$

$$+ 9\left[-104{,}036 + 12{,}978 + \frac{18.02 \times (450)^2}{2 \times 32.17 \times 778} \right]$$

$$+ (n_{O_2} - 12.5)\left[11{,}279 + \frac{32.0 \times (450)^2}{2 \times 32.17 \times 778} \right]$$

$$+ 3.76 n_{O_2}\left[10{,}651 + \frac{28.02 \times (450)^2}{2 \times 32.17 \times 778} \right]$$

$$= (-2{,}174{,}652 + 54{,}243 n_{O_2}) \text{ Btu/lb mole fuel}$$

Therefore,

$$-105{,}076 + 6355 n_{O_2} = -2{,}174{,}652 + 54{,}243 n_{O_2}$$

$$n_{O_2} = \frac{2{,}069{,}576}{47{,}888} = 43.2 \text{ moles } O_2/\text{mole fuel}$$

$$\text{moles air/mole fuel} = 4.76(43.2) = 205.8$$

$$FA_{\text{Ideal}} = \frac{114.23}{205.8 \times 28.97} = 0.0194$$

$$\eta_{\text{comb}} = \frac{0.0194}{0.0211} = 0.92 = 92\%$$

Example 13.18

In a certain steam power plant 715,000 lbm of water per hour enter the boiler at a pressure of 1850 lbf/in.^{2} and a temperature of 415 F. Steam leaves the boiler at 1320 lbf/in.^{2}, 925 F. The power output of the turbine is 81,000 kw. Coal is used at the rate of 59,000 lbm/hr, and has a higher heating value of 14,310 Btu/lbm. Determine the efficiency of steam generator and over-all thermal efficiency of the plant.

In power plants the efficiency of both the boiler and the over-all efficiency of the plant are based on the higher heating value of the fuel. The efficiency of the boiler is defined by Eq. 13.19 as

$$\eta_{\text{steam generator}} = \frac{\text{heat transferred to } H_2O/\text{lbm fuel}}{\text{higher heating value}}$$

Therefore

$$\eta_{\text{steam generator}} = \frac{715{,}000}{59{,}000} \times \frac{(1448.0 - 392.8)}{14{,}310} = 89.4\%$$

The thermal efficiency is defined by Eq. 13.20.

$$\eta_{\text{th}} = \frac{w}{\text{heating value}} = \frac{81{,}000 \times 3412}{59{,}000 \times 14{,}310} = 32.7\%$$

PROBLEMS

13.1 Butane is burned with air and a volumetric analysis of the combustion products on a dry basis yields the following composition:

CO_2	7.8%
CO	1.1
O_2	8.2
N_2	82.9

Determine the per cent of theoretical air used in this combustion process.

13.2 A hydrocarbon fuel is burned with air and the following volumetric analysis on a dry basis is obtained from the products of combustion.

CO_2	10.5%
O_2	5.3
N_2	84.2

Determine the composition of the fuel on a mass basis and the per cent of theoretical air utilized in the combustion process.

13.3 Octane is burned with the theoretical air in a constant pressure process ($P = 14.7$ lbf/in.^{2}) and the products are cooled to 80 F.

(*a*) How many lbm of water are condensed per lbm of fuel?

(*b*) Suppose the air used for combustion has a relative humidity of 90% and is at a temperature of 80 F and 14.7 lbf/in.^{2} pressure. How many lbm of water will be condensed per lbm of fuel when the products are cooled to 80 F?

13.4 The hot exhaust gas from an internal combustion engine is analyzed and found to have the following composition on a volumetric basis:

CO_2	10%
H_2O	13
CO	2
O_2	3
N_2	72

This gas is to be fed to an exhaust gas reactor and mixed with a certain amount of air, as shown in Fig. 13.9, to eliminate the CO. It is decided that a mole fraction of O_2 of 10% in the mixture at point 3 will ensure that no CO remains. What must the ratio of flows be entering the reactor?

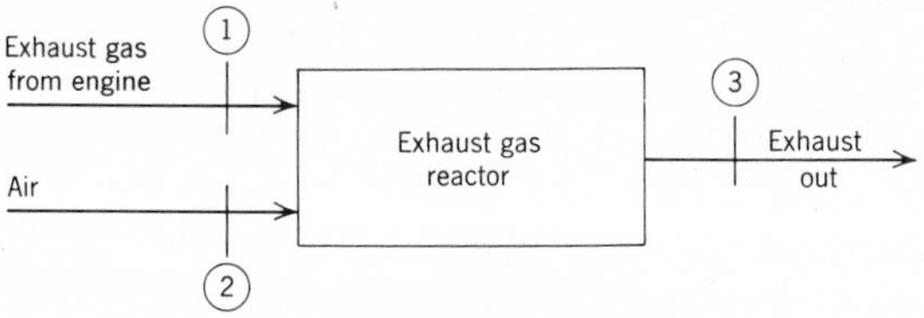

Fig. 13.9 Sketch for Problem 13.4.

13.5 In a test of rocket propellant performance, liquid hydrazine (N_2H_4) at 77 F, 1 atm, and oxygen gas at 77 F, 1 atm, are fed to a combustion chamber in the ratio 0.5 lbm O_2/lbm N_2H_4. The heat transfer from the chamber to the surroundings is estimated to be 50 Btu/lbm N_2H_4. Determine the temperature of the products, assuming only H_2O, H_2, and N_2 to be present. The enthalpy of formation of liquid N_2H_4 is +21,690 Btu/mole.

13.6 Repeat Problem 13.5, assuming that saturated liquid oxygen at 1 atm pressure is used instead of 77 F oxygen gas in the combustion process.

13.7 An internal combustion engine burns liquid octane (C_8H_{18}) and uses 125% theoretical air. The air and fuel enter at 77 F, the products leave the engine exhaust ports at 1160 F. In the engine 85% of the carbon burns to CO_2 and the remainder burns to CO. The heat transfer from this engine is just equal to the work done by the engine. Determine:

(*a*) Power output of the engine if the engine burns 20 lbm fuel/hr.

(*b*) The composition and the dew point of the products of combustion.

13.8 A natural gas consisting of 80% methane and 20% ethane (on a volume basis) is burned with 150% theoretical air in a steady-state, steady-flow process. Heat is transferred from the products of combustion until the temperature reaches 800 F. The fuel enters the combustion chamber at 77 F and the air at 260 F.

Determine the heat transfer per mole of fuel.

13.9 Liquid ethanol (C_2H_5OH) is burned with 150% theoretical oxygen in a steady-state, steady-flow process. The reactants enter the combustion chamber at 77 F, and the products are cooled and leave at 150 F, 1 atm pressure. Calculate the heat transfer per mole of ethanol. The enthalpy of formation of $C_2H_5OH(l)$ is $-119{,}441$ Btu/mole.

13.10 It has been proposed to use hydrogen (H_2) and an oxidizer consisting of 50% oxygen (O_2) and 50% fluorine (F_2), on a mole basis, as a rocket propellant combination. In a test, the hydrogen and oxidizer both enter a combustion chamber at 77 F, 1 atm, with excess H_2 being used to control the temperature of the products. It may be assumed that the products consist of H_2O, HF, and the excess H_2, and also that the heat transfer from the combustion chamber to the surroundings is 1000 Btu/mole of products. Determine the ratio of H_2 to oxidizer to be fed to the combustion chamber if the temperature of the products is 5400 R. For HF:

$$\bar{h}_f^\circ = -115{,}560 \text{ Btu/mole}; \quad \Delta\bar{h}^\circ_{537\to5400\text{ R}} = 37{,}200 \text{ Btu/mole}$$

13.11 A bomb is charged with 1 mole of CO and 2 moles of O_2. The total pressure is 1 atm and the temperature is 537 R before combustion. Combustion then occurs and the products are cooled to 2340 R. Assume complete combustion. Determine:

(*a*) The final pressure. (Note that the number of moles changes during combustion.)

(*b*) The heat transfer.

13.12 Liquid octane enters the combustion chamber of a gas turbine engine at 77 F and air enters from the compressor at 440 F. It is determined that 98% of the carbon in the fuel burns to form CO_2 and the remaining 2% burns to form CO. What amount of excess air will be required if the temperature of the products is to be limited to 1520 F?

13.13 A thermoelectric generator such as shown in Fig. 1.11 converts heat directly to electrical energy without moving parts. Such a device has been proposed for use in a portable power supply. For the purpose of thermodynamic analysis, the power supply unit may be represented as shown in Fig. 13.10.

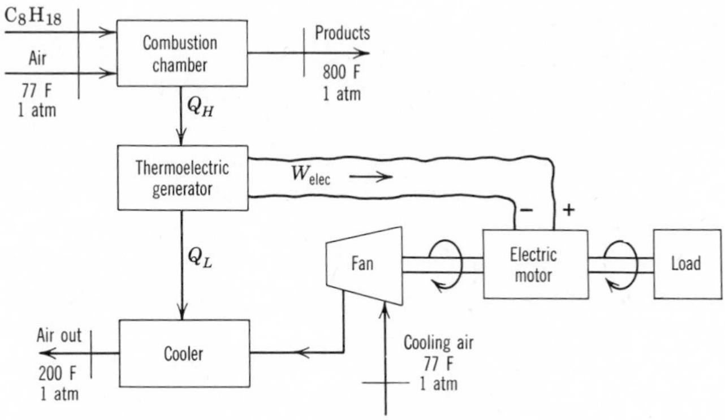

Fig. 13.10 Sketch for Problem 13.13.

Other data are as follows:

$$\eta_{\text{generator}} = \frac{W_{\text{elec}}}{Q_H} = 0.15$$

$$\eta_{\text{elec. motor}} = 0.95$$

$$\text{Electric motor output} = 2 \text{ hp}$$

$$\text{Fan requirement} = 0.1 \text{ hp}$$

Volumetric (dry) analysis of combustion products

CO_2	9.0%
CO	1.0
O_2	7.0
N_2	83.0

(*a*) Write the combustion equation per mole of octane, and determine the per cent theoretical air.

(*b*) Calculate the octane flow rate, lbm/hr.

(*c*) Calculate the volume flow rate of cooling air required (inlet conditions), ft³/min.

13.14 A jet engine is to be test run using liquid methane as a fuel with 300% theoretical air. The test conditions are as shown in Fig. 13.11. Determine the exit velocity of the products.

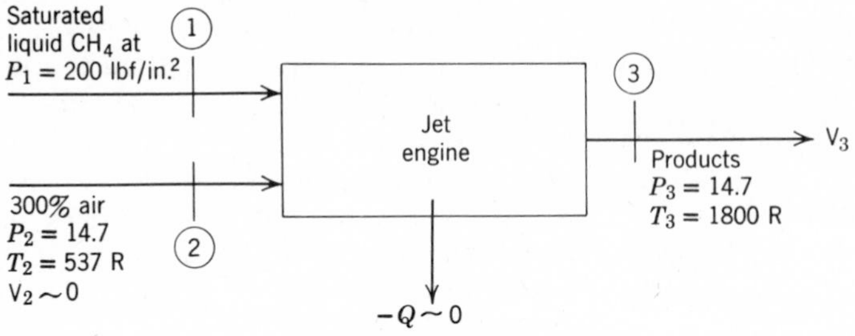

Fig. 13.11 Sketch for Problem 13.14.

13.15 Gaseous propane at 77 F is mixed with air at 300 F and burned; 300% theoretical air is used. What is the adiabatic flame temperature?

13.16 Gaseous propane at 77 F is mixed with air at 300 F and burned. What percentage of theoretical air must be used if the temperature of the products is to be 1700 F? Assume an adiabatic process and complete combustion.

13.17 How much error is introduced if the adiabatic flame temperature of Problem 13.15 is calculated assuming that all substances have constant specific heat, values from Table B.6?

13.18 The natural gas B from Table 13.2 is burned with 150% theoretical air. Calculate the adiabatic flame temperature for complete combustion, if the temperature of the reactants is 77 F.

13.19 Sulfur at 77 F is burned with 50% excess air, the air being at a temperature of 300 F. Assuming all the sulfur is burned to SO_2, calculate the adiabatic flame temperature. The enthalpy of formation of SO_2 is −127,700 Btu/lb mole. The constant-pressure specific heat of SO_2 is given by the relation

$$\bar{C}_p = 7.70 + 0.0029T - 0.26 \times 10^{-6}T^2$$

where $\bar{C}_p$ = Btu/lb mole-R
T = °R

13.20 Acetylene gas at 77 F, 1 atm pressure, is fed to the head of a cutting torch. Calculate the flame temperature if the acetylene is burned with
(*a*) One hundred per cent theoretical air at 77 F.
(*b*) One hundred per cent theoretical oxygen at 77 F.

13.21 The enthalpy of formation of magnesium oxide, MgO (s) is −143.84 k cal/gm mole at 25 C. The melting point of magnesium oxide is approximately 3000 K, and the increase in enthalpy between 298 K and 3000 K is 30.7 k cal/gm mole. The enthalpy of sublimation at 3000 K is estimated at 100 k cal/gm mole, and the specific heat of magnesium-oxide vapor above 3000 K is estimated at 8.9 cal/gm mole-K.
(*a*) Determine the enthalpy of combustion per lbm of magnesium.
(*b*) Estimate the adiabatic flame temperature when magnesium is burned with theoretical oxygen.

13.22 Consider natural gas D listed in Table 13.2. Calculate the enthalpy of combustion of this gas at 77 F, considering the products to include:
(*a*) Vapor water.
(*b*) Liquid water.

13.23 Hydrogen peroxide is sometimes used as the oxidizer in special power plants such as torpedoes and rockets. Determine the enthalpy of combustion at 77 F per lbm of reactants for the following combustion process:

$$4H_2O_2(l) + CH_4 \rightarrow 6H_2O + CO_2$$

The enthalpy of formation $H_2O_2(l)$ is −80,700 Btu/lb mole.

13.24 (*a*) Determine the enthalpy of formation of liquid benzene at 77 F.
(*b*) Benzene (l) at 77 F is burned with air at 440 F in a steady-flow process. The products of combustion are cooled to 2060 F and have the following volumetric analysis on a dry basis:

	% by volume
CO_2	10.7
CO	3.6
O_2	5.3
N_2	80.4

Calculate the heat transfer per mole of fuel during the combustion process.

13.25 One mole of carbon at 77 F, 1 atm and two moles of O_2 at 260 F, 1 atm enter a combustion chamber, and the products leave at 2060 F, 1 atm. Assuming complete combustion and a steady-state steady-flow process, determine:

(*a*) The heat transfer per mole of fuel.

(*b*) The net entropy change for the process, if the heat is transferred to the surroundings at 77 F.

13.26 Methyl alcohol (CH_3OH) is burned with 120% theoretical air at 20 lbf/in.^{2}, after which the products of combustion are passed through a heat exchanger and cooled to 120 F. Considering the process to be steady-flow, calculate the absolute entropy of the products leaving the heat exchanger per mole of alcohol burned.

13.27 The following data are taken from the test of a gas turbine on a test stand.

Fuel—C_4H_{10} (g) at 77 F and 1 atm
Air—300% theoretical air at 77 F and 1 atm
Velocity of inlet air = 200 ft/sec
Velocity of products at exit = 2200 ft/sec
Temperature and pressure of products = 1160 F and 1 atm

Assuming complete combustion, determine

(*a*) The net heat transfer per mole of fuel.

(*b*) The net increase of entropy per mole of fuel.

13.28 A fuel cell is a device which converts a portion of the energy of a chemical reaction directly into electrical energy. The overall reaction taking place may be described as a controlled-rate combustion process. Consider the fuel cell shown in Fig. 13.12

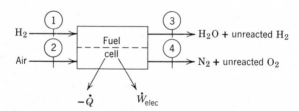

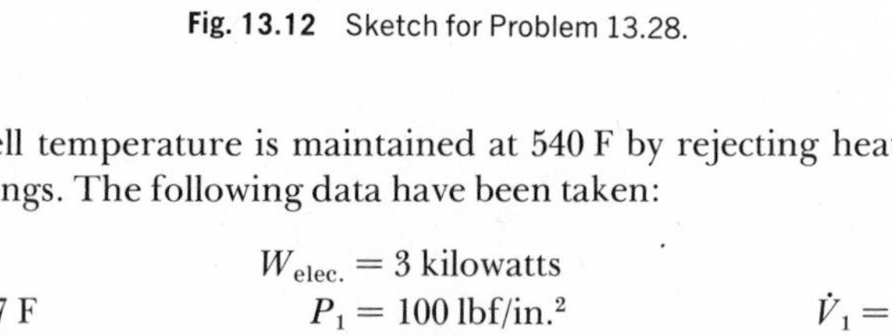

Fig. 13.12 Sketch for Problem 13.28.

The cell temperature is maintained at 540 F by rejecting heat to the surroundings. The following data have been taken:

	$W_{elec.} = 3$ kilowatts	
$T_1 = 77$ F	$P_1 = 100$ lbf/in.^{2}	$\dot{V}_1 = 0.20$ ft^3/min
$T_2 = 77$ F	$P_2 = 100$ lbf/in.^{2}	$\dot{V}_2 = 0.45$ ft^3/min
$T_3 = 540$F	$P_3 = 90$ lbf/in.^{2}	
$T_4 = 540$ F	$P_4 = 90$ lbf/in.^{2}	$(y_{O_2})_4 = 0.05$

(*a*) Write the overall reaction equation, on a basis of moles/hr for each substance.

(*b*) Calculate the heat transfer from the cell, Btu/hr.

(*c*) Calculate the entropy flow at point 4, Btu/R-hr.

13.29 Calculate the irreversibility for the process described in Problem 13.11.

13.30 Consider the combustion of gaseous propane at 14.7 lbf/in.², 77 F, with theoretical air at 14.7 lbf/in.², 77 F in a steady-state, steady-flow process. Assume that combustion is complete and that the products leave at 77 F. Determine the decrease in Gibbs function for the two following cases:

(*a*) The reactants and products are both separated into their various constituents, and each constituent is at 1 atm pressure, 77 F. This is shown schematically in Fig. 13.13.

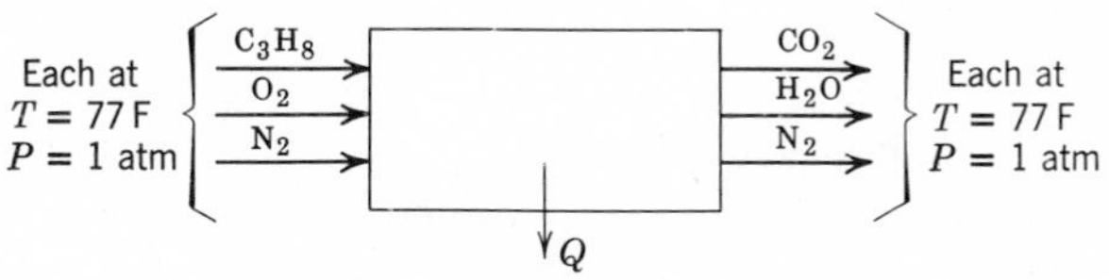

Fig. 13.13 Sketch for Problem 13.30*a*.

(*b*) The reactants and products each consist of a mixture at a total pressure of 1 atm and a temperature of 77 F. This is shown schematically in Fig. 13.14.

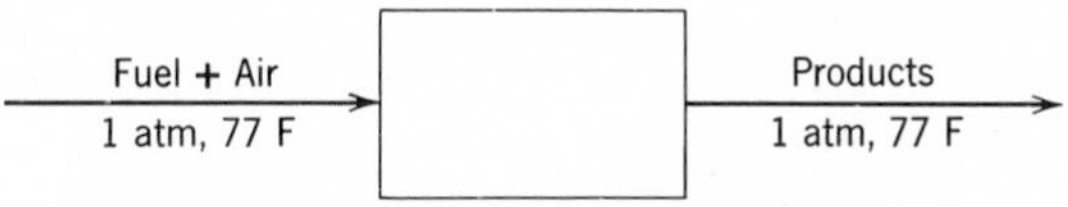

Fig. 13.14 Sketch for Problem 13.30*b*.

13.31 A mixture of butane and 150% theoretical air enters a combustion chamber at 77 F, 20 lbf/in.², and the products of combustion leave at 1340 F, 20 lbf/in.². Assuming a complete combustion, determine the heat transfer from the combustion chamber and the irreversibility for the process, both per mole of butane.

13.32 Consider one cylinder of a spark-ignition internal-combustion engine. Before the compression stroke the cylinder is filled with a mixture of air and ethane. Assume that 150% theoretical air has been used, and that the pressure is 1 atm and the temperature 77 F before compression. The compression ratio of the engine is 9 to 1.

(*a*) Determine the pressure and temperature after compression assuming a reversible adiabatic compression. Assume that the ethane behaves as an ideal gas.

(*b*) Assume further that complete combustion is achieved while the piston remains at top dead center (i.e., after the reversible adiabatic compression), and that the combustion process is adiabatic. Determine the temperature and pressure after combustion, and the increase in entropy during the combustion process.

(*c*) What is the irreversibility for this process?

13.33 A small air-cooled gasoline engine is tested, and the output is found to be 1.34 hp. The temperature of the products is measured and found to be 730 F. The products are analyzed with the following results, on a dry volumetric basis:

CO_2	11.4%
O_2	1.6
CO	2.9
N_2	84.1

The fuel used may be considered to be C_8H_{18}, and the air and fuel enter the engine at 77 F. The rate at which fuel is used is 1.20 lbm/hr.

(*a*) Determine the rate of heat transfer from the engine.

(*b*) What is the efficiency of the engine?

13.34 Consider the combustion process of Problem 13.16, in which propane at 77 F is burned with air at 300 F and the temperature of the products is 1700 F. Given these same states, assume that the combustion efficiency is 90% and that 95% of the carbon in the propane burns to form CO_2 and 5% burns to form CO. What is the heat transfer from the combustion chamber for this process?

13.35 Consider the hydrogen-oxygen fuel cell described in Section 1.2 and shown schematically in Fig. 1.7. The maximum theoretical voltage $\mathscr{E}^0$ obtainable from such a device at a given temperature is given by the expression

$$\mathscr{E}^0 = \frac{-\Delta G}{23{,}062\, n}$$

in which ΔG is the change in Gibbs function in cal for the overall chemical reaction at this temperature, and n is the number of moles of electrons involved in each half-cell reaction.

(*a*) Calculate $\mathscr{E}^0$ for this fuel cell, in which the H_2 and O_2 each enter at 25 C, 1 atm, and liquid H_2O leaves at 25 C, 1 atm.

(*b*) Repeat part *a* if the fuel cell operates on air at 25 C, 1 atm instead of on pure O_2 at this state.

14

Introduction to Phase and Chemical Equilibrium

In most of our considerations up to this point we have assumed that we are dealing either with systems that are in equilibrium or with those in which the deviation from equilibrium is infinitesimal, as in a quasi-equilibrium or reversible process. For irreversible processes we made no attempt to describe the state of the system during the process but dealt only with the initial and final states of the system, at which time we considered the system to be in equilibrium.

In this chapter we shall examine the criteria for equilibrium and from them derive certain relations which will enable us, under certain conditions, to determine the properties of a system when it is in equilibrium. The specific systems we shall consider are those involving equilibrium between phases and those involving chemical equilibrium in a single phase (homogeneous equilibrium) as well as certain related topics.

14.1 Requirements for Equilibrium

As a general requirement for equilibrium we postulate that a system is in equilibrium when there is no possibility that it can do any work when it is isolated from its surroundings. In applying this criterion to a system it is helpful to divide the system into two or more subsystems, and consider the possibility of doing work by any conceivable interaction between these two subsystems. For example, in Fig. 14.1 a system has been divided into two systems and an engine, of any conceivable variety, placed between these subsystems. A system may be so defined as to include the immediate surroundings. In this case we can let the immediate surroundings be a subsystem and thus consider the general case of the equilibrium between a system and its surroundings.

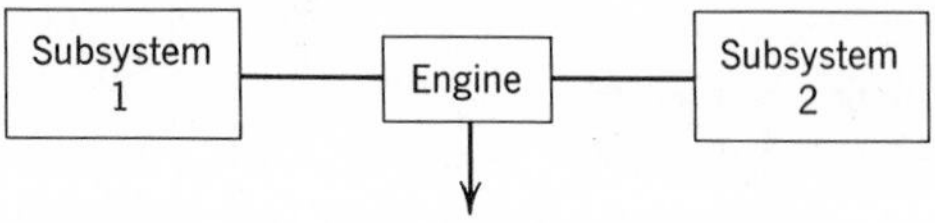

Fig. 14.1 Two subsystems that communicate through an engine.

The first requirement for equilibrium is that the two subsystems have the same temperature, for otherwise we could operate a heat engine between the two systems and do work. Thus we conclude that one requirement for equilibrium is that a system must be at a uniform temperature to be in equilibrium. It is also evident that there must be no unbalanced mechanical forces between the two systems, or else one could operate a turbine or piston engine between the two systems and do work.

However, we would like to establish general criteria for equilibrium which would apply to all simple compressible systems, including those that undergo chemical reactions. We will find that the Gibbs function is a particularly significant property in defining the criteria for equilibrium.

Let us first consider a qualitative example to illustrate this point. Consider a natural gas well that is one mile deep, and let us assume that the temperature of the gas is constant throughout the gas well. Suppose we have analyzed the composition of the gas at the top of the well, and we would like to know the composition of the gas at the bottom of the well. Further, let us assume that equilibrium conditions prevail in the well. If this is true we would expect that an engine such as shown in Fig. 14.2 (which operates on the basis of the pressure and composition change with elevation and does not involve combustion) would not be capable of doing any work.

If we consider a steady-state, steady-flow process for a control volume around this engine, we could apply Eq. 9.61

$$W_{\text{rev}} = m_i\left(g_i + \frac{\mathsf{V}_i^2}{2g_c} + Z_i\frac{g}{g_c}\right) - m_e\left(g_e + \frac{\mathsf{V}_e^2}{2g_c} + Z_e\frac{g}{g_c}\right)$$

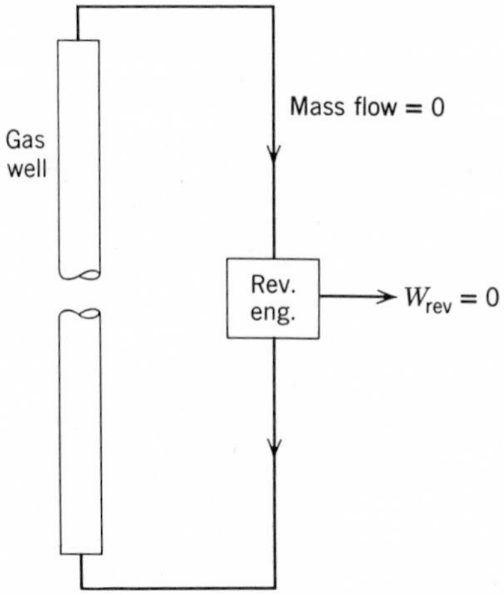

Fig. 14.2 Illustration showing the relation between reversible work and the criteria for equilibrium.

However,

$$W_{\text{rev}} = 0, \quad m_i = m_e, \quad \text{and} \quad \frac{V_i^2}{2g_c} = \frac{V_e^2}{2g_c}$$

Then we can write,

$$g_i + Z_i \frac{g}{g_c} = g_e + Z_e \frac{g}{g_c}$$

and the requirement for equilibrium in the well between two levels that are a distance dZ apart would be

$$dg_T + g/g_c dZ_T = 0$$

In contrast to a deep gas well, most of the systems that we consider are of such size that ΔZ is negligibly small, and therefore we consider the pressure in the system to be uniform.

This leads to the general statement of equilibrium that applies to simple compressible systems that may undergo a change in chemical composition, namely, that at equilibrium

$$dG_{T,P} = 0 \tag{14.1}$$

In the case of a chemical reaction, it is helpful to think of the equilibrium state as the state in which the Gibbs function is a minimum. For example, consider a system consisting initially of n_A moles of substance

A and n_B moles of substance B, which react in accordance with the relation

$$\nu_A A + \nu_B B \rightleftharpoons \nu_C C + \nu_D D$$

Let the reaction take place at constant pressure and temperature. If we plot G for this system as a function of n_A, the number of moles of A present, we would have a curve as shown in Fig. 14.3. At the minimum point on the curve, $dG_{T,P} = 0$, and this will be the equilibrium composition for this system at the given temperature and pressure. The subject of chemical equilibrium will be developed further in Section 14.6.

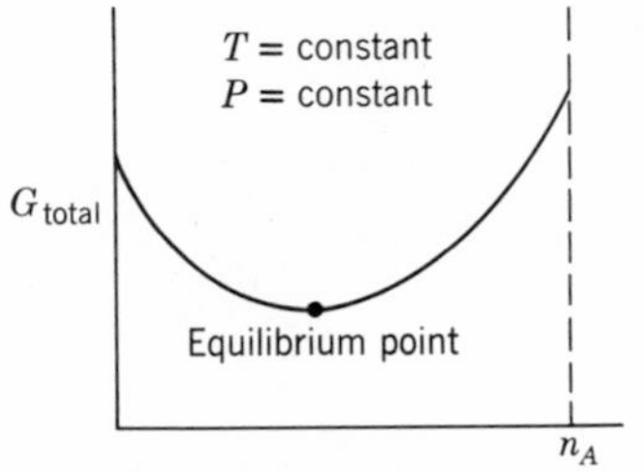

Fig. 14.3 Illustration of the requirement for chemical equilibrium.

14.2 Equilibrium Between Two Phases of a Pure Substance

As another example of this requirement for equilibrium, let us consider the equilibrium between two phases of a pure substance. Consider a system consisting of two phases of a pure substance at equilibrium. We know that under these conditions the two phases are at the same pressure and temperature. Consider the change of state associated with a transfer of dn moles from phase 1 to phase 2 while the temperature and pressure remain constant. That is,

$$dn^1 = -dn^2$$

The Gibbs function of this system is given by

$$G = f(T, P, n^1, n^2)$$

where n^1 and n^2 designate the number of moles in each phase. Therefore,

$$dG = \left(\frac{\partial G}{\partial T}\right)_{P,n^1,n^2} dT + \left(\frac{\partial G}{\partial P}\right)_{T,n^1,n^2} dP + \left(\frac{\partial G}{\partial n^1}\right)_{T,P,n^2} dn^1 + \left(\frac{\partial G}{\partial n^2}\right)_{T,P,n^1} dn^2$$

By definition,

$$\left(\frac{\partial G}{\partial n^1}\right)_{T,P,n^2} = \bar{g}^1; \quad \left(\frac{\partial G}{\partial n^2}\right)_{T,P,n^1} = \bar{g}^2$$

Therefore, at constant temperature and pressure,

$$dG = \bar{g}^1\, dn^1 + \bar{g}^2\, dn^2 = dn^1(\bar{g}^1 - \bar{g}^2)$$

Now at equilibrium (Eq. 14.1)

$$dG_{T,P} = 0$$

Therefore, at equilibrium, we have

$$\bar{g}^1 = \bar{g}^2 \tag{14.2}$$

That is, under equilibrium conditions, the Gibbs function of each phase of a pure substance is equal. Let us check this by determining the Gibbs function of saturated liquid (water) and saturated vapor (steam) at 14.7 lbf/in.^{2}. From the steam tables:
For the liquid:

$$g_f = h_f - Ts_f = 180.16 - 671.7 \times 0.3123 = -29.5 \text{ Btu/lbm}$$

For the vapor:

$$g_g = h_g - Ts_g = 1150.5 - 671.7 \times 1.7567 = -29.5 \text{ Btu/lbm}$$

Equation 14.2 can also be derived by applying the relation

$$T\, ds = dh - v\, dP$$

to the change of phase that takes place at constant pressure and temperature. For this process this relation can be integrated as follows:

$$\int_f^g T\, ds = \int_f^g dh$$

$$T(s_g - s_f) = (h_g - h_f)$$

$$h_f - Ts_f = h_g - Ts_g$$

$$g_f = g_g$$

The Clapeyron equation, which was derived in Section 12.4 can be derived by an alternate method by considering the fact that the Gibbs functions of two phases in equilibrium are equal. In Chapter 12 we considered the relation (Eq. 12.9) for a simple compressible substance,

$$dg = v\, dP - s\, dT$$

Consider a system that consists of a saturated liquid and a saturated vapor in equilibrium, and let this system undergo a change of pressure dP. The corresponding change in temperature, as determined from the vapor-pressure curve, is dT. Both phases will undergo the change in Gibbs function, dg, but since the phases always have the same value of the Gibbs function when they are in equilibrium, it follows that

$$dg_f = dg_g$$

But, from Eq. 12.9,

$$dg = v\,dP - s\,dT$$

it follows that

$$dg_f = v_f\,dP - s_f\,dT$$

$$dg_g = v_g\,dP - s_g\,dT$$

Since

$$dg_f = dg_g$$

it follows that

$$v_f\,dP - s_f\,dT = v_g\,dP - s_g\,dT$$

$$dP(v_g - v_f) = dT(s_g - s_f) \tag{14.3}$$

$$\frac{dP}{dT} = \frac{s_{fg}}{v_{fg}} = \frac{h_{fg}}{Tv_{fg}}$$

In summary, when different phases of a pure substance are in equilibrium, each phase has the same value of the Gibbs function per unit mass. This fact is relevant to different solid phases of a pure substance and is important in metallurgical applications of thermodynamics. Example 14.1 illustrates this principle.

Example 14.1

What pressure is required to make diamonds from graphite at a temperature of 25 C? The following data are given for a temperature of 25 C and a pressure of 1 atm.

	Graphite	Diamond
$\bar{g}$	0	1233 Btu/lb mole-R
v	0.00712 ft³/lbm	0.00456 ft³/lbm
β_T	3.0×10^{-6} atm^{-1}	0.16×10^{-6} atm^{-1}

The basic principle in the solution is that graphite and diamond can exist in equilibrium when they have the same value of the Gibbs function. At 1 atm pressure the Gibbs function of the diamond is greater than that of the graphite. However, the rate of increase in Gibbs function with pressure is greater for the graphite than the diamond, and therefore, at some pressure they can exist in equilibrium, and our problem is to find this pressure.

We have already considered the relation

$$dg = v\,dP - s\,dT$$

Since we are considering a process that takes place at constant temperature this reduces to

$$dg_T = v\,dP_T \tag{a}$$

Now at any pressure P and the given temperature the specific volume can be found from the following relation, which utilizes the isothermal compressibility factor.

$$v = v^\circ + \int_{P=1}^{P} \left(\frac{\partial v}{\partial P}\right)_T dP = v^\circ + \int_{P=1}^{P} \frac{v}{v}\left(\frac{\partial v}{\partial P}\right)_T dP$$

$$= v^\circ - \int_{P=1}^{P} v\beta_T\,dP \tag{b}$$

The superscript $^\circ$ will be used in this example to indicate the properties at a pressure of 1 atmosphere and a temperature of 25 C.

The specific volume changes only slightly with pressure, so that $v \approx v^\circ$. Also, we assume that β_T is constant and that we are considering a pressure of many atmospheres. With these assumptions this equation can be integrated to give

$$v = v^\circ - v^\circ\beta_T P = v^\circ(1-\beta_T P) \tag{c}$$

We can now substitute this into Eq. a to give the relation

$$dg_T = [v^\circ(1-\beta_T P)]\,dP_T$$

$$g - g^\circ = v^\circ(P - P^\circ) - v^\circ\beta_T\frac{(P^2 - P^{\circ 2})}{2} \tag{d}$$

If we assume that $P^\circ \ll P$ this reduces to

$$g - g^\circ = v^\circ\left[P - \frac{\beta_T P^2}{2}\right] \tag{e}$$

For the graphite, $g^\circ = 0$ and we can write

$$g_G = v_G^\circ\left[P - (\beta_T)_G\frac{P^2}{2}\right]$$

For the diamond, g° has a definite value and we have

$$g_D = g_D^\circ + v_D^\circ\left[P - (\beta_T)_D\frac{P^2}{2}\right]$$

But, at equilibrium the Gibbs function of the graphite and diamond are equal.

$$g_G = g_D$$

Therefore,

$$v_G^\circ\left[P - (\beta_T)_G\frac{P^2}{2}\right] = g_D^\circ + v_D^\circ\left[P - (\beta_T)_D\frac{P^2}{2}\right]$$

$$(v_G^\circ - v_D^\circ)P - [v_G^\circ(\beta_T)_G - v_D^\circ(\beta_T)_D]\frac{P^2}{2} = g_D^\circ$$

$$(0.00712 - 0.00456)P$$
$$- (0.00712 \times 3.0 \times 10^{-6} - 0.00456 \times 0.16 \times 10^{-6})\frac{P^2}{2}$$
$$= \frac{1233}{12} \times \frac{778}{14.7 \times 144}$$

Solving this for P we find

$$P = 15{,}500 \text{ atm}$$

That is, at 15,500 atm, 25 C, graphite and diamond can exist in equilibrium, and the possibility exists for conversion from graphite to diamonds.

The preceding example could also have been evaluated in terms of the fugacities instead of the Gibbs functions for the two forms of carbon. Similarly, the equilibrium requirement for liquid and vapor water discussed earlier could be expressed in terms of the fugacities of the two phases. That is,

$$f^L = f^V = f^{\text{sat}}$$

where f^{sat} is the fugacity of the substance determined at the given temperature and its saturation pressure P^{sat}.

The fugacity of a pure compressed liquid or solid phase at pressures moderately greater than the saturation pressure can be calculated with considerable accuracy as follows. For a pure substance at constant temperature,

$$dg_T = RT(d\ln f)_T = v\,dP_T$$

Integrating at constant temperature between the saturation state and the pressure P, we have

$$\int_{f^{\text{sat}}}^{f} RT(d\ln f)_T = \int_{P^{\text{sat}}}^{P} v\,dP_T$$

$$RT\ln\frac{f}{f^{\text{sat}}} = \int_{P^{\text{sat}}}^{P} v\,dP_T$$

If we assume that v is a constant, which often holds with considerable accuracy for liquids and solids, we have

$$RT\ln\frac{f}{f^{\text{sat}}} \approx v(P-P^{\text{sat}}) \tag{14.4}$$

Equation 14.4 is used to determine the fugacity coefficients of compressed liquid presented in the generalized chart given in Appendix Fig. B.10.

Since v is small for the liquid and solid phases, the quantity $v(P-P^{\text{sat}})$ is small for moderate changes of pressure. Therefore, for liquids and solids at moderate pressures we conclude that

$$f^L \approx f^{\text{sat}} \quad \text{and} \quad f^S \approx f^{\text{sat}} \tag{14.5}$$

14.3 Equilibrium of a Multicomponent, Multiphase System

As an introduction to more complicated systems, let us consider a system consisting of two components and two phases. To show the general characteristics of such a system, consider a mixture of oxygen and nitrogen at a pressure of one atmosphere and in the range of temperature where both the liquid and vapor phases are present.

Figure 14.4 shows the composition of the liquid and vapor phases which are in equilibrium as a function of temperature at a pressure of 1 atm. The upper line, marked "vapor line" gives the composition of the vapor phase, and the lower line gives the composition of the liquid phase. If we have pure nitrogen, the boiling point is 77.3 K. If we have pure oxygen, the boiling point is 90.2 K. If we have a mixture of nitrogen and oxygen with both phases present at a temperature of 84 K, the vapor will have a composition of 64 per cent N_2 and 36 per cent O_2. The liquid will have a composition of 30 per cent N_2 and 70 per cent O_2.

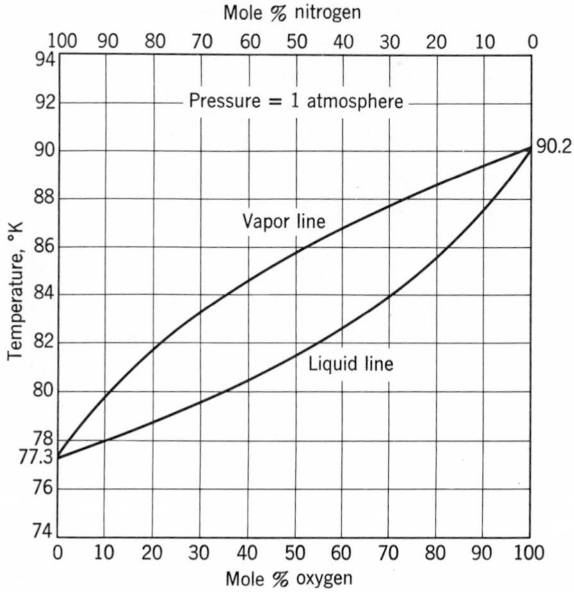

Fig. 14.4 Equilibrium diagram for liquid-vapor phases of the nitrogen-oxygen system at a pressure of 1 atm.

Consider a mixture consisting of 21 per cent oxygen and 79 per cent nitrogen (approximately the composition of air) and at a pressure of one atmosphere and an initial temperature of 74 K. At this state the mixture will be in the liquid phase. Let this liquid be slowly heated while the pressure remains constant. By referring to Fig. 14.4 we note that when the temperature reaches 78.8 K, the first bubble will form. This vapor will have a composition of approximately 6 per cent O_2 and 94 per cent N_2. As more heat is added the temperature increases and the mole fraction of oxygen in the liquid increases. When the last drop of liquid remains, the vapor will have a composition of 21 per cent O_2 and 79 per cent N_2, the temperature will be 81.9 K, and the liquid composition will be 54 per cent O_2 and 46 per cent N_2.

To understand the phase behavior of a system such as described here, it may be helpful to briefly discuss the more general phase diagram. We recall that the phase behavior for a pure substance is dictated by the saturation line on P–T coordinates, such as in Fig. 3.2. For a two-component system A and B, overall composition is now an additional independent property, and phase behavior is represented in terms of surfaces on a pressure-composition-temperature diagram, as shown in Fig. 14.5a. This diagram includes three constant-temperature planes to

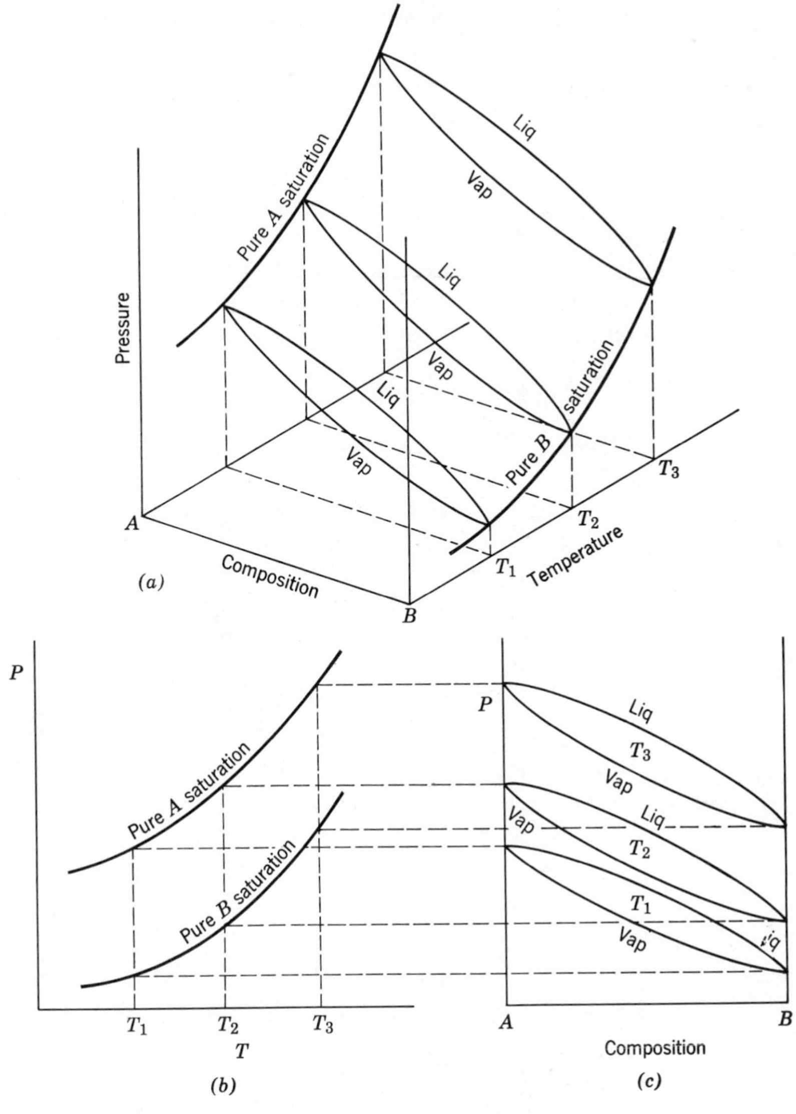

Fig. 14.5 Phase diagram for a two-component system.

indicate that there is a lower vapor surface and an upper liquid surface that together form an equilibrium envelope between the two pure substance saturation lines. Only within this envelope is it possible to have two phases present in equilibrium. Beneath the envelope the two-component system is all superheated vapor and above the envelope the system is all compressed liquid.

We realize now that a constant-pressure plane (1 atmosphere) of the three-dimensional phase diagram, as viewed from above, results in the temperature-composition diagram, Fig. 14.4, discussed in some detail previously. The other two projections of such a diagram are as shown in Figs. 14.5*b* and 14.5*c*.

In studying the characteristics of such a two-component system, we must first ask the question, "What is the requirement for phase equilibrium in this system?" We can proceed in a manner analogous to that followed in Section 14.2 for a pure substance, with superscripts 1 and 2 denoting the two phases, and in this case subscripts *A* and *B* referring to the two components. It follows from applying Eq. 12.24 to each phase that

$$dG^1 = -S^1\,dT + V^1\,dP + \mu_A{}^1\,dn_A{}^1 + \mu_B{}^1\,dn_B{}^1 \tag{14.6}$$

$$dG^2 = -S^2\,dT + V^2\,dP + \mu_A{}^2\,dn_A{}^2 + \mu_B{}^2\,dn_B{}^2$$

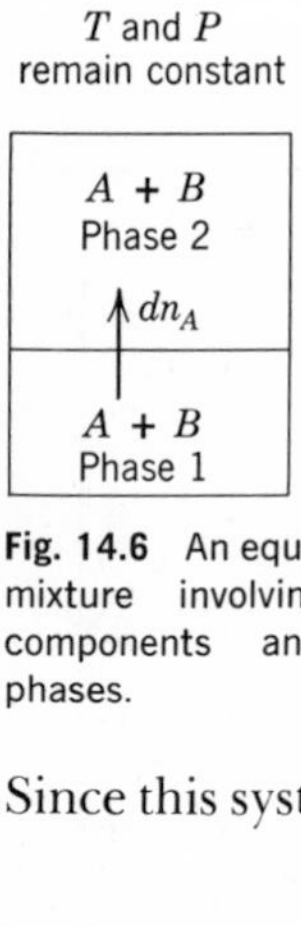

Fig. 14.6 An equilibrium mixture involving two components and two phases.

Let us consider this two-phase, two-component mixture that is in equilibrium as a system, and let each phase be considered as a subsystem. One possible change of state that might occur is for a very small amount of component *A* to be transferred from phase 1 to phase 2 while the moles of *B* in each phase and the temperature and pressure remain constant. This is shown schematically in Fig. 14.6.

For this change of state

$$dn_A{}^2 = -dn_A{}^1 \tag{14.7}$$

Since this system is in equilibrium, $dG = 0$. Therefore,

$$dG = dG^1 + dG^2 = 0 \tag{14.8}$$

Since T, P, $n_B{}^1$ and $n_B{}^2$ are all constant, it follows from Eqs. 14.6 and 14.8 that

$$\begin{aligned} dG &= \mu_A{}^1\,dn_A{}^1 + \mu_A{}^2\,dn_A{}^2 \\ &= \mu_A{}^1\,dn_A{}^1 - \mu_A{}^2\,dn_A{}^1 = dn_A{}^1(\mu_A{}^1 - \mu_A{}^2) \\ &= 0 \end{aligned}$$

Therefore, at equilibrium

$$\mu_A{}^1 = \mu_A{}^2 \tag{14.9}$$

Thus the requirement for equilibrium is that the chemical potential of each component is the same in all phases. If the chemical potential is not the same in all phases there will be a tendency for mass to pass from one phase to the other. When the chemical potential of a component is the same in both phases, there is no tendency for a net transfer of mass from one phase to the other.

This requirement for equilibrium is readily extended to multicomponent, multiphase systems, for the chemical potential of each component must be the same in all phases.

$$\begin{array}{l} \mu_A{}^1 = \mu_A{}^2 = \mu_A{}^3 = \cdots \text{ for all phases} \\ \mu_B{}^1 = \mu_B{}^2 = \mu_B{}^3 = \cdots \text{ for all phases} \\ — \quad — \quad — \\ — \quad — \quad — \\ — \quad — \quad — \\ — \quad — \quad — \\ \text{for all components} \end{array} \tag{14.10}$$

The next question we might ask is whether we can predict the equilibrium composition of a two-phase mixture from the known properties of the pure substances that make up the mixture. While we cannot do this in general, fairly accurate procedures have been developed for two-phase mixtures at moderate pressures. This involves the use of Raoult's Rule, which is developed, with statements of the simplifying assumptions on which it is based, in the next few paragraphs.

We found in Section 12.3, Eq. 12.25, that the chemical potential of component A is identical to the partial Gibbs function $\overline{G}_A$ of that component in the mixture. Thus, the requirement for equilibrium, Eq. 14.9, can also be expressed as

$$\overline{G}_A{}^1 = \overline{G}_A{}^2 \tag{14.11}$$

in which form the result is analogous to that for pure substance phase equilibrium, Eq. 14.2. It also follows from the definition of fugacity, Eq. 12.65, that the fugacity of a component in a mixture can similarly be defined in terms of its partial Gibbs function as

$$(d\overline{G}_A)_T = \overline{R}Td(\ln \bar{f}_A)_T \tag{14.12}$$

with the requirement that

$$\lim_{P\to 0} (\bar{f}_A/y_A P) = 1 \tag{14.13}$$

As a result, the requirement for phase equilibrium, Eq. 14.9, can be expressed in terms of fugacity as

$$\bar{f}_A{}^1 = \bar{f}_A{}^2 \tag{14.14}$$

That is, at equilibrium the fugacity of a given component is the same in each phase. This is perhaps the most useful formulation of the requirement for equilibrium between phases at a given temperature and pressure, as we have previous experience in evaluating fugacities. However, we have not had the occasion to examine the fugacity of a component in a mixture. Therefore, to utilize this result, it will be necessary to develop a specific model for determining the fugacities of components in a mixture.

In this text, we shall consider only one simple model for phase equilibrium, which will be discussed in terms of a two-component, liquid-vapor system. The requirement for equilibrium at temperature T and pressure P is then

$$\bar{f}_A^{\,L} = \bar{f}_A^{\,V}$$
$$\bar{f}_B^{\,L} = \bar{f}_B^{\,V} \qquad (14.15)$$

Let us first examine the behavior of the liquid phase, in which we denote molal composition by the symbol x_A, to distinquish from the vapor phase composition y_A. Our liquid model requires three separate assumptions:

1. The fugacity of a component A in the liquid can be expressed as the product of its mole fraction and the fugacity of the pure liquid A at the same pressure and temperature as the system, or

$$\bar{f}_A^{\,L} = x_A f_A^{\,L} \qquad (14.16)$$

This is termed the Lewis-Randall rule, and is derived from general thermodynamic considerations of mixtures by assuming that there is no volume change on mixing two substances at constant T and P. This assumption has been closely verified experimentally for many simple systems at low pressures.

2. The fugacity of pure liquid A at T and P of the system is equal to the fugacity of saturated A (liquid or vapor) at the same T, and its corresponding saturation pressure P_A^{sat}, or

$$f_A^{\,L} = f_A^{\text{sat}} \qquad (14.17)$$

This is equivalent to assuming that the $\int v\,dP$ correction of Eq. 14.4 is negligibly small.

3. Pure saturated vapor A at T and P_A^{sat} behaves as an ideal gas, or

$$f_A^{\text{sat}} = P_A^{\text{sat}} \qquad (14.18)$$

Combining these assumptions, Eqs. 14.16–18, we obtain

$$\bar{f}_A{}^L = x_A P_A{}^{\text{sat}} \tag{14.19}$$

which is commonly termed Raoult's Rule.

We next examine the behavior of the equilibrium vapor phase, which we assume to behave as a mixture of ideal gases. It follows, Eq. 14.13, that

$$\bar{f}_A{}^V = y_A P \tag{14.20}$$

Now, substituting Eqs. 14.19 and 14.20 for each component into Eqs. 14.15, we obtain the result for the Raoult's Rule-Ideal Gas Model:

$$x_A P_A{}^{\text{sat}} = y_A P \tag{14.21}$$

$$x_B P_B{}^{\text{sat}} = y_B P \tag{14.22}$$

which together with

$$x_A + x_B = 1 \tag{14.23}$$

$$y_A + y_B = 1 \tag{14.24}$$

forms a set of four equations in four unknowns at a given T and P ($P_A{}^{\text{sat}}$ and $P_B{}^{\text{sat}}$ depend only on T). This set of equations can then be solved to determine the equilibrium compositions in each phase.

The assumptions made in developing this model for phase equilibrium are clearly dependent on pressure and are not suitable for high-pressure or complex systems. However, the model has been found to give reasonable results in predicting the behavior of many simple systems at low pressures.

Example 14.2

Air (assumed to be 21% O_2, 79% N_2) is cooled to 80 K, 1 atmosphere pressure. Calculate the composition of the liquid and vapor phases at this condition, assuming the Raoult's Rule–Ideal Gas model, and compare the results with Fig. 14.4. At 80 K:

$$P_{N_2}^{\text{sat}} = 19.71 \text{ lbf/in.}^2, \quad P_{O_2}^{\text{sat}} = 4.36 \text{ lbf/in.}^2$$

For convenience let $N_2 = A$, and $O_2 = B$ in the calculations. Using Eqs. 14.21 and 14.22,

$$x_A P_A{}^{\text{sat}} = y_A P$$

$$x_B P_B{}^{\text{sat}} = y_B P$$

Therefore, at 80 K, 1 atm, pressure,

$$x_A(19.71) = y_A(14.7)$$

$$x_B(4.36) = y_B(14.7)$$

But,

$$x_A + x_B = 1$$

$$y_A + y_B = 1$$

Substituting for y_A, y_B, in the last equation,

$$\frac{19.71}{14.7}x_A + \frac{4.36}{14.7}x_B = 1.34x_A + 0.2965x_B = 1$$

$$4.52x_A + x_B = 3.37$$

Therefore,

$$4.52x_A + (1 - x_A) = 3.37$$

$$x_A = \frac{2.37}{3.52} = 0.674$$

$$y_A = 1.34x_A = 0.904$$

From Fig. 14.4, which is based on experimental data, we find that at 80 K, 1 atm,

$$x_A = 0.66, \quad y_A = 0.89$$

Thus, we conclude that for this system and this temperature and pressure, Raoult's Rule gives fairly accurate results.

14.4 The Gibbs Phase Rule (without Chemical Reaction)

The Gibbs phase rule, which was derived by Professor J. Willard Gibbs of Yale University in 1875, ranks among the truly significant contributions to physical science. In this section we consider the Gibbs phase rule for a system that does not involve a chemical reaction. For such a system the Gibbs phase rule is

$$\mathcal{P} + \mathcal{V} = \mathcal{C} + 2 \tag{14.25}$$

where $\mathscr{P}$ is the number of phases present, $\mathscr{V}$ is the variance, and $\mathscr{C}$ the number of components present. The term variance designates the number of intensive properties that must be specified in order to completely fix the state of the system. For example, in Example 14.2 we considered a two phase mixture of oxygen and nitrogen. For this system the number of phases $\mathscr{P}$ equals 2, the number of components $\mathscr{C}$ is 2, and therefore the variance $\mathscr{V}$ is 2. This is evident from Fig. 14.4, for this diagram is for a fixed pressure (1 atm), and the fixing of one additional intensive property, such as temperature, mole fraction of a given component in the liquid phase, or mole fraction of a given component in the vapor phase, will completely determine all other intensive properties and thus fix the state of the system (although not the relative amounts of the two phases).

Let us consider further the application of the Gibbs phase rule to a pure substance. In this case $\mathscr{C} = 1$. When we have one phase present, such as superheated vapor, $\mathscr{P} = 1$, and we conclude that $\mathscr{V} = 1 + 2 - 1 = 2$. That is, two intensive properties must be specified in order to fix the state. We are already familiar with the superheated vapor tables for a number of substances and recognize that these tables are presented with pressure and temperature as the two independent properties. Such a system is referred to as a bivariant system.

Suppose we have two phases of a pure substance in equilibrium, such as saturated liquid and saturated vapor. In this case $\mathscr{C} = 1$, $\mathscr{P} = 2$, and $\mathscr{V} = 1 + 2 - 2 = 1$. That is, one intensive property fixes the value of all other intensive properties and thus determines the state of each phase. Again, from our familiarity with tables of thermodynamic properties we recall that either pressure or temperature is used as the independent intensive property for tabulating liquid-vapor equilibrium data for a pure substance. Such a system is known as monovariant.

Consider also the triple point of a pure substance. In this case $\mathscr{C} = 1$ and $\mathscr{P} = 3$. Therefore $\mathscr{V} = 1 + 2 - 3 = 0$. That is, at the triple point all intensive properties are fixed, and if any intensive property is varied, we are no longer at the triple point. This is known as an invariant system.

It might be well at this point to again emphasize that a pure substance can have several phases in the solid state. For example, Fig. 3.6 shows the various phases of water. Note that there are several states at which three phases exist in equilibrium, and each of these is a triple point.

The application of the phase rule to a two-component, two-phase system has already been made. As a final application let us consider a two-component, three-phase system. For such a system $\mathscr{V} = 2 + 2 - 3 = 1$. That is, it is a monovariant system, and specifying one property, such as pressure or temperature fixes the state of the system.

The validity of the Gibbs phase rule can be outlined by considering a system consisting of $\mathscr{C}$ components and $\mathscr{P}$ phases in equilibrium at a given temperature and pressure. Assuming that each component exists in each phase, the state of the system could be completely specified if the concentration of each component in each phase and the temperature and pressure were specified. This would be a total of $\mathscr{C}\mathscr{P}+2$ intensive properties.

We known, however, that these are not all independent intensive properties. If we determine the number of equations we have between these intensive properties, we can subtract this from the $\mathscr{C}\mathscr{P}+2$ intensive properties and find the number of independent intensive properties, or, as it has been defined, the variance. The fact that at equilibrium the chemical potential of each component is the same in all phases gives rise to a total of $\mathscr{C}(\mathscr{P}-1)$ equations between the intensive properties of the mixture.

Further, the fact that the sum of the mole fractions equals unity in each of the $\mathscr{P}$ phases gives $\mathscr{P}$ additional equations. Therefore the variance is

$$\mathscr{V}=\mathscr{C}\mathscr{P}+2-\mathscr{P}-\mathscr{C}(\mathscr{P}-1)=\mathscr{C}+2-\mathscr{P}$$

14.5 Metastable Equilibrium

Although the limited scope of this book precludes an extensive treatment of metastable equilibrium, a brief introduction to the subject is presented in this section. Let us first consider an example of metastable equilibrium.

Consider a slightly superheated vapor, such as steam, expanding in a convergent-divergent nozzle, as shown in Fig. 14.7. Assuming the process is reversible and adiabatic, the steam will follow path 1-*a* on the *T*-*s* diagram, and at point *a* we would expect condensation to occur. However, if point *a* is reached in the divergent section of the nozzle, it is observed that no condensation occurs until point *b* is reached, and at this point the condensation occurs very abruptly in what is referred to as a condensation shock. Between points *a* and *b* the steam exists as a vapor, but the temperature is below the saturation temperature for the given pressure. This is known as a metastable state. The possibility of a metastable state exists with any phase transformation. The dotted lines on the equilibrium diagram shown in Fig. 14.8 represent possible metastable states for solid-liquid-vapor equilibrium.

The nature of a metastable state is often pictured schematically by the kind of diagram shown in Fig. 14.9. The ball is in a stable position (the "metastable state") for small displacements, but with a large displace-

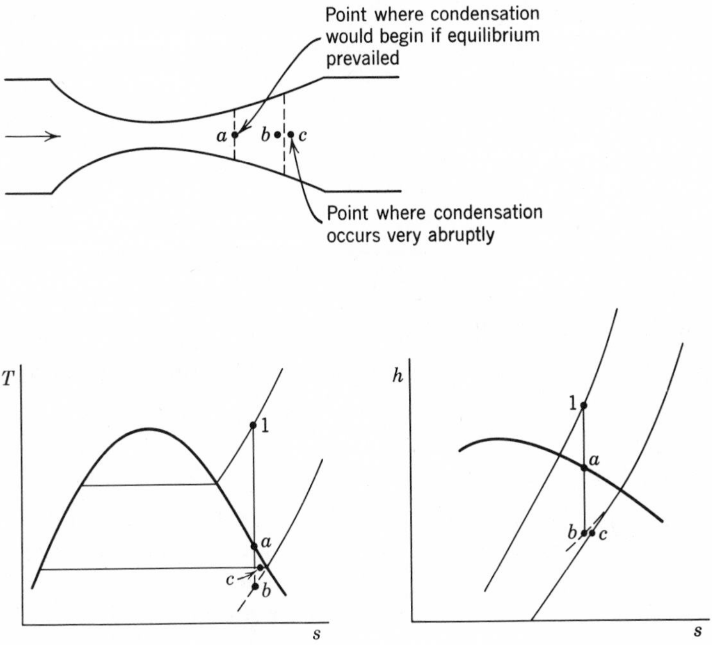

Fig. 14.7 Illustration of the phenomenon of supersaturation in a nozzle.

ment it moves to a new equilibrium position. The steam expanding in the nozzle is in a metastable state between *a* and *b*. This means that droplets smaller than a certain critical size will re-evaporate, and only when droplets larger than this critical size have formed (this corresponds to moving the ball out of the depression) will the new equilibrium state appear.

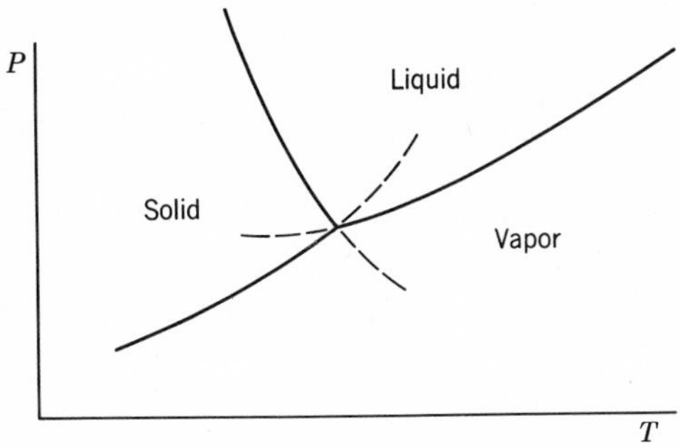

Fig. 14.8 Metastable states for solid-liquid-vapor equilibrium.

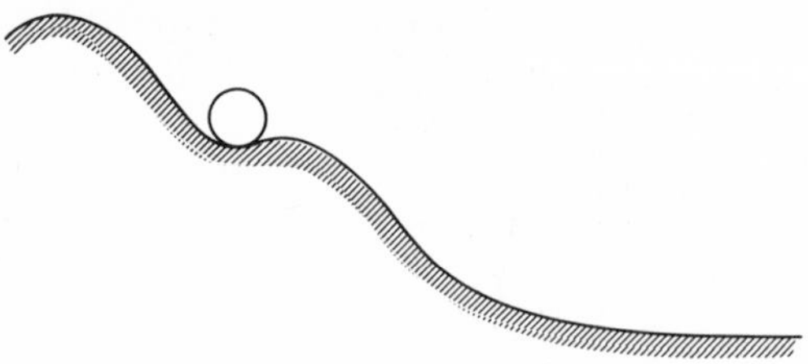

Fig. 14.9 Schematic diagram illustrating a metastable state.

14.6 Chemical Equilibrium

We now turn our attention to chemical equilibrium and consider first a chemical reaction involving only one phase. This is referred to as a homogeneous chemical reaction. It may be helpful to visualize this as a gaseous phase, but the basic considerations apply to any phase.

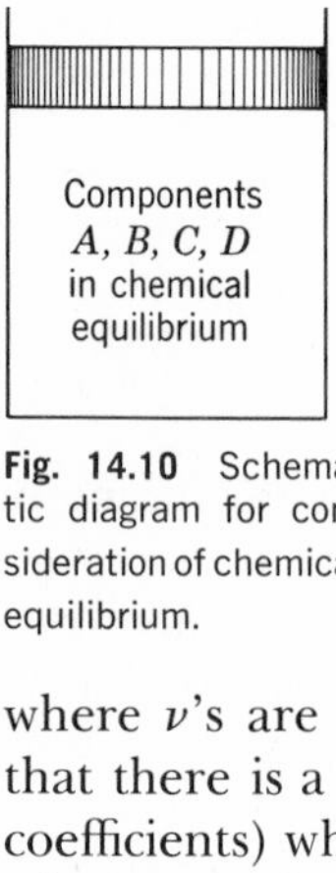

Fig. 14.10 Schematic diagram for consideration of chemical equilibrium.

Consider a vessel, Fig. 14.10, that contains four compounds, A, B, C, and D, which are in chemical equilibrium at a given pressure and temperature. For example, these might consist of CO, CO_2, H_2, and H_2O in equilibrium. Let the number of moles of each component be designated n_A, n_B, n_C, and n_D. Further, let the chemical reaction which takes place between these four constituents be

$$\nu_A A + \nu_B B \rightleftharpoons \nu_C C + \nu_D D \qquad (14.26)$$

where ν's are the stoichiometric coefficients. It should be emphasized that there is a very definite relation between the ν's (the stoichiometric coefficients) whereas the n's (the number of moles present) for any constituent can be varied simply by varying the amount of that component in the reaction vessel.

Let us now consider how the requirement for equilibrium, namely, that $dG_{T,P} = 0$ at equilibrium, applies to a homogeneous chemical reaction. When we considered phase equilibrium (Sec. 14.3) we proceeded by assuming that the two phases were in equilibrium at a given temperature and pressure, and then let a small quantity of one component be transferred from one phase to the other. In a similar manner, let us assume that the four components are in chemical equilibrium and then assume that from this equilibrium state, while the temperature and pressure remain constant, the reaction proceeds an infinitesimal amount toward the right as Eq. 14.26 is written. This results in a decrease in the moles of A and B and an increase in the moles of C and D. Let us designate the degree of reaction by ϵ, and define the degree of reaction by the relations

$$\begin{aligned} dn_A &= -\nu_A \, d\epsilon \\ dn_B &= -\nu_B \, d\epsilon \\ dn_C &= +\nu_C \, d\epsilon \\ dn_D &= +\nu_D \, d\epsilon \end{aligned} \tag{14.27}$$

That is, the change in the number of moles of any component during a chemical reaction is given by the product of the stoichiometric coefficients (the ν's) and the degree of reaction.

Let us evaluate the change in the Gibbs function associated with this chemical reaction that proceeds to the right in the amount $d\epsilon$. In doing so we use, as would be expected, the Gibbs function of each component in the mixture, namely, the partial molal Gibbs function (or its equivalent, the chemical potential)

$$dG_{T,P} = \bar{G}_C \, dn_C + \bar{G}_D \, dn_D + \bar{G}_A \, dn_A + \bar{G}_B \, dn_B$$

Substituting Eq. 14.27, we have

$$dG_{T,P} = (\nu_C \bar{G}_C + \nu_D \bar{G}_D - \nu_A \bar{G}_A - \nu_B \bar{G}_B) d\epsilon \tag{14.28}$$

At this point, we need to be able to express $\bar{G}_i$ for each component in terms of properties that can be readily evaluated. In this text we shall consider chemical equilibrium for mixtures of ideal gases only. For this case, and using as a reference state the standard-state pressure P° and the same temperature as the mixture,

$$\begin{aligned} \bar{G}_i &= \bar{H}_i - T\bar{S}_i \\ &= \bar{h}_i^\circ - T\bar{s}_i^\circ + \bar{R}T \ln \frac{y_i P}{P^\circ} \\ &= \bar{g}_i^\circ + \bar{R}T \ln \frac{y_i P}{P^\circ} \end{aligned} \tag{14.29}$$

That is, the partial Gibbs function of each component in the mixture (at T and P) is expressed in terms of the standard-state Gibbs function for the pure substance (at T and P°) plus a function of P, y_i and T. Substituting this relation into Eq. 14.28, we have

$$\begin{aligned} dG_{T,P} = \Bigg[&\nu_C \left(\bar{g}_C^\circ + \bar{R}T \ln \frac{y_C P}{P^\circ} \right) + \nu_D \left(\bar{g}_D^\circ + \bar{R}T \ln \frac{y_D P}{P^\circ} \right) \\ &- \nu_A \left(\bar{g}_A^\circ + \bar{R}T \ln \frac{y_A P}{P^\circ} \right) - \nu_B \left(\bar{g}_B^\circ + \bar{R}T \ln \frac{y_B P}{P^\circ} \right) \Bigg] d\epsilon \end{aligned} \tag{14.30}$$

Let us define ΔG° as follows,

$$\Delta G^\circ = \nu_C \bar{g}_C^\circ + \nu_D \bar{g}_D^\circ - \nu_A \bar{g}_A^\circ - \nu_B \bar{g}_B^\circ \tag{14.31}$$

That is, ΔG° is the change in the Gibbs function that would occur if the chemical reaction given by Eq. 14.26 (which involves the stoichiometric amounts of each component) proceeded completely from left to right, with the reactants A and B initially separated and each at temperature T and the standard state pressure and the products C and D finally separated and each at temperature T and the standard state pressure. Note also that ΔG° for a given reaction is a function of only the temperature. This will be most important to bear in mind as we proceed with our developments of homogeneous chemical equilibrium. Let us now digress from our development to consider an example involving the calculation of ΔG°.

Example 14.3

Determine the value of ΔG° for the reaction $H_2O \rightleftharpoons H_2 + \frac{1}{2}O_2$ at 298 K and at 2000 K, with the H_2O in the gaseous phase.

We take 1 atm as our standard state pressure, and recall that we assume further that $\bar{g}$ for all the elements is zero at one atmosphere pressure and 298 K (537 R). Therefore, at 298 K,

$$(\bar{g}_f^\circ)_{H_2} = 0; \quad (\bar{g}_f^\circ)_{O_2} = 0$$

From Table 13.3, at 298 K,

$$(\bar{g}_f^\circ)_{H_2O} = -98{,}345 \text{ Btu/lb mole.}$$

$$\Delta G^\circ = (\bar{g}_f^\circ)_{H_2} + \tfrac{1}{2}(\bar{g}_f^\circ)_{O_2} - (\bar{g}_f^\circ)_{H_2O}$$

$$= 0 + 0 - (-98{,}345) = 98{,}345 \text{ Btu}$$

At 2000 K (3600 R), from Table B.9

$$\begin{aligned}
\bar{g}_{H_2}^\circ = \bar{g}_{3600}^\circ - \bar{g}_{537}^\circ &= (\bar{h}_{3600}^\circ - \bar{h}_{537}^\circ) - (3600\bar{s}_{3600}^\circ - 536.67\bar{s}_{537}^\circ) \\
&= 22{,}772 - 3600(45.004) + 536.67(31.208) \\
&= 22{,}772 - 162{,}014 + 16{,}748 \\
&= -122{,}494 \text{ Btu/lb mole}
\end{aligned}$$

Similarly, for the O_2 at 2000 K (3600 R),

$$\bar{g}^\circ_{O_2} = (\bar{h}^\circ_{3600} - \bar{h}^\circ_{537}) - (3600\bar{s}^\circ_{3600} - 536.67\bar{s}^\circ_{537})$$

$$= 25{,}468 - 3600(64.210) + 536.67(49.004)$$

$$= 25{,}468 - 231{,}156 + 26{,}299$$

$$= -179{,}389 \text{ Btu/lb mole}$$

For the H_2O,

$$\bar{g}^\circ_{H_2O} = \bar{g}^\circ_f + \bar{g}^\circ_{3600} - \bar{g}^\circ_{537}$$

$$= -98{,}345 + 31{,}271 - [3600(63.234) - 536.67(45.106)]$$

$$= -98{,}345 + 31{,}271 - 227{,}642 + 24{,}207$$

$$= -270{,}509 \text{ Btu/lb mole}$$

Therefore, at 2000 K

$$\Delta G^\circ = -122{,}494 + \tfrac{1}{2}(-179{,}389) - (-270{,}509) = 58{,}321 \text{ Btu.}$$

The value for ΔG° at 2000 K can alternately be evaluated by using the fact that at constant T,

$$\Delta G^\circ = \Delta H^\circ - T\Delta S^\circ$$

At 2000 K,

$$\Delta H^\circ = (\bar{h}^\circ_{3600} - \bar{h}^\circ_{537})_{H_2} + \tfrac{1}{2}(\bar{h}^\circ_{3600} - \bar{h}^\circ_{537})_{O_2} - (\bar{h}^\circ_f + \bar{h}^\circ_{3600} - \bar{h}^\circ_{537})_{H_2O}$$

$$= 22{,}772 + \tfrac{1}{2}(25{,}468) - (-104{,}036 + 31{,}271)$$

$$= 108{,}271 \text{ Btu}$$

$$\Delta S^\circ = (\bar{s}^\circ_{3600})_{H_2} + \tfrac{1}{2}(\bar{s}^\circ_{3600})_{O_2} - (\bar{s}^\circ_{3600})_{H_2O}$$

$$= 45.004 + \tfrac{1}{2}(64.210) - 63.234$$

$$= 13.875 \text{ Btu/R}$$

Therefore,

$$\Delta G^\circ = 108{,}271 - 3600(13.875)$$

$$= 108{,}271 - 49{,}950 = 58{,}321 \text{ Btu}$$

Returning now to our development, substituting Eq. 14.31 into Eq. 14.30 and rearranging, we can write

$$dG_{T,P} = \left\{\Delta G^\circ + \bar{R}T \ln\left[\frac{y_C^{\nu_C} y_D^{\nu_D}}{y_A^{\nu_A} y_B^{\nu_B}}\left(\frac{P}{P^\circ}\right)^{\nu_C+\nu_D-\nu_A-\nu_B}\right]\right\} d\epsilon \tag{14.32}$$

At equilibrium $dG_{T,P} = 0$. Therefore, since $d\epsilon$ is arbitrary,

$$\ln\left[\frac{y_C^{\nu_C} y_D^{\nu_D}}{y_A^{\nu_A} y_B^{\nu_B}}\left(\frac{P}{P^\circ}\right)^{\nu_C+\nu_D-\nu_A-\nu_B}\right] = -\frac{\Delta G^\circ}{\bar{R}T} \tag{14.33}$$

For convenience, we define the equilibrium constant K as

$$\ln K = -\frac{\Delta G^\circ}{\bar{R}T} \tag{14.34}$$

which we note must be a function of temperature only for a given reaction, since ΔG° is given by Eq. 14.31 in terms of the properties of the pure substances at a given temperature and the standard-state pressure.

Example 14.4

Determine the equilibrium constant K, expressed as $\ln K$, for the reaction $H_2O \rightleftharpoons H_2 + \frac{1}{2} O_2$ at 537 R and 3600 R.

We have already found, in Example 14.3, ΔG° for this reaction at these two temperatures. Therefore, at 537 R,

$$(\ln K)_{537} = -\frac{\Delta G^\circ_{537}}{\bar{R}T} = \frac{-98{,}345}{1.987 \times 537} = -92.21$$

At 3600 R we have,

$$(\ln K)_{3600} = \frac{-\Delta G^\circ_{3600}}{\bar{R}T} = \frac{-58{,}321}{1.987 \times 3600} = -8.151$$

Table B.10 gives the values of the equilibrium constant for a number of reactions. Note again that for each reaction the value of the equilibrium constant is determined from the properties of each of the pure constituents at the standard state pressure and is a function of temperature only.

For other reaction equations, the chemical equilibrium constant can be calculated as in Example 14.4 or can be determined analytically in the following manner. Consider the general reaction Eq. 14.26. The standard-state Gibbs function for each constituent can be expressed by writing Eq. 12.9 at the constant pressure P°. For component A,

$$d\bar{g}_A^\circ = -\bar{s}_A^\circ \, dT = -\frac{\bar{h}_A^\circ - \bar{g}_A^\circ}{T} \, dT$$

or

$$\frac{T \, d\bar{g}_A^\circ - \bar{g}_A^\circ \, dT}{T^2} = -\frac{\bar{h}_A^\circ}{T^2} \, dT$$

which can also be written as

$$d\left(\frac{\bar{g}_A^\circ}{T}\right)_{P^\circ} = -\frac{\bar{h}_A^\circ}{T^2} \, dT_{P^\circ} \tag{14.35}$$

Therefore, for the reaction given by Eq. 14.26,

$$d\left(\frac{\Delta G^\circ}{T}\right)_{P^\circ} = -\frac{\Delta H^\circ}{T^2} \, dT_{P^\circ} \tag{14.36}$$

where ΔG° is defined by Eq. 14.30 and ΔH° similarly by

$$\Delta H^\circ = \nu_C \bar{h}_C^\circ + \nu_D \bar{h}_D^\circ - \nu_A \bar{h}_A^\circ - \nu_B \bar{h}_B^\circ \tag{14.37}$$

Now, substituting the definition of the equilibrium constant Eq. 14.34 into Eq. 14.36,

$$d \ln K = \frac{\Delta H^\circ}{\bar{R}T^2} \, dT_{P^\circ} \tag{14.38}$$

which is termed the van't Hoff equation. In integrating this equation, we must be careful to note that ΔH° as defined by Eq. 14.37 is a function of temperature.

Substituting the definition of equilibrium constant according to Eq. 14.34 into the equilibrium expression Eq. 14.33 yields

$$K = \frac{y_C^{\nu_C} y_D^{\nu_D}}{y_A^{\nu_A} y_B^{\nu_B}} \left(\frac{P}{P^\circ}\right)^{\nu_C + \nu_D - \nu_A - \nu_B} \tag{14.39}$$

which is the form of the chemical equilibrium equation for ideal gas reactions that we shall utilize to evaluate equilibrium composition.

Let us now consider a number of examples that illustrate the procedure for determining the equilibrium composition for a homogeneous reaction and the influence of certain variables on the equilibrium composition.

Example 14.5

One mole of carbon at 77 F and one atmosphere pressure reacts with one mole of oxygen at 77 F and one atmosphere pressure to form an equilibrium mixture of CO_2, CO, and O_2 at 3000 K, one atmosphere pressure, in a steady-flow process. Determine the equilibrium composition and the heat transfer for this process.

It is convenient to view the over-all process as though it occurred in two separate steps, a combustion process followed by a heating and dissociation of the combustion product CO_2, as indicated in Fig. 14.11.

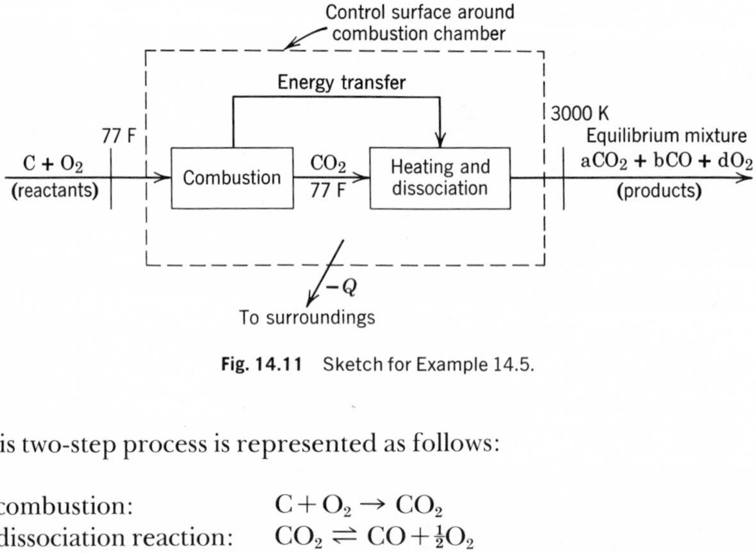

Fig. 14.11 Sketch for Example 14.5.

This two-step process is represented as follows:

combustion: $C + O_2 \rightarrow CO_2$
dissociation reaction: $CO_2 \rightleftharpoons CO + \frac{1}{2}O_2$

That is, the energy released by the combustion of C and O_2 heats the CO_2 formed to high temperature, resulting in the dissociation of part of the CO_2 to CO and O_2. Thus, the over-all reaction can be written

$$C + O_2 \rightarrow aCO_2 + bCO + dO_2$$

where the unknown coefficients a, b, and d must be found by solution of the equilibrium equation associated with the dissociation reaction. Once this is accomplished, we can write the first law for a control volume around the combustion chamber to calculate the heat transfer.

From the combustion equation we find that the initial composition for the dissociation reaction is 1 mole CO_2. Therefore, letting z be the number of moles of CO_2 dissociated, we find

	CO_2	$\rightleftharpoons CO$	$+\frac{1}{2}O_2$
Initial:	1	0	0
Change:	$-z$	$+z$	$+z/2$
At equilibrium:	$(1-z)$	z	$z/2$

Therefore, the over-all reaction is

$$C + O_2 \rightarrow (1-z)CO_2 + z\,CO + \frac{z}{2}O_2$$

and the total number of moles at equilibrium is

$$n = (1-z) + z + z/2 = 1 + z/2$$

The equilibrium mole fractions are

$$y_{CO_2} = \frac{1-z}{1+z/2}, \quad y_{CO} = \frac{z}{1+z/2}, \quad y_{O_2} = \frac{z/2}{1+z/2}$$

From Table B.10 we find that the value of the equilibrium constant at 3000 K for the dissociation reaction considered here is

$$\ln K = -1.117, \quad K = 0.328$$

Substituting these quantities along with $P = 1$ atm into Eq. 14.39, we have the equilibrium equation,

$$K = 0.328 = \frac{y_{CO}\,y_{O_2}^{1/2}}{y_{CO_2}}\left(\frac{P}{P^\circ}\right)^{1+1/2-1}$$

$$= \frac{\left(\dfrac{z}{1+z/2}\right)\left(\dfrac{z/2}{1+z/2}\right)^{1/2}(1)^{1/2}}{\left(\dfrac{1-z}{1+z/2}\right)}$$

or, in more convenient form,

$$\frac{K^2}{(P/P^\circ)} = \frac{(0.328)^2}{1} = \left(\frac{z}{1-z}\right)^2\left(\frac{z}{2+z}\right)$$

In order to obtain the physically meaningful root of this mathematical relation, we note that the number of moles of each component must be greater than zero. Thus, the root of interest to us must lie in the range

$$0 \leqslant z \leqslant 1$$

Solving the equilibrium equation by trial and error, we find

$$z = 0.436$$

Therefore, the over-all process is

$$C + O_2 \rightarrow 0.564\ CO_2 + 0.436\ CO + 0.218\ O_2$$

where the equilibrium mole fractions are

$$y_{CO_2} = \frac{0.564}{1.218} = 0.463$$

$$y_{CO} = \frac{0.436}{1.218} = 0.358$$

$$y_{O_2} = \frac{0.218}{1.218} = 0.179$$

The heat transfer from the combustion chamber to the surroundings can be calculated using the enthalpies of formation and Table B.9. For this process,

$$H_R = (\bar{h}_f^\circ)_C + (\bar{h}_f^\circ)_{O_2} = 0 + 0 = 0$$

The equilibrium products leave the chamber at 3000 K, or 5400 R. Therefore,

$$\begin{aligned} H_p &= n_{CO_2}(\bar{h}_f^\circ + \bar{h}_{5400}^\circ - \bar{h}_{537}^\circ)_{CO_2} \\ &\quad + n_{CO}(\bar{h}_f^\circ + \bar{h}_{5400}^\circ - \bar{h}_{537}^\circ)_{CO} \\ &\quad + n_{O_2}(\bar{h}_f^\circ + \bar{h}_{5400}^\circ - \bar{h}_{537}^\circ)_{O_2} \\ &= 0.564(-169{,}297 + 65{,}763) \\ &\quad + 0.436(-47{,}551 + 40{,}243) \\ &\quad + 0.218(0 + 42{,}203) \\ &= -52{,}400 \text{ Btu/mole C burned} \end{aligned}$$

Substituting into the first law,

$$\begin{aligned} Q_{\text{c.v.}} &= H_P - H_R \\ &= -52{,}400 \text{ Btu/mole C burned} \end{aligned}$$

Example 14.6

One mole of carbon at 77 F reacts with two moles of oxygen at 77 F to form an equilibrium mixture of CO_2, CO, and O_2 at 3000 K, 1 atmosphere pressure. Determine the equilibrium composition.

The over-all process can be imagined to occur in two steps as in the previous example. The combustion process is

$$C + 2O_2 \rightarrow CO_2 + O_2$$

and the subsequent dissociation reaction is

	CO_2	$\rightleftharpoons$ CO	$+\frac{1}{2}O_2$
Initial:	1	0	1
Change:	$-z$	$+z$	$+z/2$
At equilibrium:	$(1-z)$	z	$(1+z/2)$

We find that in this case the over-all process is

$$C + 2O_2 \rightarrow (1-z)CO_2 + zCO + (1+z/2)O_2$$

and the total number of moles at equilibrium is

$$n = (1-z) + z + (1+z/2) = 2 + z/2$$

The mole fractions are

$$y_{CO_2} = \frac{1-z}{2+z/2}, \quad y_{CO} = \frac{z}{2+z/2}, \quad y_{O_2} = \frac{1+z/2}{2+z/2}$$

The equilibrium constant for the reaction $CO_2 \rightleftharpoons CO + \frac{1}{2}O_2$ at 3000 K was found in Example 14.5 to be 0.328. Therefore, the equilibrium equation is, substituting these quantities and $P = 1$ atm,

$$K = 0.328 = \frac{y_{CO}y_{O_2}^{1/2}}{y_{CO_2}}\left(\frac{P}{P^\circ}\right)^{1+1/2-1}$$

$$= \frac{\left(\dfrac{z}{2+z/2}\right)\left(\dfrac{1+z/2}{2+z/2}\right)^{1/2}}{\left(\dfrac{1-z}{2+z/2}\right)}(1)^{1/2}$$

or,

$$\frac{K^2}{(P/P^\circ)} = \frac{(0.328)^2}{1} = \left(\frac{z}{1-z}\right)_2\left(\frac{2+z}{4+z}\right)$$

We note that in order for the number of moles of each component to be greater than zero,

$$0 \leqslant z \leqslant 1$$

Solving the equilibrium equation for z, we find

$$z = 0.310$$

so that the over-all process is

$$C + 2O_2 \rightarrow 0.69\ CO_2 + 0.31\ CO + 1.155\ O_2$$

The mole fractions of the components in the equilibrium mixture are

$$y_{CO_2} = \frac{0.69}{2.155} = 0.320$$

$$y_{CO} = \frac{0.31}{2.155} = 0.144$$

$$y_{O_2} = \frac{1.155}{2.155} = 0.536$$

The heat transferred from the chamber in this process could be found by the same procedure followed in Example 14.5, considering the over-all process as expressed above.

14.7 Simultaneous Reactions

In developing the equilibrium equation and equilibrium constant expressions of Section 14.6, it was assumed that there was only a single chemical reaction equation relating the substances present in the system. To demonstrate the more general situation in which there is more than one chemical reaction, we will now analyze a system involving two simultaneous reactions by a procedure analogous to that followed in Section 14.6. These results are then readily extended to systems involving several simultaneous reactions.

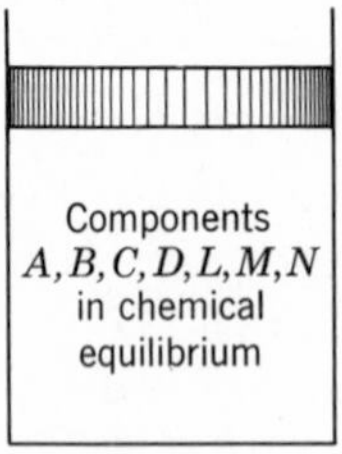

Fig. 14.12 Sketch demonstrating simultaneous reactions.

Consider a mixture of substances $A, B, C, D, L, M,$ and N as indicated in Fig. 14.12. These substances are assumed to exist at a condition of chemical equilibrium at temperature T and pressure P, and are related by the two independent reactions

$$(1)\ \nu_{A_1} A + \nu_B B \rightleftharpoons \nu_C C + \nu_D D \qquad (14.40)$$

$$(2)\ \nu_{A_2} A + \nu_L L \rightleftharpoons \nu_M M + \nu_N N \qquad (14.41)$$

We have considered the situation where one of the components (substance A) is involved in each of the reactions in order to demonstrate the effect of this condition on the resulting equations. As in the previous section, the changes in amounts of the components are related by the various stoichiometric coefficients (which are not the same as the number of moles of each substance present in the vessel). We also realize that the coefficients ν_{A_1}, and ν_{A_2} are not necessarily the same. That is, substance A does not in general take part in each of the reactions to the same extent.

Development of the requirement for equilibrium is completely analogous to that of Section 14.6. We consider that each reaction proceeds an infinitesimal amount toward the right side. This results in a decrease in the number of moles of A, B, and L, and an increase in the moles of C, D, M, and N. Letting the degrees of reaction be ϵ_1 and ϵ_2 for reactions 1 and 2, respectively, the changes in the number of moles are, for infinitesimal shifts from the equilibrium composition,

$$\begin{aligned} dn_A &= -\nu_{A_1} d\epsilon_1 - \nu_{A_2} d\epsilon_2 \\ dn_B &= -\nu_B d\epsilon_1 \\ dn_L &= -\nu_L d\epsilon_2 \\ dn_C &= +\nu_C d\epsilon_1 \\ dn_D &= +\nu_D d\epsilon_1 \\ dn_M &= +\nu_M d\epsilon_2 \\ dn_N &= +\nu_N d\epsilon_2 \end{aligned} \qquad (14.42)$$

The change in Gibbs function for the mixture in the vessel at constant temperature and pressure is

$$dG_{T,P} = \bar{G}_A dn_A + \bar{G}_B dn_B + \bar{G}_C dn_C + \bar{G}_D dn_D + \bar{G}_L dn_L + \bar{G}_M dn_M + \bar{G}_N dn_N$$

Substituting the expressions of Eq. 14.42 and collecting terms,

$$\begin{aligned} dG_{T,P} = {} & (\nu_C \bar{G}_C + \nu_D \bar{G}_D - \nu_{A_1} \bar{G}_A - \nu_B \bar{G}_B) d\epsilon_1 \\ & + (\nu_M \bar{G}_M + \nu_N \bar{G}_N - \nu_{A_2} \bar{G}_A - \nu_L \bar{G}_L) d\epsilon_2 \end{aligned} \qquad (14.43)$$

Assuming an ideal gas mixture, it is convenient to again express each of the partial molal Gibbs functions as

$$\bar{G}_i = \bar{g}_i^\circ + \bar{R}T \ln y_i \frac{P}{P^\circ}$$

Equation 14.43 written in this form becomes

$$dG_{T,P} = \left\{\Delta G_1^\circ + \bar{R}T \ln\left[\frac{y_C^{\nu_C} y_D^{\nu_D}}{y_A^{\nu_{A_1}} y_B^{\nu_B}}\left(\frac{P}{P^\circ}\right)^{\nu_C+\nu_D-\nu_{A_1}-\nu_B}\right]\right\} d\epsilon_1$$

$$+\left\{\Delta G_2^\circ + \bar{R}T \ln\left[\frac{y_M^{\nu_M} y_N^{\nu_N}}{y_A^{\nu_{A_2}} y_L^{\nu_L}}\left(\frac{P}{P^\circ}\right)^{\nu_M+\nu_N-\nu_{A_2}-\nu_L}\right]\right\} d\epsilon_2 \qquad (14.44)$$

In this equation the standard-state change in Gibbs function for each reaction is defined as

$$\Delta G_1^\circ = \nu \bar{g}_C^\circ + \nu_D \bar{g}_D^\circ - \nu_{A_1} \bar{g}_A^\circ - \nu_B \bar{g}_B^\circ \qquad (14.45)$$

$$\Delta G_2^\circ = \nu_M \bar{g}_M^\circ + \nu_N \bar{g}_N^\circ - \nu_{A_2} \bar{g}^\circ - \nu_L \bar{g}_L^\circ \qquad (14.46)$$

Equation 14.44 expresses the change in Gibbs function of the system at constant T, P, for infinitesimal degrees of reaction of both reactions 1 and 2, Eqs. 14.40 and 14.41. The requirement for equilibrium is that $dG_{T,P} = 0$. Therefore, since reactions 1 and 2 are independent, $d\epsilon_1$ and $d\epsilon_2$ can be independently varied. It follows that at equilibrium each of the bracketed terms of Eq. 14.44 must be zero. Defining equilibrium constants for the two reactions by

$$\ln K_1 = -\frac{\Delta G_1^\circ}{\bar{R}T} \qquad (14.47)$$

and

$$\ln K_2 = -\frac{\Delta G_2^\circ}{\bar{R}T} \qquad (14.48)$$

we find that, at equilibrium

$$K_1 = \frac{y_C^{\nu_C} y_D^{\nu_D}}{y_A^{\nu_{A_1}} y_B^{\nu_B}}\left(\frac{P}{P^\circ}\right)^{\nu_C+\nu_D-\nu_{A_1}-\nu_B} \qquad (14.49)$$

and

$$K_2 = \frac{y_M^{\nu_M} y_N^{\nu_N}}{y_A^{\nu_{A_2}} y_L^{\nu_L}}\left(\frac{P}{P^\circ}\right)^{\nu_M+\nu_N-\nu_{A_2}-\nu_L} \qquad (14.50)$$

These expressions must be solved simultaneously for the equilibrium composition of the mixture. The following example is presented to demonstrate and clarify this procedure.

Example 14.7

One mole of H_2O vapor is heated to 3000 K, 1 atm pressure. Determine the equilibrium composition, assuming that H_2O, H_2, O_2, and OH are present.

These are two independent reactions relating the four components of the mixture at equilibrium. These can be written as:

$$(1)\ H_2O \rightleftharpoons H_2 + \tfrac{1}{2}O_2$$

$$(2)\ H_2O \rightleftharpoons \tfrac{1}{2}H_2 + OH$$

Let a be the number of moles of H_2O dissociating according to reaction (1) during the heating, and b the number of moles of H_2O dissociating according to reaction (2). Since the initial composition is 1 mole H_2O, the changes according to the two reactions are

$$\begin{array}{lccc} & (1)\ H_2O & \rightleftharpoons H_2 & +\tfrac{1}{2}O_2 \\ \text{Change:} & -a & +a & +\tfrac{1}{2}a \end{array}$$

$$\begin{array}{lccc} & (2)\ H_2O & \rightleftharpoons \tfrac{1}{2}H_2 & +OH \\ \text{Change:} & -b & +\tfrac{1}{2}b & +b \end{array}$$

Therefore, the number of moles of each component at equilibrium is its initial number plus the change, so that at equilibrium

$$\begin{aligned} n_{H_2O} &= 1 - a - b \\ n_{H_2} &= a + \tfrac{1}{2}b \\ n_{O_2} &= \tfrac{1}{2}a \\ n_{OH} &= b \\ \hline n &= 1 + \tfrac{1}{2}a + \tfrac{1}{2}b \end{aligned}$$

The overall chemical reaction that occurs during the heating process can be written

$$H_2O \rightarrow (1-a-b)H_2O + (a+\tfrac{1}{2}b)H_2 + \tfrac{1}{2}aO_2 + bOH$$

The right-hand side of this expression is the equilibrium composition of the system. Since the number of moles of each substance must necessarily be greater than zero, we find that the possible values of a and b are restricted to

$$\begin{aligned} a &\geqslant 0 \\ b &\geqslant 0 \\ (a+b) &\leqslant 1 \end{aligned}$$

The two equilibrium equations are,

$$K_1 = \frac{y_{H_2} y_{O_2}^{1/2}}{y_{H_2O}} \left(\frac{P}{P^\circ}\right)^{1+1/2-1}$$

$$K_2 = \frac{y_{H_2}^{1/2} y_{OH}}{y_{H_2O}} \left(\frac{P}{P^\circ}\right)^{1/2+1-1}$$

Since the mole fraction of each component is the ratio of the number of moles of the component to the total number of moles of the mixture, these equations can be written in the form,

$$K_1 = \frac{\left(\dfrac{a+\frac{1}{2}b}{1+\frac{1}{2}a+\frac{1}{2}b}\right)\left(\dfrac{\frac{1}{2}a}{1+\frac{1}{2}a+\frac{1}{2}b}\right)^{1/2}}{\left(\dfrac{1-a-b}{1+\frac{1}{2}a+\frac{1}{2}b}\right)} \left(\frac{P}{P^\circ}\right)^{1/2}$$

$$= \left(\frac{a+\frac{1}{2}b}{1-a-b}\right)\left(\frac{\frac{1}{2}a}{1+\frac{1}{2}a+\frac{1}{2}b}\right)^{1/2}\left(\frac{P}{P^\circ}\right)^{1/2}.$$

and

$$K_2 = \frac{\left(\dfrac{a+\frac{1}{2}b}{1+\frac{1}{2}a+\frac{1}{2}b}\right)^{1/2}\left(\dfrac{b}{1+\frac{1}{2}a+\frac{1}{2}b}\right)}{\left(\dfrac{1-a-b}{1+\frac{1}{2}a+\frac{1}{2}b}\right)} \left(\frac{P}{P^\circ}\right)^{1/2}$$

$$= \left(\frac{a+\frac{1}{2}b}{1+\frac{1}{2}a+\frac{1}{2}b}\right)^{1/2}\left(\frac{b}{1-a-b}\right)\left(\frac{P}{P^\circ}\right)^{1/2}$$

giving two equations in the two unknowns a and b, since $P = 1$ atmosphere and the values of K_1, K_2 are known. From Table B.10 at 3000 K, we find

$$K_1 = 0.0457, \; K_2 = 0.0528$$

Therefore, the equations can be solved simultaneously for a and b. The values satisfying the equations are

$$a = 0.1082, b = 0.1080$$

Substituting these values into the expressions for the number of moles of each component and of the mixture, we find the equilibrium mole fractions to be

$$y_{H_2O} = 0.7070$$
$$y_{H_2} = 0.1465$$
$$y_{O_2} = 0.0489$$
$$y_{OH} = 0.0976$$

The methods used in this section can readily be extended to equilibrium systems involving more than two independent reactions. In each case, the number of simultaneous equilibrium equations is equal to the number of independent reactions. Solution of a large set of nonlinear simultaneous equations naturally becomes quite difficult, however, and is not easily accomplished by hand calculations. Instead, solution of these problems is normally made using iterative procedures on a digital computer.

14.8 Ionization

In the final section of this chapter, we shall consider the equilibrium of systems involving ionized gases, or plasmas, a field that has found increasing application and study in recent years. In previous sections we have discussed chemical equilibrium, with a particular emphasis on molecular dissociation, as for example the reaction

$$N_2 \rightleftharpoons 2N$$

which occurs to an appreciable extent for most molecules only at high temperature (of the order of magnitude 3000–10,000 K). At still higher temperatures, such as found in electric arcs, the gas becomes ionized. That is, some of the atoms lose an electron, according to the reaction

$$N \rightleftharpoons N^+ + e^-$$

where N^+ denotes a singly-ionized nitrogen atom (one that has lost one electron and consequently has a positive charge) and e^- represents the free electron. As temperature is increased still further, many of the ionized atoms lose another electron, according to the reaction

$$N^+ \rightleftharpoons N^{++} + e^-$$

and thus become doubly-ionized. As the temperature is increased further, the process continues until a temperature is reached at which all the electrons have been stripped from the nucleus.

Ionization generally is appreciable only at high temperature. However, dissociation and ionization both tend to occur to greater extents at low pressure, and consequently dissociation and ionization may be appreciable in such environments as the upper atmosphere, even at moderate temperature. Other effects such as radiation will also cause ionization, but these effects are not considered here.

The problems involved with analyzing the composition in a plasma become much more difficult than for an ordinary chemical reaction, since in an electric field the free electrons in the mixture do not exchange energy with the positive ions and neutral atoms at the same rate as they do with the field. Consequently, in a plasma in an electric field, the electron gas is not at exactly the same temperature as the heavy particles. However, for moderate fields, it is a reasonable approximation to assume a condition of thermal equilibrium in the plasma, at least for preliminary calculations. Under this condition, we can treat the ionization equilibrium in exactly the same manner as an ordinary chemical equilibrium analysis.

At these extremely high temperatures, we may assume that the plasma behaves as an ideal gas mixture of neutral atoms, positive ions, and electron gas. Thus, for the ionization of some atomic species A,

$$A \rightleftharpoons A^{+} + e^{-} \tag{14.51}$$

we may write the ionization equilibrium equation in the form

$$K = \frac{y_{A^+} y_{e^-}}{y_A} \left(\frac{P}{P^\circ} \right)^{1+1-1} \tag{14.52}$$

The ionization-equilibrium constant K is defined in the ordinary manner as

$$\ln K = -\frac{\Delta G^\circ}{\bar{R}T} \tag{14.53}$$

and is a function of temperature only. The standard-state Gibbs function change for reaction 14.51 is found from

$$\Delta G^\circ = \bar{g}^\circ_{A^+} + \bar{g}^\circ_{e^-} - \bar{g}^\circ_{A} \tag{14.54}$$

The standard-state Gibbs function for each component at the given plasma temperature can be calculated using the procedures of statistical thermodynamics, so that ionization-equilibrium constants can be tabulated as functions of temperature.

Solution of the ionization-equilibrium equation, Eq. 14.52, is then accomplished in the same manner as for an ordinary chemical-reaction equilibrium.

Example 14.8

Calculate the equilibrium composition if argon gas is heated in an arc to 10,000 K, 0.01 atm, assuming the plasma to consist of Ar, Ar^+, e^-. The ionization-equilibrium constant for the reaction

$$Ar \rightleftharpoons Ar^+ + e^-$$

at this temperature is 0.00042.

Consider an initial composition of 1 mole neutral argon, and let z be the number of moles ionized during the heating process. Therefore,

	Ar	$\rightleftharpoons$	Ar^+	$+$	e^-
Initial:	1		0		0
Change:	$-z$		$+z$		$+z$
Equilibrium:	$(1-z)$		z		z

and

$$n = (1-z) + z + z = 1 + z$$

Since the number of moles of each component must be positive, the variable z is restricted to the range

$$0 \leqslant z \leqslant 1$$

The equilibrium mole fractions are

$$y_{Ar} = \frac{n_{Ar}}{n} = \frac{1-z}{1+z}$$

$$y_{Ar^+} = \frac{n_{Ar^+}}{n} = \frac{z}{1+z}$$

$$y_{e^-} = \frac{n_{e^-}}{n} = \frac{z}{1+z}$$

The equilibrium equation is

$$K = \frac{y_{Ar^+} y_{e^-}}{y_{Ar}} \left(\frac{P}{P^\circ}\right)^{1+1-1} = \frac{\left(\dfrac{z}{1+z}\right)\left(\dfrac{z}{1+z}\right)}{\left(\dfrac{1-z}{1+z}\right)} \left(\frac{P}{P^\circ}\right)$$

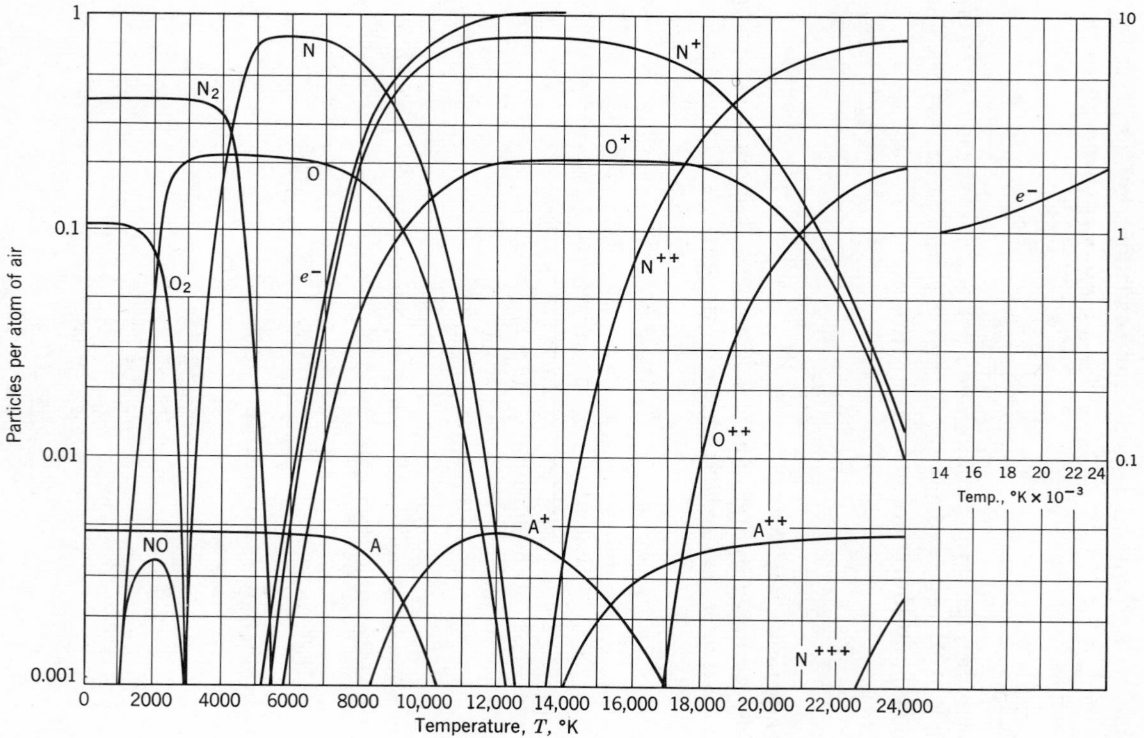

$$\frac{\rho}{\rho_0} = 10^{-6} \left(\rho_0 = 0.0807 \ \frac{\text{lbm}}{\text{ft}^3} \right)$$

Fig. 14.13 Equilibrium composition of air (W. E. Moeckel & K. C. Weston, NACA TN 4265 (1958)).

so that, at 10,000 K, 0.01 atm,

$$0.00042 = \left(\frac{z^2}{1-z^2}\right)(0.01)$$

Solving,

$$z = 0.201$$

and the composition is found to be

$$y_{\mathrm{Ar}} = 0.666$$
$$y_{\mathrm{Ar}^+} = 0.167$$
$$y_{e^-} = 0.167$$

Simultaneous reactions, such as simultaneous molecular dissociation and ionization reactions or multiple ionization reactions, can be analyzed in the same manner as the ordinary simultaneous chemical reactions of Section 14.7. In doing so, we again make the assumption of thermal equilibrium in the plasma, which as mentioned before is, in many cases, a reasonable approximation. Figure 14.13 shows the equilibrium composition of air at high temperature and low density, and indicates the overlapping regions of the various dissociation and ionization processes.

PROBLEMS

14.1 Show that at the triple point of water all three phases have the same value of the Gibbs function.

14.2 Consider a gas well one mile deep filled with pure ethane. The temperature may be assumed to be uniform throughout the well at 100 F. The pressure is measured at the top of the well and found to be 800 lbf/in.^{2}. What is the pressure at the bottom of the well?

14.3 Consider a gas well one mile deep containing a mixture of methane and ethane at a uniform temperature of 100 F. At the top of the well the pressure is 1000 lbf/in.^{2} and the composition is 80% CH_4, 20% C_2H_6 on a mole basis. Assuming the mixture to behave as an ideal gas, determine the pressure and composition at the bottom of the well.

14.4 Determine the phase compositions in the liquid-vapor equilibrium system nitrogen-oxygen at 110 K, 10 atm pressure, and compare the results with the experimental values (50.0% N_2 in the liquid phase, 68.5% N_2 in the vapor phase). At 110 K, the pure-substance saturation pressures are 214 lbf/in.^{2} for N_2 and 78.8 lbf/in.^{2} for O_2.

14.5 It is desired to maintain a liquid-vapor equilibrium mixture of argon and nitrogen at 100 K and a pressure at which the vapor will contain 50 mole per cent of each component. What is this pressure? The saturation pressures for pure argon and nitrogen are 45.7 lbf/in.^{2} and 112.8 lbf/in.^{2} at 100 K.

14.6 A spacecraft breathing mixture of oxygen and helium is to be prepared by the process shown in Fig. 14.14. Helium gas is bubbled through a liquid oxygen storage tank such that a saturated gas mixture of He and O_2 is in equilibrium with the pure liquid O_2 at 200 R and pressure P. The gas is withdrawn, throttled to 10 lbf/in.^{2} and heated to 60 F. Assume that the tank temperature is uniform and constant at 200 R. What are the required tank pressure P and the heat transfer in the heater? The saturation pressure of oxygen at 200 R is 85.3 lbf/in.^{2}.

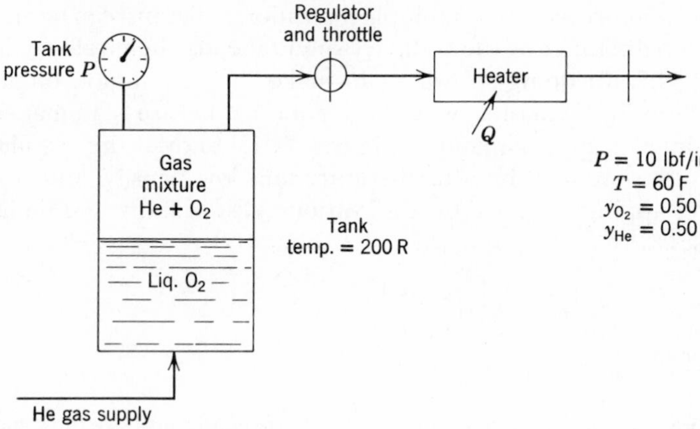

Fig. 14.14 Sketch for Problem 14.6.

14.7 Consider the hydrogen-oxygen fuel cell described in Problem 13.35, in which it was assumed that pure liquid water exits at 25 C, 1 atm.

(*a*) Determine the maximum EMF of the fuel cell operating at 25 C, assuming pure ideal gas water exits at 25 C, 1 atm. What is the significance of this result?

(*b*) Repeat part *a*, considering the fuel cell to operate at a temperature of 600 K. The hydrogen and oxygen each enter at 600 K, 1 atm, and ideal gas water exits at 600 K, 1 atm.

14.8 Calculate the equilibrium constant for the reaction

$$N_2 \rightleftharpoons 2N$$

at temperatures of 537 R and 9000 R.

14.9 Plot to scale the values of $\ln K$ vs. $1/T$ for the reaction $CO_2 \rightleftharpoons CO + \frac{1}{2}O_2$. Write an equation for the $\ln K$ as a function of temperature.

14.10 (*a*) One mole of nitrogen at 77 F, 1 atm pressure is heated to 5000 K in a constant-pressure process. Determine the equilibrium composition at 5000 K, 1 atm pressure, assuming that N_2 and N are the constituents present.
(*b*) Repeat part *a* for a pressure of 0.01 atm.

14.11 Oxygen is heated from room temperature in a steady-flow process at a constant pressure of 10 atm. At what temperature will the mole fraction of O_2 in the mixture of O_2 and O be 40%? (That is, the mole fraction of the O_2 is 40% and of the O, 60%.)

14.12 Catalytic gas generators are frequently used to decompose a liquid, providing a desired gas mixture (spacecraft control systems, fuel cell gas supply, and so forth). Consider pure hydrazine N_2H_4 fed to a gas generator, which yields an equilibrium mixture of N_2, H_2, NH_3 at 200 F, 50 lbf/in.^{2}. Calculate the per cent composition of the mixture. For the reaction:

$$\tfrac{1}{2}N_2 + \tfrac{3}{2}H_2 \rightleftharpoons NH_3$$

$$\ln K = 3.22 \text{ at } 200 \text{ F}$$

14.13 A mixture of 1 mole CO_2, $\frac{1}{2}$ mole CO, and 2 moles O_2 at 77 F and 20 lbf/in.^{2} pressure is heated in a constant pressure steady-flow process to 3000 K. Determine the equilibrium composition at 3000 K on a per cent basis, assuming the reaction equation to be

$$CO_2 \rightleftharpoons CO + \tfrac{1}{2}O_2$$

14.14 Repeat Problem 14.13 for the case where the initial mixture also includes 2 moles N_2, which does not dissociate during the process.

14.15 Calculate the equilibrium constant at 3000 K for the reaction

$$2CO_2 \rightleftharpoons 2CO + O_2.$$

Repeat Problem 14.13 using this reaction equation instead of the one specified in Problem 14.13.

14.16 Hydrogen fuel cells are in operation at the present time, for certain special applications. For most purposes, however, it would be more practical to be able to use a hydrocarbon fuel. One means of doing this is to "reform" the hydrocarbon to obtain H_2, which is then fed to the fuel cell. As a preliminary step in the analysis of such a procedure, consider the reforming reaction

$$CH_4 + H_2O \rightleftharpoons 3H_2 + CO$$

(*a*) Determine the equilibrium constant for this reaction at 980 F.
(*b*) One mole CH_4 and 4 moles H_2O are fed to a catalytic reformer. Assume that an equilibrium mixture of CH_4, H_2O, H_2, and CO leaves at 980 F, 14.7 lbf/in.^{2}. Find the equilibrium composition of this mixture.

14.17 (*a*) One mole of hydrogen at 537 R, 30 lbf/in.², is heated to 5400 R at constant pressure in a steady-state, steady-flow process.

Determine the equilibrium composition at the outlet and the heat transfer per mole of H_2 entering.

(*b*) Plot the equilibrium composition of hydrogen (H_2 and H) as a function of temperature for 1 atm and for 0.01 atm pressure.

14.18 Diatomic fluorine gas (F_2) enters a heat exchanger at 77 F, 100 lbf/in.² at the rate of 1 mole/min. The gas is heated to 3000 R, during which process some of the fluorine dissociates according to the reaction

$$F_2 \rightleftharpoons 2F$$

Calculate:

(*a*) The equilibrium constant for the reaction at 3000 R.

(*b*) The composition of the mixture leaving the heat exchanger at 3000 R, 100 lbf/in.².

(*c*) The heat transfer rate, Btu/min.

	$(\bar{h}_f^\circ)_{537}\left(\frac{\text{Btu}}{\text{mole}}\right)$	$\bar{s}_{537}^\circ\left(\frac{\text{Btu}}{\text{mole-R}}\right)$	$\bar{C}_{po}\left(\frac{\text{Btu}}{\text{mole-R}}\right)$
F_2	0	48.45	$6.5+5.5\times 10^{-4}T$
F	34020	37.92	4.968

14.19 The presence of even small amounts of the various oxides of nitrogen in combustion products is an important factor in the air pollution problem. Consider a mixture comprised of the following basic products of combustion:

CO_2	11 volume %
H_2O	12
O_2	4
N_2	73

At a condition of 2000 K, 2 atm pressure, it is believed that NO (but not NO_2) will also be present in the mixture. Determine the per cent composition of the mixture at this state without using Table B.10.

The following data are known at 2000 K:

$$\tfrac{1}{2}N_2 + O_2 \rightleftharpoons NO_2, \ \Delta G^\circ = 34{,}070 \text{ cal}$$

$$NO + \tfrac{1}{2}O_2 \rightleftharpoons NO_2, \ \Delta G^\circ = 18{,}450 \text{ cal}$$

14.20 A mixture of 40% CO_2, 40% O_2, 20% N_2 (by volume) at 77 F, 25 lbf/in.², is fed to a steady-flow heater at the rate of 1 mole/min. An equilibrium mixture of CO_2, CO, O_2, N_2 leaves the heater at 4500 F, 25 lbf/in.². Determine

(*a*) The composition of the mixture leaving the heater.

(*b*) Heat transfer to the heater/min.

14.21 One mole of carbon monoxide at 1 atm pressure, 1000 K, reacts with one mole of oxygen at 1 atm pressure, 1000 K to form an equilibrium mixture at 2800 K and a total pressure of 1 atm.

(*a*) Determine the heat transfer for this process, assuming that the reaction takes place in a steady-state, steady-flow process.

(*b*) Calculate the irreversibility for this process.

14.22 Methane is to be burned with oxygen in an adiabatic steady-flow process. Both enter the combustion chamber at 537 R, and the products leave at 10 lbf/in.^{2}. What per cent excess O_2 should be used if the flame temperature is to be 5040 R? Assume that some of the CO_2 formed dissociates to CO and O_2, such that the products leaving the chamber consist of CO_2, CO, H_2O, and O_2 at equilibrium.

14.23 One mole of hydrogen at 77 F and 1 atm pressure reacts with 2 moles of oxygen at 77 F and 1 atm pressure in a steady-flow adiabatic process. The products are at 1 atm pressure and consist of an equilibrium mixture of H_2, O_2, and H_2O. Determine the adiabatic flame temperature.

14.24 Calculate the net entropy change for the process described in Problem 14.23.

14.25 Many combustion processes involve products of combustion at temperatures not greater than 2000 F. A gas turbine is a typical example. Usually the effects of dissociation are ignored at these temperatures. Is this justified for CO_2 and H_2O?

14.26 One mole of carbon (solid) and b moles of O_2 are contained in a closed tank. The initial conditions are $P_1 = 25$ lbf/in.^{2}, $T_1 = 80$ F. The carbon is then ignited, and the final state is an equilibrium mixture of CO_2, O_2, and CO at $T_2 = 5300$ F, and the mole fraction of CO_2 is 0.20. Calculate the initial number of moles of O_2, b, and the final pressure, P_2.

14.27 A mixture of 2 moles CO_2, 3 moles O_2, and b moles N_2 at room temperature and 50 lbf/in.^{2} is heated to 3500 F in a steady-flow process. Consider that at the outlet the mixture consists of CO_2, O_2, N_2, CO in equilibrium. Determine the minimum value for b if the mole fraction of CO at the outlet is to be kept below 0.001.

14.28 An important step in the manufacture of chemical fertilizer is the production of ammonia, according to the reaction

$$N_2 + 3H_2 \rightleftharpoons 2NH_3$$

(*a*) Using the data given below, calculate the equilibrium constant for this reaction at 750 R.

(*b*) For an initial composition of 25% N_2, 75% H_2, calculate the equilibrium composition at 750 R, 50 atm pressure.

$$\text{For } NH_3\text{: } (\bar{h}_f^\circ)_{537} = -19{,}872 \frac{\text{Btu}}{\text{mole}}; \ (\bar{g}_f^\circ)_{537} = -7157 \frac{\text{Btu}}{\text{mole}}$$

$$\bar{C}_{po} = 6.086 + 8.812 \times 10^{-3}\text{T} - 1.506 \times 10^{-6}\text{T}^2 \frac{\text{Btu}}{\text{mole-R}}$$

14.29 A gas mixture at 77 F, 15 lbf/in.^{2}, which consists of 50% CO_2 and 50% CO (by volume) is fed to a chemical reactor at the rate of 2 moles/min. Steam from a line at 400 F, 200 lbf/in.^{2} is also fed to the reactor at the rate of 2 moles/min. In the reactor the gases are heated in the presence of a catalyst such that the mixture leaving the reactor at 1340 F, 15 lbf/in.^{2} may be assumed to be in equilibrium according to the reaction

$$CO + H_2O \rightleftharpoons CO_2 + H_2$$

Assume that all the constituents are ideal gases except for the inlet steam. Calculate:

(*a*) The equilibrium constant for this reaction at 1340 F using the values from Table B.10.

(*b*) Enthalpy of the inlet steam relative to the elements at 77 F, 1 atm.

(*c*) Composition (on a mole basis) of the mixture leaving the reactor.

(*d*) Heat transfer to the reactor per minute.

14.30 A mixture of one mole H_2O and one mole O_2 at 400 K is heated to 3000 K, 2 atm pressure, in a steady-state, steady-flow process. Determine the equilibrium composition at the outlet of the heat exchanger, assuming that H_2O, H_2, O_2, and OH are present.

14.31 One lb mole of air (assume 78% N_2, 21% O_2, 1% Ar) at room temperature is heated to 7200 R, 30 lbf/in.^{2} Find the equilibrium composition at this temperature and pressure, assuming that only N_2, O_2, NO, O, and Ar are present.

14.32 Operation of an MHD converter requires an electrically conducting gas. It is proposed to utilize helium gas "seeded" with 1.0 mole per cent cesium, as shown in Fig. 14.15. The cesium is partly ionized ($Cs \rightleftharpoons Cs^+ + e^-$) by heating the mixture to 2500 K, 10 atm in a nuclear reactor, in order to provide free electrons. No helium is ionized in this process, so that the mixture entering the converter consists of He, Cs, Cs^+, e^-. In order to analyze the converter process, it is necessary to know precisely the mole fraction of electrons in the mixture. Determine this fraction. At 2500 K, $\ln K = -13.4$ for the reaction given above.

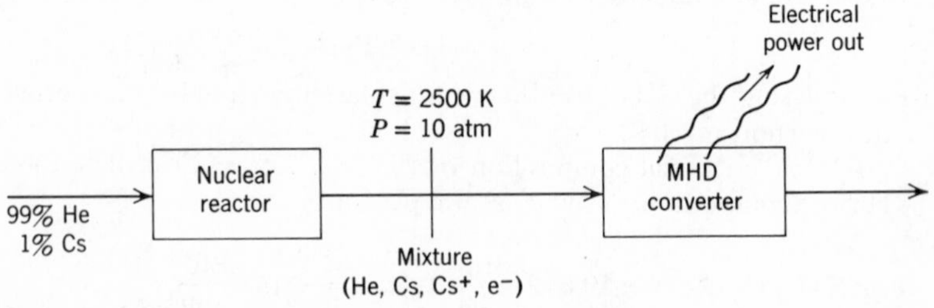

Fig. 14.15 Sketch for Problem 14.32.

14.33 One gm mole of argon at room temperature is heated to 20,000 K, 1 atm pressure. Assume that the plasma at this condition consists of an equilibrium mixture of Ar, Ar^+, A^{++}, e^-, according to the simultaneous reactions

(1) $Ar \rightleftharpoons Ar^+ + e^-$
(2) $Ar^+ \rightleftharpoons Ar^{++} + e^-$

The ionization-equilibrium constants for these reactions at 20,000 K have been calculated from spectroscopic data as: $\ln K_1 = 3.11$; $\ln K_2 = -4.92$. Determine the equilibrium composition of the plasma.

14.34 Plot the equilibrium composition of nitrogen at 0.1 atm pressure (assume N_2, N, N^+, e^{-1} present) from 5000 K to 15,000 K. For the reaction

$$N \rightleftharpoons N^+ + e^-$$

the ionization-equilibrium constant K has been calculated from spectroscopic data as

T(°K)	K
10,000	6.26×10^{-4}
12,000	1.51×10^{-2}
14,000	0.151
16,000	0.92

14.35 Write a computer program to solve the following problem. One mole of carbon at 25 C is to be burned with b moles of oxygen in a constant-pressure, adiabatic process. The products consist of an equilibrium mixture of CO_2, CO, and O_2. It is desired to determine the flame temperature for various combinations of b and the pressure P, assuming variable specific heat for the constituents.

15

Molecular Distributions and Models

The developments and discussions of thermodynamics to this point in the text have been principally from the macroscopic or bulk point of view. In the remaining chapters we reconsider these same fundamental concepts and laws within the framework of the microscopic or molecular viewpoint. Such a task requires an introduction to the fundamentals of mathematical probability and statistics, the concepts of distributions of particles among available states, and the basic statistical models of thermodynamic analysis, which are the topics covered in the present chapter.

15.1 Introduction

In adopting the microscopic viewpoint of thermodynamic systems, we recall from the discussions of Section 2.6 that this necessarily involves a detailed consideration of the energy associated with the particles comprising the system. In that section, we discussed the various modes or general types of energy storage by molecules, without regard for specific values of energy allowable to the molecules in any of these modes. At this point, we must concern ourselves with this question of allowable values of energy.

As discussed in Section 2.6, we shall concern ourselves in this analysis with systems of independent particles (the particles to be analyzed may be electrons, atoms, molecules, photons, etc.). Thus, we may speak of the energy of any individual particle without concern for the influence of other particles in the system. The fundamental assumption of our analysis from this viewpoint is the existence of discrete quantum states available for occupancy by the individual particles. These quantum states should be thought of as being associated with the physical system, and not as properties of the particles. In general, the number of quantum states available is much larger than the number of particles, such that at any instant of time the majority of states are not occupied. Each quantum state is identifiable by a unique set of quantum numbers, and has associated with it a specific value of energy. By this we mean that any particle occupying a certain quantum state must necessarily possess the energy and other characteristics of that state. We may for the present regard each quantum state and its energy as including all possible modes of storage discussed in Section 2.6.

The energy and other characteristics of a quantum state are fixed by the appropriate boundary parameters, those variables restricting the manner and extent of movement of the particle. For example, in the simple case of translation of a particle inside a box, the boundary parameters fixing the energy states available to the particle are the dimensions of the box, since the sole restriction on the movement is that of the walls confining the particle to the interior. If the boundary parameters are changed, the energy value of each quantum state changes accordingly, quite irrespective of whether the state is occupied by a particle. This dependency of energy states on boundary parameters will be developed quantitatively in Chapter 18; for the present it is sufficient to be aware of the existence of this relation.

The basic problem to be analyzed in thermodynamics concerns the manner in which a fixed number of particles with a given total energy tend to distribute themselves among the available quantum states. In general, many of these quantum states have common values of associated energy, such that they can be conveniently grouped according to energy levels. The number of quantum states at each energy value or level is then termed the degeneracy of that level. Thus, the analysis of particle distributions is commonly made over degenerate energy levels instead of over the individual quantum states. Such analysis becomes extremely complex, because in general we are concerned with systems containing billions of particles and even larger numbers of quantum states. As a result, we must rely on certain statistical and averaging techniques in our calculations. In order to develop the appropriate analytical models and techniques, it is necessary for us to introduce the fundamental concepts and terminology of mathematical probability and statistics, and particularly to examine a number of basic combinatorial problems. These topics are considered in the following three sections.

15.2 Mathematical Probability

The subject of mathematical probability has a long history, having been developed formally over the past 300 years. While this subject is one that we all have an intuitive feeling for, there are many seemingly simple problems in which intuition may lead us astray. This is commonly due to a misinterpretation of the possible outcomes of a test or experiment. For example, suppose we toss a pair of fairly weighted coins simultaneously and inquire about the probability that both will be heads. It is perhaps a natural mistake to list the possible outcomes of this experiment as follows: 2 heads; 2 tails; 1 head and 1 tail. Having done so, we would be led to assign a probability of $\frac{1}{3}$ to each of these outcomes. If this experiment is actually performed a number of times, however, it will soon become apparent that the relative frequencies of occurrence of the outcomes listed above approach the values $\frac{1}{4}$, $\frac{1}{4}$, $\frac{1}{2}$. The reason for this apparent paradox is quite simple—the true outcomes of the experiment are four in number: H-H, T-T, H-T, T-H. Having no information to lead us to believe that these are not equally likely, we assign a probability of $\frac{1}{4}$ to each (the last two correspond to 1 head and 1 tail, so that this probability becomes $\frac{1}{2}$).

This simple example leads us to a formal definition of mathematical probability of an event as the limiting value of the relative frequency of occurrence of that event when the experiment is repeated a large number of times. Frequently, it is possible to calculate probabilities *a priori* (by recognizing all the possible outcomes), so that it is unnecessary to actually perform the experiment. For such cases, it is convenient to state the definition of probability as follows.

If there are N mutually exclusive, equally likely outcomes of an experiment, M of which result in a certain event A, then the probability of A is given by the ratio M/N. That is,

$$P(A) = M/N \tag{15.1}$$

The term mutually exclusive means that no two of the N outcomes can occur simultaneously, and will be discussed later in this section.

This probability can be represented pictorially by setting up a sample space as shown in Fig. 15.1. Each of the possible outcomes of an experiment is represented by a point (or unit area) in the space S, which is set up with a uniform point-density. Those outcomes favorable to event A fall inside area A and those resulting in other events fall elsewhere in the total area S. Thus, in accordance with our definition, the probability of event A is the ratio of area A to the total area S.

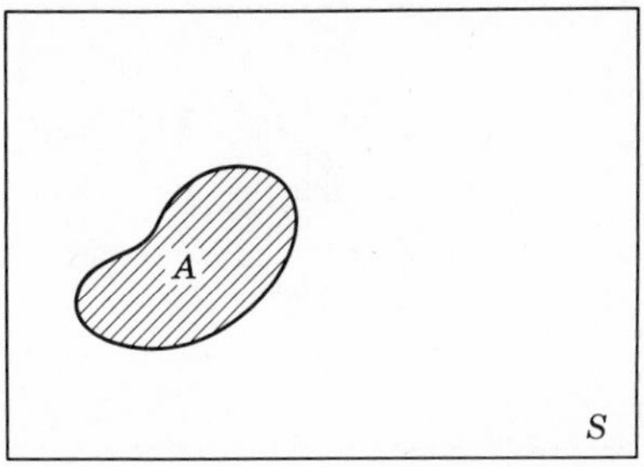

Fig. 15.1 Outcomes of an experiment favorable to event A.

Consistent with our definition of probability are the following:

$$\begin{aligned} P(A) &= 0 \text{ constitutes an impossibility} \\ P(A) &= 1 \text{ constitutes a certainty} \end{aligned} \tag{15.2}$$

so that in general, for any random event A,

$$0 \leqslant P(A) \leqslant 1 \tag{15.3}$$

The discussion above concerns only simple probability, and does not include situations in which we have information about the experiment before it is conducted. This information, which will have some influence on our probability predictions, may possibly be in the form of a restriction placed on the outcome, or it may concern results of trials already completed.

Let us assume that event B depends in some manner on another event A. Of importance in such a case is the conditional probability of B, denoted by the symbol $P(B/A)$, which is to be read as the probability of B given the fact that A has occurred. In order to clarify the difference between simple and conditional probabilities, consider the possible results of an experiment as shown in Fig. 15.2. It is apparent that

$$P(A) = \frac{\text{Area } A}{\text{Area } S}, \qquad P(B) = \frac{\text{Area } B}{\text{Area } S}$$

If we are given the information that A has occurred, then it is known that the result is a point somewhere inside the boundaries of A. The only portion of this area also resulting in event B is the intersection, labeled AB. Thus, we conclude that the conditional probability is given by

$$P(B/A) = \frac{\text{Area } AB}{\text{Area } A}$$

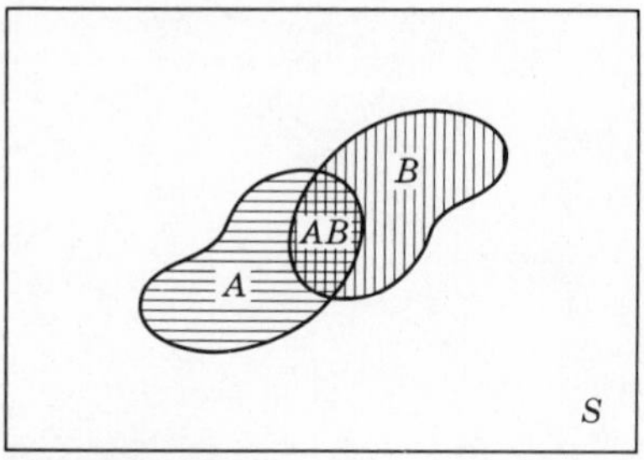

Fig. 15.2 Representation of the intersection of two events.

That is, the information conveyed to us has enabled us to re-evaluate the probability that B will occur, which can be seen by comparison of the area ratios.

Related to this discussion is the topic of the compound probability AB, that is, the probability that both A and B will occur. From Fig. 15.2 we note that the area of intersection is AB, so that in accordance with the definition of probability

$$P(AB) = \frac{\text{Area } AB}{\text{Area } S}$$

It follows from our discussion above and from the products of area ratios that

$$P(AB) = P(A) \times P(B/A) \tag{15.4}$$

and it is similarly apparent that

$$P(AB) = P(B) \times P(A/B) \tag{15.5}$$

Two other definitions are of interest at this point, those of independent events and mutually exclusive events. Event B is said to be independent of event A if

$$P(B/A) = P(B) \tag{15.6}$$

It follows from Eqs. 15.4 and 15.5 that this also requires that event A be independent of B according to a similar equation. Two events A and B are said to be mutually exclusive if they do not intersect, that is, if

$$P(AB) = 0 \tag{15.7}$$

This requirement was used in connection with the definition of probability according to Eq. 15.1.

Example 15.1

A standard deck of playing cards consists of 13 cards in each of 4 suits, 2 of which are red (hearts and diamonds) and 2 of which are black (clubs and spades). Each suit contains the cards 2, 3, 4, . . ., 10, J, Q, K, A. Two cards are drawn from a well-shuffled deck and placed face down on the table.

1. Without looking at either card, determine the probability that the second card drawn is an ace.
2. Suppose the first card drawn is turned over and found to be an ace. How does this knowledge influence the prediction of part 1?
3. What is the probability that both cards drawn are aces?

The solution to this problem is as follows.

1. We are concerned here only with the second card drawn from the deck. The fact that one other card has already been placed apart from the remainder of the deck has no influence on our prediction regarding the second, since we do not know what the first card is. That is, there was no information conveyed by the first draw. Thus, the second card could equally well be any of the cards in the deck, so $N = 52$. Of these, four would result in the desired event B, the draw of an ace on the second draw, giving

$$P(B) = P(\text{ace}) = M/N = 4/52 = 1/13$$

2. In this case, we have received the information that the card of interest to us cannot possibly be the ace drawn first, but could be any of the remaining 51. The number resulting in the desired event B has also been reduced by 1, as one of the four aces is gone. Let event A be the draw of an ace on the first draw; then event B is still the draw of an ace on the second draw. Event A conveys to us the information that there are only 51 possibilities, of which 3 are favorable to event B. Then the probability of event B given event A, $P(B/A)$, is

$$P(B/A) = P(\text{ace/one ace already drawn}) = 3/51 = 1/17$$

3. Defining events A and B as in part 2, we note that

$$P(A) = 4/52 = 1/13$$
$$P(B/A) = 3/51 = 1/17$$

Therefore, from Eq. 15.4, the probability that both cards are aces is

$$P(AB) = P(A) \times P(B/A) = 1/13 \times 1/17 = 1/221$$

15.3 Permutations, Combinations, and Repeated Trials

In the preceding section, we have calculated probabilities in cases for which it was not difficult to determine the number of possible outcomes of an experiment either through intuitive reasoning or by direct counting. It is important at this point to develop analytical means for evaluating more complex situations.

First, let us define a permutation of a group of objects as any specific arrangement or ordering of the members of the group. For example, consider the three letters ABC. There are six arrangements or permutations of the three letters:

$$\begin{matrix} ABC & BAC & CAB \\ ACB & BCA & CBA \end{matrix}$$

In such a problem, it is a simple matter to list all the possibilities, but suppose we asked for the number of permutations of all the letters in the alphabet? It would be an extremely tedious, if not a hopeless, task to attempt to list all the possibilities.

In an effort to develop formulas for calculating permutations, let us analyze what has been done in listing the arrangements of the three letters ABC. In writing down any arrangement, there are three choices for the first letter, either A, B, or C. After this decision has been made, however, there are only two letters remaining, so that there are two possible choices for the second member of the arrangement. Once that has been selected, there is but one letter remaining and therefore only a single choice for the third letter in the arrangement. Thus we find that there are

$$3 \times 2 \times 1 = 6$$

possible arrangements or permutations of the group of three letters. Looking at this problem in terms of sampling, we might alternatively say that it is possible to pick six distinguishable samples of three letters from this group.

By extending this analysis to increasingly more complex examples, we conclude that, for N objects, there are N choices for the first member of any arrangement, $(N-1)$ choices for the second, . . . , and one for the Nth, so that the number of permutations of N objects considered as a group, ${}_N\mathscr{P}_N$, is

$${}_N\mathscr{P}_N = N(N-1)(N-2) \cdots 3 \times 2 \times 1 = N! \tag{15.8}$$

In the symbol used for the number of permutations, the subscript N before the symbol $\mathscr{P}$ denotes the number of objects in the group. The subscript N after the symbol specifies that all N objects were considered together, or we might say that the sample consisted of N objects. We have also introduced the factorial symbol $N!$ defined as the product of all integers from 1 to N. It is of interest to note that

$$N \times (N-1)! = N! \tag{15.9}$$

Consequently, Eq. 15.9 for $N = 1$ defines

$$0! = 1 \tag{15.10}$$

Up to this point, our concern has been only with the possible arrangements of all the objects of which the group is comprised. Suppose we consider now the possibility of picking only a portion of the group. For example, the number of permutations of two letters from the group of four letters $ABCD$ is given by

$$\begin{array}{cccc} AB & BA & CA & DA \\ AC & BC & CB & DB \\ AD & BD & CD & DC \end{array}$$

or a total of 12. Analyzing the process, we find that there are four choices for the first letter and three for the second, giving the total number of possible arrangements

$$4 \times 3 = 12$$

as we found by direct listing.

Extending this reasoning to the general case, we may think of this process as the number of ways of picking a sample of M objects out of a group of N without replacement ($M \leqslant N$), or, in other words, the number of permutations of N objects taken M at a time without repetition. There are N choices for the first object, $(N-1)$ choices for the second, . . ., and $(N-M+1)$ choices for the Mth object, so that

$${}_N\mathscr{P}_M = N(N-1)(N-2) \cdots (N-M+2)(N-M+1) \tag{15.11}$$

where the symbol ${}_N\mathscr{P}_M$ is read as the number of permutations of N objects taken M at a time.

Equation 15.11 can be converted to a more useful form by using the factorial expression

$$N! = N(N-1) \cdots (N-M+1) \times (N-M)! \tag{15.12}$$

Comparing Eq. 15.11 and Eq. 15.12

$$ {}_{N}\mathscr{P}_{M} = \frac{N!}{(N-M)!} \qquad (15.13) $$

We see that, for $M = N$, Eq. 15.13 reduces to Eq. 15.8 (since $0! = 1$), and the two are found to be consistent.

Equation 15.13 may be viewed in the following manner: Of the $N!$ permutations of N objects, there are $(N-M)!$ permutations of the $(N-M)$ objects remaining after M have been selected. Therefore the number of permutations when only M are selected from the group of N, ${}_{N}\mathscr{P}_{M}$, is given by

$$ (N-M)! \times {}_{N}\mathscr{P}_{M} = {}_{N}\mathscr{P}_{N} $$

In these considerations, we have not allowed repetition of symbols, or, in terms of sampling, the sampling has been made without replacement. Suppose that repetition (or replacement) had been permitted. Now, for the problem of selecting two letters from the group *ABCD*, there would be four choices for the first, and also four for the second, or

$$ 4 \times 4 = 4^2 = 16 $$

possible ways of choosing two of the four letters. In addition to the twelve given previously, there would now also be the four doubles *AA*, *BB*, *CC*, *DD*. In general, the number of ways of picking a sample of M objects from a group of N with replacement, or the number of permutations of N objects taken M at a time with repetition, is

$$ {}_{N}\mathscr{P}_{M(\text{rep})} = N^M \qquad (15.14) $$

instead of that given by Eq. 15.13.

There are other aspects to be considered. Suppose that, in the consideration of selecting a sample of M objects from a group of N, we are not interested in the order of the objects within the sample drawn. For example, in the problem of selecting two letters from the group *ABCD*, if order is not considered there are only the six results

AB *BC* *CD*
AC *BD*
AD

instead of the 12 results given previously. That is, all such results as *AB* and *BA* are no longer considered as being different. Since we are now interested only in selections of objects rather than arrangements, we use the terminology "number of combinations" instead of "number of permutations." For the preceding problem the number of combinations of four objects taken two at a time is 6. In a manner similar to our explanation following Eq. 15.13, we conclude that the number of combinations of four letters taken two at a time should be the number of permutations of four objects taken two at a time divided by the 2! permutations of the two objects selected, so as to discount the order within the sample. Then the number of combinations is given by

$$\frac{(4!/2!)}{2!} = \frac{4!}{2!2!} = 6$$

as found above by counting.

The number of combinations of N objects taken M at a time (i.e., the number of ways of choosing a sample of M objects from a group of N objects where the order within the sample is not to be considered) is given by

$$_N\mathscr{C}_M = \frac{_N\mathscr{P}_M}{M!} = \frac{N!}{M!(N-M)!} \tag{15.15}$$

Since the order within the sample of M is not to be considered, the number of combinations is given by the number of permutations of N objects taken M at a time divided by $M!$, the number of permutations of the M objects within the sample.

Example 15.2

In how many ways can two cards be drawn from a standard deck of 52, if (1) order is not considered and (2) order is considered?

1. If the order is not considered, then, from Eq. 15.15, there are

$$_{52}\mathscr{C}_2 = \frac{52!}{2!(52-2)!} = \frac{52 \times 51}{2} = 1326$$

different pairs that can be drawn.

2. If order is to be considered, then, from Eq. 15.13, there are

$$_{52}\mathscr{P}_2 = \frac{52!}{(52-2)!} = 52 \times 51 = 2652$$

possible outcomes, exactly twice that of part (1). Each of the results of part (1) must in this case be considered as two separate results to account for ordering of the pair. That is, 10H − 6C is different from 6C − 10H, where H and C refer to hearts and clubs.

The preceding example must have raised questions about which quantity is of more practical interest, permutations or combinations. This is a problem that must be faced in every analysis and one that should not be taken lightly. The only answer to be given is that each case is individual, and each time we must ask ourselves whether or not ordering is of importance. For example, in one form of the card game of poker, each player is dealt five cards before the game begins. In such a case, the order in which the five cards in a hand were dealt to a player is of no significance, the only thing of importance being the resulting group. On the other hand, in another form of the same game, each player is dealt only two cards at the beginning, and receives the third, fourth, and fifth individually at later times during the course of the game itself. In this case, the order in which the cards are received is certainly of concern to a player, for, as he receives additional information with each card, he attempts to re-evaluate his chances for eventually winning the game. At the end of the game, the ordering has no effect on determining who is the winner, but order does influence the players during the game. A player who did not look at any of his cards until he had received all five would obviously be at a tremendous disadvantage in the long run. We can all, no doubt, think of many additional illustrations in which ordering within a sample or group is either of primary concern or of no importance whatsoever. There are also, of course, many complex situations in which such a decision is not at all obvious.

A combinatorial problem that will be of particular interest to us in analyzing molecular distributions follows from Eq. 15.15. Suppose that we are concerned with dividing a group of N distinguishable objects not into groups of M and (N-M) but into a large number of groups, with M_1 objects in group 1, M_2 objects in group 2, . . . M_k objects in group k, where

$$\sum_{k=1}^{k} M_k = N$$

and where the order of objects within each of the k groups is not to be considered. The number of combinations ${}_N\mathscr{C}_{M_1,\, M_2 \ldots, M_k}$ can be deduced from a repeated application of Eq. 15.14 to each of the groups 1, 2, . . . k-1, with the result

$${}_N\mathscr{C}_{M_1, M_2, \ldots, M_k} = \frac{N!}{M_1!M_2! \ldots M_k!} = \frac{N!}{\prod_k (M_k!)} \tag{15.16}$$

The symbol $\prod_k$ in Eq. 15.16 will be used extensively to denote the continued product over all values of k.

Let us now utilize the results of this section in evaluating probabilities in repeated trials. We consider an experiment with two outcomes, success or failure. A success may be the appearance of heads in the toss of a coin, for example; in this case the appearance of tails constitutes a failure. Or, in tossing a die, the appearance of a 3 may be the success, any of the five other results being considered a failure. For any trial, let the probability of success be p. Then the probability of failure would be $(1-p)$. For N trials, the probability of achieving N successes is p^N, since the trials are independent. Similarly, the probability that all N trials will result in failures is $(1-p)^N$. It follows that the probability of achieving M successes immediately followed by $(N-M)$ failures is

$$p^M \times (1-p)^{N\text{-}M}$$

The probability for M successes and $(N-M)$ failures in any specific order is also given by this expression, since the outcome of each of the N trials is specified as being a success or a failure. However, it may be possible to achieve M successes in N trials in more than one way, and commonly we are not interested in the order in which the success appeared during the experiment, but only in the final number of successes. Thus we should multiply the probability of achieving M successes one way by the number of ways M successes out of N can occur without regard for order (the number of combinations of N objects taken M at a time). Consequently, it follows from Eq. 15.15 that the probability $P(M)$ of achieving M successes in any unspecified order out of N trials is

$$P(M) = \frac{N!}{M!(N-M)!}(p)^M(1-p)^{N-M} \tag{15.17}$$

Example 15.3

A fairly weighted coin is tossed six times in succession. What is the most probable number of heads? What is the probability of the most probable number?

The probability for obtaining heads on any single toss is 1/2. Then, from Eq. 15.17 for $N=6$,

$$P(M) = \frac{6!}{M!(6-M)!}\left(\frac{1}{2}\right)^M\left(\frac{1}{2}\right)^{6-M}$$

where M is the number of heads. If all the possible values of M are evaluated, the probabilities are

M	$P(M)$
0	0.015625
1	0.09375
2	0.234375
3	0.3125
4	0.234375
5	0.09375
6	0.015625

The most probable number of heads is seen to be 3, which has a probability of 0.3125.

15.4 Distribution Functions, Mean Values, and Deviations

Suppose that we analyze the results of the preceding section, and in particular discuss Example 15.3 in detail. The number of heads, M, resulting in the six-coin-toss experiment may be viewed as a variable, one that may take on only the discrete values 0, 1, 2, . . . , 6. If we performed that experiment a great many times, we would expect the results to be distributed with a relative frequency close to the calculated values of $P(M)$. If these values are plotted on a histogram, the probabilities are seen to be symmetric about the most probable result, 3 heads, as shown in Fig. 15.3.

The function $P(M)$ may also be called a distribution function, and the distribution specified by Eq. 15.17 is called the binomial distribution. When the probability function is used for this purpose, the distribution function is said to be normalized, since

$$\sum_M P(M) = 1 \tag{15.18}$$

Suppose that the six-coin-toss experiment is performed R times, where R is a very large number. The number of times that M heads will appear is $R \times P(M)$. The average number of heads occurring per trial, $\langle M \rangle$ which is also called the arithmetic mean value, is the sum of $M \times R \times P(M)$ over all values of M, with the total divided by the number of experiments R. That is,

$$\langle M \rangle = \frac{1}{R} \sum_M M \times R \times P(M) = \sum_M M \times P(M) \tag{15.19}$$

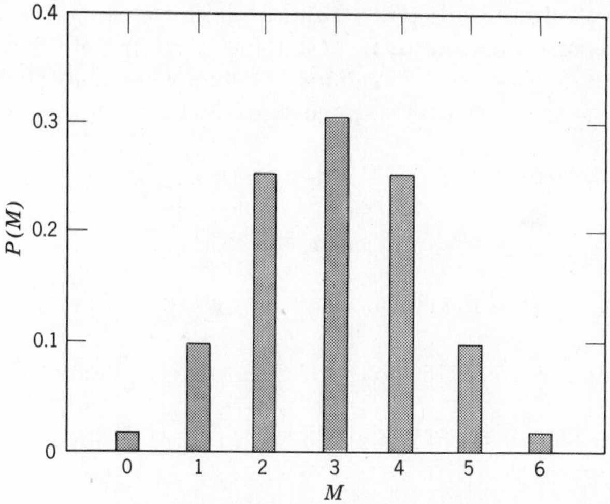

Fig. 15.3 Histogram showing the results of Example 15.3.

so that the mean value of M is independent of the number of trials, assuming that the number is sufficiently large. It can be shown by using Eq. 15.19 that the mean value of M for Example 15.3 is 3.0, in this case the same as the most probable result.

In a similar manner, we could determine the mean value of the quantity M^2, which is designated $\langle M^2 \rangle$, and is found to be

$$\langle M^2 \rangle = \sum_M M^2 \times P(M) \tag{15.20}$$

a quantity that will be found to be of interest to us. Note that the brackets indicates the mean value of the entire quantity inside it. For Example 15.3, the mean value of M^2 is found by Eq. 15.20 to be 10.50, which we note is not the same as the quantity $\langle M \rangle^2$.

Mean values such as those given by Eq. 15.19 or Eq. 15.20 supply some information but do not indicate the shape of the distribution. For this reason, it is necessary to have some knowledge about deviations from the mean value. We consider two such deviations here, first the average deviation from the mean value, which is given by

$$\langle M - \langle M \rangle \rangle = \sum_M (M - \langle M \rangle) \times P(M) = \sum_M M \times P(M) - \langle M \rangle \times \sum_M P(M)$$

Using Eq. 15.18 and Eq. 15.19 and the fact that $\langle M \rangle$ is a constant, this reduces to

$$\langle M - \langle M \rangle \rangle = \langle M \rangle - \langle M \rangle = 0 \tag{15.21}$$

which is consistent with the definition of the mean value. Thus, this deviation is of no help to us in describing the shape of the distribution. The second deviation to be examined is the mean-square deviation, also called the variance, σ^2, which can be expressed as

$$\sigma^2 = \langle (M - \langle M \rangle)^2 \rangle = \sum_M (M - \langle M \rangle)^2 \times P(M)$$
$$= \sum_M (M^2 - 2M\langle M \rangle + \langle M \rangle^2) \times P(M)$$
$$= \sum_M M^2 \times P(M) - 2\langle M \rangle \sum_M M \times P(M) + \langle M \rangle^2 \sum_M P(M)$$

Substituting Eqs. 15.18, 15.19, and 15.20 into this equation yields

$$\sigma^2 = \langle (M - \langle M \rangle)^2 \rangle = \langle M^2 \rangle - 2\langle M \rangle \times \langle M \rangle + \langle M \rangle^2 = \langle M^2 \rangle - \langle M \rangle^2 \tag{15.22}$$

A term commonly used in statistics is the standard deviation, σ, the square root of the variance given in Eq. 15.22. For Example 15.3, the variance is found from Eq. 15.22 to be

$$\sigma^2 = 10.50 - (3.0)^2 = 1.50$$

and the standard deviation is the square root of this number.

Figure 15.4 demonstrates the significance and utility of the standard deviation, σ. Both distributions have the same mean value $\langle M \rangle$. However, the distribution in Fig. 15.4*a* has a small value of σ, while that shown in 15.4*b* has a large σ. Thus the standard deviation (or variance),

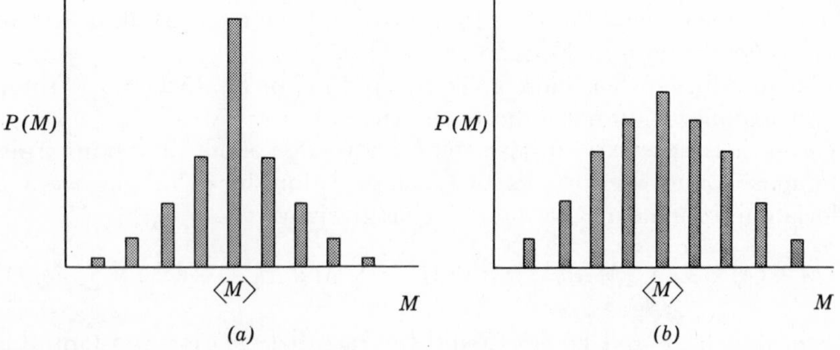

Fig. 15.4 Histograms for distributions with different standard deviations. (*a*) Small value of σ; (*b*) large value of σ.

together with the mean value, gives a meaningful description of the distribution in a variable.

We must keep in mind that, in the majority of practical applications, the approach is necessarily the opposite of that considered so far in this section. Commonly, the *a priori* probabilities and distribution functions are not known, and the role of statistics thus lies in the analysis of a set of experimental data, in terms of calculated mean values and standard deviations. In such analyses, the probabilities in Eqs. 15.19, 15.20, and 15.22 are simply the observed relative frequencies of occurrence, in accordance with our original definition of probability.

A case of particular interest to us is one in which the over-all range of the discrete variable M is very large compared to the unit change in the variable, and in which the probability distribution function $P(M)$ changes by only a small amount for a unit change in the variable. In this case, we may approximate M very closely by a continuous variable x and $P(M)$ by a continuous function $P(x)$. This representation, shown in Fig. 15.5, may be viewed as the limiting case of a histogram like that of Fig. 15.3.

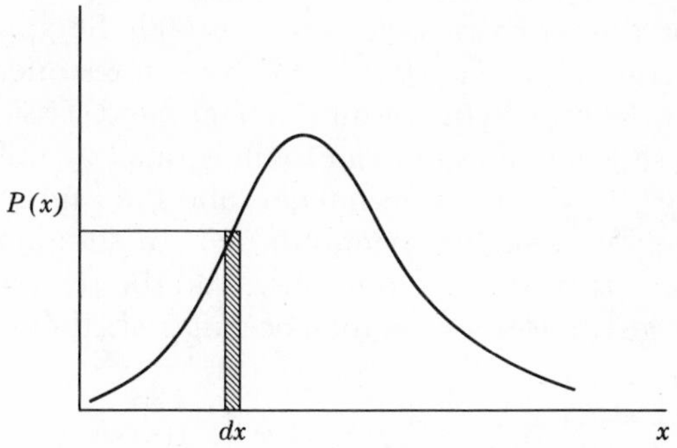

Fig. 15.5 Representation of a continuous distribution function.

Now, for a continuous variable, the probability that the variable will have a value in the small interval between x and $x + dx$ is, as shown in Fig. 15.5,

$$P(x)\, dx$$

and, since the probability is used as the distribution function, the distribution is said to be normalized. That is, when it is integrated over all possible values of x,

$$\int P(x)\, dx = 1 \tag{15.23}$$

Equation 15.23 may be viewed as the limiting case of Eq. 15.18 as the intervals between discrete values of the variable become progressively smaller (relative to the range) and the distribution function may be assumed continuous. In a similar manner, the limiting case of Eq. 15.19 for the mean value of the variable is given by

$$\langle x \rangle = \int xP(x)\,dx \tag{15.24}$$

and, similarly, the subsequent three expressions are

$$\langle x^2 \rangle = \int x^2 P(x)\,dx \tag{15.25}$$

$$\langle x - \langle x \rangle \rangle = 0 \tag{15.26}$$

$$\sigma^2 = \langle (x - \langle x \rangle)^2 \rangle = \langle x^2 \rangle - \langle x \rangle^2 \tag{15.27}$$

These expressions are all analogous to those previously derived, of course, but in this case, we have the obvious mathematical advantage of being able to work with integrals instead of summations. It will be very important for us to understand the conditions under which summations of functions of discrete variables can accurately be approximated by integration of continuous functions (this constitutes one of the fundamental differences between quantum mechanics and classical mechanics). For our purposes, this question can be illustrated as follows: consider some function $f(j)$, where j is an integer, and the function is such that $f(j)$ changes only very slightly from j to $j+1$. In addition, the range of summation over j is from zero to infinity, and the function is such that $f(j) \to 0$ for large j. The question then becomes whether the approximation

$$\sum_{j=0}^{\infty} f(j) = \sum_{j=0}^{\infty} f(j)\Delta j \approx \int_0^{\infty} f(x)\,dx \tag{15.28}$$

is valid. A powerful mathematical tool in providing an answer to this question is the Euler-Maclaurin summation theorem, which is developed in Appendix A.3, and is stated by Eq. A.28.

15.5 Molecular Distributions

As was mentioned briefly in Section 15.1, the basic problem in statistical thermodynamics is the analysis of molecular distributions among degenerate energy levels (quantum states grouped according to values of energy) at conditions of fixed total number of particles and system energy. This is a difficult problem, and is further complicated by the question of an appropriate molecular model. To avoid encountering all the complicating factors at once, let us begin by analyzing an extremely simplified system.

Consider a system comprised of four identical particles that may be distributed among only four possible non-degenerate energy levels. These levels have the relative values of energy ϵ_j of 0, 1, 2, 3 (subscript j used to denote the level) as fixed by the significant boundary parameters; the total energy of the system in the same units is equal to 3. Let us assume that although identical, the particles are distinguishable from one another, and may for our purposes be labeled A, B, C, D. Let us also make the assumption that the individual quantum states are equally likely for any given particle, which is a fundamental hypothesis of statistical mechanics. Since we have already specified non-degenerate energy levels, that is, a single quantum state per level, this requires that the energy levels are equally likely in our example.

We now define a system macrostate as a given distribution of "number of particles" among energy levels, without concern for further detail at the microscopic level (for example, we do not care which particle is in which level). Thus, a macrostate is specified solely by a list of values N_j; in our example this means a set of four numbers N_0, N_1, N_2, N_3. The first question to be resolved, then, concerns the number of macrostates possible in this system, recognizing that the only macrostates allowable are those that satisfy the two system constraints

$$N = \sum_j N_j = 4$$

$$U = \sum_j N_j \epsilon_j = 3$$

It is easily found that there are only 3 possible macrostates, these being labeled I, II, and III and shown in Table 15.1.

Table 15.1
Number of Particles in Each Energy Level

Energy Level ϵ_j	Macrostate I	Macrostate II	Macrostate III
3	1	0	0
2	0	1	0
1	0	1	3
0	3	2	1

If we were to observe the system at any instant of time, it would necessarily be found in one of these three macrostates because of the particle and energy restrictions that were imposed. The next question to be resolved, then, concerns the relative probabilities of each of the three macrostates. To answer this we must examine the system in greater detail at a particle or microscopic level. In macrostate I, for example, one of the particles is in energy level 3 and the remaining three particles are all in level 0. However, since the particles are assumed to be distinguishable from one another, this macrostate can arise from four different microscopic configurations: $A0$, $B0$, $C0$, $D3$ (to be read as particle A in level 0, particle B in level 0, particle C in level 0, particle D in level 3); $A0, B0, D0, C3$; $A0, C0, D0, B3$; and $B0, C0, D0, A3$.

Consequently, it is useful to define another term; a microstate is an exact specification of the microscopic distribution of particles among available quantum states. Thus, a microstate involves total and complete knowledge of the particle configuration, consistent with the model to be used. Each of the four specifications above is therefore a microstate corresponding to macrostate I. We shall in general use the symbol w to denote the number of microstates consistent with a given macrostate. Thus, in our example,

$$\mathrm{w_I} = 4$$

In a similar manner, we can determine all the microstates corresponding to macrostates II and III. These results are listed in Table 15.2.

Table 15.2
All Microstates Corresponding to the Three Macrostates

Macrostate I	Macrostate II	Macrostate III
$A0, B0, C0, D3$	$A0, B0, C1, D2$	$A0, B1, C1, D1$
$A0, B0, D0, C3$	$A0, B0, D1, C2$	$B0, A1, C1, D1$
$A0, C0, D0, B3$	$A0, C0, B1, D2$	$C0, A1, B1, D1$
$B0, C0, D0, A3$	$A0, C0, D1, B2$	$D0, A1, B1, C1$
	$A0, D0, B1, C2$	
	$A0, D0, C1, B2$	
	$B0, C0, A1, D2$	
	$B0, C0, D1, A2$	
	$B0, D0, A1, C2$	
	$B0, D0, C1, A2$	
	$C0, D0, A1, B2$	
	$C0, D0, B1, A2$	

From these results, it is found that

$$w_{II} = 12$$
$$w_{III} = 4$$

Listing all the possible microstates as in Table 15.2 is helpful in understanding what constitutes a microstate, but it is certainly not a practical thing to do in a more complicated example. We recognize, however, that the number of microstates consistent with each macrostate in our problem can be found analytically using Eq. 15.16,

$$w_I = {}_4\mathscr{C}_{3,0,0,1} = \frac{4}{3!0!0!1!} = 4$$

$$w_{II} = {}_4\mathscr{C}_{2,1,1,0} = \frac{4!}{2!1!1!0!} = 12$$

$$w_{III} = {}_4\mathscr{C}_{1,3,0,0} = \frac{4!}{1!3!0!0!} = 4$$

which is in agreement with the values found by direct listing.

The assumption of equal probabilities for individual quantum states for a given particle is consistent with assuming that the individual system microstates are equally likely to be observed at any given instant of time.

In this context, w is a measure of the relative probabilities of the various macrostates, and for this reason is commonly termed the thermodynamic probability of the macrostate. Since w_{II} is larger than either w_I or w_{III}, we conclude that macrostate II is the most probable macrostate for this system at the given conditions. The mathematical probabilities of the three macrostates can be found by normalization over all microstates as

$$P(\text{I}) = \frac{w_I}{w_{TOTAL}} = \frac{4}{20}$$

$$P(\text{II}) = \frac{w_{II}}{w_{TOTAL}} = \frac{12}{20}$$

$$P(\text{III}) = \frac{w_{III}}{w_{TOTAL}} = \frac{4}{20}$$

where w_{TOTAL} represents the total number of microstates for all possible macrostates.

Let us now add a degree of complexity to our problem, and make it somewhat more realistic, by removing the restriction of non-degenerate levels of energy. Suppose that there are three quantum states at energy level 0, three at level 1, four at level 2, and four at level 3. This may be represented schematically as in Fig. 15.6, with boxes at different heights representing the individual quantum states at different levels of energy. Using the symbol g_j to denote the degeneracy of each energy level ϵ_j, we have $g_j = 3, 3, 4, 4$ for $\epsilon_j = 0, 1, 2, 3$, respectively.

Degeneracy g_j	Energy Level ϵ_j	Quantum States			
4	3	31	32	33	34
4	2	21	22	23	24
3	1	11	12	13	
3	0	01	02	03	

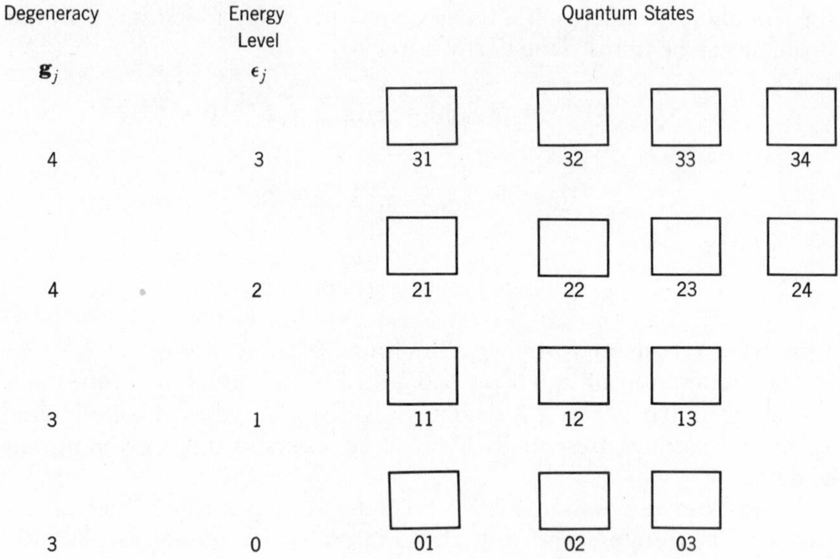

Fig. 15.6 Representation of quantum states for example problem.

In evaluating the molecular distribution problem for this situation, we recognize that allowing the energy levels to be degenerate has no effect on the possible macrostates, which are the same as listed in Table 15.1, I, II and III. However, there is an important effect on the possible microstates since, by definition, a microstate involves an exact specification of particle distribution among quantum states. For example, macrostate I previously could arise from four microstates, one of which was $A0$, $B0$, $C0$, $D3$. Referring to Fig. 15.6, we find that this description no longer gives us a complete specification of the particle configuration and must itself be subdivided according to individual states. One such new microstate is $A01$, $B01$, $C01$, $D31$; others are $A01$, $B01$, $C01$, $D32$; $A01$, $B01$, $C02$, $D31$; etc. It is apparent that there are now a tremendous number of microstates corresponding to what we previously termed microstate $A0$, $B0$, $C0$, $D3$ — so many that we would not care to have to list them. However, this number can be found from applying our combinatorial analysis from Section 15.3 in the following manner. Corresponding to $A0$, $B0$, $C0$, $D3$ there are simultaneously three choices (boxes) for A at level 0, three choices for B at level 0, three for C at level 0, and four for D at level 3, or a total of

$$3 \times 3 \times 3 \times 4 = 108$$

microstates. By the same reasoning, there are 108 microstates each for $A0$, $B0$, $D0$, $C3$; $A0$, $C0$, $D0$, $B3$; and $B0$, $C0$, $D0$, $A3$; our three other microstates of the previous example. Thus, the total number of microstates consistent with macrostate I in this example is

$$w_{\mathrm{I}} = 4 \times 108 = 432$$

We may therefore speak of two types of contributions to microstates for a given macrostate: (1) those associated with which distinguishable particle is in which level, and (2) those associated with the number of ways the particles at each level can be distributed among the quantum states at that level.

By a similar analysis, for each of the 12 previous microstates of macrostate II, there are now simultaneously three choices each for the two specific particles at level 0, three for the specific particle at level 1, and four for the specific particle at level 2, or

$$3 \times 3 \times 3 \times 4 = 108$$

microstates. Consequently, there are now a total of

$$w_{\mathrm{II}} = 12 \times 108 = 1296$$

microstates resulting in macrostate II. In the same manner, we find that for macrostate III, each of the four microstates of the previous example is now subdivided into

$$3 \times 3 \times 3 \times 3 = 81$$

so that macrostate III now results from a total of

$$w_{\mathrm{III}} = 4 \times 81 = 324$$

microstates.

It is found that macrostate II is again the most probable, having the largest value of thermodynamic probability. We also note that macrostates I and III are not equally likely, as was the case previously. The mathematical probabilities for each of these could be found by normalization using w_{TOTAL}, which is 2052 total microstates for all three macrostates. In most problems in thermodynamics, however, the mathematical probabilities of particular macrostates are not of as much interest as are the thermodynamic probabilities and the total number of microstates w_{TOTAL}. In regard to the latter quantity, it is perhaps a subtle point that this system having 2052 possible microstates is a tremendously more complex system than the one having 20 possible microstates. Consequently, our knowledge of the true microscopic configuration in this system at any instant of time is much less. We might say that the degree of microscopic disorder in this system is correspondingly larger, since the system may possibly be in any of its microstates. The quantitative relation of microscopic disorder to the second law of thermodynamics is one of the most important results of a study of thermodynamics from the microscopic viewpoint, as will become evident in succeeding chapters.

In our first example of this section, we found it relatively simple to identify all the macrostates and even all the microstates by direct listing. In the second example, the number of microstates became sufficiently large that we would not care to list all 2052, although it would certainly not be an impossible task. We, of course, realize that a real thermodynamic problem to be analyzed does not involve a system of 4 particles with 4 energy levels each having a degeneracy of 3 to 4. These numbers are each commonly of the order of magnitude of 10^{20} or larger. Thus it becomes unthinkable to consider approaching the problem in the above manner. For this reason, it is necessary to develop a formal molecular model and a formal mathematical approach to the problems of molecular distributions and thermodynamic analysis. The question of molecular models is discussed in the following section, and the formal approach to large-scale systems is developed in Chapter 16.

15.6 Molecular Models

In the preceding section, we analyzed two very simple distribution problems. In these examples we counted microstates by assuming that identical particles are distinguishable (*A*, *B*, *C*, *D*) from one another. This assumption is consistent with the hypotheses of classical mechanics, but not with those of quantum mechanics, in which identical particles are assumed to be fundamentally indistinguishable from one another.

In our examples we also permitted microstates in which two or more particles simultaneously occupy the same quantum state. This is consistent with the hypotheses of classical mechaincs and also in quantum mechanics for some types of particles, but not for others.

It will therefore be necessary for us to develop three separate statistical models for the analysis of particle distributions: one for classical mechanics assuming distinguishable particles and no restriction on the number occupying a quantum state (Boltzmann statistics); and two models for quantum mechanics, assuming indistinguishable particles, one with a maximum of one particle per quantum state (Fermi-Dirac statistics) and the other with no restriction on the number of particles per quantum state (Bose-Einstein statistics). In each case it is presumed that a set of energy levels ϵ_j, with degeneracies g_j, is available to the particles comprising the system. There are N particles in the system, which has an energy of U.

Boltzmann statistics

Since the second example of the preceding section was analyzed using assumptions consistent with Boltzmann statistics (distinguishable particles with no limit on the number of particles per quantum state), it may be helpful for the reader to keep that example and also the schematic representation of Fig. 15.6 in mind.

We consider some particular system macrostate, as specified by a set of numbers N_j, which gives the distribution of particles, in number, among the available energy levels. In seeking the number of microstates consistent with this macrostate, we must consider two factors. First the number of ways in which N distinguishable particles can be divided into groups (levels) with N_1 particles in group 1, N_2 in group 2, . . . , is, according to Eq. 15.16,

$$\frac{N!}{\prod_j (N_j!)}$$

which has already been used in analyzing our previous examples.

We must also consider each energy level ϵ_j, with its number of quantum states g_j and number of particles N_j. Since there are g_j choices for the location of each particle, the number of ways that the N_j particles at a level can be placed in the g_j states of the level is, from Eq. 15.14,

$$g_j^{N_j}$$

Therefore the number of microstates consistent with a given macrostate, or in other words, the thermodynamic probability of the macrostate, is the product of this factor over all levels times the factor for interchanges between levels; i.e., for the Boltzmann statistical model,

$$w = \left(\frac{N!}{\prod_j (N_j!)}\right)\left(\prod_j g_j^{N_j}\right) = N!\prod_j \left(\frac{g_j^{N_j}}{N_j!}\right) \qquad (15.29)$$

Fermi-Dirac statistics

This model assumes that the identical particles comprising the system are indistinguishable from one another, and also that there is a maximum of one particle per quantum state (thus for any macrostate, $N_j \leq g_j$ for each energy level). Of the N particles in the system, for a given macrostate there are N_1 particles in level 1, N_2 in level 2, etc. Since we cannot distinguish which particles are in which levels but only the numbers in each, there is no factor contributing to microstates associated with interchanges of particles between levels, such as was found for Boltzmann statistics by Eq. 15.16.

The only factor to be considered, then, is the number of ways that the N_j particles at each level can be distributed among the g_j states of that level, where $N_j \leq g_j$. The indistinguishability of the particles complicates the analysis somewhat. If the particles were distinguishable, the number of ways of placing N_j particles in the g_j states would be given by Eq. 15.13 as

$$\frac{g_j!}{(g_j - N_j)!}$$

since there are g_j choices for the first particle, $(g_j - 1)$ choices for the second, . . . , $(g_j - N_j + 1)$ choices for the N_j^{th}, as in the analysis of that combinatorial problem. We can use this result if we divide it by the $N_j!$ permutations of the N_j particles to effectively remove their distinguishability from one another, with the result

$$\frac{g_j!}{(g_j - N_j)!N_j!}$$

Therefore, the thermodynamic probability for a given macrostate is the product of this factor over all levels; i.e., for the Fermi-Dirac statistical model,

$$w = \prod_j \left(\frac{g_j!}{(g_j - N_j)!N_j!}\right) \qquad (15.30)$$

Bose-Einstein statistics

The third statistical model of interest to us also assumes, as does Fermi-Dirac statistics, that the particles are indistinguishable; but in this case there is no restriction on the number of particles per quantum state. As in the previous case, we are not concerned with interchanges of particles among levels as contributing to microstates for a given macrostate, but only with the number of ways that the N_j particles at each level can be distributed among the g_j states of that level. This turns out to be a very difficult question to answer for the assumed behavior, and can best be handled by resorting to a schematic example.

Let us consider a group of g_j boxes represented by the symbols B_1, $B_2, B_3, \ldots, B_{g_j}$, and a group of N_j objects to be placed in the boxes. For the moment we assume the objects to be distinguishable from one another and represent them accordingly as $0_1, 0_2, 0_3, \ldots, 0_{N_j}$. Now consider any arrangement of the list of these (g_j+N_j) symbols that is completely arbitrary except that the list begins with the symbol B_1, for example,

$$B_1\ 0_2\ 0_7\ 0_5\ B_5\ B_3\ 0_4\ B_6\ 0_1\ 0_9 \ldots.$$

Let us read this arrangement of symbols as specifying the following particular distribution of objects in boxes: Box 1 contains objects 2, 7, 5 (in that order); box 5 is empty; box 3 contains object 4; box 6 contains objects 1, 9; etc. After the symbol B_1, there are $(g_j+N_j-1)!$ permutations or arrangements of the (g_j+N_j-1) symbols, and consequently that number of possible lists beginning with B_1. However, $(g_j-1)!$ of these lists correspond to the same numbered objects in the same order placed in the same numbered boxes, if the relative positions of the (g_j-1) B's (with their following 0's) are shifted. One such shift would be

$$B_1\ 0_2\ 0_7\ 0_5\ B_3\ 0_4\ B_5\ B_6\ 0_1\ 0_9 \ldots.$$

which gives exactly the same distribution as the preceding list. Therefore, the $(g_j+N_j-1)!$ lists are reduced by the factor $(g_j-1)!$ Now, recall that the objects are in reality identical and indistinguishable, so that all we are concerned with is that there are any 3 objects in no specific order in box 1, no objects in box 5, any single object in 3, etc., so that the number of lists of interest to us must further be reduced by the factor $N_j!$ permutations of the N_j objects. Consequently, the number of ways of placing the N_j indistinguishable objects in the g_j boxes with no restriction on the number per box is

$$\frac{(g_j+N_j-1)!}{(g_j-1)!N_j!}$$

This factor represents the number of ways of placing the N_j particles at each energy level in the g_j states at that level for the assumptions of this model. Consequently, the thermodynamic probability of a given macrostate is the product over all levels; i.e., for the Bose-Einstein statistical model,

$$w = \prod_j \left(\frac{(g_j + N_j - 1)!}{(g_j - 1)!N_j!} \right) \tag{15.31}$$

Example 15.4

A system comprised of four particles has four energy levels with relative energy values of 0, 1, 2, 3. The total energy of the system is 3, and the statistical weights of the four levels are 3, 3, 4, 4, respectively. Determine the thermodynamic probability for each of the possible macrostates, assuming (1) Boltzmann statistics, (2) Fermi-Dirac statistics, and (3) Bose-Einstein statistics.

The possible distributions of particles among energy levels resulting in a total energy of 3 are the same as in the examples considered in Section 15.5. These three macrostates I, II, and III are listed in Table 15.1.

1. For Boltzmann statistics, using Eq. 15.29 for distinguishable particles,

$$w = N! \prod_j \left(\frac{g_j^{N_j}}{N_j!} \right)$$

For the three possible macrostates

$$w_I = 4! \left(\frac{3^3}{3!} \right) \left(\frac{3^0}{0!} \right) \left(\frac{4^0}{0!} \right) \left(\frac{4^1}{1!} \right) = 432$$

$$w_{II} = 4! \left(\frac{3^2}{2!} \right) \left(\frac{3^1}{1!} \right) \left(\frac{4^1}{1!} \right) \left(\frac{4^0}{0!} \right) = 1296$$

$$w_{III} = 4! \left(\frac{3^1}{1!} \right) \left(\frac{3^3}{3!} \right) \left(\frac{4^0}{0!} \right) \left(\frac{4^0}{0!} \right) = 324$$

the same results as were found informally in the analysis of Section 15.5.

2. For Fermi-Dirac statistics, from Eq. 15.30 for indistinguishable particles with a maximum of one per quantum state,

$$w = \prod_j \left(\frac{g_j!}{(g_j - N_j)!N_j!} \right)$$

$$w_I = \left(\frac{3!}{0!3!} \right) \left(\frac{3!}{3!0!} \right) \left(\frac{4!}{4!0!} \right) \left(\frac{4!}{3!1!} \right) = 4$$

$$w_{II} = \left(\frac{3!}{1!2!}\right)\left(\frac{3!}{2!1!}\right)\left(\frac{4!}{3!1!}\right)\left(\frac{4!}{4!0!}\right) = 36$$

$$w_{III} = \left(\frac{3!}{2!1!}\right)\left(\frac{3!}{0!3!}\right)\left(\frac{4!}{4!0!}\right)\left(\frac{4!}{4!0!}\right) = 3$$

3. For Bose-Einstein statistics, from Eq. 15.31 for indistinguishable particles,

$$w = \prod_j \left(\frac{(g_j + N_j - 1)!}{(g_j - 1)!N_j!}\right)$$

$$w_I = \left(\frac{5!}{2!3!}\right)\left(\frac{2!}{2!0!}\right)\left(\frac{3!}{3!0!}\right)\left(\frac{4!}{3!1!}\right) = 40$$

$$w_{II} = \left(\frac{4!}{2!2!}\right)\left(\frac{3!}{2!1!}\right)\left(\frac{4!}{3!1!}\right)\left(\frac{3!}{3!0!}\right) = 72$$

$$w_{III} = \left(\frac{3!}{2!1!}\right)\left(\frac{5!}{2!3!}\right)\left(\frac{3!}{3!0!}\right)\left(\frac{3!}{3!0!}\right) = 30$$

It is found that macrostate II is the most probable according to each of the three statistical models, since that macrostate has the largest thermodynamic probability in each case. It is worth noting that the three models do not predict the same values for either thermodynamic or mathematical probabilities for the allowed macrostates, and consequently must predict a different type of behavior at the molecular level and perhaps at the macroscopic level as well. This must certainly raise the question of whether one of these models is correct and the others incorrect, the answer to which can only come from applying the models to the analysis of thermodynamic systems and comparing the results with experimental observations. This question will be further explored in subsequent chapters.

PROBLEMS

15.1 A box contains 5 white balls and 3 black balls. What is the probability that the second ball drawn from the box is white, if:

(*a*) The color of the first is not known?

(*b*) The first is white?

15.2 There are 10 coins in a box, 4 pennies, 3 nickels, 2 dimes, and 1 quarter. One coin is then drawn out, and its denomination is not disclosed.

(*a*) What is the probability that the second coin drawn will be a nickel?

(*b*) If it is now learned that the first coin drawn is not a penny, what is the probability that the second will be a nickel?

15.3 In a certain manufacturing process it is known from previous experience that 70% of the finished parts are of acceptable quality.

(*a*) If 2 parts are selected at random, what is the probability that at least 1 will be acceptable?

(*b*) How many parts must one select in order to be 99% sure that at least 1 of them will be acceptable?

15.4 A system is comprised of electrical components, each of which is 99.99% reliable. If any component fails, the system fails. Find the probability that the system will operate properly, if the system is comprised of:

(*a*) 10 components.

(*b*) 500 components. (Use the binomial series.)

15.5 In examining trials involving two events A and B, it is frequently of interest to determine the probability that either A or B or both A and B will occur. Show that this probability is given by the expression

$$P(A \text{ and/or } B) = P(A) + P(B) - P(AB)$$

15.6 Two types of missiles are fired at a target. Type A has a probability of 0.3 of scoring a hit, and type B a probability of 0.6. What is the probability that the target will be hit?

15.7 (*a*) How many 7-digit telephone numbers are there, if the only restriction is that the first digit cannot be a 0 or a 1?

(*b*) How many license plates can be made using 2 letters followed by 4 digits? By using 3 letters followed by 3 digits? By using 6 digits?

15.8 How many 3-letter words can be constructed from the letters in the alphabet if no repetition of letters is allowed? How many combinations of letters taken 3 at a time are there? Explain why these results are different.

15.9 Find the number of ways that 4 pennies, 3 nickels, 2 dimes, and 1 quarter can be arranged in a line.

15.10 A shipment of 50 manufactured parts has been received from the factory. It has been decided that, if 10% are defective, the shipment should be returned. Instead of testing all the parts, however, a sample will be taken, and, if every part in the sample is reliable, the entire shipment will be accepted without further tests. Evaluate this testing procedure if the sample consists of:

(*a*) 5 parts.

(*b*) 10 parts.

15.11 A fairly weighted coin is tossed 10 times. Calculate the probabilities for each of the possible numbers of heads. How are these probabilities affected if the coin is weighted to land heads 60% of the time?

15.12 Five distinguishable balls are dropped at random into 3 boxes. What is the probability that 3 balls will be in the first box, 2 in the second, and 0 in the third, if there is no limit on the number per box?

15.13 The experiment of Problem 15.12 is repeated with indistinguishable balls. Is the probability the same?

15.14 As a simple preliminary to the problem of mixing two substances, consider the following experiment. Box I contains 2 white balls, and box II contains 2 black balls. A ball is taken at random from box I and dropped into II, resulting in an "intermediate result" of 1 ball in I and 3 in II. The trial is completed by taking a ball at random from II and dropping it in I. This two-step trial is to be repeated many times.

(*a*) Calculate the probabilities for each of the possible intermediate and final configurations for the first trial.

(*b*) Repeat part (*a*) for the second and third trials.

(*c*) What limits are approached as the number of trials becomes large?

15.15 A system consisting of two particles has available energy levels of 0, 1, 2 units, each level with a degeneracy of 2. The system energy is fixed at 2 units.

(*a*) List the possible macrostates for this system.

(*b*) List the possible microstates for each macrostate, for the three statistical molecular models discussed in Section 15.6.

(*c*) Calculate the values of part *b* from Eqs. 15.29–31.

15.16 Repeat Example 15.4 for the case in which the energy level degeneracies g_j are all equal to 6, instead of the values given in the example.

15.17 Consider a system having non-degenerate ($g_j = 1$) energy levels of 0, 1, 2, 3 units, as in the system analyzed in Section 15.5. This system, however, contains 8 particles and has a total system energy of 6 units. Determine the thermodynamic probability for each of the possible macrostates, assuming distinguishable particles.

15.18 Repeat Problem 15.17 for Boltzmann, Bose-Einstein, and Fermi-Dirac statistics, assuming that each of the energy levels has a degeneracy of 6.

15.19 Repeat Problem 15.17 for the case in which the available non-degenerate energy levels are 0, 1, 2, 3, 4, 5, 6 units.

15.20 Repeat Problem 15.17 for the case in which the system contains 12 particles and has a total system energy of 9 units.

16

Statistical Mechanics

In the preceding chapter, the concept of molecular distributions was developed by dealing with very simplified systems, comprised of a small number of independent particles and having a small number of available quantum states. Even so, it became necessary to develop analytical statistical models for the purpose of calculating the number of microstates corresponding to each macrostate of the system. In this chapter, these results will be extended to the analysis of large-scale systems, for the ultimate purpose of determining thermodynamic properties and behavior. The basic concepts of classical mechanics are introduced, and problems related to molecular velocity distribution and thermal effusion are also considered.

16.1 The Thermodynamic Equilibrium State

The basic problem in statistical thermodynamics is that of specification of the thermodynamic equilibrium state in terms of the appropriate independent variables. Consider as a thermodynamic system a gas enclosed in a rigid vessel, as shown in Fig. 16.1, with independent

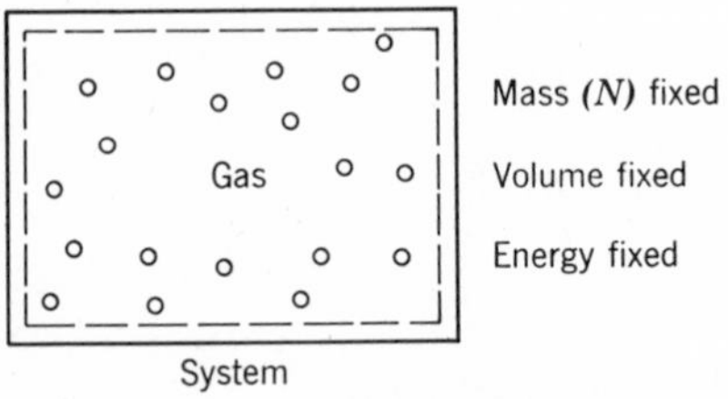

Fig. 16.1 A thermodynamic system.

variables chosen to be U, V, and N. It is assumed that the gas is a pure substance, such that the particles are identical, and also that the density is low, such that the particles may be considered as independent. Specification of the volume V fixes the set of available quantum states, that is, a set of energy levels ϵ_j with corresponding degeneracies g_j. Specification of the system mass N and energy U then places two constraints on the system macrostates (which are given by a set of numbers N_j). Only those macrostates are possible that are consistent with the relations

$$N = \sum_j N_j \tag{16.1}$$

$$U = \sum_j N_j \epsilon_j \tag{16.2}$$

The concept of thermodynamic equilibrium presumes that all the macrostates (and microstates) consistent with these system constraints are accessible to the system, and it is a basic assumption of statistical mechanics that every microstate is equally likely. This means, then, that the macrostates are not equally likely. Since observed behavior at the macroscopic level in general varies with the macrostates, it is of interest to determine the relative probabilities of finding the system in its various macrostates. For this purpose, it is necessary to use an appropriate statistical model and the corresponding expression for thermodynamic probability, as developed in Section 15.6.

We note that this problem is the same as that considered in Example 15.4, although here we presume a large-scale system, such that the numbers of particles N and energy levels ϵ_j (and possibly their degeneracies g_j as well) are now extremely large. As a result, it is no longer possible to directly list the system macrostates, as was done in that example. However, by analyzing simple problems where macrostates can be listed, it is found that the thermodynamic probabilities of a small number of macrostates tend to dominate as the numbers N, ϵ_j, and g_j are made progressively larger, as is depicted schematically in Fig. 16.2. In statistical terminology, the distribution becomes very sharply peaked (a small standard deviation) about the mathematically most-probable macrostate. For the sizes of numbers corresponding to ordinary thermodymanic systems, this sharpness of distribution becomes extreme.

As a result of this behavior we are able to draw a most important conclusion regarding the nature of thermodynamic equilibrium. While presuming that all macrostates are accessible to the system, we conclude that a small number of these, clustered about the most probable, occur in such larger numbers of ways microscopically than all the others that

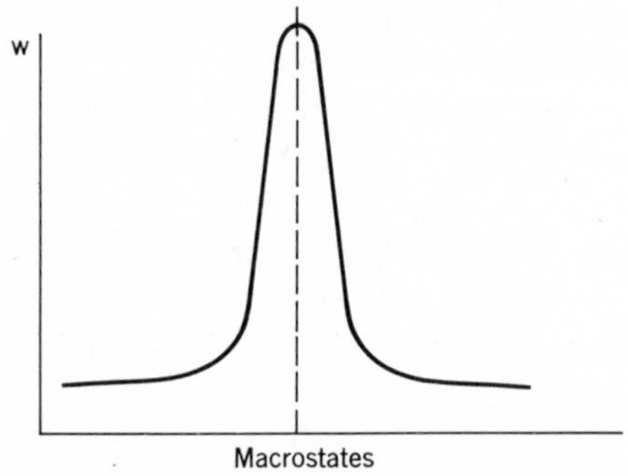

Fig. 16.2 The thermodynamic probability for a large-scale system.

the others are negligible by comparison. In other words, although equilibrium is recognized to be of a dynamic nature, with the system continually passing from one microstate to another, this characteristic is not observed at the macroscopic level, since essentially all the microstates assumed by the system correspond to macrostates that are nearly identical (within experimental limits). Small, short-term fluctuations from this group may be expected, but large or long-term fluctuations would be expected to be extremely unlikely. Thus, in calculating the properties and behavior of a system at a condition of thermodynamic equilibrium, we may utilize the single mathematically most probable macrostate as representative of this small group of macrostates that actually occur.

For a given statistical molecular model (expressing thermodynamic probability for any given macrostate) the method of Lagrange undetermined multipliers enables us to analytically determine the most probable macrostate. Since the system is subject to two mathematical constraints, Eqs. 16.1 and 16.2, this result will be specified in terms of two undetermined multipliers, which must be evaluated by relation of the mathematical results to the physical problem. At this point, we shall simply determine the most probable macrostate, in terms of the undetermined multipliers, for the three molecular models developed in Section 15.6.

Boltzmann statistics

First, for Boltzmann statistics, let us, for convenience, rewrite the distribution equation 15.29 in logarithmic form:

$$\ln \mathrm{w} = \ln N! + \ln \left(\prod_j \frac{\mathrm{g}_j^{N_j}}{N_j!} \right)$$

Since the logarithm of a product of quantities is equal to the sum of the logarithms, it follows that

$$\ln \text{w} = \ln N! + \sum_j (N_j \ln \text{g}_j - \ln N_j!) \tag{16.3}$$

Now, assuming that N is very large and also that the N_j's (for those levels that contribute significantly to w) are very large, the factorials in Eq. 16.3 may be accurately represented by Stirling's formula (Appendix A.1, Eq. A.9). Thus, Eq. 16.3 becomes

$$\ln \text{w} = N \ln N - N + \sum_j (N_j \ln \text{g}_j - N_j \ln N_j + N_j) \tag{16.4}$$

and using Eq. 16.1,

$$\ln \text{w} = N \ln N + \sum_j (N_j \ln \text{g}_j - N_j \ln N_j) \tag{16.5}$$

To find the most probable macrostate, the maximum of w, we differentiate Eq. 16.5 with respect to N_j and set $d(\ln \text{w})$ equal to zero. Since N and all the g_j's are constant, it follows that

$$d(\ln \text{w}) = \sum_j (\ln \text{g}_j \, dN_j - \ln N_j \, dN_j - dN_j) = 0 \tag{16.6}$$

But

$$\sum_j dN_j = dN = 0 \tag{16.7}$$

Therefore, for the maximum of w,

$$d(\ln \text{w}) = \sum_j \ln \left(\frac{\text{g}_j}{N_j}\right) dN_j = 0 \tag{16.8}$$

Furthermore, for a given internal energy of the system,

$$dU = \sum_j \epsilon_j \, dN_j = 0 \tag{16.9}$$

The most probable macrostate for Boltzmann statistics is given by Eq. 16.8 subject to the two constraints, constant N and U, that is, Eqs. 16.7 and 16.9. Consequently, the expression giving the maximum of w can be found by the method of Lagrange multipliers, which is discussed in detail in Appendix A.2. Let us multiply Eq. 16.7 by some arbitrary value α, and Eq. 16.9 by some other constant β. Subtracting (for convenience) Eq. 16.8 from the sum of these results in the expression

$$\sum_j \left(\ln \frac{N_j}{\text{g}_j} + \alpha + \beta\epsilon_j\right) dN_j = 0 \tag{16.10}$$

The sum in Eq. 16.10 is taken over all energy levels, for example

$$j = 1, 2, 3, \ldots, (J-1), J$$

or a total of J levels. However, only $(J-2)$ of the N_j's are independent, because of the two restrictions. If the first two terms of the series in Eq. 16.10 are taken as the dependent terms, the multipliers α, β can then be selected such that the coefficients of dN_1 and dN_2 are identically zero. That is,

$$\ln \frac{N_1}{g_1} + \alpha + \beta\epsilon_1 = 0 \tag{16.11}$$

$$\ln \frac{N_2}{g_2} + \alpha + \beta\epsilon_2 = 0 \tag{16.12}$$

Since the remaining $(J-2)$ variables are independent, dN_3, dN_4, . . . , dN_J can all be varied arbitrarily. Thus the only way that Eq. 16.10 can be satisfied is for each of the coefficients to be identically zero, or

$$\ln \frac{N_j}{g_j} + \alpha + \beta\epsilon_j = 0 \tag{16.13}$$

for all values of j. This is more conveniently written in the form

$$N_j = g_j e^{-\alpha} e^{-\beta\epsilon_j} \tag{16.14}$$

This equation is called the Boltzmann distribution law, which specifies the most probable distribution of particles among energy levels. The distribution according to Eq. 16.14 is seen to be of an exponential form. Therefore, given values of the multipliers α, β, which remain to be determined, the general form of the distribution is that shown in Fig. 16.3. The shape of the curve depends considerably upon the magnitudes of α and β.

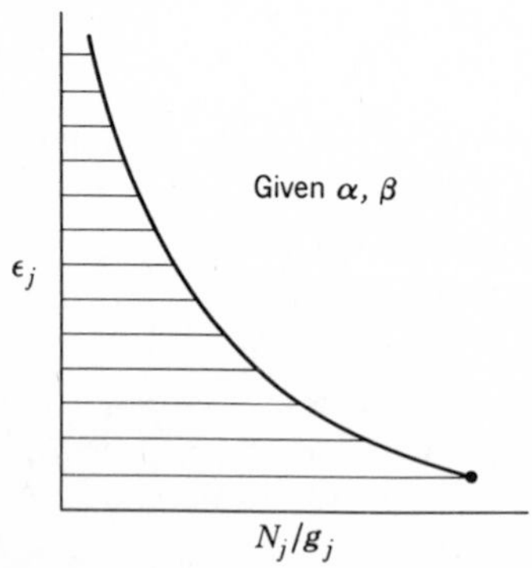

Fig. 16.3 Exponential nature of the Boltzmann distribution function.

Bose-Einstein statistics

For indistinguishable particles, the Bose-Einstein expression for thermodynamic probability w is given by Eq. 15.31. However, it will be found that, for systems of interest,

$$g_j \gg 1 \tag{16.15}$$

so that the 1's appearing in Eq. 15.31 can be neglected. The expression is then written in the logarithmic form

$$\begin{aligned}\ln w &= \ln\left[\prod_j \frac{(g_j+N_j)!}{g_j!N_j!}\right]\\ &= \sum_j\,[\ln\,(g_j+N_j)!-\ln g_j!-\ln N_j!]\end{aligned} \tag{16.16}$$

If it is also assumed that both g_j and N_j are sufficiently large (for those levels that contribute appreciably to w), then Stirling's formula can be used to eliminate the factorials, giving

$$\begin{aligned}\ln w &= \sum_j\,[(g_j+N_j)\ln\,(g_j+N_j)-(g_j+N_j)\\ &\quad -g_j\ln g_j+g_j-N_j\ln N_j+N_j]\\ &= \sum_j\,[(g_j+N_j)\ln\,(g_j+N_j)-g_j\ln g_j-N_j\ln N_j]\end{aligned} \tag{16.17}$$

To determine the most probable macrostate we must find the maximum of Eq. 16.17 subject to the two constraints, fixed N and U, or Eqs. 16.7 and 16.9. Therefore, differentiating Eq. 16.17 and setting $d(\ln w)$ equal to zero, we obtain

$$\begin{aligned}d(\ln w) &= \sum_j\,[dN_j+\ln\,(g_j+N_j)\,dN_j-dN_j-\ln N_j\,dN_j]\\ &= \sum \ln\left(\frac{g_j+N_j}{N_j}\right)dN_j=0\end{aligned} \tag{16.18}$$

Now, if we multiply Eq. 16.7 by some value α and Eq. 16.9 by β, and subtract Eq. 16.18 from their sum, we find

$$\sum_j\left[-\ln\left(\frac{g_j+N_j}{N_j}\right)+\alpha+\beta\epsilon_j\right]dN_j=0 \tag{16.19}$$

As with Boltzmann statistics, two of the N_j's are not independent, but their coefficients can be made identically zero by the proper choice of the multipliers α and β. Since all the remaining dN_j's are arbitrary, the coefficient of each must be zero in order that Eq. 16.19 be satisfied. Therefore

$$-\ln\left(\frac{g_j}{N_j}+1\right)+\alpha+\beta\epsilon_j=0 \tag{16.20}$$

for all values of j, or

$$N_j=\frac{g_j}{e^{\alpha}e^{\beta\epsilon_j}-1} \tag{16.21}$$

for the most probable macrostate. Equation 16.21 is called the Bose-Einstein distribution law and is seen to differ from Eq. 16.14 for the Boltzmann distribution only by the 1 in the denominator.

Fermi-Dirac statistics

Our third statistical model is that for indistinguishable particles with a maximum of one particle per quantum state, the Fermi-Dirac model. Equation 15.30 for Fermi-Dirac statistics is, in logarithmic form,

$$\begin{aligned}\ln w &= \ln\left[\prod_j \frac{g_j!}{(g_j-N_j)!N_j!}\right]\\ &= \sum_j [\ln g_j! - \ln (g_j-N_j)! - \ln N_j!]\end{aligned} \tag{16.22}$$

But, using Stirling's formula, we obtain

$$\begin{aligned}\ln w &= \sum_j [g_j \ln g_j - g_j - (g_j-N_j)\ln (g_j-N_j)\\ &\qquad + (g_j-N_j) - N_j \ln N_j + N_j]\\ &= \sum_j [g_j \ln g_j - (g_j-N_j)\ln (g_j-N_j) - N_j\ln N_j]\end{aligned} \tag{16.23}$$

To find the maximum of w as given by Eq. 16.23, at fixed N, U,

$$\begin{aligned}d(\ln w) &= \sum_j [dN_j + \ln (g_j-N_j)\,dN_j - dN_j - \ln N_j\,dN_j]\\ &= \sum_j \ln\left(\frac{g_j-N_j}{N_j}\right)dN_j = 0\end{aligned} \tag{16.24}$$

As in the previous cases, we multiply Eq. 16.7 by some α and Eq. 16.9 by

β, and subtract Eq. 16.24 from their sum. The result is

$$\sum_j \left[-\ln\left(\frac{g_j - N_j}{N_j}\right) + \alpha + \beta\epsilon_j\right] dN_j = 0 \tag{16.25}$$

The multipliers α and β are so chosen as to make the coefficients of the two dependent N_j's zero. Then, to satisfy Eq. 16.25, the coefficients of all the independent N_j's must also be zero, and

$$-\ln\left(\frac{g_j}{N_j} - 1\right) + \alpha + \beta\epsilon_j = 0 \tag{16.26}$$

for all values of j, or

$$N_j = \frac{g_j}{e^{\alpha}e^{\beta\epsilon_j} + 1} \tag{16.27}$$

This equation specifies the most probable macrostate for Fermi-Dirac statistics.

16.2 Corrected Boltzmann Statistics

In the preceding section, we considered a thermodynamic system of independent particles at given U, V, N, and developed analytical expressions for the most probable macrostate according to our three molecular models. If we accept quantum mechanics as being correct, then the proper model would be either that of Bose-Einstein or Fermi-Dirac, depending on the appropriateness of the one particle per quantum state restriction. Consequently, with this assumption, the Boltzmann model, based on the distinguishable particle assumption of classical mechanics, cannot be valid and would seem to be of no interest to us. However, we shall find shortly that for most gases at low to moderate density (relatively low pressure or high temperature), the number of quantum states available at any level is much larger than the number of particles in that level. That is,

$$\frac{N_j}{g_j} \ll 1 \tag{16.28}$$

for all j. From the distribution equations, 16.21 or 16.27, this is seen to be equivalent to the statement

$$e^{\alpha}e^{\beta\epsilon_j} \gg 1 \tag{16.29}$$

Therefore, under these special conditions, the -1 in the denominator of the distribution equation for Bose-Einstein statistics and the $+1$ for Fermi-Dirac statistics are negligible, and both models predict essentially the same distribution of particles among energy levels as does Boltzmann statistics.

Before concluding that our three models are identical when Eq. 16.28 is valid, however, let us see what each predicts for the thermodynamic probability under these conditions. If Eqs. 16.15 and 16.28 are reasonable, then

$$\frac{(g_j+N_j-1)!}{(g_j-1)!} = g_j(g_j+1)(g_j+2)\cdots(g_j+N_j-1) \approx g_j^{N_j} \tag{16.30}$$

and the thermodynamic probability for a given distribution according to Bose-Einstein statistics given in general by Eq. 15.31 reduces to the form

$$w \approx \prod_j \frac{g_j^{N_j}}{N_j!} \tag{16.31}$$

Similarly, assuming Eqs. 16.15 and 16.28, and also making use of Eq. 15.12,we obtain

$$\frac{g_j!}{(g_j-N_j)!} = g_j(g_j-1)\cdots(g_j-N_j+1) \approx g_j^{N_j} \tag{16.32}$$

Substituting Eq. 16.32 into Eq. 15.30, we see that the thermodynamic probability for a given distribution according to Fermi-Dirac statistics also reduces to Eq. 16.31 under these conditions, as did that for Bose-Einstein statistics.

Comparing Eq. 16.31 with Eq. 15.29 for Boltzmann statistics, we find that, whereas the three models predict essentially the same distribution when Eq. 16.28 is valid, the thermodynamic probability for any given macrostate is different by $N!$ for the Boltzmann statistics, or

$$w_{\text{B-E}} \approx w_{\text{F-D}} \approx \left(\frac{w}{N!}\right)_{\text{Boltz}} \tag{16.33}$$

Of course, dividing the value of w for Boltzmann statistics, which assumed distinguishable particles, by $N!$ has the effect of discounting the distinguishability of the N particles, a procedure which has been discussed in Chapter 15.

We find that, when the assumption of Eq. 16.28 is valid, both Bose-Einstein and Fermi-Dirac statistics reduce to a common statistical model with a distribution given by Eq. 16.14 and a thermodynamic probability given by Eq. 16.31. This model, which will be the one of principal interest to us in the remainder of this text, is termed "corrected" Boltzmann statistics, the name resulting from the comparison in Eq. 16.33.

The molecular distribution for the most probable macrostate has been expressed in terms of the unknown multipliers α and β, according to Eq. 16.14. At this point, it is convenient to examine the relationship between these parameters. If we substitute the distribution equation 16.14 into the system mass constraint, Eq. 16.1,

$$N = \sum_j N_j = e^{-\alpha} \sum_j \mathsf{g}_j e^{-\beta \epsilon_j} \tag{16.34}$$

we note that the resulting expression can be used to eliminate α in terms of β and the basic energy state parameters ϵ_j, g_j. Thus, the summation on the right side of Eq. 16.34 becomes of fundamental importance in statistical thermodynamics, and is termed the partition function Z,

$$Z = \sum_j \mathsf{g}_j e^{-\beta \epsilon_j} \tag{16.35}$$

In words, the partition function is the summation over all energy levels j of the Boltzmann factor $e^{-\beta \epsilon_j}$ times the number of quantum states g_j at that level. Thus, the partition function could alternately be expressed as the summation of the Boltzmann factor over all individual quantum states i

$$Z = \sum_{i(\text{states})} e^{-\beta \epsilon_i} \tag{16.36}$$

if it is found more convenient to sum over individual states than over degenerate energy levels. It will also prove beneficial to our subsequent developments to recognize that the partition function is a fundamental thermodynamic property, expressable as

$$Z = Z(\beta, V) \tag{16.37}$$

since the available energy states are fixed and specified by the system boundary parameters (volume, for a simple compressible substance).

In terms of the definition of the partition function, Eq. 16.34 can be written as

$$e^{-\alpha} = \frac{N}{Z} \tag{16.38}$$

and correspondingly, the distribution equation for the most probable macrostate, Eq. 16.14, can be written in the form

$$N_j = \frac{N}{Z} g_j e^{-\beta \epsilon_j} \tag{16.39}$$

If this result is substituted into the system energy constraint, Eq. 16.2, we obtain

$$U = \sum_j N_j \epsilon_j = \frac{N}{Z} \sum_j g_j \epsilon_j e^{-\beta \epsilon_j} \tag{16.40}$$

which includes a new summation that appears similar to that defining Z. Let us now differentiate Eq. 16.35, keeping in mind the functional relationship given by Eq. 16.37. It is found that

$$\left(\frac{\partial Z}{\partial \beta}\right)_V = -\sum_j g_j \epsilon_j e^{-\beta \epsilon_j} \tag{16.41}$$

Substituting this into Eq. 16.40, we have

$$U = -\frac{N}{Z}\left(\frac{\partial Z}{\partial \beta}\right)_V \tag{16.42}$$

This expression demonstrates the fundamental nature and importance of the partition function in statistical thermodynamics. We find that it is possible to express an important thermodynamic property U in terms of Z, such that evaluation of Z permits a direct calculation of the energy of a system at a given state.

Equation 16.42 is also useful at this point in identifying three areas of statistical thermodynamics that we wish to examine in the remainder of this text. These include: (1) expression of other thermodynamic properties of interest in terms of the partition function; (2) identification of the multiplier β; and (3) evaluation of the partition function from the fundamental energy state parameters (which in general requires a knowledge of quantum mechanics). However, before proceeding to a general discussion of these problems, we shall, in the remainder of this chapter, consider the simple special case of the classical monatomic gas. We note that for this case, the last two of these three areas can be analyzed rather easily.

16.3 The Classical Monatomic Gas

Consider a thermodynamic system as shown in Figure 16.1, in which the state is specified in terms of fixed N, V, and U. We wish to analyze the simplest possible substance, and for that purpose consider the gas to be ideal and monatomic, ground electronic level only, and described by corrected Boltzmann statistics and classical mechanics. The monatomic ideal gas with all atoms in the ground electronic level has only translational energy, which according to classical mechanics, is the familiar kinetic energy, $\frac{1}{2}m\mathsf{V}^2$. (On the atomic and molecular scale, we shall not include the unit conversion constant g_c discussed in Section 2.5, but keep in mind that units must always be consistent.)

For our model, the translational energy is conveniently expressed in terms of velocity components as

$$\epsilon_{\mathsf{V}_x\mathsf{V}_y\mathsf{V}_z} = \tfrac{1}{2}m(\mathsf{V}_x^2 + \mathsf{V}_y^2 + \mathsf{V}_z^2) \qquad (16.43)$$

which is the energy corresponding to a particular set of velocity components V_x, V_y, and V_z. The partition function, Eq. 16.35, can then be rewritten in terms of these variables as

$$Z = \sum_{\text{all}\,\mathsf{V}_x,\mathsf{V}_y,\mathsf{V}_z} \mathsf{g}_{\mathsf{V}_x\mathsf{V}_y\mathsf{V}_z}\, e^{-(\beta m/2)(\mathsf{V}_x^2+\mathsf{V}_y^2+\mathsf{V}_z^2)} \qquad (16.44)$$

In examining this expression, we encounter the real problem at hand. In order to evaluate Eq. 16.44, it is necessary to know what the increments of velocity components are to be, and associated with this, the number of quantum states in the increment. Therefore, we should more properly state that the energy expressed by Eq. 16.43 is that for a specified interval V_x, V_y, V_z to $\mathsf{V}_x + \Delta\mathsf{V}_x$, $\mathsf{V}_y + \Delta\mathsf{V}_y$, $\mathsf{V}_z + \Delta\mathsf{V}_z$, and that $\mathsf{g}_{\mathsf{V}_x\mathsf{V}_y\mathsf{V}_z}$ is the number of quantum states in this interval $\Delta\mathsf{V}_x$, $\Delta\mathsf{V}_y$, $\Delta\mathsf{V}_z$.

For this purpose, we must discuss the concept of a phase space. A single point in this phase space specifies the state, both location and momentum, of a particle. We simplify this discussion by considering only one dimension x, for which the phase space would be two dimensions, x and $m\mathsf{V}_x$. We speak of a point in this two-dimensional space, but realize that we must in fact deal with a unit area $\Delta x \Delta m\mathsf{V}_x$ of some minimum size, in accordance with the Heisenberg Uncertainty Principle. This principle, to be discussed in detail in Chapter 18, establishes a minimum fundamental uncertainty in any simultaneous measurement of position and momentum, as

$$(\Delta x \Delta m\mathsf{V}_x)_{\text{min.}} \sim h \qquad (16.45)$$

in which h is Planck's constant. This then fixes the unit cell size in our phase space, such that a given interval V_x to $\mathsf{V}_x + \Delta\mathsf{V}_x$ is described as shown in Fig. 16.4.

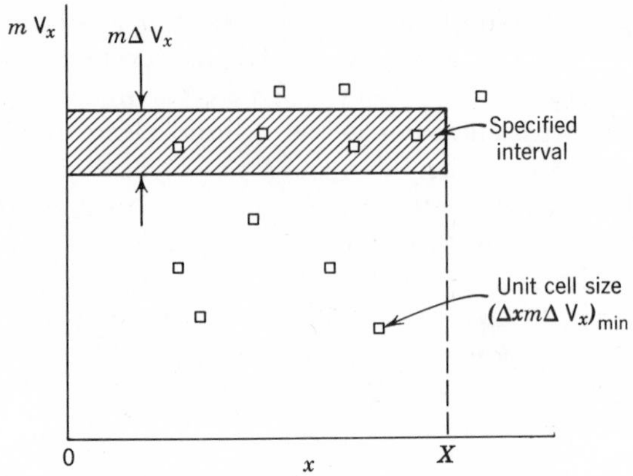

Fig. 16.4 A two-dimensional phase space.

Our interest is in the number of cells in the specified interval V_x to $V_x + \Delta V_x$ without regard for the location of the particle within the tank, i.e., $\Delta x \rightarrow X$. Thus, the number of cells or quantum states in this range is the area of this region of the phase space divided by the unit area, or

$$g_{V_x} = \frac{mX\Delta V_x}{h} \tag{16.46}$$

It follows readily that in our three-dimensional problem, a similar argument involving a six-dimensional phase space with unit cell volume of h^3 results in the conclusion that

$$g_{V_xV_yV_z} = \frac{m^3V}{h^3}\Delta V_x \Delta V_y \Delta V_z \tag{16.47}$$

and the expression for the partition function now becomes

$$Z = \frac{m^3V}{h^3}\sum_{V_x}\sum_{V_y}\sum_{V_z} e^{-(\beta m/2)(V_x^2+V_y^2+V_z^2)}\,\Delta V_x \Delta V_y \Delta V_z \tag{16.48}$$

If we now compare the magnitude of h with typical values for X and m, we see from Eq. 16.45 that our specified interval ΔV_x can be made an extremely small number. Therefore, Eq. 16.48 is, for all practical purposes,

$$Z = \frac{m^3V}{h^3}\int_{V_x}\int_{V_y}\int_{V_z} e^{-(\beta m/2)(V_x^2+V_y^2+V_z^2)}\,dV_x\,dV_y\,dV_z \tag{16.49}$$

in which the integration is to be performed over all possible values, positive and negative, for the components. In addition, it is certainly reasonable to presume that the x, y, z directions are independent, so that Eq. 16.49 becomes

$$
\begin{aligned}
Z &= \frac{m^3 V}{h^3}\left[\int_{\mathsf{V}_s=-\infty}^{+\infty} e^{-(\beta m/2)\mathsf{V}_s^2}\, d\mathsf{V}_s\right]^3 \\
&= \frac{m^3 V}{h^3}\left[\sqrt{\frac{2\pi}{\beta m}}\right]^3 \\
&= V\left(\frac{2\pi m}{\beta h^2}\right)^{3/2} \qquad (16.50)
\end{aligned}
$$

Based on this result, we are now able to evaluate the internal energy for the classical monatomic gas by using Eq. 16.42. Differentiating Eq. 16.50,

$$\left(\frac{\partial Z}{\partial \beta}\right)_V = -\frac{3}{2}V\left(\frac{2\pi m}{h^2}\right)^{3/2}\beta^{-5/2} = -\frac{3}{2}\frac{Z}{\beta}$$

such that from Eq. 16.42,

$$U = -\frac{N}{Z}\left(-\frac{3}{2}\frac{Z}{\beta}\right) = \frac{3N}{2\beta} \qquad (16.51)$$

which establishes the relation of energy to the multiplier β.

We shall now show, in a similar manner, that the internal energy of a monatomic ideal gas is a very simple function of the temperature of the system, and consequently we shall relate the multiplier β to the temperature. Consider a monatomic ideal gas contained in a box of dimensions X, Y, Z, as shown in Fig. 16.5. Consider the atoms striking the wall perpendicular to the x-axis at distance $x = X$ during an interval of time Δt. An atom with a given x-component of velocity V_x and any V_y, V_z must lie within a distance $\mathsf{V}_x\Delta t$ of the wall at the beginning of the time interval (and consequently within a fraction $\mathsf{V}_x\Delta t/X$ of the total volume) in order to strike that wall during the interval. The number of atoms in the entire box with this particular x-component velocity is denoted by N_{V_x}, and the number of these striking the X wall during a unit interval is

$$\frac{\mathsf{V}_x\Delta t}{X}\times N_{\mathsf{V}_x}\frac{1}{\Delta t}$$

at which time each undergoes a change in x-direction momentum of $2m\mathsf{V}_x$. Thus the contribution to the force exerted on the wall by atoms of

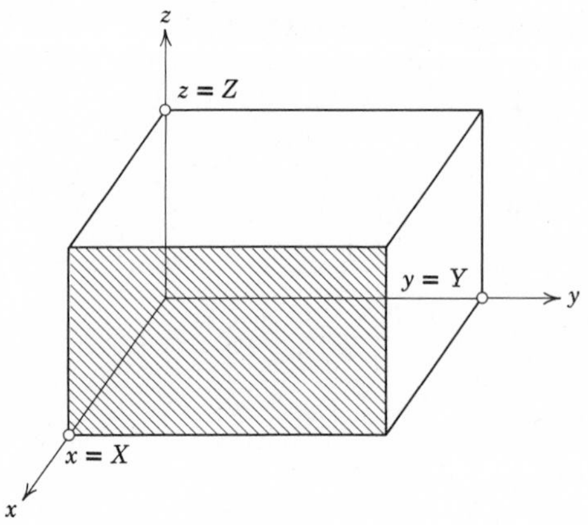

Fig. 16.5 A box of dimensions X, Y, Z confining a monatomic ideal gas.

a particular x-component velocity V_x is

$$\frac{2m\mathsf{V}_x^{\,2}}{X} \times N_{\mathsf{V}_x}$$

and the contribution to pressure is

$$\frac{2m\mathsf{V}_x^{\,2}}{X(YZ)} \times N_{\mathsf{V}_x} = \frac{2m\mathsf{V}_x^{\,2}}{V} N_{\mathsf{V}_x}$$

where V is the volume of the box. The pressure of the gas in the container is that due to the contributions of all V_x greater than zero, and assuming, V_x to be essentially continuous,

$$P = \frac{2m}{V} \int_0^{\infty} \mathsf{V}_x^{2} N_{\mathsf{V}_x}\, d\mathsf{V}_x \tag{16.52}$$

If we assume the gas to be isotropic, that is, that the $+x$- and $-x$-directions are equally probable, then

$$N_{\mathsf{V}_x} = N_{-\mathsf{V}_x} \tag{16.53}$$

and

$$P = \frac{m}{V} \int_{-\infty}^{+\infty} \mathsf{V}_x^{2} N_{\mathsf{V}_x}\, d\mathsf{V}_x \tag{16.54}$$

In accordance with our discussions of Chapter 15 the mean value of the square of the x-component velocity is given by

$$\langle \mathsf{V}_x{}^2 \rangle = \frac{1}{N} \int_{-\infty}^{+\infty} \mathsf{V}_x{}^2 N_{\mathsf{V}_x} d\mathsf{V}_x \tag{16.55}$$

so that Eq. 16.54 can be conveniently written as

$$P = \frac{m}{V} N \langle \mathsf{V}_x{}^2 \rangle \tag{16.56}$$

Furthermore, since the system is isotropic, the x-, y-, and z-directions are all equally probable, and

$$\langle \mathsf{V}_x{}^2 \rangle = \langle \mathsf{V}_y{}^2 \rangle = \langle \mathsf{V}_z{}^2 \rangle = \tfrac{1}{3} \langle \mathsf{V}^2 \rangle \tag{16.57}$$

so that

$$PV = \tfrac{1}{3} m N \langle \mathsf{V}^2 \rangle = \tfrac{2}{3} N \left(\tfrac{1}{2} m \langle \mathsf{V}^2 \rangle \right) = \tfrac{2}{3} U \tag{16.58}$$

The equation of state for an ideal gas is, from Chapter 3,

$$PV = NkT \tag{16.59}$$

where k is the gas constant per atom or molecule, the Boltzmann constant. Combining the last two equations yields

$$U = \tfrac{3}{2} NkT \tag{16.60}$$

which relates the internal energy of a monatomic ideal gas to the temperature.

Now, if we compare the two relations of Eqs. 16.51 and 16.60, we conclude that the multiplier β is inversely proportional to the temperature, and is given by the expression

$$\beta = \frac{1}{kT} \tag{16.61}$$

Although we have shown that Eq. 16.61 is valid for a classical, monatomic ideal gas, it is also possible to demonstrate this relation between β and T under more general conditions. This general case will be developed in Chapter 17, but until that time we shall assume that Eq. 16.61 is valid in general.

From the relation of β to temperature demonstrated above, we may now express the partition function, Eq. 16.50, as

$$\mathrm{Z} = V \left(\frac{2\pi mkT}{h^2} \right)^{3/2} \tag{16.62}$$

or, using the equation of state, Eq. 16.59,

$$Z = N\left(\frac{2\pi m}{h^2}\right)^{3/2}\frac{(kT)^{5/2}}{P} \tag{16.63}$$

Similarly, the distribution equation for the most probable macrostate, Eq. 16.39, becomes

$$N_j = \frac{N}{Z} g_j e^{-\epsilon_j/kT} \tag{16.64}$$

for the classical monatomic gas.

16.4 The Maxwell-Boltzmann Velocity Distribution

Since we have just utilized the assumptions and procedures of classical mechanics to evaluate the partition function for a monatomic ideal gas, let us now extend our consideration to find the corresponding distribution of molecular velocities. These results will have numerous useful applications.

Consider a system comprised of N particles at T, V. We once again consider the most elementary case: corrected Boltzmann statistics, monatomic ideal gas, and classical mechanics. Of the N atoms in the system, a number, dN_{V_x,V_y,V_z}, have velocity components in the interval V_x to $V_x + dV_x$, V_y to $V_y + dV_y$, V_z to $V_z + dV_z$. The statistical weight for the velocity element $dV_x\,dV_y\,dV_z$ was given in Section 16.3, Eq. 16.47:

$$g_{V_x,V_y,V_z} = \left(\frac{m^3V}{h^3}\right) dV_x\,dV_y\,dV_z$$

for the same set of assumptions considered here. Therefore the fraction of atoms, $dN_{V_x,V_y,V_z}/N$ in this velocity element is given by the equilibrium distribution equation (16.64) as

$$\frac{dN_{V_x,V_y,V_z}}{N} = \frac{(m^3V/h^3)\exp\left[-\frac{m}{2kT}(V_x^2+V_y^2+V_z^2)\right]dV_x dV_y dV_z}{Z} \tag{16.65}$$

But, substituting Eq. 16.62 for Z, which was evaluated under the same assumptions as Eq. 16.65, this becomes

$$\frac{dN_{V_x,V_y,V_z}}{N} = \left(\frac{m}{2\pi kT}\right)^{3/2} \exp\left[-\frac{m}{2kT}(V_x^2+V_y^2+V_z^2)\right]dV_x dV_y dV_z \tag{16.66}$$

which is termed the Maxwell-Boltzmann velocity distribution equation.

Instead of using the vector velocity components, it is often more convenient to use the scalar magnitude of velocity or speed of the particles, given by

$$\mathsf{V} = |(\mathsf{V}_x^2 + \mathsf{V}_y^2 + \mathsf{V}_z^2)^{1/2}| \tag{16.67}$$

It is helpful to think of the components V_x, V_y, V_z as the cartesian coordinates of a point on the surface of a sphere of radius V, as shown in Fig. 16.6. Using spherical coordinates, the velocity components can be expressed by

$$\begin{aligned} \mathsf{V}_x &= \mathsf{V} \sin\theta \cos\phi \\ \mathsf{V}_y &= \mathsf{V} \sin\theta \sin\phi \\ \mathsf{V}_z &= \mathsf{V} \cos\theta \end{aligned} \tag{16.68}$$

It is important to keep in mind that, while the components V_x, V_y, V_z can have values from $-\infty$ to $+\infty$, the scalar V can have values only from 0 to $+\infty$. To write Eq. 16.66 in terms of V, we note that an integral over $d\mathsf{V}_x d\mathsf{V}_y d\mathsf{V}_z$ in cartesian coordinates is replaced in spherical coordinates by the integral over $\mathsf{V}^2 \sin\theta\, d\theta d\phi d\mathsf{V}$. Consequently, Eq. 16.66 can be written as the fraction of particles in the interval V, θ, ϕ to $\mathsf{V}+d\mathsf{V}$, $\theta+d\theta$, $\phi+d\phi$

as

$$\frac{dN_{\mathsf{V}\theta\phi}}{N} = \left(\frac{m}{2\pi kT}\right)^{3/2} \mathsf{V}^2 e^{-m\mathsf{V}^2/2kT} \sin\theta\, d\theta d\phi d\mathsf{V} \tag{16.69}$$

which is an alternative form of the Maxwell-Boltzmann velocity distribution equation.

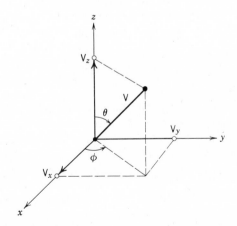

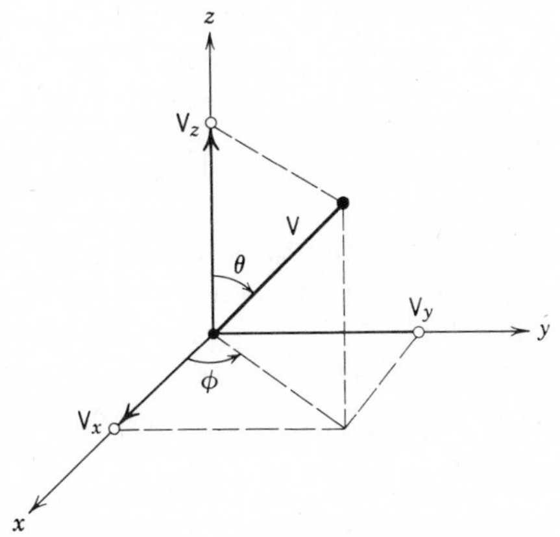

Fig. 16.6 The cartesian and spherical coordinate systems.

Frequently, we are concerned with molecular speeds but not with orientation. That is, we would like to know the fraction of particles in the interval V to V+dV and with any θ, ϕ. This distribution is obtained by integrating Eq. 16.69 over all values of θ and ϕ, and results in

$$\frac{dN_{\mathrm{V}}}{N} = \int_{\phi}\int_{\theta} \frac{dN_{\mathrm{V}\theta\phi}}{N}$$

$$= \left(\frac{m}{2\pi kT}\right)^{3/2} \mathrm{V}^2 e^{-m\mathrm{V}^2/2kT} d\mathrm{V} \int_{\phi=0}^{2\pi}\int_{\theta=0}^{\pi} \sin\theta\, d\theta\, d\phi$$

$$= 4\pi\left(\frac{m}{2\pi kT}\right)^{3/2} \mathrm{V}^2 e^{-m\mathrm{V}^2/2kT}\, d\mathrm{V} \qquad (16.70)$$

A number of useful results can be obtained through integration of this expression, or of Eq. 16.65.

The fraction of particles having an x-component velocity between some V_x and $\mathrm{V}_x + d\mathrm{V}_x$ and any value of V_y, V_z can be found by integrating Eq. 16.65 over all possible values of V_y and V_z. This yields

$$\frac{dN_{\mathrm{V}_x}}{N} = \int_{\mathrm{V}_y}\int_{\mathrm{V}_z} \frac{dN_{\mathrm{V}_x\mathrm{V}_y\mathrm{V}_z}}{N}$$

$$= \left(\frac{m}{2\pi kT}\right)^{3/2}\left[\int_{-\infty}^{+\infty}\int_{-\infty}^{+\infty} \exp\left[-\frac{m}{2kT}(\mathrm{V}_x^2+\mathrm{V}_y^2+\mathrm{V}_z^2)\right] d\mathrm{V}_y\, d\mathrm{V}_z\right] d\mathrm{V}_x$$

$$= \left(\frac{m}{2\pi kT}\right)^{1/2} \exp\left[-\frac{m}{2kT}\mathrm{V}_x^2\right] d\mathrm{V}_x \qquad (16.71)$$

A plot of this function is shown in dimensionless form in Fig. 16.7. It is apparent that the component velocity is symmetrical in the $\pm x$-direction with a maximum at zero. The average value of V_x is given by

$$\langle \mathrm{V}_x \rangle = \left(\frac{m}{2\pi kT}\right)^{1/2} \int_{-\infty}^{+\infty} \mathrm{V}_x \exp\left[-\frac{m}{2kT}\mathrm{V}_x^2\right] d\mathrm{V}_x = 0 \qquad (16.72)$$

The integral is zero because of the symmetry displayed in Fig. 16.7. This symmetry of component velocity and the result of Eq. 16.72 are in agreement with the assumption of Section 16.3 that the gas is isotropic.

Let us return now to the velocity distribution as expressed by Eq. 16.70. The most probable speed V_{mp} is that corresponding to the maximum of that function found from

$$\frac{d}{d\mathrm{V}}\left(\frac{dN_{\mathrm{V}}}{N\, d\mathrm{V}}\right) = 4\pi\left(\frac{m}{2\pi kT}\right)^{3/2} \mathrm{V} \exp\left(-\frac{m\mathrm{V}^2}{2kT}\right)\left[2 - \frac{m\mathrm{V}^2}{kT}\right] = 0 \quad (16.73)$$

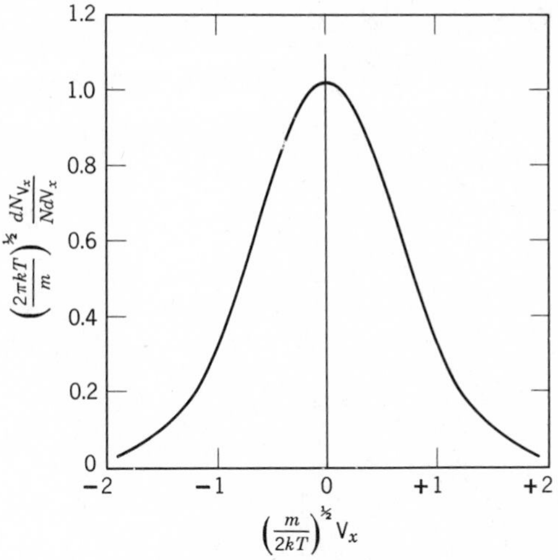

Fig. 16.7 *x*-component Maxwell-Boltzmann velocity distribution.

so that

$$V_{mp} = \left(\frac{2kT}{m}\right)^{1/2} \tag{16.74}$$

On the other hand, the mean speed is given by

$$\begin{aligned} \langle V \rangle &= \int_0^\infty V\left(\frac{dN_V}{N\,dV}\right) dV = 4\pi\left(\frac{m}{2\pi kT}\right)^{3/2} \int_0^\infty V^3 \exp\left(-\frac{mV^2}{2kT}\right) dV \\ &= 4\pi\left(\frac{m}{2\pi kT}\right)^{3/2}\left[2\left(\frac{kT}{m}\right)^2\right] = \left(\frac{8kT}{\pi m}\right)^{1/2} \end{aligned} \tag{16.75}$$

or

$$\langle V \rangle = 1.1284 V_{mp} \tag{16.76}$$

Similarly, the mean-square speed is

$$\begin{aligned} \langle V^2 \rangle &= \int_0^\infty V^2\left(\frac{dN_V}{N\,dV}\right) dV = 4\pi\left(\frac{m}{2\pi kT}\right)^{3/2} \int_0^\infty V^4 \exp\left(-\frac{mV^2}{2kT}\right) dV \\ &= 4\pi\left(\frac{m}{2\pi kT}\right)^{3/2}\left[\frac{3\sqrt{\pi}}{8}\left(\frac{2kT}{m}\right)^{5/2}\right] = \frac{3kT}{m} \end{aligned} \tag{16.77}$$

Thus the mean kinetic energy and, consequently, internal energy are in agreement with that found from Eq. 16.60. Using Eqs. 16.77 and 16.74, the root-mean-square speed is

$$\langle V^2 \rangle^{1/2} = 1.2248 V_{mp} \qquad (16.78)$$

If we now plot the speed distribution of Eq. 16.70, as in Fig. 16.8, we see that the distribution is asymmetric, tending to zero for large V. The values given in Eqs. 16.74, 16.76, 16.78 are also pointed out in this figure.

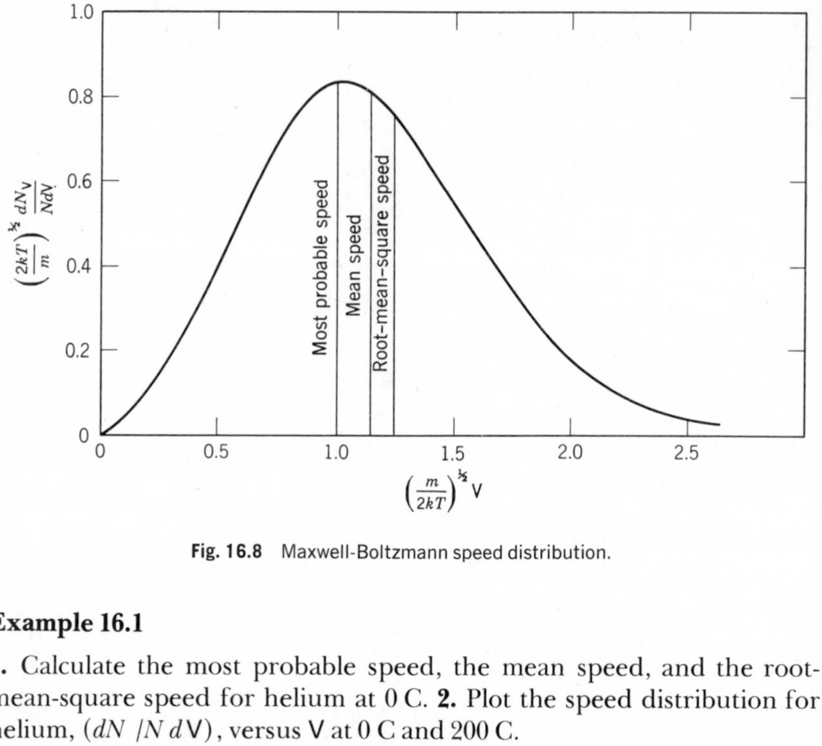

Fig. 16.8 Maxwell-Boltzmann speed distribution.

Example 16.1

1. Calculate the most probable speed, the mean speed, and the root-mean-square speed for helium at 0 C. **2.** Plot the speed distribution for helium, $(dN/N\,dV)$, versus V at 0 C and 200 C.

1. From Table B.11, the mass of the helium atom is

$$m = 1.66 \times 10^{-24} \times (4.003) = 6.645 \times 10^{-24} \text{ gm}$$

From Eq. 16.74, the most probable speed at 0 C is

$$V_{mp} = \left(\frac{2kT}{m}\right)^{1/2} = \left[\frac{2(1.38044 \times 10^{-16})273.15}{6.645 \times 10^{-24}}\right]^{1/2}$$

$$= 10.64 \times 10^4 \text{ cm/sec} = 1064 \text{ meters/sec}$$

From Eq. 16.77, the mean or average speed is

$$\langle \mathsf{V} \rangle = 1.1284 \times (1064) = 1201 \text{ meters/sec}$$

and, from Eq. 16.78, the root-mean-square speed is

$$\langle \mathsf{V}^2 \rangle^{1/2} = 1.2248 \times (1064) = 1303 \text{ meters/sec}$$

2. From Eqs. 16.70 and 16.74,

$$\frac{dN_{\mathsf{V}}}{N\,d\mathsf{V}} = \frac{4}{\sqrt{\pi}\mathsf{V}_{\text{mp}}}\left(\frac{\mathsf{V}}{\mathsf{V}_{\text{mp}}}\right)^2 \exp -\left(\frac{\mathsf{V}}{\mathsf{V}_{\text{mp}}}\right)^2$$

At 0 C,

$$\frac{dN_{\mathsf{V}}}{N\,d\mathsf{V}} = 0.00212\left(\frac{\mathsf{V}}{1064}\right)^2 \exp -\left(\frac{\mathsf{V}}{1064}\right)^2$$

At 200 C,

$$\mathsf{V}_{\text{mp}} = 1064 \times \left(\frac{473.15}{273.15}\right)^{1/2} = 1400 \text{ meters/sec}$$

and

$$\frac{dN_{\mathsf{V}}}{N\,d\mathsf{V}} = 0.00161\left(\frac{\mathsf{V}}{1400}\right)^2 \exp -\left(\frac{\mathsf{V}}{1400}\right)^2$$

Plots of these distributions are shown in Fig. 16.9. We note that, as

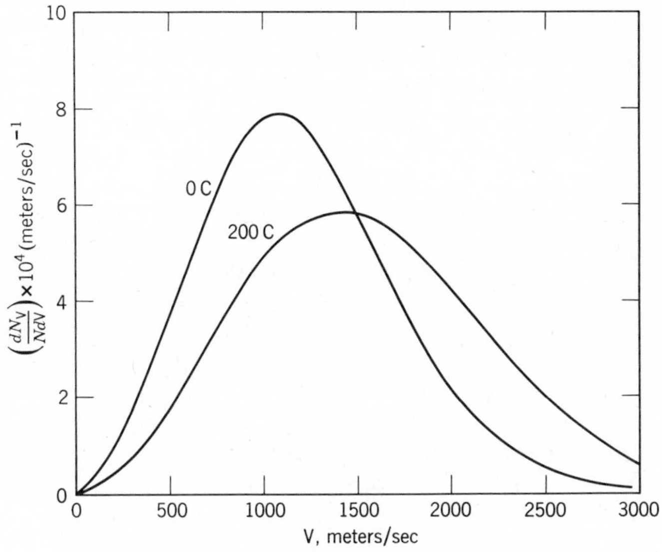

Fig. 16.9 Speed distributions for Example 16.1.

the temperature is increased, the distribution becomes broader, more atoms shifting to the higher energy levels (higher velocities). There is, of course, a resulting increase in internal energy of the gas.

16.5 Free Molecular Flow

In this section, we examine a class of problems involving molecular flow that is readily analyzed using the techniques developed in the present chapter.

Let us consider a vessel containing a gas in a certain thermodynamic state (N, V, T). One wall of the vessel contains a hole, such that molecules of the gas may flow out of the vessel. If this hole is large relative to molecular dimensions, then a certain portion of the molecules striking the hole can be expected to collide with other molecules and be deflected. If, however, the hole is sufficiently small, then it may be expected that nearly every molecule striking the hole will pass freely through. The characteristics of these two types of flow can be expected to be quite different.

Our interest here is in the second case, the free molecular flow, for which we must focus our attention on two questions, namely, how many particles strike the hole in a given time, and what constitutes a sufficiently small hole for the assumption of free molecular flow to be reasonable? To answer the first question, we may develop an analysis similar to that leading to the concept of pressure in Section 16.3. Referring to a gas contained in a vessel as shown in Fig. 16.5 of that development, we let dN_{V_x} be the number of molecules having an x-component velocity in the interval V_x to $\mathsf{V}_x + d\mathsf{V}_x$, and any value of V_y and V_z. In this case, we are concerned not with the number of particles of this component striking the entire X-wall (area YZ) but only a certain area A on that wall. Therefore, this number is

$$\frac{\text{No. of } \mathsf{V}_x \text{ striking } A}{\text{sec.}} = \frac{\mathsf{V}_x \Delta t}{X} dN_{\mathsf{V}_x}, \frac{1}{\Delta t}\frac{A}{YZ}$$

$$= \frac{\mathsf{V}_x A}{V} dN_{\mathsf{V}_x},$$

If we now assume that the Maxwell-Boltzmann velocity distribution is valid, and substitute Eq. 16.71 for dN_{V_x},

$$\frac{\text{No. of } \mathsf{V}_x \text{ striking } A}{\text{sec.}} = \frac{NA}{V}\left(\frac{m}{2\pi kT}\right)^{1/2} \mathsf{V}_x e^{-m\mathsf{V}_x^2/2kT} d\mathsf{V}_x$$

and integrate over all values of V_x with which molecules can strike this area from inside,

$$\frac{\text{Total No. striking } A}{\text{sec.}} = \frac{NA}{V}\left(\frac{m}{2\pi kT}\right)^{1/2} \int_0^\infty V_x e^{-mV_x^2/2kT} dV_x$$

$$= \tfrac{1}{4}\frac{N}{V}A\left(\frac{8kT}{\pi m}\right)^{1/2} = \tfrac{1}{4}\frac{N}{V}A\langle V\rangle \qquad (16.79)$$

We find that the number of particles striking the hole depends on the number density N/V of particles inside the vessel and on their mean speed $\langle V\rangle$ as given by Eq. 16.75, in addition to the size of the hole.

In regard to the second question, we must determine the mean free path Λ of a molecule, its average distance traveled between collisions. First, we recognize that during a time Δt, a molecule travels an average distance equal to $\langle V\rangle\Delta t$. Thus, the problem becomes one of finding the average number of collisions suffered by a molecule during Δt. Since only an order of magnitude of Λ is required for our purposes, we shall not analyze this problem in detail, but merely attempt to justify the result. If we consider the molecules as rigid spheres of diameter σ, then each molecule presents a target size, or collision cross-section of $\pi\sigma^2$ to other molecules of the same diameter. The average number of collisions should be proportional to this value, and also to the number density of particles and to their average speeds. Therefore, the mean free path becomes

$$\Lambda = \frac{\langle V\rangle\Delta t}{C\frac{N}{V}\langle V\rangle\pi\sigma^2\Delta t} = \frac{1}{C\frac{N}{V}\pi\sigma^2} \qquad (16.80)$$

Using the Maxwell-Boltzmann velocity distribution, Eq. 16.69, and analyzing the characteristics of relative motion of particles in a system, the value of the constant in Eq. 16.80 is found to be $C = \sqrt{2}$.

Example 16.2

Calculate the mean free path for N_2 at 30 C, 1 atm pressure. Use a molecular diameter $\sigma = 4$ Å.

From Eq. 16.59,

$$\frac{N}{V} = \frac{P}{kT} = \frac{1.013 \times 10^6}{1.38 \times 10^{-16} \times 303.15} = 2.43 \times 10^{19}\frac{1}{\text{cm}^3}$$

$$\pi\sigma^2 = \pi(4 \times 10^{-8})^2 = 50.3 \times 10^{-16}\ \text{cm}^2$$

Therefore, from Eq. 16.80,

$$\Lambda = \frac{1}{\sqrt{2}\frac{N}{V}\pi\sigma^2} = \frac{1}{\sqrt{2}(2.43\times 10^{19})(50.3\times 10^{-16})}$$

$$= 5.8\times 10^{-6}\ \text{cm} = 580\ \text{Å}$$

We are now in a position to state that if the hole in the vessel wall is considerably larger than the mean free path of the molecules inside the vessel, the flow may be expected to be of the ordinary macroscopic, or bulk flow type. If the hole is considerably smaller, say an order of magnitude, than Λ, the flow may be expected to be of the free molecule type, as given by Eq. 16.79. Between these extremes the problem becomes more difficult, and, for the most part, empirical in nature.

PROBLEMS

16.1 Stirling's approximation for $N!$ (Eq. A.9) is used in the determination of the most probable macrostate in Section 16.1. Compare the value given by this expression with the correct value, for $N = 10$ and for $N = 20$ (note $20! = 2.432\times 10^{17}$).

16.2 The method of Lagrange undetermined multipliers is used to determine the most probable macrostate in Section 16.1. As a simple example of this technique, consider the following problem: A rectangle has a base x and height y, where $x + y = 8$. Use the method of Lagrange multipliers to find the values of x and y giving the maximum area.

16.3 A system contains 3000 particles and has an energy of 4100 units. There are three non-degenerate energy levels, of energies 1, 2, and 3 units, respectively.

(*a*) One macrostate consistent with the given constraints is $N_1 = 2000$, $N_2 = 900$, $N_3 = 100$. Shifts of one particle from level 2 to 1 and one from 2 to 3 or shifts of one particle from level 1 to 2 and one from 3 to 2 represent small changes in both directions from the given macrostate. Calculate the ratio of thermodynamic probabilities for each of these small changes. Is the stated macrostate the equilibrium distribution that is consistent with the constraints?

(*b*) Determine analytically the equilibrium distribution for this system. Then use the particle-shifting procedure of part (*a*) to demonstrate that it is the equilibrium distribution.

16.4 Develop the expression for the most probable macrostate for Bose-Einstein statistics, for the case in which the simplification of Eq. 16.15 is not valid.

16.5 A two-dimensional ideal gas is a useful model for certain applications (for example, in analyzing the behavior of atoms adsorbed on a surface but still free to move in two directions). Determine the "pressure", force per unit length, for such a two-dimensional gas in terms of the energy and area.

16.6 A monatomic ideal gas is contained in a cylinder fitted with a piston, as shown in Fig. 16.10. The piston is then moved very slowly toward the right at velocity V_p. An atom with velocity V striking the moving piston at angle θ rebounds with velocity V′ and at angle θ'. Determine the relations between these quantities. Then show that the rate of decrease in energy of the particle can be expressed as $(2m\mathsf{V}\cos\theta)\mathsf{V}_p$, and that for the system as $P(dV/dt)$.

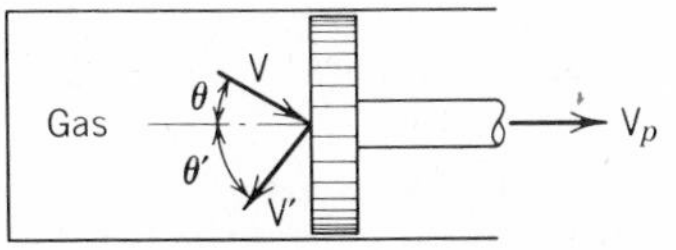

Fig. 16.10 Sketch for Problem 16.6.

16.7 The behavior of helium is accurately represented by Bose-Einstein statistics. As an indication of the applicability of the "corrected" Boltzmann model at different temperatures, evaluate the quantity e^{α} for helium at 1 atm pressure and 5, 50, and 500 K, assuming classical mechanics. What conclusions can be drawn from the results?

16.8 Consider a monatomic gas comprised of electrons (which is properly represented by Fermi-Dirac statistics). In certain cases, the behavior of the electron gas can be closely approximated by corrected Boltzmann statistics. Is this a reasonable assumption for a temperature of 1000 K and a number density $N/V = 10^{10}$ electrons/cm^3? Explain your answer.

16.9 Assuming a Maxwell-Boltzmann velocity distribution, compare the average speeds of helium and xenon at 300 K and at 3000 K.

16.10 Consider a monatomic ideal gas. Show that the average positive x-component of velocity is

$$\langle \mathsf{V}_x \rangle = \left(\frac{kT}{2\pi m}\right)^{1/2} \qquad (\mathsf{V}_x \geqslant 0 \text{ only})$$

and also that

$$\langle \mathsf{V}_x^2 \rangle = \frac{kT}{m}$$

16.11 Consider a monatomic gas having a Maxwell-Boltzmann velocity distribution.

(*a*) Show that the corresponding energy distribution is given by

$$\frac{dN_\epsilon}{N} = \frac{2}{\sqrt{\pi}(kT)^{3/2}} \epsilon^{1/2} e^{-\epsilon/kT}\, d\epsilon$$

where dN_ϵ is the number of particles having energy in the interval ϵ to $\epsilon + d\epsilon$.

(*b*) For a system having the energy distribution of part (*a*), calculate the most probable energy and the average energy.

16.12 (*a*) For a gas with a Maxwell-Boltzmann velocity distribution, show that the fraction of atoms in the system having a speed between 0 and V can be expressed as

$$\frac{N_{0\text{ to }\mathsf{V}}}{N} = \text{erf}\,(x) - \frac{2}{\sqrt{\pi}} x e^{-x^2}$$

where $x = \mathsf{V}/\mathsf{V}_{mp}$.

(*b*) Calculate the fraction of atoms in argon gas at 300 K with speeds less than 500 meters/sec at any instant of time.

16.13 At what temperature will 70% of the atoms in a helium-filled tank have speeds greater than 2000 meters/sec?

16.14 Show that it is not necessary to assume a Maxwell-Boltzmann velocity distribution in the development of Eq. 16.79. (The result of Problem 16.10 may prove helpful.)

16.15 Consider a wall of a tank containing a monatomic gas. Taking x, y coordinates in the plane of the wall and z as the inward-directed normal, show that the distribution of particles striking the surface can be expressed in spherical coordinates as

$$\frac{\text{number }(\mathsf{V}, \theta, \phi)}{\text{cm}^2\text{-sec}} = \frac{N}{V}\left(\frac{m}{2\pi kT}\right)^{3/2} \mathsf{V}^3 e^{-(m/2kT)\mathsf{V}^2} \cos\theta \sin\theta\, d\theta\, d\phi\, d\mathsf{V}$$

16.16 (*a*) Consider a gas contained in volume V and having an arbitrary speed distribution dN_V (the number of particles in an interval V to V$+d$V, and any θ, ϕ). Using the result of Problem 16.15, show that the rate at which particles of speed V strike a unit area of wall surface is given by

$$\frac{1}{4}\frac{\mathsf{V}}{V} dN_\mathsf{V}$$

(*b*) Derive eq. 16.79 using the result of part *a*.

16.17 Neon gas is contained in a cubic one-liter tank at 25 C, 1 atm pressure.

(*a*) Calculate the rate at which neon atoms strike a 1 cm² area of the wall surface.

(*b*) How many times per second, on the average, does an atom strike a wall of the tank?

16.18 A very small pinhole develops in the wall of a vessel, and every atom of the gas inside that strikes the hole escapes from the tank. Show that the average energy per particle escaping is $2kT$. Why is this different from the average energy of the gas inside the tank?

16.19 A classical experiment for verifying the Maxwell-Boltzmann velocity distribution involves the apparatus shown schematically in Fig. 16.11. A

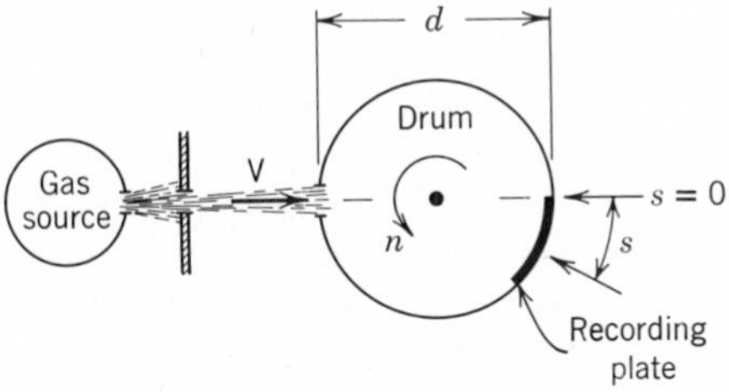

Fig. 16.11 Sketch for Problem 16.19.

unidirectional molecular beam (produced by collimating a stream of atoms escaping from a small hole in a vessel) enters a slit in a drum of diameter d. The drum rotates at high speed, n revolutions/sec, and these atoms are distributed along the recording plate according to their speeds. Develop an expression for the intensity of deposition of atoms as a function of displacement s and other physical parameters. Sketch the relation between deposition intensity and displacement.

16.20 The equilibrium of a liquid with its vapor is of a dynamic character. At equilibrium, molecules of liquid evaporate at the same rate at which molecules of vapor condense, so that there is no net change in the amount of each phase present. Consider mercury at 0 C, at which temperature the vapor pressure is 1.85×10^{-4} mm Hg. If every vapor molecule striking the liquid surface condenses, what are the rates at which condensation and evaporation occur at equilibrium?

16.21 A tank of volume V contains a gas (molecular weight M) at conditions P_0, T, as shown in Fig. 16.12. The tank is located in a large environment of the same gas at conditions P_A, T, where $P_A > P_0$. A very small hole of area A

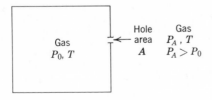

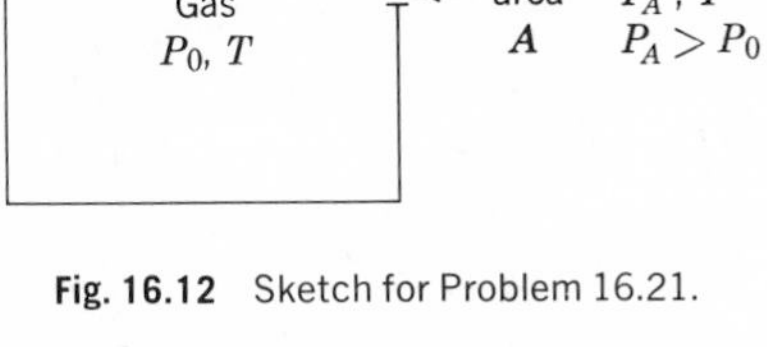

Fig. 16.12 Sketch for Problem 16.21.

develops in a wall of the tank, such that every particle striking the hole from either side passes through freely. Assuming that T is constant, derive an expression for the pressure inside the tank as a function of time.

16.22 A 1-liter vacuum container is maintained at 1 micron (10^{-3} mm Hg) pressure. Air at 1 atm pressure, 25 C, surrounds the container. A pinhole of 10^{-6} cm diameter then develops in the wall.

(*a*) Find the initial rate at which molecules pass through the hole in each direction, treating air as a pure substance with $M = 28.97$.

(*b*) What is the pressure inside the container after 1 hour?

16.23 A "getter" is a substance used to remove gas molecules from a closed volume (by adsorption or chemical combination); it finds considerable application in high vacuum technology. As an example, consider a 100 cc volume containing air that is initially pumped to 1 mm Hg pressure and then sealed. Inside this volume is a "getter" of 1 cm² surface, which is assumed to remove 0.1% of the molecules striking the surface. Determine the pressure inside the volume as a function of time, assuming that the temperature remains constant at 25 C.

16.24 A "potential barrier" model is useful for many applications, especially in the field of electronics. Consider a monatomic ideal gas having Maxwell-Boltzmann velocity distribution and confined in an enclosure, one "wall" of which consists of a potential barrier φ (perhaps due to an electric field). Thus all particles that reach this wall with a normal component of kinetic energy greater than φ escape, while all those with energy component less than φ are reflected. Develop an expression for the flux of particles through this barrier in terms of the significant parameters.

16.25 A monatomic gas is contained in a vessel of volume V at initial conditions P_0, T_0. The gas slowly escapes through a small hole of area A, under conditions of free molecule flow. There is no flow into the vessel from outside. The vessel is well insulated, such that there is no heat transfer with the surroundings, and the specific heat of the walls can be neglected.

(*a*) Find the relation between P and T inside the vessel at any point during the process.

(*b*) Compare the result of part *a* with the case in which the hole area is much larger than the mean free path of the gas.

16.26 Determine the expression for temperature inside the vessel as a function of time for Problem 16.25.

16.27 A tank of volume V contains a gas at initial conditions T_0, P_0. The gas escapes slowly through a hole of area A, under conditions of free molecule flow. There is no flow into the tank from outside. At any time during the process, it can be assumed that the temperature and pressure in the tank are related by the expression

$$\frac{T}{T_0} = \left(\frac{P}{P_0}\right)^c$$

where c is a constant. Develop an expression for temperature T as a function of time.

16.28 Consider the same situation as described in Problem 16.21, except that the gas outside the tank is not the same as that inside. The small hole now develops in the wall, and the pressure inside begins to drop, even though the outside pressure is larger. Show under what conditions this may happen.

17

Statistical Thermodynamics

In this chapter, we consider the first and second laws of thermodynamics, as well as various thermodynamic properties, from a microscopic point of view. In so doing, we shall relate the definitions of the properties and statements of the laws that are given in classical thermodynamics to the concepts and definitions of statistical thermodynamics, and thus demonstrate the added insight into the nature of physical processes that a microscopic point of view provides. In addition to discussing these quantities, we shall develop expressions for the thermodynamic properties of interest in terms of the partition function, so that the subsequent problem of evaluating properties will be focused upon the determination of that quantity.

Since the developments of this chapter are based on the preceding chapter, we must limit the discussion to systems of independent particles, and we shall concentrate primarily on systems following corrected Boltzmann statistics. However, the relevance of Bose-Einstein and Fermi-Dirac statistics at temperatures approaching absolute zero will also be demonstrated.

17.1 The First Law of Thermodynamics

Consider a thermodynamic system, for which the first law of thermodynamics is given by Eq. 5.7. If kinetic and potential energy changes of the system as a whole are negligible, this reduces to

$$dU = \delta Q - \delta W \tag{17.1}$$

where the change in internal energy is given in terms of the two macroscopic energy transfer quantities, namely heat transfer and work.

Let us now examine the change in internal energy of this system from a microscopic viewpoint. The energy of the system is given by Eq. 16.2,

$$U = \sum_j N_j \epsilon_j$$

A fundamental assumption in the development and discussion of the equilibrium state in Chapter 16 was that the molecular energy values ϵ_j are fixed by specifying the system boundary parameters (V for a simple compressible substance). Therefore, we may write Eq. 16.2 in differential form as

$$\begin{aligned} dU &= \sum_j \epsilon_j \, dN_j + \sum_j N_j \, d\epsilon_j \\ &= \sum_j \epsilon_j \, dN_j + \sum_j N_j \left(\frac{d\epsilon_j}{dV}\right) dV \end{aligned} \tag{17.2}$$

where the second form emphasizes the dependency of ϵ_j on the boundary parameter V. We find that from the microscopic viewpoint the change in internal energy, Eq. 17.2, is expressed as the sum of two contributions, as in the macroscopic point of view, Eq. 17.1. The first is the change due to a net redistribution of particles among the available energy levels. The second is the change resulting from a shift in magnitude of energy associated with each of the levels or groups of quantum states, this shift being caused by a change in the system boundary parameters.

Let us now try to associate these two distinct contributions of Eq. 17.2 with the macroscopic quantities, work and heat transfer. For example, consider the work done when the system boundary of a simple compressible substance is allowed to move, as was analyzed in Section 4.3. If the driving force is only infinitesimally greater than that resisting the movement, then the process will proceed very slowly, and the system will pass through a series of equilibrium states. In such a reversible process, the work done at the moving boundary is given by Eq. 4.2,

$$\delta W_r = P \, dV$$

Note that the second term of Eq. 17.2 gives the change of energy associated with the volume change. We conclude that the second term of Eq. 17.2 and the macroscopic work both represent the energy change associated with the change in volume, and are therefore equivalent.

Equating the two, we have

$$\begin{aligned}\delta W_r &= -\sum_j N_j d\epsilon_j \\ &= -\sum_j N_j\left(\frac{d\epsilon_j}{dV}\right) dV \end{aligned} \tag{17.3}$$

In the general case, other thermodynamic driving forces and corresponding boundary parameters may be of significance. Analysis of this possibility for the macroscopic viewpoint leads to Eq. 4.13, while from the microscopic viewpoint we recognize that the energy levels depend on whatever boundary parameters are important to the problem at hand. Therefore, in the general case, we again reach the conclusion that the identity of Eq. 17.3 is valid, although the second equality would be modified to account for dependence of ϵ_j on all the boundary parameters.

Now by comparing Eqs. 17.1, 17.2 and 17.3, we conclude that the first term of Eq. 17.2 and the macroscopic heat transfer both correspond to the same effect, for a reversible process. That is,

$$\delta Q_r = \sum_j \epsilon_j \, dN_j \tag{17.4}$$

so that heat transfer can be visualized from the microscopic viewpoint as that energy change resulting in a net redistribution of particles among the available fixed energy levels.

To aid in the visualization of the quantities work and heat transfer, let us represent the change in energy resulting from two sequential processes as shown in Fig. 17.1. Figure 17.1*a* represents the equilibrium

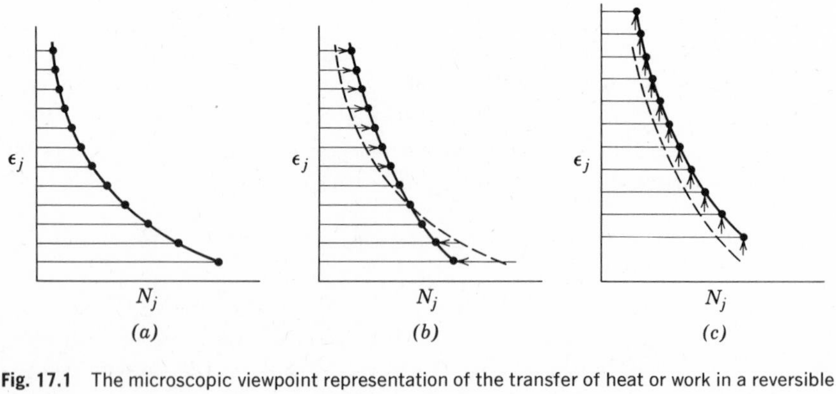

Fig. 17.1 The microscopic viewpoint representation of the transfer of heat or work in a reversible process.

distribution in a system of some particular energy U. Consider first a reversible heat transfer to the system during a process in which there is no work; that is, the various boundary parameters such as volume are held constant. Therefore the energies of the various levels remain fixed, and the increase in the energy of the system arises solely from a net shifting of particles from lower levels to higher levels, as shown in Fig. 17.1*b*. Next, consider a reversible adiabatic process in which work is put into the system, as for example by boundary movement. From the first law, we conclude that the internal energy is increased. Taking a microscopic viewpoint, we see that the change in boundary parameters causes a shift in energy of the various levels, as shown in Fig. 17.1*c*. The increase in energy of the system results in this case from the shifting of the energy levels, and not from a net redistribution of particles among levels.

Although our example is an extremely simple one, the ability to visualize a heat transfer or work on a microscopic scale should give us a further insight into the nature of these quantities, even if only for a reversible process. The nature of an irreversible process will be discussed in conjunction with the second law of thermodynamics in Section 17.3.

17.2 Entropy

Let us consider as a thermodynamic system two metal blocks that initially are at different temperatures. If these blocks are brought into thermal communication, they will gradually come to the same temperature, at a value intermediate to the two initial temperatures, by means of a spontaneous heat transfer. We say that in the final state the blocks have reached thermal equilibrium. The change occurred because of the initial imbalance in temperature, the driving force or thermodynamic potential. If we also include other thermodynamic potentials (pressure, chemical potential, etc.), we say that a condition of thermodynamic equilibrium exists in a system when the system is at equilibrium with respect to all potentials. That is, there is no imbalance in any driving force, and consequently no tendency for any spontaneous change to occur. From the methods of classical thermodynamics, the equilibrium state of a system of a given mass, volume, and internal energy is the state of maximum entropy, as can be seen from the property relation, Eq. 8.5. Mathematically, this is stated as

$$dS_{U,V,m} = 0 \quad \text{for an infinitesimal change}$$

$$\Delta S_{U,V,m} < 0 \quad \text{for a finite change}$$

Let us now view the system from a microscopic point of view. A particular macrostate, or distribution d of particles among the various energy levels can occur in w_d ways, the thermodynamic probability for that macrostate. For an equilibrium state of the system of given V, N, U, all macrostates consistent with the system constraints are accessible or available to the system. Thus the total number of microstates available to the system, w_{tot}, is the summation of w_d over all possible macrostates, or

$$w_{tot} = \sum_{d} w_d$$

On the other hand, we can also consider an inhibited state, one in which the system is restrained in some manner, at least temporarily, from reaching all possible macrostates (and microstates), so that w_{tot} for such a state is some smaller number than that given above.

Comparing this microscopic viewpoint of the equilibrium state at fixed V, N, U with the classical concept, we conclude that the entropy should be directly related to the total number of microstates available to the system. However, we must keep in mind that entropy is an extensive thermodynamic property; this statement means that, if two independent systems are combined, their entropies are additive. If the entropy is directly proportional to $\ln w_{tot}$ it will have this behavior, since the total thermodynamic probabilities of two systems are multiplicative when the systems are combined. Finally, the constant of proportionality must have the units of energy per degree.

With these considerations in mind, we define entropy from the microscopic viewpoint by the expression

$$S = k \ln w_{tot} \tag{17.5}$$

where k is the Boltzmann constant.

The definition of S according to Eq. 17.5 has the required characteristics of entropy discussed above. It remains to be shown that this definition is completely consistent with the classical definition of entropy from the second law of thermodynamics. Before proceeding to this demonstration, however, let us make a number of observations regarding Eq. 17.5. We note that there is an increase in entropy associated with a change from a given macroscopic state to a state that has a greater number of microscopic ways of occurring. In this sense, entropy can be considered as a measure of disorder, randomness, or lack of information about the microscopic configuration of the particles of which the system is comprised. A perfectly ordered system, with w_{tot} equal to unity, corresponds to zero entropy and implies a complete knowledge of the microscopic state of the system.

Our principal concern in this text is with equilibrium states. While all possible macrostates consistent with the constraints are available to the system, the most probable macrostate has a thermodynamic probability tremendously larger than all macrostates differing from it by a distinguishable amount. Thus the thermodynamic equilibrium state may be represented for all practical purposes by the most probable macrostate, that is,

$$\ln \mathrm{w}_{\mathrm{mp}} \approx \ln \mathrm{w}_{\mathrm{tot}}$$

so that the entropy for an equilibrium state can be calculated from the expression

$$S = k \ln \mathrm{w}_{\mathrm{mp}} \tag{17.6}$$

In our consideration of equilibrium states, let us now express Eq. 17.6 in a more readily usable form. For corrected Boltzmann statistics, the thermodynamic probability for any given macrostate is given by Eq. 16.31, which, in logarithmic form and with Stirling's formula for $N!$ is

$$\ln \mathrm{w} = \sum_j N_j \left[\ln\left(\frac{g_j}{N_j}\right) + 1\right] \tag{17.7}$$

However, for the most probable macrostate, the ratio (g_j/N_j) can be eliminated by introducing the equilibrium distribution equation, Eq. 16.39. Substituting this into Eq. 17.7, we have

$$\begin{aligned} \ln \mathrm{w}_{\mathrm{mp}} &= \sum_j N_j \left[\ln\left(\frac{Z}{N} e^{\beta\epsilon_j}\right) + 1\right] \\ &= \ln\left(\frac{Z}{N}\right) \sum_j N_j + \beta \sum_j N_j\epsilon_j + \sum N_j \\ &= N \ln\left(\frac{Z}{N}\right) + \beta U + N \end{aligned} \tag{17.8}$$

Therefore, the entropy of a system in an equilibrium state, Eq. 17.6, can be written in the form

$$S = k \ln \mathrm{w}_{\mathrm{mp}} = Nk\left[\ln\left(\frac{Z}{N}\right) + 1\right] + k\beta U \tag{17.9}$$

Recognizing the functional dependencies of U and Z according to Eqs. 16.42 and 16.37, we note that Eq. 17.9 is of the form

$$S = S(N, \beta, V) \tag{17.10}$$

We have previously found in Section 16.3 that for the special case of the classical monatomic gas, the multiplier β is inversely proportional to temperature, as given by Eq. 16.61. We shall now proceed to demonstrate that this inverse relation of β to T is of general validity for corrected Boltzmann statistics. The procedure followed in this proof is to compare the definition of entropy according to classical thermodynamics with that for statistical thermodynamics. From the classical approach, entropy was defined by Eq. 8.2,

$$dS = \left(\frac{\delta Q}{T}\right)_{\text{rev}}$$

Let us now derive an expression for dS from the statistical point of view, by differentiating Eq. 17.9 at fixed N,

$$dS = k\left(\frac{N}{Z}dZ + \beta dU + Ud\beta\right) \tag{17.11}$$

But, from the definition of partition function, Eq. 16.35,

$$\frac{N}{Z}dZ = \frac{N}{Z}(-\beta)\sum_j g_j e^{-\beta\epsilon_j} d\epsilon_j - \frac{Nd\beta}{Z}\sum_j g_j \epsilon_j e^{-\beta\epsilon_j} \tag{17.12}$$

However, if we assume the change to be reversible, then the system passes through a sequence of equilibrium states, such that the equilibrium distribution equation, Eq. 16.39, must be valid at any point along the path. For this case, substituting Eq. 16.39 into Eq. 17.12, we have, for the reversible process,

$$\frac{N}{Z}dZ = -\beta\sum_j N_{j\,\text{mp}}\, d\epsilon_j - d\beta\sum_j N_{j\,\text{mp}}\epsilon_j$$

Substituting Eqs. 16.2 and 17.3, we have

$$\frac{N}{Z}dZ = \beta\delta W_r - Ud\beta \tag{17.13}$$

Now, substituting Eq. 17.13 into Eq. 17.11 and using the first law, Eq. 17.1,

$$\begin{aligned} dS &= k(\beta\delta W_r - Ud\beta + \beta dU + Ud\beta) \\ &= k\beta(\delta W_r + dU) \\ &= k\beta(\delta Q_r) \end{aligned} \tag{17.14}$$

Comparing this result with the classical definition of entropy, we conclude that the relationship of Eq. 16.61,

$$\beta = 1/kT$$

is a general relation for corrected Boltzmann statistics.

17.3 The Second Law of Thermodynamics

We have discussed the quantities heat and work and have in the process examined the first law from a microscopic viewpoint. We now proceed to consider the second law of thermodynamics from a microscopic point of view, and again consider first the reversible process. Following this, we shall examine the nature of irreversible processes from the microscopic viewpoint.

Suppose that a gas is contained in an insulated cylinder fitted with a frictionless piston, as shown in Fig. 17.2. The system initially has some

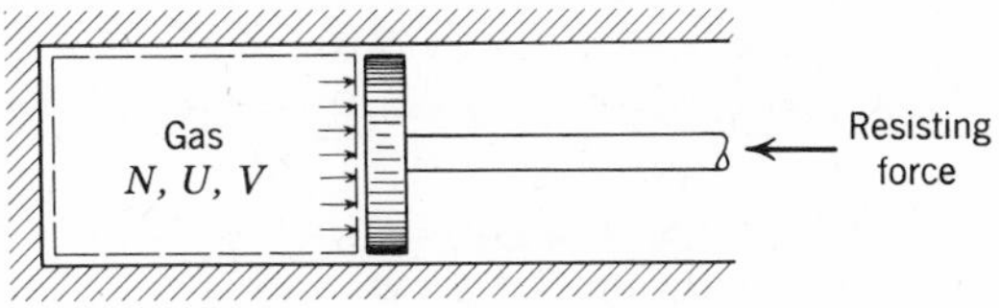

Fig. 17.2 The reversible expansion of a gas.

known number of particles N, internal energy U, and volume V, and is assumed to exist at a condition of equilibrium. The pressure of the gas on the face of the piston is exactly balanced by the proper resisting force (including the contribution of ambient pressure on the back face of the piston). Now let the resisting force be reduced by an infinitesimal amount, allowing the piston to move very slowly toward the right. The gas will expand, and since the piston is moving slowly, the system will pass through a series of equilibrium states and undergo a reversible expansion. Since the cylinder is well insulated, we assume that there is no heat transfer during the process.

From our discussion of the first law, we know that the energy change that takes place in this system during the process described is due solely to a shifting of the energy levels and not to any net redistribution of particles among the levels. We note from Eq. 17.7 that, for the succession of equilibrium states that the system passes through the thermodynamic probability remains constant because there was no net redistribution of particles. Consequently, in accordance with our definition of entropy in Eq. 17.6 the entropy of the system remains constant during the process. The same conclusion is, of course, reached on the basis of classical thermodynamics, but the molecular point of view now gives us another way of visualizing the concept of an isentropic process.

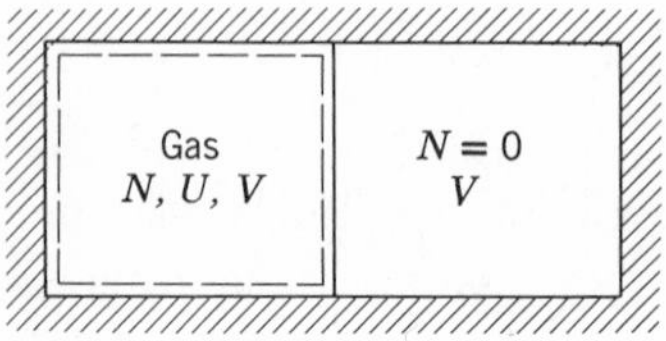

Fig. 17.3 An irreversible expansion of a gas.

Let us consider the opposite extreme, an irreversible expansion in which no work is done, as represented in Fig. 17.3. The tank is insulated, and a gas on the left side is restrained by a thin partition. The initial state of the gas is identical with that of the previous example: N particles, an internal energy U, and volume V. In this case, an equal volume on the other side of the partition is completely evacuated. If the partition is broken, the gas will expand to fill the entire volume. This expansion is irreversible and is certainly very different in nature from the reversible expansion examined previously.

If we analyze this process from a classical viewpoint, we are limited to a description of the system before the process and after the gas has again come to equilibrium at the end of the process. We do not know exactly what happened between these two end states, since the changes occurred very rapidly and the system was not at equilibrium. However, as in the first example, there is no heat transfer during the process because the tank is well insulated. Furthermore, in this process no work is done, since there was no resistance to the expansion (assuming the hole in the partition to be reasonably large compared to molecular dimensions). We therefore conclude from the first law that there is no change in internal energy during the process.

When we examine this process from a microscopic viewpoint, we cannot describe the series of states that the system passes through; this is the same difficulty that is encountered in an analysis from the macroscopic point of view. We are, however, able to make some interesting observations about the process. There is a change in volume, and consequently there will be a shift in the energy levels. Since no work was done during the process, Eq. 17.3, which applies to a reversible process, cannot in general be valid for the irreversible process as well. Also, since the energy of the system remained constant, we conclude from Eq. 17.2 that there must have been some net redistribution of particles among levels during the process. Because there was no heat transfer, we conclude also that Eq. 17.4 is not in general valid for an irreversible process.

(The heat transfer is given, however, by Eq. 17.4 for a process in which all the boundary parameters are fixed.) Consequently, our description of an irreversible process is quite limited in scope. We conclude that there is a general shifting of levels and of particles among levels, but we do not know the path followed—only the end states. However, the microscopic viewpoint gives us another means of viewing an irreversible process, and in that respect is very useful.

The microscopic point of view is even more useful and informative when we consider the second law for this process. Since the system is isolated from its surroundings (no energy transfer), from a classical viewpoint the second law of thermodynamics can be written, Eq. 8.16,

$$\Delta S_{\substack{\text{isolated}\\ \text{system}}} \geq 0$$

That is, the entropy change for an isolated system undergoing a spontaneous process will increase, or, in the limit, for a completely reversible process, remain constant.

For an analysis from the microscopic viewpoint, let us imagine the system described in Fig. 17.3 at the instant of time after the partition has been broken, before the gas has begun to expand. It is, at least in theory, possible that at the moment that the partition is ruptured, none of the molecules in the vicinity of the hole is moving toward the partition, and consequently all the gas will remain on the left side of the container for a period of time. If the hole is extremely small, of molecular dimensions, this might seem reasonable, but it certainly is unacceptable to us if the hole is large, as we assumed previously. That is, although there is a probability that all the gas will remain on the left side, that probability is so small that we do not expect to observe this behavior in nature. Thus it is overwhelmingly more probable that the gas will expand to fill the entire volume as this additional number of microstates becomes available to the system. Similarly, once the gas has filled the volume, there always exists the minute possibility that at some later time all the molecules will be found on the left side of the tank, but it is equally unlikely that this condition will exist for any finite period of time as it is that the gas would have remained on the left side when the partition ruptured, since the number of microscopic states resulting in such a macroscopic configuration is infinitesimal compared to the total number of states available.

Since the total thermodynamic probability for the state associated with the gas filling the volume is much greater than that for the initial state of the gas on the left side of the tank, we find from Eq. 17.6 that the spontaneous process occurring when the partition breaks corresponds to an increase in entropy, which is, of course, in agreement with the classical statement of the second law. The microscopic approach is also very informative, however, because it gives a different viewpoint from which to examine the second law of thermodynamics and, in particular, a more meaningful concept of the thermodynamic equilibrium state and the property entropy. For any change of state of an isolated system (during which there is an increase in entropy), a larger number of microstates is made available to the system, as for example by the removal of some internal inhibition or restriction on the system.

We should always bear in mind that the second law of thermodynamics is our attempt to describe systematically certain aspects of phenomena that we observe in nature. However, our description from the microscopic point of view involves a perspective quite different from that in our description from a macroscopic point of view, and provides very significant additional insight into the fundamental nature of the processes that give rise to the second law.

Before proceeding, we should mention that, although we have discussed only two specific examples here, a completely reversible expansion and an irreversible one with no work output, there are any number of degrees of irreversibility between these extremes. For example, in the first case, if the resisting force on the piston is reduced by a finite amount, then the process will proceed at a finite rate, and the system will not pass through a series of equilibrium states. The resulting process is irreversible to a certain extent, and the work done by the system on its surroundings will be some amount less than that for the reversible expansion. In such a case, the work and heat transfer are not in general given by Eqs. 17.3 and 17.4 since the process is not reversible, and the entropy will increase, although not as much as in the case of the completely irreversible expansion.

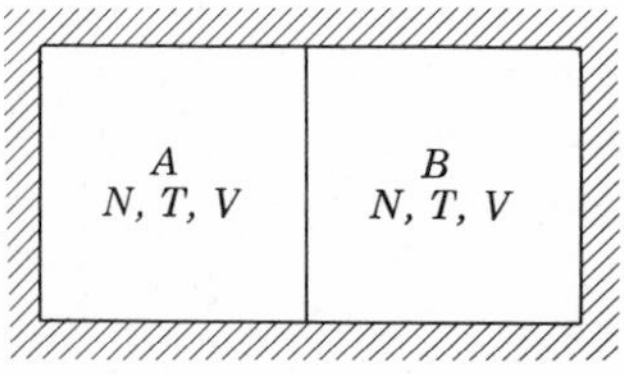

Fig. 17.4 The mixing of two gases.

Another factor causing irreversibilities is the mixing of two different substances. Consider an insulated tank containing two ideal gases A and B, initially separated by a partition as shown in Fig. 17.4. The number of molecules, N, of each gas is the same, and each gas initially occupies a volume V and is at temperature T. Gases A and B considered together constitute an isolated system, the entropy of which at the initial state 1 is

$$S_1 = S_{A_1} + S_{B_1} = k \ln \mathrm{w}_{\mathrm{mp}_{A_1}} + k \ln \mathrm{w}_{\mathrm{mp}_{B_1}} = k \ln (\mathrm{w}_{\mathrm{mp}_{A_1}} \times \mathrm{w}_{\mathrm{mp}_{B_1}}) \quad (17.15)$$

If the partition is now removed or broken, the two gases will gradually mix, and after a period of time the entire tank will be filled with a uniform mixture of A and B at a condition of equilibrium. The number of microstates consistent with the most probable macrostate in the mixture at state 2 will be the product of those for the two components as they exist in the mixture, and the entropy of the mixture is

$$S_2 = k \ln \mathrm{w}_{\mathrm{mp}_{AB2}} = k \ln (\mathrm{w}_{\mathrm{mp}_{A2}} \times \mathrm{w}_{\mathrm{mp}_{B2}}) \tag{17.16}$$

To determine the change in thermodynamic probability for A and B as a result of the mixing process, let us visualize the process in a manner similar to that used before. Immediately after the partition is removed, all the molecules of substance A are still on the left side of the tank and those of B on the right. Now there is certainly a possibility that A and B will remain for a period of time in this separated state, but we realize that, owing to the random motions of the molecules, the probability of the two gases mixing is much greater. That is,

$$\begin{aligned} \mathrm{w}_{\mathrm{mp}_{A2}} &> \mathrm{w}_{\mathrm{mp}_{A1}} \\ \mathrm{w}_{\mathrm{mp}_{B2}} &> \mathrm{w}_{\mathrm{mp}_{B1}} \end{aligned} \tag{17.17}$$

Using Eq. 17.17 in comparing Eqs. 17.15 and 17.16, we find that the mixing process constitutes an increase in entropy, or

$$S_2 - S_1 > 0 \tag{17.18}$$

which is the result obtained from a classical analysis.

For this problem, it is important to realize that, if the same gas is initially on both sides of the partition, the preceding analysis is not valid. Once the partition is removed, we are no longer able to distinguish which molecules were initially on the left side and which were on the right. Consequently, the thermodynamic probability for the system after the partition is broken cannot be split into that associated with the molecules that started on the left and right sides, respectively. It equals the product of the initial values of w for the two parts, and there is no change in entropy.

In this section we have discussed irreversibility resulting from unrestrained expansion (boundary movement at a finite rate resulting from a finite pressure difference) and that due to mixing of different substances. In general, any change driven by a finite imbalance in a thermodynamic potential is irreversible. Since the force resisting the change is a finite amount less than the driving force, the process occurs at a finite rate. Consequently, the system does not pass through a series of equilibrium states and the process is not reversible. Other examples of irreversibility, as discussed in Chapter 7, are processes involving friction, heat transfer through a finite temperature difference, and flow of electrical current through a finite resistance, which can similarly be examined from the microscopic point of view.

17.4 Thermodynamic Properties

In earlier sections, we have developed a number of general equilibrium state equations for a system obeying the corrected Boltzmann statistical model. These include Eq. 16.35 for the partition function, Eq. 16.39 for the equilibrium distribution and Eq. 16.42 for internal energy, all from Section 16.2, and Eq. 17.9 for entropy in Section 17.2. In that section we also demonstrated the general relationship between β and temperature, as given previously by Eq. 16.61 for a special case. Because of this general validity of Eq. 16.61, we may now rewrite Eq. 16.35 as

$$Z = \sum_j g_j e^{-\epsilon_j/kT} \tag{17.19}$$

Eq. 16.39 as

$$N_j = \frac{N}{Z} g_j e^{-\epsilon_j/kT} \tag{17.20}$$

Eq. 16.42 as

$$U = \frac{NkT^2}{Z}\left(\frac{\partial Z}{\partial T}\right)_V = NkT^2\left(\frac{\partial \ln Z}{\partial T}\right)_V \tag{17.21}$$

and Eq. 17.9 as

$$S = Nk\left[\ln\left(\frac{Z}{N}\right) + 1\right] + \frac{U}{T} \tag{17.22}$$

These four expressions now constitute the basic set of relations from which we will develop equations for other thermodynamic properties for corrected Boltzmann systems.

The Helmholtz function A has been defined in Eq. 9.56, in connection with the second law of thermodynamics. We find from Eq. 17.22 that

$$A = U - TS = -NkT\left[\ln\left(\frac{Z}{N}\right) + 1\right] \tag{17.23}$$

Using Eqs. 12.21 and 12.25, it can be shown that the thermodynamic property relation for a pure substance of varying mass can be expressed in terms of Helmholtz function as

$$dA = -S\,dT - P\,dV + \bar{g}\,dn \tag{17.24}$$

where $\bar{g}$, is the Gibbs function per mole. Since this represents a functional dependence of the form

$$A = A(T, V, n)$$

it follows that

$$\bar{g} = \left(\frac{\partial A}{\partial n}\right)_{T,V} \tag{17.25}$$

Now, since

$$n = \frac{N}{N_0} \tag{17.26}$$

the Gibbs function can be found from Eqs. 17.23 and 17.25,

$$\begin{aligned} G = n\bar{g} = N\left(\frac{\partial A}{\partial N}\right)_{T,V} &= N\left\{-kT\left[\ln\left(\frac{Z}{N}\right)+1\right] - NkT\left(-\frac{1}{N}\right)\right\} \\ &= -NkT\ln\left(\frac{Z}{N}\right) \end{aligned} \tag{17.27}$$

Therefore, from the definitions of Gibbs and Helmholtz functions and substituting Eqs. 17.23 and 17.27,

$$PV = G - A = NkT \tag{17.28}$$

which is the equation of state for an ideal gas. It is interesting to observe that the assumption of independent particles and corrected Boltzmann statistics leads to this result, Eq. 17.28, and that ideal gas behavior has not been assumed independently in our development.

Using the definition of enthalpy along with Eqs. 17.21 and 17.28, we can express that property as

$$H = U + PV = NkT\left[T\left(\frac{\partial \ln Z}{\partial T}\right)_V + 1\right] \tag{17.29}$$

There is another development of interest in regard to the equation of state. Suppose we consider a system in which the only significant boundary parameter is the volume. Then, for a small change occurring along a reversible path,

$$\delta W_r = P\,dV = -\sum_j N_j\,d\epsilon_j \tag{17.30}$$

But, since the system passes through a series of equilibrium states, the equilibrium distribution equation, Eq. 17.20, may be substituted into Eq. 17.30, giving

$$P\,dV = -\frac{N}{Z}\sum_j g_j\,e^{-\epsilon_j/kT}\,d\epsilon_j \tag{17.31}$$

Note that the summation in this equation is the same as that obtained by differentiating Z at constant temperature. If we do this,

$$dZ = -\frac{1}{kT}\sum_j g_j\,e^{-\epsilon_j/kT}\,d\epsilon_j \tag{17.32}$$

we find that, at constant temperature,

$$P\,dV = \frac{NkT}{Z}\,dZ = NkT\,d\ln Z \tag{17.33}$$

Therefore, for a system at equilibrium,

$$P = NkT\left(\frac{\partial \ln Z}{\partial V}\right)_T \tag{17.34}$$

Since this development has been made for a system of independent particles, it follows that Eq. 17.34 must be identical with Eq. 17.28. A comparison of the two indicates that, for an ideal gas,

$$Z = V \times f(T) \tag{17.35}$$

This is, in fact, the result that was found in Eq. 16.62,

$$Z = V\left(\frac{2\pi mkT}{h^2}\right)^{3/2}$$

when the partition function was evaluated for the special case of a classical, monatomic ideal gas.

An important consequence of Eq. 17.35 becomes evident if we use Eq. 17.35 in the evaluation of the internal energy from Eq. 17.21. Writing

Eq. 17.35 in logarithmic form, we have

$$\ln Z = \ln V + \ln [f(T)]$$

It follows that

$$\left(\frac{\partial \ln Z}{\partial T}\right)_V = \frac{d \ln [f(T)]}{dT}$$

Substituting this result into Eq. 17.21, we find

$$\begin{aligned} U &= NkT^2\left(\frac{\partial \ln Z}{\partial T}\right)_V \\ &= NkT^2\left\{\frac{d \ln [f(T)]}{dT}\right\} \\ &= U(N, T) \end{aligned} \tag{17.36}$$

We conclude that internal energy per unit mass will be expressed as some function of temperature only, and will not depend on volume (or pressure).

In practical applications, especially in tabulating the properties of a substance for use in thermodynamic calculations, it is convenient to write the equations for the various thermodynamic properties on a unit mole basis. Using Eq. 17.26 and the relation

$$\bar{R} = N_0 k \tag{17.37}$$

we can write Eq. 17.28 as

$$P\bar{v} = \bar{R}T \tag{17.38}$$

which is, of course, the same as Eq. 3.1. Similarly Eq. 17.21 becomes

$$\bar{u} = \frac{U}{n} = \bar{R}T^2\left(\frac{\partial \ln Z}{\partial T}\right)_V \tag{17.39}$$

We could also rewrite Eq. 17.29 for enthalpy on a mole basis. However, in practice, enthalpy is more easily calculated from its definition, and Eq. 17.38, by

$$\bar{h} = \bar{u} + P\bar{v} = \bar{u} + \bar{R}T \tag{17.40}$$

In a similar manner, the specific heats $\bar{C}_v$ and $\bar{C}_p$ could be expressed in terms of the partition function, but are more readily evaluated from their definitions. For the constant-volume specific heat, from Eqs. 5.14 and 17.36,

$$\bar{C}_{vo} = \left(\frac{\partial \bar{u}}{\partial T}\right)_v = \left(\frac{d\bar{u}}{dT}\right) \tag{17.41}$$

For the constant-pressure specific heat, from Eqs. 5.15, 17.36 and 17.40,

$$\bar{C}_{po} = \left(\frac{\partial \bar{h}}{\partial T}\right)_P = \left(\frac{d\bar{h}}{dT}\right) \tag{17.42}$$

Equations for other thermodynamic properties discussed in this section are written on a mole basis, using Eqs. 17.26 and 17.37, as

$$\bar{s} = \frac{S}{n} = \bar{R}\left[\ln\left(\frac{Z}{N}\right) + 1\right] + \frac{\bar{u}}{T} \tag{17.43}$$

$$\bar{a} = \frac{A}{n} = -\bar{R}T\left[\ln\left(\frac{Z}{N}\right) + 1\right] \tag{17.44}$$

$$\bar{g} = \frac{G}{n} = -\bar{R}T\ln\left(\frac{Z}{N}\right) \tag{17.45}$$

17.5 Bose-Einstein and Fermi-Dirac Statistics

While considering the laws and functions of thermodynamics from a microscopic point of view, it is of interest to develop briefly the more general Bose-Einstein and Fermi-Dirac statistics. We shall not devote much time to these models here because we have found that, in the majority of cases of interest to us, each reduces to the much simpler common form that we have termed corrected Boltzmann statistics. However, since at very low temperatures the assumption that $g_j \gg N_j$ is not reasonable, it is important to determine the behavior of these more general models near absolute zero. We shall also find that the expressions for thermodynamic properties for these models are not easily simplified, as are those for corrected Boltzmann statistics.

For Bose-Einstein statistics, the equilibrium distribution equation obtained by maximizing the thermodynamic probability for given values of N, U, V was found to be (Eq. 16.21)

$$N_j = \frac{g_j}{e^{\alpha}e^{\beta\epsilon_j} - 1}$$

It was shown in Chapter 16 that in the special case of corrected Boltzmann statistics the multiplier β is equal to $1/kT$, and in the present chapter it was found that entropy defined in terms of w and using the relation $\beta = 1/kT$ is consistent with the entropy definition of classical thermodynamics from the second law. Therefore let us assume that this identification of β with the temperature according to Eq. 16.61 is valid in general, so that the Bose-Einstein distribution equation can be expressed as

$$N_j = \frac{g_j}{e^{\alpha}e^{\epsilon_j/kT} - 1} \tag{17.46}$$

To determine the entropy from Bose-Einstein statistics, we use the expression for thermodynamic probability given in Eq. 16.17:

$$\ln w = \sum_j [(g_j + N_j) \ln (g_j + N_j) - g_j \ln g_j - N_j \ln N_j]$$

It is more convenient for our purposes to write the equation in the form

$$\ln w = \sum_j \left[g_j \ln \left(1 + \frac{N_j}{g_j}\right) + N_j \ln \left(1 + \frac{g_j}{N_j}\right)\right] \tag{17.47}$$

For the most probable distribution, or macroscopic equilibrium state, we substitute Eq. 17.46 into Eq. 17.47 and obtain

$$\ln w_{mp} = \sum_j \left[g_j \ln (1 - e^{-\alpha}e^{-\epsilon_j/kT})^{-1} + N_j\left(\alpha + \frac{\epsilon_j}{kT}\right)\right]$$

$$= -\sum_j g_j \ln (1 - e^{-\alpha}e^{-\epsilon_j/kT}) + \alpha N + \frac{U}{kT} \tag{17.48}$$

Using Eq. 17.48, we find that the entropy is

$$S = k \ln w_{mp} = -k \sum_j g_j \ln (1 - e^{-\alpha}e^{-\epsilon_j/kT}) + \alpha Nk + \frac{U}{T} \tag{17.49}$$

and, from the definition of the Helmholtz function,

$$A = U - TS = kT \sum_j g_j \ln (1 - e^{-\alpha}e^{-\epsilon_j/kT}) - \alpha NkT \tag{17.50}$$

We find that the expressions for these properties are not given in terms of the partition function as defined by Eq. 17.19. By using the logarithmic series, it is easy to show that, when the ratio N_j/g_j is much less than unity, the equations above reduce to those found previously for corrected Boltzmann statistics. This demonstration is left as an exercise.

To find the Gibbs function, we approach the problem as before by making use of the thermodynamic property relation. From Eqs. 17.25 and 17.26,

$$G = N\left(\frac{\partial A}{\partial N}\right)_{T,V} \tag{17.51}$$

Before differentiating the equation for the Helmholtz function, however, we recall that for corrected Boltzmann statistics it was found that α is given by Eq. 16.38, so that

$$\alpha = \ln\frac{Z}{N} = f(N, T, V) \tag{17.52}$$

Therefore we assume that the multiplier α is some function of N, T, V for Bose-Einstein statistics also. Now, from Eqs. 17.50 and 17.51,

$$\begin{aligned} G &= N\left[kT\sum_j \frac{g_j}{e^{\alpha}e^{\epsilon_j/kT}-1}\left(\frac{\partial \alpha}{\partial N}\right)_{T,V} - NkT\left(\frac{\partial \alpha}{\partial N}\right)_{T,V} - \alpha kT\right] \\ &= N\left[kT\sum_j N_j\left(\frac{\partial \alpha}{\partial N}\right)_{T,V} - NkT\left(\frac{\partial \alpha}{\partial N}\right)_{T,V} - \alpha kT\right] \\ &= -\alpha NkT \end{aligned} \tag{17.53}$$

Therefore

$$\alpha = -\frac{G}{NkT} \tag{17.54}$$

Previously, the Gibbs function for corrected Boltzmann statistics was found to be given by Eq. 17.27. From that equation and Eq. 17.52,

$$G = -NkT\ln\frac{Z}{N} = -NkT\alpha \tag{17.55}$$

which is identical with Eq. 17.54 for Bose-Einstein statistics.

Since it is desirable to eliminate the multiplier α from our distribution equation, Eq. 17.54 can be substituted into Eq. 17.46, and the Bose-Einstein distribution equation becomes

$$N_j = \frac{g_j}{e^{(\epsilon_j - G/N)/kT} - 1} \tag{17.56}$$

The development for the general Fermi-Dirac statistics will not be discussed here, because the method is entirely analogous to that for the Bose-Einstein statistics. The details are left as an exercise. When the development is carried through, it is found that the entropy is expressed as

$$S = k \sum_j g_j \ln (1 + e^{-\alpha} e^{-\epsilon_j/kT}) + \alpha N k + \frac{U}{T} \tag{17.57}$$

and also that α is given by Eq. 17.54, as was found for Bose-Einstein statistics (and the corrected Boltzmann model as well). Consequently, the distribution equation for Fermi-Dirac statistics is conveniently written in the form

$$N_j = \frac{g_j}{e^{(\epsilon_j - G/N)/kT} + 1} \tag{17.58}$$

Equations for other thermodynamic properties are not presented here since, as for the entropy, they do not reduce to a convenient form. Evaluation of properties for the general Bose-Einstein and Fermi-Dirac statistical models is difficult for this reason.

As mentioned above, it is of interest to examine the two general models as the temperature approaches absolute zero. For Bose-Einstein statistics, let us take $\epsilon_0 = 0$ and find the ratio of the number of particles in any level j to that in the ground level from the distribution equation in the form of Eq. 17.46. For any value of $j > 0$,

$$\frac{N_j}{N_0} = \frac{g_j}{g_0} \left[\frac{e^{\alpha} - 1}{e^{\alpha} e^{\epsilon_j/kT} - 1} \right] \tag{17.59}$$

If e^{α} approaches unity at low temperature, which is known to be the case for corrected Boltzmann statistics, then, since the ratio g_j/g_0 is finite, it is apparent from Eq. 17.59 that

$$\lim_{T \to 0} \left(\frac{N_j}{N_0} \right) = 0 \tag{17.60}$$

for any $j > 0$. That is, at the limiting temperature of absolute zero, all the particles would be in the ground level of energy. That e^{α} approaches unity as T approaches zero is in fact reasonable for Bose-Einstein statistics is seen by writing the distribution equation for the ground energy level $\epsilon_0 = 0$.

$$N_0 = \frac{g_0}{e^{\alpha} - 1} \tag{17.61}$$

The value of g_0, the statistical weight of the ground level, is known to be a small number (perhaps equal to unity). As the temperature is decreased, N_0 becomes increasingly larger, such that e^{α} must approach unity at very low temperature.

To determine the behavior of entropy for a Bose-Einstein system as the temperature goes to zero, let us examine the original expression for w for this model, Eq. 15.31. At the limit of T equal to zero, there is only one possible macrostate, namely that all the particles are in the ground level, and, for this macrostate,

$$\mathrm{w} = \frac{(\mathrm{g}_0 + N - 1)!}{(\mathrm{g}_0 - 1)!N!} \tag{17.62}$$

If the ground level is non-degenerate, then g_0 is equal to unity, w is identically one, and the entropy is exactly zero according to Eq. 17.5. Even if this is not exact, it is nevertheless true that g_0 is a small number, so that w is small, in which case the entropy is still very nearly equal to zero. This is in agreement with the statement of the third law of thermodynamics, which is discussed in Chapter 13.

For a system obeying Fermi-Dirac statistics, the behavior at temperatures approaching absolute zero is entirely different from the Bose-Einstein system, since in this case there can be no more than one particle per quantum state. The ground energy level will be completely filled by g_0 particles, the first level above that by g_1 particles, etc. Therefore, in this system at absolute zero, the particles would assume the lowest possible energy states, but could not all crowd into the ground level. This behavior can be seen by examining the distribution equation, Eq. 17.58, at low temperature. As the temperature approaches absolute zero, the Gibbs function per particle G/N, also called the chemical potential per particle, approaches a zero-point value μ_0. For all the energy levels for which ϵ_j is less than μ_0,

$$\lim_{T \to 0} \left(e^{(\epsilon_j - \mu_0)/kT}\right) = e^{-\infty} = 0 \tag{17.63}$$

so, that for these levels,

$$\lim_{T \to 0} \left(\frac{N_j}{\mathrm{g}_j}\right) = 1 \tag{17.64}$$

That is, all these levels are completely filled. On the other hand, for those levels for which ϵ_j is greater than μ_0,

$$\lim_{T \to 0} \left(e^{(\epsilon_j - \mu_0)/kT}\right) = e^{+\infty} \tag{17.65}$$

and

$$\lim_{T \to 0} \left(\frac{N_j}{\mathrm{g}_j}\right) = 0 \tag{17.66}$$

That is, all energy levels above an energy equal to μ_0 are empty at absolute zero. This limiting distribution for Fermi-Dirac statistics at absolute zero temperature is depicted in Fig. 17.5. The separating value of energy μ_0, below which all levels are filled and above which all levels are empty at absolute zero temperature, is known as the Fermi energy.

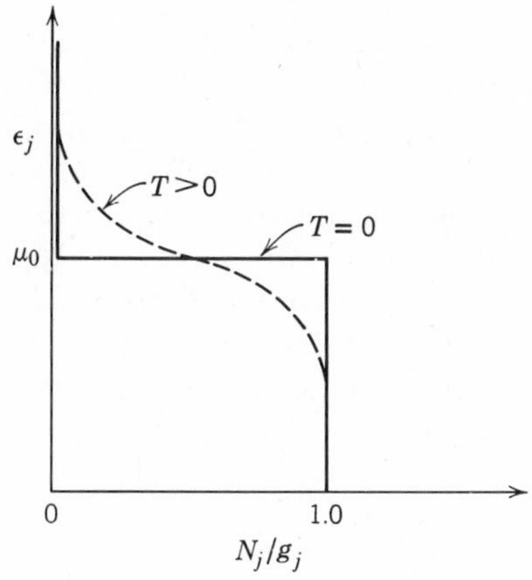

Fig. 17.5 The Fermi-Dirac distribution.

We also note that, as the temperature is increased, some particles shift into the higher energy levels, and the distribution changes to that shown by the broken line in Fig. 17.5. As the temperature continues to increase, the upper portion of the distribution curve approaches the exponential function predicted by corrected Boltzmann statistics.

The internal energy and Gibbs function at absolute zero can be found,[1] since

$$\sum_j g_j \geqslant \sum_j N_j = N \tag{17.67}$$

where the summation extends over those levels containing particles. Since N and the various g_j's are known, Eq. 17.67 specifies the maximum value of j to be considered. Note that we have used the $>$ sign because it is possible that the uppermost level containing particles at zero temperature ($\epsilon_j = G/N$) will not be completely filled. This is certainly of minor importance, however, since the numbers become very large.

[1]This problem will be considered in Section 20.4, in terms of the behavior of the conduction electrons in a metal.

To evaluate the entropy at absolute zero, we use the equation for thermodynamic probability for Fermi-Dirac statistics, Eq. 15.30,

$$w = \prod_j \frac{g_j!}{(g_j - N_j)!N_j!}$$

For all levels of energy less than μ_0, N_j equals g_j and the corresponding term in Eq. 15.30 is unity. For energy levels of magnitude greater than μ_0, N_j equals zero, so that all these terms are also equal to one. Now, if the uppermost level containing particles is full, then w for the distribution at absolute zero is identically one and the entropy is zero as found before for the Bose-Einstein case. Or, if there are not enough particles to fill the level $\epsilon_j = \mu_0$ completely, then for this level $N_j < g_j$, and the contribution to w for that level and consequently w for the entire distribution will be some number slightly greater than unity. However, the product of the logarithm of this number and the Boltzmann constant will be extremely small, so that the entropy even in this case is very nearly zero.

PROBLEMS

17.1 Assuming Boltzmann statistics, what is the entropy of the system described in Example 15.4?

17.2 The entropy of one mole of carbon dioxide is 51.072 cal/K at 25 C, 1 atm pressure. Determine the thermodynamic probability of the most probable macrostate.

17.3 The classical, monatomic gas was analyzed in Section 16.3.
(*a*) Develop an expression for $\ln w_{mp}$ as a function of Z and N for this gas.
(*b*) Determine w_{mp} for one mole of helium at 25 C, 1 atm pressure.

17.4 Consider one mole of a gas at room conditions, for which a typical value for $\ln w_{mp}$ is about 10^{25}. In developing Eq. 17.6 from Eq. 17.5, it is recognized that the most probable macrostate is not the only macrostate with a large thermodynamic probability, but is instead representative of a group of very similar macrostates. Suppose that there were 10^{10} macrostates with a value of w nearly as large as the most probable macrostate. Would it still be reasonable to use Eq. 17.6 for entropy?

17.5 The most probable macrostate in a thermodynamic system has been presumed to be extremely more likely than any macrostate differing from it by a distinguishable amount. For a volume V containing N particles at energy U, consider a macrostate d for which $N_j = N_{j_{mp}} + \delta N_j$, where $|\delta N_j| \ll N_{j_{mp}}$. Starting with Eq. 17.7, show that the entropy for such a macrostate must be less than that corresponding to the most probable macrostate.

17.6 Consider 1 mole of gas having a macrostate d for which

$$\frac{|\delta N_j|}{N_{j_{\text{mp}}}} = 0.001\%$$

for every level j. Calculate the ratio of thermodynamic probabilities w_{mp}/w_d, and find the corresponding difference in entropies.

17.7 Consider a classical monatomic gas contained in one side of a tank as shown in Fig. 17.3. The partition is broken, and the gas is allowed to fill the entire volume. Using the expression for Z given by Eq. 16.62, analyze this process from the microscopic viewpoint. Show that the expression for the net entropy change is of the same form as the classical result.

17.8 Consider the classical monatomic gas, discussed in Section 16.3.

(*a*) Using the expression for Z given by Eq. 16.62, determine the internal energy and enthalpy per mole.

(*b*) Determine the constant-volume and constant-pressure specific heats, and also the ratio $\bar{C}_{po}/\bar{C}_{vo}$.

17.9 Calculate the entropy per mole of monatomic nitrogen at 25 C, 1 atm pressure, and compare the result with the value given in Table B.9.

17.10 Develop the entropy equation for a Fermi-Dirac system, Eq. 17.57, and demonstrate that Eq. 17.54 is a valid relation for such a system.

17.11 In the development of thermodynamic properties for Bose-Einstein statistics in Section 17.4, it was assumed that $\beta = 1/kT$. Using a procedure equivalent to that of Section 17.2, show that this assumption is correct.

17.12 It is desired to show that a general Bose-Einstein gas of independent particles does not obey the ideal gas equation of state. By series expansion, show that the equation of state is instead

$$\frac{PV}{NkT} = 1 - \frac{1}{2N} e^{-2\alpha} \sum_j g_j e^{-2\epsilon_j/kT} \ldots$$

(Hint: Begin by combining Eqs. 17.50 and 17.54)

17.13 Consider a Bose-Einstein gas with no internal modes of energy. Using the result of Problem 17.12, show that the equation of state becomes,

$$\frac{PV}{NkT} = 1 - \frac{C}{2^{5/2}}\left(\frac{N}{V}\right) - \cdots$$

and find C in terms of m and T.

17.14 Consider a Bose-Einstein system at very low temperature. Show that the limiting relation, Eq. 17.60, is valid without requiring that the ground level energy ϵ_0 is equal to zero.

17.15 Suppose that a Bose-Einstein system has a ground level degeneracy g_0 of 10. Is the entropy of this system equal to zero at absolute zero temperature? Calculate the value at zero temperature for a mole of this substance.

18

Quantum Mechanics

In the three preceding chapters, we have developed the basic concepts of statistical mechanics, utilizing the microscopic viewpoint and certain molecular models. These developments have led to the expression of thermodynamic properties in terms of the partition function, the evaluation of which depends upon a knowledge of the energy levels as determined from quantum mechanics. We must therefore develop the mathematical aspects of quantum mechanics, at least to the extent required for the determination of the energy levels, and this is the purpose of this chapter.

18.1 The Bohr Theory of the Atom

As an introduction to the subject of quantum mechanics, we describe the Bohr theory of the atom, which was presented originally by Niels Bohr in 1913 and was frequently modified during the succeeding decade until its general replacement by the new quantum mechanics.

Bohr's theory is fundamentally a combination of classical mechanics and Planck's quantum theory of radiation.[1] The atomic model consists of electrons orbiting around a central nucleus. We consider here the simplest case, namely, the hydrogen atom, which consists of a single electron revolving about a nucleus comprised of one proton. The force exerted on the electron by the nucleus is the Coulomb potential $C\mathscr{Z}e^2/r^2$, where $\mathscr{Z}$ is the atomic number of the substance (i.e., the charge on the nucleus, which is equal to unity for hydrogen), e is the electron charge, and r is the radius of the orbit. Bohr postulated that only discrete orbits

[1]The topic of thermal radiation will be treated in some detail in Section 19.10 as an example of a Bose-Einstein system.

are permissible, these being specified by the requirement that the angular momentum of the electron be equal to $\mathbf{n}(h/2\pi)$, where $\mathbf{n}$ is the principal quantum number, which may have only integral values 1, 2, 3, ... For an electron to jump from one orbit to a higher one, it must receive a quantum of radiation (energy according to Planck's theory) of exactly the appropriate frequency. This absorption of energy results in a discrete line in the spectrum observed for hydrogen. The various wavelengths at which hydrogen and other atoms absorb or emit energy will be discussed in Section 18.8. With Bohr's hypothesis and using the Coulomb potential, it is found from Newton's second law of motion that the radii of these allowed orbits are directly proportional to $\mathbf{n}^2$.

This basic theory is far too simple to explain the observed spectral lines for even the hydrogen atom, much less those for more complex atoms. Modifications to the theory allow for elliptical orbits as well as circular, as specified by the azimuthal quantum number $\mathbf{l}$, which may take on only the values $\mathbf{l} = 0, 1, 2, \ldots (\mathbf{n} - 1)$. Figure 18.1 shows the Bohr orbits for the first few values of the quantum numbers $\mathbf{n}$ and $\mathbf{l}$.

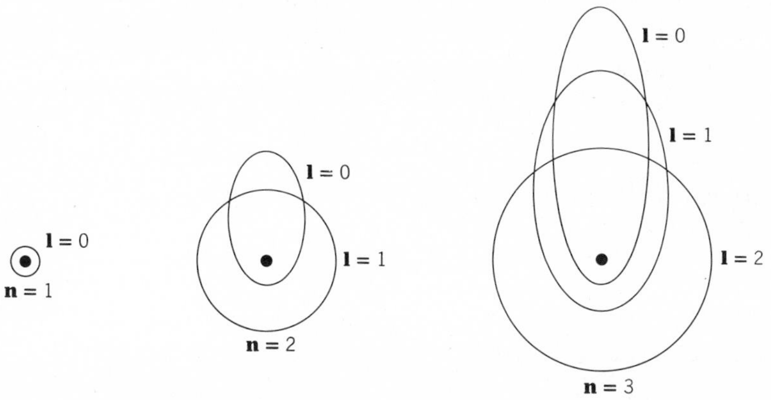

Fig. 18.1 Electron orbits for the Bohr theory of the hydrogen atom.

The energy is dependent primarily upon the principal quantum number $\mathbf{n}$, and energy differences for different values of $\mathbf{l}$ are very small for the hydrogen atom. The magnetic quantum number $\mathbf{m}_l$, with allowed values $\mathbf{m}_l = 0, \pm 1, \pm 2, \ldots, \pm \mathbf{l}$, was introduced to account for the observed splitting of spectral lines in a magnetic field, and the electron spin quantum number $\mathbf{m}_s = \pm\frac{1}{2}$ was introduced in an attempt to account for the behavior of helium and other more complex substances.

In spite of its growing complexity, the Bohr theory was still unable to account for or explain many of the observed phenomena, and it was therefore discarded in favor of the new quantum mechanics. The model is still useful today, however, as a tool for many calculations and especially as a convenient pictorial representation of the atom.

18.2 Wave Characteristics of Electrons and the Heisenberg Uncertainty Principle

The introduction of Planck's quantum theory of radiation began a great controversy about the fundamental nature of light and electromagnetic radiation in general. Radiation having a corpuscular nature could not easily explain the problems of dispersion, reflection, interference, diffraction, and others handled so readily by the wave theory. On the other hand, using the quantum theory viewpoint, one could rather simply explain the problem of energy exchange—the photoelectric effect in particular—and other difficulties that had plagued the wave theory for years. Furthermore, it seemed for a time that Bohr's atomic model, which incorporated the quantum theory, would provide a basis for the explanation of spectra and atomic behavior in general.

Let us consider a photon of electromagnetic radiation. From Planck's equation for energy exchange,

$$\epsilon = h\nu \tag{18.1}$$

where ν is the frequency, and from Einstein's equation,

$$\epsilon = mc^2 \tag{18.2}$$

we find that, although a photon or quantum of radiation has no mass, it does have associated with it a momentum (mc),

$$(mc) = \frac{h\nu}{c} = \frac{h}{\lambda} \tag{18.3}$$

in which the wavelength λ is given by

$$\lambda = \frac{c}{\nu} \tag{18.4}$$

We view radiation, then, as having a dualistic nature—the wave properties of frequency and wavelength and the particle or corpuscular property of momentum.

In 1923, when the controversy over the nature of radiation had reached essentially an impasse with both viewpoints capable of only partial success, de Broglie postulated that matter may also possess a similar dualistic nature. His hypothesis that a wavelength given by the relation

$$\lambda = \frac{h}{m\mathsf{V}} \tag{18.5}$$

can be associated with an electron of mass m and velocity V was first confirmed by Davisson and Germer in 1925, and later by others. It was found that passing a narrow beam of electrons through a thin metal foil produces a set of concentric diffraction rings, similar to the optical rings of light diffraction, a phenomenon explained by wave interference. The electron ring spacings could also be predicted according to Eq. 18.5. From the magnitude of Planck's constant h, it is clear that the de Broglie matter wavelengths for large macroscopic bodies are so small as to result in a totally negligible influence, but for electrons and other atomic-scale considerations, a clear distinction between particle and wave is not always possible.

The Heisenberg uncertainty principle can be demonstrated by the following thought experiment. Suppose that we wish to observe an electron through some powerful microscope. It is impossible, even in theory, to determine the exact position of the electron at any instant of time, inasmuch as some source of radiation must be used to illuminate the electron. Consequently, the minimum uncertainty Δx in the position of the electron is of the order of magnitude of the wavelength of the illuminating radiation. That is,

$$\Delta x \sim \lambda$$

We conclude that, in order to minimize the uncertainty in position of the electron, it would be desirable to have a microscope using radiation of a very short wavelength. However, in making this measurement or observation, at least one photon of radiation must strike the electron and be reflected to the observer. This photon possesses a momentum h/λ, at least part of which is transferred to the electron as they collide. Consequently, there is a resulting uncertainty about the momentum of the electron of the order of magnitude h/λ, or

$$\Delta(m\mathsf{V}_x) \sim \frac{h}{\lambda}$$

with the opposite dependency on λ. We find that radiation having a short wavelength results in a small uncertainty in position of the electron but a large uncertainty in its momentum. Conversely, by using radiation of long wavelength we could determine the momentum of an electron quite accurately, but we would then be very uncertain about its position. The product of these effects,

$$\Delta x \times \Delta(m\mathsf{V}_x) \sim h \tag{18.6}$$

is an approximate statement of Heisenberg's uncertainty principle, one of the fundamental equations of quantum mechanics. It should be pointed out that this basic uncertainty principle is not observed or of consequence in ordinary macroscopic measurements, because of the extremely small magnitude of Planck's constant. However, it is of prime importance on the atomic scale, where we are concerned with small masses and distances.

18.3 The Schrödinger Wave Equation

As a preliminary to a consideration of the Schrödinger wave equation for matter waves, we first examine the partial differential equation representing a vibrating string as developed in classical physics. Letting the variable y denote the amplitude of vibration at time t for any distance x from the fixed end point, this equation is found to be, from Newton's second law,

$$\left(\frac{\partial^2 y}{\partial x^2}\right) = \frac{1}{\mathsf{V}_p{}^2}\left(\frac{\partial^2 y}{\partial t^2}\right) \tag{18.7}$$

in which V_p is the phase velocity.

In general, the amplitude y is a function of both x and t. Let us assume that this dependency is separable into the form

$$y(x,t) = f(x)g(t) \tag{18.8}$$

Taking the partial derivatives of Eq. 18.8 and substituting into Eq. 18.7, we obtain

$$\frac{1}{f}\left(\frac{d^2 f}{dx^2}\right) = \frac{1}{g\mathsf{V}_p{}^2}\left(\frac{d^2 g}{dt^2}\right) = -\alpha^2 \tag{18.9}$$

We note that one side of this equation is a function only of x and the other a function only of t. Inasmuch as x and t are independent variables, Eq. 18.9 can be valid only if both sides are constant, which for convenience we have set equal to $-\alpha^2$.

The second equality of Eq. 18.9 can be written as

$$\frac{d^2g}{dt^2}+\alpha^2 V_p^{\,2} g=0 \qquad (18.10)$$

The solution to this ordinary differential equation is

$$g=C_1 \cos(\alpha V_p t)+C_2 \sin(\alpha V_p t) \qquad (18.11)$$

from which we conclude that g is a periodic function of time having the frequency

$$\nu=\frac{\alpha V_p}{2\pi} \qquad (18.12)$$

Therefore the constant α can be expressed as

$$\alpha=\frac{2\pi\nu}{V_p}=\frac{2\pi}{\lambda} \qquad (18.13)$$

where λ is the wavelength.

Using Eq. 18.13 and the equality involving f and α in Eq. 18.7, it follows that

$$\frac{d^2f}{dx^2}+\frac{4\pi^2}{\lambda^2}f=0 \qquad (18.14)$$

The function f then represents the amplitude of the standing wave at any position x for a vibration of associated wavelength λ.

In 1926, Schrödinger applied the differential wave equation to the matter waves of de Broglie. Using the symbol Ψ_x for the x-direction amplitude or wave function (the physical interpretation of which will be discussed shortly), and the de Broglie equation (Eq. 18.5), the differential equation is

$$\frac{d^2\Psi_x}{dx^2}+\frac{4\pi^2(mV_x)^2}{h^2}\Psi_x=0 \qquad (18.15)$$

It is more convenient to use the total energy ϵ_x, where

$$\epsilon_x=(\mathrm{KE})_x+(\mathrm{PE})_x=\tfrac{1}{2}mV_x^{\,2}+\Phi_x \qquad (18.16)$$

where Φ_x is the potential energy.

The one-dimensional wave equation then becomes

$$\frac{d^2\Psi_x}{dx^2}+\frac{8\pi^2 m}{h^2}(\epsilon_x-\Phi_x)\Psi_x=0 \qquad (18.17)$$

In order to generalize the equation to three dimensions, we define a wave function Ψ as

$$\Psi = \Psi_x \Psi_y \Psi_z \tag{18.18}$$

and recognize that

$$\epsilon = \epsilon_x + \epsilon_y + \epsilon_z \tag{18.19}$$

and

$$\Phi = \Phi_x + \Phi_y + \Phi_z \tag{18.20}$$

so that, in three dimensions,

$$\nabla^2 \Psi + \frac{8\pi^2 m}{h^2} (\epsilon - \Phi) \Psi = 0 \tag{18.21}$$

where the operator ∇^2 is

$$\nabla^2 = \frac{\partial^2}{\partial x^2} + \frac{\partial^2}{\partial y^2} + \frac{\partial^2}{\partial z^2} \tag{18.22}$$

Equation 18.21 is known as the time-independent Schrödinger wave equation. In general, as for the vibrating string, the wave function Ψ is dependent upon time as well as position and may even be a complex function having real and imaginary parts. Consequently, there is a corresponding general time-dependent Schrödinger wave equation, but for stationary states of the system (standing waves), the case of interest to us here, we need consider only the special form, Eq. 18.21.

A basic hypothesis of wave mechanics is that the wave function Ψ has the characteristic of a probability function. Since Ψ may in general be complex while a probability must be real, the product of Ψ and its complex conjugate Ψ^* is taken as representing probability. Furthermore, a probability function must be continuous and single-valued and must approach zero for large values of distance. It is found that a Ψ function having the required behavior is obtained from solutions of the Schrödinger wave equation (Eq. 18.21) only for discrete energy states of the system. When the function Ψ is real, the product $\Psi\Psi^*$ reduces to Ψ^2, so that for stationary states, where the particle being represented is presumed to have a definite energy ϵ_i, the quantity

$$\Psi_i^2 \, dx \, dy \, dz$$

becomes the probability of finding the particle of energy ϵ_i in the element of space x to $x + dx$, y to $y + dy$, z to $z + dz$.

18.4 Translation

In applying the Schrödinger wave equation to the molecules of a gas, we assume that the various energy modes are separable. We consider first the translational energy, which would constitute the entire energy of a monatomic ideal gas with all the atoms in the electronic ground level.

The gas is confined in a box of dimensions X, Y, Z, but to simplify the problem let us first consider a one-dimensional case, in which the particles move only in the x-direction, between 0 and X. Applying the Schrödinger equation to a single particle, for the x-direction only, yields

$$\frac{d^2\Psi_x}{dx^2}+\frac{8\pi^2 m}{h^2}(\epsilon_x-\Phi_x)\Psi_x=0 \tag{18.23}$$

However, since the particle must remain inside the box, the probability, and consequently the wave function Ψ_x, must be zero for all x outside this range. This physical requirement is satisfied by making the potential energy $\Phi_x=\infty$ for $x<0$ and $x>X$. We also note that, for an ideal gas, $\Phi_x=0$ inside the box. Therefore, for the range

$$0<x<X \tag{18.24}$$

the Schrödinger wave equation is

$$\frac{d^2\Psi_x}{dx^2}=\frac{8\pi^2 m}{h^2}\epsilon_x\Psi_x=0 \tag{18.25}$$

with the boundary conditions

$$\Psi_x(0)=0 \tag{18.26}$$

$$\Psi_x(X)=0 \tag{18.27}$$

The solution to Eq. 18.25 can be written in the form

$$\Psi_x=C_1\sin\left[x\sqrt{\frac{8\pi^2 m\epsilon_x}{h^2}}\right]+C_2\cos\left[x\sqrt{\frac{8\pi^2 m\epsilon_x}{h^2}}\right] \tag{18.28}$$

From the boundary condition at $x=0$ (Eq. 18.26), it follows that

$$0=C_1(0)+C_2(1)$$

and we conclude that

$$C_2=0 \tag{18.29}$$

From the boundary condition at $x = X$ (Eq. 18.27),

$$0 = C_1 \sin\left[X\sqrt{\frac{8\pi^2 m\epsilon_x}{h^2}}\right] \tag{18.30}$$

Now the possibility that $C_1 = 0$ would result in the physically impossible solution that $\Psi_x = 0$ for all values of x. The only alternative is that

$$\sin\left[X\sqrt{\frac{8\pi^2 m\epsilon_x}{h^2}}\right] = 0$$

or

$$X\sqrt{\frac{8\pi^2 m\epsilon_x}{h^2}} = \mathbf{k}_x \pi \tag{18.31}$$

with

$$\mathbf{k}_x = 1, 2, 3, \ldots$$

The value $\mathbf{k}_x = 0$ results in a trivial solution ($\Psi_x = 0$ for all x) and is therefore not permissible.

It is seen from Eq. 18.31 that only discrete energy states are allowed, these states being given in terms of the x-direction translational quantum number $\mathbf{k}_x$ as

$$\epsilon_{\mathbf{k}_x} = \frac{h^2}{8mX^2}\mathbf{k}_x^{\,2} \qquad \mathbf{k}_x = 1, 2, 3, \ldots \tag{18.32}$$

For each energy state $\epsilon_{\mathbf{k}_x}$ as specified by the quantum number $\mathbf{k}_x$, there is a corresponding x-direction wave function $\Psi_{\mathbf{k}_x}$, which from Eqs. 18.28, 18.29, and 18.31 can be expressed in the form

$$\Psi_{\mathbf{k}_x}(x) = C_1 \sin\left(\mathbf{k}_x \pi \frac{x}{X}\right) \tag{18.33}$$

The constant C_1 in this equation can be evaluated by normalization, because the square of the wave function represents the probability. Therefore

$$\int_{x=0}^{X} \Psi_{\mathbf{k}_x}^{\,2}(x)\, dx = 1 \tag{18.34}$$

since for each $\mathbf{k}_x$, a particle having the corresponding energy $\epsilon_{\mathbf{k}_x}$ must be found somewhere in the box, according to Eq. 18.24. If we substitute Eq. 18.33 into Eq. 18.34,

$$C_1^{\,2} \int_0^X \sin^2\left(\mathbf{k}_x \pi \frac{x}{X}\right) dx = 1 \tag{18.35}$$

and integrate, we find

$$C_1 = \sqrt{\frac{2}{X}} \tag{18.36}$$

Therefore each quantum state as specified by $\mathbf{k}_x$ has associated with it the wave function

$$\Psi_{\mathbf{k}_x}(x) = \sqrt{\frac{2}{X}} \sin\left(\mathbf{k}_x \pi \frac{x}{X}\right) \tag{18.37}$$

The diagrams of Fig. 18.2 show the wave functions and square of the wave functions for the first few quantum states for the particle in the one-dimensional box.

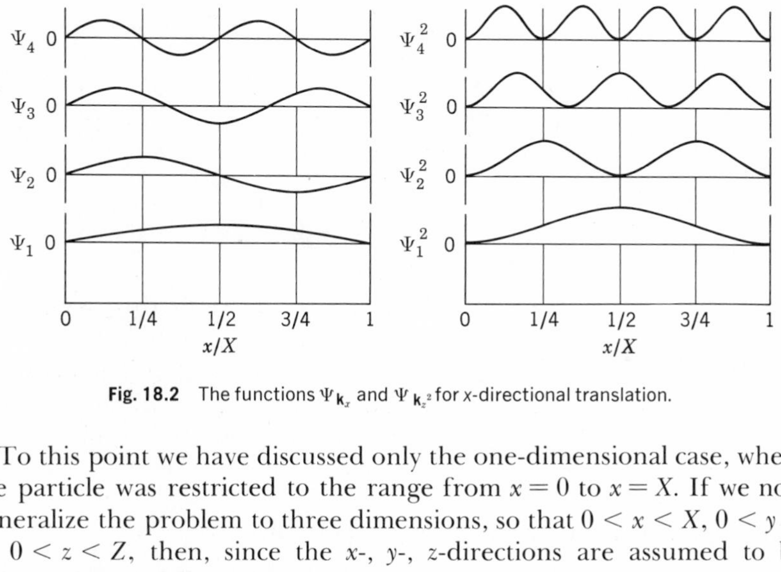

Fig. 18.2 The functions $\Psi_{\mathbf{k}_x}$ and $\Psi_{\mathbf{k}_z^2}$ for x-directional translation.

To this point we have discussed only the one-dimensional case, where the particle was restricted to the range from $x = 0$ to $x = X$. If we now generalize the problem to three dimensions, so that $0 < x < X$, $0 < y < Y$, $0 < z < Z$, then, since the x-, y-, z-directions are assumed to be independent, it follows that

$$\epsilon = \epsilon_x + \epsilon_y + \epsilon_z \tag{18.38}$$

From the Schrödinger time-independent equation, Eq. 18.21 and the boundary conditions, it is found that only discrete energy states are permissible; these energies are given in terms of three quantum numbers for translation by the expression

$$\epsilon_{\mathbf{k}_x,\mathbf{k}_y,\mathbf{k}_z} = \frac{h^2}{8m}\left(\frac{\mathbf{k}_x^2}{X^2} + \frac{\mathbf{k}_y^2}{Y^2} + \frac{\mathbf{k}_z^2}{Z^2}\right) \tag{18.39}$$

in which each of the quantum numbers $\mathbf{k}_x$, $\mathbf{k}_y$, $\mathbf{k}_z$ may have any integral value 1, 2, 3, ... A particular quantum state is then specified by the values of each of these numbers. It is important to realize that the states (5, 2, 3), (5, 3, 2), (2, 5, 3), etc., are all different quantum states, although all have exactly the same magnitude of energy. This point should help to clarify the distinction between energy (or quantum) states and energy levels. Each of the energy states $\epsilon_{\mathbf{k}_x,\mathbf{k}_y,\mathbf{k}_z}$, has associated with it a wave function

$$\Psi_{\mathbf{k}_x\mathbf{k}_y\mathbf{k}_z}(x, y, z) = \sqrt{\frac{8}{XYZ}} \sin\left(\mathbf{k}_x\pi \frac{x}{X}\right) \sin\left(\mathbf{k}_y\pi \frac{y}{Y}\right) \sin\left(\mathbf{k}_z\pi \frac{z}{Z}\right) \qquad (18.40)$$

where the constant has been evaluated by normalizing the wave function, as in the one-dimensional problem.

Very commonly, we are concerned with a system for which

$$X = Y = Z = V^{1/3} \qquad (18.41)$$

so that Eq. 18.39 can be simplified to the form

$$\epsilon_{\mathbf{k}_x\mathbf{k}_y\mathbf{k}_z} = \frac{h^2}{8mV^{2/3}} (\mathbf{k}_x^{\,2} + \mathbf{k}_y^{\,2} + \mathbf{k}_z^{\,2}) \qquad (18.42)$$

with $\mathbf{k}_x$, $\mathbf{k}_y$, $\mathbf{k}_z = 1, 2, 3, \ldots$. In this form, we see that the energy associated with a given quantum state is dependent upon the volume of the system, as stated earlier in Chapters 16 and 17; this statement was especially important in relating changes in energy to volume changes and ultimately to boundary movement work for a reversible process involving a simple compressible substance.

18.5 Application of the Wave Equation to Molecules

In the preceding section, we applied the Schrödinger wave equation to the problem of translation, the only energy mode for a monatomic ideal gas with all the atoms in the electronic ground level. In proceeding to the more general case of a molecule comprised of more than a single atom, the problem becomes relatively more complex because there may be several modes of energy storage. Let us consider a system of independent particles (an ideal gas), so that there is no system energy resulting from intermolecular forces and we may speak of the energy per molecule.

We assume first that the energy of the molecule can be separated into two parts: the translational energy of the molecule as a whole, expressed in terms of the motion of the center of mass of the molecule; and the energy internal to the molecule itself, that is, energy with respect to a coordinate system having its origin at the center of mass of the molecule. For the energy of the molecule, we may write

$$\epsilon = \epsilon_t + \epsilon_{\text{int}} \tag{18.43}$$

and, for the wave function,

$$\Psi = \Psi_t \Psi_{\text{int}} \tag{18.44}$$

We note also that, since we are concerned with independent particles, any potential energy is associated with the internal modes of energy. The Schrödinger wave equation can then be separated into two parts, one dependent solely upon the motion of the center of mass, and the other only upon the energy of the component atoms with respect to the center of mass. The first of these

$$\left[\frac{\partial^2 \Psi_t}{\partial x^2_{\text{C.M.}}} + \frac{\partial^2 \Psi_t}{\partial y^2_{\text{C.M.}}} + \frac{\partial^2 \Psi_t}{\partial z^2_{\text{C.M.}}}\right] + \frac{8\pi^2 m}{h^2} \epsilon_t \Psi_t = 0 \tag{18.45}$$

has been evaluated for translational energies and wave functions in the previous section for a monatomic ideal gas. The same results apply for the translation of a molecule consisting of more than one atom, provided that the x, y, z in Eq. 18.45 are the coordinates of the center of mass of the molecule (relative to an external coordinate frame) and that m is the mass of the molecule, i.e., the sum of the masses of the component atoms.

The second equation, dependent only upon the internal modes, becomes

$$\left(\frac{\partial^2 \Psi_{\text{int}}}{\partial x^2} + \frac{\partial^2 \Psi_{\text{int}}}{\partial y^2} + \frac{\partial^2 \Psi_{\text{int}}}{\partial z^2}\right) + \frac{8\pi^2 m}{h^2} (\epsilon_{\text{int}} - \Phi_{\text{int}})\, \Psi_{\text{int}} = 0 \tag{18.46}$$

In this equation, the coordinate system x, y, z is taken with respect to the center of mass of the molecule. The energy may include contributions from rotational, vibrational, and electronic states, as discussed in Chapter 2. These energy modes are not strictly independent. However, for most substances, the energy of the first excited electronic level is so high that essentially all the molecules will be found in the ground level. In addition, the vibrational energy levels will be found to be relatively large (compared to those for rotation), so that, at moderate temperature, most molecules in a system will be found in the lowest few vibrational levels at any time. Under such conditions, there will not be an appreciable interaction between rotation and vibration, and the vibration can reasonably be assumed to be harmonic.

Our simple model for internal energy modes consists therefore of a rigid rotator, harmonic oscillator, and ground electronic level. For the energy contribution of the internal modes, we can write

$$\epsilon_{\text{int}} = \epsilon_r + \epsilon_v + \epsilon_{e0} \tag{18.47}$$

and, for the internal wave function,

$$\Psi_{\text{int}} = \Psi_r \Psi_v \Psi_{e0} \tag{18.48}$$

We find that Eqs. 18.47 and 18.48 permit a separation of the wave equation, Eq. 18.46, into two parts, one for rotation and the other for vibration (ϵ_{e0} is a constant). In the next section, we consider that portion of the wave equation for a rigid rotator, and in Section 18.7 that for a harmonic oscillator.

At high temperatures, this simple internal model becomes inaccurate, and corrections for rotational centrifugal stretching, anharmonicity of the vibration, and rotation-vibration coupling must be included. For substances in which excited electronic levels are significant (O_2 and NO, for example), the rotation and vibration must be considered for each electronic level, and evaluation becomes extremely tedious.

18.6 Rigid Rotator

In the simplified model for internal modes of energy, interaction between rotation and vibration is assumed to be negligible, and we consider the molecule to be rigid insofar as rotation is concerned. In evaluating the rotational contribution to energy, we find it convenient to work in spherical polar coordinates, taking the origin at the center of mass of the molecule, as shown for a diatomic molecule in Fig. 18.3. The molecule is comprised of two atoms, 1 and 2, of masses m_1 and m_2, respectively. The position of atom 1 with respect to the center of mass is specified by the radius r_1, the angle with the z-axis, θ, and the angle of the projection of r_1 in the xy-plane with the x-axis, ϕ. Angle θ is restricted to $0 \leqslant \theta \leqslant \pi$, while, for ϕ, $0 \leqslant \phi \leqslant 2\pi$. Therefore, for atom 2, the angle θ is replaced by $(\pi - \theta)$, and the angle ϕ by $(\pi + \phi)$.

From Fig. 18.3, it is seen that the relations between the two coordinate systems are

$$\begin{aligned} x_1 &= r_1 \sin\theta \cos\phi \\ y_1 &= r_1 \sin\theta \sin\phi \\ z_1 &= r_1 \cos\theta \end{aligned} \tag{18.49}$$

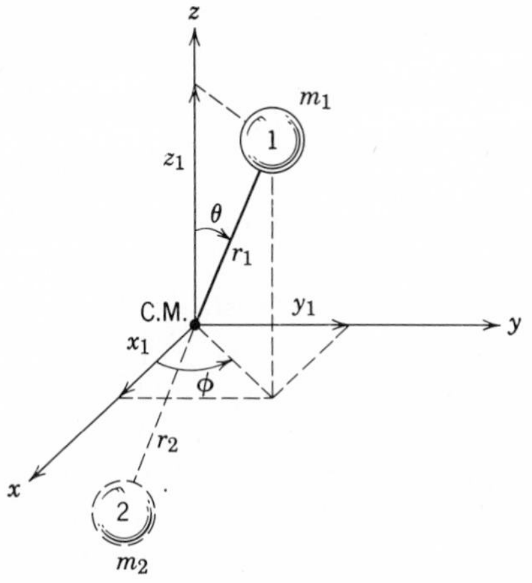

Fig. 18.3 The spherical coordinate system for a diatomic molecule.

Thus, the total kinetic energy of the rotating molecule is given by

$$\mathrm{KE}_{m_1+m_2} = \frac{I_\varepsilon}{2}\left[\left(\frac{\partial\theta}{\partial t}\right)^2 + \sin^2\theta\left(\frac{\partial\phi}{\partial t}\right)^2\right] \tag{18.50}$$

where the moment of inertia I_ε of the rigid molecule is defined by

$$I_\varepsilon = \sum_i m_i r_i^2 = m_1 r_1^2 + m_2 r_2^2 \tag{18.51}$$

We find it convenient to use the total distance between atoms (centers) making up the molecule, instead of the distances r_1 and r_2. As shown in Fig. 18.4, the equilibrium distance r_ε is defined as

$$r_\varepsilon = r_1 + r_2 \tag{18.52}$$

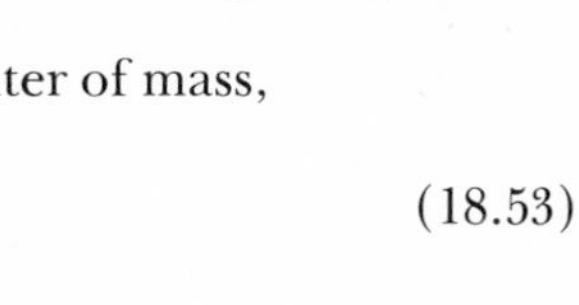

Fig. 18.4 The coordinates of a diatomic rigid rotator.

From a balance of moments taken about the center of mass,

$$m_1 r_1 = m_2 r_2 \tag{18.53}$$

so that Eq. 18.51 is conveniently written in the form

$$I_\varepsilon = \left(\frac{m_1 m_2}{m_1 + m_2}\right) r_\varepsilon^2 \doteq m_r r_\varepsilon^2 \tag{18.54}$$

where the quantity m_r is defined as

$$m_r = \frac{m_1 m_2}{m_1 + m_2} \tag{18.55}$$

and is called the reduced mass of the molecule.

We find that the total kinetic energy of rotation is equivalent to that of a single particle of mass m_r moving on a sphere of radius r_ε. Thus, to determine the energy levels and wave functions for a rigid diatomic rotator, we apply the Schrödinger wave equation to such a particle. The wave equation for internal modes, Eq. 18.46, is separated according to the expressions given by Eqs. 18.47 and 18.48. The potential energy Φ is assumed to be a function of r only, and as such will be considered in the vibrational energy evaluation. After separation, the partial differential equation for rotation as applied to a particle of mass m_r and constant dimension r_ε is

$$\nabla^2 \Psi_r + \frac{8\pi^2 m_r}{h^2} \epsilon_r \Psi_r = 0 \tag{18.56}$$

The Laplacian operator ∇^2 (Eq. 18.22), written in spherical polar coordinates, becomes

$$\nabla^2 = \frac{1}{r^2}\frac{\partial}{\partial r}\left(r^2 \frac{\partial}{\partial r}\right) + \frac{1}{r^2 \sin\theta}\frac{\partial}{\partial\theta}\left(\sin\theta\frac{\partial}{\partial\theta}\right) + \frac{1}{r^2\sin^2\theta}\frac{\partial^2}{\partial\phi^2} \tag{18.57}$$

However, for our application, $r = r_\varepsilon =$ constant, so that the first term in Eq. 18.57 equals zero. Substituting this relation into Eq. 18.56, and noting from Eq. 18.54 that $I_\varepsilon = m_r r_\varepsilon^2$, the wave equation for the rigid rotator is

$$\frac{1}{\sin\theta}\frac{\partial}{\partial\theta}\left(\sin\theta\frac{\partial\Psi_r}{\partial\theta}\right) + \frac{1}{\sin^2\theta}\frac{\partial^2\Psi_r}{\partial\phi^2} + \frac{8\pi^2 I_\varepsilon \epsilon_r}{h^2}\Psi_r = 0 \tag{18.58}$$

Even for our simplified model of rigid rotation, the wave equation, Eq. 18.58, is a relatively complex partial differential equation in θ and ϕ and does not readily yield a solution. Suppose that we assume a product solution, similar to the procedure used in solving the vibrating string equation in Section 18.3. We assume in this case that the wave function for rotation can be expressed as a product of two functions, one dependent only upon θ and the other only upon ϕ, or

$$\Psi_r(\theta, \phi) = f(\theta)\, g(\phi) \tag{18.59}$$

Taking the required derivatives of Eq. 18.59 and substituting into the wave equation, Eq. 18.58, the result is easily rearranged to the form

$$\frac{\sin\theta}{f}\frac{d}{d\theta}\left[\sin\theta\left(\frac{df}{d\theta}\right)\right]+\frac{8\pi^2 I_\varepsilon \epsilon_r}{h^2}\sin^2\theta=-\frac{1}{g}\frac{d^2g}{d\phi^2}=\lambda^2 \tag{18.60}$$

where λ^2 must be a constant, since one side of the equation depends only on θ and the other only on ϕ. Thus, as was the case previously, we have two ordinary differential equations, which can be written as

$$\frac{d^2g}{d\phi^2}+\lambda^2 g=0 \tag{18.61}$$

$$\frac{d^2f}{d\theta^2}+\frac{\cos\theta}{\sin\theta}\left(\frac{df}{d\theta}\right)+\left[\frac{8\pi^2 I_\varepsilon \epsilon_r}{h^2}-\frac{\lambda^2}{\sin^2\theta}\right]f=0 \tag{18.62}$$

Solution of Eq. 18.61 yields the conclusion that g is a periodic function of ϕ. However, in order that Ψ^2 have the necessary characteristics of a probability density, g must be finite and single-valued. This imposes the restriction that the constant λ be integral. The remaining differential equation, Eq. 18.62, can be converted into a standard form, called the associated Legendre equation, through a change of variable. The solution to this equation can then be found in terms of finite polynomials called associated Legendre functions. This solution imposes two further restrictions, however. One limits the energy term in Eq. 18.62 to certain discrete values, according to

$$\frac{8\pi^2 I_\varepsilon \epsilon_r}{h^2}=\mathbf{j}(\mathbf{j}+1);\quad \mathbf{j}=0,1,2,\ldots \tag{18.63}$$

and the other restricts the maximum value of λ to the value $\mathbf{j}$.

Substituting the two solutions into Eq. 18.59, the rotational wave function is

$$\Psi_r(\theta,\phi)=[C_1\sin\lambda\phi+C_2\cos\lambda\phi]\,[P_{\mathbf{j}}^{|\lambda|}(\cos\theta)] \tag{18.64}$$

with

$$\lambda=0,\pm1,\pm2,\ldots,\pm(\mathbf{j}-1),\pm\mathbf{j} \tag{18.65}$$

The second term in Eq. 18.64 is the associated Legendre function, depending on both rotational quantum numbers $\mathbf{j}$ and λ. The discrete rotational energies are, from Eq. 18.63,

$$\epsilon_r=\frac{h^2}{8\pi^2 I_\varepsilon}\mathbf{j}(\mathbf{j}+1);\quad \mathbf{j}=0,1,2,3,\ldots \tag{18.66}$$

One very important result follows from a close examination of these three expressions, Eqs. 18.64, 18.65 and 18.66. From the first two, we note that for each value of **j** there are (2**j** + 1) values of λ and therefore (2**j** + 1) different wave functions (discrete rotational quantum states). However, since there is only one value of rotational energy ϵ_r for each **j**, we conclude that the rotational energy levels given by Eq. 18.66 are degenerate levels, the degeneracy g_r being given by

$$g_r = (2\mathbf{j} + 1) \tag{18.67}$$

18.7 Harmonic Oscillator

In evaluating the energy and wave function for our simple vibrational model, the harmonic oscillator, we consider again a diatomic molecule, as shown in Fig. 18.5. At some instant of time, atoms 1 and 2 are at a total distance of separation r as shown in the diagram, which is, in general, different from the equilibrium distance r_ε (that at which the force between atoms changes sign) used for the rigid rotator analysis. We assume in our model that the vibration is harmonic; this means that the restoring force is directly proportional to $(r - r_\varepsilon)$, the displacement of the atoms from their equilibrium separation. Thus, this displacement x,

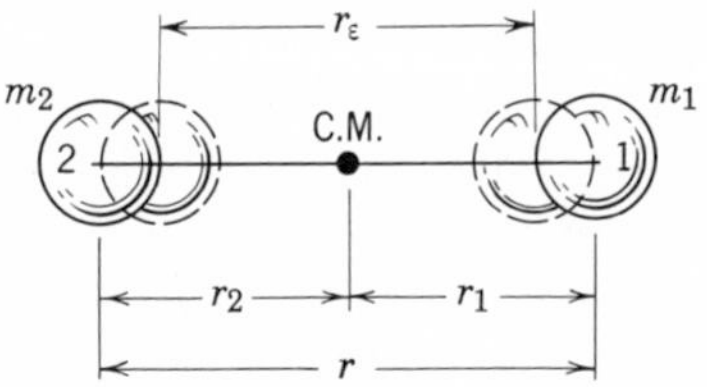

Fig. 18.5 The coordinates of a harmonic oscillator.

$$x = r - r_\varepsilon \tag{18.68}$$

becomes the significant dimension in the motion. The assumption of direct proportionality between force and displacement is reasonable if the displacement remains relatively small.

Using Newton's second law, it is found that the vibrational potential energy $\Phi(x)$ for any x is given by

$$\Phi(x) = -\int_0^x F\,dx = 2\pi^2 \nu_\varepsilon{}^2 m_r x^2 \tag{18.69}$$

in which ν_ε is the frequency of the vibration, and m_r is the molecular reduced mass as defined according to Eq. 18.55.

For our simple internal model, separation of the wave equation for internal modes (Eq. 18.46) into rotational and vibrational components has been discussed in Section 18.5. Therefore, we write the vibrational portion of Eq. 18.46 for a hypothetical particle of mass m_r which moves in the single direction x and substitute Eq. 18.69 for the potential energy, $\Phi(x)$. The resulting wave equation for the harmonic oscillator is

$$\frac{d^2\Psi_v}{dx^2}+\frac{8\pi^2 m_r}{h^2}(\epsilon_v-2\pi^2\nu_\varepsilon{}^2 m_r x^2)\Psi_v=0 \tag{18.70}$$

This differential equation can be converted into a standard form, called Hermite's equation by performing changes in both independent and dependent variables. It is then found that acceptable solutions (in terms of the required characteristics of Ψ^2) are possible only for the discrete values of vibrational energy

$$\epsilon_v = h\nu_\varepsilon(\mathbf{v}+\tfrac{1}{2}) \qquad \mathbf{v}=0,1,2,3,\ldots \tag{18.71}$$

Frequently, it is more convenient to use the vibrational wave number ω_ε, defined by

$$\omega_\varepsilon = \frac{\nu_\varepsilon}{c} \tag{18.72}$$

where c is the velocity of light. In this form, the vibrational energy is written

$$\epsilon_v = hc\omega_\varepsilon(\mathbf{v}+\tfrac{1}{2}); \qquad \mathbf{v}=0,1,2,3,\ldots \tag{18.73}$$

It is also of interest to note here that for the ground state $\mathbf{v}=0$ there is a residual or zero-point energy contribution of $\frac{1}{2}h\nu_\varepsilon$ or $\frac{1}{2}hc\omega_\varepsilon$.

The solution to the wave equation is given in terms of finite Hermite polynomials as

$$\Psi_v(z) = Ce^{-z^2/2}H_{\mathbf{v}}(z) \tag{18.74}$$

where

$$z=\sqrt{\frac{4\pi^2 c\omega_\varepsilon m_r}{h}}(r-r_\varepsilon) \tag{18.75}$$

These polynomials depend only on the single vibrational quantum number ν. Since the vibrational energies also depend only on this number, according to Eq. 18.73, we conclude that these are energies of single states, not degenerate levels as was the case for rotation.

The constant in Eq. 18.74 is determined by normalization such that $\Psi_v{}^2(z)$ represents the probability of finding a molecule of energy ϵ_v at position z. The details of carrying out this process need not concern us here, but the results present a very informative picture, as shown in Fig. 18.6. For each energy level, the probability distribution as represented

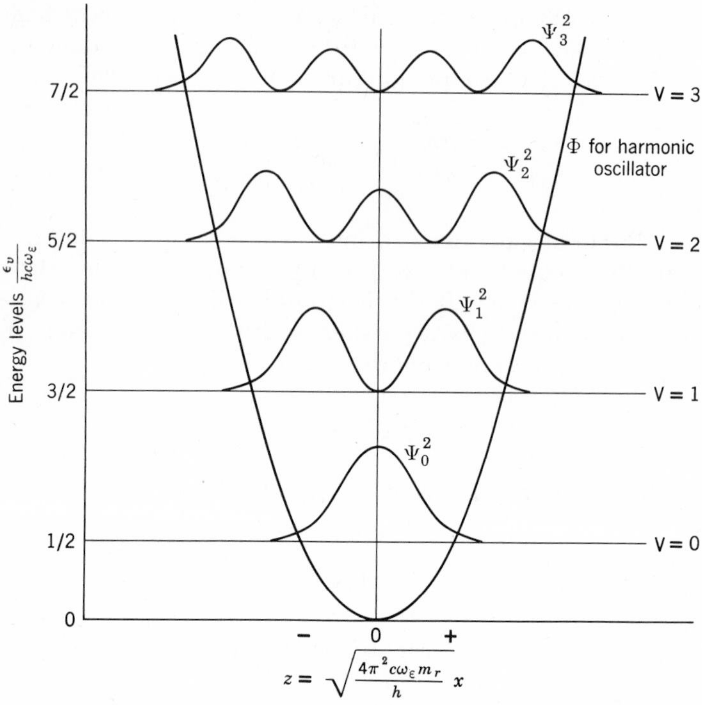

Fig. 18.6 The probability function and energy levels of a harmonic oscillator.

by Ψ_v^2 has been superimposed on the diagram of energy versus distance from the equilibrium point.

18.8 Electronic States of Atoms and Molecules

In the application of the Schrödinger wave equation to internal energy modes of a molecule, we have analyzed the simple internal model—rigid rotator, harmonic oscillator, ground electronic level—but we have not as yet discussed the electronic states for atoms or molecules. For our purposes, a detailed investigation of this subject is not necessary, because for most substances in the temperature range of interest the electronic energy levels are so widely spaced that essentially all the particles are found in the ground state at any instant of time. Nevertheless, we should be familiar enough with the subject to be able to recognize the exceptions to this statement. A second and still more important reason for at least a qualitative discussion of electronic states is that the ground level of an atom or molecule may be degenerate, resulting in a contribution to the entropy of the substance.

Let us consider the most elementary case, the hydrogen atom, as we did at the beginning of the chapter in our discussion of Bohr's theory. We apply the Schrödinger wave equation, Eq. 18.21, to the atom's single electron, which is under the influence of the nucleus according to the Coulomb potential. It is again convenient to work in spherical polar coordinates, so that the Laplacian operator is given by Eq. 18.57, which was used for the molecular rigid rotator. In direct contrast to that application, however, the distance r in this case is not constant, and the first term of Eq. 18.57 is not equal to zero. To achieve a solution of the differential equation, we assume that the wave function can be expressed as the product

$$\Psi(r, \theta, \phi) = \Psi_r(r)\,\Psi_\theta(\theta)\,\Psi_\phi(\phi) \tag{18.76}$$

where each of the three functions is dependent upon only one of the coordinates. Substitution of Eq. 18.76 into the wave equation results in a separation of the differential equation into three terms, each in terms of ordinary derivatives. Since the variables r, θ, ϕ are independent, we conclude that each term must be a constant. Solution of the θ and ϕ terms then follows directly in exactly the same manner as for the molecular rigid rotator. From this solution we obtain the pair of quantum numbers $\mathbf{l}$, a non-negative integer analogous to the $\mathbf{j}$ for a molecule, and $\mathbf{m}_l$, where

$$\mathbf{m}_l = 0, \pm 1, \pm 2, \ldots, \pm \mathbf{l} \tag{18.77}$$

and is analogous to the λ for a molecule.

The remaining second-order differential equation in r then includes the potential energy resulting from the Coulomb force and also the term in $\mathbf{l}(\mathbf{l}+1)$. We find that a solution of this difficult equation is valid only if the energy quantized according to the relation

$$\epsilon = \left(\frac{2\pi^2 C^2 \mathscr{Z}^2 e^4 m}{h^2}\right)\frac{1}{\mathbf{n}^2} \tag{18.78}$$

where the quantum number $\mathbf{n}$, called the principal quantum number, can assume only positive integral values,

$$\mathbf{n} = 1, 2, 3, \ldots. \tag{18.79}$$

For an acceptable solution it is also required that the azimuthal number $\mathbf{l}$ be restricted to the values

$$\mathbf{l} = 0, 1, 2, \ldots, (\mathbf{n}-1) \tag{18.80}$$

while the values of $\mathbf{m}_l$, the magnetic quantum number, are as given by Eq. 18.77.

For our solution, we find that the energy levels for atomic hydrogen are dependent only upon the principal quantum number $\mathbf{n}$, as given by Eq. 18.78 and shown in Fig. 18.7. Energies are given in units of electron volts and also reciprocal centimeters (ϵ/hc) with respect to the ground state. The limiting value for very large $\mathbf{n}$ is 13.595 volts (this is called the ionization potential), under which condition the electron leaves the influence of the nucleus. The vertical lines in the diagram show some of the transitions observed in the spectra of the hydrogen atom; the length of the line between any two levels is proportional to the wave number of the observed spectral line and the width is proportional to its intensity. Since the energy levels depend only upon $\mathbf{n}$, the levels must be degenerate according to the number of allowed values of the quantum numbers $\mathbf{l}$ and $\mathbf{m}_l$. Actually, from a more general solution which includes the effects of relativity and electron spin, it is found that there are very slight differences in energy for different values of $\mathbf{l}$ and $\mathbf{m}_l$; this is in agreement with very precise spectral observations. The energy levels for atomic hydrogen, including the azimuthal number $\mathbf{l}$, is shown in Fig. 18.8. The dashed lines indicate so-called forbidden transitions, which ordinarily are not observed in the spectra.

The results discussed here for the hydrogen atom are the same as those found from Bohr's theory. We find very striking differences between the theories, however, even for hydrogen. In the new quantum or wave mechanics, the results follow directly from the sole assumption of the validity of the wave equation. Another important point that should be emphasized is that, although the energy is sharply defined according to Eq. 18.78, there has been no mention of the nature of orbits as assumed by Bohr. Indeed, if we evaluate the wave function of Eq. 18.76 in terms of r, θ, ϕ and normalize in the three directions, the square of the result then designates the probability of locating the electron in a certain position with respect to the nucleus. Let us consider here the radial direction r only. For given values of $\mathbf{n}$ (and energy) and $\mathbf{l}$, the quantity Ψ^2 represents the probability of locating the electron within an element of volume. Since the volume element is represented by $4\pi^2\,dr$, the probability that the electron will be found between r and $r+dr$ and with any values of θ, ϕ is

$$4\pi\,\Psi^2 r^2\,dr$$

The nature of the function $\Psi^2 r^2$ is depicted in Fig. 18.9 for the first few values of $\mathbf{n}$ and $\mathbf{l}$. It is of particular interest to compare the probability functions of Fig. 18.9 with the corresponding orbits of the Bohr theory

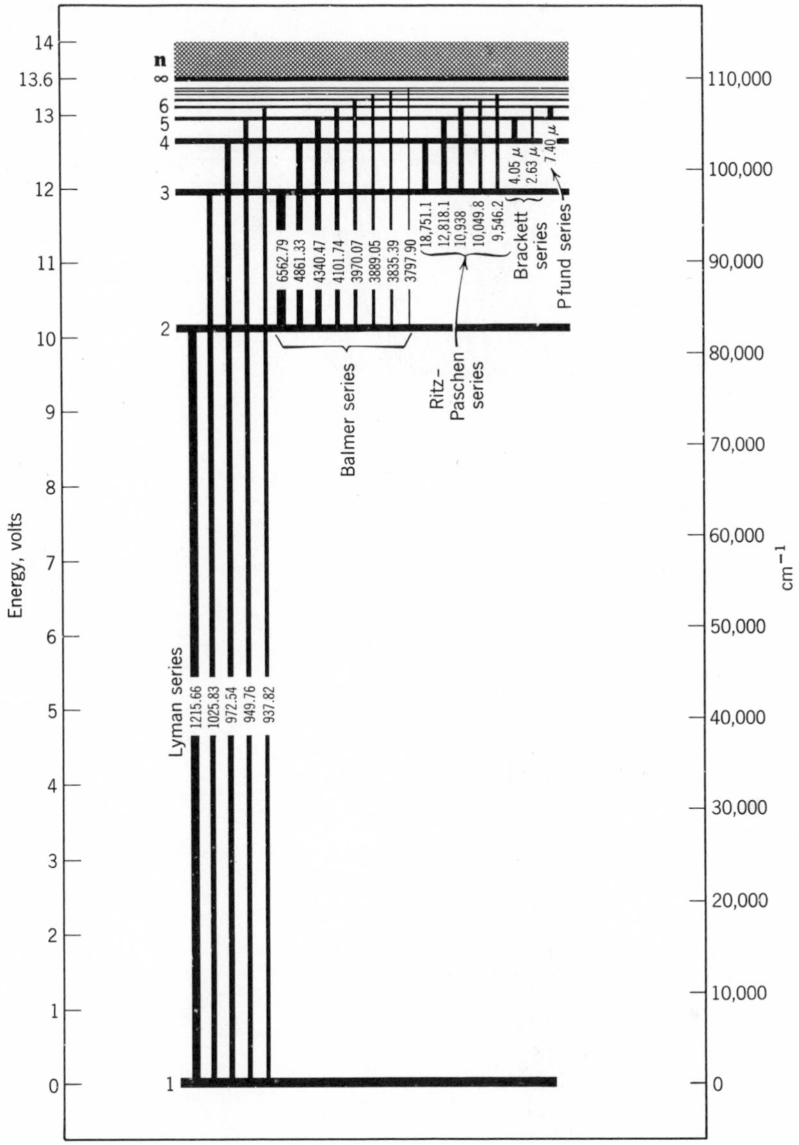

Fig. 18.7 Electronic energy levels of the hydrogen atom (from G. Herzberg, *Atomic Spectra and Atomic Structure*, Dover Publications, New York, 1944, p. 24).

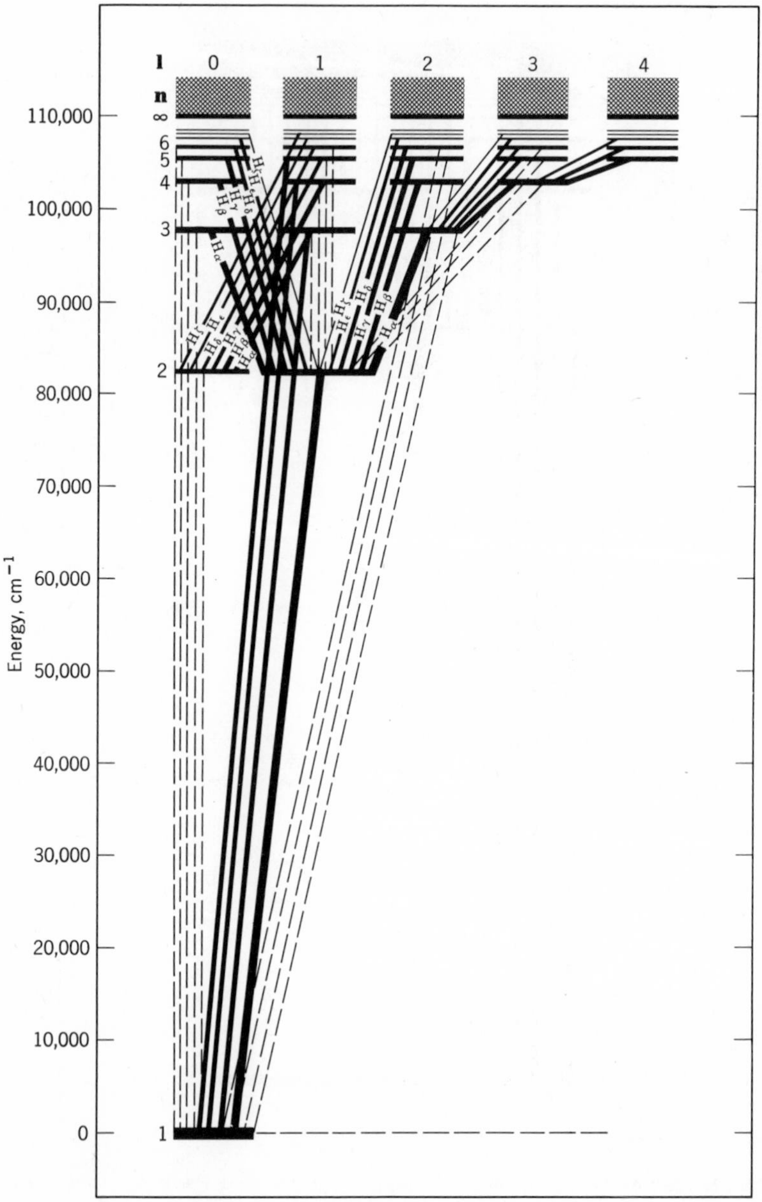

Fig. 18.8 Electronic energy levels of the hydrogen atom (from G. Herzberg, *Atomic Spectra and Atomic Structure*, Dover Publications, New York, 1944, p. 26).

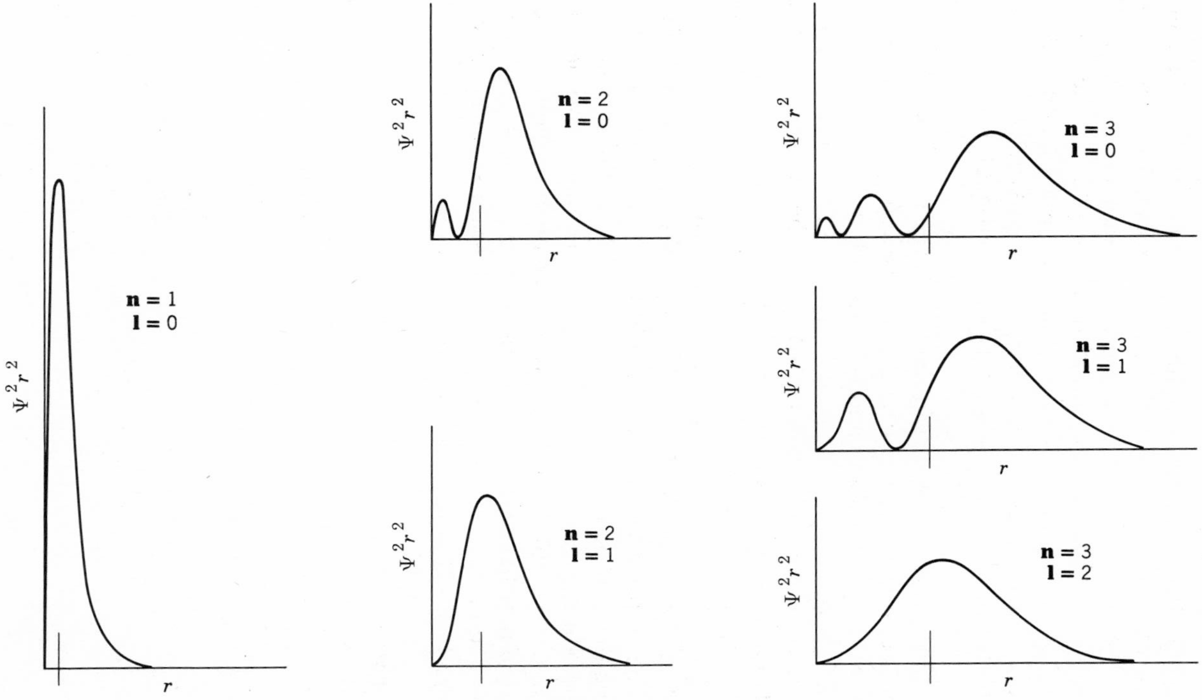

Fig. 18.9 The radial probability function for the electron of a hydrogen atom.

shown in Fig. 18.1. The marks on the r-axes in Fig. 18.9 indicate the radii of the corresponding circular Bohr orbits, which are seen to coincide with the maxima of the probability function whenever $\mathbf{l} = \mathbf{n} - 1$. Whenever $\mathbf{l} \neq \mathbf{n} - 1$, there is more than one peak in the probability function, which corresponds to the elliptical orbits of the old theory. From this comparison, we can understand why Bohr's theory was able to predict certain behavior as well as it did although it failed in other cases.

It should also be noted that in this model the values of the azimuthal and magnetic quantum numbers $\mathbf{l}$ and $\mathbf{m}_l$ give a shape to the atom as they influence the wave function and consequently the probability in the θ and ϕ directions.

The fourth quantum number of the atom, the electron spin number $\mathbf{m}_s$, did not appear in our solution of the wave equation, but it does result logically in the more general solution. We can justify the existence of such a number here in the same way as it was done before the development of wave mechanics. It is observed that the spectral lines of certain atoms split into even-numbered multiples under the influence of a magnetic field, while the splitting of the states of different $\mathbf{m}_l$ can only result in odd numbers 1, 3, 5, ... Therefore there must be a fourth quantum number to account for this observed phenomenon. Considering an electron to behave as a tiny spinning magnet with $\mathbf{m}_s = \pm\frac{1}{2}$ (indicating the orientation of the spin angular momentum vector) resolves this problem. Table 18.1 lists the first few quantum states for the single

Table 18.1

Electron State	n	l	m_l	m_s
1*s*	1	0	0	$\pm\frac{1}{2}$
2*s*	2	0	0	$\pm\frac{1}{2}$
2*p*		1	−1	$\pm\frac{1}{2}$
			0	$\pm\frac{1}{2}$
			+1	$\pm\frac{1}{2}$
3*s*	3	0	0	$\pm\frac{1}{2}$
3*p*		1	−1	$\pm\frac{1}{2}$
			0	$\pm\frac{1}{2}$
			+1	$\pm\frac{1}{2}$
3*d*		2	−2	$\pm\frac{1}{2}$
			−1	$\pm\frac{1}{2}$
			0	$\pm\frac{1}{2}$
			+1	$\pm\frac{1}{2}$
			+2	$\pm\frac{1}{2}$

electron of the hydrogen atom and shows the restrictions on the four quantum numbers. For the hydrogen atom in the absence of a magnetic field, the energy is primarily a function of the principal quantum number. Thus there are two states of the same energy for the ground level ($\mathbf{n} = 1$), eight states of approximately the same energy for the first excited level ($\mathbf{n} = 2$), eighteen for the second excited level ($\mathbf{n} = 3$), etc. Since the energy actually does depend slightly upon $\mathbf{l}$ and $\mathbf{m}_l$ even for hydrogen, we must be careful not to say that these states have exactly the same energy.

The spectroscopic notation used to designate electron configuration is also listed in Table 18.1. In this notation, the number refers to the value of $\mathbf{n}$ and the letter following denotes the azimuthal number $\mathbf{l}$. Values of $\mathbf{l}$ equal to 0, 1, 2, 3, . . . are designated by the letters *s*, *p*, *d*, *f*, . . . , in this notation. For atoms having more than one electron, a superscript after the letter gives the number of electrons with those values of $\mathbf{n}$ and $\mathbf{l}$. For example, the symbol $2p^4$ means that four electrons have the numbers $\mathbf{n} = 2, \mathbf{l} = 1$.

So far we have discussed only the hydrogen atom, because of its simplicity. The first logical extension of the theory would be to the He^+ ion, which has a nucleus of two protons and two neutrons and a single electron. Evaluation of this problem is the same as for the H atom, except for a different Coulomb potential between the nucleus and electron. The complexity grows rapidly when we consider the helium atom, however, because now there are two electrons instead of a single one. There will thus be a potential between each electron and the nucleus, and another between the pair of electrons. However, the latter potential is relatively small, and this is neglected in order to achieve a solution to the problem. The wave equation is then written for each electron, and after conbination it is found that valid solutions exist only for sets of the quantum numbers discussed previously and as given in Table 18.1. There is an additional consideration now, namely that of the Pauli exclusion principle, which will be discussed in the following section. For the present, let us say that one statement of the exclusion principle is that no two electrons in a single atom may simultaneously have an identical set of quantum numbers. Therefore, again referring to Table 18.1, we conclude that, for the ground state of the helium atom, one electron has the numbers $1, 0, 0, +\frac{1}{2}$, and the other has $1, 0, 0, -\frac{1}{2}$. That is, if the numbers $\mathbf{n}$, $\mathbf{l}$, $\mathbf{m}_l$ are all the same for two electrons, then their spins must be in opposite directions.

For more complex atoms with greater numbers of electrons, we find that, according to Pauli's exclusion principle, additional electrons must fall in successively higher states because the lower ones are already filled.

As an example, the ground level of the oxygen atom has the electron configuration $1s^2 2s^2 2p^4$.

The periodic table of the elements is also readily understood as a result of this theory. We shall consider two examples here, the inert gases and the halogens. The inert gases—helium (2 electrons), neon (10), argon (18), krypton (36), zenon (54)—all have closed or filled outer shells or subshells and are therefore very stable and chemically inactive (for Xe, there are no $4f$ electrons, which would have a higher energy than $5s$ and $5p$ electrons; hence the closed outer shell is $5p^6$). On the other hand, the halogens—fluorine (9 electrons), chlorine (17), bromine (35), iodine (53)—all lack a single electron to complete a shell or subshell and are consequently very active chemically (the situation with I is like that for Xe, the outer shell being $5p^5$). Each of the halogens has a low-lying electronic state. The first few electronic states observed for several simple atoms are given in Table 18.2.

Table 18.2
Atomic Energy Levels

Substance	Electron Configuration	Term Symbol	Energy, cm^{-1}
H (1 electron)	$1s$	$^2S_{1/2}$	0.00
	$2p$	$^2P_{1/2}$	82258.907
	$2s$	$^2S_{1/2}$	82258.942
	$2p$	$^2P_{3/2}$	82259.272
	$3p$	$^2P_{1/2}$	97492.198
He (2)	$1s^2$	$^1S_0$	0
	$1s2s$	$^3S_1$	159850.318
		$^1S_0$	166271.70
N (7)	$1s^2 2s^2 2p^3$	$^4S_{3/2}$	0
		$^2D_{5/2}$	19223.9
		$^2D_{3/2}$	19233.1
O (8)	$1s^2 2s^2 2p^4$	$^3P_2$	0
		$^3P_1$	158.5
		$^3P_0$	226.5
		$^1D_2$	15867.7
F (9)	$1s^2 2s^2 2p^5$	$^2P_{3/2}$	0
		$^2P_{1/2}$	404.0
	$1s^2 2s^2 2p^4 3s$	$^4P_{5/2}$	102406.5

From C. E. Moore, "Atomic Energy Levels," *National Bureau of Standards Circular* **467**, Vols. I–III (1949, 1952, 1958).

For the majority of substances the energy of the first excited electronic level is so high that at moderate temperature all but a negligible fraction of the atoms are in the ground level at any time. Even for such substances, it is necessary that the degeneracy of the ground level be known, and this can be found from the term symbol for the level.

The letters used for the atomic term symbols are analogous to those lower case letters used to indicate electron configuration. Of interest to us here, however, is the right-hand subscript on the term symbol, which gives the total angular momentum number **J** for the atom.

For each term there are $2\mathbf{J}+1$ components of the total angular momentum along a chosen axis, each with essentially the same energy in the absence of a magnetic field. Therefore there are $(2\mathbf{J}+1)$ wave functions for each **J**, and the degeneracy of each term is given by that quantity $(2\mathbf{J}+1)$. For example, the ground electronic level of monatomic oxygen is $^3P_2$, and the first two excited levels are $^3P_1$, $^3P_0$. Consequently, the corresponding degeneracies are 5, 3, 1, respectively. For hydrogen we find that only the ground level is important at moderate temperature since the first excited level is so high, but we draw the most important conclusion that the ground level degeneracy is equal to 2. We note also that the degeneracies of the next three terms are 2, 2, 4, for a total of eight states all having very nearly the same observed energy, which of course is in agreement with our previous discussion and Table 18.1, where there were eight states of $\mathbf{n} = 2$.

A discussion of the electronic states for molecules is necessarily even more complex than for atoms. Fortunately, for most molecules even the first excited level is sufficiently high in energy that it is satisfactory to assume that all molecules are in the ground level at any given time. This means that our only problem is that of determining the degeneracy of the electronic ground level, which again is found from the molecular term symbol, for which Greek letters are used. In this case, the degeneracy of a term equals the left-hand superscript for a Σ term and equals twice that superscript for all other terms. The reason for this involves the possible combinations of angular and spin momentum orientations, and does not concern us here.

While most molecules do not have low-lying electronic levels, diatomic oxygen is an exception, having several terms of not extremely different energies, as shown in Fig. 18.10. For molecules with such low-lying terms, the problem of evaluation of the partition function becomes extremely tedious, because the rotational and vibrational constants are in general somewhat different for the different electronic levels. Thus each electronic level must be evaluated separately, as each has its own rotation and vibration.

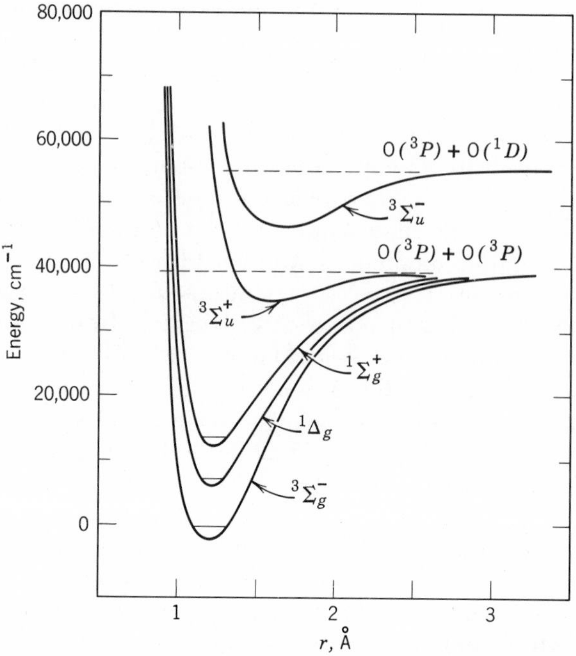

Fig. 18.10 The potential energy functions for diatomic oxygen (from G. Herzberg, *Molecular Spectra and Molecular Structure, I, Spectra of Diatomic Molecules*, 2nd edition, D. Van Nostrand Co., Princeton, N.J., 1950, p. 446).

18.9 The Pauli Exclusion Principle

In this section we discuss the symmetry nature of wave functions, the Pauli exclusion principle, and the very important point—whether Fermi-Dirac statistics or Bose-Einstein statistics represent the behavior of a given system of particles.

For simplicity, let us consider two quantum states i and j, having energies ϵ_i and ϵ_j, respectively. Now suppose that there are two identical particles 1 and 2, one in each of the states i, j. This pair may be two electrons, two protons, etc. The total energy of the system, then, is

$$\epsilon = \epsilon_i + \epsilon_j$$

and we have a situation referred to as exchange degeneracy, since the same energy results if particle 1 is in state i and particle 2 is in state j, or vice versa. The system wave functions describing these two possibilities can be written as the product of the single-particle wave functions, neglecting interactions, or

$$\Psi_i(1)\Psi_j(2)$$

or

$$\Psi_i(2)\Psi_j(1)$$

The first of these is the product of the single-particle functions (including spin coordinates) for particle 1 in state i, particle 2 in state j. The second is the product of the functions for particle 2 in state i, particle 1 in state j. Obviously, both must satisfy the combined or system wave equation describing the two-particle system. We have a problem, though, in that either of these system functions considers that the two particles can be distinguished from one another, such that we can tell which particle is in which quantum state. Such a function would, of course, be unacceptable, as a basic hypothesis of quantum mechanics is that the particles are identical and indistinguishable. This difficulty is resolved when we realize that linear combinations of these functions must also satisfy the wave equation. We have two choices, then, for a system wave function, namely a symmetric function

$$\Psi_S = \Psi_i(1)\Psi_j(2) + \Psi_i(2)\Psi_j(1) \tag{18.81}$$

or an antisymmetric function

$$\Psi_A = \Psi_i(1)\Psi_j(2) - \Psi_i(2)\Psi_j(1) \tag{18.82}$$

We note that the sign of the symmetric system wave function remains the same upon interchange of the coordinates of the two identical particles, but the antisymmetric system function changes sign if the particles are interchanged.

For any system of a pair of identical particles, either Eq. 18.81 or Eq. 18.82 must be valid (neglecting interaction). The important point is that both cannot be valid because then their linear combination would also necessarily be valid, and we would be back to our original, unacceptable functions.

From the experimentally observed behavior, it is found that particles with half-integral spin—electrons, protons, nuclei of odd mass number, for example—have only antisymmetric states and wave functions. Particles with integral spin—photons, nuclei of even mass number—have only symmetric states and wave functions.

It is not difficult to construct symmetric or antisymmetric system functions from single-particle functions for a system of more than two quantum states and more than two identical particles. We realize immediately that it is impossible to construct an antisymmetric function from a collection of single-particle functions if any two of the single-particle functions (i.e., quantum states) are identical. In such a case those two states could be exchanged with no effect on the system function. We recognize this statement as one form of the Pauli exclusion principle presented earlier: no two electrons in an atom can simultaneously have an identical set of quantum numbers. Thus we now find that this statement is equivalent to the wave mechanical statement that the system wave function is antisymmetric. This, then, is the Pauli exclusion principle. For the electrons in an atom, it is commonly stated as: the wave function for any atom must be antisymmetric in all the electrons, since interchange of the coordinates of any pair must result in a sign change of the system function. We also note that the Pauli exclusion principle says nothing about any two particles that are not identical, as, for example, a proton and an electron, or about particles having symmetric wave functions.

We are now in a position to make the necessary connection between quantum mechanics and the types of statistics developed earlier. We find that the particles with antisymmetric wave functions discussed above are subject to the exclusion principle and consequently can have no two identical single-particle states occupied. These particles therefore obey the Fermi-Dirac statistics, developed under the assumption of no more than one particle per state. Conversely, for particles having symmetric wave functions there is no such restriction; hence these particles obey the Bose-Einstein statistics, for which it is possible to have more than one particle per state.

18.10 Band Theory of Solids

In Section 18.8, we discussed the electronic states of atoms, which were assumed to be independent of one another, that is, at infinite separation. In this, the final section of this chapter, we consider, in a qualitative sense, the effects on the energy levels of bringing such atoms into close proximity with one another—which is the situation in the lattice structure of a solid. The result of this interaction is shown schematically in Fig. 18.11, where, at infinite separation ($1/r = 0$), the energy levels are those for the individual atoms. As the distance between atoms is progressively decreased, these levels split and change as indicated on the diagram. A solid is comprised of a large number of atoms packed in a crystal lattice, for which the original single-atom levels have blended into energy bands, each comprised of a large, but finite, number of electron

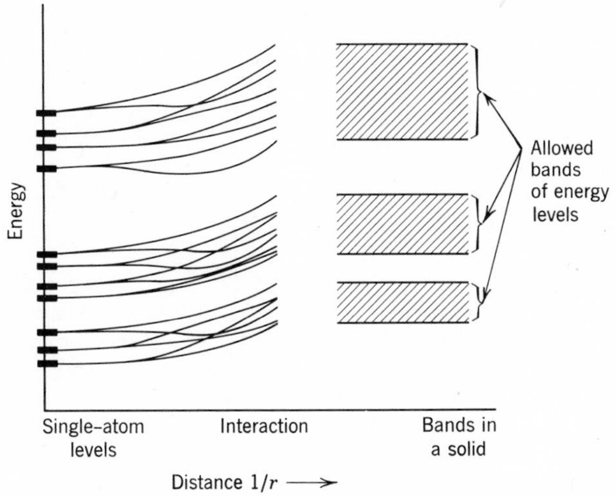

Fig. 18.11 Energy bands in a solid.

levels. Between these allowed energy bands are forbidden ranges in which there are no states available to the electrons.

In Section 18.9 we determined from the exclusion principle that electrons behave according to Fermi-Dirac statistics. From the discussion of Fermi-Dirac statistics in Section 17.5, we conclude that, at absolute zero temperature, electrons fill the lower energy levels (and therefore bands) up to an energy μ_0, the Fermi level, above which the energy levels are empty. For an electrical insulator, this Fermi level is such that there is a filled energy band above which there is a large energy gap (5–10 ev) to the next allowed band, which is empty. This situation is depicted in Fig. 18.12*a*. The highest filled band is termed the valence band, and the next band above this (empty) is called the conduction band. For a metal, an electrical conductor, the situation at absolute zero temperature is different; the Fermi level occurs in the middle of an allowed band, as shown in Fig. 18.12*b*. Thus there are electrons in the conduction band at this condition.

As the temperature is increased, to room temperature for example, because of the large energy gap only a very few electrons in an insulator have received sufficient energy to move into the conduction band. Consequently, the electrons in an insulator are rather immobile, and the material is a poor conductor. Conversely, the electrons in a metal are very easily excited into the levels of slightly higher energies within the same band, and they are quite mobile. The material is therefore an excellent conductor.

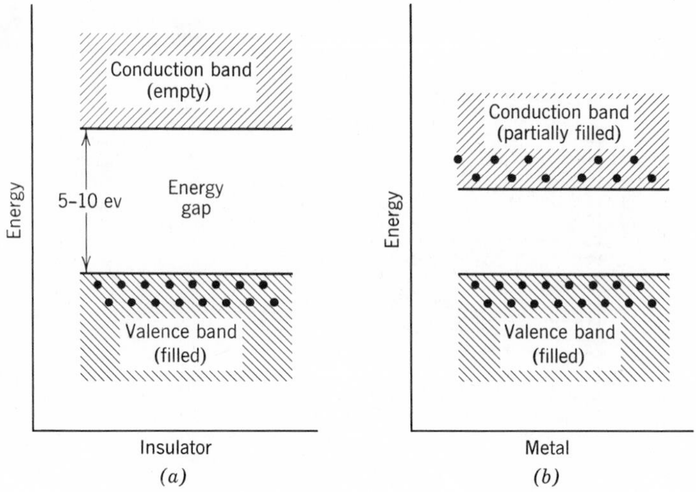

Fig. 18.12 Band characteristics of insulators and metals.

A semiconductor has basically the same characteristics as an insulator, except that the energy gap is considerably smaller, on the order of 1 ev as compared with 5–10 ev for the insulator. Consider the intrinsic semiconductor germanium, which has an atomic number of 32 and consequently a valence of 4, i.e., four valence electrons in the outer ring beyond the closed shells of 2, 8, and 18 electrons (see Section 18.8). Therefore, in a crystal structure, each germanium atom shares these four valence electrons with one from each of four neighboring atoms to form the stable structure of an effective eight-electron outer ring. The energy gap for this pure material is approximately 0.7 ev.

A different situation results when an intrinsic semiconductor is "doped" with controlled small amounts of certain impurities. For example, consider the addition of a small amount of arsenic (atomic number 33, and a valence of 5) to germanium. At the lattice sites now occupied by arsenic atoms, sharing of electrons with neighboring germanium atoms occurs as in pure germanium, except that now there is one extra electron at each of these sites and it is relatively mobile. In terms of the energy bands, the addition of the arsenic atoms therefore corresponds to the addition of an "impurity level" containing electrons. The amount of arsenic can be controlled to locate this impurity level just below the conduction band ($\sim$ 0.01 ev) as shown in Fig. 18.13*a*. Thus only small amounts of energy are required to excite electrons from the impurity, or donor, level into the conduction band. Semiconductors of

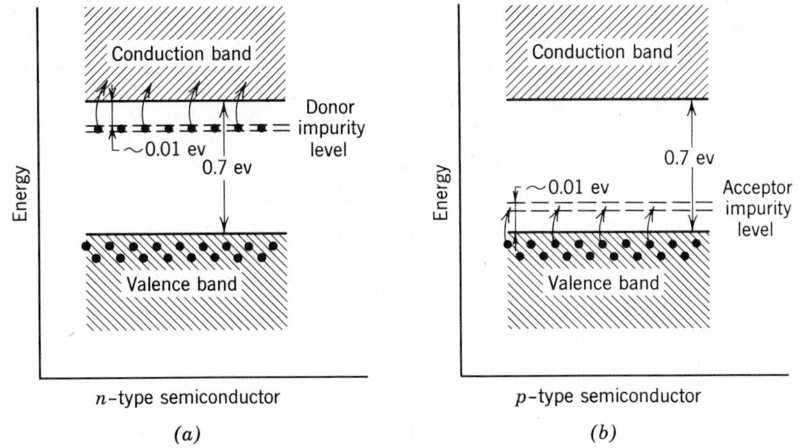

Fig. 18.13 Band characteristics of doped semiconductors.

this type are called *n*-type semiconductors because of their negative (electron) charge carriers.

Finally, consider the addition of a small amount of gallium (atomic number 31, and a valence of 3) to germanium. At the lattice sites now occupied by the gallium atoms, the sharing of valence electrons with four neighboring germanium atoms results in an electron deficiency of one, or a positive "hole." Consequently, an electron from another germanium atom jumps into the hole to complete the bond, leaving a hole in that atom. The process continues, resulting in the creation of mobile holes, or positive charge carriers in the material. In terms of band theory, the addition of gallium creates an impurity level of holes, which can be located just above the valence band of pure germanium, as in Fig. 18.13*b*. Only small amounts of energy are required to excite electrons from the valence band into the holes in this impurity, or acceptor, level, thereby leaving behind mobile holes in the valence band. Semiconductors of this type are called *p*-type semiconductors, because of their positive (hole) charge carriers.

PROBLEMS

18.1 Consider the Bohr theory of the hydrogen atom. The proportionality constant C in the Coulomb potential has the value 8.98×10^{18} dyne-cm^2/coulomb^2.

(*a*) Show that the radii of the Bohr orbits are proportional to $\mathbf{n}^2$. Calculate the radius of the ground level orbit.

(*b*) Develop an expression for the change in energy corresponding to any change $\mathbf{n}$ to $\mathbf{n}'$.

18.2 (*a*) In 1885, Balmer correlated hydrogen spectral data from the visible and near ultraviolet region according to the expression

$$\nu = c\mathscr{R}\left(\frac{1}{2^2} - \frac{1}{\mathbf{n}^2}\right); \quad \mathbf{n} = 3, 4, 5, \ldots$$

where $\mathscr{R}$ (called the Rydberg constant) has the value 109,678 cm^{-1}. Using the results of Problem 18.1 and Planck's relation, calculate the theoretical value of $\mathscr{R}$ according to Bohr's theory.

(*b*) The Lyman series for H in the far ultraviolet region is expressed by the relation

$$\nu = c\mathscr{R}\left(\frac{1}{1^2} - \frac{1}{\mathbf{n}^2}\right); \quad \mathbf{n} = 2, 3, 4, \ldots$$

Calculate the energy levels for H for both the Lyman and the Balmer series, and compare the results with Fig. 18.7. What is the ionization energy for atomic hydrogen?

18.3 A certain spectral series for singly ionized helium (He^+) is correlated by the formula

$$\nu = 4c\mathscr{R}\left(\frac{1}{(\mathbf{n}')^2} - \frac{1}{\mathbf{n}^2}\right)$$

where $\mathscr{R}$ has the value 109,722 cm^{-1}, nearly the same as the Rydberg constant of Problem 18.2. Discuss this expression in terms of the Bohr theory.

18.4 Explanation of the photoelectric effect was one of the early successes of quantum mechanics. For a given metal, it is found that light of less than a certain threshold frequency will not stimulate electron emission from the surface. Above this threshold frequency, the maximum kinetic energy of emitted electrons is directly proportional to frequency and independent of the intensity of the incident light. As explained by the quantum theory, the maximum kinetic energy of an emitted electron is the energy received from a photon of radiation minus an amount φ expended in escaping from the metal. For cesium, φ, called the work function, has the value 1.9 ev. Show that light in the visible region of the spectra (4000–7000 Å) stimulates photoelectron emission in cesium.

18.5 Consider an α particle (the nucleus of a He^4 atom) traveling with a kinetic energy of 6 Mev. What is the deBroglie wavelength associated with this particle?

18.6 Suppose that the position of an electron moving in the *x*-direction is known to within 0.01 mm. What is the order of magnitude of uncertainty in the velocity of the electron? If the particle is an α particle instead of an electron, how is the result changed?

18.7 Verify that Eq. 18.37 is a solution of the Schrödinger wave equation for translational energy, Eq. 18.25.

18.8 Consider an electron gas confined in a volume of 1 cm^3. Determine the spacing of adjacent translational energy states at the level where all three quantum numbers are 10 and also at the level where all three quantum numbers are 1000.

18.9 Repeat Problem 18.8 for
(*a*) Diatomic hydrogen.
(*b*) Diatomic chlorine.

18.10 Assuming that the contributions of translational and internal modes are independent (Eqs. 18.43 and 18.44), show that the Schrödinger wave equation applied to the molecule can be separated into the two parts, Eqs. 18.45 and 18.46.

18.11 Show that Eq. 18.50 is the representation of the combined rotational kinetic energy for the two atoms comprising a rigid diatomic molecule.

18.12 For rotational energy, show that Eq. 18.58 separates to the form of Eq. 18.60 by using Eq. 18.59.

18.13 Consider the rotational differential equation, Eq. 18.62.
(*a*) Use the variable $x = \cos\theta$ and Eq. 18.63 to obtain the associated Legendre differential equation.
(*b*) Verify that the associated Legendre functions $P_{\mathbf{j}}^{|\lambda|}(x)$ are solutions of the result of part *a*. These are defined as

$$P_{\mathbf{j}}^{|\lambda|}(x) = \frac{1}{2^{\mathbf{j}}\mathbf{j}!}(1-x^2)^{|\lambda|/2}\frac{d^{|\lambda|+\mathbf{j}}}{dx^{|\lambda|+\mathbf{j}}}[(x^2-1)^{\mathbf{j}}]$$

18.14 (*a*) From the definition given in Problem 18.13, why is the maximum value of λ restricted to that of $\mathbf{j}$?
(*b*) Calculate the associated Legendre functions for values of $\mathbf{j} = 0$, 1, and 2.

18.15 (*a*) Calculate the spacing of rotational energy levels for diatomic hydrogen for the first five values of $\mathbf{j}$. Compare the results with the spacings of Problem 18.9. (Use molecular constants from Table B.14)
(*b*) Repeat the calculation for diatomic chlorine.

18.16 The solution to the vibrational wave equation is given by Eq. 18.74 in terms of Hermite polynomials, defined by

$$H_{\mathbf{v}}(z) = (-1)^{\mathbf{v}}e^{z^2}\frac{d^{\mathbf{v}}}{dz^{\mathbf{v}}}[e^{-z^2}]$$

Verify that Eq. 18.74 is a solution of the differential equation, Eq. 18.70.

18.17 (*a*) Calculate the spacing of vibrational levels for diatomic hydrogen. Compare the result with Problems 18.9 and 18.15.
(*b*) Repeat the calculation for diatomic chlorine.

18.18 (*a*) The first excited electronic level for diatomic hydrogen is 90,171 cm^{-1} above the ground level. Compare this spacing with those of the rotational and vibrational levels.

(*b*) Repeat the calculation for diatomic chlorine, for which the first electronic level is 18,147 cm^{-1} above the ground level.

18.19 Discuss the first 18 elements of the periodic table (hydrogen through argon) in terms of their respective ground level electron configurations.

18.20 In this chapter we have been concerned with the stationary states of particles, and consequently we have considered the time-independent Schrödinger wave equation. The general time-dependent Schrödinger equation is

$$\left(\frac{h^2}{8\pi^2 m}\right)\nabla^2\Psi - \Phi\Psi = \left(\frac{h}{2\pi i}\right)\frac{\partial\Psi}{\partial t}$$

Consider this equation applied to the *x*-direction only, for convenience. Using the procedures followed in Section 18.3, show that, if the potential energy Φ is not a function of time, the general wave function Ψ is separable to

$$\Psi(x, t) = \Psi(x)\varphi(t)$$

with two resulting expressions, one the relation

$$\varphi(t) = e^{-(2\pi i/h)\epsilon t}$$

and the other the time-independent Schrödinger equation, Eq. 18.17.

19

The Properties of Gases

In Chapter 17, certain thermodynamic properties of interest were expressed as functions of the partition function, which is defined in terms of the energy levels and their degeneracies. This partition function development is valid for systems obeying corrected Boltzmann statistics, the approximation to both Bose-Einstein and Fermi-Dirac systems. Then, in Chapter 18, we evaluated the energies and degeneracies for the various energy modes in terms of the appropriate sets of quantum numbers. We now proceed to use these results to evaluate the partition function and, therefore, the thermodynamic properties. In this chapter, we limit our considerations to the properties of gases, after first discussing general separation of the partition function into translational and internal mode components. Three special problems, the representation of electromagnetic radiation by a photon gas, the behavior of mixtures, and chemical reaction and equilibrium are also introduced.

19.1 Contributions to the Partition Function and to Properties

For corrected Boltzmann statistics, the partition function has been expressed by Eq. 17.19:

$$Z = \sum_j g_j e^{-\epsilon_j/kT}$$

Since the summation is to be made over all energy levels accessible to the molecule, it must include all the various modes of energy. It would prove convenient to separate this into summations over the individual energy modes. The degree to which such a separation can be carried depends upon several factors (the substance and the range of temperature, for example), but in all cases we can separate the energy into translational and internal contributions in accordance with Eq. 18.43,

$$\epsilon = \epsilon_t + \epsilon_{\text{int}}$$

Since the corresponding wave functions Ψ (representing quantum states) are in such a case multiplicative, as given by Eq. 18.44,

$$\Psi = \Psi_t \Psi_{\text{int}}$$

the component energy level degeneracies must also be multiplicative, or

$$\mathsf{g} = \mathsf{g}_t \mathsf{g}_{\text{int}} \tag{19.1}$$

Substitution of Eqs. 18.43 and 19.1 into the partition function Eq. 17.19, results in the separation

$$\begin{aligned} Z &= \sum_{\text{all } j} \mathsf{g}_{j_t} \mathsf{g}_{j_{\text{int}}} \exp\left[-\frac{\epsilon_{j_t} + \epsilon_{j_{\text{int}}}}{kT}\right] \\ &= \left[\sum_{j_t} \mathsf{g}_{j_t} \exp\left(-\frac{\epsilon_{j_t}}{kT}\right)\right]\left[\sum_{j_{\text{int}}} \mathsf{g}_{j_{\text{int}}} \exp\left(-\frac{\epsilon_{j_{\text{int}}}}{kT}\right)\right] \\ &= Z_t Z_{\text{int}} \end{aligned} \tag{19.2}$$

where, by definition,

$$Z_t = \sum_{j_t} \mathsf{g}_{j_t} \exp\left(-\frac{\epsilon_{j_t}}{kT}\right) \tag{19.3}$$

$$Z_{\text{int}} = \sum_{j_{\text{int}}} \mathsf{g}_{j_{\text{int}}} \exp\left(-\frac{\epsilon_{j_{\text{int}}}}{kT}\right) \tag{19.4}$$

Thus the partition function can be written as the product of the translational partition function, Eq. 19.3, and the internal partition function, Eq. 19.4. The first of these can readily be evaluated by using the expression for translational energy found from quantum mechanics in the preceding chapter. The evaluation of the second usually proves more difficult, depending upon the number of modes contributing internally (in general, rotational, vibrational, electronic) and the degree to which they can be separated from one another.

Having found that the partition function can be separated into a product involving translational and internal modes of energy, let us now substitute Eq. 19.4 into our equations for the thermodynamic properties. From Eq. 17.39,

$$\begin{aligned}\bar{u} &= \bar{R}T^2\left[\frac{\partial \ln (Z_t Z_{\text{int}})}{\partial T}\right]_V \\ &= \bar{R}T^2\left(\frac{\partial \ln Z_t}{\partial T}\right)_V + \bar{R}T^2\left(\frac{\partial \ln Z_{\text{int}}}{\partial T}\right)_V \\ &= \bar{u}_t + \bar{u}_{\text{int}} \qquad (19.5)\end{aligned}$$

We find that the internal energy can be calculated as a sum of contributions, that due to translation

$$\bar{u}_t = \bar{R}T^2\left(\frac{\partial \ln Z_t}{\partial T}\right)_V \qquad (19.6)$$

and that resulting from the internal modes

$$\bar{u}_{\text{int}} = \bar{R}T^2\left(\frac{\partial \ln Z_{\text{int}}}{\partial T}\right)_V \qquad (19.7)$$

The enthalpy of an ideal gas can now be found from Eq. 17.40 as

$$\bar{h} = \bar{u} + P\bar{v} = \bar{u} + \bar{R}T = \bar{u}_t + \bar{u}_{\text{int}} + \bar{R}T$$

This can also be expressed as

$$\bar{h} = \bar{h}_t + \bar{h}_{\text{int}} \qquad (19.8)$$

if the translational enthalpy is defined as

$$\bar{h}_t = \bar{u}_t + \bar{R}T \qquad (19.9)$$

It is then noted that the internal contribution to enthalpy and internal energy are the same, namely,

$$\bar{h}_{\text{int}} = \bar{u}_{\text{int}} \qquad (19.10)$$

The $\bar{R}T$ term has been associated with the translation in Eq. 19.9. This can be justified from experimental observation for substances having no internal contributions. We might alternatively justify this by using the expression developed for pressure in Chapter 17, Eq. 17.34,

$$P = NkT\left(\frac{\partial \ln Z}{\partial V}\right)_T$$

We recall from the results of quantum mechanics that the internal modes are not functions of volume and consequently do not contribute anything to the pressure (or to $P\bar{v}$). From the definition of enthalpy, we then arrive at Eq. 19.10.

Expressions for specific heats can also be developed, but these quantities are most readily evaluated from their definitions, Eqs. 17.41 and 17.42, using Eqs. 19.5 to 19.10.

From Eqs. 17.43, 19.2, and 19.5, we find that the entropy can also be expressed as a sum of the contributions of translational and internal modes,

$$\bar{s} = \bar{R}\left[\ln\left(\frac{Z_t Z_{\text{int}}}{N}\right) + 1\right] + \frac{\bar{u}_t + \bar{u}_{\text{int}}}{T}$$

This can also be written as

$$\bar{s} = \bar{s}_t + \bar{s}_{\text{int}} \tag{19.11}$$

if the translational entropy is defined as

$$\bar{s}_t = \bar{R}\left[\ln\left(\frac{Z_t}{N}\right) + 1\right] + \frac{\bar{u}_t}{T} \tag{19.12}$$

and the internal contribution to this property is defined as

$$\bar{s}_{\text{int}} = \bar{R} \ln Z_{\text{int}} + \frac{\bar{u}_{\text{int}}}{T} \tag{19.13}$$

The factors N and unity are associated with the translational contribution, which is in agreement with observed data for substances lacking internal contributions.

In a similar manner, the Helmholtz function is given by

$$\bar{a} = \bar{a}_t + \bar{a}_{\text{int}} \tag{19.14}$$

with

$$\bar{a}_t = -\bar{R}T\left[\ln\left(\frac{Z_t}{N}\right) + 1\right] \tag{19.15}$$

$$\bar{a}_{\text{int}} = -\bar{R}T \ln Z_{\text{int}} \tag{19.16}$$

The Gibbs function is given by

$$\bar{g} = \bar{g}_t + \bar{g}_{\text{int}} \tag{19.17}$$

where

$$\bar{g}_t = -\bar{R}T \ln\left(\frac{Z_t}{N}\right) \tag{19.18}$$

$$\bar{g}_{\text{int}} = \bar{a}_{\text{int}} = -\bar{R}T \ln Z_{\text{int}} \tag{19.19}$$

19.2 Translation

In the preceding section, we performed a general separation of the partition function into translational and internal mode components. For the remainder of this chapter, we shall discuss the thermodynamic properties for the various classes of gases, namely monatomic, diatomic and polyatomic. Each of these classes differs in its number and types of internal modes, as mentioned in Chapter 2, and will be considered in turn in later sections of this chapter. However, the expression for the translational energy as given by Eq. 18.42,

$$\epsilon_{\mathbf{k}_x,\mathbf{k}_y,\mathbf{k}_z} = \frac{h^2}{8mV^{2/3}}(\mathbf{k}_x{}^2 + \mathbf{k}_y{}^2 + \mathbf{k}_z{}^2)$$

is the same for each of these types of molecules. Thus, the results of the present section will apply to all the classes of molecules. We need only be careful to use the total mass of the molecule (as found by summing the masses of the component atoms for molecules comprised of two or more atoms) for m in Eq. 18.42.

In order to use Eq. 18.42 to evaluate the translational partition function, we first recall that this expression gives the energy of a particular quantum state and not a degenerate level of energy. Thus, the translational partition function can be calculated most directly by summing over all quantum or energy states, according to Eq. 16.36, instead of over all levels of energy as implied in Eq. 19.3. That is,

$$Z_t = \sum_{j_t \text{ levels}} \mathrm{g}_{j_t} \exp\left[-\frac{\epsilon_{j_t}}{kT}\right]$$

$$= \sum_{\mathbf{k}_x=1}^{\infty} \sum_{\mathbf{k}_y=1}^{\infty} \sum_{\mathbf{k}_z=1}^{\infty} \exp\left[-\frac{\epsilon_{\mathbf{k}_x,\mathbf{k}_y,\mathbf{k}_z}}{kT}\right] \tag{19.20}$$

Now, substituting Eq. 18.48, which gives the energy of a state, into Eq. 19.20, we have

$$Z_t = \sum_{\mathbf{k}_x=1}^{\infty} \sum_{\mathbf{k}_y=1}^{\infty} \sum_{\mathbf{k}_z=1}^{\infty} \exp\left[-\frac{h^2}{8mV^{2/3}kT}(\mathbf{k}_x{}^2+\mathbf{k}_y{}^2+\mathbf{k}_z{}^2)\right]$$

Since the three functions and summations in $\mathbf{k}_x$, $\mathbf{k}_y$, $\mathbf{k}_z$ are identical, an arbitrary subscript i can be used for each, so that

$$Z_t = \left\{\sum_{\mathbf{k}_i=1}^{\infty} \exp\left[-\frac{h^2}{8mV^{2/3}kT}\mathbf{k}_i{}^2\right]\right\}^3 \tag{19.21}$$

The summation inside the brackets in this equation is of the form

$$\sum_{\mathbf{k}=1}^{\infty} e^{-r\mathbf{k}^2} = \sum_{\mathbf{k}=0}^{\infty} e^{-r\mathbf{k}^2} - 1$$

with r a constant. The summation on the right side here is evaluated in terms of an integral using the Euler-Maclaurin summation theorem, Eq. A.28,

$$\sum_{\mathbf{k}=0}^{\infty} e^{-r\mathbf{k}^2} = \int_0^{\infty} e^{-r\mathbf{k}^2}\,d\mathbf{k} + \tfrac{1}{2}(1) - \tfrac{1}{12}(0) + \tfrac{1}{720}(0) + \cdots$$

Therefore only the first two terms are different from zero, and, on combining this expression with the previous one, we have

$$\sum_{\mathbf{k}=1}^{\infty} e^{-r\mathbf{k}^2} = \int_0^{\infty} e^{-r\mathbf{k}^2}\,d\mathbf{k} - \tfrac{1}{2}$$

To apply this result to Eq. 19.21, we anticipate the result that Z_t will be a large number. The difference of $-\frac{1}{2}$ between the summation and integral will then be negligible. In other words, we are making the assumption that the translational energy levels are so closely spaced that the summation of Eq. 19.21 can be directly replaced by an integral. This will be found to be of general validity for translational energy in our applications, but the Euler-Maclaurin theorem gives us a means for justifying the approximation. Therefore

$$Z_t = \left\{\int_0^{\infty} \exp\left[-\frac{h^2}{8mV^{2/3}kT}\mathbf{k}_i{}^2\right] d\mathbf{k}_i\right\}^3$$

On performing this integration, we find

$$Z_t = \left[\tfrac{1}{2}\sqrt{\pi}\sqrt{\frac{8mV^{2/3}kT}{h^2}}\right]^3 = V\left(\frac{2\pi mkT}{h^2}\right)^{3/2} \tag{19.22}$$

This result is identical with that found for a monatomic ideal gas assuming classical mechanics in Section 16.3. We note that the assumption discussed above (neglecting the factor of $\frac{1}{2}$) which led to Eq. 19.22 is equivalent to the assumption made in the analysis of Section 16.3.

For the translational contributions to thermodynamic properties, it is convenient to write Eq. 19.22 in logarithmic form:

$$\ln Z_t = \ln V + \tfrac{3}{2}\ln\left(\frac{2\pi mk}{h^2}\right) + \tfrac{3}{2}\ln T \tag{19.23}$$

Now, using Eq. 19.23 to evaluate Eq. 17.34,

$$P = NkT\left(\frac{\partial \ln Z}{\partial V}\right)_T$$

we obtain

$$P = NkT\left(\frac{1}{V}\right) = \frac{n\bar{R}T}{V} = \frac{\bar{R}T}{\bar{v}} \tag{19.24}$$

as found previously in Eq. 17.38. Translation provides the entire contribution to pressure, since the internal modes are not functions of volume. From Eq. 19.6,

$$\bar{u}_t = \bar{R}T^2\left(\frac{\partial \ln Z_t}{\partial T}\right)_V$$

the internal energy is

$$\bar{u}_t = \bar{R}T^2\left(\frac{3}{2T}\right) = \tfrac{3}{2}\bar{R}T \tag{19.25}$$

and the enthalpy is found from Eq. 19.9 to be

$$\bar{h}_t = \tfrac{3}{2}\bar{R}T + \bar{R}T = \tfrac{5}{2}\bar{R}T \tag{19.26}$$

The translational contribution to the constant-volume specific heat is, from Eqs. 17.41 and 19.25,

$$\bar{C}_{v_t} = \left(\frac{d\bar{u}_t}{dT}\right) = \tfrac{3}{2}\bar{R} \tag{19.27}$$

while that to the constant-pressure specific heat is, from Eqs. 17.42 and 19.26,

$$\bar{C}_{p_t} = \left(\frac{d\bar{h}_t}{dT}\right) = \tfrac{5}{2}\bar{R} \tag{19.28}$$

The translational entropy is found from Eq. 19.12 to be

$$\bar{s}_t = \bar{R}\left[\ln\left(\frac{Z_t}{N}\right) + 1\right] + \tfrac{3}{2}\bar{R} = \bar{R}\left[\ln\left(\frac{Z_t}{N}\right) + \tfrac{5}{2}\right] \tag{19.29}$$

From Eqs. 19.22 and 19.24,

$$\frac{Z_t}{N} = \frac{V}{N}\left(\frac{2\pi mkT}{h^2}\right)^{3/2} = \left(\frac{2\pi m}{h^2}\right)^{3/2} \frac{(kT)^{5/2}}{P} \tag{19.30}$$

Similarly, the translational contributions to the Helmholtz and Gibbs functions are found from Eqs. 19.15 and 19.18, respectively, using Eq. 19.30 for the quantity (Z_t/N).

Example 19.1

Calculate the translational contributions to enthalpy, entropy, and Gibbs function for helium at 25 C, 1 atm pressure.

For He:

$$m = 1.66 \times 10^{-24}(4.003) = 6.645 \times 10^{-24}\ \text{gm}$$

From Eq. 19.26,

$$\bar{h}_t = \tfrac{5}{2}\bar{R}T = \tfrac{5}{2}(1.987)(298.15) = 1481\ \text{cal/mole}$$

From Eq. 19.30,

$$\frac{Z_t}{N} = \left(\frac{2\pi m}{h^2}\right)^{3/2} \frac{(kT)^{5/2}}{P}$$

$$= \left(\frac{2\pi \times 6.645 \times 10^{-24}}{(6.625 \times 10^{-27})^2}\right)^{3/2} \frac{(1.3804 \times 10^{-16} \times 298.15)^{5/2}}{1.01325 \times 10^6}$$

$$= 3.10 \times 10^5$$

From Eq. 19.29,

$$\bar{s}_t = \bar{R}\left[\ln\left(\frac{Z_t}{N}\right) + \tfrac{5}{2}\right] = 1.987[12.66 + 2.5] = 30.1\ \text{cal/mole-K}$$

Therefore, from Eq. 19.18,

$$\bar{g}_t = -\bar{R}T \ln\left(\frac{Z_t}{N}\right)$$

$$= -1.987(298.15)(12.66)$$

$$= -7500\ \text{cal/mole}$$

Alternatively, using the definition of the Gibbs function,

$$\bar{g}_t = \bar{h}_t - T\bar{s}_t = 1481 - 298.15(30.1) = -7500 \text{ cal/mole}$$

19.3 An Alternative Evaluation of the Translational Partition Function

It was noted in Section 16.2 that the partition function defined by Eq. 16.35 in terms of a summation over degenerate levels of energy can alternatively be expressed as a sum over all individual quantum states, so that

$$Z = \sum_{j \text{ levels}} g_j e^{-\epsilon_j/kT} = \sum_{i \text{ states}} e^{-\epsilon_i/kT}$$

The latter approach was used in evaluating the partition function for translation in the previous section, since the translational energy as given by Eq. 18.42 is the energy of a quantum state, and not that of a degenerate level of energy. It is also desirable, however, to analyze this problem from the stand-point of energy levels, and to do so we must develop an expression for energy level degeneracy. That is, the number of individual quantum states (each specified by a set of integers $\mathbf{k}_x$, $\mathbf{k}_y$, $\mathbf{k}_z$) having energy in an interval ϵ to $\epsilon + d\epsilon$ must be found.

Let us consider a three-dimensional space of coordinates $\mathbf{k}_x$, $\mathbf{k}_y$, $\mathbf{k}_z$, a two-dimensional cross section of which is shown in Fig. 19.1. Each of the

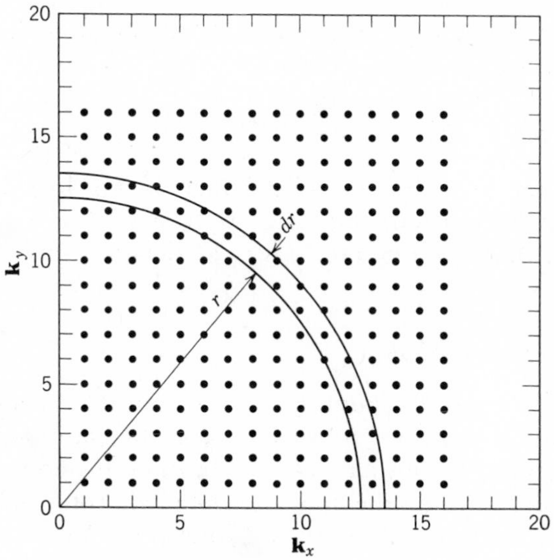

Fig. 19.1 Cross section of the three-dimensional $\mathbf{k}_x$, $\mathbf{k}_y$, $\mathbf{k}_z$ space.

three variables is restricted to positive integral values, and we note that a point in this $\mathbf{k}_x$, $\mathbf{k}_y$, $\mathbf{k}_z$ space represents a translational quantum state. The distance r of any point from the origin is given by

$$r = (\mathbf{k}_x{}^2 + \mathbf{k}_y{}^2 + \mathbf{k}_z{}^2)^{1/2} \tag{19.31}$$

We now restrict the discussion to those states (sets of $\mathbf{k}_x$, $\mathbf{k}_y$, $\mathbf{k}_z$) for which r is large compared to unity, so that r can be treated as essentially continuous. (Translational states for which r is small constitute only a negligible fraction of the total number and can therefore be neglected here.) Thus the number of states g_r in the spherical shell from r to $r + dr$ is

$$\mathsf{g}_r = \tfrac{1}{8}(4\pi r^2\, dr) \tag{19.32}$$

where the factor $\frac{1}{8}$ accounts for the restriction of $\mathbf{k}_x$, $\mathbf{k}_y$, $\mathbf{k}_z$ to all positive values only.

The quantity of interest to us, however, is the number of states in an interval of energy ϵ to $\epsilon + d\epsilon$. From Eqs. 18.42 and 19.31,

$$\epsilon = \frac{h^2}{8mV^{2/3}} r^2 \tag{19.33}$$

and

$$d\epsilon = \frac{h^2}{8mV^{2/3}} 2r\, dr \tag{19.34}$$

Using Eqs. 19.33 and 19.34 in Eq. 19.32, we find that the number of states in the interval ϵ to $\epsilon + d\epsilon$ is

$$\mathsf{g}_\epsilon = \frac{4\pi mV}{h^3} (2m\epsilon)^{1/2}\, d\epsilon \tag{19.35}$$

If we use Eq. 19.35 (for the degeneracy) in Eq. 19.3 (for the translational partition function) and integrate from $\epsilon = 0$ to $\epsilon = \infty$, the result is once again that given by Eq. 19.22. The details of the evaluation are left as an exercise.

19.4 The Monatomic Gas

For any ideal gas, the contributions to partition function and to properties are those from translation and the internal modes. The translational contributions have already been evaluated in Section 19.2, and for a monatomic gas the only internal mode contribution is that of the electronic levels of the atom. By convention we take the ground electronic level energy as the reference state, that is,

$$\epsilon_{e_0} = 0 \tag{19.36}$$

The internal partition function then becomes

$$Z_{\text{int}} = Z_e = g_{e_0} + g_{e_1} \exp\left[-\frac{\epsilon_{e_1}}{kT}\right] + g_{e_2} \exp\left[-\frac{\epsilon_{e_2}}{kT}\right] + \cdots \quad (19.37)$$

This can be evaluated by direct summation over the electronic terms as necessary. Using the definitions

$$Z_e' = T\left(\frac{dZ_e}{dT}\right) = \sum_j g_{e_j}\left(\frac{\epsilon_{e_j}}{kT}\right) \exp\left[-\frac{\epsilon_{e_j}}{kT}\right] \quad (19.38)$$

$$Z_e'' = T\left(\frac{dZ_e'}{dT}\right) = \sum_j g_{e_j}\left(\frac{\epsilon_{e_j}}{kT}\right)^2 \exp\left[-\frac{\epsilon_{e_j}}{kT}\right] - Z_e' \quad (19.39)$$

it is found from Eqs. 19.7, 19.10, 19.13, and 19.19 that the electronic contributions to properties are given by the following relations.

$$\bar{h}_e = \bar{u}_e = \bar{R}T\frac{Z_e'}{Z_e} \quad (19.40)$$

$$\bar{s}_e = \bar{R}\left[\ln Z_e + \frac{Z_e'}{Z_e}\right] \quad (19.41)$$

$$\bar{g}_e = \bar{a}_e = -\bar{R}T \ln Z_e \quad (19.42)$$

Differentiating Eq. 19.40, we find

$$\bar{C}_{p_e} = \bar{C}_{v_e} = \bar{R}\left[\frac{Z_e'' + Z_e'}{Z_e} - \left(\frac{Z_e'}{Z_e}\right)^2\right] \quad (19.43)$$

Example 19.2

1. Using values from Table 18.2, determine the fraction of atoms in each of the first three electronic levels for monatomic fluorine at 1000 K.
2. Calculate the internal energy, constant-volume specific heat, and entropy of monatomic fluorine at 1000 K, 1 atm pressure.

1. From the data for fluorine in Table 18.2, we have the following:

Term	ϵ/hc, cm^{-1}	$g = 2J + 1$
$^2P_{3/2}$	0	4
$^2P_{1/2}$	404.0	2
$^4P_{5/2}$	102,406.5	6

Therefore, at 1000 K,

$$\frac{\epsilon_{e1}}{kT} = \left(\frac{\epsilon_{e1}}{hc}\right)\left(\frac{hc}{k}\right)\frac{1}{T} = 404.0(1.4388)\left(\tfrac{1}{1000}\right) = 0.5813$$

Similarly,

$$\frac{\epsilon_{e2}}{kT} = 102406.5(1.4388)\left(\tfrac{1}{1000}\right) = 147.2$$

so that the electronic partition function is, from Eq. 19.37,

$$Z_e = 4 + 2 \times e^{-0.5813} + 6 \times e^{-147.2} = 5.118$$

From the distribution equation, Eq. 17.20

$$\frac{N_{e_j}}{N} = \frac{g_{e_j} \exp\left(-\epsilon_{e_j}/kT\right)}{Z_e}$$

the fraction of particles in the electronic ground level is

$$\frac{N_{e_0}}{N} = \frac{g_{e_0}(1)}{Z_e} = \frac{4}{5.118} = 0.782$$

Similarly, the fraction of particles in the first excited level is

$$\frac{N_{e_1}}{N} = \frac{g_{e_1} \exp\left(-\epsilon_{e_1}/kT\right)}{Z_e} = \frac{2 \times e^{-0.5813}}{5.118} = 0.218$$

For the second excited level, we find

$$\frac{N_{e_2}}{N} = \frac{g_{e_2} \exp\left(-\epsilon_{e_2}/kT\right)}{Z_e} = \frac{6 \times e^{-147.2}}{5.118} \approx 0$$

2. A negligible fraction of the fluorine atoms are in the second and higher excited levels, so we need consider only the ground and first excited levels as contributing to properties. Let us first evaluate the translational contributions. From Eq. 19.25,

$$\bar{u}_t = \tfrac{3}{2}\bar{R}T = \tfrac{3}{2}(1.987)(1000) = 2981 \text{ cal/mole}$$

From Eq. 19.27,

$$\bar{C}_{v_t} = \tfrac{3}{2}\bar{R} = \tfrac{3}{2}(1.987) = 2.981 \text{ cal/mole}$$

For monatomic fluorine, the atomic mass is

$$m = 1.66 \times 10^{-24}(19) = 31.54 \times 10^{-24} \text{ gm}$$

so that, using Eq. 19.30,

$$\ln \frac{Z_t}{N} = \ln \left[\left(\frac{2\pi m}{h^2} \right)^{3/2} \frac{(kT)^{5/2}}{P} \right]$$

$$= \ln \left[\left(\frac{2\pi \times 31.54 \times 10^{-24}}{(6.625 \times 10^{-27})^2} \right)^{3/2} \frac{(1.3804 \times 10^{-16} \times 1000)^{5/2}}{1.01325 \times 10^6} \right]$$

$$= 18.019$$

Therefore, from Eq. 19.29,

$$\bar{s}_t = \bar{R}\left[\ln \left(\frac{Z_t}{N} \right) + \tfrac{5}{2} \right] = 1.987[18.019 + 2.5] = 40.7 \text{ cal/mole-K}$$

For the electronic contributions, it is convenient to construct a table like the accompanying one, using

$$y_j = \frac{\epsilon_{ej}}{kT}$$

ϵ_{e_j} cm^{-1}	g_{e_j}	y_j	e^{-y_j}	$g_{e_j} e^{-y_j}$	$g_{e_j} y_j e^{-y_j}$	$g_{e_j} y_j^2 e^{-y_j}$
0	4	0	1	4.0	0	0
404.0	2	0.5813	0.559	1.118	0.650	0.378
102406.5	6	147.2	~0	~0	~0	~0
				$Z_e = 5.118$	$Z_e' = 0.650$	$Z_e'' + Z_e' = 0.378$

The summations for Z_e, Z_e', $Z_e'' + Z_e'$ follow from Eqs. 19.37 to 19.39. From Eq. 19.40,

$$\bar{u}_e = \bar{R}T \frac{Z_e'}{Z_e} = 1.987(1000)\left(\frac{0.650}{5.118} \right) = 252 \text{ cal/mole}$$

From Eq. 19.43,

$$\bar{C}_{v_e} = \bar{R}\left[\frac{Z_e'' + Z_e'}{Z_e} - \left(\frac{Z_e'}{Z_e} \right)^2 \right]$$

$$= 1.987\left[\frac{0.378}{5.118} - \left(\frac{0.650}{5.118} \right)^2 \right] = 0.115 \text{ cal/mole-K}$$

From Eq. 19.41,

$$\bar{s}_e = \bar{R}\left[\ln Z_e + \frac{Z_e'}{Z_e}\right]$$

$$= 1.987\left[\ln 5.118 + \frac{0.650}{5.118}\right] = 3.50 \text{ cal/mole-K}$$

Therefore the values of the properties to be determined are

$$\bar{u} = \bar{u}_t + \bar{u}_e = 2981 + 252 = 3233 \text{ cal/mole}$$
$$\bar{C}_{v_o} = \bar{C}_{v_t} + \bar{C}_{v_e} = 2.981 + 0.115 = 3.096 \text{ cal/mole-K}$$
$$\bar{s} = \bar{s}_t + \bar{s}_e = 40.70 + 3.50 = 44.20 \text{ cal/mole-K}$$

It is of interest to note that, for the very common case in which only the ground electronic level is of significance, the partition function Z_e reduces to the degeneracy of the ground level, and both Z_e' and Z_e'' are zero. Consequently, the electronic level contributions $\bar{h}_e$, $\bar{u}_e$, $\bar{C}_{p_e}$, $\bar{C}_{v_e}$ are all zero, while $\bar{s}_e$, $\bar{g}_e$, $\bar{a}_e$, are zero only if the ground level degeneracy is unity.

19.5 The Diatomic Gas—Simple Internal Model

The evaluation of thermodynamic properties of diatomic gases is approached in the same manner as that utilized for monatomic gases. The translational contributions to the thermodynamic properties are again given by the equations developed in Section 19.2. Evaluation of the internal contributions for diatomic gases is much more difficult than for monatomic gases, however, since a diatomic molecule has rotational and vibrational energies in addition to the electronic states.

In this section, we consider the simple internal model, consisting of rigid rotator, harmonic oscillator, ground electronic level, with the energy modes separated as outlined in Chapter 18. With this assumption, the component energies are additive, as indicated in Eq. 18.47

$$\epsilon_{\text{int}} = \epsilon_r + \epsilon_v + \epsilon_{e0}$$

The component wave functions (and, therefore, energy level degeneracies) for this model are multiplicative, as in Eq. 18.48. By convention, the energy of the ground electronic level, ϵ_{e0}, is taken to be zero, but we include the term here as a reminder that the degeneracy of this level must be included in our calculations.

Before we proceed, we must consider the reference state or "zero-of-energy." We have already noted that the vibrational energy is written with respect to the minimum point on the potential curve, which is shown in Fig. 19.2 (in the vicinity of the minimum point, the assumption of harmonic oscillator is a good one). Thus, when $\mathbf{v} = 0$, the ground vibrational state energy ϵ_{vo}, is given by the equation

$$\epsilon_{vo} = \tfrac{1}{2} h\nu_{\mathscr{E}} \tag{19.44}$$

Since it is more convenient to calculate vibrational contributions with respect to the ground vibrational state, we write Eq. 18.47 in the form

$$\epsilon_{\text{int}} = \epsilon_r + (\epsilon_v - \epsilon_{vo}) + \epsilon_{vo} + \epsilon_{eo} \tag{19.45}$$

One other point to be considered in selecting the energy reference state is that the reference must be made consistent with that for monatomic gases. This is particularly important in the analysis of chemical reactions,

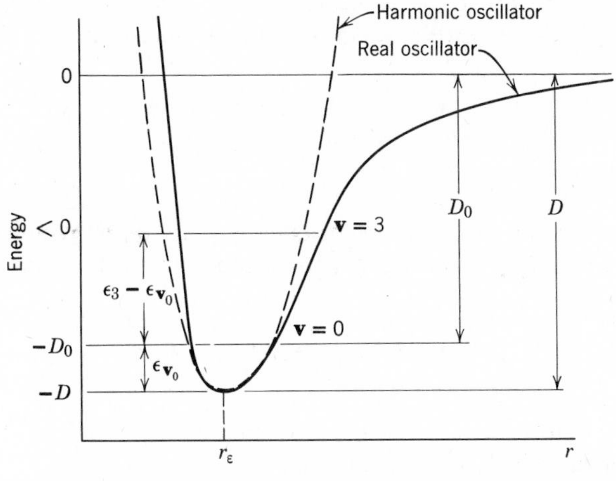

Fig. 19.2 The zero-energy reference for a diatomic molecule.

as will be found in Section 19.12. In order that the two be consistent, we should select for the zero-of-energy of the diatomic molecule that energy at which the molecule is dissociated to the monatomic form, in which the separation r of the atoms is so large that they are effectively no longer joined together. Therefore, as is evident from Fig. 19.2, ϵ_{int} relative to this base is given by the expression

$$\epsilon_{\text{int}} = \epsilon_r + (\epsilon_v - \epsilon_{vo}) + \epsilon_{vo} + \epsilon_{eo} - D = \epsilon_r + (\epsilon_v - \epsilon_{vo}) + \epsilon_{eo} - D_0 \tag{19.46}$$

in terms of the observed dissociation energy of the molecule, D_0.

Equation 19.4 gives an expression for the internal partition function in terms of the various internal modes of energy. Substituting Eq. 19.46 into Eq. 19.4, we have

$$Z_{int} = \sum_{\text{all } j} g_{r_j} g_{v_j} g_{eo} \exp\left[\frac{-(\epsilon_{r_j} + (\epsilon_{v_j} - \epsilon_{vo}) + \epsilon_{eo} - D_0)}{kT}\right]$$

$$= Z_r Z_v Z_{eo} Z_{chem} \tag{19.47}$$

The rotational partition function Z_r in Eq. 19.47 is

$$Z_r = \sum_{r_j} g_{r_j} \exp\left[\frac{-\epsilon_{r_j}}{kT}\right] \tag{19.48}$$

It was found in Chapter 18 that the vibrational energy levels are nondegenerate. Therefore g_v is unity for all levels, and the vibrational partition function Z_v with respect to the ground vibrational state energy ϵ_{vo} is

$$Z_v = \sum_{v_j} \exp\left[\frac{-(\epsilon_{v_j} - \epsilon_{vo})}{kT}\right] \tag{19.49}$$

Since the electronic ground level energy ϵ_{eo} is by convention taken as zero, the partition function Z_{eo} for the electronic ground state is

$$Z_{eo} = g_{eo} \exp\left[\frac{-\epsilon_{eo}}{kT}\right] = g_{eo} \tag{19.50}$$

Finally, we define the chemical partition function Z_{chem} in terms of the dissociation energy D_0 by the relation

$$Z_{chem} = \exp\left[\frac{D_0}{kT}\right] \tag{19.51}$$

If Eq. 19.47 is written in logarithmic form, we have

$$\ln Z_{int} = \ln Z_r + \ln Z_v + \ln Z_{eo} + \ln Z_{chem} \tag{19.52}$$

From this equation, it is apparent that each of the four terms has a contribution to the thermodynamic properties as given by Eqs. 19.7, 19.10, 19.13, 19.16, and 19.19 for $\bar{u}$, $\bar{h}$, $\bar{s}$, $\bar{a}$, $\bar{g}$, respectively. These contributions will now be evaluated in the following three sections.

19.6 Rotation

We may now utilize the results of quantum mechanics to evaluate the rotational partition function, as defined by Eq. 19.48, and subsequently the rotational contributions to thermodynamic properties. Substituting the expressions for energy (Eq. 18.66) and rotational level degeneracy (Eq. 18.67) into Eq. 19.48, we have

$$Z_r = \sum_{\mathbf{j}=0}^{\infty} (2\mathbf{j}+1) \exp\left[-\frac{h^2}{8\pi^2 I_\varepsilon kT}\mathbf{j}(\mathbf{j}+1)\right] \tag{19.53}$$

Let us for the moment assume that the rotational energy levels are spaced sufficiently close together that the summation in this expression can be replaced directly by an integral. If we also change variable according to the relation

$$y^2 = \mathbf{j}(\mathbf{j}+1)$$

Eq. 19.53 becomes

$$Z_r = \int_0^{\infty} 2y \exp\left[-\frac{h^2}{8\pi^2 I_\varepsilon kT}y^2\right] dy = \frac{8\pi^2 I_\varepsilon kT}{h^2} = \frac{T}{\theta_r} \tag{19.54}$$

where the constant θ_r is defined by

$$\theta_r = \frac{h^2}{8\pi^2 I_\varepsilon k} \tag{19.55}$$

and, having the units of temperature, is termed the characteristic rotational temperature.

There is one additional factor that must be considered in the evaluation of the rotational partition function for homonuclear molecules (for example, N_2, O_2, H_2, etc.). Since the two atoms of which the molecule is comprised are identical, the molecule can be rotated through 180° and the new configuration will be indistinguishable from the original one. To correct for this double counting, we introduce a symmetry number σ into the result, so that the correct expression for the rotational partition function is

$$Z_r = \frac{T}{\sigma\theta_r} \tag{19.56}$$

in which $\sigma = 1$ for heteronuclear molecules, $\sigma = 2$ for homonuclear molecules. The argument used here in introducing the symmetry number is admittedly somewhat artificial and incomplete. The complete explanation for the justification of this number is based on the symmetry characteristics of the molecular wave function and, being a rather complete and advanced subject in itself, will not be examined in this text.

For the rotational contributions to thermodynamic properties, we rewrite Eq. 19.56 in logarithmic form:

$$\ln Z_r = \ln T - \ln \sigma\theta_r \tag{19.57}$$

Then, on applying Eq. 19.7,

$$\bar{u}_{\text{int}} = \bar{R}T^2\left(\frac{\partial \ln Z_{\text{int}}}{\partial T}\right)_V$$

to the rotation, we find that

$$\bar{u}_r = \bar{R}T^2\left(\frac{1}{T}\right) = \bar{R}T \tag{19.58}$$

Similarly, it follows from Eq. 19.10 that

$$\bar{h}_r = \bar{u}_r = \bar{R}T \tag{19.59}$$

When Eqs. 19.58 and 19.59 are differentiated with respect to temperature, it follows from the definitions of the specific heats that

$$\bar{C}_{v_r} = \bar{C}_{p_r} = \bar{R} \tag{19.60}$$

The rotational contribution to entropy can be found by applying Eq. 19.13,

$$\bar{s}_{\text{int}} = \bar{R} \ln Z_{\text{int}} + \frac{\bar{u}_{\text{int}}}{T}$$

to the rotation. The result is

$$\bar{s}_r = \bar{R}\left[\ln\left(\frac{T}{\sigma\theta_r}\right) + 1\right] \tag{19.61}$$

The rotational contributions to Helmholtz and Gibbs functions are, from Eqs. 19.16 and 19.19,

$$\bar{a}_r = \bar{g}_r = -\bar{R}T \ln\left(\frac{T}{\sigma\theta_r}\right) \tag{19.62}$$

Example 19.3

Calculate the rotational partition function and the rotational contributions to internal energy and entropy for diatomic nitrogen at 500 K.

From Table B.14, for N_2,

$$r_\varepsilon = 1.0976 \text{ Å}$$

The mass of a single nitrogen atom is

$$m_{\text{N}} = 1.66 \times 10^{-24}(14.008) = 23.25 \times 10^{-24} \text{ gm}$$

or half the mass of the diatomic molecule. The reduced mass is found from Eq. 18.55:

$$m_r = \frac{m_{\text{N}} m_{\text{N}}}{m_{\text{N}} + m_{\text{N}}} = \frac{m_{\text{N}}}{2} = 11.625 \times 10^{-24} \text{ gm}$$

Then, from Eq. 18.54,

$$I_\varepsilon = m_r r_\varepsilon^2 = 11.625 \times 10^{-24}(1.0976 \times 10^{-8})^2 = 14.0 \times 10^{-40} \text{ gm cm}^2$$

The characteristic rotational temperature θ_r is found from Eq. 19.55,

$$\theta_r = \frac{h^2}{8\pi^2 I_\varepsilon k} = \frac{(6.625 \times 10^{-27})^2}{8\pi^2(14.0 \times 10^{-40})(1.3804 \times 10^{-16})} = 2.875 \text{ K}$$

Since diatomic nitrogen is a homonuclear molecule (two N atoms), the symmetry number σ is 2. The rotational partition function can be found from Eq. 19.56,

$$Z_r = \frac{T}{\sigma\theta_r} = \frac{500}{2(2.875)} = 87.0$$

The internal energy of rotation is, from Eq. 19.58,

$$\bar{u}_r = \bar{R}T = 1.987(500) = 994 \text{ cal/mole}$$

The rotational entropy is calculated from Eq. 19.61:

$$\bar{s}_r = \bar{R}[\ln Z_r + 1] = 1.987[\ln 87.0 + 1] = 10.85 \text{ cal/mole-K}$$

The expression for the rotational partition function for a rigid rotator as given by Eq. 19.56 was obtained under the assumption that the rotational energy levels are very closely spaced. To determine the error resulting from this approximation, we must evaluate the summation of Eq. 19.53, using the Euler-Maclaurin summation theorem, as we did for translational energy in Section 19.2. The procedure here is straightforward but involves the rather tedious job of taking the derivatives of the function in that equation; this will be left as an exercise. The result, after symmetry number σ has been included can be written in the form[1]

$$Z_r = \frac{T}{\sigma\theta_r}\left[1 + \frac{1}{3}\left(\frac{\theta_r}{T}\right) + \frac{1}{15}\left(\frac{\theta_r}{T}\right)^2 + \frac{4}{315}\left(\frac{\theta_r}{T}\right)^3 + \cdots\right] \qquad (19.63)$$

Note that in this equation Z_r is given in terms of the simplified result plus a series of correction terms. We find from Eq. 19.63 that the simplification made in Eq. 19.54 is reasonable only for $T \gg \theta_r$ for, in this case, the correction terms are negligibly small. When correction terms are of importance, it is necessary to evaluate the rotational contributions to properties directly from Eq. 19.63 instead of determining properties from Eqs. 19.58 to 19.62.

If all the terms of Eq. 19.63 are used in expressing the rotational partition function, the following relations for various thermodynamic properties result.

$$\bar{h}_r = \bar{u}_r = \bar{R}T\left(\frac{Z_r'}{Z_r}\right) \qquad (19.64)$$

$$\bar{s}_r = \bar{R}\left[\ln Z_r + \left(\frac{Z_r'}{Z_r}\right)\right] \qquad (19.65)$$

$$\bar{g}_r = \bar{a}_r = -\bar{R}T \ln Z_r \qquad (19.66)$$

$$\bar{C}_{p_r} = \bar{C}_{v_r} = \bar{R}\left[\frac{Z_r'' + Z_r'}{Z_r} - \left(\frac{Z_r'}{Z_r}\right)^2\right] \qquad (19.67)$$

We note that this set of expressions is identical in form with Eqs. 19.40 to 19.43, developed for electronic energy contributions of a monatomic gas. In this case, the quantities Z_r' and Z_r'' are

$$Z_r' = T\left(\frac{dZ_r}{dT}\right) = \frac{T}{\sigma\theta_r}\left[1 - \frac{1}{15}\left(\frac{\theta_r}{T}\right)^2 - \frac{8}{315}\left(\frac{\theta_r}{T}\right)^3\right] \qquad (19.68)$$

$$Z_r'' = T\left(\frac{dZ_r'}{dT}\right) = \frac{T}{\sigma\theta_r}\left[1 + \frac{1}{15}\left(\frac{\theta_r}{T}\right)^2 + \frac{16}{315}\left(\frac{\theta_r}{T}\right)^3\right] \qquad (19.69)$$

Example 19.4

Calculate the rotational contributions to internal energy and specific heat $\bar{C}_{v0}$ for hydrogen fluoride, HF, at 100 K.

From Table B.14 for HF,

$$r_\varepsilon = 0.9168 \text{ Å}$$

[1]For $T/\theta_r > 1$; otherwise, the rotational partition function can be evaluated by direct expansion of Eq. 19.53.

The atomic masses are

$$m_{\mathrm{H}} = 1.66 \times 10^{-24}(1.008) = 1.673 \times 10^{-24}\ \mathrm{gm}$$
$$m_{\mathrm{F}} = 1.66 \times 10^{-24}(19) = 31.54 \times 10^{-24}\ \mathrm{gm}$$

From Eqs. 18.55 and 18.54,

$$m_r = \frac{m_{\mathrm{H}} m_{\mathrm{F}}}{m_{\mathrm{H}} + m_{\mathrm{F}}} = \frac{1.673(31.54)}{33.21} \times 10^{-24} = 1.59 \times 10^{-24}\ \mathrm{gm}$$

$$I_\varepsilon = m_r r_\varepsilon^{\,2} = 1.59 \times 10^{-24}(0.9168 \times 10^{-8})^2 = 1.34 \times 10^{-40}\ \mathrm{gm\ cm^2}$$

Thus, from Eq. 19.55,

$$\theta_r = \frac{h^2}{8\pi^2 I_\varepsilon k} = \frac{(6.625 \times 10^{-27})^2}{8\pi^2(1.34 \times 10^{-40})(1.3804 \times 10^{-16})} = 30.3\ \mathrm{K}$$

We note that, at 100 K, the ratio

$$\frac{T}{\theta_r} = \frac{100}{30.3} = 3.30$$

is not a large number. Therefore the rotational partition function should be evaluated from the more general expression of Eq. 19.63, as the use of Eq. 19.56 would introduce a significant error. From Eq. 19.63, with $\sigma = 1$ for the HF molecule, we have

$$Z_r = \frac{T}{\sigma\theta_r}\left[1 + \frac{1}{3}\left(\frac{\theta_r}{T}\right) + \frac{1}{15}\left(\frac{\theta_r}{T}\right)^2 + \frac{4}{315}\left(\frac{\theta_r}{T}\right)^3 + \cdots\right]$$
$$= 3.30[1 + \tfrac{1}{3}(0.303) + \tfrac{1}{15}(0.303)^2 + \tfrac{4}{315}(0.303)^3]$$
$$= 3.30[1 + 0.101 + 0.0061 + 0.0004]$$
$$= 3.30[1.1075] = 3.655$$

From Eq. 19.68,

$$Z_r' = \frac{T}{\sigma\theta_r}\left[1 - \frac{1}{15}\left(\frac{\theta_r}{T}\right)^2 - \frac{8}{315}\left(\frac{\theta_r}{T}\right)^3\right]$$
$$= 3.30[1 - \tfrac{1}{15}(0.303)^2 - \tfrac{8}{315}(0.303)^3] = 3.277$$

Therefore, from Eq. 19.64,

$$\bar{u}_r = \bar{R}T\left(\frac{Z_r'}{Z_r}\right) = 1.987(100)\left(\frac{3.277}{3.655}\right) = 178.0\ \mathrm{cal/mole}$$

Had the simplified expression, Eq. 19.58, been used, the result would have been

$$\bar{u}_r = \bar{R}T = 198.7 \text{ cal/mole}$$

which is in error by nearly 12%. From Eq. 19.69,

$$Z_r'' = \frac{T}{\sigma\theta_r}\left[1 + \frac{1}{15}\left(\frac{\theta_r}{T}\right)^2 + \frac{16}{315}\left(\frac{\theta_r}{T}\right)^3\right]$$
$$= 3.30[1 + \tfrac{1}{15}(0.303)^2 + \tfrac{16}{315}(0.303)^3] = 3.325$$

Therefore, using Eq. 19.67, we find

$$\bar{C}_{v_r} = \bar{R}\left[\frac{Z_r'' + Z_r'}{Z_r} - \left(\frac{Z_r'}{Z_r}\right)^2\right]$$
$$= 1.987\left[\left(\frac{3.325 + 3.277}{3.655}\right) - \left(\frac{3.277}{3.655}\right)^2\right]$$
$$= 1.999 \text{ cal/mole-K}$$

as compared with the value 1.987 cal/mole-K from the simplified expression, Eq. 19.60.

19.7 Vibration

For the vibrational partition function in our simple internal model, we consider the molecule to behave as a harmonic oscillator, the energy being given by Eq. 18.71. In the vibrational partition function (Eq. 19.49), the energy considered is that with respect to the ground state $\mathbf{v} = 0$, which is

$$\epsilon_v - \epsilon_{v0} = h\nu_\varepsilon \mathbf{v} \tag{19.70}$$

where $\mathbf{v}$ is the vibrational quantum number. Substituting Eq. 19.70 into the vibrational partition function (Eq. 19.49), we obtain

$$Z_v = \sum_{\mathbf{v}=0}^{\infty} e^{-(h\nu_\varepsilon/kT)\mathbf{v}} \tag{19.71}$$

Unlike the translational and rotational energies, the vibrational energy levels are so widely spaced that it is not possible to replace the summation in Eq. 19.70 by an integral, even if the Euler-Maclaurin summation theorem is used. It is noted, however, that this expression is given by

$$Z_v = \sum_{\mathbf{v}=0}^{\infty} e^{-(h\nu_\varepsilon/kT)\mathbf{v}} = \frac{1}{1 - e^{-\theta_v/T}} \tag{19.72}$$

where the characteristic vibrational temperature θ_v, is defined as

$$\theta_v = \frac{h\nu_\varepsilon}{k} = \frac{hc\omega_\varepsilon}{k} \tag{19.73}$$

In the evaluation of the vibrational contributions to thermodynamic properties, it is convenient to express Eq. 19.72 in the logarithmic form

$$\ln Z_v = -\ln\,(1 - e^{-\theta_v/T}) \tag{19.74}$$

The internal energy contribution is found from Eq. 19.7,

$$\bar{u}_{\text{int}} = \bar{R}T^2 \left(\frac{\partial \ln Z_{\text{int}}}{\partial T} \right)_V$$

However, since the partition function Z_v to be used in this expression is evaluated with respect to the ground vibrational state energy ϵ_{vo}, it follows that the molal energy calculated from the equation will be that with respect to the molal ground state energy $\bar{u}_{vo}$, which equals $N_0\epsilon_{vo}$. Therefore, from Eqs. 19.7 and 19.74 we find

$$(\bar{u}_v - \bar{u}_{vo}) = \bar{R}T^2 \left[-\frac{e^{-\theta_v/T}}{1 - e^{-\theta_v/T}} \left(-\frac{\theta_v}{T^2} \right) \right] = \frac{\bar{R}\theta_v}{e^{\theta_v/T} - 1} \tag{19.75}$$

Similarly, using Eq. 19.10, we have

$$(\bar{h}_v - \bar{u}_{vo}) = (\bar{u}_v - \bar{u}_{vo}) = \frac{\bar{R}\theta_v}{e^{\theta_v/T} - 1} \tag{19.76}$$

Differentiating Eqs. 19.75 and 19.76, we obtain

$$\bar{C}_{v_v} = \bar{C}_{p_v} = \frac{\bar{R}(\theta_v/T)^2 e^{\theta_v/T}}{(e^{\theta_v/T} - 1)^2} \tag{19.77}$$

The entropy contribution is found directly from Eq. 19.13,

$$\bar{s}_{\text{int}} = \bar{R} \ln Z_{\text{int}} + \frac{\bar{u}_{\text{int}}}{T}$$

It is easily shown that this expression is independent of the zero-point energy $\bar{u}_{vo}$. Therefore, substituting Eqs. 19.74 and 19.75 into Eq. 19.13, we have

$$\bar{s}_v = \bar{R} \left[-\ln\,(1 - e^{-\theta_v/T}) + \frac{\theta_v/T}{e^{\theta_v/T} - 1} \right] \tag{19.78}$$

It follows also from Eqs. 19.16 and 19.19 that

$$(\bar{a}_v - \bar{u}_{vo}) = (\bar{g}_v - \bar{u}_{vo}) = \bar{R}T \ln (1 - e^{-\theta_v/T}) \tag{19.79}$$

Example 19.5

Calculate the specific heat $\bar{C}_{po}$ for diatomic iodine at 0 C.

The translational, rotational, and vibrational contributions to the specific heat must all be considered. The translational contribution is found from Eq. 19.28,

$$\bar{C}_{p_t} = \tfrac{5}{2}\bar{R} = \tfrac{5}{2}(1.987) = 4.968 \text{ cal/mole-K}$$

For rotation, it is found from the values in Table B.14 that, for diatomic iodine,

$$\theta_r = 0.054 \text{ K}$$

Therefore, at $T = 273.15$ K, we find that $T \gg \theta_r$; hence the rotational partition function is accurately given by Eq. 19.56 and, correspondingly, the specific heat contribution is given by Eq. 19.60,

$$\bar{C}_{p_r} = \bar{R} = 1.987 \text{ cal/mole-K}$$

To determine the vibrational contribution, we use the value of ω_ε from Table B.14 to calculate the characteristic vibrational temperature from Eq. 19.73:

$$\theta_v = \left(\frac{hc}{k}\right)\omega_\varepsilon = 1.4388(214.5) = 308.8 \text{ K}$$

Therefore

$$\frac{\theta_v}{T} = \frac{308.8}{273.15} = 1.131$$

The value for $\bar{C}_{p_v}$ can now be found from Eq. 19.77:

$$\bar{C}_{p_v} = \frac{\bar{R}(\theta_v/T)^2 e^{\theta_v/T}}{(e^{\theta_v/T} - 1)^2} = 1.987(0.8999) = 1.788 \text{ cal/mole-K}$$

Adding the translational, rotational, and vibrational contributions, we find, for the specific heat $\bar{C}_{po}$,

$$\bar{C}_{po} = \bar{C}_{p_t} + \bar{C}_{p_r} + \bar{C}_{p_v} = 4.968 + 1.987 + 1.788 = 8.743 \text{ cal/mole-K}$$

19.8 Electronic Ground Level and Chemical Energy

Since the electronic energy in the ground electronic level is by convention assigned the value zero, only the ground level degeneracy is significant for electronic contributions in the simple internal model for the diatomic gas. In logarithmic form, the electronic partition function (Eq. 19.50) is

$$\ln Z_e = \ln g_{e0} = \text{constant} \tag{19.80}$$

It therefore follows from Eqs. 19.7, 19.10, and the definitions of $\bar{C}_v$ and $\bar{C}_p$ that

$$\bar{u}_e = \bar{h}_e = \bar{C}_{v_e} = \bar{C}_{p_e} = 0 \tag{19.81}$$

However, there is an electronic contribution to entropy as given by Eq. 19.13, from which we find

$$\bar{s}_e = \bar{R} \ln g_{e0} \tag{19.82}$$

For the Helmholtz and Gibbs functions we have, from Eq. 19.19,

$$\bar{a}_e = \bar{g}_e = -\bar{R}T \ln g_{e0} \tag{19.83}$$

Note that $\bar{s}_e$, $\bar{a}_e$, and $\bar{g}_e$ are zero only if the ground electronic level degeneracy is unity.

Finally, we must determine the contribution of the chemical energy to thermodynamic properties. As noted in Section 19.5, we have selected the zero-of-energy for the molecule as the dissociated state, so that the reference state will be consistent with that chosen previously for a monatomic substance. This selection introduces a constant value of $-D_0$ into the internal mode energy of the molecule, with a resulting contribution to the thermodynamic properties. Therefore, in accordance with Eq. 19.51,

$$\ln Z_{\text{chem}} = \frac{D_0}{kT} = \frac{N_0 D_0}{\bar{R}T} = -\frac{\bar{h}^0_{f0}}{\bar{R}T} \tag{19.84}$$

where $\bar{h}^0_{f0}$ is the enthalpy of formation of the molecule at absolute zero temperature and is by definition

$$\bar{h}^0_{f0} = -N_0 D_0 \tag{19.85}$$

With the molecule in the ground electronic level, the enthalpy of formation is a negative number, since the energy of the dissociated molecule has been selected as zero, as discussed with reference to Fig. 19.2.

From Eqs. 19.7 and 19.10,

$$\bar{h}_{\text{chem}} = \bar{u}_{\text{chem}} = \bar{R}T^2\left(+\frac{\bar{h}^0_{f0}}{\bar{R}T^2}\right) = \bar{h}^0_{f0} \tag{19.86}$$

which is a constant. Therefore

$$\bar{C}_{v_{\text{chem}}} = \bar{C}_{p_{\text{chem}}} = 0 \tag{19.87}$$

It also follows from Eq. 19.13 that

$$\bar{s}_{\text{chem}} = \bar{R}\left(-\frac{\bar{h}^0_{f0}}{\bar{R}T}\right) + \frac{\bar{u}_{\text{chem}}}{T} = 0 \tag{19.88}$$

Since $\bar{s}_{\text{chem}}$ is zero, we find from Eqs. 19.19 and 19.86 that

$$\bar{a}_{\text{chem}} = \bar{g}_{\text{chem}} = \bar{h}^0_{f0} \tag{19.89}$$

The relation between the enthalpy of formation as defined by Eq. 19.85 and that utilized in Chapters 13 and 14 in the study of chemical reactions is examined in Section 19.12, and in the problems at the end of this chapter.

19.9 The Polyatomic Gas

Translational contributions to properties for polyatomic gases are evaluated as for the monatomic and diatomic species, using the total mass of the molecule and the results of Section 19.2. The evaluation of internal modes proves to be a difficult task, however, even though we assume the simple internal model: rigid rotator, harmonic oscillator, ground electronic level. The first question to be faced is that concerning the number of degrees of freedom for the molecule. Consider a molecule comprised of a atoms, for which the total number of degrees of freedom of motion must equal $3a$. Since each molecule has three directions of translational motion, this leaves a total of $(3a-3)$ degrees of freedom for the rotational and vibrational modes. We must now divide our discussion of poly-atomic molecules into two classes according to structure, linear and non-linear, for which the behavior, and consequently, the thermodynamic properties are different.

As examples of linear molecules, consider CO_2 and N_2O, the structures of which are as shown in Fig. 19.3. In either case, the moment of inertia about the axis joining the three nuclei is negligibly small, while the remaining two moments are equal, which is the same as for diatomic

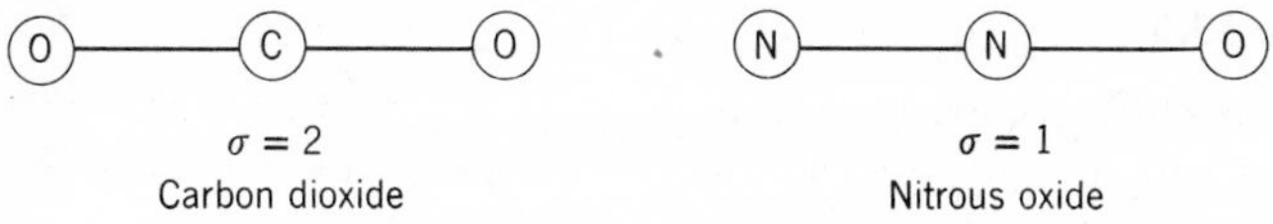

Fig. 19.3 The alignment of two linear polyatomic molecules.

molecules. Therefore, the rotational behavior of a linear polyatomic molecule is the same as determined in Section 19.6. The partition function is found from Eq. 19.56, with the characteristic rotational temperature given by Eq. 19.55. Rotational contributions to properties are found from Eqs. 19.58 to 19.62, as before. Evaluation of the equilibrium-point moment of inertia I_ε, which is required in Eq. 19.55, is somewhat more complicated for the polyatomic gas. For the linear polyatomic molecule, I_ε is given by the relation

$$I_\varepsilon = \sum_{i=1}^{a} m_i r_i^2 \tag{19.90}$$

where r_i is the equilibrium distance of atom i from the center of mass of the molecule. This equilibrium distance r_i is defined by the expression

$$\sum_{i=1}^{a} m_i r_i = 0 \tag{19.91}$$

It is also necessary to know the symmetry number σ for the molecule. This requires a knowledge of the structure of the molecule, as is seen from Fig. 19.3.

Since the linear molecule has but two rotational degrees of freedom, there are $(3a-5)$ degrees of freedom remaining for vibration of the molecule. The contributions of these $(3a-5)$ vibrational modes can be considered as being independent in the simple model, and the vibration can be assumed harmonic for each mode. Thus we have $(3a-5)$ vibrational partition function contributions of the form of Eq. 19.72. The $(3a-5)$ characteristic vibrational temperatures θ_v are given in terms of the normal frequencies or wave numbers, as in Eq. 19.73. Two or more of these modes may have the same frequency, depending upon the molecule. Contributions to properties from these modes are given by Eqs. 19.75 to 19.79 for each of the $(3a-5)$ vibrational modes of the molecule. We should keep in mind that there is a $\frac{1}{2}h\nu_\varepsilon$ zero-point contribution for each of the vibrational modes, which by convention is included in the chemical energy term. These zero-point terms add a constant value to $\bar{u}, \bar{h}, \bar{a}, \bar{g}$, but of course not to $\bar{s}, \bar{C}_{p0}, \bar{C}_{v0}$ for the molecule.

Electronic ground state contributions to properties are found, as for diatomic substances, by using Eqs. 19.81 to 19.83.

Example 19.6

Calculate the entropy of carbon dioxide at 1200 K, 1 atm pressure.

From Table B.14 for CO_2 the C—O distances are 1.160 Å, and the vibrational wave numbers are

$$\omega_1 = 1342.9 \text{ cm}^{-1}$$
$$\omega_2 = 667.3 \text{ (2 modes)}$$
$$\omega_3 = 2349.3$$

For the ground electronic level, $g_{e_0} = 1$. The mass of the CO_2 molecule is

$$m = 1.66 \times 10^{-24}(44.01) = 73.05 \times 10^{-24} \text{ gm}$$

In solving this problem, we consider in turn the translational, rotational, vibrational, and electronic modes, and find the entropy contribution of each. We consider first the translational contribution to entropy. From Eq. 19.30, at 1200 K, 1 atm,

$$\frac{Z_t}{N} = \left(\frac{2\pi m}{h^2}\right)^{3/2} \frac{(kT)^{5/2}}{P}$$

$$= \left(\frac{2\pi \times 73.05 \times 10^{-24}}{(6.625 \times 10^{-27})^2}\right)^{3/2} \frac{(1.3804 \times 10^{-16} \times 1200)^{5/2}}{1.01325 \times 10^6}$$

$$= 3.70 \times 10^8$$

The translational entropy is, from Eq. 19.29,

$$\bar{s}_t = \bar{R}\left[\ln \frac{Z_t}{N} + \tfrac{5}{2}\right] = 1.987[19.73 + 2.5] = 44.15 \text{ cal/mole-K}$$

Next we consider the rotational contribution. From Eq. 19.91, the center of mass is at the center of the C atom. Therefore $r_C = 0$, $r_O = \pm 1.160$ Å, and, using Eq. 19.90, we have

$$I_\varepsilon = \sum_i m_i r_i^2 = 2(16.0 \times 1.66 \times 10^{-24})(1.160 \times 10^{-8})^2$$

$$= 71.5 \times 10^{-40} \text{ gm-cm}^2$$

From Eq. 19.55,

$$\theta_r = \frac{h^2}{8\pi^2 I_\varepsilon k} = \frac{(6.625 \times 10^{-27})^2}{8\pi^2 \times 71.5 \times 10^{-40} \times 1.3804 \times 10^{-16}} = 0.562 \text{ K}$$

The rotational partition function is found from Eq. 19.56:

$$Z_r = \frac{T}{\sigma\theta_r} = \frac{1200}{2(0.562)} = 1067$$

The resulting contribution to entropy is, from Eq. 19.61,

$$\bar{s}_r = \bar{R}[\ln Z_r + 1] = 1.987[\ln 1067 + 1] = 15.80 \text{ cal/mole-K}$$

As regards the vibrational contribution, there are four vibrational degrees of freedom, two of which have the same frequency. Thus

$$\theta_{v_1} = \frac{hc}{k}\,\omega_1 = 1.4388(1342.9) = 1932 \text{ K}$$
$$\theta_{v_2} = 1.4388(667.3) = 960 \text{ K (double)}$$
$$\theta_{v_3} = 1.4388(2349.3) = 3380 \text{ K}$$

and

$$\frac{\theta_{v_1}}{T} = \frac{1932}{1200} = 1.61$$
$$\frac{\theta_{v_2}}{T} = \frac{960}{1200} = 0.80 \text{ (double)}$$
$$\frac{\theta_{v_3}}{T} = \frac{3380}{1200} = 2.82$$

Using Eq. 19.78 for each of the four vibrational modes, the vibrational entropy is found to be

$$\bar{s}_v = \bar{R}\sum_i \left\{-\ln\left[1-\exp\left(-\frac{\theta_{v_i}}{T}\right)\right] + \frac{\theta_{v_i}/T}{\exp(\theta_{v_i}/T)-1}\right\}$$
$$= 1.987[(0.2230+0.4022)+2(0.5966+0.6528)+(0.0615+0.1787)]$$
$$= 6.70 \text{ cal/mole-K}$$

From Eq. 19.82, the electronic contribution is

$$\bar{s}_e = \bar{R}\ln g_{e0} = 0$$

Therefore the entropy of CO_2 at 1200 K, 1 atm, is found by summing the various contributions. The result is

$$\bar{s} = \bar{s}_t + \bar{s}_r + \bar{s}_v + \bar{s}_e = 44.15 + 15.80 + 6.70 + 0$$
$$= 66.65 \text{ cal/mole-K}$$

An example of a non-linear polyatomic molecule is H_2O, as discussed briefly in Section 2.6. This type of molecule has three degrees of freedom for rotation, and therefore only $(3a-6)$ for vibrational modes. We consider again only the simple internal model, and assume the rotational levels to be closely spaced so that the classical approximation is valid. However, the rotational energy for a non-linear molecule is not the same as that for a linear one. It was pointed out above that, for a linear molecule, the moments of inertia about two of the mutually perpendicular axes are equal while that about the axis joining the nuclei is negligible. For the general non-linear molecule this is not the case, and the development of rotational energy and partition function expressions becomes extremely complex, even for the classical rigid rotator. We shall not derive these expressions here but will simply present the result, which is given in terms of the three principal moments of inertia, I_x, I_y, I_z, and the rotational symmetry number σ. The relation is

$$\begin{aligned} Z_r &= \frac{\pi^{1/2}}{\sigma}\left(\frac{8\pi^2 I_x kT}{h^2}\right)^{1/2}\left(\frac{8\pi^2 I_y kT}{h^2}\right)^{1/2}\left(\frac{8\pi^2 I_z kT}{h^2}\right)^{1/2} \\ &= \frac{8\pi^2}{\sigma h^3}(I_x I_y I_z)^{1/2}(2\pi kT)^{3/2} \end{aligned} \tag{19.92}$$

One should note the similarity between this result and that for the linear molecule, which has only two moments, and these are equal in magnitude.

The three principal moments of inertia for the molecule must be found in the following manner. Let us consider first any set of mutually perpendicular coordinate axes x, y, z, having their origin at the center of mass of the molecule. The center of mass is defined by the expressions

$$\sum_{i=1}^{a} m_i x_i = \sum_{i=1}^{a} m_i y_i = \sum_{i=1}^{a} m_i z_i = 0 \tag{19.93}$$

The moments of inertia corresponding to the selected coordinate directions are given by

$$\begin{aligned} I_{xx} &= \sum_{i=1}^{a} m_i (y_i^2 + z_i^2) \\ I_{yy} &= \sum_{i=1}^{a} m_i (z_i^2 + x_i^2) \\ I_{zz} &= \sum_{i=1}^{a} m_i (x_i^2 + y_i^2) \end{aligned} \tag{19.94}$$

The products of inertia for this coordinate system are

$$I_{xy} = I_{yx} = \sum_{i=1}^{a} m_i x_i y_i$$
$$I_{yz} = I_{zy} = \sum_{i=1}^{a} m_i y_i z_i \qquad (19.95)$$
$$I_{zx} = I_{xz} = \sum_{i=1}^{a} m_i z_i x_i$$

The six values given by Eqs. 19.94 and 19.95 depend on the selection of coordinate axes and are changed by rotation of the coordinate system about the origin. The particular set of axes for which the products of inertia (Eq. 19.95) are all zero is called the principal set of axes. The corresponding moments of Eq. 19.94 are called the principal moments of inertia, designated simply by I_x, I_y, I_z, and are the values to be used in calculating the partition function from Eq. 19.92. The principal set of axes can usually be selected quite easily by letting any one of the coordinate directions x, y, z coincide with a line of symmetry in the molecule.

Non-linear molecules are classified according to the relative values of their three principal moments of inertia. A molecule for which all three principal moments are equal is referred to as a spherical top molecule. One for which two of the three principal moments are the same is called a symmetric top molecule, and a molecule with all three moments different is termed an asymmetric top molecule.

The rotational symmetry number for non-linear molecules can be very difficult to determine, especially for complex substances. The methane molecule is shown in Fig. 19.4. It has the four H atoms arranged in the shape of a tetrahedron with the single C atom at the center. To determine σ, let us label the H atoms as shown in the figure, and refer to the plane formed by atoms 1, 2, and 3 in the diagram as the base plane. Successive rotations of 120° about the z-axis result in three indistinguishable configurations of the molecule. However, the molecule could

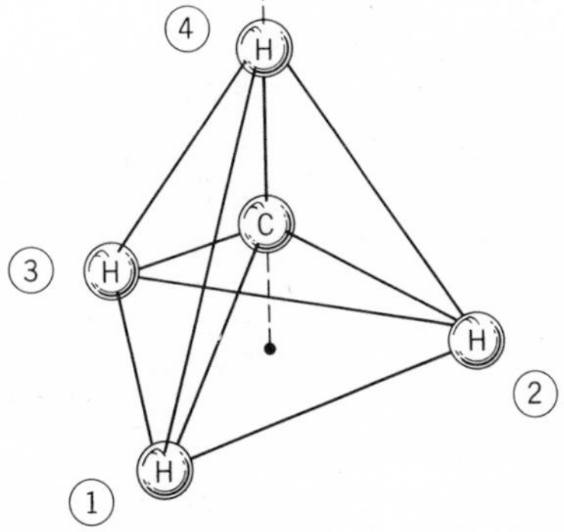

Fig. 19.4 The configuration of the methane molecule.

also be rotated about an axis normal to z, putting 1, 2, 4, for example, in the base plane. Subsequently 120° rotations about the z-axis result in three additional indistinguishable positions. After similar rotations to put either 1, 3, 4 or 2, 3, 4 in the base plane, each with three positions about the z-axis, we conclude that there is a total of twelve indistinguishable configurations for the CH_4 molecule and that $\sigma = 12$. For many molecules, the determination of rotational symmetry number is even more involved.

The rotational contributions to thermodynamic properties are found from the partition function expression (Eq. 19.92). Using Eqs. 19.7 and 19.10 gives

$$\bar{h}_r = \bar{u}_r = \bar{R}T^2\left(\frac{d \ln Z_r}{dT}\right) = \bar{R}T^2\left(\frac{3}{2T}\right) = \tfrac{3}{2}\bar{R}T \tag{19.96}$$

Differentiating this result with respect to temperature, we find

$$\bar{C}_{p_r} = \bar{C}_{v_r} = \tfrac{3}{2}\bar{R} \tag{19.97}$$

From Eqs. 19.13 and 19.96,

$$\bar{s}_r = \bar{R} \ln Z_r + \frac{\bar{u}_r}{T} = \bar{R}\left[\ln Z_r + \tfrac{3}{2}\right] \tag{19.98}$$

while, from Eqs. 19.16 and 19.19,

$$\bar{g}_r = \bar{a}_r = -\bar{R}T \ln Z_r \tag{19.99}$$

Example 19.7

Calculate the rotational contribution to Gibbs function for water at 1000 K.

From Table B.14, for H_2O the O—H distance is 0.9584 Å and the H—O—H angle 104.45°, as shown in Fig. 19.5. It is first necessary to find

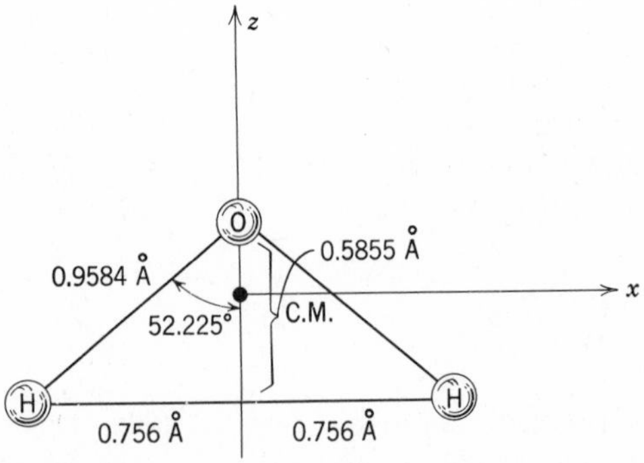

Fig. 19.5 Diagram of the water molecule for Example 19.7.

the center of mass and the principal moments for the molecule. We select a coordinate system with the z-axis coinciding with the line of symmetry and the x—z plane in the plane of the molecule as shown in the diagram. The y-axis is then normal to the page.

The atomic masses are

$$m_O = 1.66 \times 10^{-24}(16.0) = 26.55 \times 10^{-24}\text{ gm}$$
$$m_H = 1.66 \times 10^{-24}(1.008) = 1.673 \times 10^{-24}\text{ gm}$$

For this coordinate system,

$$\begin{aligned} x_O &= 0 \\ x_H &= \pm 0.9584 \sin 52.225° = \pm 0.756\text{ Å} \\ y_O &= y_H = 0 \\ z_O - z_H &= 0.9584 \cos 52.225° = 0.5855 \end{aligned}$$

From Eq. 19.92,

$$\sum_i m_i z_i = 26.55 \times 10^{-24} \times z_O + 2 \times 1.673 \times 10^{-24}\,(z_O - 0.5855) = 0$$

Therefore

$$\begin{aligned} z_O &= +0.0655\text{ Å} \\ z_H &= -0.5200\text{ Å} \end{aligned}$$

Since $y_O = y_H = 0$, it is apparent that

$$I_{xy} = I_{yz} = 0$$

and

$$\begin{aligned} I_{zx} &= \sum_i m_i z_i x_i \\ &= 26.55 \times 10^{-24}(0.0655 \times 10^{-8})(0) \\ &\quad + 1.673 \times 10^{-24}(-0.5200 \times 10^{-8})(0.756 \times 10^{-8}) \\ &\quad + 1.673 \times 10^{-24}(-0.5200 \times 10^{-8})(-0.756 \times 10^{-8}) = 0 \end{aligned}$$

so that the axes selected are the principal set, and Eq. 19.94 therefore gives the principal moments. These are found to be

$$\begin{aligned} I_x &= \sum_i m_i(y_i^2 + z_i^2) \\ &= 26.55 \times 10^{-24}[0 + (0.0655 \times 10^{-8})^2] \\ &\quad + 2 \times 1.673 \times 10^{-24}[0 + (-0.5200 \times 10^{-8})^2] \\ &= 1.019 \times 10^{-40}\text{ gm-cm}^2 \end{aligned}$$

Similarly,

$$I_y = \sum_i m_i(z_i^2 + x_i^2) = 2.932 \times 10^{-40} \text{ gm-cm}^2$$

$$I_z = \sum_i m_i(x_i^2 + y_i^2) = 1.913 \times 10^{-40} \text{ gm-cm}^2$$

The water molecule is consequently found to be an asymmetric top. The symmetry number of the molecule is 2, since the molecule could be rotated 180° about the z-axis into an indistinguishable position. Therefore, from Eq. 19.92, using $\sigma = 2$, we find

$$\begin{aligned} Z_r &= \frac{8\pi^2}{\sigma h^3}(I_x I_y I_z)^{1/2}(2\pi k T)^{3/2} \\ &= \frac{8\pi^2}{2(6.625 \times 10^{-27})^3} \\ &\quad \times (1.019 \times 10^{-40} \times 2.932 \times 10^{-40} \times 1.913 \times 10^{-40})^{1/2} \\ &\quad \times (2\pi \times 1.3804 \times 10^{-16} \times 1000)^{3/2} \\ &= 263 \end{aligned}$$

Then, from Eq. 19.99, the rotational Gibbs function for H_2O at 1000 K is

$$\bar{g}_r = -\bar{R}T \ln Z_r = -1.987(1000) \ln 263 = -11{,}070 \text{ cal/mole}$$

The vibrational contributions for the non-linear molecule can be handled in exactly the same manner as for the linear polyatomic molecule. The sole difference is that in this case there are only $(3a-6)$ vibrational modes and separate contributions to properties, instead of $(3a-5)$ as for the linear molecule. Each of the modes contributes to properties according to the set of equations 19.75 to 19.79. The electronic ground level contributions are found from Eqs. 19.81 to 19.83 as before.

19.10 The Photon Gas

In this section, we consider electromagnetic radiation as represented by a gas of photons, an application of statistical thermodynamics in which corrected Boltzmann statistics is not a reasonable model. It was found in Section 16.2 that, if the multiplier e^α is very large, then both the quantum statistical models, Bose-Einstein and Fermi-Dirac, reduce to a common approximate form, which we call the corrected Boltzmann model. Whenever e^α is not large compared with unity, however, the appropriate quantum statistical model must be employed in its general form.

The atoms comprising a system change from one electronic state to a higher or lower energy state as a result of absorbing or emitting photons of radiation of the appropriate frequency, as discussed in Chapter 18. Since these photons are strictly independent of one another, we can reasonably treat a collection of photons constituting an amount of radiant energy as an ideal gas. In doing so, we realize that photons behave as elementary particles of zero spin. Therefore the Pauli exclusion principle is not applicable, and the Bose-Einstein model is the appropriate quantum statistical model for such a system.

The behavior of a body emitting or absorbing radiation at some steady state temperature T can be analyzed in terms of the radiation field or photon gas in equilibrium with the body. Let us consider, then, a simple model consisting of a box of sides $x = y = z = l$, having perfectly reflecting walls and containing a quantity of radiant energy U. We shall proceed to determine the equilibrium distribution of photons comprising this radiation among the various available energy levels (actually frequencies). This problem is in many respects identical with those considered previously, namely, determination of the equilibrium distribution of particles in a system by maximizing the entropy subject to the system constraints.

Let us assume for the moment a set of discrete energy levels ϵ_j, with corresponding degeneracies g_j. A particular distribution is specified by the number of photons, N_j, in the various levels of energy, such that the system energy is given by

$$U = \sum_j N_j \epsilon_j \tag{19.100}$$

The problem at hand is different from previous ones in one very important aspect. In earlier developments, the total number of particles in the system was always fixed, providing a second constraint on the system variables. Photons, however, can be emitted or absorbed at the walls of the system in such a manner that the number of photons in the system is not necessarily conserved. For example, at some instant of time, two photons of energy ϵ_j could be absorbed at the wall and another of energy $2\epsilon_j$ emitted, with the system energy U remaining constant as required.

Thus the equilibrium distribution in this system is determined through the maximization of the thermodynamic probability for Bose-Einstein statistics, which from Eq. 16.18 is

$$d \ln \mathrm{w} = \sum_j \ln \left(\frac{\mathsf{g}_j + N_j}{N_j} \right) dN_j = 0 \tag{19.101}$$

subject to the single constraint

$$dU = \sum_j \epsilon_j \, dN_j = 0 \qquad (19.102)$$

Multiplying Eq. 19.102 by β and adding to the negative of Eq. 19.101, we obtain

$$\sum_j \left[-\ln \left(\frac{g_j + N_j}{N_j} \right) + \beta \epsilon_j \right] dN_j = 0 \qquad (19.103)$$

By the method of undetermined multipliers, we conclude that the coefficients of all the dN_j must be zero. Therefore the equilibrium distribution of photons is specified by

$$N_j = \frac{g_j}{e^{\beta \epsilon_j} - 1} \qquad (19.104)$$

As in earlier developments,

$$\beta = \frac{1}{kT}$$

so that

$$N_j = \frac{g_j}{e^{\epsilon_j / kT} - 1} \qquad (19.105)$$

This result is the Bose-Einstein equilibrium distribution equation for ordinary particles, Eq. 17.46, with the multiplier α taken as zero.

Having developed the distribution equation, Eq. 19.105, we now turn to the problem of evaluating the energy levels ϵ_j and degeneracies g_j. From Planck's equation, Eq. 18.1,

$$\epsilon = h\nu$$

we recognize that such a level ϵ_j corresponds to a small range of frequencies ν_j to $\nu_j + \Delta\nu$, and also that the degeneracy g_j represents the number of quantum states in this interval $\Delta\nu$.

The differential wave equation describing the photons is identical in form with that for the translation of a particle in a box analyzed in Section 18.4. As in the solution to that equation, the equilibrium condition is specified by standing waves having nodes at the boundaries of the container. The solution is that given by Eq. 18.40 in terms of the set of positive integers $\mathbf{k}_x$, $\mathbf{k}_y$, $\mathbf{k}_z$,

$$\Psi_{\mathbf{k}_x, \mathbf{k}_y, \mathbf{k}_z}(x, y, z) = C \sin\left(\mathbf{k}_x \pi \frac{x}{l} \right) \sin\left(\mathbf{k}_y \pi \frac{y}{l} \right) \sin\left(\mathbf{k}_z \pi \frac{z}{l} \right)$$

In order that the wave nodes coincide with the wall of the container (the wave function equals zero at the walls), the wave number (reciprocal of wavelength λ) components are restricted to the values

$$\omega_x = \frac{\mathbf{k}_x}{2l}, \quad \omega_y = \frac{\mathbf{k}_y}{2l}, \quad \omega_z = \frac{\mathbf{k}_z}{2l} \tag{19.106}$$

Therefore the allowed wave numbers ω can be represented as

$$\omega = \frac{1}{\lambda} = (\omega_x^{\,2} + \omega_y^{\,2} + \omega_z^{\,2})^{1/2} = \frac{r}{2l} \tag{19.107}$$

where

$$r = (\mathbf{k}_x^{\,2} + \mathbf{k}_y^{\,2} + \mathbf{k}_z^{\,2})^{1/2} \tag{19.108}$$

which is the same as the vector r defined in Section 19.3 for translational states. Proceeding in the same manner as followed in that section, we find that the number of sets of all positive $\mathbf{k}_x$, $\mathbf{k}_y$, $\mathbf{k}_z$ in a spherical shell between r and $r + dr$ is

$$\tfrac{1}{8}(4\pi r^2\, dr)$$

This can alternatively be expressed in terms of the wavelength λ using Eq. 19.107, or in terms of the frequency ν, where

$$\nu = \frac{c}{\lambda} = \frac{cr}{2l}$$

Thus the number of sets of $\mathbf{k}_x$, $\mathbf{k}_y$, $\mathbf{k}_z$ can be written as

$$\tfrac{1}{8}(4\pi)\left(\frac{2l\nu}{c}\right)^2\left(\frac{2l}{c}\right) d\nu = \frac{4\pi V}{c^3}\nu^2\, d\nu$$

For any direction of wave propagation, the radiation has two planes of polarization normal to the direction of propagation. Consequently, this result must be multiplied by a factor of 2 in order to give the total number of quantum states, g_ν, for photons in the frequency band ν to $\nu + d\nu$. Therefore we have

$$\mathsf{g}_\nu = \frac{8\pi V}{c^3}\nu^2\, d\nu \tag{19.109}$$

corresponding to an energy ϵ_ν given by Eq. 18.1 or by

$$\epsilon_\nu = h\nu = \frac{hcr}{2l} = \frac{hc}{2V^{1/3}}(\mathbf{k}_x^{\,2} + \mathbf{k}_y^{\,2} + \mathbf{k}_z^{\,2})^{1/2} \tag{19.110}$$

If we now substitute Eqs. 19.109 and 18.1 into Eq. 19.105, the equilibrium distribution can be expressed in terms of the number of photons, dN_ν, in the frequency band ν to $\nu + d\nu$ as

$$dN_\nu = \frac{8\pi V}{c^3} \frac{\nu^2}{e^{h\nu/kT} - 1} d\nu \tag{19.111}$$

Using this result, we see that the radiant energy per unit volume, or the energy density, for the frequency band ν to $\nu + d\nu$ is

$$\frac{\epsilon_\nu \, dN_\nu}{V} = \frac{8\pi h}{c^3} \frac{\nu^3}{e^{h\nu/kT} - 1} d\nu \tag{19.112}$$

which is called Planck's distribution equation for a black body (ideal) radiator. If the energy density according to Eq. 19.112 is plotted versus frequency at a given temperature, the distribution is as indicated in Fig. 19.6. We note from this figure that the maximum in energy density

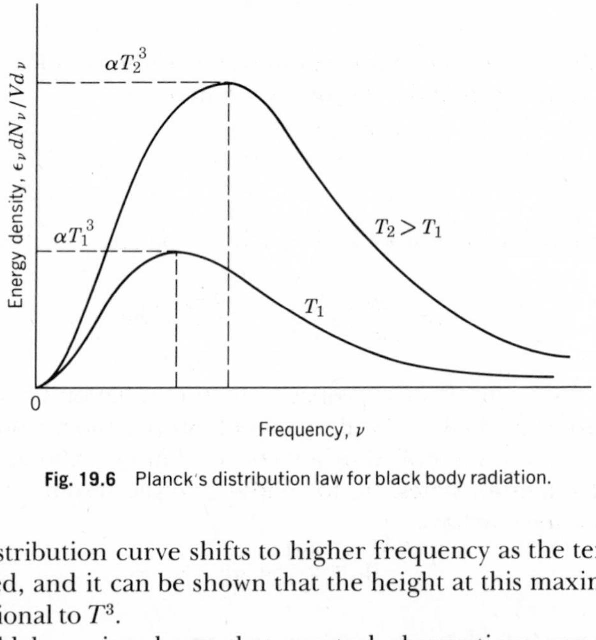

Fig. 19.6 Planck's distribution law for black body radiation.

for the distribution curve shifts to higher frequency as the temperature is increased, and it can be shown that the height at this maximum point is proportional to T^3.

It should be pointed out that spectral observations are commonly made in terms of the wavelength λ instead of the frequency. For this purpose, the Planck distribution equation can easily be rewritten in terms of λ, which is left to the problems at the end of this chapter.

It is of interest to examine two special cases of the distribution equation written in the form of Eq. 19.112, the very low and the very high frequency ranges, respectively. If, at a given temperature, we consider only the very low frequency end, so that

$$\frac{h\nu}{kT} \ll 1$$

then

$$e^{h\nu/kT} - 1 \approx \frac{h\nu}{kT}$$

With this approximation, the radiant energy distribution, Eq. 19.112, reduces to

$$\frac{\epsilon_\nu \, dN_\nu}{V} = \frac{8\pi\nu^2 kT}{c^3} d\nu \tag{19.113}$$

This is the Rayleigh-Jeans distribution equation for low frequency radiation, derived originally from classical mechanics. This expression predicts a continually increasing energy density with frequency, which obviously cannot be reasonable at high frequencies.

The other special case to be analyzed is that of very high frequency, such that

$$\frac{h\nu}{kT} \gg 1$$

for which

$$e^{h\nu/kT} - 1 \approx e^{h\nu/kT}$$

We note that neglecting unity (as compared with the exponential term) is the approximation made in corrected Boltzmann statistics. With this approximation, the energy density (Eq. 19.112) reduces to

$$\frac{\epsilon_\nu \, dN_\nu}{V} = \frac{8\pi h}{c^3} \nu^3 e^{-h\nu/kT} \, d\nu \tag{19.114}$$

This is called Wien's formula, the form of which was proposed originally as an empirical correlation of high frequency distributions.

Our primary interest in considering these two special cases, the Rayleigh-Jeans and Wien formulas, is that historically both preceded the correct expression, Eq. 19.112. Planck originally proposed this relation as a means for interpolating between the low frequency (Rayleigh-Jeans) and high frequency (Wien) equations. Of course, his development was not based on Bose-Einstein statistics, but it did incorporate the concept of photons and proved to be the foundation of the science of quantum mechanics.

Our final analysis in treating the photon gas as a representative model of radiation is the determination of the radiant energy and entropy, from which other thermodynamic properties of interest can then be found. The total energy per unit volume, or energy density, at any temperature is found as the area beneath the distribution curve of Fig. 19.6. Using Eq. 19.112, this is found to be

$$\frac{U}{V}=\int_0^\infty \left(\frac{\epsilon_\nu \, dN_\nu}{V\, d\nu}\right) d\nu = \frac{8\pi h}{c^3}\int_0^\infty \frac{\nu^3}{e^{h\nu/kT}-1}\, d\nu \qquad (19.115)$$

Substituting $x = h\nu/kT$, and integrating Eq. 19.115 (Table B.13),

$$\frac{U}{V}=\frac{8\pi (kT)^4}{(hc)^3}\int_0^\infty \frac{x^3}{e^x-1}\, dx = \frac{8\pi^5}{15}\frac{(kT)^4}{(hc)^3} \qquad (19.116)$$

The energy density at a given temperature is therefore found to be proportional to T^4. If we consider a small rate of flow of radiant energy (through a small pinhole in the wall), the rate of flow is found to be proportional to the energy density inside the box and consequently also proportional to the fourth power of the temperature. This relation is the well-known Stefan-Boltzmann radiation law.

The entropy of a photon gas is that for a Bose-Einstein system with α taken to be zero. From Eq. 17.49, this is found to be

$$S = (k \ln \text{w}_{\text{mp}})_{\text{B-E}} = -k\sum_j \text{g}_j \ln(1-e^{-\epsilon_j/kT}) + \frac{U}{T} \qquad (19.117)$$

Using Eqs. 18.1, 19.109, and 19.116, the photon gas entropy density is found to be

$$\begin{aligned}\frac{S}{V} &= -k\frac{8\pi}{c^3}\int_0^\infty \nu^2 \ln(1-e^{-h\nu/kT})\, d\nu + \frac{U/V}{T}\\ &= -8\pi k^4\left(\frac{T}{hc}\right)^3\int_0^\infty x^2\ln(1-e^{-x})\, dx + \frac{U/V}{T}\\ &= \frac{32\pi^5}{45}k^4\left(\frac{T}{hc}\right)^3 \end{aligned} \qquad (19.118)$$

As mentioned above, other thermodynamic properties of interest can be found from their definitions and the classical relations.

19.11 Mixtures of Gases

This section is related to the discussion of mixtures of ideal gases from the classical viewpoint in Section 11.1, and is intended to provide another means for viewing the behavior of such a system. We consider a mixture of two components A and B, and assume a system of independent particles obeying corrected Boltzmann statistics.

The first problem in analyzing the behavior and properties of a mixture is the determination of the equilibrium state, the development of which is analogous to that for a pure gas in Section 16.1. We consider a system of volume V containing a fixed number of particles N_A of species A and a fixed number N_B of species B, with the system having a given internal energy U. Since the particles are independent, we assume that specification of the system boundary parameter V fixes independent sets of quantum states available to species A and to species B. Thus, each species has its own set of energy levels, which for component A are

$$\epsilon_{A_1}, \epsilon_{A_2}, \ldots, \epsilon_{A_j}$$

with degeneracies

$$g_{A_1}, g_{A_2}, \ldots, g_{A_j}$$

and for component B the levels are

$$\epsilon_{B_1}, \epsilon_{B_2}, \ldots, \epsilon_{B_j}$$

with degeneracies

$$g_{B_1}, g_{B_2}, \ldots, g_{B_j}$$

At any instant of time, the N_A particles of A are distributed in some manner among the energy levels for A, and the N_B particles of B are distributed in some manner among the energy levels for B. Any given system macrostate is then specified in terms of the number of particles of each species that are found in each of its levels, that is, by the sets

$$N_{A_1}, N_{A_2}, \ldots, N_{A_j}$$

$$N_{B_1}, N_{B_2}, \ldots, N_{B_j}$$

The possible macrostates are, of course, subject to the restrictions

$$N_A = \sum_{A_j} N_{A_j} \tag{19.119}$$

$$N_B = \sum_{B_j} N_{B_j} \tag{19.120}$$

and also, since the system energy U is given,

$$U = U_A + U_B = \sum_{Aj} N_{Aj}\epsilon_{Aj} + \sum_{Bj} N_{Bj}\epsilon_{Bj} \qquad (19.121)$$

The thermodynamic probabilities w_A, w_B for the components are multiplicative, since any of the microstates of component A can exist in combination with any of those for component B. Thus the thermodynamic probability w for the mixture is, using the corrected Boltzmann expression Eq. 16.31 for A and B,

$$w = w_A w_B = \left(\prod_{Aj} \frac{g_{Aj}^{N_{Aj}}}{N_{Aj}!}\right)\left(\prod_{Bj} \frac{g_{Bj}^{N_{Bj}}}{N_{Bj}!}\right) \qquad (19.122)$$

If we rewrite Eq. 19.122 in logarithmic form and expand each term by Stirling's formula, the expression for w becomes

$$\ln w = N_A + \sum_{Aj} N_{Aj} \ln \frac{g_{Aj}}{N_{Aj}} + N_B + \sum_{Bj} N_{Bj} \ln \frac{g_{Bj}}{N_{Bj}} \qquad (19.123)$$

We seek the most probable distribution of the sets of particles N_A, N_B, among their respective sets of energy levels; that is, the macrostate having the maximum value of w. If we differentiate Eq. 19.123 and set $d \ln w = 0$ as in the previous developments, we find that

$$\sum_{Aj} \ln \frac{g_{Aj}}{N_{Aj}} dN_{Aj} + \sum_{Bj} \ln \frac{g_{Bj}}{N_{Bj}} dN_{Bj} = 0 \qquad (19.124)$$

This relation specifies the most probable macrostate, subject to the constraints

$$dN_A = \sum_{Aj} dN_{Aj} = 0 \qquad (19.125)$$

$$dN_B = \sum_{Aj} dN_{Bj} = 0 \qquad (19.126)$$

and

$$dU = \sum_{Aj} \epsilon_{Aj} dN_{Aj} + \sum_{Bj} \epsilon_{Bj} dN_{Bj} = 0 \qquad (19.127)$$

Using the method of undetermined multipliers, we multiply Eq. 19.125 by α_A, Eq. 19.126 by α_B, and Eq. 19.127 by β. Adding the three resulting expressions to the negative of Eq. 19.124 we have

$$\sum_{Aj} \left(\ln \frac{N_{Aj}}{g_{Aj}} + \alpha_A + \beta\epsilon_{Aj}\right) dN_{Aj} + \sum_{Bj} \left(\ln \frac{N_{Bj}}{g_{Bj}} + \alpha_B + \beta\epsilon_{Bj}\right) dN_{Bj} = 0 \qquad (19.128)$$

Since components A and B are independent, we conclude from Eq. 19.128 that, for the most probable macrostate,

$$N_{A_j} = g_{A_j} e^{-\alpha_A} e^{-\epsilon_{Aj}/kT} \tag{19.129}$$

for all values of A_j, and

$$N_{B_j} = g_{B_j} e^{-\alpha_B} e^{-\epsilon_{Bj}/kT} \tag{19.130}$$

for all values of B_j. In both expressions the relation $\beta = 1/kT$ has been assumed, as was previously found to be the case for corrected Boltzmann statistics.

If we now sum Eq. 19.129 over all A_j's,

$$N_A = e^{-\alpha_A} \sum_{A_j} g_{A_j} e^{-\epsilon_{A_j}/kT} = e^{-\alpha_A} Z_A \tag{19.131}$$

where, according to Eq. 17.19, Z_A is the partition function for A,

$$Z_A = \sum_{A_j} g_{A_j} e^{-\epsilon_{A_j}/kT} \tag{19.132}$$

Similarly, for component B, from Eq. 19.130,

$$N_B = e^{-\alpha_B} \sum_{B_j} g_{B_j} e^{-\epsilon_{B_j}/kT} = e^{-\alpha_B} Z_B \tag{19.133}$$

where

$$Z_B = \sum_{B_j} g_{B_j} e^{-\epsilon_{B_j}/kT} \tag{19.134}$$

Thus, eliminating the multipliers α_A, α_B, the equilibrium distribution for the binary mixture whose components obey corrected Boltzmann statistics can be expressed as

$$N_{A_j} = \frac{N_A g_{A_j} e^{-\epsilon_{A_j}/kT}}{Z_A} \quad \text{for all } A_j \tag{19.135}$$

$$N_{B_j} = \frac{N_B g_{B_j} e^{-\epsilon_{B_j}/kT}}{Z_B} \quad \text{for all } B_j \tag{19.136}$$

It is important to note that the component partition functions Z_A, Z_B, are to be evaluated for A and B, respectively, each at the temperature and volume of the system. Thus we conclude that each of the components of the mixture behaves as though it existed alone in the total volume V at the temperature of the mixture. We find, then, that the assumption of the existence of independent sets of energy levels at the microscopic level, as specified by the boundary parameter, leads logically to a result for the equilibrium state that is identical to a mixture model assumed in the classical development in Chapter 11.

Let us now examine the thermodynamic properties for this mixture of gases. Substituting Eqs. 19.135 and 19.136 for N_{A_j} and N_{B_j}, and by the same procedure as in Section 16.2, we have for U_A

$$U_A = \sum_{A_j} N_{A_j} \epsilon_{A_j} = \frac{N_A}{Z_A} \sum_{A_j} g_{A_j} \epsilon_{A_j} e^{-\epsilon_{A_j}/kT}$$

$$= \frac{N_A k T^2}{Z_A} \left(\frac{\partial Z_A}{\partial T} \right)_V = n_A \bar{R} T^2 \left(\frac{\partial \ln Z_A}{\partial T} \right)_V \quad (19.137)$$

and similarly for U_B,

$$U_B = \sum_{B_j} N_{B_j} \epsilon_{B_j} = n_B \bar{R} T^2 \left(\frac{\partial \ln Z_B}{\partial T} \right)_V \quad (19.138)$$

Substituting these relations into Eq. 19.121, we find that the internal energy of the system is

$$U = n_A \bar{R} T^2 \left(\frac{\partial \ln Z_A}{\partial T} \right)_V + n_B \bar{R} T^2 \left(\frac{\partial \ln Z_B}{\partial T} \right)_V = n_A \bar{u}_A + n_B \bar{u}_B \quad (19.139)$$

where $\bar{u}_A$ is the molal internal energy for substance A as though it existed alone in volume V at temperature T, while $\bar{u}_B$ is the corresponding value for substance B. This conclusion is in accord with the corresponding analysis made for an ideal gas mixture from the viewpoint of classical thermodynamics in Chapter 11.

The entropy of a system at equilibrium is defined by Eq. 17.6,

$$S = k \ln w_{mp}$$

Therefore, substituting Eqs. 19.135 and 19.136 for the ratios g_{A_j}/N_{A_j} and g_{B_j}/N_{B_j}, which correspond to the most probable distributions, into Eq. 19.123, we obtain

$$\ln w_{mp} = N_A + \sum_{A_j} N_{A_j} \ln \left(\frac{Z_A}{N_A} e^{+\epsilon_{A_j}/kT} \right) + N_B + \sum_{B_j} N_{B_j} \ln \left(\frac{Z_B}{N_B} e^{+\epsilon_{B_j}/kT} \right)$$

$$= N_A + \ln \frac{Z_A}{N_A} \sum_{A_j} N_{A_j} + \frac{1}{kT} \sum_{A_j} N_{A_j} \epsilon_{A_j}$$

$$+ N_B + \ln \frac{Z_B}{N_B} \sum_{B_j} N_{B_j} + \frac{1}{kT} \sum_{B_j} N_{B_j} \epsilon_{B_j}$$

$$= N_A + N_A \ln\frac{Z_A}{N_A} + \frac{U_A}{kT} + N_B + N_B \ln\frac{Z_B}{N_B} + \frac{U_B}{kT}$$

$$= N_A\left[\ln\left(\frac{Z_A}{N_A}\right) + 1\right] + \frac{U_A}{kT} + N_B\left[\ln\left(\frac{Z_B}{N_B}\right) + 1\right] + \frac{U_B}{kT}$$

If we now substitute this result into Eq. 17.6, we obtain for the entropy of the mixture

$$S = N_A k\left[\ln\left(\frac{Z_A}{N_A}\right) + 1\right] + \frac{U_A}{T} + N_B k\left[\ln\left(\frac{Z_B}{N_B}\right) + 1\right] + \frac{U_B}{T}$$

$$= n_A\left\{\bar{R}\left[\ln\left(\frac{Z_A}{N_A}\right) + 1\right] + \frac{\bar{u}_A}{T}\right\} + n_B\left\{\bar{R}\left[\ln\left(\frac{Z_B}{N_B}\right) + 1\right] + \frac{\bar{u}_B}{T}\right\} \quad (19.140)$$

or,

$$\bar{s} = n_A\bar{s}_A + n_B\bar{s}_B \quad (19.141)$$

In Eq. 19.141, $\bar{s}_A$ is the entropy of substance A evaluated as though it exists alone in the system volume V at temperature T, and $\bar{s}_B$ is the corresponding value for substance B. We find again that the result is in accord with an analysis of an ideal gas mixture made from the classical viewpoint.

Expressions for the other thermodynamic properties of interest for ideal gas mixtures, $\bar{C}_{p0}$, $\bar{C}_{v0}$, H, A, G, follow readily from their definitions and the relations for internal energy, entropy, and the ideal gas equation of state.

19.12 Chemical Reaction and Equilibrium

This final section of the present chapter is concerned with the special topics of chemical reaction and equilibrium, and is intended as supplementary material to that covered in Chapters 13 and 14. It is not intended to cover these subjects in detail here, but rather to examine two particular topics, that of enthalpy of formation and the concept of chemical equilibrium from the molecular viewpoint.

To briefly summarize the classical presentation of enthalpy of formation in Chapter 13, we recall that the subject was approached from the calorimetric viewpoint, that of measuring the heat transfer associated with a chemical reaction. These reactions were specified to occur at a constant temperature of 25 C. It was then noted that since one element cannot be converted into another in a chemical reaction, we are free to choose reference values for enthalpy of the various elements independently of one another. For convenience, we choose these reference values for the stable forms of the elements (for example C, N_2, H_2) as zero at the specified temperature of 25 C. Then, measurement of the heat transfer for the various chemical reactions establishes the enthalpy of formation of other forms of these elements and also of compounds relative to this arbitrarily selected reference.

While the above argument presents a logical picture for the establishment of the concept of enthalpy of formation, the numerical values as presented in Table 13.3 are not determined by calorimetric procedures, as implied. Instead, it is more precise to determine the fundamental reference value D_0, as shown in Fig. 19.2, from spectroscopic measurements on the molecular species. The enthalpy of formation (25 C reference) can then be calculated from this value and the properties of the gases. The procedure for such a calculation is most easily demonstrated by an example.

Example 19.8

Calculate the enthalpy of formation of monatomic nitrogen relative to the 25 C chemical reference base.

First, for the monatomic species, N, at 298.15 K:

$$(\bar{h})_{\mathrm{N}} = \bar{h}_t + \bar{h}_e + \bar{h}_{\mathrm{chem}}$$

From Eq. 19.26,

$$\bar{h}_t = \tfrac{5}{2}\bar{R}T = \tfrac{5}{2} \times 1.987 \times 298.15 = 1481 \text{ cal/mole}$$

From the electronic level data in Table 18.2 and Eqs. 19.37, 19.38 and 19.40,

$$\bar{h}_e = \bar{R}T \frac{Z_e'}{Z_e} \approx 0$$

The spectroscopic reference base is taken as zero at 0K for the monatomic species, as indicated in Fig. 19.2. Therefore,

$$\bar{h}_{\mathrm{chem}} = 0$$

and

$$(\bar{h})_{\mathrm{N}} = 1481 + 0 + 0 = 1481 \text{ cal/mole}$$

Now, for the diatomic species, N_2, at 298.15 K:

$$(\bar{h})_{\mathrm{N_2}} = \bar{h}_t + \bar{h}_r + (\bar{h}_v - \bar{u}_{v0}) + \bar{h}_e + \bar{h}_{\mathrm{chem}}$$

As for the monatomic form,

$$\bar{h}_t = \tfrac{5}{2}\bar{R}T = 1481 \text{ cal/mole}$$

For the simple internal model, from Eq. 19.59,

$$\bar{h}_r = \bar{R}T = 1.987 \times 298.15 = 592 \text{ cal/mole}$$

From the data in Table B.14 and Eq. 19.73,

$$\theta = \frac{h_\varepsilon \omega_\varepsilon}{k} = 1.4388 \times 2357.6 = 3395 \text{ K}$$

so that at 298.15 K

$$x = \frac{\theta_v}{T} = \frac{3395}{298.15} = 11.37$$

and from Eq. 19.76,

$$(\bar{h}_v - \bar{u}_{v0}) = \bar{R}T\left(\frac{x}{e^x - 1}\right) \approx 0$$

For the simple internal model, from Eq. 19.81,

$$\bar{h}_e = 0$$

and from Table B.14 and Eq. 19.85,

$$\begin{aligned}\bar{h}_{\text{chem}} &= -N_0 D_0 \\ &= -6.023 \times 10^{23} \times 9.757 \times 1.6021 \times 10^{-12} \times \frac{1}{4.184 \times 10^7} \\ &= -225{,}040 \text{ cal/mole}\end{aligned}$$

Thus,

$$\begin{aligned}(\bar{h})_{N_2} &= 1481 + 592 + 0 + 0 - 225{,}040 \\ &= -222{,}967 \text{ cal/mole}\end{aligned}$$

We now have the enthalpies of both species at 298.15 K relative to the spectroscopic reference base. Consequently, for the chemical reaction

$$\tfrac{1}{2}N_2 \rightarrow N$$

$$\begin{aligned}\Delta H^\circ_{298} &= (\bar{h})_N - \tfrac{1}{2}(\bar{h})_{N_2} \\ &= 1481 - \tfrac{1}{2}(-222{,}967) \\ &= +112{,}965 \text{ cal}\end{aligned}$$

which is equivalent to the calorimetric heat transfer for the constant-temperature reaction at 25 C. Therefore, relative to the 25 C chemical reference value

$$(\bar{h}^\circ_f)_{N_2} = 0$$

We find

$$(\bar{h}_f^\circ)_{\rm N} = \Delta H_{298}^\circ + \tfrac{1}{2}(\bar{h}_f^\circ)_{\rm N_2}$$
$$= +112{,}965 \text{ cal/mole of N}$$

which agrees with the value listed for N in Table B.9.

The second topic that we wish to examine in this section from the molecular viewpoint is that of the concept of chemical equilibrium. Let us consider the dissociation of a diatomic species AB to the two monatomic species A and B. The reaction equation is written as

$$AB \rightleftharpoons A + B \tag{19.142}$$

We assume that the equilibrium mixture behaves as an ideal gas mixture of the three components, AB, A and B. The analysis of this mixture from the molecular viewpoint is analogous to that in Section 19.11 for a binary mixture, except now we have three distinguishable components in volume V and therefore three independent sets of energy levels, one for each species.

Let us consider first a mixture comprised of a given total number of particles of each species N_{AB}, N_A, N_B (not necessarily the number of each at a condition of chemical equilibrium) contained in volume V at some temperature T. These particles are distributed in some manner among the various energy levels, the macrostate being specified by the number of particles of each species in each of its energy levels,

$$\begin{gathered} N_{AB_1},\ N_{AB_2},\ \ldots,\ N_{AB_j} \\ N_{A_1},\ N_{A_2},\ \ldots,\ N_{A_j} \\ N_{B_1},\ N_{B_2},\ \ldots,\ N_{B_j} \end{gathered} \tag{19.143}$$

with the restrictions

$$\begin{gathered} N_{AB} = \sum_{AB_j} N_{AB_j} \\ N_A = \sum_{A_j} N_{A_j} \\ N_B = \sum_{B_j} N_{B_j} \end{gathered} \tag{19.144}$$

For any given macrostate, the thermodynamic probability w for the mixture is

$$\begin{aligned} \mathrm{w} &= (\mathrm{w}_{AB})(\mathrm{w}_A)(\mathrm{w}_B) \\ &= \left(\prod_{AB_j} \frac{g_{AB_j}^{N_{AB_j}}}{N_{AB_j}!}\right)\left(\prod_{A_j} \frac{g_{A_j}^{N_{A_j}}}{N_{A_j}!}\right)\left(\prod_{B_j} \frac{g_{B_j}^{N_{B_j}}}{N_{B_j}!}\right) \end{aligned} \tag{19.145}$$

As discussed in Section 19.11, each state for component A can be associated with any state of B or of AB, etc. Therefore the value of w for the mixture is the product of those for the individual constituents, as indicated in Eq. 19.145. The energy corresponding to this given macrostate is expressed by

$$U=\sum_{AB_j} N_{AB_j}\epsilon_{AB_j}+\sum_{A_j} N_{A_j}\epsilon_{A_j}+\sum_{B_j} N_{B_j}\epsilon_{B_j} \tag{19.146}$$

We seek the most probable macrostate for this system, subject to the given constraints, Eqs. 19.144 and 19.146, a fixed number of particles of each type, and fixed energy of the system. This is found by maximizing Eq. 19.145 subject to these constraints. From the results of Section 19.11, Eqs. 19.135 and 19.136, it is easily seen that this result is

$$N_{AB_j}=\frac{N_{AB}}{Z_{AB}}\,g_{AB_j}e^{-\epsilon_{AB_j}/kT},\quad \text{for all } AB_j \tag{19.147}$$

$$N_{A_j}=\frac{N_A}{Z_A}\,g_{A_j}e^{-\epsilon_{A_j}/kT},\quad \text{for all } A_j \tag{19.148}$$

$$N_{B_j}=\frac{N_B}{Z_B}\,g_{B_j}e^{-\epsilon_{B_j}/kT},\quad \text{for all } B_j \tag{19.149}$$

This is the most probable macrostate for the given values N_{AB}, N_A, N_B, and U. The thermodynamic probability corresponding to this most probable macrostate is, using Eq. 19.139 for three components,

$$\ln \mathrm{w}_{\mathrm{mp}}=N_{AB}\left[\ln\left(\frac{Z_{AB}}{N_{AB}}\right)+1\right]+N_A\left[\ln\left(\frac{Z_A}{N_A}\right)+1\right]+N_B\left[\ln\left(\frac{Z_B}{N_B}\right)+1\right]+\frac{U}{kT} \tag{19.150}$$

To review briefly, we began our development by considering a mixture comprised of given N_{AB}, N_A, and N_B, but not necessarily the proportions for chemical equilibrium, and we found that the thermodynamic probability for the most probable macrostate in the mixture of these proportions is given by Eq. 19.150. There will be an equation of the same form for each mixture of different proportions (different values of N_{AB}, N_A, N_B), but of course the value of w_{mp} will be different for each different mixture. The particular mixture with the largest value of w_{mp} is the chemical equilibrium mixture and can be found by determining the maximum of Eq. 19.150 while varying N_{AB}, N_A, N_B. The possible values of N_{AB}, N_A, N_B are, however, restricted according to the initial composition in the system. The total number of atoms of A must remain the same, whether in the diatomic form AB or in the monatomic form, and the total number of atoms of B must similarly remain constant. Since there is one atom of A and one of B in AB,

$$\text{total atoms of type } A = N_{AB} + N_A = \text{constant} \tag{19.151}$$
$$\text{total atoms of type } B = N_{AB} + N_B = \text{constant} \tag{19.152}$$

To find the maximum in Eq. 19.150 subject to these interrelations, we differentiate Eq. 19.150 with respect to N_{AB}, N_A, N_B, at fixed U, V and set the result equal to zero. Upon performing this differentiation, it is found that, for the chemical equilibrium composition,

$$d \ln w_{mp} = \ln \frac{Z_{AB}}{N_{AB}} dN_{AB} + \ln \frac{Z_A}{N_A} dN_A + \ln \frac{Z_B}{N_B} dN_B = 0 \tag{19.153}$$

subject to the constraints

$$dN_{AB} + dN_A = 0 \tag{19.154}$$
$$dN_{AB} + dN_B = 0 \tag{19.155}$$

We now multiply Eq. 19.154 by γ and Eq. 19.155 by δ and add the result to Eq. 19.153. By the method of undetermined multipliers, we conclude that, for the equilibrium composition N_{AB_e}, N_{A_e}, N_{B_e},

$$\gamma + \delta + \ln \frac{Z_{AB}}{N_{AB_e}} = 0 \tag{19.156}$$

$$\gamma + \ln \frac{Z_A}{N_{A_e}} = 0 \tag{19.157}$$

$$\delta + \ln \frac{Z_B}{N_{B_e}} = 0 \tag{19.158}$$

Eliminating the multipliers γ and δ from these equations, we have

$$\frac{N_{A_e} N_{B_e}}{N_{AB_e}} = \frac{Z_A Z_B}{Z_{AB}} \tag{19.159}$$

which is the equilibrium equation and is analogous to the expression (Eq. 14.39) developed from the classical viewpoint. To demonstrate this analogy, recall that the mole fraction of component i is

$$y_i = \frac{n_i}{n} = \frac{N_i}{N} \tag{19.160}$$

and also that the partition function, including all contributions, can be written for component i as

$$Z_i = V \times f_i(T) \qquad (19.161)$$

Therefore, substituting these relations into Eq. 19.159 for the three species, we have

$$\frac{y_A y_B}{y_{AB}} = \frac{V}{N} f(T) = \frac{kT}{P} f(T) = K\left(\frac{P^\circ}{P}\right) \qquad (19.162)$$

which is the classical result (Eq. 14.39 applied to the reaction Eq. 19.142) in terms of the equilibrium constant K.

PROBLEMS

19.1 In Section 19.2, the translational partition function was determined for a system of dimensions $X = Y = Z$. Repeat this development for a system in which X, Y, and Z are not equal.

19.2 Consider the evaluation of the translational partition function, Eq. 19.22, from the quantum mechanical energy expression, Eq. 19.21. For argon at −100 C, 1 atm pressure, estimate the error introduced by neglecting the 1/2 term in this development.

19.3 A two-dimensional ideal gas has been considered in Problem 16.5. Consider such a gas confined to a surface of dimensions $X = Y$ and possessing only translational energy.

(*a*) Show that the energy of a quantum state is given by

$$\epsilon_{\mathbf{k}_x, \mathbf{k}_y} = \frac{h^2}{8m\mathscr{A}}(\mathbf{k}_x^{\,2} + \mathbf{k}_y^{\,2})$$

where $\mathscr{A}$ is the surface area.

(*b*) Find expressions for the internal energy and entropy of this gas.

(*c*) Using the thermodynamic relation

$$P = -\left(\frac{\partial A}{\partial \mathscr{A}}\right)_T$$

show that P is given by the result found in Problem 16.5.

19.4 The equation for translational entropy of an ideal gas can be written in the form

$$s = R[C_1 \ln T + C_2 \ln M - \ln P + C_3]$$

which is known as the Sackur-Tetrode equation. If T is in degrees K, M is the molecular weight, and P is in atmospheres, determine the values of the constants C_1, C_2, C_3.

19.5 Evaluate the translational partition function using the approach of Section 19.3, beginning with Eq. 19.35 for the number of translational states in an energy interval ϵ to $\epsilon + d\epsilon$.

19.6 Consider a 1 cm³ sample of O_2 at 100 C, 1 atm pressure. Using Eq. 19.35, calculate the number of translational quantum states in the interval of energy kT to $1.001kT$.

19.7 Considering translational energy only, use Eq. 19.35 and the result of Problem 16.11 to find an expression for the fractional occupancy dN_ϵ/g_ϵ at any level of energy ϵ to $\epsilon + d\epsilon$. Is the result a surprising one? Plot this expression versus ϵ/kT.

19.8 Calculate the entropy per mole of argon at 300 K, 1 atm pressure.

19.9 Calculate the entropy per mole of monatomic hydrogen at 300 K, 1 atm pressure and compare the result with the value given in Table B.9.

19.10 Calculate the entropy of singly-ionized nitrogen (N atoms minus one electron) at 2000 K, 2 atm pressure.
Observed electronic terms for N^+:

Term	(ϵ/hc), cm^{-1}
$^3P_0$	0
$^3P_1$	49.1
$^3P_2$	131.3
$^1D_2$	15315.7

19.11 Calculate the values of $\bar{h}$, $\bar{C}_{po}$, and $\bar{s}$ for monatomic oxygen at 100 K and 1000 K, 1 atm pressure. Compare the enthalpy and entropy values with those given in Table B.9.

19.12 Plot the electronic contribution to specific heat versus temperature for monatomic oxygen.

19.13 The following are the first several electronic terms observed for monatomic chlorine gas:

Term	(ϵ/hc), cm^{-1}
$^2P_{3/2}$	0
$^2P_{1/2}$	881
$^4P_{5/2}$	71954
$^4P_{3/2}$	72484
$^4P_{1/2}$	72823

Sketch the specific heat $\bar{C}_{po}$ as a function of temperature. Include estimates of any temperatures at which unusual behavior occurs.

19.14 Consider a monatomic gas having two non-degenerate electronic states $\epsilon_{e0} = 0$ and ϵ_{e1}.

(*a*) Plot the electronic contribution to specific heat versus "reduced" temperature, kT/ϵ_{e1}.

(*b*) Determine analytically the value of reduced temperature at which the specific heat contribution is a maximum.

19.15 Plot the distribution of molecules among the various rotational levels for carbon monoxide at 100 K and at 300 K.

19.16 For HCl, find the temperature at which exactly 50% of the molecules have a rotational quantum number greater than 3.

19.17 (*a*) Using the Euler-Maclaurin summation theorem, show that Eq. 19.63 is valid.

(*b*) Estimate the error in calculating the rotational partition function for carbon monoxide at 50 K, 100 K, and 500 K by the integrated expression, Eq. 19.56.

19.18 When T/θ_r is less than unity, the rotational partition function for a heteronuclear molecule can be evaluated by direct expansion of Eq. 19.53.

(*a*) For such a region, develop an expression for the rotational internal energy.

(*b*) Plot the rotational internal energy versus T for HF at temperatures below 100 K, using the appropriate method for determining u_r.

19.19 Repeat Problem 19.18 for the rotational specific heat of HF, and discuss the results of the two problems.

19.20 Hydrogen deuteride (HD), the diatomic molecule comprised of a hydrogen atom and a deuteron (1 proton and 1 neutron in the nucleus), has a characteristic rotational temperature $\theta_r = 63$ K. Calculate the rotational entropy of HD at 20, 100, and 500 K, using whatever method is appropriate at each temperature.

19.21 Prove that the vibrational entropy is independent of the selection for the ground level reference of energy, u_{v_0}.

19.22 For diatomic nitrogen at 2000 K, calculate the probability that a molecule will have a vibrational quantum number less than 3.

19.23 Plot the distribution of molecules among the various vibrational levels for diatomic bromine at 300 K and 3000 K.

19.24 For diatomic fluorine at 1 atm pressure, compare the relative magnitudes of the various contributions to $\bar{C}_{p_0}$ and $\bar{s}$ at 300 K and also at 3000 K.

19.25 Calculate the entropy of diatomic bromine at 800 K, 5 atm pressure. Clearly specify any assumptions made in the calculation.

19.26 Calculate the entropy of nitric oxide (NO) at 300 K, 1 atm. The electronic term is $^2\pi$, a total of 4 states—two of which lie 121.1 cm^{-1} above the other two. Is the simple integrated result valid for the rotational partition function at this temperature?

19.27 Diatomic chlorine at 1000 K, 1 atm, is cooled in a constant-pressure, steady flow process to 300 K. Calculate the heat transfer and entropy change per mole during this process.

19.28 Atoms from a monatomic gas are adsorbed onto a surface of area $\mathscr{A}$ until an equilibrium condition is established (Gibbs function per mole of the adsorbed atoms equal to that of the free gas). Assume ground electronic state for each. For the adsorbed atoms, assume a two-dimensional translation on the adsorbing surface plus a vibrational degree of freedom normal to the surface. (The vibration is represented by θ_v and a ground-point energy ϵ_0). Derive an expression for the concentration of adsorbed atoms on the surface (atoms/cm^2) in terms of the free gas number density (atoms/cm^3) and other relevant parameters.

19.29 The first three electronic levels of diatomic oxygen are:

Term	ϵ/hc, cm^{-1}
$^3\Sigma_g^-$	0
$^1\Delta_g$	7,882
$^1\Sigma_g^+$	13,121

Assume a rigid rotator-harmonic oscillator for which the constants r_ε and ω_ε are the same for each of the electronic levels and no higher electronic levels are populated.

(*a*) Calculate the fraction of molecules in each of the three electronic levels at 5000 K.

(*b*) Determine the constant-pressure specific heat of O_2 at 5000 K.

19.30 Calculate the enthalpy and the entropy of diatomic oxygen at 5000 K, 1 atm pressure, and compare the results with the values given by Table B.9.

19.31 Calculate the Gibbs function with respect to ground state energy for nitrous oxide (N_2O) at 800 K, 1 atm pressure.

19.32 Carbon dioxide at 300 K, 1 atm pressure, is heated in a constant-pressure, steady-flow process to 1200 K.

(*a*) Calculate the heat transfer for the process and the entropy change of the carbon dioxide.

(*b*) Repeat part *a* using the Gas Tables, B.9.

19.33 Calculate the entropy and the constant-pressure specific heat for water at 1000 K, 1 atm pressure, and compare the value of entropy with that in Table B.9.

19.34 (*a*) Calculate the entropy of ideal gas water at 25 C, 1 atm pressure, and compare the value with that of Table B.9. Ideal gas water is a hypothetical state at this condition, but is useful as a reference state for other calculations.

(*b*) Using the result of part *a* and the steam tables, calculate the entropy of liquid water at 25 C, 1 atm pressure, and compare the result with the value of Table 13.3.

19.35 Determine the principal moments of inertia for the methane (CH_4) molecule, and calculate the rotational entropy at 400 K.

19.36 Consider the black body radiation distribution equation, Eq. 19.112. At any given temperature, find the frequency at which the energy density is a maximum. Show that this maximum energy density is proportional to T^3.

19.37 Spectral observations of thermal radiation are made in terms of the wavelength λ instead of frequency, as was mentioned in Section 19.10. Thus, in analyzing the data, it is necessary to consider the Planck distribution wirtten in terms of λ, where $\epsilon_\lambda \, dN_\lambda = \epsilon_\nu \, dN_\nu$.

(*a*) Write the Planck distribution equation in terms of λ, and sketch the distribution as a function of λ.

(*b*) For any temperature T, find the wavelength at which the function $(\epsilon_\lambda \, dN_\lambda / V d\lambda)$ is a maximum.

19.38 Plot the black body distribution versus frequency for a temperature of 1000 K. Compare the result with the Rayleigh-Jeans and Wien equations.

19.39 Consider a photon gas as representative of thermal radiation in equilibrium with its container. Using the thermodynamic relation

$$P = -\left(\frac{\partial A}{\partial V}\right)_T$$

calculate the radiation pressure at 1000 K.

19.40 In Section 16.5 an expression for the flux of particles escaping a tank was developed. An analogous result can be found for the energy flux of radiant energy.

(*a*) Show that the radiant energy flux through a small hole (such that the equilibrium distribution at temperature T inside the container is not disturbed) is given by the relation

$$\frac{\text{Energy}}{\text{cm}^2\text{-sec}} = \frac{1}{4}\left(\frac{U}{V}\right)c$$

in terms of the equilibrium energy density and speed of photons inside the container.

(*b*) The radiant energy flux found in part *a* is also equal to σT^4, where σ is the Stefan-Boltzmann constant. Determine the numerical value of σ.

19.41 The intensity of solar radiation is a maximum at approximately 5000 Å.

(*a*) Assuming the sun to be an ideal radiator, calculate its surface temperature. (Use the result of Problem 19.37.)

(*b*) Use the results of part *a* and Problem 19.40 to find the flux of solar radiant energy.

19.42 A large tank contains a mixture of two gases A and B at temperature T and pressure P. The mole fractions are y_A and y_B, respectively. A very small hole is then opened in the wall of the container; thus every particle striking the hole may be assumed to escape. Does the escaping stream of particles have the same composition as the mixture in the tank? How do the relative proportions depend on molecular weights?

19.43 A 10-liter tank contains a mixture of 50% helium, 50% nitrogen at 30 C, 0.1 atm. One end of the tank contains a porous plug consisting of very fine holes, through which the gas is to be withdrawn. The total hole area is estimated to be 10^{-4} cm^2. Determine the pressure and composition inside the tank as a function of time.

19.44 Two dissimilar ideal gases A and B are contained in an insulated tank separated by a wall. There are N_A molecules of A in volume V_A and N_B molecules of B in volume V_B. Both are at the same pressure and temperature. The wall is then broken and the gases are allowed to mix.

(*a*) Use the partition functions to show that the entropy change due to mixing is given by

$$-\bar{R} \sum_i n_i \ln y_i$$

(*b*) If gases A and B are identical, show that there is no change in entropy when the wall is broken.

19.45 Calculate the entropy per mole for a mixture of 50% helium, 50% neon, at 500 C, 10 atm.

19.46 A 1-liter tank containing helium initially at 0.5 atm pressure is located in a large environment of argon at 1 atm pressure. The temperature is uniform and constant at 20 C. The tank wall contains a covered porous plug consisting of very small holes, a total hole area of 10^{-5} cm^2. The cover is then removed from the porous plug for 20 minutes. Assume that any helium atoms leaving the tank do not reenter, but argon atoms entering from outside may flow out again. Calculate the pressure and composition inside the tank at the end of the 20-minute period, and explain why the values come out as they do.

19.47 To demonstrate the necessity of including the constant ground electronic level contribution to entropy, consider the following process: one mole of HF is dissociated to form one mole of H and one mole of F. Assuming each substance to be in its ground electronic level, calculate the electronic contribution to the change in entropy for the process.

19.48 Calculate the enthalpy of formation of liquid water at 25 C, 1 atm, relative to the "chemical" reference base (zero for ideal gas H_2 and O_2 at 25 C). Whatever models, assumptions or approximations are used must be clearly stated and justified. Use the value $D_0 = 9.51$ ev for H_2O.

19.49 Using a value $D_0 = 16.55$ ev for CO_2, calculate the equilibrium constant at 3000 K for the reaction

$$CO_2 \rightleftharpoons CO + \tfrac{1}{2}O_2$$

and compare the result with the value given in Table B.10.

19.50 Calculate the equilibrium constant for the reaction $N_2 \rightleftharpoons 2N$ at 2000 K and at 6000 K, and compare the results with the values listed in Table B.10.

19.51 At temperatures above 10,000 K, nitrogen is essentially completely dissociated to the monatomic species and ionization begins to be of significance.

(*a*) Determine the ionization equilibrium constant at 15,000 K for the reaction

$$\mathrm{N} \rightleftharpoons \mathrm{N}^+ + \mathrm{e}^-$$

For monatomic N, use Table 18.2 for the electronic levels, and an ionization energy (ground level of N^+ above N) of 14.54 ev. For the positive ion N^+, use the electronic term data given in Problem 19.10.

(*b*) For an initial composition of pure N_2 at room temperature, what is the equilibrium composition at 15,000 K, 0.1 atm?

19.52 Show that, if Eq. 19.150 is maximized at constant U, V, the result is Eq. 19.153.

19.53 Carry out a maximization at constant T, V, analogous to that of Eqs. 19.150 to 19.153. (In terms of the independent variables T, V, it is necessary to utilize the Helmholtz function instead of entropy.)

19.54 Consider the molecular dissociation $A_2 \rightleftharpoons 2A$, where A is an arbitrary substance. Develop an expression for the chemical equilibrium constant in terms of the significant fundamental parameters. Assume ground electronic levels only, with degeneracies $\mathrm{g}_{e_{A_1}}$, $\mathrm{g}_{e_{A_2}}$, and use the simple internal model for A_2.

19.55 Write a computer program to solve the following problem. For one of the gases listed in both Tables B.9 and B.14, it is desired to compare the values for enthalpy and entropy as calculated by the methods of this chapter with the values given in Table B.9, over the entire range of that table.

20

The Properties of Solids

Up to this point in our development and application of statistical thermodynamics, we have been concerned with the analysis of systems of independent particles. In these systems, the forces between particles are negligibly small, and we may therefore correctly speak of single-particle states and the energy per particle. The characteristic structure of a solid, however, is that of atoms which are held tightly together in a crystal lattice by strong interatomic forces, as discussed in Section 18.10. The individual atoms possess energy as a result of vibrations about their equilibrium positions in the lattice. These vibrations become stronger with respect to one another as energy is added to the crystal. Certainly, the behavior of each atom in the lattice is strongly influenced by its neighboring atoms in the lattice structure. Nevertheless, it is found to be reasonable to treat the crystal as a collection of independent or semi-independent harmonic oscillators, and to calculate the energy and other properties of the solid phase on the basis of this assumption. In this chapter, we apply the results of Section 19.7, the properties of the individual harmonic oscillator, to two models for the solid phase, for which expressions for various thermodynamic properties of solids are derived. Finally, the electronic contributions to thermodynamic properties are evaluated on the basis of the development of the Fermi-Dirac statistical model.

20.1 The Einstein Solid

In Einstein's simple model of the solid phase, it is assumed that each of the N atoms in a crystal vibrates harmonically and independently in three directions about its equilibrium position in the crystal lattice. The system is therefore assumed to consist of $3N$ independent harmonic oscillators, all of the same fundamental frequency ν_ε, since the vibrational modes are independent of one another. The properties of such a system are then $3N$ times those for a single oscillator. In Section 19.7, the thermodynamic properties were expressed on a molal basis (N_0 oscillators), so that for an Einstein solid the properties per mole are simply three times those expressed by Eqs. 19.75 to 19.79.

Of particular interest is the specific heat, which can be compared with experimentally determined values. It follows from Eq. 19.77 that, for the Einstein solid, the specific heat $\bar{C}_{v_E}$ is

$$\bar{C}_{v_E} = \frac{3\bar{R}(\theta_E/T)^2 e^{\theta_E/T}}{(e^{\theta_E/T}-1)^2} \tag{20.1}$$

where θ_E, which is called the Einstein temperature, has been used instead of θ_v. Figure 20.1 shows a plot of the Einstein specific heat as given by Eq. 20.1 versus the dimensionless temperature T/θ_E.

At high temperature, $\bar{C}_{v_E}$ approaches a limit of $3R$, which is in agreement with most observed data. This high temperature limiting value of $3\bar{R}$ was observed experimentally by Dulong and Petit in the early nineteenth century; this is the analytical result that one obtains from classical mechanics for all temperatures. It should be pointed out that for some solids the specific heat does not approach this value, because of anharmonicity or additional contributions from electronic states.

At low temperature, the Einstein specific heat expression, Eq. 20.1, reduces to approximately.

$$\bar{C}_{v_E} \cong 3\bar{R}\left(\frac{\theta_E}{T}\right)^2 e^{-\theta_E/T} \tag{20.2}$$

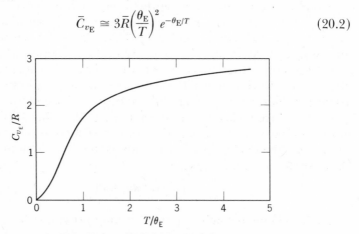

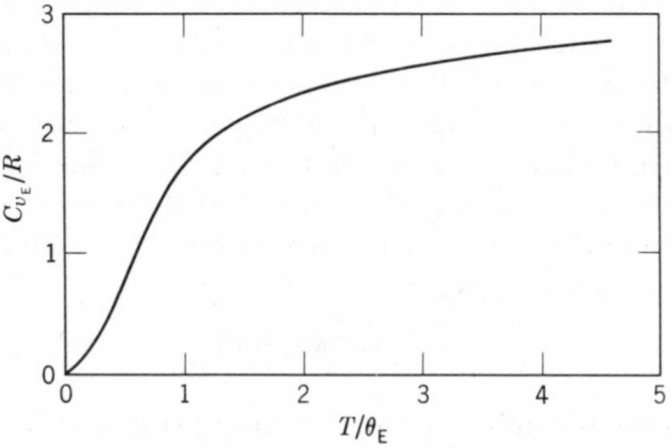

Fig. 20.1 The specific heat of an Einstein solid.

and approaches zero as the temperature approaches absolute zero. This zero limit is also in agreement with experimental observation, but the quantitative behavior of the Einstein model at low temperature is not particularly good. It is found experimentally that the specific heat of a crystal at low temperature is very closely proportional to T^3. If the value for θ_E is so selected that reasonable agreement with experimental observations is obtained over a wide range of temperature, then the calculated specific heats at low temperature are considerably smaller than the experimental values. One concludes that the Einstein model, which assumes $3N$ harmonic oscillators of the same frequency, is certainly a vast oversimplification of the problem. Nevertheless, it demonstrates a qualitatively correct behavior, and was a tremendous advance beyond classical mechanics, which predicts a constant value of $3\bar{R}$ for the specific heat.

20.2 The Debye Solid

The failure of the Einstein model of the solid phase to provide accurate correlation with experimental specific heat data, especially at low temperature, led Debye, in 1912, to improve the mathematical model. Einstein assumed that the vibrational modes in the crystal are independent of one another, and consequently that all the modes are of the same frequency. It is only reasonable to expect, however, that the atoms in a crystal vibrate with many different frequencies, and also that low frequency vibrations are of particular importance in analyzing low temperature behavior of the crystal.

Debye therefore assumed that there are a large number of very closely spaced frequencies of vibration, such that a continuous distribution function $f(\nu)$ can be used in representing the number of normal vibrations in any interval ν to $\nu + d\nu$. From the point of view of a distribution function, the Einstein model having $3N$ oscillators of frequency ν_ε is as shown in Fig. 20.2*a*, and Debye distribution of frequencies is shown in Fig. 20.2*b*. Since the atoms are held in a lattice having certain atomic distances and orientations, there is some limiting high frequency value corresponding to a wavelength of the order of magnitude of the interatomic distances. This maximum value ν_D is called the Debye frequency. It follows that the area under the Debye distribution curve of Fig. 20.2*b* must equal the total number of vibrational modes, $3N$, or in terms of the distribution function,

$$\int_0^{\nu_D} f(\nu)\, d\nu = 3N \tag{20.3}$$

To determine the nature of the distribution $f(\nu)$, Debye assumed that the vibrations in the solid can be treated as elastic waves in a continuous

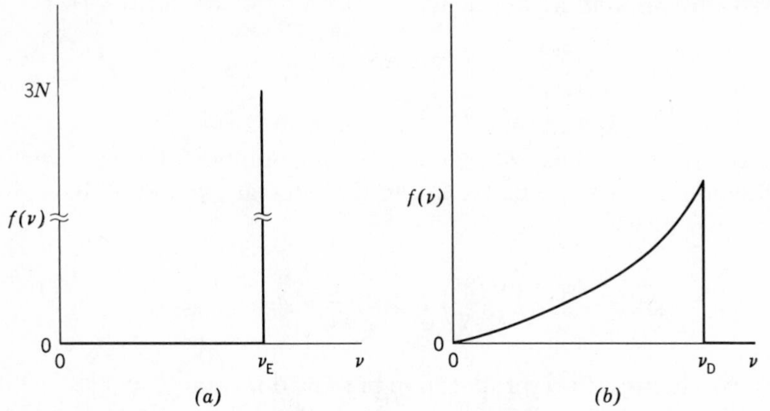

Fig. 20.2 The distribution function for (*a*) the Einstein and (*b*) the Debye models.

medium. The normal vibrational modes are represented by the various standing waves, with nodes coinciding with the crystal boundary. Since this is the same mathematical problem as that for the translation of a particle in a box (and also for the photon gas representing radiation), the solution is given by Eq. 18.40. For a crystal of side l, the wave function is, in terms of the three positive integers $\mathbf{k}_x$, $\mathbf{k}_y$, $\mathbf{k}_z$,

$$\Psi_{\mathbf{k}_x,\mathbf{k}_y,\mathbf{k}_z}(x, y, z) = C \sin\left(\mathbf{k}_x\pi\frac{x}{l}\right) \sin\left(\mathbf{k}_y\pi\frac{y}{l}\right) \sin\left(\mathbf{k}_z\pi\frac{z}{l}\right) \tag{20.4}$$

There is, consequently, a wave function, Eq. 20.4, for each set of integers $\mathbf{k}_x$, $\mathbf{k}_y$, $\mathbf{k}_z$. We now define

$$r = (\mathbf{k}_x^2 + \mathbf{k}_y^2 + \mathbf{k}_z^2)^{1/2}$$

where r is the radius of a sphere in a space of cartesian coordinates $\mathbf{k}_x$, $\mathbf{k}_y$, $\mathbf{k}_z$. The number of sets of these positive integers, $N_{\mathbf{k}_x,\mathbf{k}_y,\mathbf{k}_z}$, in the interval r to $r + dr$ is then given by

$$N_{\mathbf{k}_x,\mathbf{k}_y,\mathbf{k}_z} = \tfrac{1}{8}(4\pi r^2\, dr) \tag{20.5}$$

The factor 1/8 is introduced into Eq. 20.5 because only 1/8 of the surface of the sphere of radius r corresponds to all positive $\mathbf{k}_x$, $\mathbf{k}_y$, $\mathbf{k}_z$. This analysis is the same as that discussed for translational states in Section 19.3.

For wave nodes coincidental with the crystal boundaries, the allowed wavelength λ must be, as in Eq. 19.107 for the photon gas,

$$\lambda = 2\frac{l}{r} \tag{20.6}$$

and the corresponding frequency ν for a given wave velocity V is

$$\nu = \frac{\mathsf{V}}{\lambda} = \frac{\mathsf{V}r}{2l} \tag{20.7}$$

Therefore, from Eqs. 20.5 and 20.7, the number of frequencies of oscillation, N_ν (sets of $\mathbf{k}_x$, $\mathbf{k}_y$, $\mathbf{k}_z$), in the interval ν to $\nu + d\nu$ for a given wave velocity V is

$$N_\nu = \frac{4\pi}{8}\left(\frac{2l\nu}{\mathsf{V}}\right)^2\left(\frac{2l}{\mathsf{V}}\right) d\nu = 4\pi\frac{l^3}{\mathsf{V}^3}\nu^2\, d\nu = 4\pi\frac{V}{\mathsf{V}^3}\nu^2\, d\nu \tag{20.8}$$

In considering wave propagation in some direction of an elastic medium, we must include both the longitudinal waves of velocity V_l and the simultaneous transverse waves of velocity V_t, the latter having two normal directions of polarization. Therefore the total number of vibrational modes in the interval ν to $\nu + d\nu$ is, using Eq. 20.8,

$$f(\nu)\, d\nu = \left(\frac{1}{\mathsf{V}_l^{\,3}} + \frac{2}{\mathsf{V}_t^{\,3}}\right) 4\pi V \nu^2\, d\nu = \frac{12\pi V}{\langle \mathsf{V} \rangle^3}\nu^2\, d\nu \tag{20.9}$$

where the average wave velocity $\langle \mathsf{V} \rangle$ has been defined by the expression

$$\frac{3}{\langle \mathsf{V} \rangle^3} = \frac{1}{\mathsf{V}_l^3} + \frac{2}{\mathsf{V}_t^3}. \tag{20.10}$$

Having determined the distribution function $f(\nu)$ in terms of Eq. 20.9, we can now integrate Eq. 20.3. The result is

$$3N = \int_0^{\nu_D} \frac{12\pi V}{\langle \mathsf{V} \rangle^3}\nu^2\, d\nu = \frac{4\pi V}{\langle \mathsf{V} \rangle^3}\nu_D{}^3$$

or

$$\nu_D{}^3 = \frac{3N\langle \mathsf{V} \rangle^3}{4\pi V} \tag{20.11}$$

The distribution function $f(\nu)$ can alternatively be expressed in terms of the Debye frequency ν_D as

$$f(\nu)\, d\nu = \frac{9N}{\nu_D{}^3}\nu^2\, d\nu \tag{20.12}$$

for all values of ν between zero and ν_D.

The internal energy with respect to the ground state for a mole of oscillators of frequency ν is given by Eq. 19.75. Thus, for a Debye solid, using Eq. 20.12 we have, on a mole basis,

$$(\bar{u}-\bar{u}_{v0}) = \int_0^{\nu_D}\left[\frac{k(h\nu/k)}{e^{h\nu/kT}-1}\right]f(\nu)\,d\nu = \frac{9N_0kT}{\nu_D{}^3}\int_0^{\nu_D}\left[\frac{h\nu/kT}{e^{h\nu/kT}-1}\right]\nu^2\,d\nu \tag{20.13}$$

If we let

$$x = \frac{h\nu}{kT} = \frac{\theta}{T} \tag{20.14}$$

this relation can be rewritten in the form

$$(\bar{u}-\bar{u}_{v0}) = 3\bar{R}T\left[\frac{3}{x_D{}^3}\int_0^{x_D}\frac{x^3}{e^x-1}\,dx\right] = 3\bar{R}T[D(x_D)] \tag{20.15}$$

where

$$x_D = \frac{h\nu_D}{kT} = \frac{\theta_D}{T} \tag{20.16}$$

The quantity θ_D is called the Debye temperature, and is defined in terms of ν_D by Eq. 20.16.

Table 20.1

$x_D = \theta_D/T$	$D(x_D)$	$x_D = \theta_D/T$	$D(x_D)$
0	1.0000	3	0.2836
0.1	0.9630	4	0.1817
0.2	0.9270	5	0.1176
0.5	0.8250	10	0.0193
1	0.6744	20	0.0024
2	0.4411	∞	0

The function $D(x_D)$ defined by Eq. 20.15 is called the Debye function; this must be evaluated numerically as a function of x_D or θ_D/T. A few values of the Debye function are given in Table 20.1.

The specific heat for the Debye solid is found by differentiating Eq. 20.15:

$$\bar{C}_v = \left(\frac{\partial\bar{u}}{\partial T}\right)_V = d\frac{(\bar{u}-\bar{u}_{v0})}{dT} = 3\bar{R}\left[4\times D(x_D) - \frac{3x_D}{e^{x_D}-1}\right] \tag{20.17}$$

At high temperature, the Debye function approaches unity and the specific heat $\bar{C}_v$ approaches a limiting value of $3\bar{R}$, which is identical with the result obtained for the Einstein model. As mentioned previously, this is in excellent agreement with observed specific heat data for most monatomic solids. At very low temperature, the Debye function reduces to the form

$$D(x_D) = \frac{\pi^4}{5}(x_D)^{-3} \tag{20.18}$$

so that, at low temperature,

$$\bar{C}_v = \frac{12\pi^4}{5}\bar{R}\left(\frac{T}{\theta_D}\right)^3 \tag{20.19}$$

From this expression, we note the proportionality of $\bar{C}_v$ to T^3 at temperatures near absolute zero, which is in good agreement with experimental values. We also note that $\bar{C}_v$ goes to zero at absolute zero temperature.

Over the entire range of temperature, the shape of the specific heat from the Debye theory is qualitatively the same as that for the simpler Einstein model shown in Fig. 20.1. The Debye specific heat function is shown in Fig. 20.3 and is compared with a few experimental points for lead, silver, zinc, and aluminum.

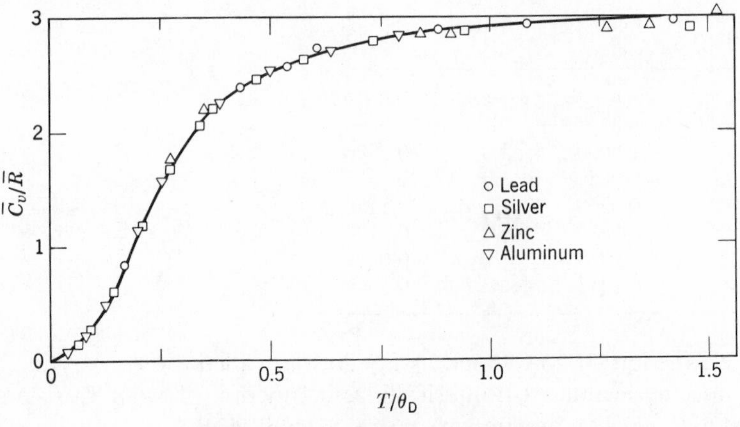

Fig. 20.3 The specific heat of a Debye solid.

Quantitatively, the Debye model is excellent and is superior to the Einstein model, particularly at low temperatures. If the theory were exact, it would be possible to select a Debye temperature θ_D that would correlate the experimental data over a very wide temperature range. For many solids this can be done, but for some materials the Debye temperature varies somewhat with the range of temperature in which the data are to be correlated. Average values for the parameter θ_D for several substances are listed in Table 20.2.

Table 20.2

Element	θ_D, °K	Element	θ_D, °K
Pb	86	Zn	240
Hg	90	Cu	315
K	99	Al	398
Bi	111	Cr	405
Na	160	Co	445
Sn	165	Fe	453
Ag	215	Ni	456
Pt	225	Be	980
Ca	230	C (diamond)	1860

The entropy for a Debye solid is, of course, independent of the arbitrary zero reference for energy. Equation 19.78 gives the entropy for N_0 vibrational modes of a given frequency ν. Therefore, using the Debye frequency distribution function, Eq. 20.12, the molal entropy is found to be

$$\bar{s} = k \int_0^{\nu_D} \left[\frac{x}{e^x - 1} - \ln\,(1 - e^{-x}) \right] f(\nu)\, d\nu$$

$$= \frac{9\bar{R}}{x_D{}^3} \int_0^{x_D} \left[\frac{x}{e^x - 1} - \ln\,(1 - e^{-x}) \right] x^2\, dx$$

$$= \bar{R}[4 \times D(x_D) - 3 \ln\,(1 - e^{-x_D})] \tag{20.20}$$

At low temperature, Eq. 20.20 reduces to the form

$$\bar{s} = \frac{4\pi^4}{5} \bar{R} \left(\frac{T}{\theta_D} \right)^3 \tag{20.21}$$

and is proportional to T^3. The entropy then approaches zero as temperature approaches absolute zero; this, as noted earlier, is in agreement with the reference state for entropy. On a physical basis, this is seen to be a result of each of the $3N$ vibrational modes dropping into its nondegenerate ground state as the temperature becomes close to absolute zero.

Example 20.1

Calculate the specific heat and entropy of aluminum at 100 K. From Table 20.2, for aluminum,

$$\theta_D = 398 \text{ K}$$

Therefore, at 100 K,

$$x_D = \frac{\theta_D}{T} = \frac{398}{100} = 3.98$$

Interpolating in Table 20.1, the Debye function at 100 K is found to be

$$D(x_D) = 0.1837$$

For $x_D = 3.98$,

$$\frac{x_D}{e^{x_D} - 1} = 0.07580, \quad -\ln(1 - e^{-x_D}) = 0.01887$$

Therefore, the specific heat is, from Eq. 20.17,

$$\bar{C}_v = 3(1.987)[4(0.1837) - 3(0.07580)] = 3.02 \text{ cal/mole-K}$$

From Eq. 20.20, the entropy is

$$\bar{s}_v = 1.987[4(0.1837) + 3(0.01887)] = 1.570 \text{ cal/mole-K}$$

The Debye model of the solid phase has been found to give good quantitative agreement with experimental observations for many substances. More complex models of the solid have been developed to explain a number of problems and features of solid behavior for which the Debye model is inadequate, but these models will not be discussed here.

20.3 The Electron Gas in a Metal

In previous sections of this chapter, we have discussed the behavior of solids and have noted that the Debye model, which considers vibration of the atoms in the lattice sites, gives quite accurate representation of the thermodynamic properties of crystals, even for many metals. Such a model has, of course, not included the contributions to properties of the conduction, or valence, electrons in metals. As discussed in Section 18.10, the electrons in the conduction band of a metal are quite mobile. Because of the large zero-point energy of a Fermi-Dirac system, the conduction electrons have sufficient energy to move quite freely through the lattice of positive ions. Since forces between the electrons are relatively small, we assume that it is reasonable to treat the collection of free electrons in a metal as a system of independent particles, which we term the electron gas.

The Fermi-Dirac model must be used to represent this system, for which the equilibrium distribution of electrons among energy levels is (Eq. 17.58),

$$N_j = \frac{g_j}{e^{(\epsilon_j - \mu)/kT} + 1}$$

in which the particle chemical potential μ has been substituted for G/N. It was found in Section 17.5 that, at absolute zero temperature, all energy levels below the Fermi level μ_0 are filled, while all levels above μ_0 are empty, resulting in the distribution shown in Fig. 17.5.

If we wish to evaluate the Fermi energy μ_0, it will therefore be necessary to determine the number of quantum states available at all levels of energy less than μ_0. The number of translational states in an interval of energy ϵ to $\epsilon + d\epsilon$ has been given previously by Eq. 19.35,

$$g_{\epsilon\ \text{trans}} = \frac{4\pi mV}{h^3}(2m\epsilon)^{1/2}\, d\epsilon$$

Since this expression considers translation only, this result must be multiplied by a factor of 2 for the electron gas to account for electron spin degeneracy. That is, an electron in any translational state can have a spin of either $+\frac{1}{2}$ or $-\frac{1}{2}$. Thus the total number of quantum states available to the electron gas in an interval of energy ϵ to $\epsilon + d\epsilon$ is

$$g_\epsilon = \frac{8\pi mV}{h^3}(2m\epsilon)^{1/2}\, d\epsilon \tag{20.22}$$

At absolute zero temperature, Eq. 20.22 also represents the number of particles, dN_ϵ, in the energy range ϵ to $\epsilon + d\epsilon$ for all energies less than the Fermi level μ_0, since all quantum states below μ_0 are filled at this condition. Therefore the total number of electrons, N, in the system is found to be

$$N = \int^{N_{\epsilon=\mu_0}} dN_\epsilon = \frac{8\pi mV}{h^3}(2m)^{1/2}\int_0^{\mu_0} \epsilon^{1/2}\, d\epsilon = \frac{8\pi V}{3h^3}(2m\mu_0)^{3/2} \tag{20.23}$$

The Fermi energy μ_0, which is also the chemical potential G/N of the system at zero temperature, is found from Eq. 20.23 to be

$$\mu_0 = \frac{h^2}{8m}\left(\frac{3N}{\pi V}\right)^{2/3} \tag{20.24}$$

The total energy U_0 of the system at absolute zero temperature is

$$U_0 = \int_{N_\epsilon=0}^{N_\epsilon=\mu_0} \epsilon \, dN_\epsilon = \frac{8\pi mV}{h^3}(2m)^{1/2}\int_0^{\mu_0} \epsilon^{3/2}\, d\epsilon$$

$$= \tfrac{3}{5}N\mu_0 \tag{20.25}$$

We note that the average energy per electron, U_0/N, at absolute zero is equal to three-fifths of the maximum energy μ_0.

It was determined in Section 17.5 that the entropy of a Fermi-Dirac system at absolute zero is zero, or

$$S_0 = 0 \tag{20.26}$$

The zero-point Helmholtz function A_0 is seen to be

$$A_0 = U_0 - T_0 S_0 = U_0 \tag{20.27}$$

and the Gibbs function is

$$G_0 = A_0 + P_0 V = U_0 + P_0 V \tag{20.28}$$

However, from Eq. 20.25,

$$G_0 = N\mu_0 = \tfrac{5}{3}U_0 \tag{20.29}$$

so that, for the electron gas,

$$P_0 V = \tfrac{2}{3}U_0 \tag{20.30}$$

which is of the same form as the relation found for the ordinary classical Boltzmann gas, Eq. 16.58, at normal temperatures.

A representative calculation of the Fermi energy for the electron gas in a metal by using Eq. 20.24 indicates that μ_0 is an extremely large number, of the order of magnitude of $10^5 \times k$. Consequently, it is reasonable to treat the behavior of the electron gas at ordinary temperatures, room temperature for example, as merely small perturbations from the zero temperature behavior. This procedure involves a number of series expansions and approximations and is quite complex mathematically. It is found that the chemical potential for the electron gas in a metal is, in terms of the zero-point value,

$$\mu = \mu_0\left[1 - \frac{\pi^2}{12}\left(\frac{kT}{\mu_0}\right)^2 + \cdots\right] \tag{20.31}$$

Similarly, the energy is

$$U = U_0\left[1 + \frac{5\pi^2}{12}\left(\frac{kT}{\mu_0}\right)^2 - \cdots\right] \tag{20.32}$$

The specific heat contribution is therefore

$$\overline{C}_v = \left[\frac{\partial (U/n)}{\partial T}\right]_V = \frac{N_0\pi^2k^2T}{2\mu_0} - \cdots \tag{20.33}$$

The entropy of the electron gas is readily found from the classical expression, as the integral of $(\overline{C}_v/T)\,dT$ from 0 to T.

Because of the large value of μ_0 mentioned above, it is seen that the specific heat contribution of the electron gas is quite small at ordinary temperature, of the order of magnitude of $10^{-2} \times \overline{R}$. This is in basic agreement with the conclusions reached in studying the results of the Debye model of solids, which includes only the lattice contribution to specific heat. Of course, if the electron gas had been treated strictly as an ordinary Boltzmann gas, the contribution to specific heat would be $\frac{3}{2}\overline{R}$, which is not at all in agreement with experimental data. As a final observation on the electron contribution to specific heat, we note that, at low temperature, only the first term of Eq. 20.33 is significant, and the specific heat contribution is proportional to T. The lattice contribution has been found previously to be proportional to T^3 at low temperature. Thus it should be possible to distinguish between these contributions at temperatures close to absolute zero; this behavior has been confirmed experimentally.

PROBLEMS

20.1 The specific heat of graphitic carbon at 300 K is 2.06 cal/mole-K. Calculate the characteristic Einstein temperature from this value, and use it to compute the specific heat at 50 K and at 100 K, assuming an Einstein solid. Compare the results with the experimental values 0.12 and 0.40 cal/mole-K.

20.2 Repeat Problem 20.1, assuming the Debye model of the solid phase.

20.3 Calculate the specific heat and the Helmholtz function for lead at the boiling point temperature of hydrogen, 20.4 K.

20.4 Experimental values for the constant-volume specific heat of solid argon are given in the accompanying table.

T, K	C_v, cal/mole-K
10	0.80
15	1.88
20	2.94
30	4.25
40	4.61
60	5.12
80	5.74

Calculate a characteristic Debye temperature from each of these data points, and plot the results versus temperature. If the Debye theory were exact, θ_D would be a constant, the same value from each experimental point. Is this approximately true for argon?

20.5 Recently measured values of the specific heat of copper are listed below.

T, K	C, joules/gm-K
40	0.060
60	0.137
100	0.254
180	0.346
300	0.386

Find values for the characteristic Debye and Einstein temperatures from each of these values. Which theory results in a better correlation of the data?

20.6 Helium gas is contained in a 1-liter aluminum tank at 300 K, 10 atm pressure. This system (tank and gas) is then cooled using liquid nitrogen to 77.3 K. If the mass of the aluminum tank is 1500 gm, how much heat is rejected to the liquid nitrogen during this cooling process?

20.7 A vertical nickel cylinder and piston contains a diatomic gas at 1 atm pressure. The initial gas volume is 1000 cm^3, and the total mass of nickel is 300 gm. Calculate the heat transfer if the system (gas plus metal) is allowed to warm from −90 C to −45 C. For the diatomic gas, it may be assumed that in this temperature range

$$\theta_r \ll T \ll \theta_v$$

20.8 Consider a monatomic solid in equilibrium with its vapor at very low temperature. Using the requirement for equilibrium between two phases of a pure substance (equal chemical potentials), show that the sublimation pressure for the solid is expressed as

$$\ln P_{\text{sub}} = \frac{C_1}{T} + C_2 \ln T + C_3 T^3 + C_4$$

Neglect the specific volume of the solid and any electronic contribution.

20.9 (*a*) Using the Clapeyron equation,

$$\frac{dP_{\text{sub}}}{dT} = \frac{\Delta h_{\text{sub}}}{T \Delta v_{\text{sub}}}$$

and the result of Problem 20.8, find an expression for the enthalpy of sublimation for a solid at low temperature.

(*b*) Calculate the sublimation pressure and Δh_{sub} for solid argon at 4.2 K, using $\theta_D = 83.5$ K. The triple-point of argon is 83.7 K, 0.68 atm.

20.10 The coefficient of thermal expansion, α, is defined, Eq. 12.47, as

$$\alpha = \frac{1}{V}\left(\frac{\partial V}{\partial T}\right)_P$$

Using the Maxwell relation

$$\left(\frac{\partial V}{\partial T}\right)_P = -\left(\frac{\partial S}{\partial P}\right)_T$$

show that α for a Debye solid can be expressed as

$$\alpha = \frac{nC_v}{V}\left(\frac{\partial \ln \theta_D}{\partial P}\right)_T$$

20.11 Calculate the Fermi energy μ_0 and the pressure exerted by the electron gas in a metal at 0 K. Assume one conduction electron per atom of metal and a metal density of 0.1 mole/cm³.

20.12 Consider the electron gas in a metal.

(*a*) Using the thermodynamic relation

$$P = -\left(\frac{\partial A}{\partial V}\right)_T$$

show that, at room temperature,

$$PV = \tfrac{2}{3}U$$

(*b*) Using the data of Problem 20.11, calculate μ and the electron gas pressure at 300 K.

20.13 (*a*) Consider the electron gas in tungsten. Assume a metal density of 19.3 gm/cm³ and two conduction electrons per atom. Find the number density N/V of the electron gas, the Fermi energy μ_0, and a characteristic temperature μ_0/k.

(*b*) Now consider the electrons in a vacuum tube as a Fermi-Dirac gas with a number density N/V of 10^{11} per cm³. Calculate the Fermi energy μ_0 and μ_0/k for this gas, and compare the values with those of part *a*. What is the significance of this comparison?

20.14 Calculate the lattice and electronic contributions to the entropy of sodium at 25 C. Use a metal density of 0.97 gm/cm³, and 1 conduction electron per atom.

20.15 Using a metal density of 10.5 gm/cm³ and assuming one conduction electron per atom, calculate the lattice and electronic contributions to the specific heat of silver at 30 K and 300 K.

20.16 Recent experimental specific heat data for copper are correlated by the expression

$$C_p = 30.6\left(\frac{T}{344.5}\right)^3 + 10.8 \times 10^{-6} T$$

at temperatures below 10 K. Neglecting the difference between C_p and C_v, discuss this expression and the numerical constants in terms of the Debye theory—electron gas model.

20.17 A certain metal has a density of 0.1 mole/cm³, 2 conduction electrons per atom of metal, and a Debye temperature of 250 K. Find 2 temperatures at which the electronic contribution to C_v is 10% of the lattice contribution.

20.18 (*a*) Using Eqs. 17.58 and 20.22 for an electron gas, show that the number of electrons, dN_ϵ, in an interval of energy ϵ to $\epsilon + d\epsilon$ is

$$dN_\epsilon = \frac{4\pi V}{h^3} \frac{(2m)^{3/2} \epsilon^{1/2}}{e^{(\epsilon-\mu)/kT} + 1} d\epsilon$$

(*b*) Plot the function $dN_\epsilon/d\epsilon$ vs. ϵ for the electron gas of a 1 cm³ sample of copper at 0 K and 300 K. Use a density of 8.96 gm/cm³ and one conduction electron per atom for copper.

20.19 (*a*) Use the result of Problem 20.18*a* to find the speed distribution in an electron gas. Show that the number of electrons, dN_V in an interval V to $\mathsf{V} + d\mathsf{V}$ is

$$dN = \frac{8\pi m^3 V}{h^3} \frac{\mathsf{V}^2}{e^{(m\mathsf{V}^2/2-\mu)/kT} + 1} d\mathsf{V}$$

(*b*) Find an expression for the average speed of an electron gas at 0 K. What is the order of magnitude of this value for a typical metal?

20.20 Substitute the relation $d\mathsf{V}_x\, d\mathsf{V}_y\, d\mathsf{V}_z = 4\pi \mathsf{V}^2\, d\mathsf{V}$ into the result of Problem 20.19*a* to obtain a relation for the distribution of component velocities in an electron gas. Using this result, show that the number of electrons having an x-component in the interval V_x to $\mathsf{V}_x + d\mathsf{V}_x$ and any V_y, V_z is

$$dN_{\mathsf{V}_x} = \frac{4\pi V m^2 kT}{h^3} \ln\left[e^{(\mu-\epsilon_x)/kT} + 1\right] d\mathsf{V}_x$$

(Transform V_y, V_z, to polar coordinates to perform the necessary integration.)

20.21 The problem of particles meeting a potential barrier at a surface was analyzed for Boltzmann statistics in Problem 16.24. To analyze the phenomenon of thermionic emission, the emission of electrons from an incandescent cathode, however, it is necessary to utilize Fermi-Dirac statistics for the electron gas in the cathode metal.

(*a*) Show that the rate at which electrons of a given V_x strike a unit surface (from the interior) is, as was found in Chapter 16,

$$\frac{\text{No. of } \mathsf{V}_x}{\text{cm}^2\text{-sec}} = \frac{\mathsf{V}_x}{V}\, dN_{\mathsf{V}_x}$$

where dN_{V_x} is the number of electrons in the metal in the interval V_x to $\mathsf{V}_x + d\mathsf{V}_x$, and x is taken normal to the surface.

(*b*) If ϵ_m is the minimum kinetic energy normal to the surface with which an electron can escape the surface, show that the total rate of electron flow per unit area of cathode is given by

$$\frac{\text{No.}}{\text{cm}^2\text{-sec}} = CT^2 e^{-\varphi/kT}$$

where $\varphi = (\epsilon_m - \mu)$ is called the work function of the cathode. Compare this result with that of Problem 16.24.

(*c*) Each electron escaping from the cathode surface carries the charge e. Therefore, the electrical current density J_{sat}/A is given by

$$\frac{J_{\text{sat}}}{A} = CeT^2 e^{-\varphi/kT}$$

which is termed the Richardson-Dushman equation for thermionic emission. Using the value $\varphi = 4.5$ ev and the theoretical value for C, calculate J_{sat}/A for tungsten at 3000 K. What assumption is implied in using the theoretical value of C in such a calculation?

Appendix A

Mathematical Developments

Appendix A is concerned with the development of three mathematical techniques that are required in the formulation and evaluation of the subject of statistical thermodynamics. Stirling's Formula and the Lagrange Undetermined Multipliers are used in the analysis of the thermodynamic equilibrium state and expression of properties in Chapters 16 and 17. The Euler-Maclaurin Summation Formula is used in the evaluation of partition function summations in Chapter 19.

A.1 Stirling's Formula

Stirling's Formula is concerned with an approximation to $N!$ that is valid for large values of N.

It follows from an integration by parts that

$$\int_0^\infty x^N e^{-x}\,dx = N\int_0^\infty x^{N-1}e^{-x}\,dx \tag{A.1}$$

From the similarity of the two integrals, we note that the integral on the right side of Eq. A.1 could in turn be expressed using this same rule. When we repeat the integration by the same rule over and over, it soon becomes apparent that, if N is an integer, the integral on the left side of Eq. A.1 is equal to $N!$ Let us approximate this integral for large N. For the variable x, substitute the expression

$$x = N + \sqrt{N}y \tag{A.2}$$

Now, since

$$x^N = e^{N \ln x} \tag{A.3}$$

the integrand can be written

$$x^N e^{-x}\, dx = \sqrt{N}\, e^{-N} e^{N \ln (N+\sqrt{N}y) - \sqrt{N}y}\, dy \tag{A.4}$$

Expanding the logarithm, for large N,

$$\ln (N+\sqrt{N}\,y) = \ln N + \ln\left(1+\frac{y}{\sqrt{N}}\right) = \ln N + \frac{y}{\sqrt{N}} - \frac{y^2}{2N} + \cdots \tag{A.5}$$

If we substitute Eq. A.5 into Eq. A.4 and integrate, the result should be approximately $N!$ for large values of N. Therefore

$$N! \cong \sqrt{N}\, e^{-N} N^N \int_{-\sqrt{N}}^{+\infty} e^{-y^2/2}\, dy \tag{A.6}$$

But, if N is large, an examination of the integrand of Eq. A.6 shows that the contribution from $-\infty$ to $-\sqrt{N}$ is negligible, so that

$$N! \cong \sqrt{N}\, e^{-N} N^N \int_{-\infty}^{+\infty} e^{-y^2/2}\, dy \cong N^N e^{-N} \sqrt{2\pi N} \tag{A.7}$$

For our purposes, the logarithm of $N!$ is of interest. Equation A.7 can be written in the form

$$\ln N! \cong N \ln N - N + \tfrac{1}{2} \ln 2\pi N \tag{A.8}$$

But, for large values of N,

$$\tfrac{1}{2} \ln 2\pi N \ll N$$

and

$$\ln N! \cong N \ln N - N \tag{A.9}$$

This equation is known as Stirling's formula.

A.2 Lagrange Multipliers

Lagrange's Method of Undetermined Multipliers is a formal mathematical procedure for determining a maximum point in a continuous function subject to one or more constraints.

We consider first some function $f(x, y, z)$, a continuous function of the three independent variables x, y, z. Since the function is continuous, the total differential may be expressed as

$$df = \left(\frac{\partial f}{\partial x}\right) dx + \left(\frac{\partial f}{\partial y}\right) dy + \left(\frac{\partial f}{\partial z}\right) dz \tag{A.10}$$

At the maximum (or minimum) of the function, it is necessary that

$$df = \left(\frac{\partial f}{\partial x}\right) dx + \left(\frac{\partial f}{\partial y}\right) dy + \left(\frac{\partial f}{\partial z}\right) dz = 0 \tag{A.11}$$

But, since the variables x, y, z are independent, it follows that changes dx, dy, dz are also independent. Thus the only way that Eq. A.11 can be satisfied for such arbitrary changes is for all of the coefficients to be identically zero at the maximum. That is,

$$\frac{\partial f}{\partial x} = \frac{\partial f}{\partial y} = \frac{\partial f}{\partial z} = 0 \tag{A.12}$$

Now suppose that, instead, the variables x, y, z are not all independent but are restricted by some interrelation

$$g(x, y, z) = 0 \tag{A.13}$$

so that any two of the three variables can be chosen as the independent variables, the third being dependent according to Eq. A.13. In this case, the argument leading to Eq. A.12 is no longer valid, although a maximum is still given by Eq. A.11. However, we now have a second relation

$$\left(\frac{\partial g}{\partial x}\right) dx + \left(\frac{\partial g}{\partial y}\right) dy + \left(\frac{\partial g}{\partial z}\right) dz = 0 \tag{A.14}$$

at the maximum point. Suppose we multiply the last equation by some arbitrary constant λ and add the result to Eq. A.11. This gives the expression

$$\left(\frac{\partial f}{\partial x} + \lambda \frac{\partial g}{\partial x}\right) dx + \left(\frac{\partial f}{\partial y} + \lambda \frac{\partial g}{\partial y}\right) dy + \left(\frac{\partial f}{\partial z} + \lambda \frac{\partial g}{\partial z}\right) dz = 0 \tag{A.15}$$

Since any two of the variables can be selected as independent, let us vary dy and dz arbitrarily. This, of course, requires that the change dx will be dependent. Therefore we choose the value of λ such that

$$\left(\frac{\partial f}{\partial x}+\lambda\frac{\partial g}{\partial x}\right)=0 \tag{A.16}$$

at the maximum. Having selected such a value for λ, the only way that Eq. A.15 can be identically zero for any arbitrary dy and dz is for their coefficients to be identically zero, or

$$\left(\frac{\partial f}{\partial y}+\lambda\frac{\partial g}{\partial y}\right)=0 \tag{A.17}$$

$$\left(\frac{\partial f}{\partial z}+\lambda\frac{\partial g}{\partial z}\right)=0 \tag{A.18}$$

Similarly, any two variables can be taken as independent; then the undetermined multiplier λ is so chosen as to make the coefficient of the dependent variable equal to zero. The subsequent argument is then analogous, and the same expressions, Eqs. A.16, A.17, and A.18 result.

We have considered here the simple case of a function subject to but one constraint and have consequently found it necessary to introduce one undetermined multiplier. This method can very easily be extended to determine the maximum of a function of any number of variables under any lesser number of constraints. By exactly the same procedure and entirely analogous arguments, the resulting number of undetermined multipliers is found to be equal to the number of restraining equations.

A.3 Euler-Maclaurin Summation Formula

The Euler-Maclaurin Summation Formula is concerned with the replacement of summations by integrals, and the errors involved with doing so.

Consider some function $f(j)$, where j is an integer, and the function is such that $f(j)$ changes only very slightly from j to $j+1$. If in addition the function is such that $f(j)\rightarrow 0$ for large j, then

$$\sum_{j=0}^{\infty} f(j)=\sum_{j=0}^{\infty} f(j)\Delta j\approx\int_{x=0}^{\infty} f(x)\,dx \tag{A.19}$$

If we assume also that the derivatives of $f(x)\rightarrow 0$ for large x, then it may be more accurate to represent the summation of Eq. A.19 by a series including values of $f(x)$ and its derivatives at the lower limit 0, such as

$$\sum_{j=0}^{\infty} f(j) = \sum_{j=0}^{\infty} f(j)\Delta j = \int_{x=0}^{\infty} f(x)\,dx + C_0 f(0) + C_1 f^{1}(0)$$
$$+ C_2 f^{11}(0) + C_3 f^{111}(0) + C_4 f^{1V}(0) + C_5 f^{V}(0) + \cdots \qquad \text{(A.20)}$$

To evaluate the constants in Eq. A.20, let us consider a simple function that meets our requirements, such as the function

$$f(j) = e^{-mj} \qquad \text{(A.21)}$$

summed from 0 to ∞. At the lower limit,

$$\begin{aligned} f(0) &= 1 \\ f^{1}(0) &= -m \\ f^{11}(0) &= +m^2 \\ f^{111}(0) &= -m^3 \\ &\cdots\cdots\cdots \end{aligned} \qquad \text{(A.22)}$$

Furthermore, we know that

$$\int_0^{\infty} e^{-mx}\,dx = \frac{1}{m} \qquad \text{(A.23)}$$

and

$$\sum_{j=0}^{\infty} e^{-mj} = \frac{1}{1-e^{-m}} \qquad \text{(A.24)}$$

Now, if we expand e^{-m} in a series,

$$e^{-m} = 1 - m + \frac{m^2}{2} - \frac{m^3}{6} + \frac{m^4}{24} + \cdots \qquad \text{(A.25)}$$

and substitute the series into Eq. A.24 and carry out the division,

$$\sum_{j=0}^{\infty} e^{-mj} = \frac{1}{m} + \frac{1}{2} + \frac{m}{12} - \frac{m^3}{720} + \frac{m^5}{30{,}240} + \cdots \qquad \text{(A.26)}$$

Substituting Eqs. A.26, A.23, and the set A.22 into Eq. A.20, and comparing coefficients, we find that the constants are

$$
\begin{aligned}
C_0 &= +\tfrac{1}{2} \\
C_1 &= -\tfrac{1}{12} \\
C_2 &= 0 \\
C_3 &= \tfrac{1}{720} \\
C_4 &= 0 \\
C_5 &= -\tfrac{1}{30,240} \\
&\cdots\cdots\cdots\cdots\cdots
\end{aligned}
\tag{A.27}
$$

so that Eq. A.20 can be written

$$
\begin{aligned}
\sum_{j=0}^{\infty} f(j) = \int_0^{\infty} f(x)\,dx + \tfrac{1}{2} f(0) \\
- \tfrac{1}{12} f^{1}(0) + \tfrac{1}{720} f^{111}(0) - \tfrac{1}{30,240} f^{V}(0) + \cdots
\end{aligned}
\tag{A.28}
$$

Of course, having considered only one particular function, we should not conclude that Eq. A.28 is valid in general. If this formula is tested against other functions, however, it is in fact found to be a general expression for functions satisfying the specified requirements. Equation A.28 is a special case of the Euler-Maclaurin summation formula where the upper limit is infinite and the lower limit is zero.

Appendix B

Figures, Tables, and Charts

Table B.1
Thermodynamic Properties of Steam[a]
Table B.1.1
Saturated Steam: Temperature Table

Temp. Fahr. T	Press. Lbf/Sq.In. P	Specific Volume		Internal Energy			Enthalpy			Entropy		
		Sat. Liquid v_f	Sat. Vapor v_g	Sat. Liquid u_f	Evap. u_{fg}	Sat. Vapor u_g	Sat. Liquid h_f	Evap. h_{fg}	Sat. Vapor h_g	Sat. Liquid s_f	Evap. s_{fg}	Sat. Vapor s_g
32.018	0.08866	0.016022	3302	0.00	1021.2	1021.2	0.01	1075.4	1075.4	0.00000	2.1869	2.1869
35	0.09992	0.016021	2948	2.99	1019.2	1022.2	3.00	1073.7	1076.7	0.00607	2.1704	2.1764
40	0.12166	0.016020	2445	8.02	1015.8	1023.9	8.02	1070.9	1078.9	0.01617	2.1430	2.1592
45	0.14748	0.016021	2037	13.04	1012.5	1025.5	13.04	1068.1	1081.1	0.02618	2.1162	2.1423
50	0.17803	0.016024	1704.2	18.06	1009.1	1027.2	18.06	1065.2	1083.3	0.03607	2.0899	2.1259
60	0.2563	0.016035	1206.9	28.08	1002.4	1030.4	28.08	1059.6	1087.7	0.05555	2.0388	2.0943
70	0.3632	0.016051	867.7	38.09	995.6	1033.7	38.09	1054.0	1092.0	0.07463	1.9896	2.0642
80	0.5073	0.016073	632.8	48.08	988.9	1037.0	48.09	1048.3	1096.4	0.09332	1.9423	2.0356
90	0.6988	0.016099	467.7	58.07	982.2	1040.2	58.07	1042.7	1100.7	0.11165	1.8966	2.0083
100	0.9503	0.016130	350.0	68.04	975.4	1043.5	68.05	1037.0	1105.0	0.12963	1.8526	1.9822
110	1.2763	0.016166	265.1	78.02	968.7	1046.7	78.02	1031.3	1109.3	0.14730	1.8101	1.9574
120	1.6945	0.016205	203.0	87.99	961.9	1049.9	88.00	1025.5	1113.5	0.16465	1.7690	1.9336
130	2.225	0.016247	157.17	97.97	955.1	1053.0	97.98	1019.8	1117.8	0.18172	1.7292	1.9109
140	2.892	0.016293	122.88	107.95	948.2	1056.2	107.96	1014.0	1121.9	0.19851	1.6907	1.8892
150	3.722	0.016343	96.99	117.95	941.3	1059.3	117.96	1008.1	1126.1	0.21503	1.6533	1.8684
160	4.745	0.016395	77.23	127.94	934.4	1062.3	127.96	1002.2	1130.1	0.23130	1.6171	1.8484
170	5.996	0.016450	62.02	137.95	927.4	1065.4	137.97	996.2	1134.2	0.24732	1.5819	1.8293
180	7.515	0.016509	50.20	147.97	920.4	1068.3	147.99	990.2	1138.2	0.26311	1.5478	1.8109

[a]Abridged from Joseph H. Keenan, Frederick G. Keyes, Philip G. Hill, and Joan G. Moore, *Steam Tables*, (New York: John Wiley & Sons, Inc., 1969).

Table B.1.1 (Continued)
Saturated Steam: Temperature Table

Temp. Fahr. T	Press. Lbf/ Sq. In. P	Specific Volume Sat. Liquid v_f	Specific Volume Sat. Vapor v_g	Internal Energy Sat. Liquid u_f	Internal Energy Evap. u_{fg}	Internal Energy Sat. Vapor u_g	Enthalpy Sat. Liquid h_f	Enthalpy Evap. h_{fg}	Enthalpy Sat. Vapor h_g	Entropy Sat. Liquid s_f	Entropy Evap. s_{fg}	Entropy Sat. Vapor s_g
190	9.343	0.016570	40.95	158.00	913.3	1071.3	158.03	984.1	1142.1	0.27866	1.5146	1.7932
200	11.529	0.016634	33.63	168.04	906.2	1074.2	168.07	977.9	1145.9	0.29400	1.4822	1.7762
210	14.125	0.016702	27.82	178.10	898.9	1077.0	178.14	971.6	1149.7	0.30913	1.4508	1.7599
212	14.698	0.016716	26.80	180.11	897.5	1077.6	180.16	970.3	1150.5	0.31213	1.4446	1.7567
220	17.188	0.016772	23.15	188.17	891.7	1079.8	188.22	965.3	1153.5	0.32406	1.4201	1.7441
230	20.78	0.016845	19.386	198.26	884.3	1082.6	198.32	958.8	1157.1	0.33880	1.3901	1.7289
240	24.97	0.016922	16.327	208.36	876.9	1085.3	208.44	952.3	1160.7	0.35335	1.3609	1.7143
250	29.82	0.017001	13.826	218.49	869.4	1087.9	218.59	945.6	1164.2	0.36772	1.3324	1.7001
260	35.42	0.017084	11.768	228.64	861.8	1090.5	228.76	938.8	1167.6	0.38193	1.3044	1.6864
270	41.85	0.017170	10.066	238.82	854.1	1093.0	238.95	932.0	1170.9	0.39597	1.2771	1.6731
280	49.18	0.017259	8.650	249.02	846.3	1095.4	249.18	924.9	1174.1	0.40986	1.2504	1.6602
290	57.53	0.017352	7.467	259.25	838.5	1097.7	259.44	917.8	1177.2	0.42360	1.2241	1.6477
300	66.98	0.017448	6.472	269.52	830.5	1100.0	269.73	910.4	1180.2	0.43720	1.1984	1.6356
310	77.64	0.017548	5.632	279.81	822.3	1102.1	280.06	903.0	1183.0	0.45067	1.1731	1.6238
320	89.60	0.017652	4.919	290.14	814.1	1104.2	290.43	895.3	1185.8	0.46400	1.1483	1.6123
330	103.00	0.017760	4.312	300.51	805.7	1106.2	300.84	887.5	1188.4	0.47722	1.1238	1.6010
340	117.93	0.017872	3.792	310.91	797.1	1108.0	311.30	879.5	1190.8	0.49031	1.0997	1.5901
350	134.53	0.017988	3.346	321.35	788.4	1109.8	321.80	871.3	1193.1	0.50329	1.0760	1.5793
360	152.92	0.018108	2.961	331.84	779.6	1111.4	332.35	862.9	1195.2	0.51617	1.0526	1.5688
370	173.23	0.018233	2.628	342.37	770.6	1112.9	342.96	854.2	1197.2	0.52894	1.0295	1.5585
380	195.60	0.018363	2.339	352.95	761.4	1114.3	353.62	845.4	1199.0	0.54163	1.0067	1.5483
390	220.2	0.018498	2.087	363.58	752.0	1115.6	364.34	836.2	1200.6	0.55422	0.9841	1.5383
400	247.1	0.018638	1.8661	374.27	742.4	1116.6	375.12	826.8	1202.0	0.56672	0.9617	1.5284

410	276.5	0.018784	1.6726	385.01	732.6	1117.6	385.97	817.2	1203.1	0.57916	0.9395	1.5187
420	308.5	0.018936	1.5024	395.81	722.5	1118.3	396.89	807.2	1204.1	0.59152	0.9175	1.5091
430	343.3	0.019094	1.3521	406.68	712.2	1118.9	407.89	796.9	1204.8	0.60381	0.8957	1.4995
440	381.2	0.019260	1.2192	417.62	701.7	1119.3	418.98	786.3	1205.3	0.61605	0.8740	1.4900
450	422.1	0.019433	1.1011	428.6	690.9	1119.5	430.2	775.4	1205.6	0.6282	0.8523	1.4806
460	466.3	0.019614	0.9961	439.7	679.8	1119.6	441.4	764.1	1205.5	0.6404	0.8308	1.4712
470	514.1	0.019803	0.9025	450.9	668.4	1119.4	452.8	752.4	1205.2	0.6525	0.8093	1.4618
480	565.5	0.020002	0.8187	462.2	656.7	1118.9	464.3	740.3	1204.6	0.6646	0.7878	1.4524
490	620.7	0.020211	0.7436	473.6	644.7	1118.3	475.9	727.8	1203.7	0.6767	0.7663	1.4430
500	680.0	0.02043	0.6761	485.1	632.3	1117.4	487.7	714.8	1202.5	0.6888	0.7448	1.4335
520	811.4	0.02091	0.5605	508.5	606.2	1114.8	511.7	687.3	1198.9	0.7130	0.7015	1.4145
540	961.5	0.02145	0.4658	532.6	578.4	1111.0	536.4	657.5	1193.8	0.7374	0.6576	1.3950
560	1131.8	0.02207	0.3877	557.4	548.4	1105.8	562.0	625.0	1187.0	0.7620	0.6129	1.3749
580	1324.3	0.02278	0.3225	583.1	515.9	1098.9	588.6	589.3	1178.0	0.7872	0.5668	1.3540
600	1541.0	0.02363	0.2677	609.9	480.1	1090.0	616.7	549.7	1166.4	0.8130	0.5187	1.3317
620	1784.4	0.02465	0.2209	638.3	440.2	1078.5	646.4	505.0	1151.4	0.8398	0.4677	1.3075
640	2057.1	0.02593	0.1805	668.7	394.5	1063.2	678.6	453.4	1131.9	0.8681	0.4122	1.2803
660	2362	0.02767	0.14459	702.3	340.0	1042.3	714.4	391.1	1105.5	0.8990	0.3493	1.2483
680	2705	0.03032	0.11127	741.7	269.3	1011.0	756.9	309.8	1066.7	0.9350	0.2718	1.2068
700	3090	0.03666	0.07438	801.7	145.9	947.7	822.7	167.5	990.2	0.9902	0.1444	1.1346
705.44	3204	0.05053	0.05053	872.6	0	872.6	902.5	0	902.5	1.0580	0	1.0580

Table B.1.2
Saturated Steam: Pressure Table

Press. Lbf/ Sq. In. P	Temp. Fahr. T	Specific Volume: Sat. Liquid v_f	Specific Volume: Sat. Vapor v_g	Internal Energy: Sat. Liquid u_f	Internal Energy: Evap. u_{fg}	Internal Energy: Sat. Vapor u_g	Enthalpy: Sat. Liquid h_f	Enthalpy: Evap. h_{fg}	Enthalpy: Sat. Vapor h_g	Entropy: Sat. Liquid s_f	Entropy: Evap. s_{fg}	Entropy: Sat. Vapor s_g
1.0	101.70	0.016136	333.6	69.74	974.3	1044.0	69.74	1036.0	1105.8	0.13266	1.8453	1.9779
2.0	126.04	0.016230	173.75	94.02	957.8	1051.8	94.02	1022.1	1116.1	0.17499	1.7448	1.9198
3.0	141.43	0.016300	118.72	109.38	947.2	1056.6	109.39	1013.1	1122.5	0.20089	1.6852	1.8861
4.0	152.93	0.016358	90.64	120.88	939.3	1060.2	120.89	1006.4	1127.3	0.21983	1.6426	1.8624
5.0	162.21	0.016407	73.53	130.15	932.9	1063.0	130.17	1000.9	1131.0	0.23486	1.6093	1.8441
6.0	170.03	0.016451	61.98	137.98	927.4	1065.4	138.00	996.2	1134.2	0.24736	1.5819	1.8292
8.0	182.84	0.016526	47.35	150.81	918.4	1069.2	150.84	988.4	1139.3	0.26754	1.5383	1.8058
10	193.19	0.016590	38.42	161.20	911.0	1072.2	161.23	982.1	1143.3	0.28358	1.5041	1.7877
14.696	211.99	0.016715	26.80	180.10	897.5	1077.6	180.15	970.4	1150.5	0.31212	1.4446	1.7567
15	213.03	0.016723	26.29	181.14	896.8	1077.9	181.19	969.7	1150.9	0.31367	1.4414	1.7551
20	227.96	0.016830	20.09	196.19	885.8	1082.0	196.26	960.1	1156.4	0.33580	1.3962	1.7320
25	240.08	0.016922	16.306	208.44	876.9	1085.3	208.52	952.2	1160.7	0.35345	1.3607	1.7142
30	250.34	0.017004	13.748	218.84	869.2	1088.0	218.93	945.4	1164.3	0.36821	1.3314	1.6996
35	259.30	0.017073	11.900	227.93	862.4	1090.3	228.04	939.3	1167.4	0.38093	1.3064	1.6873
40	267.26	0.017146	10.501	236.03	856.2	1092.3	236.16	933.8	1170.0	0.39214	1.2845	1.6767
45	274.46	0.017209	9.403	243.37	850.7	1094.0	243.51	928.8	1172.3	0.40218	1.2651	1.6673
50	281.03	0.017269	8.518	250.08	845.5	1095.6	250.24	924.2	1174.4	0.41129	1.2476	1.6589
55	287.10	0.017325	7.789	256.28	840.8	1097.0	256.46	919.9	1176.3	0.41963	1.2317	1.6513
60	292.73	0.017378	7.177	262.06	836.3	1098.3	262.25	915.8	1178.0	0.42733	1.2170	1.6444
65	298.00	0.017429	6.657	267.46	832.1	1099.5	267.67	911.9	1179.6	0.43450	1.2035	1.6380
70	302.96	0.017478	6.209	272.56	828.1	1100.6	272.79	908.3	1181.0	0.44120	1.1909	1.6321
75	307.63	0.017524	5.818	277.37	824.3	1101.6	277.61	904.8	1182.4	0.44749	1.1790	1.6265
80	312.07	0.017570	5.474	281.95	820.6	1102.6	282.21	901.4	1183.6	0.45344	1.1679	1.6214
85	316.29	0.017613	5.170	286.30	817.1	1103.5	286.58	898.2	1184.8	0.45907	1.1574	1.6165

90	320.31	0.017655	4.898	290.46	813.8	1104.3	290.76	895.1	1185.9	0.46442	1.1475	1.6119
95	324.16	0.017696	4.654	294.45	810.6	1105.0	294.76	892.1	1186.9	0.46952	1.1380	1.6076
100	327.86	0.017736	4.434	298.28	807.5	1105.8	298.61	889.2	1187.8	0.47439	1.1290	1.6034
110	334.82	0.017813	4.051	305.52	801.6	1107.1	305.88	883.7	1189.6	0.48355	1.1122	1.5957
120	341.30	0.017886	3.730	312.27	796.0	1108.3	312.67	878.5	1191.1	0.49201	1.0966	1.5886
130	347.37	0.017957	3.457	318.61	790.7	1109.4	319.04	873.5	1192.5	0.49989	1.0822	1.5821
140	353.08	0.018024	3.221	324.58	785.7	1110.3	325.05	868.7	1193.8	0.50727	1.0688	1.5761
150	358.48	0.018089	3.016	330.24	781.0	1111.2	330.75	864.2	1194.9	0.51422	1.0562	1.5704
160	363.60	0.018152	2.836	335.63	776.4	1112.0	336.16	859.8	1196.0	0.52078	1.0443	1.5651
170	368.47	0.018214	2.676	340.76	772.0	1112.7	341.33	855.6	1196.9	0.52700	1.0330	1.5600
180	373.13	0.018273	2.533	345.68	767.7	1113.4	346.29	851.5	1197.8	0.53292	1.0223	1.5553
190	377.59	0.018331	2.405	350.39	763.6	1114.0	351.04	847.5	1198.6	0.53857	1.0122	1.5507
200	381.86	0.018387	2.289	354.9	759.6	1114.6	355.6	843.7	1199.3	0.5440	1.0025	1.5464
250	401.04	0.018653	1.8448	375.4	741.4	1116.7	376.2	825.8	1202.1	0.5680	0.9594	1.5274
300	417.43	0.018896	1.5442	393.0	725.1	1118.2	394.1	809.8	1203.9	0.5883	0.9232	1.5115
350	431.82	0.019124	1.3267	408.7	710.3	1119.0	409.9	795.0	1204.9	0.6060	0.8917	1.4978
400	444.70	0.019340	1.1620	422.8	696.7	1119.5	424.2	781.2	1205.5	0.6218	0.8638	1.4856
450	456.39	0.019547	1.0326	435.7	683.9	1119.6	437.4	768.2	1205.6	0.6360	0.8385	1.4746
500	467.13	0.019748	0.9283	447.7	671.7	1119.4	449.5	755.8	1205.3	0.6490	0.8154	1.4645
550	477.07	0.019943	0.8423	458.9	660.2	1119.1	460.9	743.9	1204.8	0.6611	0.7941	1.4551
600	486.33	0.02013	0.7702	469.4	649.1	1118.6	471.7	732.4	1204.1	0.6723	0.7742	1.4464
700	503.23	0.02051	0.6558	488.9	628.2	1117.0	491.5	710.5	1202.0	0.6927	0.7378	1.4305
800	518.36	0.02087	0.5691	506.6	608.4	1115.0	509.7	689.6	1199.3	0.7110	0.7050	1.4160
900	532.12	0.02123	0.5009	523.0	589.6	1112.6	526.6	669.5	1196.0	0.7277	0.6750	1.4027
1000	544.75	0.02159	0.4459	538.4	571.5	1109.9	542.4	650.0	1192.4	0.7432	0.6471	1.3903
1200	567.37	0.02232	0.3623	566.7	536.8	1103.5	571.7	612.3	1183.9	0.7712	0.5961	1.3673
1400	587.25	0.02307	0.3016	592.7	503.3	1096.0	598.6	575.5	1174.1	0.7964	0.5497	1.3461
1600	605.06	0.02386	0.2552	616.9	470.5	1087.4	624.0	538.9	1162.9	0.8196	0.5062	1.3258
1800	621.21	0.02472	0.2183	640.0	437.6	1077.7	648.3	502.1	1150.4	0.8414	0.4645	1.3060
2000	636.00	0.02565	0.18813	662.4	404.2	1066.6	671.9	464.4	1136.3	0.8623	0.4238	1.2861
2500	668.31	0.02860	0.13059	717.7	313.4	1031.0	730.9	360.5	1091.4	0.9131	0.3196	1.2327
3000	695.52	0.03431	0.08404	783.4	185.4	968.8	802.5	213.0	1015.5	0.9732	0.1843	1.1575
3203.6	705.44	0.05053	0.05053	872.6	0	872.6	902.5	0	902.5	1.0580	0	1.0580

Table B.1.3
Superheated Vapor

°F	v	u	h	s	v	u	h	s	v	u	h	s
	$P = 1.0 (101.70)$				$P = 5.0 (162.21)$				$P = 10.0 (193.19)$			
Sat	333.6	1044.0	1105.8	1.9779	73.53	1063.0	1131.0	1.8441	38.42	1072.2	1143.3	1.7877
200	392.5	1077.5	1150.1	2.0508	78.15	1076.3	1148.6	1.8715	38.85	1074.7	1146.6	1.7927
240	416.4	1091.2	1168.3	2.0775	83.00	1090.3	1167.1	1.8987	41.32	1089.0	1165.5	1.8205
280	440.3	1105.0	1186.5	2.1028	87.83	1104.3	1185.5	1.9244	43.77	1103.3	1184.3	1.8467
320	464.2	1118.9	1204.8	2.1269	92.64	1118.3	1204.0	1.9487	46.20	1117.6	1203.1	1.8714
360	488.1	1132.9	1223.2	2.1500	97.45	1132.4	1222.6	1.9719	48.62	1131.8	1221.8	1.8948
400	511.9	1147.0	1241.8	2.1720	102.24	1146.6	1241.2	1.9941	51.03	1146.1	1240.5	1.9171
440	535.8	1161.2	1260.4	2.1932	107.03	1160.9	1259.9	2.0154	53.44	1160.5	1259.3	1.9385
500	571.5	1182.8	1288.5	2.2235	114.20	1182.5	1288.2	2.0458	57.04	1182.2	1287.7	1.9690
600	631.1	1219.3	1336.1	2.2706	126.15	1219.1	1335.8	2.0930	63.03	1218.9	1335.5	2.0164
700	690.7	1256.7	1384.5	2.3142	138.08	1256.5	1384.3	2.1367	69.01	1256.3	1384.0	2.0601
800	750.3	1294.9	1433.7	2.3550	150.01	1294.7	1433.5	2.1775	74.98	1294.6	1433.3	2.1009
1000	869.5	1373.9	1534.8	2.4294	173.86	1373.9	1534.7	2.2520	86.91	1373.8	1534.6	2.1755
1200	988.6	1456.7	1639.6	2.4967	197.70	1456.6	1639.5	2.3192	98.84	1456.5	1639.4	2.2428
1400	1107.7	1543.1	1748.1	2.5584	221.54	1543.1	1748.1	2.3810	110.76	1543.0	1748.0	2.3045
	$P = 14.696 (211.99)$				$P = 20 (227.96)$				$P = 40 (267.26)$			
Sat	26.80	1077.6	1150.5	1.7567	20.09	1082.0	1156.4	1.7320	10.501	1092.3	1170.0	1.6767
240	28.00	1087.9	1164.0	1.7764	20.47	1086.5	1162.3	1.7405				
280	29.69	1102.4	1183.1	1.8030	21.73	1101.4	1181.8	1.7676	10.711	1097.3	1176.6	1.6857
320	31.36	1116.8	1202.1	1.8280	22.98	1116.0	1201.0	1.7930	11.360	1112.8	1196.9	1.7124
360	33.02	1131.2	1221.0	1.8516	24.21	1130.6	1220.1	1.8168	11.996	1128.0	1216.8	1.7373
400	34.67	1145.6	1239.9	1.8741	25.43	1145.1	1239.2	1.8395	12.623	1143.0	1236.4	1.7606
440	36.31	1160.1	1258.8	1.8956	26.64	1159.6	1258.2	1.8611	13.243	1157.8	1255.8	1.7828

500	38.77	1181.8	1287.3	1.9263	28.46	1181.5	1286.8	1.8919	14.164	1180.1	1284.9	1.8140
600	42.86	1218.6	1335.2	1.9737	31.47	1218.4	1334.8	1.9395	15.685	1217.3	1333.4	1.8621
700	46.93	1256.1	1383.8	2.0175	34.47	1255.9	1383.5	1.9834	17.196	1255.1	1382.4	1.9063
800	51.00	1294.4	1433.1	2.0584	37.46	1294.3	1432.9	2.0243	18.701	1293.7	1432.1	1.9474
1000	59.13	1373.7	1534.5	2.1330	43.44	1373.5	1534.3	2.0989	21.70	1373.1	1533.8	2.0223
1200	67.25	1456.5	1639.3	2.2003	49.41	1456.4	1639.2	2.1663	24.69	1456.1	1638.9	2.0897
1400	75.36	1543.0	1747.9	2.2621	55.37	1542.9	1747.9	2.2281	27.68	1542.7	1747.6	2.1515
1600	83.47	1633.2	1860.2	2.3194	61.33	1633.2	1860.1	2.2854	30.66	1633.0	1859.9	2.2089
	$P = 60 (292.73)$				$P = 80 (312.07)$				$P = 100 (327.86)$			
Sat	7.177	1098.3	1178.0	1.6444	5.474	1102.6	1183.6	1.6214	4.434	1105.8	1187.8	1.6034
320	7.485	1109.5	1192.6	1.6634	5.544	1106.0	1188.0	1.6271				
360	7.924	1125.3	1213.3	1.6893	5.886	1122.5	1209.7	1.6541	4.662	1119.7	1205.9	1.6259
400	8.353	1140.8	1233.5	1.7134	6.217	1138.5	1230.6	1.6790	4.934	1136.2	1227.5	1.6517
440	8.775	1156.0	1253.4	1.7360	6.541	1154.2	1251.0	1.7022	5.199	1152.3	1248.5	1.6755
500	9.399	1178.6	1283.0	1.7678	7.017	1177.2	1281.1	1.7346	5.587	1175.7	1279.1	1.7085
600	10.425	1216.3	1332.1	1.8165	7.794	1215.3	1330.7	1.7838	6.216	1214.2	1329.3	1.7582
700	11.440	1254.4	1381.4	1.8609	8.561	1253.6	1380.3	1.8285	6.834	1252.8	1379.2	1.8033
800	12.448	1293.0	1431.2	1.9022	9.321	1292.4	1430.4	1.8700	7.445	1291.8	1429.6	1.8449
1000	14.454	1372.7	1533.2	1.9773	10.831	1372.3	1532.6	1.9453	8.657	1371.9	1532.1	1.9204
1200	16.452	1455.8	1638.5	2.0448	12.333	1455.5	1638.1	2.0130	9.861	1455.2	1637.7	1.9882
1400	18.445	1542.5	1747.3	2.1067	13.830	1542.3	1747.0	2.0749	11.060	1542.0	1746.7	2.0502
1600	20.44	1632.8	1859.7	2.1641	15.324	1632.6	1859.5	2.1323	12.257	1632.4	1859.3	2.1076
1800	22.43	1726.7	1975.7	2.2179	16.818	1726.5	1975.5	2.1861	13.452	1726.4	1975.3	2.1614
2000	24.41	1824.0	2095.1	2.2685	18.310	1823.9	2094.9	2.2367	14.647	1823.7	2094.8	2.2121
	$P = 120 (341.30)$				$P = 140 (353.08)$				$P = 160 (363.60)$			
Sat	3.730	1108.3	1191.1	1.5886	3.221	1110.3	1193.8	1.5761	2.836	1112.0	1196.0	1.5651
360	3.844	1116.7	1202.0	1.6021	3.259	1113.5	1198.0	1.5812				
400	4.079	1133.8	1224.4	1.6288	3.466	1131.4	1221.2	1.6088	3.007	1128.8	1217.8	1.5911
450	4.360	1154.3	1251.2	1.6590	3.713	1152.4	1248.6	1.6399	3.228	1150.5	1246.1	1.6230

Table B.1.3 (Continued)
Superheated Vapor

°F	v	u	h	s	v	u	h	s	v	u	h	s
	$P = 120(341.30)$				$P = 140(353.08)$				$P = 160(363.60)$			
500	4.633	1174.2	1277.1	1.6868	3.952	1172.7	1275.1	1.6682	3.440	1171.2	1273.0	1.6518
550	4.900	1193.8	1302.6	1.7127	4.184	1192.6	1300.9	1.6944	3.646	1191.3	1299.2	1.6784
600	5.164	1213.2	1327.8	1.7371	4.412	1212.1	1326.4	1.7191	3.848	1211.1	1325.0	1.7034
700	5.682	1252.0	1378.2	1.7825	4.860	1251.2	1377.1	1.7648	4.243	1250.4	1376.0	1.7494
800	6.195	1291.2	1428.7	1.8243	5.301	1290.5	1427.9	1.8068	4.631	1289.9	1427.0	1.7916
1000	7.208	1371.5	1531.5	1.9000	6.173	1371.0	1531.0	1.8827	5.397	1370.6	1530.4	1.8677
1200	8.213	1454.9	1637.3	1.9679	7.036	1454.6	1636.9	1.9507	6.154	1454.3	1636.5	1.9358
1400	9.214	1541.8	1746.4	2.0300	7.895	1541.6	1746.1	2.0129	6.906	1541.4	1745.9	1.9980
1600	10.212	1632.3	1859.0	2.0875	8.752	1632.1	1858.8	2.0704	7.656	1631.9	1858.6	2.0556
1800	11.209	1726.2	1975.1	2.1413	9.607	1726.1	1975.0	2.1242	8.405	1725.9	1974.8	2.1094
2000	12.205	1823.6	2094.6	2.1919	10.461	1823.5	2094.5	2.1749	9.153	1823.3	2094.3	2.1601
	$P = 180(373.13)$				$P = 200(381.86)$				$P = 225(391.87)$			
Sat	2.533	1113.4	1197.8	1.5553	2.289	1114.6	1199.3	1.5464	2.043	1115.8	1200.8	1.5365
400	2.648	1126.2	1214.4	1.5749	2.361	1123.5	1210.8	1.5600	2.073	1119.9	1206.2	1.5427
450	2.850	1148.5	1243.4	1.6078	2.548	1146.4	1240.7	1.5938	2.245	1143.8	1237.3	1.5779
500	3.042	1169.6	1270.9	1.6372	2.724	1168.0	1268.8	1.6239	2.405	1165.9	1266.1	1.6087
550	3.228	1190.0	1297.5	1.6642	2.893	1188.7	1295.7	1.6512	2.558	1187.0	1293.5	1.6366
600	3.409	1210.0	1323.5	1.6893	3.058	1208.9	1322.1	1.6767	2.707	1207.5	1320.2	1.6624
700	3.763	1249.6	1374.9	1.7357	3.379	1248.8	1373.8	1.7234	2.995	1247.7	1372.4	1.7095
800	4.110	1289.3	1426.2	1.7781	3.693	1288.6	1425.3	1.7660	3.276	1287.8	1424.2	1.7523
900	4.453	1329.4	1477.7	1.8175	4.003	1328.9	1477.1	1.8055	3.553	1328.3	1476.2	1.7920
1000	4.793	1370.2	1529.8	1.8545	4.310	1369.8	1529.3	1.8425	3.827	1369.3	1528.6	1.8292
1200	5.467	1454.0	1636.1	1.9227	4.918	1453.7	1635.7	1.9109	4.369	1453.4	1635.3	1.8977

Continued from previous page

Temp	v	u	h	s
1400	6.137	1541.2	1745.6	1.9849
1600	6.804	1631.7	1858.4	2.0425
1800	7.470	1725.8	1974.6	2.0964
2000	8.135	1823.2	2094.2	2.1470

Temp	v	u	h	s
1400	5.521	1540.9	1745.3	1.9732
1600	6.123	1631.6	1858.2	2.0308
1800	6.722	1725.6	1974.4	2.0847
2000	7.321	1823.0	2094.0	2.1354

Temp	v	u	h	s
1400	4.906	1540.7	1744.9	1.9600
1600	5.441	1631.3	1857.9	2.0177
1800	5.975	1725.4	1974.2	2.0716
2000	6.507	1822.9	2093.8	2.1223

$P = 250 \ (401.04)$

Temp	v	u	h	s
Sat	1.8448	1116.7	1202.1	1.5274
450	2.002	1141.1	1233.7	1.5632
500	2.150	1163.8	1263.3	1.5948
550	2.290	1185.3	1291.3	1.6233
600	2.426	1206.1	1318.3	1.6494
650	2.558	1226.5	1344.9	1.6739
700	2.688	1246.5	1371.1	1.6970
800	2.943	1287.0	1423.2	1.7401
900	3.193	1327.6	1475.3	1.7799
1000	3.440	1368.7	1527.9	1.8172
1200	3.929	1453.0	1634.8	1.8858
1400	4.414	1540.4	1744.6	1.9483
1600	4.896	1631.1	1857.6	2.0060
1800	5.376	1725.2	1974.0	2.0599
2000	5.856	1822.7	2093.6	2.1106

$P = 275 \ (409.52)$

Temp	v	u	h	s
Sat	1.6813	1117.5	1203.1	1.5192
450	1.8026	1138.3	1230.0	1.5495
500	1.9407	1161.7	1260.4	1.5820
550	2.071	1183.6	1289.0	1.6110
600	2.196	1204.7	1316.4	1.6376
650	2.317	1225.3	1343.2	1.6623
700	2.436	1245.7	1369.7	1.6856
800	2.670	1286.2	1422.1	1.7289
900	2.898	1327.0	1474.5	1.7689
1000	3.124	1368.2	1527.2	1.8064
1200	3.570	1452.6	1634.3	1.8751
1400	4.011	1540.1	1744.2	1.9376
1600	4.450	1630.9	1857.3	1.9954
1800	4.887	1725.0	1973.7	2.0493
2000	5.323	1822.5	2093.4	2.1000

$P = 300 \ (417.43)$

Temp	v	u	h	s
Sat	1.5442	1118.2	1203.9	1.5115
450	1.6361	1135.4	1226.2	1.5365
500	1.7662	1159.5	1257.5	1.5701
550	1.8878	1181.9	1286.7	1.5997
600	2.004	1203.2	1314.5	1.6266
650	2.117	1224.1	1341.6	1.6516
700	2.227	1244.6	1368.3	1.6751
800	2.442	1285.4	1421.0	1.7187
900	2.653	1326.3	1473.6	1.7589
1000	2.860	1367.7	1526.5	1.7964
1200	3.270	1452.2	1633.8	1.8653
1400	3.675	1539.8	1743.8	1.9279
1600	4.078	1630.7	1857.0	1.9857
1800	4.479	1724.9	1973.5	2.0396
2000	4.879	1822.3	2093.2	2.0904

$P = 350 \ (431.82)$

Temp	v	u	h	s
Sat	1.3267	1119.0	1204.9	1.4978
450	1.3733	1129.2	1218.2	1.5125
500	1.4913	1154.9	1251.5	1.5482
550	1.5998	1178.3	1281.9	1.5790
600	1.7025	1200.3	1310.6	1.6068
650	1.8013	1221.6	1338.3	1.6323
700	1.8975	1242.5	1365.4	1.6562
800	2.085	1283.8	1418.8	1.7004

$P = 400 \ (444.70)$

Temp	v	u	h	s
Sat	1.1620	1119.5	1205.5	1.4856
450	1.1745	1122.6	1209.6	1.4901
500	1.2843	1150.1	1245.2	1.5282
550	1.3833	1174.6	1277.0	1.5605
600	1.4760	1197.3	1306.6	1.5892
650	1.5645	1219.1	1334.9	1.6153
700	1.6503	1240.4	1362.5	1.6397
800	1.8163	1282.1	1416.6	1.6844

$P = 450 \ (456.39)$

Temp	v	u	h	s
Sat	1.0326	1119.6	1205.6	1.4746
450				
500	1.1226	1145.1	1238.5	1.5097
550	1.2146	1170.7	1271.9	1.5436
600	1.2996	1194.3	1302.5	1.5732
650	1.3803	1216.6	1331.5	1.6000
700	1.4580	1238.2	1359.6	1.6248
800	1.6077	1280.5	1414.4	1.6701

Table B.1.3 (Continued)
Superheated Vapor

°F	v	u	h	s	v	u	h	s	v	u	h	s
	$P = 350 (431.82)$				$P = 400 (444.70)$				$P = 450 (456.39)$			
900	2.267	1325.0	1471.8	1.7409	1.9776	1323.7	1470.1	1.7252	1.7524	1322.4	1468.3	1.7113
1000	2.446	1366.6	1525.0	1.7787	2.136	1365.5	1523.6	1.7632	1.8941	1364.4	1522.2	1.7495
1200	2.799	1451.5	1632.8	1.8478	2.446	1450.7	1631.8	1.8327	2.172	1450.0	1630.8	1.8192
1400	3.148	1539.3	1743.1	1.9106	2.752	1538.7	1742.4	1.8956	2.444	1538.1	1741.7	1.8823
1600	3.494	1630.2	1856.5	1.9685	3.055	1629.8	1855.9	1.9535	2.715	1629.3	1855.4	1.9403
1800	3.838	1724.5	1973.1	2.0225	3.357	1724.1	1972.6	2.0076	2.983	1723.7	1972.1	1.9944
2000	4.182	1822.0	2092.8	2.0733	3.658	1821.6	2092.4	2.0584	3.251	1821.3	2092.0	2.0453
	$P = 500 (467.13)$				$P = 600 (486.33)$				$P = 700 (503.23)$			
Sat	0.9283	1119.4	1205.3	1.4645	0.7702	1118.6	1204.1	1.4464	0.6558	1117.0	1202.0	1.4305
500	0.9924	1139.7	1231.5	1.4923	0.7947	1128.0	1216.2	1.4592				
550	1.0792	1166.7	1266.6	1.5279	0.8749	1158.2	1255.4	1.4990	0.7275	1149.0	1243.2	1.4723
600	1.1583	1191.1	1298.3	1.5585	0.9456	1184.5	1289.5	1.5320	0.7929	1177.5	1280.2	1.5081
650	1.2327	1214.0	1328.0	1.5860	1.0109	1208.6	1320.9	1.5609	0.8520	1203.1	1313.4	1.5387
700	1.3040	1236.0	1356.7	1.6112	1.0727	1231.5	1350.6	1.5872	0.9073	1226.9	1344.4	1.5661
800	1.4407	1278.8	1412.1	1.6571	1.1900	1275.4	1407.6	1.6343	1.0109	1272.0	1402.9	1.6145
900	1.5723	1321.0	1466.5	1.6987	1.3021	1318.4	1462.9	1.6766	1.1089	1315.6	1459.3	1.6576
1000	1.7008	1363.3	1520.7	1.7371	1.4108	1361.2	1517.8	1.7155	1.2036	1358.9	1514.9	1.6970
1100	1.8271	1406.0	1575.1	1.7731	1.5173	1404.2	1572.7	1.7519	1.2960	1402.4	1570.2	1.7337
1200	1.9518	1449.2	1629.8	1.8072	1.6222	1447.7	1627.8	1.7861	1.3868	1446.2	1625.8	1.7682
1400	2.198	1537.6	1741.0	1.8704	1.8289	1536.5	1739.5	1.8497	1.5652	1535.3	1738.1	1.8321
1600	2.442	1628.9	1854.8	1.9285	2.033	1628.0	1853.7	1.9080	1.7409	1627.1	1852.6	1.8906
1800	2.684	1723.3	1971.7	1.9827	2.236	1722.6	1970.8	1.9622	1.9152	1721.8	1969.9	1.9449
2000	2.926	1820.9	2091.6	2.0335	2.438	1820.2	2090.8	2.0131	2.0887	1819.5	2090.1	1.9958

	$P = 800(518.36)$				$P = 1000(544.75)$				$P = 1250(572.56)$			
Sat	0.5691	1115.0	1199.3	1.4160	0.4459	1109.9	1192.4	1.3903	0.3454	1101.7	1181.6	1.3619
550	0.6154	1138.8	1229.9	1.4469	0.4534	1114.8	1198.7	1.3966				
600	0.6776	1170.1	1270.4	1.4861	0.5140	1153.7	1248.8	1.4450	0.3786	1129.0	1216.6	1.3954
650	0.7324	1197.2	1305.6	1.5186	0.5637	1184.7	1289.1	1.4822	0.4267	1167.2	1266.0	1.4410
700	0.7829	1222.1	1338.0	1.5471	0.6080	1212.0	1324.6	1.5135	0.4670	1198.4	1306.4	1.4767
750	0.8306	1245.7	1368.6	1.5730	0.6490	1237.2	1357.3	1.5412	0.5030	1226.1	1342.4	1.5070
800	0.8764	1268.5	1398.2	1.5969	0.6878	1261.2	1388.5	1.5664	0.5364	1251.8	1375.8	1.5341
900	0.9640	1312.9	1455.6	1.6408	0.7610	1307.3	1448.1	1.6120	0.5984	1300.0	1438.4	1.5820
1000	1.0482	1356.7	1511.9	1.6807	0.8305	1352.2	1505.9	1.6530	0.6563	1346.4	1498.2	1.6244
1100	1.1300	1400.5	1567.8	1.7178	0.8976	1396.8	1562.9	1.6908	0.7116	1392.0	1556.6	1.6631
1200	1.2102	1444.6	1623.8	1.7526	0.9630	1441.5	1619.7	1.7261	0.7652	1437.5	1614.5	1.6991
1400	1.3674	1534.2	1736.6	1.8167	1.0905	1531.9	1733.7	1.7909	0.8689	1529.0	1730.0	1.7648
1600	1.5218	1626.2	1851.5	1.8754	1.2152	1624.4	1849.3	1.8499	0.9699	1622.2	1846.5	1.8243
1800	1.6749	1721.0	1969.0	1.9298	1.3384	1719.5	1967.2	1.9046	1.0693	1717.6	1965.0	1.8791
2000	1.8271	1818.8	2089.3	1.9808	1.4608	1817.4	2087.7	1.9557	1.1678	1815.7	2085.8	1.9304

	$P = 1500(596.39)$				$P = 1750(617.31)$				$P = 2000(636.00)$			
Sat	0.2769	1091.8	1168.7	1.3359	0.2268	1080.2	1153.7	1.3109	0.18813	1066.6	1136.3	1.2861
600	0.2816	1096.6	1174.8	1.3416								
650	0.3329	1147.0	1239.4	1.4012	0.2627	1122.5	1207.6	1.3603	0.2057	1091.1	1167.2	1.3141
700	0.3716	1183.4	1286.6	1.4429	0.3022	1166.7	1264.6	1.4106	0.2487	1147.7	1239.8	1.3782
750	0.4049	1214.1	1326.5	1.4767	0.3341	1201.3	1309.5	1.4485	0.2803	1187.3	1291.1	1.4216
800	0.4350	1241.8	1362.5	1.5058	0.3622	1231.3	1348.6	1.4802	0.3071	1220.1	1333.8	1.4562
850	0.4631	1267.7	1396.2	1.5320	0.3878	1258.8	1384.4	1.5081	0.3312	1249.5	1372.0	1.4860
900	0.4897	1292.5	1428.5	1.5562	0.4119	1284.8	1418.2	1.5334	0.3534	1276.8	1407.6	1.5126
1000	0.5400	1340.4	1490.3	1.6001	0.4569	1334.3	1482.3	1.5789	0.3945	1328.1	1474.1	1.5598
1100	0.5876	1387.2	1550.3	1.6399	0.4990	1382.2	1543.8	1.6197	0.4325	1377.2	1537.2	1.6017
1200	0.6334	1433.5	1609.3	1.6765	0.5392	1429.4	1604.0	1.6571	0.4685	1425.2	1598.6	1.6398
1400	0.7213	1526.1	1726.3	1.7431	0.6158	1523.1	1722.6	1.7245	0.5368	1520.2	1718.8	1.7082
1600	0.8064	1619.9	1843.7	1.8031	0.6896	1617.6	1841.0	1.7850	0.6020	1615.4	1838.2	1.7692

Table B.1.3 (Continued)
Superheated Vapor

°F	v	u	h	s	v	u	h	s	v	u	h	s
	$P = 1500(596.39)$				$P = 1750(617.31)$				$P = 2000(636.00)$			
1800	0.8899	1715.7	1962.7	1.8582	0.7617	1713.9	1960.5	1.8404	0.6656	1712.0	1958.3	1.8249
2000	0.9725	1814.0	2083.9	1.9096	0.8330	1812.3	2082.0	1.8919	0.7284	1810.6	2080.2	1.8765
	$P = 2500(668.31)$				$P = 3000(695.52)$				$P = 3500$			
Sat	0.13059	1031.0	1091.4	1.2327	0.08404	968.8	1015.5	1.1575				
650									0.02491	663.5	679.7	0.8630
700	0.16839	1098.7	1176.6	1.3073	0.09771	1003.9	1058.1	1.1944	0.03058	759.5	779.3	0.9506
750	0.2030	1155.2	1249.1	1.3686	0.14831	1114.7	1197.1	1.3122	0.10460	1058.4	1126.1	1.2440
800	0.2291	1195.7	1301.7	1.4112	0.17572	1167.6	1265.2	1.3675	0.13626	1134.7	1223.0	1.3226
850	0.2513	1229.5	1345.8	1.4456	0.19731	1207.7	1317.2	1.4080	0.15818	1183.4	1285.9	1.3716
900	0.2712	1259.9	1385.4	1.4752	0.2160	1241.8	1361.7	1.4414	0.17625	1222.4	1336.5	1.4096
950	0.2896	1288.2	1422.2	1.5018	0.2328	1272.7	1402.0	1.4705	0.19214	1256.4	1380.8	1.4416
1000	0.3069	1315.2	1457.2	1.5262	0.2485	1301.7	1439.6	1.4967	0.2066	1287.6	1421.4	1.4699
1100	0.3393	1366.8	1523.8	1.5704	0.2772	1356.2	1510.1	1.5434	0.2328	1345.2	1496.0	1.5193
1200	0.3696	1416.7	1587.7	1.6101	0.3036	1408.0	1576.6	1.5848	0.2566	1399.2	1565.3	1.5624
1400	0.4261	1514.2	1711.3	1.6804	0.3524	1508.1	1703.7	1.6571	0.2997	1501.9	1696.1	1.6368
1600	0.4795	1610.8	1832.6	1.7424	0.3978	1606.3	1827.1	1.7201	0.3395	1601.7	1821.6	1.7010
1800	0.5312	1708.2	1954.0	1.7986	0.4416	1704.5	1949.6	1.7769	0.3776	1700.8	1945.4	1.7583
2000	0.5820	1807.2	2076.4	1.8506	0.4844	1803.9	2072.8	1.8291	0.4147	1800.6	2069.2	1.8108
	$P = 4000$				$P = 5000$				$P = 6000$			
650	0.02447	657.7	675.8	0.8574	0.02377	648.0	670.0	0.8482	0.02322	640.0	665.8	0.8405
700	0.02867	742.1	763.4	0.9345	0.02676	721.8	746.6	0.9156	0.02563	708.1	736.5	0.9028

750	0.06331	960.7	1007.5	1.1395	0.03364	821.4	852.6	1.0049	0.02978	788.6	821.7	0.9746
800	0.10522	1095.0	1172.9	1.2740	0.05932	987.2	1042.1	1.1583	0.03942	896.9	940.7	1.0708
850	0.12833	1156.5	1251.5	1.3352	0.08556	1092.7	1171.9	1.2596	0.05818	1018.8	1083.4	1.1820
900	0.14622	1201.5	1309.7	1.3789	0.10385	1155.1	1251.1	1.3190	0.07588	1102.9	1187.2	1.2599
950	0.16151	1239.2	1358.8	1.4144	0.11853	1202.2	1311.9	1.3629	0.09008	1162.0	1262.0	1.3140
1000	0.17520	1272.9	1402.6	1.4449	0.13120	1242.0	1363.4	1.3988	0.10207	1209.1	1322.4	1.3561
1100	0.19954	1333.9	1481.6	1.4973	0.15302	1310.6	1452.2	1.4577	0.12218	1286.4	1422.1	1.4222
1200	0.2213	1390.1	1553.9	1.5423	0.17199	1371.6	1530.8	1.5066	0.13927	1352.7	1507.3	1.4752
1300	0.2414	1443.7	1622.4	1.5823	0.18918	1428.6	1603.7	1.5493	0.15453	1413.3	1584.9	1.5206
1400	0.2603	1495.7	1688.4	1.6188	0.20517	1483.2	1673.0	1.5876	0.16854	1470.5	1657.6	1.5608
1600	0.2959	1597.1	1816.1	1.6841	0.2348	1587.9	1805.2	1.6551	0.19420	1578.7	1794.3	1.6307
1800	0.3296	1697.1	1941.1	1.7420	0.2626	1689.8	1932.7	1.7142	0.21801	1682.4	1924.5	1.6910
2000	0.3625	1797.3	2065.6	1.7948	0.2895	1790.8	2058.6	1.7676	0.24087	1784.3	2051.7	1.7450

Table B.1.4
Compressed Liquid

T	v	u	h	s	v	u	h	s	v	u	h	s
	$P = 500(467.13)$				$P = 1000(544.75)$				$P = 1500(596.39)$			
Sat	0.019748	447.70	449.53	0.64904	0.021591	538.39	542.38	0.74320	0.023461	604.97	611.48	0.80824
32	0.015994	0.00	1.49	0.00000	0.015967	0.03	2.99	0.00005	0.015939	0.05	4.47	0.00007
50	0.015998	18.02	19.50	0.03599	0.015972	17.99	20.94	0.03592	0.015946	17.95	22.38	0.03584
100	0.016106	67.87	69.36	0.12932	0.016082	67.70	70.68	0.12901	0.016058	67.53	71.99	0.12870
150	0.016318	117.66	119.17	0.21457	0.016293	117.38	120.40	0.21410	0.016268	117.10	121.62	0.21364
200	0.016608	167.65	169.19	0.29341	0.016580	167.26	170.32	0.29281	0.016554	166.87	171.46	0.29221
250	0.016972	217.99	219.56	0.36702	0.016941	217.47	220.61	0.36628	0.016910	216.96	221.65	0.36554
300	0.017416	268.92	270.53	0.43641	0.017379	268.24	271.46	0.43552	0.017343	267.58	272.39	0.43463
350	0.017954	320.71	322.37	0.50249	0.017909	319.83	323.15	0.50140	0.017865	318.98	323.94	0.50034
400	0.018608	373.68	375.40	0.56604	0.018550	372.55	375.98	0.56472	0.018493	371.45	376.59	0.56343
450	0.019420	428.40	430.19	0.62798	0.019340	426.89	430.47	0.62632	0.019264	425.44	430.79	0.62470
500					0.02036	483.8	487.5	0.6874	0.02024	481.8	487.4	0.6853
550									0.02158	542.1	548.1	0.7469
	$P = 2000(636.00)$				$P = 3000(695.52)$				$P = 5000$			
Sat	0.025649	662.40	671.89	0.86227	0.034310	783.45	802.50	0.97320				
32	0.015912	0.06	5.95	0.00008	0.015859	0.09	8.90	0.00009	0.015755	0.11	14.70	−0.00001
50	0.015920	17.91	23.81	0.03575	0.015870	17.84	26.65	0.03555	0.015773	17.67	32.26	0.03508
100	0.016034	67.37	73.30	0.12839	0.015987	67.04	75.91	0.12777	0.015897	66.40	81.11	0.12651
200	0.016527	166.49	172.60	0.29162	0.016476	165.74	174.89	0.29046	0.016376	164.32	179.47	0.28818
300	0.017308	266.93	273.33	0.43376	0.017240	265.66	275.23	0.43205	0.017110	263.25	279.08	0.42875
400	0.018439	370.38	377.21	0.56216	0.018334	368.32	378.50	0.55970	0.018141	364.47	381.25	0.55506
450	0.019191	424.04	431.14	0.62313	0.019053	421.36	431.93	0.62011	0.018803	416.44	433.84	0.61451
500	0.02014	479.8	487.3	0.6832	0.019944	476.2	487.3	0.6794	0.019603	469.8	487.9	0.6724
560	0.02172	551.8	559.8	0.7565	0.021382	546.2	558.0	0.7508	0.020835	536.7	556.0	0.7411

600	0.02330	605.4	614.0	0.8086	0.02274	597.0	609.6	0.8004	0.02191	584.0	604.2	0.7876
640					0.02475	654.3	668.0	0.8545	0.02334	634.6	656.2	0.8357
680					0.02879	728.4	744.3	0.9226	0.02535	690.6	714.1	0.8873
700									0.02676	721.8	746.6	0.9156

Table B.1.5
Saturated Solid–Vapor

	Specific Volume			Internal Energy			Enthalpy			Entropy		
Temp. Fahr. T	Press. lbf/in² P	Sat. Solid v_i	Sat. Vapor $v_g \times 10^{-3}$	Sat. Solid u_i	Subl. u_{ig}	Sat. Vapor u_g	Sat. Solid h_i	Subl. h_{ig}	Sat. Vapor h_g	Sat. Solid s_i	Subl. s_{ig}	Sat. Vapor s_g
32.018	0.0887	0.01747	3.302	−143.34	1164.6	1021.2	−143.34	1218.7	1075.4	−0.292	2.479	2.187
32	0.0886	0.01747	3.305	−143.35	1164.6	1021.2	−143.35	1218.7	1075.4	−0.292	2.479	2.187
30	0.0808	0.01747	3.607	−144.35	1164.9	1020.5	−144.35	1218.9	1074.5	−0.294	2.489	2.195
25	0.0641	0.01746	4.506	−146.84	1165.7	1018.9	−146.84	1219.1	1072.3	−0.299	2.515	2.216
20	0.0505	0.01745	5.655	−149.31	1166.5	1017.2	−149.31	1219.4	1070.1	−0.304	2.542	2.238
15	0.0396	0.01745	7.13	−151.75	1167.3	1015.5	−151.75	1219.7	1067.9	−0.309	2.569	2.260
10	0.0309	0.01744	9.04	−154.17	1168.1	1013.9	−154.17	1219.9	1065.7	−0.314	2.597	2.283
5	0.0240	0.01743	11.52	−156.56	1168.8	1012.2	−156.56	1220.1	1063.5	−0.320	2.626	2.306
0	0.0185	0.01743	14.77	−158.93	1169.5	1010.6	−158.93	1220.2	1061.2	−0.325	2.655	2.330
−5	0.0142	0.01742	19.03	−161.27	1170.2	1008.9	−161.27	1220.3	1059.0	−0.330	2.684	2.354
−10	0.0109	0.01741	24.66	−163.59	1170.9	1007.3	−163.59	1220.4	1056.8	−0.335	2.714	2.379
−15	0.0082	0.01740	32.2	−165.89	1171.5	1005.6	−165.89	1220.5	1054.6	−0.340	2.745	2.405
−20	0.0062	0.01740	42.2	−168.16	1172.1	1003.9	−168.16	1220.6	1052.4	−0.345	2.776	2.431
−25	0.0046	0.01739	55.7	−170.40	1172.7	1002.3	−170.40	1220.6	1050.2	−0.351	2.808	2.457
−30	0.0035	0.01738	74.1	−172.63	1173.2	1000.6	−172.63	1220.6	1048.0	−0.356	2.841	2.485
−35	0.0026	0.01737	99.2	−174.82	1173.8	998.9	−174.82	1220.6	1045.8	−0.361	2.874	2.513
−40	0.0019	0.01737	133.8	−177.00	1174.3	997.3	−177.00	1220.6	1043.6	−0.366	2.908	2.542

Table B.2
Thermodynamic Properties of Ammonia[a]
Table B.2.1
Saturated Ammonia

Temp. F	Abs. Press. lbf/in.² P	Specific Volume ft³/lbm			Enthalpy Btu/lbm			Entropy Btu/lbm R		
		Sat. Liquid v_f	Evap. v_{fg}	Sat. Vapor v_g	Sat. Liquid h_f	Evap. h_{fg}	Sat. Vapor h_g	Sat. Liquid s_f	Evap. s_{fg}	Sat. Vapor s_g
−60	5.55	0.0228	44.707	44.73	−21.2	610.8	589.6	−0.0517	1.5286	1.4769
−55	6.54	0.0229	38.357	38.38	−15.9	607.5	591.6	−0.0386	1.5017	1.4631
−50	7.67	0.0230	33.057	33.08	−10.6	604.3	593.7	−0.0256	1.4753	1.4497
−45	8.95	0.0231	28.597	28.62	−5.3	600.9	595.6	−0.0127	1.4495	1.4368
−40	10.41	0.02322	24.837	24.86	0	597.6	597.6	0.000	1.4242	1.4242
−35	12.05	0.02333	21.657	21.68	5.3	594.2	599.5	0.0126	1.3994	1.4120
−30	13.90	0.0235	18.947	18.97	10.7	590.7	601.4	0.0250	1.3751	1.4001
−25	15.98	0.0236	16.636	16.66	16.0	587.2	603.2	0.0374	1.3512	1.3886
−20	18.30	0.0237	14.656	14.68	21.4	583.6	605.0	0.0497	1.3277	1.3774
−15	20.88	0.02381	12.946	12.97	26.7	580.0	606.7	0.0618	1.3044	1.3664
−10	23.74	0.02393	11.476	11.50	32.1	576.4	608.5	0.0738	1.2820	1.3558
−5	26.92	0.02406	10.206	10.23	37.5	572.6	610.1	0.0857	1.2597	1.3454
0	30.42	0.02419	9.092	9.116	42.9	568.9	611.8	0.0975	1.2377	1.3352
5	34.27	0.02432	8.1257	8.150	48.3	565.0	613.3	0.1092	1.2161	1.3253
10	38.51	0.02446	7.2795	7.304	53.8	561.1	614.9	0.1208	1.1949	1.3157
15	43.14	0.02460	6.5374	6.562	59.2	557.1	616.3	0.1323	1.1739	1.3062
20	48.21	0.02474	5.8853	5.910	64.7	553.1	617.8	0.1437	1.1532	1.2969
25	53.73	0.02488	5.3091	5.334	70.2	548.9	619.1	0.1551	1.1328	1.2879
30	59.74	0.02503	4.8000	4.825	75.7	544.8	620.5	0.1663	1.1127	1.2790
35	66.26	0.02518	4.3478	4.373	81.2	540.5	621.7	0.1775	1.0929	1.2704

40	73.32	0.02533	3.9457	3.971	86.8	536.2	623.0	0.1885	1.0733	1.2618
45	80.96	0.02548	3.5885	3.614	92.3	531.8	624.1	0.1996	1.0539	1.2535
50	89.19	0.02564	3.2684	3.294	97.9	527.3	625.2	0.2105	1.0348	1.2453
55	98.06	0.02581	2.9822	3.008	103.5	522.8	626.3	0.2214	1.0159	1.2373
60	107.6	0.02597	2.7250	2.751	109.2	518.1	627.3	0.2322	0.9972	1.2294
65	117.8	0.02614	2.4939	2.520	114.8	513.4	628.2	0.2430	0.9786	1.2216
70	128.8	0.02632	2.2857	2.312	120.5	508.6	629.1	0.2537	0.9603	1.2140
75	140.5	0.02650	2.0985	2.125	126.2	503.7	629.9	0.2643	0.9422	1.2065
80	153.0	0.02668	1.9283	1.955	132.0	498.7	630.7	0.2749	0.9242	1.1991
85	166.4	0.02687	1.7741	1.801	137.8	493.6	631.4	0.2854	0.9064	1.1918
90	180.6	0.02707	1.6339	1.661	143.5	488.5	632.0	0.2958	0.8888	1.1846
95	195.8	0.02727	1.5067	1.534	149.4	483.2	632.6	0.3062	0.8713	1.1775
100	211.9	0.02747	1.3915	1.419	155.2	477.8	633.0	0.3166	0.8539	1.1705
105	228.9	0.02769	1.2853	1.313	161.1	472.3	633.4	0.3269	0.8366	1.1635
110	247.0	0.02790	1.1891	1.217	167.0	466.7	633.7	0.3372	0.8194	1.1566
115	266.2	0.02813	1.0999	1.128	173.0	460.9	633.9	0.3474	0.8023	1.1497
120	286.4	0.02836	1.0186	1.047	179.0	455.0	634.0	0.3576	0.7851	1.1427
125	307.8	0.02860	0.9444	0.973	185.1	448.9	634.0	0.3679	0.7679	1.1358

[a]Reprinted by permission from National Bureau of Standards Circular No. 142, *Tables of Thermodynamic Properties of Ammonia.*

Table B.2.2
Superheated Ammonia

Abs. Press. lbf/in.² (Sat. Temp.)		Temperature, F 0	20	40	60	80	100	120	140	160	180	200	220
	v	28.58	29.90	31.20	32.49	33.78	35.07	36.35	37.62	38.90	40.17	41.45	
10	h	618.9	629.1	639.3	649.5	659.7	670.0	680.3	690.6	701.1	711.6	722.2	
(−41.34)	s	1.477	1.499	1.520	1.540	1.559	1.578	1.596	1.614	1.631	1.647	1.664	
	v	18.92	19.82	20.70	21.58	22.44	23.31	24.17	25.03	25.88	26.74	27.59	
15	h	617.2	627.8	638.2	648.5	658.9	669.2	679.6	690.0	700.5	711.1	721.7	
(−27.29)	s	1.427	1.450	1.471	1.491	1.511	1.529	1.548	1.566	1.583	1.599	1.616	
	v	14.09	14.78	15.45	16.12	16.78	17.43	18.08	18.73	19.37	20.02	20.66	21.3
20	h	615.5	626.4	637.0	647.5	658.0	668.5	678.9	689.4	700.0	710.6	721.2	732.0
(−16.64)	s	1.391	1.414	1.436	1.456	1.476	1.495	1.513	1.531	1.549	1.565	1.582	1.598
	v	11.19	11.75	12.30	12.84	13.37	13.90	14.43	14.95	15.47	15.99	16.50	17.02
25	h	613.8	625.0	635.8	646.5	657.1	667.7	678.2	688.8	699.4	710.1	720.8	731.6
(−7.96)	s	1.362	1.386	1.408	1.429	1.449	1.468	1.486	1.504	1.522	1.539	1.555	1.571
	v	9.25	9.731	10.20	10.65	11.10	11.55	11.99	12.43	12.87	13.30	13.73	14.16
30	h	611.9	623.5	634.6	645.5	656.2	666.9	677.5	688.2	698.8	709.6	720.3	731.1
(−.57)	s	1.337	1.362	1.385	1.406	1.426	1.446	1.464	1.482	1.500	1.517	1.533	1.550
	v		8.287	8.695	9.093	9.484	9.869	10.25	10.63	11.00	11.38	11.75	12.12
35	h		622.0	633.4	644.4	655.3	666.1	676.8	687.6	698.3	709.1	719.9	730.7
(5.89)	s		1.341	1.365	1.386	1.407	1.427	1.445	1.464	1.481	1.498	1.515	1.531
	v		7.203	7.568	7.922	8.268	8.609	8.945	9.278	9.609	9.938	10.27	10.59
40	h		620.4	632.1	643.4	654.4	665.3	676.1	686.9	697.7	708.5	719.4	730.3
(11.66)	s		1.323	1.347	1.369	1.390	1.410	1.429	1.447	1.465	1.482	1.499	1.515
	v		6.363	6.694	7.014	7.326	7.632	7.934	8.232	8.528	8.822	9.115	9.406
45	h		618.8	630.8	642.3	653.5	664.6	675.5	686.3	697.2	708.0	718.9	729.9
(16.87)	s		1.307	1.331	1.354	1.375	1.395	1.414	1.433	1.450	1.468	1.485	1.501

	v			5.988	6.280	6.564	6.843	7.117	7.387	7.655	7.921	8.185	8.448
50	*h*			629.5	641.2	652.6	663.7	674.7	685.7	696.6	707.5	718.5	729.4
(21.67)	*s*			1.317	1.340	1.361	1.382	1.401	1.420	1.437	1.455	1.472	1.488
	v			4.933	5.184	5.428	5.665	5.897	6.126	6.352	6.576	6.798	7.019
60	*h*			626.8	639.0	650.7	662.1	673.3	684.4	695.5	706.5	717.5	728.6
(30.21)	*s*			1.291	1.315	1.337	1.358	1.378	1.397	1.415	1.432	1.449	1.466

		60	80	100	120	140	160	180	200	240	280	320	360
	v	4.401	4.615	4.822	5.025	5.224	5.420	5.615	5.807	6.187	6.563		
70	*h*	636.6	648.7	660.4	671.8	683.1	694.3	705.5	716.6	738.9	761.4		
(37.7)	*s*	1.294	1.317	1.338	1.358	1.377	1.395	1.413	1.430	1.463	1.494		
	v	3.812	4.005	4.190	4.371	4.548	4.722	4.893	5.063	5.398	5.73		
80	*h*	634.3	646.7	658.7	670.4	681.8	693.2	704.4	715.6	738.1	760.7		
(44.4)	*s*	1.275	1.298	1.320	1.340	1.360	1.378	1.396	1.414	1.447	1.478		
	v	3.353	3.529	3.698	3.862	4.021	4.178	4.332	4.484	4.785	5.081		
90	*h*	631.8	644.7	657.0	668.9	680.5	692.0	703.4	714.7	737.3	760.0		
(50.47)	*s*	1.257	1.281	1.304	1.325	1.344	1.363	1.381	1.400	1.432	1.464		
	v	2.985	3.149	3.304	3.454	3.600	3.743	3.883	4.021	4.294	4.562		
100	*h*	629.3	642.6	655.2	667.3	679.2	690.8	702.3	713.7	736.5	759.4		
(56.05)	*s*	1.241	1.266	1.289	1.310	1.331	1.349	1.368	1.385	1.419	1.451		
	v		2.166	2.288	2.404	2.515	2.622	2.727	2.830	3.030	3.227	3.420	
140	*h*		633.8	647.8	661.1	673.7	686.0	698.0	709.9	733.3	756.7	780.0	
(74.79)	*s*		1.214	1.240	1.263	1.284	1.305	1.324	1.342	1.376	1.409	1.440	
	v			1.720	1.818	1.910	1.999	2.084	2.167	2.328	2.484	2.637	
180	*h*			639.9	654.4	668.0	681.0	693.6	705.9	730.1	753.9	777.7	
(89.78)	*s*			1.199	1.225	1.248	1.269	1.289	1.308	1.344	1.377	1.408	
	v				1.443	1.525	1.601	1.675	1.745	1.881	2.012	2.140	2.265
220	*h*				647.3	662.0	675.8	689.1	701.9	726.8	751.1	775.3	799.5
(102.42)	*s*				1.192	1.217	1.239	1.260	1.280	1.317	1.351	1.383	1.413
	v				1.302	1.380	1.452	1.521	1.587	1.714	1.835	1.954	2.069
240	*h*				643.5	658.8	673.1	686.7	699.8	725.1	749.8	774.1	798.4
(108.09)	*s*				1.176	1.203	1.226	1.248	1.268	1.305	1.339	1.371	1.402

Table B.2.2 (Continued)
Superheated Ammonia

Abs. Press. lbf/in.^{2} (Sat. Temp.)		Temperature, F											
		60	80	100	120	140	160	180	200	240	280	320	360
	v				1.182	1.257	1.326	1.391	1.453	1.572	1.686	1.796	1.904
260	h				639.5	655.6	670.4	684.4	697.7	723.4	748.4	772.9	797.4
(113.42)	s				1.162	1.189	1.213	1.235	1.256	1.294	1.329	1.361	1.391
	v				1.078	1.151	1.217	1.279	1.339	1.451	1.558	1.661	1.762
280	h				635.4	652.2	667.6	681.9	695.6	721.8	747.0	771.7	796.3
(118.45)	s				1.147	1.176	1.201	1.224	1.245	1.283	1.318	1.351	1.382

Table B.3
Thermodynamic Properties of Freon-12 (Dichlorodifluoromethane)[a]
Table B.3.1
Saturated Freon-12

		Specific Volume ft³/lbm			Enthalpy Btu/lbm			Entropy Btu/lbm R		
Temp F T	Abs. Press. lbf/in.² P	Sat. Liquid v_f	Evap. v_{fg}	Sat. Vapor v_g	Sat. Liquid h_f	Evap. h_{fg}	Sat. Vapor h_g	Sat. Liquid s_f	Evap. s_{fg}	Sat. Vapor s_g
−130	0.41224	0.009736	70.7203	70.730	−18.609	81.577	62.968	−0.04983	0.24743	0.19760
−120	0.64190	0.009816	46.7312	46.741	−16.565	80.617	64.052	−0.04372	0.23731	0.19359
−110	0.97034	0.009899	31.7671	31.777	−14.518	79.663	65.145	−0.03779	0.22780	0.19002
−100	1.4280	0.009985	21.1541	22.164	−12.466	78.714	66.248	−0.03200	0.21883	0.18683
−90	2.0509	0.010073	15.8109	15.821	−10.409	77.764	67.355	−0.02637	0.21034	0.18398
−80	2.8807	0.010164	11.5228	11.533	−8.3451	76.812	68.467	−0.02086	0.20229	0.18143
−70	3.9651	0.010259	8.5584	8.5687	−6.2730	75.853	69.580	−0.01548	0.19464	0.17916
−60	5.3575	0.010357	6.4670	6.4774	−4.1919	74.885	70.693	−0.01021	0.18716	0.17714
−50	7.1168	0.010459	4.9637	4.9742	−2.1011	73.906	71.805	−0.00506	0.18038	0.17533
−40	9.3076	0.010564	3.8644	3.8750	0	72.913	72.913	0	0.17373	0.17373
−30	11.999	0.010674	3.0478	3.0585	2.1120	71.903	74.015	0.00496	0.16733	0.17229
−20	15.267	0.010788	2.4321	2.4429	4.2357	70.874	75.110	0.00983	0.16119	0.17102
−10	19.189	0.010906	1.9628	1.9727	6.3716	69.824	76.196	0.01462	0.15527	0.16989
0	23.849	0.011030	1.5979	1.6089	8.5207	68.750	77.271	0.01932	0.14956	0.16888
10	29.335	0.011160	1.3129	1.3241	10.684	67.651	78.335	0.02395	0.14403	0.16798
20	35.736	0.011296	1.0875	1.0988	12.863	66.522	79.385	0.02852	0.13867	0.16719
30	43.148	0.011438	0.90736	0.91880	15.058	65.361	80.419	0.03301	0.13347	0.16648
40	51.667	0.011588	0.76198	0.77357	17.273	64.163	81.436	0.03745	0.12841	0.16586
50	61.394	0.011746	0.64362	0.65537	19.507	62.926	82.433	0.04184	0.12346	0.16530
60	72.433	0.011913	0.54648	0.55839	21.766	61.643	83.409	0.04618	0.11861	0.16479

Table B.3.1 (Continued)
Saturated Freon-12

Temp F T	Abs. Press. lbf/in.² P	Specific Volume ft³/lbm			Enthalpy Btu/lbm			Entropy Btu/lbm R		
		Sat. Liquid v_f	Evap. v_{fg}	Sat. Vapor v_g	Sat. Liquid h_f	Evap. h_{fg}	Sat. Vapor h_g	Sat. Liquid s_f	Evap. s_{fg}	Sat. Vapor s_g
70	84.888	0.012089	0.46609	0.47818	24.050	60.309	84.359	0.05048	0.11386	0.16434
80	98.870	0.012277	0.39907	0.41135	26.365	58.917	85.282	0.05475	0.10917	0.16392
90	114.49	0.012478	0.34281	0.35529	28.713	57.461	86.174	0.05900	0.10453	0.16353
100	131.86	0.012693	0.29525	0.30794	31.100	55.929	87.029	0.06323	0.09992	0.16315
110	151.11	0.012924	0.25577	0.26769	33.531	54.313	87.844	0.06745	0.09534	0.16279
120	172.35	0.013174	0.22019	0.23326	36.013	52.597	88.610	0.07168	0.09073	0.16241
130	195.71	0.013447	0.19019	0.20364	38.553	50.768	89.321	0.07583	0.08609	0.16202
140	221.32	0.013746	0.16424	0.17799	41.162	48.805	89.967	0.08021	0.08138	0.16159
150	249.31	0.014078	0.14156	0.15564	43.850	46.684	90.534	0.08453	0.07657	0.16110
160	279.82	0.014449	0.12159	0.13604	46.633	44.373	91.006	0.08893	0.07260	0.16053
170	313.00	0.014871	0.10386	0.11873	49.529	41.830	91.359	0.09342	0.06643	0.15985
180	349.00	0.015360	0.08794	0.10330	52.562	38.999	91.561	0.09804	0.06096	0.15900
190	387.98	0.015942	0.073476	0.089418	55.769	35.792	91.561	0.10284	0.05511	0.15793
200	430.09	0.016659	0.060069	0.076728	59.203	32.075	91.278	0.10789	0.04862	0.15651
210	475.52	0.017601	0.047242	0.064843	62.959	27.599	90.558	0.11332	0.03921	0.15453
220	524.43	0.018986	0.035154	0.053140	67.246	21.790	89.036	0.11943	0.03206	0.15149
230	577.03	0.021854	0.017581	0.039435	72.893	12.229	85.122	0.12739	0.01773	0.14512
233.6 (critical)	596.9	0.02870	0	0.02870	78.86	0	78.86	0.1359	0	0.1359

Table B.3.2
Superheated Freon-12

Temp. F	v	h	s	v	h	s	v	h	s
	5 lbf/in.^{2}			10 lbf/in.^{2}			15 lbf/in.^{2}		
0	8.0611	78.582	0.19663	3.9809	78.246	0.18471	2.6201	77.902	0.17751
20	8.4265	81.309	0.20244	4.1691	81.014	0.19061	2.7494	80.712	0.18349
40	8.7903	84.090	0.20812	4.3556	83.828	0.19635	2.8770	83.561	0.18931
60	9.1528	86.922	0.21367	4.5408	86.689	0.20197	3.0031	86.451	0.19498
80	9.5142	89.806	0.21912	4.7248	89.596	0.20746	3.1281	89.383	0.20051
100	9.8747	92.738	0.22445	4.9079	92.548	0.21283	3.2521	92.357	0.20593
120	10.234	95.717	0.22968	5.0903	95.546	0.21809	3.3754	95.373	0.21122
140	10.594	98.743	0.23481	5.2720	98.586	0.22325	3.4981	98.429	0.21640
160	10.952	101.812	0.23985	5.4533	101.669	0.22830	3.6202	101.525	0.22148
180	11.311	104.925	0.24479	5.6341	104.793	0.23326	3.7419	104.661	0.22646
200	11.668	108.079	0.24964	5.8145	107.957	0.23813	3.8632	107.835	0.23135
220	12.026	111.272	0.25441	5.9946	111.159	0.24291	3.9841	111.046	0.23614
	20 lbf/in.^{2}			25 lbf/in.^{2}			30 lbf/in.^{2}		
20	2.0391	80.403	0.17829	1.6125	80.088	0.17414	1.3278	79.765	0.17065
40	2.1373	83.289	0.18419	1.6932	83.012	0.18012	1.3969	82.730	0.17671
60	2.2340	86.210	0.18992	1.7723	85.965	0.18591	1.4644	85.716	0.18257
80	2.3295	89.168	0.19550	1.8502	88.950	0.19155	1.5306	88.729	0.18826
100	2.4241	92.164	0.20095	1.9271	91.968	0.19704	1.5957	91.770	0.19379
120	2.5179	95.198	0.20628	2.0032	95.021	0.20240	1.6600	94.843	0.19918
140	2.6110	98.270	0.21149	2.0786	98.110	0.20763	1.7237	97.948	0.20445
160	2.7036	101.380	0.21659	2.1535	101.234	0.21276	1.7868	101.086	0.20960
180	2.7957	104.528	0.22159	2.2279	104.393	0.21778	1.8494	104.258	0.21463
200	2.8874	107.712	0.22649	2.3019	107.588	0.22269	1.9116	107.464	0.21957
220	2.9789	110.932	0.23130	2.3756	110.817	0.22752	1.9735	110.702	0.22440
240	3.0700	114.186	0.23602	2.4491	114.080	0.23225	2.0351	113.973	0.22915

Table B.3.2 (Continued)
Superheated Freon-12

Temp. F	v	h	s	v	h	s	v	h	s
	35 lbf/in.^{2}			40 lbf/in.^{2}			50 lbf/in.^{2}		
40	1.1850	82.442	0.17375	1.0258	82.148	0.17112	0.80248	81.540	0.16655
60	1.2442	85.463	0.17968	1.0789	85.206	0.17712	0.84713	84.676	0.17271
80	1.3021	88.504	0.18542	1.1306	88.277	0.18292	0.89025	87.811	0.17862
100	1.3589	91.570	0.19100	1.1812	91.367	0.18854	0.93216	90.953	0.18434
120	1.4148	94.663	0.19643	1.2309	94.480	0.19401	0.97313	94.110	0.18988
140	1.4701	97.785	0.20172	1.2798	97.620	0.19933	1.0133	97.286	0.19527
160	1.5248	100.938	0.20689	1.3282	100.788	0.20453	1.0529	100.485	0.20051
180	1.5789	104.122	0.21195	1.3761	103.985	0.20961	1.0920	103.708	0.20563
200	1.6327	107.338	0.21690	1.4236	107.212	0.21457	1.1307	106.958	0.21064
220	1.6862	110.586	0.22175	1.4707	110.469	0.21944	1.1690	110.235	0.21553
240	1.7394	113.865	0.22651	1.5176	113.757	0.22420	1.2070	113.539	0.22032
260	1.7923	117.175	0.23117	1.5642	117.074	0.22888	1.2447	116.871	0.22502
	60 lbf/in.^{2}			70 lbf/in.^{2}			80 lbf/in.^{2}		
60	0.69210	84.126	0.16892	0.58088	83.552	0.16556	. . .	. . .	. . .
80	0.72964	87.330	0.17497	0.61458	86.832	0.17175	0.52795	86.316	0.16885
100	0.76588	90.528	0.18079	0.64685	90.091	0.17768	0.55734	89.640	0.17489
120	0.80110	93.731	0.18641	0.67803	93.343	0.18339	0.58556	92.945	0.18070
140	0.83551	96.945	0.19186	0.70836	96.597	0.18891	0.61286	96.242	0.18629
160	0.86928	100.776	0.19716	0.73800	99.862	0.19427	0.63943	99.542	0.19170
180	0.90252	103.427	0.20233	0.76708	103.141	0.19948	0.66543	102.851	0.19696
200	0.93531	106.700	0.20736	0.79571	106.439	0.20455	0.69095	106.174	0.20207
220	0.96775	109.997	0.21229	0.82397	109.756	0.20951	0.71609	109.513	0.20706
240	0.99988	113.319	0.21710	0.85191	113.096	0.21435	0.74090	112.872	0.21193
260	1.0318	116.666	0.22182	0.87959	116.459	0.21909	0.76544	116.251	0.21669
280	1.0634	120.039	0.22644	0.90705	119.846	0.22373	0.78975	119.652	0.22135

90 lbf/in.²

100	0.48749	89.175	0.17234
120	0.51346	92.536	0.17824
140	0.53845	95.879	0.18391
160	0.56268	99.216	0.18938
180	0.58629	102.557	0.19469
200	0.60941	105.905	0.19984
220	0.63213	109.267	0.20486
240	0.65451	112.644	0.20976
260	0.67662	116.040	0.21455
280	0.69849	119.456	0.21923
300	0.72016	122.892	0.22381
320	0.74166	126.349	0.22830

100 lbf/in.²

100	0.43138	88.694	0.16996
120	0.45562	92.116	0.17597
140	0.47881	95.507	0.18172
160	0.50118	98.884	0.18726
180	0.52291	102.257	0.19262
200	0.54413	105.633	0.19782
220	0.56492	109.018	0.20287
240	0.58538	112.415	0.20780
260	0.60554	115.828	0.21261
280	0.62546	119.258	0.21731
300	0.64518	122.707	0.22191
320	0.66472	126.176	0.22641

125 lbf/in.²

100	0.32943	87.407	0.16455
120	0.35086	91.008	0.17087
140	0.37098	94.537	0.17686
160	0.39015	98.023	0.18258
180	0.40857	101.484	0.18807
200	0.42642	104.934	0.19338
220	0.44380	108.380	0.19853
240	0.46081	111.829	0.20353
260	0.47750	115.287	0.20840
280	0.49394	118.756	0.21316
300	0.51016	122.238	0.21780
320	0.52619	125.737	0.22235

150 lbf/in.²

120	0.28007	89.800	0.16629
140	0.29845	93.498	0.17256
160	0.31566	97.112	0.17849
180	0.33200	100.675	0.18415
200	0.34769	104.206	0.18958
220	0.36285	107.720	0.19483
240	0.37761	111.226	0.19992
260	0.39203	114.732	0.20485
280	0.40617	118.242	0.20967
300	0.42008	121.761	0.21436
320	0.43379	125.290	0.21894
340	0.44733	128.833	0.22343

175 lbf/in.²

120	· · ·	· · ·	· · ·
140	0.24595	92.373	0.16859
160	0.26198	96.142	0.17478
180	0.27697	99.823	0.18062
200	0.29120	103.447	0.18620
220	0.30485	107.036	0.19156
240	0.31804	110.605	0.19674
260	0.33087	114.162	0.20175
280	0.34339	117.717	0.20662
300	0.35567	121.273	0.21137
320	0.36773	124.835	0.21599
340	0.37963	128.407	0.22052

200 lbf/in.²

120	· · ·	· · ·	· · ·
140	· · ·	· · ·	· · ·
160	0.20579	91.137	0.16480
180	0.22121	95.100	0.17130
200	0.23535	98.921	0.17737
220	0.24860	102.652	0.18311
240	0.26117	106.325	0.18860
260	0.27323	109.962	0.19387
280	0.28489	113.576	0.19896
300	0.29623	117.178	0.20390
320	0.30730	120.775	0.20870
340	0.31815	124.373	0.21337
	0.32881	127.974	0.21793

250 lbf/in.²

160	0.16249	92.717	0.16462
180	0.17605	96.925	0.17130
200	0.18824	100.930	0.17747
220	0.19952	104.809	0.18326

300 lbf/in.²

160	· · ·	· · ·	· · ·
180	0.13482	94.556	0.16537
200	0.14697	98.975	0.17217
220	0.15774	103.136	0.17838

400 lbf/in.²

160	· · ·	· · ·	· · ·
180	· · ·	· · ·	· · ·
200	0.091005	93.718	0.16092
220	0.10316	99.046	0.16888

Table B.3.2 (Continued)
Superheated Freon-12

Temp. F	*v*	*h*	*s*	*v*	*h*	*s*	*v*	*h*	*s*
	250 lbf/in.^{2}			300 lbf/in.^{2}			400 lbf/in.^{2}		
240	0.21014	108.607	0.18877	0.16761	107.140	0.18419	0.11300	103.735	0.17568
260	0.22027	112.351	0.19404	0.17685	111.043	0.18969	0.12163	108.105	0.18183
280	0.23001	116.060	0.19913	0.18562	114.879	0.19495	0.12949	112.286	0.18756
300	0.23944	119.747	0.20405	0.19402	118.670	0.20000	0.13680	116.343	0.19298
320	0.24862	123.420	0.20882	0.20214	122.430	0.20489	0.14372	120.318	0.19814
340	0.25759	127.088	0.21346	0.21002	126.171	0.20963	0.15032	124.235	0.20310
360	0.26639	130.754	0.21799	0.21770	129.900	0.21423	0.15668	128.112	0.20789
380	0.27504	134.423	0.22241	0.22522	133.624	0.21872	0.16285	131.961	0.21258
	500 lbf/in.^{2}			600 lbf/in.^{2}					
220	0.064207	92.397	0.15683	...	...	...			
240	0.077620	99.218	0.16672	0.047488	91.024	0.15335			
260	0.087054	104.526	0.17421	0.061922	99.741	0.16566			
280	0.094923	109.277	0.18072	0.070859	105.637	0.17374			
300	0.10190	113.729	0.18666	0.078059	110.729	0.18053			
320	0.10829	117.997	0.19221	0.084333	115.420	0.18663			
340	0.11426	122.143	0.19746	0.090017	119.871	0.19227			
360	0.11992	126.205	0.20247	0.095289	124.167	0.19757			
380	0.12533	130.207	0.20730	0.10025	128.355	0.20262			
400	0.13054	134.166	0.21196	0.10498	132.466	0.20746			
420	0.13559	138.096	0.21648	0.10952	136.523	0.21213			
440	0.14051	142.004	0.22087	0.11391	140.539	0.21664			

Table B.4
Thermodynamic Properties of Nitrogen[a]
Table B.4.1
Saturated Nitrogen

Temp. °R	Abs. Press. lbf/in.² P	Specific Volume, ft³/lbm			Enthalpy, Btu/lbm			Entropy, Btu/lbm-R		
		Sat. Liquid v_f	Evap. v_{fg}	Sat. Vapor v_g	Sat. Liquid h_f	Evap. h_{fg}	Sat. Vapor h_g	Sat. Liquid s_f	Evap. s_{fg}	Sat. Vapor s_g
113.670	1.813	0.01845	23.793	23.812	0.000	92.891	92.891	0.00000	0.81720	0.81720
120.000	3.337	0.01875	13.570	13.589	3.113	91.224	94.337	0.02661	0.76020	0.78681
130.000	7.654	0.01929	6.3208	6.3401	8.062	88.432	96.494	0.06610	0.68025	0.74634
139.255	14.696	0.01984	3.4592	3.4791	12.639	85.668	98.306	0.09992	0.61518	0.71510
140.000	15.425	0.01989	3.3072	3.3271	13.006	85.436	98.443	0.10253	0.61026	0.71279
150.000	28.120	0.02056	1.8865	1.9071	17.945	82.179	100.124	0.13628	0.54786	0.68414
160.000	47.383	0.02132	1.1469	1.1682	22.928	78.548	101.476	0.16795	0.49093	0.65888
170.000	74.991	0.02219	0.7299	0.7521	28.045	74.383	102.427	0.19829	0.43754	0.63584
180.000	112.808	0.02323	0.4789	0.5021	33.411	69.478	102.889	0.22805	0.38599	0.61404
190.000	162.761	0.02449	0.3190	0.3435	39.153	63.582	102.735	0.25789	0.33464	0.59254
200.000	226.853	0.02613	0.2119	0.2380	45.283	56.474	101.757	0.28780	0.28237	0.57017
210.000	307.276	0.02845	0.1354	0.1639	52.061	47.474	99.536	0.31894	0.22607	0.54501
220.000	406.739	0.03249	0.0750	0.1075	60.336	34.536	94.872	0.35494	0.15698	0.51192
226.000	477.104	0.03806	0.0374	0.0755	68.123	20.423	88.546	0.38789	0.09037	0.47826

[a] Abstracted from National Bureau of Standards Technical Note 129A, The Thermodynamic Properties of Nitrogen from 114 to 540 R between 1.0 and 3000 PSIA. Supplement A (British Units) by Thomas R. Strobridge.

Table B.4.2
Superheated Nitrogen

Temp. °R	v ft³/lbm	h Btu/lbm	s Btu/lbm-R	v ft³/lbm	h Btu/lbm	s Btu/lbm-R	v ft³/lbm	h Btu/lbm	s Btu/lbm-R
	14.7 lbf/in.²			20 lbf/in.²			50 lbf/in.²		
150	3.7782	101.086	0.7343	2.7395	100.715	0.7109			
200	5.1366	113.849	0.8078	3.7538	113.625	0.7852	1.4534	112.315	0.7159
250	6.4680	126.443	0.8640	4.7397	126.293	0.8418	1.8663	125.432	0.7744
300	7.7876	138.958	0.9096	5.7138	138.850	0.8875	2.2662	138.239	0.8212
350	9.1015	151.432	0.9481	6.6820	151.351	0.9261	2.6599	150.896	0.8602
400	10.412	163.882	0.9814	7.6469	163.821	0.9594	3.0502	163.471	0.8938
450	11.721	176.319	1.0107	8.6098	176.271	0.9887	3.4385	175.997	0.9233
500	13.028	188.748	1.0368	9.5714	188.710	1.0149	3.8255	188.492	0.9496
540	14.073	198.690	1.0560	10.340	198.657	1.0341	4.1344	198.474	0.9688

Temp. °R	v ft³/lbm	h Btu/lbm	s Btu/lbm-R	v ft³/lbm	h Btu/lbm	s Btu/lbm-R	v ft³/lbm	h Btu/lbm	s Btu/lbm-R
	100 lbf/in.²			200 lbf/in.²			500 lbf/in.²		
200	0.6834	109.931	0.6585	0.2884	103.911	0.5875	0.1321	108.378	0.5608
250	0.9078	123.948	0.7212	0.4272	120.763	0.6631	0.1966	128.168	0.6335
300	1.1169	137.205	0.7696	0.5420	135.076	0.7153	0.2473	143.838	0.6819
350	1.3192	150.133	0.8094	0.6490	148.589	0.7570	0.2932	158.205	0.7202
400	1.5181	162.888	0.8435	0.7522	161.718	0.7921	0.3368	171.933	0.7526
450	1.7149	175.540	0.8733	0.8532	174.630	0.8225	0.3790	185.292	0.7807
500	1.9103	188.129	0.8998	0.9529	187.408	0.8494	0.4120	195.807	0.8010
540	2.0660	198.170	0.9192	1.0319	197.567	0.8690			

	1000 lbf/in.^{2}			2000 lbf/in.^{2}			3000 lbf/in.^{2}		
250	0.0384	78.126	0.4145	0.0286	70.290	0.3596	0.0261	69.719	0.3371
300	0.0828	115.224	0.5514	0.0398	97.820	0.4599	0.0321	93.216	0.4228
350	0.1150	135.789	0.6150	0.0552	122.614	0.5366	0.0403	116.066	0.4933
400	0.1417	152.487	0.6597	0.0699	142.869	0.5908	0.0493	136.883	0.5490
450	0.1659	167.637	0.6954	0.0833	160.406	0.6321	0.0582	155.522	0.5930
500	0.1887	181.969	0.7256	0.0958	176.411	0.6659	0.0667	172.551	0.6289
540	0.2063	193.069	0.7470	0.1053	188.526	0.6892	0.0732	185.361	0.6535

Table B.5
Critical Constants[a]

Substance	Formula	Molecular Weight	Temperature K	Temperature R	Pressure atm	Pressure lbf/in.^{2}	Volume, ft^3/lb-mole
Ammonia	NH_3	17.03	405.5	729.8	111.3	1636	1.16
Argon	Ar	39.944	151	272	48.0	705	1.20
Bromine	Br_2	159.832	584	1052	102	1500	2.17
Carbon dioxide	CO_2	44.01	304.2	547.5	72.9	1071	1.51
Carbon monoxide	CO	28.01	133	240	34.5	507	1.49
Chlorine	Cl_2	70.914	417	751	76.1	1120	1.99
Deuterium (Normal)	D_2	4.00	38.4	69.1	16.4	241	...
Helium	He	4.003	5.3	9.5	2.26	33.2	0.926
Helium^3	He	3.00	3.34	6.01	1.15	16.9	...
Hydrogen (Normal)	H_2	2.016	33.3	59.9	12.8	188.1	1.04
Krypton	Kr	83.7	209.4	376.9	54.3	798	1.48
Neon	Ne	20.183	44.5	80.1	26.9	395	0.668
Nitrogen	N_2	28.016	126.2	227.1	33.5	492	1.44
Nitrous oxide	N_2O	44.02	309.7	557.4	71.7	1054	1.54
Oxygen	O_2	32.00	154.8	278.6	50.1	736	1.25
Sulfur dioxide	SO_2	64.06	430.7	775.2	77.8	1143	1.95
Water	H_2O	18.016	647.4	1165.3	218.3	3204	0.90
Xenon	Xe	131.3	289.75	521.55	58.0	852	1.90
Benzene	C_6H_6	78.11	562	1012	48.6	714	4.17
n-Butane	C_4H_{10}	58.120	425.2	765.2	37.5	551	4.08
Carbon tetrachloride	CCl_4	153.84	556.4	1001.5	45.0	661	4.42
Chloroform	$CHCl_3$	119.39	536.6	965.8	54.0	794	3.85
Dichlorodifluoromethane	CCl_2F_2	120.92	384.7	692.4	39.6	582	3.49
Dichlorofluoromethane	$CHCl_2F$	102.93	451.7	813.0	51.0	749	3.16
Ethane	C_2H_6	30.068	305.5	549.8	48.2	708	2.37
Ethyl alcohol	C_2H_5OH	46.07	516.0	929.0	63.0	926	2.68
Ethylene	C_2H_4	28.052	282.4	508.3	50.5	742	1.99
n-Hexane	C_6H_{14}	86.172	507.9	914.2	29.9	439	5.89
Methane	CH_4	16.042	191.1	343.9	45.8	673	1.59
Methyl alcohol	CH_3OH	32.04	513.2	923.7	78.5	1154	1.89
Methyl chloride	CH_3Cl	50.49	416.3	749.3	65.9	968	2.29
Propane	C_3H_8	44.094	370.0	665.9	42.0	617	3.20
Propene	C_3H_6	42.078	365.0	656.9	45.6	670	2.90
Propyne	C_3H_4	40.062	401	722	52.8	776	...
Trichlorofluoromethane	CCl_3F	137.38	471.2	848.1	43.2	635	3.97

[a] K. A. Kobe and R. E. Lynn, Jr., *Chem. Rev.*, **52**: 117–236 (1953).

Table B.6

Zero-Pressure Properties of Gases

$(C_{po}, C_{vo}, \text{and } k \text{ are at } 80 \text{ F})$

Gas	Chemical Formula	Molecular Weight	R ft-lbf/ lbm R	C_{po} Btu/ lbm R	C_{vo} Btu/ lbm R	k
Air	$\cdots$	28.97	53.34	0.240	0.171	1.400
Argon	Ar	39.94	38.66	0.1253	0.0756	1.667
Carbon Dioxide	CO_2	44.01	35.10	0.203	0.158	1.285
Carbon Monoxide	CO	28.01	55.16	0.249	0.178	1.399
Helium	He	4.003	386.0	1.25	0.753	1.667
Hydrogen	H_2	2.016	766.4	3.43	2.44	1.404
Methane	CH_4	16.04	96.35	0.532	0.403	1.32
Nitrogen	N_2	28.016	55.15	0.248	0.177	1.400
Oxygen	O_2	32.000	48.28	0.219	0.157	1.395
Steam	H_2O	18.016	85.76	0.445	0.335	1.329

Table B.7

Constant-Pressure Specific Heats of Various Substances at Zero Pressure[a]

Gas or Vapor	Equation, $\bar{C}_{po}$ in Btu/lb mole-R T in degrees Rankine	Range R	Max. Error %
O_2	$\bar{C}_{po} = 11.515 - \frac{172}{\sqrt{T}} + \frac{1530}{T}$	540–5000	1.1
	$= 11.515 - \frac{172}{\sqrt{T}} + \frac{1530}{T} + \frac{0.05}{1000}(T - 4000)$	5000–9000	0.3
N_2	$\bar{C}_{po} = 9.47 - \frac{3.47 \times 10^3}{T} + \frac{1.16 \times 10^6}{T^2}$	540–9000	1.7
CO	$\bar{C}_{po} = 9.46 - \frac{3.29 \times 10^3}{T} + \frac{1.07 \times 10^6}{T^2}$	540–9000	1.1
H_2	$\bar{C}_{po} = 5.76 + \frac{0.578}{1000}T + \frac{20}{\sqrt{T}}$	540–4000	0.8
	$= 5.76 + \frac{0.578}{1000}T + \frac{20}{\sqrt{T}} - \frac{0.33}{1000}(T - 4000)$	4000–9000	1.4
H_2O	$\bar{C}_{po} = 19.86 - \frac{597}{\sqrt{T}} + \frac{7500}{T}$	540–5400	1.8
CO_2	$\bar{C}_{po} = 16.2 - \frac{6.53 \times 10^3}{T} + \frac{1.41 \times 10^6}{T^2}$	540–6300	0.8
CH_4	$\bar{C}_{po} = 4.52 + 0.00737T$	540–1500	1.2
C_2H_4	$\bar{C}_{po} = 4.23 + 0.01177T$	350–1100	1.5
C_2H_6	$\bar{C}_{po} = 4.01 + 0.01636T$	400–1100	1.5
C_3H_8	$\bar{C}_{po} = 2.258 + 0.0320T - 5.43 \times 10^{-6}T^2$	415–2700	1.8
C_4H_{10}	$\bar{C}_{po} = 4.36 + 0.0403T - 6.83 \times 10^{-6}T^2$	540–2700	1.7
C_8H_{18}	$\bar{C}_{po} = 7.92 + 0.0601T$	400–1100	est. 4
$C_{12}H_{26}$	$\bar{C}_{po} = 8.68 + 0.0889T$	400–1100	est. 4

[a]From *Bulletin No. 2,* Ga. School of Technology, by R. L. Sweigert and M. W. Beardsley, 1938, except C_3H_8 and C_4H_{10}, which are from H. M. Spencer, *J. of Am. Chem. Soc.*, **67**: 1859 (1945).

Table B.8
Thermodynamic Properties of Air at Low Pressure[a]

T R	h Btu/lbm	u Btu/lbm	$s°$ Btu/lbm R	T R	h Btu/lbm	u Btu/lbm	$s°$ Btu/lbm R	T R	h Btu/lbm	u Btu/lbm	$s°$ Btu/lbm R
200	47.67	33.96	0.36303	700	167.56	119.58	0.66321	1200	291.30	209.05	0.79628
220	52.46	37.38	0.38584	720	172.39	123.04	0.67002	1220	296.41	212.78	0.80050
240	57.25	40.80	0.40666	740	177.23	126.51	0.67665	1240	301.52	216.53	0.80466
260	62.03	44.21	0.42582	760	182.08	129.99	0.68312	1260	306.65	220.28	0.80876
280	66.82	47.63	0.44356	780	186.94	133.47	0.68942	1280	311.79	224.05	0.81280
300	71.61	51.04	0.46007	800	191.81	136.97	0.69558	1300	316.94	227.83	0.81680
320	76.40	54.46	0.47550	820	196.69	140.47	0.70160	1320	322.11	231.63	0.82075
340	81.18	57.87	0.49002	840	201.56	143.98	0.70747	1340	327.29	235.43	0.82464
360	85.97	61.29	0.50369	860	206.46	147.50	0.71323	1360	332.48	239.25	0.82848
380	90.75	64.70	0.51663	880	211.35	151.02	0.71886	1380	337.68	243.08	0.83229
400	95.53	68.11	0.52890	900	216.26	154.57	0.72438	1400	342.90	246.93	0.83604
420	100.32	71.52	0.54058	920	221.18	158.12	0.72979	1420	348.14	250.79	0.83975
440	105.11	74.93	0.55172	940	226.11	161.68	0.73509	1440	353.37	254.66	0.84341
460	109.90	78.36	0.56235	960	231.06	165.26	0.74030	1460	358.63	258.54	0.84704
480	114.69	81.77	0.57255	980	236.02	168.83	0.74540	1480	363.89	262.44	0.85062
500	119.48	85.20	0.58233	1000	240.98	172.43	0.75042	1500	369.17	266.34	0.85416
520	124.27	88.62	0.59173	1020	245.97	176.04	0.75536	1520	374.47	270.26	0.85767
540	129.06	92.04	0.60078	1040	250.95	179.66	0.76019	1540	379.77	274.20	0.86113
560	133.86	95.47	0.60950	1060	255.96	183.29	0.76496	1560	385.08	278.13	0.86456
580	138.66	98.90	0.61793	1080	260.97	186.93	0.76964	1580	390.40	282.09	0.86794
600	143.47	102.34	0.62607	1100	265.99	190.58	0.77426	1600	395.74	286.06	0.87130
620	148.28	105.78	0.63395	1120	271.03	194.25	0.77880	1620	401.09	290.04	0.87462
640	153.09	109.21	0.64159	1140	276.08	197.94	0.78326	1640	406.45	294.03	0.87791
660	157.92	112.67	0.64902	1160	281.14	201.63	0.78767	1660	411.82	298.02	0.88116
680	162.73	116.12	0.65621	1180	286.21	205.33	0.79201	1680	417.20	302.04	0.88439

Table B.8 (Continued)
Thermodynamic Properties of Air at Low Pressure

T R	h Btu/lbm	u Btu/lbm	$s°$ Btu/lbm R	T R	h Btu/lbm	u Btu/lbm	$s°$ Btu/lbm R	T R	h Btu/lbm	u Btu/lbm	$s°$ Btu/lbm R
1700	422.59	306.06	0.88758	1940	488.12	355.12	0.92362	2180	554.97	405.53	0.95611
1720	428.00	310.09	0.89074	1960	493.64	359.28	0.92645	2200	560.59	409.78	0.95868
1740	433.41	314.13	0.89387	1980	499.17	363.43	0.92926	2220	566.23	414.05	0.96123
1760	438.83	318.18	0.89697	2000	504.71	367.61	0.93205	2240	571.86	418.31	0.96376
1780	444.26	322.24	0.90003	2020	510.26	371.79	0.93481	2260	577.51	422.59	0.96626
1800	449.71	326.32	0.90308	2040	515.82	375.98	0.93756	2280	583.16	426.87	0.96876
1820	455.17	330.40	0.90609	2060	521.39	380.18	0.94026	2300	588.82	431.16	0.97123
1840	460.63	334.50	0.90908	2080	526.97	384.39	0.94296	2320	594.49	435.46	0.97369
1860	466.12	338.61	0.91203	2100	532.55	388.60	0.94564	2340	600.16	439.76	0.97611
1880	471.60	342.73	0.91497	2120	538.15	392.83	0.94829	2360	605.84	444.07	0.97853
1900	477.09	346.85	0.91788	2140	543.74	397.05	0.95092	2380	611.53	448.38	0.98092
1920	482.60	350.98	0.92076	2160	549.35	401.29	0.95352	2400	617.22	452.70	0.98331

[a]Abridged from Table 1 in *Gas Tables*, by Joseph H. Keenan and Joseph Kaye. Copyright 1948, by Joseph H. Keenan and Joseph Kaye. Published by John Wiley & Sons, Inc., New York.

Table B.9[a]

Enthalpy of Formation at 25 C, Ideal Gas Enthalpy, and Absolute Entropy at One Atmosphere Pressure

		Nitrogen, Diatomic (N_2) (Sept. 30, 1965)			Nitrogen, Monatomic (N) (Mar. 31, 1961)		
		$(\bar{h}_f^\circ)_{298} = 0$ cal/gm mole $= 0$ Btu/lb mole $M = 28.016$			$(\bar{h}_f^\circ)_{298} = 112{,}965$ cal/gm mole $= 203{,}337$ Btu/lb mole $M = 14.008$		
Temp °K	Temp °R	$(\bar{h}^\circ - \bar{h}^\circ_{298})$ cal/gm mole	$\bar{s}^\circ$: cal/gm mole-K Btu/lb mole-R	$(\bar{h}^\circ - \bar{h}^\circ_{537})$ Btu/lb mole	$(\bar{h}^\circ - \bar{h}^\circ_{298})$ cal/gm mole	$\bar{s}^\circ$: cal/gm mole-K Btu/lb mole-R	$(\bar{h}^\circ - \bar{h}^\circ_{537})$ Btu/lb mole
0	0	−2,072	0	−3,730	−1,481	0	−2,666
100	180	−1,379	38.170	−2,483	−984	31.187	−1,771
200	360	−683	42.992	−1,229	−488	34.631	−878
298	537	0	45.770	0	0	36.614	0
300	540	13	45.813	23	9	36.645	16
400	720	710	47.818	1,278	506	38.074	911
500	900	1,413	49.386	2,543	1,003	39.183	1,805
600	1,080	2,125	50.685	3,825	1,500	40.089	2,700
700	1,260	2,853	51.806	5,135	1,996	40.855	3,593
800	1,440	3,596	52.798	6,473	2,493	41.518	4,487
900	1,620	4,355	53.692	7,839	2,990	42.103	5,382
1000	1,800	5,129	54.507	9,232	3,487	42.627	6,277
1100	1,980	5,917	55.258	10,651	3,984	43.100	7,171
1200	2,160	6,718	55.955	12,092	4,481	43.532	8,066
1300	2,340	7,529	56.604	13,552	4,977	43.930	8,959
1400	2,520	8,350	57.212	15,030	5,474	44.298	9,853
1500	2,700	9,179	57.784	16,522	5,971	44.641	10,748
1600	2,880	10,015	58.324	18,027	6,468	44.962	11,642
1700	3,060	10,858	58.835	19,544	6,965	45.263	12,537
1800	3,240	11,707	59.320	21,073	7,461	45.547	13,430

Table B.9 (Continued)
Enthalpy of Formation at 25 C, Ideal Gas Enthalpy, and Absolute Entropy at One Atmosphere Pressure

		Nitrogen, Diatomic (N_2) (Sept. 30, 1965)			Nitrogen, Monatomic (N) (Mar. 31, 1961)		
		$(\bar{h}_f^\circ)_{298} = 0$ cal/gm mole $= 0$ Btu/lb mole $M = 28.016$			$(\bar{h}_f^\circ)_{298} = 112{,}965$ cal/gm mole $= 203{,}337$ Btu/lb mole $M = 14.008$		
Temp °K	Temp °R	$(\bar{h}^\circ - \bar{h}_{298}^\circ)$ cal/gm mole	$\bar{s}^\circ$: cal/gm mole-K Btu/lb mole-R	$(\bar{h}^\circ - \bar{h}_{537}^\circ)$ Btu/lb mole	$(\bar{h}^\circ - \bar{h}_{298}^\circ)$ cal/gm mole	$\bar{s}^\circ$: cal/gm mole-K Btu/lb mole-R	$(\bar{h}^\circ - \bar{h}_{537}^\circ)$ Btu/lb mole
1900	3,420	12,560	59.782	22,608	7,958	45.815	14,324
2000	3,600	13,418	60.222	24,152	8,455	46.070	15,219
2100	3,780	14,280	60.642	25,704	8,952	46.313	16,114
2200	3,960	15,146	61.045	27,263	9,449	46.544	17,008
2300	4,140	16,015	61.431	28,827	9,946	46.765	17,903
2400	4,320	16,886	61.802	30,395	10,444	46.977	18,799
2500	4,500	17,761	62.159	31,970	10,941	47.180	19,694
2600	4,680	18,638	62.503	33,548	11,439	47.375	20,590
2700	4,860	19,517	62.835	35,131	11,938	47.563	21,488
2800	5,040	20,398	63.155	36,716	12,437	47.745	22,387
2900	5,220	21,280	63.465	38,304	12,936	47.920	23,285
3000	5,400	22,165	63.765	39,897	13,437	48.090	24,187
3200	5,760	23,939	64.337	43,090	14,441	48.414	25,994
3400	6,120	25,719	64.877	46,294	15,451	48.720	27,812
3600	6,480	27,505	65.387	49,509	16,469	49.011	29,644
3800	6,840	29,295	65.871	52,731	17,495	49.288	31,491
4000	7,200	31,089	66.331	55,960	18,531	49.554	33,356
4200	7,560	32,888	66.770	59,198	19,580	49.810	35,244
4400	7,920	34,690	67.189	62,442	20,643	50.057	37,157
4600	8,280	36,496	67.591	65,693	21,721	50.297	39,098

4800	8,640	38,306	67.976	68,951	22,816	50.530	41,069
5000	9,000	40,119	68.346	72,214	23,928	50.757	43,070
5200	9,360	41,935	68.702	75,483	25,059	50.978	45,106
5400	9,720	43,755	69.045	78,759	26,210	51.195	47,178
5600	10,180	45,579	69.377	82,042	27,380	51.408	49,284
5800	10,540	47,406	69.698	85,331	28,570	51.617	51,426
6000	10,800	49,237	70.008	88,627	29,780	51.822	53,604

[a]The thermochemical data in Table B.9 are from the JANAF Thermochemical Tables, Thermal Research Laboratory, The Dow Chemical Company, Midland, Michigan. The date each table was issued is indicated.

Table B.9(Continued)

Enthalpy of Formation at 25 C, Ideal Gas Enthalpy, and Absolute Entropy at One Atmosphere Pressure

		Oxygen, Diatomic (O_2) (Sept. 30, 1965)			Oxygen, Monatomic(O) (June 30, 1962)		
		$(\bar{h}_f^\circ)_{298} = 0$ cal/gm mole $= 0$ Btu/lb mole $M = 32.00$			$(\bar{h}_f^\circ)_{298} = 59{,}559$ cal/gm mole $= 107{,}206$ Btu/lb mole $M = 16.00$		
Temp °K	Temp °R	$(\bar{h}^\circ - \bar{h}_{298}^\circ)$ cal/gm mole	$\bar{s}^\circ$: cal/gm mole-K Btu/lb mole-R	$(\bar{h}^\circ - \bar{h}_{537}^\circ)$ Btu/lb mole	$(\bar{h}^\circ - \bar{h}_{298}^\circ)$ cal/gm mole	$\bar{s}^\circ$: cal/gm mole-K Btu/lb mole-R	$(h^\circ - h_{537}^\circ)$ Btu/lb mole
0	0	−2,075	0	−3,735	−1,608	0	−2,894
100	180	−1,381	41.395	−2,486	−1,080	32.466	−1,944
200	360	−685	46.218	−1,233	−523	36.340	−941
298	537	0	49.004	0	0	38.468	0
300	540	13	49.047	23	10	38.501	18
400	720	724	51.091	1,303	528	39.991	950
500	900	1,455	52.722	2,619	1,038	41.131	1,868
600	1,080	2,210	54.098	3,978	1,544	42.054	2,779
700	1,260	2,988	55.297	5,378	2,048	42.831	3,686
800	1,440	3,786	56.361	6,815	2,550	43.501	4,590
900	1,620	4,600	57.320	8,280	3,052	44.092	5,494
1000	1,800	5,427	58.192	9,769	3,552	44.619	6,394
1100	1,980	6,266	58.991	11,279	4,051	45.095	7,292
1200	2,160	7,114	59.729	12,805	4,551	45.529	8,192
1300	2,340	7,971	60.415	14,348	5,049	45.928	9,088
1400	2,520	8,835	61.055	15,903	5,548	46.298	9,986
1500	2,700	9,706	61.656	17,471	6,046	46.642	10,883
1600	2,880	10,583	62.222	19,049	6,544	46.963	11,779
1700	3,060	11,465	62.757	20,637	7,042	47.265	12,676

1800	3,240	12,354	63.265	22,237	7,540	47.550	13,572
1900	3,420	13,249	63.749	23,848	8,038	47.819	14,468
2000	3,600	14,149	64.210	25,468	8,536	48.074	15,365
2100	3,780	15,054	64.652	27,097	9,034	48.317	16,261
2200	3,960	15,966	65.076	28,739	9,532	48.549	17,158
2300	4,140	16,882	65.483	30.388	10,029	48.770	18,052
2400	4,320	17,804	65.876	32.047	10,527	48.982	18,949
2500	4,500	18,732	66.254	33,718	11,026	49.185	19,847
2600	4,680	19,664	66.620	35,395	11,524	49.381	20,743
2700	4,860	20,602	66.974	37,084	12,023	49.569	21,641
2800	5,040	21,545	67.317	38,781	12,522	49.751	22,540
2900	5,220	22,493	67.650	40,487	13,022	49.926	23,440
3000	5,400	23,446	67.973	42,203	13,522	50.096	24,340
3200	5,760	25,365	68.592	45,657	14,524	50.419	26,143
3400	6,120	27,302	69.179	49,144	15,529	50.724	27,952
3600	6,480	29,254	69.737	52,657	16,537	51.012	29,767
3800	6,840	31,221	70.269	56,198	17,549	51.285	31,588
4000	7,200	33,201	70.776	59,762	18,565	51.546	33,417
4200	7,560	35,193	71.262	63,347	19,586	51.795	35,255
4400	7,920	37,196	71.728	66,953	20,611	52.033	37,100
4600	8,280	39,208	72.176	70,574	21,641	52.262	38,954
4800	8,640	41,229	72.606	74,212	22,676	52.482	40,817
5000	9,000	43,257	73.019	77,863	23,715	52.695	42,687
5200	9,360	45,292	73.418	81,526	24,760	52.899	44,568
5400	9,720	47,332	73.803	85,198	25,809	53.097	46,456
5600	10,180	49,377	74.175	88,879	26,863	53.289	48,353
5800	10,540	51,426	74.535	92,567	27,921	53.475	50,258
6000	10,800	53,479	74.883	96,262	28,984	53.655	25,171

Table B.9(Continued)
Enthalpy of Formation at 25 C, Ideal Gas Enthalpy, and Absolute Entropy at One Atmosphere Pressure

		Carbon Dioxide (CO_2) (Sept. 30, 1965) $(\bar{h}_f^\circ)_{298} = -94{,}054$ cal/gm mole $= -169{,}297$ Btu/lb mole; $M = 44.011$			Carbon Monoxide (CO) (Sept. 30, 1965) $(\bar{h}_f^\circ)_{298} = -26{,}417$ cal/gm mole $= -47{,}551$ Btu/lb mole; $M = 28.011$		
Temp °K	Temp °R	$(\bar{h}^\circ - \bar{h}_{298}^\circ)$ cal/gm mole	$\bar{s}^\circ$: cal/gm mole-K Btu/lb mole-R	$(\bar{h}^\circ - \bar{h}_{537}^\circ)$ Btu/lb mole	$(\bar{h}^\circ - \bar{h}_{298}^\circ)$ cal/gm mole	$\bar{s}^\circ$: cal/gm mole-K Btu/lb mole-R	$(\bar{h}^\circ - \bar{h}_{537}^\circ)$ Btu/lb mole
0	0	−2,238	0	−4,028	−2,072	0	−3,730
100	180	−1,543	42.758	−2,777	−1,379	39.613	−2,483
200	360	−816	47.769	−1,469	−683	44.435	−1,229
298	537	0	51.072	0	0	47.214	0
300	540	16	51.127	29	13	47.257	23
400	720	958	53.830	1,724	711	49.265	1,280
500	900	1,987	56.122	3,577	1,417	50.841	2,551
600	1,080	3,087	58.126	5,557	2,137	52.152	3,847
700	1,260	4,245	59.910	7,641	2,873	53.287	5,171
800	1,440	5,453	61.522	9,815	3,627	54.293	6,529
900	1,620	6,702	62.992	12,064	4,397	55.200	7,915
1000	1,800	7,984	64.344	14,371	5,183	56.028	9,329
1100	1,980	9,296	65.594	16,733	5,983	56.790	10,769
1200	2,160	10,632	66.756	19,138	6,794	57.496	12,229
1300	2,340	11,988	67.841	21,578	7,616	58.154	13,709
1400	2,520	13,362	68.859	24,052	8,446	58.769	15,203
1500	2,700	14,750	69.817	26,550	9,285	59.348	16,713
1600	2,880	16,152	70.722	29,074	10,130	59.893	18,234
1700	3,060	17,565	71.578	31,617	10,980	60.409	19,764
1800	3,240	18,987	72.391	34,177	11,836	60.898	21,305
1900	3,420	20,418	73.165	36,752	12,697	61.363	22,855

2000	3,600	21,857	73.903	39,343	13,561	61.807	24,410
2100	3,780	23,303	74.608	41,945	14,430	62.230	25,974
2200	3,960	24,755	75.284	44,559	15,301	62.635	27,542
2300	4,140	26,212	75.931	47,182	16,175	63.024	29,115
2400	4,320	27,674	76.554	49,813	17,052	63.397	30,694
2500	4,500	29,141	77.153	52,454	17,931	63.756	32,276
2600	4,680	30,613	77.730	55,103	18,813	64.102	33,863
2700	4,860	32,088	78.286	57,758	19,696	64.435	35,453
2800	5,040	33,567	78.824	60,421	20,582	64.757	37,048
2900	5,220	35,049	79.344	63,088	21,469	65.069	38,644
3000	5,400	36,535	79.848	65,763	22,357	65.370	40,243
3200	5,760	39,515	80.810	71,127	24,139	65.945	43,450
3400	6,120	42,507	81.717	76,513	25,927	66.487	46,669
3600	6,480	45,508	82.574	81,914	27,719	66.999	49,894
3800	6,840	48,518	83.388	87,332	29,516	67.485	53,129
4000	7,200	51,538	84.162	92,768	31,316	67.946	56,369
4200	7,560	54,566	84.901	98,219	33,121	68.387	59,618
4400	7,920	57,601	85.607	103,682	34,930	68.807	62,874
4600	8,280	60,644	86.284	109,159	36,741	69.210	66,134
4800	8,640	63,695	86.933	114,651	38,557	69.596	69,403
5000	9,000	66,753	87.557	120,155	40,375	69.967	72,675
5200	9,360	69,819	88.158	125,674	42,196	70.325	75,953
5400	9,720	72,893	88.738	131,207	44,021	70.669	79,238
5600	10,180	75,976	89.299	136,757	45,849	71.001	82,528
5800	10,540	79,068	89.841	142,322	47,679	71.322	85,822
6000	10,800	82,168	90.367	147,902	49,513	71.633	89,123

Table B.9(Continued)

Enthalpy of Formation at 25 C, Ideal Gas Enthalpy, and Absolute Entropy at One Atmosphere Pressure

		Water (H_2O) (Mar. 31, 1961)			Hydroxyl (OH) (Mar. 31, 1966)		
		$(\bar{h}_f^\circ)_{298} = -57{,}798$ cal/gm mole $= -104{,}036$ Btu/lb mole $M = 18.016$			$(\bar{h}_f^\circ)_{298} = 9432$ cal/gm mole $= 16{,}978$ Btu/lb mole $M = 17.008$		
Temp °K	Temp °R	$(\bar{h}^\circ - \bar{h}_{298}^\circ)$ cal/gm mole	$\bar{s}^\circ$: cal/gm mole-K Btu/lb mole-R	$(\bar{h}^\circ - \bar{h}_{537}^\circ)$ Btu/lb mole	$(\bar{h}^\circ - \bar{h}_{298}^\circ)$ cal/gm mole	$\bar{s}^\circ$: cal/gm mole-K Btu/lb mole-R	$(\bar{h}^\circ - \bar{h}_{537}^\circ)$ Btu/lb mole
0	0	−2,367	0	−4,261	−2,192	0	−3,946
100	180	−1,581	36.396	−2,846	−1,467	35.726	−2,641
200	360	−784	41.916	−1,411	−711	40.985	−1,280
298	537	0	45.106	0	0	43.880	0
300	540	15	45.155	27	13	43.925	23
400	720	825	47.484	1,485	725	45.974	1,305
500	900	1,654	49.334	2,977	1,432	47.551	2,578
600	1,080	2,509	50.891	4,516	2,137	48.837	3,847
700	1,260	3,390	52.249	6,102	2,845	49.927	5,121
800	1,440	4,300	53.464	7,740	3,556	50.877	6,401
900	1,620	5,240	54.570	9,432	4,275	51.724	7,695
1000	1,800	6,209	55.592	11,176	5,003	52.491	9,005
1100	1,980	7,210	56.545	12,978	5,742	53.195	10,336
1200	2,160	8,240	57.441	14,832	6,491	53.847	11,684
1300	2,340	9,298	58.288	16,736	7,252	54.455	13,054
1400	2,520	10,384	59.092	18,691	8,023	55.027	14,441
1500	2,700	11,495	59.859	20,691	8,805	55.566	15,849
1600	2,880	12,630	60.591	22,734	9,596	56.077	17,273
1700	3,060	13,787	61.293	24,817	10,397	56.563	18,715
1800	3,240	14,964	61.965	26,935	11,207	57.025	20,173
1900	3,420	16,160	62.612	29,088	12,024	57.467	21,643

2000	3,600	17,373	63.234	31,271	12,849	57.891	23,128
2100	3,780	18,602	63.834	33,484	13,681	58.296	24,626
2200	3,960	19,846	64.412	35,723	14,520	58.686	26,136
2300	4,140	21,103	64.971	37,985	15,364	59.062	27,655
2400	4,320	22,372	65.511	40,270	16,214	59.424	29,185
2500	4,500	23,653	66.034	42,575	17,069	59.773	30,724
2600	4,680	24,945	66.541	44,901	17,929	60.110	32,272
2700	4,860	26,246	67.032	47,243	18,794	60.436	33,829
2800	5,040	27,556	67.508	49,601	19,662	60.752	35,392
2900	5,220	28,875	67.971	51,975	20,535	61.058	36,963
3000	5,400	30,201	68.421	54,362	21,411	61.355	38,540
3200	5,760	32,876	69.284	59,177	23,174	61.924	41,713
3400	6,120	35,577	70.102	64,039	24,949	62.462	44,908
3600	6,480	38,300	70.881	68,940	26,735	62.973	48,123
3800	6,840	41,043	71.622	73,877	28,532	63.458	51,358
4000	7,200	43,805	72.331	78,849	30,338	63.922	54,608
4200	7,560	46,583	73.008	83,849	32,153	64.364	57,875
4400	7,920	49,375	73.658	88,875	33,976	64.788	61,157
4600	8,280	52,181	74.281	93,926	35,807	65.195	64,453
4800	8,640	55,000	74.881	99,000	37,644	65.586	67,759
5000	9,000	57,829	75.459	104,092	39,489	65.963	71,080
5200	9,360	60,669	76.016	109,204	41,340	66.326	74,412
5400	9,720	63,520	76.553	114,336	43,197	66.676	77,755
5600	10,180	66,381	77.074	119,486	45,060	67.015	81,108
5800	10,540	69,251	77.577	124,652	46,929	67.343	84,472
6000	10,800	72,131	78.065	129,836	48,803	67.661	87,845

Table B.9(Continued)

Enthalpy of Formation at 25 C, Ideal Gas Enthalpy, and Absolute Entropy at One Atmosphere Pressure

		Hydrogen, Diatomic (H_2) (Mar. 31, 1961)			Hydrogen, Monatomic (H) (Sept. 30, 1965)		
		$(\bar{h}_f^\circ)_{298} = 0$ cal/gm mole = 0 Btu/lb mole $M = 2.016$			$(\bar{h}_f^\circ) = 52{,}100$ cal/gm mole = 93,780 Btu/lb mole $M = 1.008$		
Temp °K	Temp °R	$(\bar{h}^\circ - \bar{h}_{298}^\circ)$ cal/gm mole	$\bar{s}^\circ$: cal/gm mole-K Btu/lb mole-R	$(\bar{h}^\circ - \bar{h}_{537}^\circ)$ Btu/lb mole	$(\bar{h}^\circ - \bar{h}_{298}^\circ)$ cal/gm mole	$\bar{s}^\circ$: cal/gm mole-K Btu/lb mole-R	$(\bar{h}^\circ - \bar{h}_{537}^\circ)$ Btu/lb mole
0	0	−2,024	0	−3,643	−1,481	0	−2,666
100	180	−1,265	24.387	−2,277	−984	21.965	−1,771
200	360	−662	28.520	−1,192	−488	25.408	−878
298	537	0	31.208	0	0	27.392	0
300	540	13	31.251	23	9	27.423	16
400	720	707	33.247	1.273	506	28.852	911
500	900	1,406	34.806	2,531	1,003	29.961	1,805
600	1,080	2,106	36.082	3,791	1,500	30.867	2,700
700	1,260	2,808	37.165	5,054	1,996	31.632	3,593
800	1,440	3,514	38.107	6,325	2,493	32.296	4,487
900	1,620	4,226	38.946	7,607	2,990	32.881	5,382
1000	1,800	4,944	39.702	8,899	3,487	33.404	6,277
1100	1,980	5,670	40.394	10,206	3,984	33.878	7,171
1200	2,160	6,404	41.033	11,527	4,481	34.310	8,066
1300	2,340	7,148	41.628	12,866	4,977	34.708	8,959
1400	2,520	7,902	42.187	14,224	5,474	35.076	9,853
1500	2,700	8,668	42.716	15,602	5,971	35.419	10,748
1600	2,880	9,446	43.217	17,003	6,468	35.739	11,642
1700	3,060	10,233	43.695	18,419	6,965	36.041	12,537
1800	3,240	11,030	44.150	19,854	7,461	36.325	13,430
1900	3,420	11,836	44.586	21,305	7,958	36.593	14,324

2000	3,600	12,651	45.004	22,772	8,455	36.848	15,219
2100	3,780	13,475	45.406	24,255	8,952	37.090	16,114
2200	3,960	14,307	45.793	25,753	9,449	37.322	17,008
2300	4,140	15,146	46.166	27,263	9,945	37.542	17,901
2400	4,320	15,993	46.527	28,787	10,442	37.754	18,796
2500	4,500	16,848	46.875	30,326	10,939	37.957	19,690
2600	4,680	17,708	47.213	31,874	11,436	38.152	20,585
2700	4,860	18,575	47.540	33,435	11,933	38.339	21,479
2800	5,040	19,448	47.857	35,006	12,430	38.520	22,374
2900	5,220	20,326	48.166	36,587	12,926	38.694	23,267
3000	5,400	21,210	48.465	38,178	13,423	38.862	24,161
3200	5,760	22,992	49.040	41,386	14,417	39.183	25,951
3400	6,120	24,794	49.586	44,629	15,410	39.484	27,738
3600	6,400	26,616	50.107	47,909	16,404	39.768	29,527
3800	6,840	28,457	50.605	51,223	17,398	40.037	31,316
4000	7,200	30,317	51.082	54,571	18,391	40.292	33,104
4200	7,560	32,194	51.540	57,949	19,385	40.534	34,893
4400	7,920	34,088	51.980	61,358	20,379	40.765	36,682
4600	8,280	35,999	52.405	64,798	21,372	40.986	38,470
4800	8,640	37,926	52.815	68,267	22,366	41.198	40,259
5000	9,000	39,868	53.211	71,762	23,359	41.400	42,046
5200	9,360	41,825	53.595	75,285	24,353	41.595	43,835
5400	9,720	43,797	53.967	78,835	25,347	41.783	45,625
5600	10,180	45,783	54.328	82,409	26,340	41.963	47,412
5800	10,540	47,783	54.679	86,009	27,334	42.138	49,201
6000	10,800	49,796	55.020	89,633	28,328	42.306	50,990

Table B.10

Logarithms to the Base *e* of the Equilibrium Constant *K*

For the reaction $\nu_A A + \nu_B B \rightleftharpoons \nu_C C + \nu_D D$ the equilibrium constant K is defined as

$$K = \frac{y_C^{\nu_C} y_D^{\nu_D}}{y_A^{\nu_A} y_B^{\nu_B}} \left(\frac{P}{P^\circ}\right)^{\nu_C+\nu_D-\nu_A-\nu_B}$$

Based on thermodynamic data given in the JANAF Thermochemical Tables, Thermal Research Laboratory, The Dow Chemical Company, Midland, Michigan.

Temperature °K	$H_2 \rightleftharpoons 2H$	$O_2 \rightleftharpoons 2O$	$N_2 \rightleftharpoons 2N$	$H_2O \rightleftharpoons H_2 + \frac{1}{2}O_2$	$H_2O \rightleftharpoons \frac{1}{2}H_2 + OH$	$CO_2 \rightleftharpoons CO + \frac{1}{2}O_2$	$\frac{1}{2}N_2 + \frac{1}{2}O_2 \rightleftharpoons NO$
298	−164.018	−186.988	−367.493	−92.214	−106.214	−103.768	−35.052
500	−92.840	−105.643	−213.385	−52.697	−60.287	−57.622	−20.295
1000	−39.816	−45.163	−99.140	−23.169	−26.040	−23.535	−9.388
1200	−30.887	−35.018	−80.024	−18.188	−20.289	−17.877	−7.569
1400	−24.476	−27.755	−66.342	−14.615	−16.105	−13.848	−6.270
1600	−19.650	−22.298	−56.068	−11.927	−13.072	−10.836	−5.294
1800	−15.879	−18.043	−48.064	−9.832	−10.663	−8.503	−4.536
2000	−12.853	−14.635	−41.658	−8.151	−8.734	−6.641	−3.931
2200	−10.366	−11.840	−36.404	−6.774	−7.154	−5.126	−3.433
2400	−8.289	−9.510	−32.024	−5.625	−5.838	−3.866	−3.019
2600	−6.530	−7.534	−28.317	−4.654	−4.725	−2.807	−2.671
2800	−5.015	−5.839	−25.130	−3.818	−3.769	−1.900	−2.372
3000	−3.698	−4.370	−22.372	−3.092	−2.943	−1.117	−2.114
3200	−2.547	−3.085	−19.950	−2.457	−2.218	−0.435	−1.888
3400	−1.529	−1.948	−17.813	−1.897	−1.582	0.163	−1.690
3600	−0.622	−0.939	−15.911	−1.398	−1.014	0.695	−1.513
3800	0.189	−0.032	−14.212	−0.951	−0.507	1.170	−1.356

4000	0.921	0.783	−12.673	−0.548	−0.050	1.593	−1.216
4500	2.473	2.500	−9.427	0.306	0.914	2.484	−0.921
5000	3.712	3.882	−6.820	0.990	1.683	3.191	−0.686
5500	4.730	5.010	−4.679	1.554	2.312	3.765	−0.497
6000	5.577	5.950	−2.878	2.026	2.837	4.239	−0.341

Table B.11
Atomic Weights

For the sake of completeness all known elements are included in the list. Several of those more recently discovered are represented only by the unstable isotopes. The values in parenthesis in the atomic weight column is, in each case, the mass number of the most stable isotope.[a]

Name	Symbol	At. No.	International Atomic Weight 1966	International Atomic Weight 1959	Valence
Actinium	Ac	89	—	(227)	—
Aluminum	Al	13	26.9815	26.98	3
Americium	Am	95	—	(243)	3, 4, 5, 6
Antimony, stibium	Sb	51	121.75	121.76	3, 5
Argon	Ar	18	39.948	39.944	0
Arsenic	As	33	74.9216	74.92	3, 5
Astatine	At	85	—	(210)	1, 3, 5, 7
Barium	Ba	56	137.34	137.36	2
Berkelium	Bk	97	—	(247)	3, 4
Beryllium	Be	4	9.0122	9.013	2
Bismuth	Bi	83	208.980	208.99	3, 5
Boron	B	5	10.811	10.82	3
Bromine	Br	35	79.904$^{(1)}$	79.916	1, 3, 5, 7
Cadmium	Cd	48	112.40	112.41	2
Calcium	Ca	20	40.08	40.08	2
Californium	Cf	98	—	(251)	—
Carbon	C	6	12.01115	12.011	2, 4
Cerium	Ce	58	140.12	140.13	3, 4
Cesium	Cs	55	132.905	132.91	1
Chlorine	Cl	17	35.453	35.457	1, 3, 5, 7
Chromium	Cr	24	51.996	52.01	2, 3, 6
Cobalt	Co	27	58.9332	58.94	2, 3
Columbium, see *Niobium*					
Copper	Cu	29	63.546$^{(2)}$	63.54	1, 2
Curium	Cm	96	—	(247)	3

Name	Symbol	At. No.	International Atomic Weight 1966	International Atomic Weight 1959	Valence
Neodymium	Nd	60	144.24	144.27	3
Neon	Ne	10	20.183	20.183	0
Neptunium	Np	93	—	(237)	4, 5, 6
Nickel	Ni	28	58.71	58.71	2, 3
Niobium (columbium)	Nb	41	92.906	92.91	3, 5
Nitrogen	N	7	14.0067	14.008	3, 5
Nobelium	No	102	—	(254)	—
Osmium	Os	76	190.2	190.2	2, 3, 4, 8
Oxygen	O	8	15.9994	16.000	2
Palladium	Pd	46	106.4	106.4	2, 4, 6
Phosphorus	P	15	30.9738	30.975	3, 5
Platinum	Pt	78	195.09	195.09	2, 4
Plutonium	Pu	94	—	(244)	3, 4, 5, 6
Polonium	Po	84	—	(209)	—
Potassium, kalium	K	19	39.102	39.100	1
Praseodymium	Pr	59	140.907	140.92	3
Promethium	Pm	61	—	(145)	3
Protactinium	Pa	91	—	(231)	—
Radium	Ra	88	—	(226)	2
Radon	Rn	86	—	(222)	0
Rhenium	Re	75	186.2	186.22	—
Rhodium	Rh	45	102.905	102.91	3
Rubidium	Rb	37	85.47	85.48	1
Ruthenium	Ru	44	101.07	101.1	3, 4, 6, 8
Samarium	Sm	62	150.35	150.35	2, 3

Dysprosium	Dy	66	162.50	162.51	3
Einsteinium	Es	99	–	(254)	–
Erbium	Er	68	167.26	167.27	3
Europium	Eu	63	151.96	152.0	2, 3
Fermium	Fm	100	–	(257)	–
Fluorine	F	9	18.9984	19.00	1
Francium	Fr	87	–	(223)	1
Gadolinium	Gd	64	157.25	157.26	3
Gallium	Ga	31	69.72	69.72	2, 3
Germanium	Ge	32	72.59	72.60	4
Gold, aurum	Au	79	196.967	197.0	1, 3
Hafnium	Hf	72	178.49	178.50	4
Helium	He	2	4.0026	4.003	0
Holmium	Ho	67	164.930	164.94	3
Hydrogen	H	1	1.00797	1.0080	1
Indium	In	49	114.82	114.82	3
Iodine	I	53	126.9044	126.91	1, 3, 5, 7
Iridium	Ir	77	192.2	192.2	3, 4
Iron, ferrum	Fe	26	55.847	55.85	2, 3
Krypton	Kr	36	83.80	83.80	0
Lanthanum	La	57	138.91	138.92	3
Lawrencium	Lr	103	(257)		
Lead, plumbum	Pb	82	207.19	207.21	2, 4
Lithium	Li	3	6.939	6.940	1
Lutetium	Lu	71	174.97	174.99	3
Magnesium	Mg	12	24.312	24.32	2
Manganese	Mn	25	54.9380	54.94	2, 3, 4, 6, 7
Mendelevium	Md	101	–	(256)	–
Mercury, hydrargyrum	Hg	80	200.59	200.61	1, 2
Molybdenum	Mo	42	95.94	95.95	3, 4, 6
Scandium	Sc	21	44.956	44.96	3
Selenium	Se	34	78.96	78.96	2, 4, 6
Silicon	Si	14	28.086	28.09	4
Silver, argentum	Ag	47	107.868$^{(3)}$	107.873	1
Sodium, natrium	Na	11	22.9898	22.991	1
Strontium	Sr	38	87.62	87.63	2
Sulfur	S	16	32.064	32.066[b]	2, 4, 6
Tantalum	Ta	73	180.948	180.95	5
Technetium	Tc	43	–	(97)	6, 7
Tellurium	Te	52	127.60	127.61	2, 4, 6
Terbium	Tb	65	158.924	158.93	3
Thallium	Tl	81	204.37	204.39	1, 3
Thorium	Th	90	232.038	(232)	4
Thulium	Tm	69	168.934	168.94	3
Tin, stannum	Sn	50	118.69	118.70	2, 4
Titanium	Ti	22	47.90	47.90	3, 4
Tungsten (wolfram)	W	74	183.85	183.86	6
Uranium	U	92	238.03	238.07	4, 6
Vanadium	V	23	50.942	50.95	3, 5
Xenon	Xe	54	131.30	131.30	0
Ytterbium	Yb	70	173.04	173.04	2, 3
Yttrium	Y	39	88.905	88.91	3
Zinc	Zn	30	65.37	65.38	2
Zirconium	Zr	40	91.22	91.22	4

[a]The 1959 atomic weights are based on O = 16.000 whereas those of 1966 are based on the isotope C^{12}.
[b]Because of natural variations in the relative abundances of the isotopes of sulfur the atomic weight of this element has a range of ±0.003.

1. ±0.002
2. ±0.001
3. ±0.001

Reprinted with permission from the *Handbook of Chemistry and Physics*, 51st ed., R. C. Weast, editor, The Chemical Rubber Co., Cleveland, Ohio, 1970.

Table B.12
Metric System Constants and Conversion Factors

Universal gas Constant $\bar{R} = 1.98726$ cal/gm mole-K
$= 0.082055$ lit-atm/gm mole-K
Avogadro's number $N_0 = 6.0232 \times 10^{23}$ 1/gm mole
(Chemical $O_2 = 16.00$ scale)
Boltzmann's constant $k = \bar{R}/N_0 = 1.3804 \times 10^{-16}$ erg/K
Planck's constant $h = 6.62517 \times 10^{-27}$ erg-sec
Speed of light $c = 2.998 \times 10^{10}$ cm/sec
$hc/k = 1.4388$ cm-K
Mass per atomic weight $m = 1.66 \times 10^{-24}$ gm
Electron mass $m_e = 9.1086 \times 10^{-28}$ gm
1 Angstrom (Å) = 10^{-8} cm
1 dyne = 1 gm-cm/sec^2
1 liter = 1000.028 cm^3
1 erg = 1 dyne-cm
1 electron volt (ev) = 1.6021×10^{-12} erg
1 joule = 10^7 ergs
1 calorie (cal) = 4.184 joules
1 atmosphere (atm) = 1.01325×10^6 dynes/cm^2 = 760 mm Hg
$T(°K) = T(°C) + 273.15$

2.54 cm = 1 inch
453.59 gm = 1 lbm
252 cal = 1 Btu
1 atm = 14.696 lbf/in.^{2}
1 °K = 1.8 °R
1 cal/gm mole-K = 1 Btu/lb mole-R

Table B.13
Series and Integrals

$$e^x = 1+\frac{x}{1!}+\frac{x^2}{2!}+\frac{x^3}{3!}+\cdots$$

$$\ln(1+x) = x-\frac{x^2}{2}+\frac{x^3}{3}-\frac{x^4}{4}+\cdots$$

$$(x+y)^n = x^n + nx^{n-1}y + \frac{n(n-1)}{2!}x^{n-2}y^2+\cdots \quad (y^2 < x^2)$$

n	$\int_0^\infty x^n e^{-ax^2}\,dx$	n	$\int_0^\infty x^n e^{-ax^2}\,dx$
0	$\frac{1}{2}\left(\frac{\pi}{a}\right)^{1/2}$	3	$\frac{1}{2a^2}$
1	$\frac{1}{2a}$	4	$\frac{3}{8}\left(\frac{\pi}{a^5}\right)^{1/2}$
2	$\frac{1}{4}\left(\frac{\pi}{a^3}\right)^{1/2}$	5	$\frac{1}{a^3}$

$$\int_{-\infty}^{+\infty} x^n e^{-ax^2}\,dx = 0 \qquad (n = 1, 3, 5, \ldots)$$

$$= 2\int_0^\infty x^n e^{-ax^2}\,dx \quad (n = 0, 2, 4, \ldots)$$

$$\int_0^\infty \frac{x^3}{e^x-1}\,dx = \frac{\pi^4}{15}$$

$$\int_0^\infty x^2 \ln(1-e^{-x})\,dx = -\frac{\pi^4}{45}$$

Error function: $\operatorname{erf}(x) = \frac{2}{\sqrt{\pi}}\int_0^x e^{-x^2}\,dx$

x	erf (x)	x	erf (x)	x	erf (x)
0	0	0.7	0.6778	1.4	0.9523
0.1	0.1125	0.8	0.7421	1.5	0.9661
0.2	0.2227	0.9	0.7969	1.6	0.9764
0.3	0.3286	1.0	0.8427	1.7	0.9838
0.4	0.4284	1.1	0.8802	1.8	0.9891
0.5	0.5205	1.2	0.9103	1.9	0.9928
0.6	0.6039	1.3	0.9340	2.0	0.9953

Table B.14
Molecular Constants
Table B.14.1
Constants for Diatomic Molecules

Date	Substance	Ground Level	r_ε, Å	ω_ε, cm^{-1}	D_0, ev
12/31/61	Br_2	$^1\Sigma_g^+$	2.284	323.2	1.971
9/30/65	Cl_2	$^1\Sigma_g^+$	1.986	561.1	2.473
9/30/65	F_2	$^1\Sigma$	1.409	923.1	1.592
3/31/61	H_2	$^1\Sigma_g^+$	0.7417	4405.3	4.477
9/30/61	I_2	$^1\Sigma_g^+$	2.667	214.5	1.544
9/30/65	N_2	$^1\Sigma_g^+$	1.0976	2357.6	9.757
9/30/65	O_2	$^3\Sigma_g^-$	1.2074	1580.2	5.115
9/30/65	CO	$^1\Sigma^+$	1.1281	2169.5	11.089
9/30/65	HBr	$^1\Sigma^+$	1.414	2649.2	3.164
9/30/64	HCl	$^1\Sigma^+$	1.2746	2989.6	4.430
12/31/68	HF	$^1\Sigma^+$	0.9168	4138.3	5.858
9/30/61	HI	$^1\Sigma^+$	1.604	2309.1	3.054
6/30/63	NO	$^2\Pi$[a]	1.1508	1903.6	6.487
3/31/66	OH	$^2\Pi$[b]	0.9706	3735.2	4.393

[a]2 states at 121.1 cm^{-1} above the 2 ground states.
[b]2 states at 139.7 cm^{-1} above the 2 ground states.

Table B.14.2
Constants for Polyatomic Molecules

Date	Substance	Bond Distance, Å	Bond Angle, deg	ω_ε, cm^{-1}
9/30/65	CO_2	C—O: 1.160	O—C—O: 180	1342.9
				667.3(2)
				2349.3
6/30/61	CS_2	C—S: 1.553	S—C—S: 180	658
				396.8(2)
				1532.5
12/31/64	N_2O	N—N: 1.1282	N—N—O: 180	1276.5
		N—O: 1.1842		589.2(2)
				2223.7
3/31/61	H_2O	O—H: 0.9584	H—O—H: 104.45	3657.1
				1594.6
				3755.8
12/31/65	H_2S	S—H: 1.328	H—S—H: 92.3	2614.6
				1182.7
				2627.5
12/31/65	NH_2	N—H: 1.024	H—N—H: 103	3173
				1499
				3220
9/30/64	NO_2	N—O: 1.197	O—N—O: 134.25	1357.8
				756.8
				1665.5
3/31/61	CH_4	C—H: 1.091	H—C—H: 109.47	2916.5
				1534 (2)
				3018.7(3)
				1306 (3)

From *Janaf Thermochemical Tables*, The Dow Chemical Co., Midland, Mich. (Constants represent the naturally occurring isotopic mixture.)

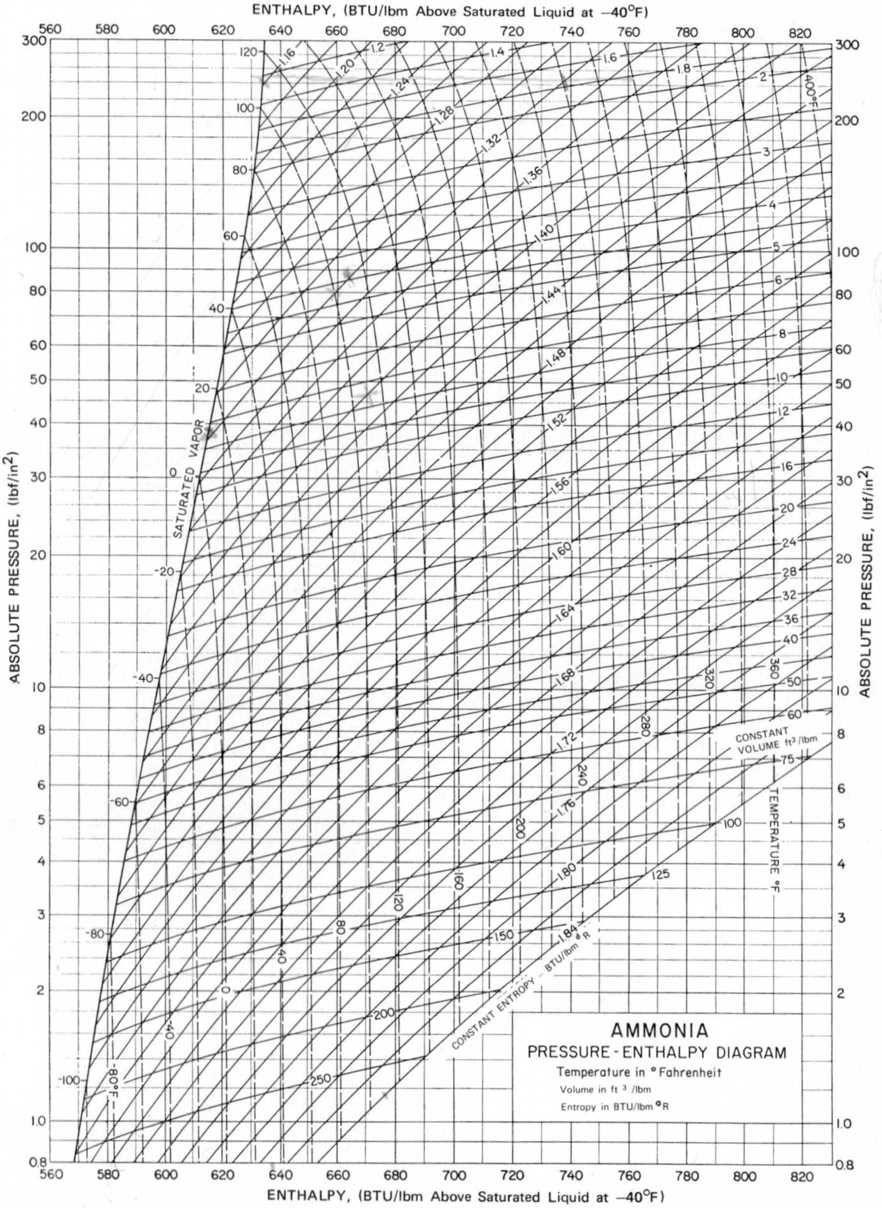

Fig. B.2 Pressure-enthalpy diagram for ammonia (Reprinted by permission from ASHRAE Handbook of Fundamentals 1967).

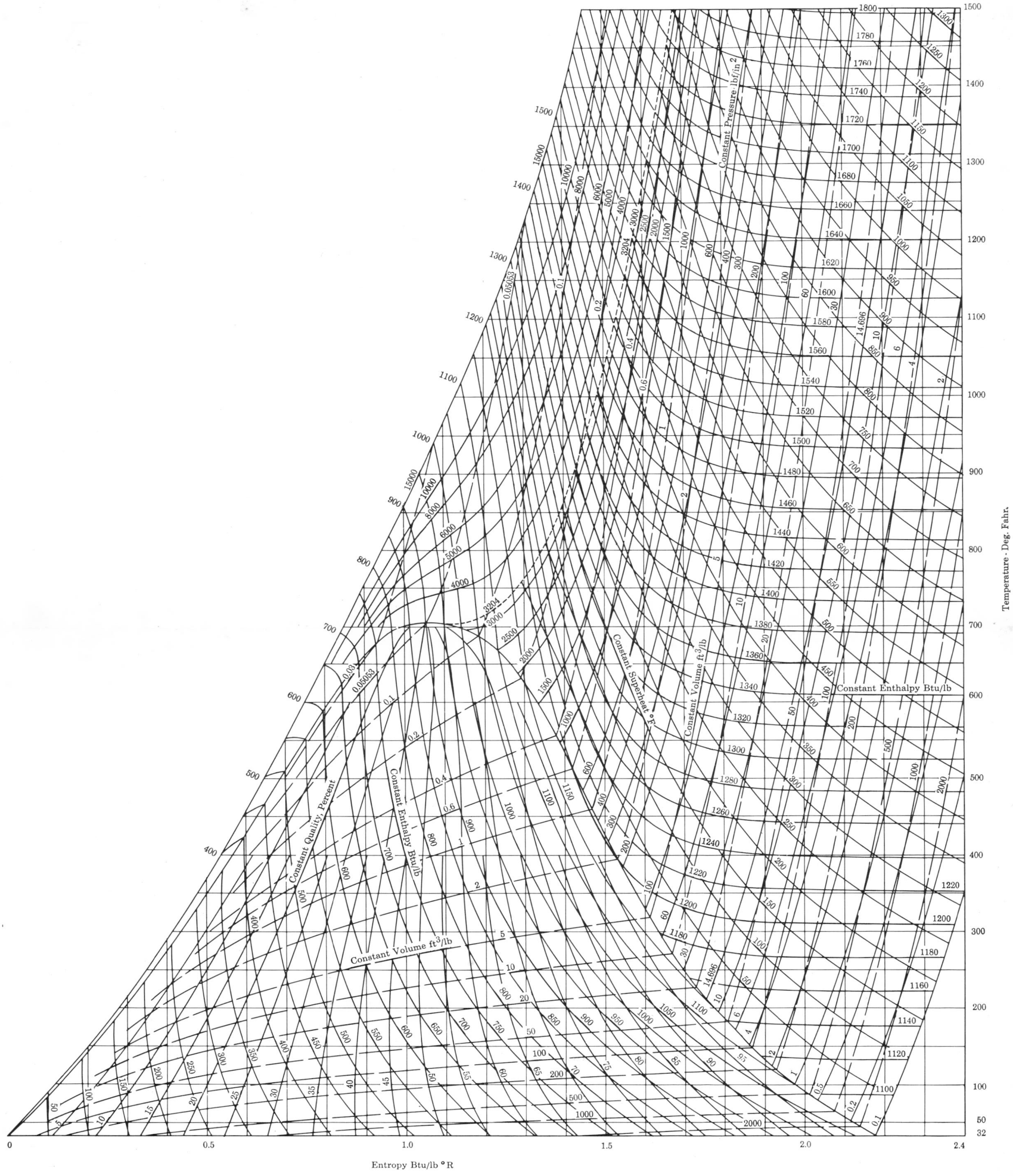

Fig. B.1 Temperature–entropy chart (from *Steam Tables* by Keenan, Keyes, Hill, & Moore Copyright © 1969, John Wiley & Sons, Inc).

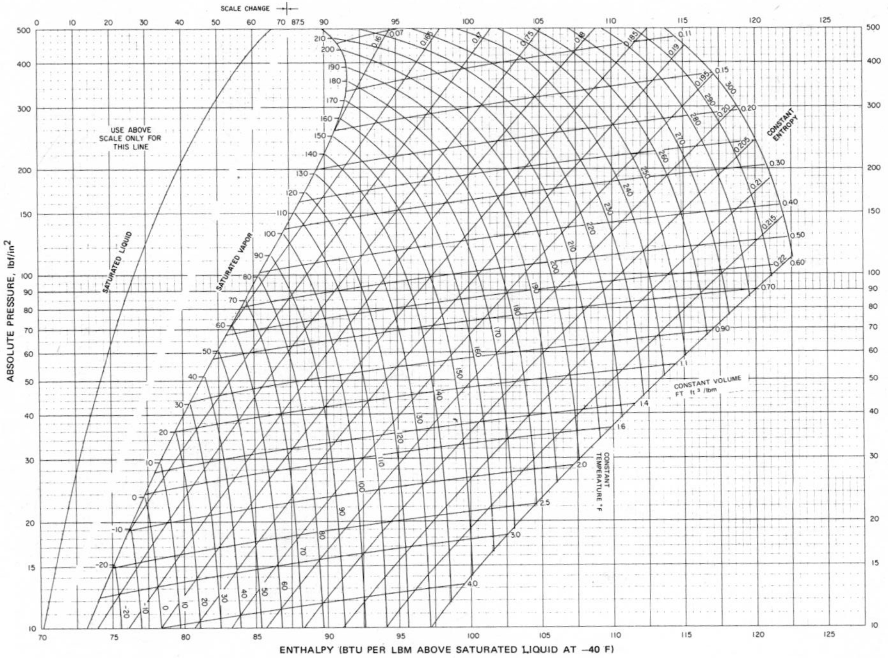

Fig. B.3 Pressure-enthalpy diagram for Freon-12 (Copyright 1967 by E. I. duPont de Nemours and Co., Reprinted by permission).

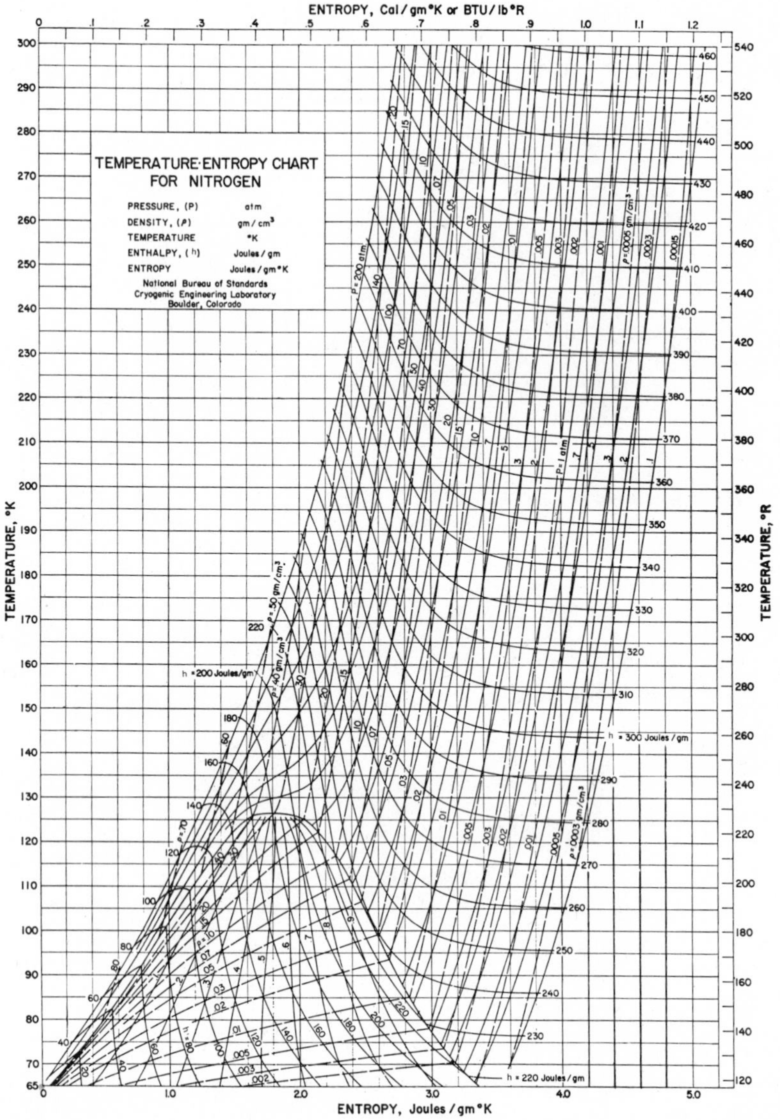

Fig. B.4 Temperature-entropy diagram for nitrogen (Reprinted by permission from ASHRAE Handbook of Fundamentals 1967).

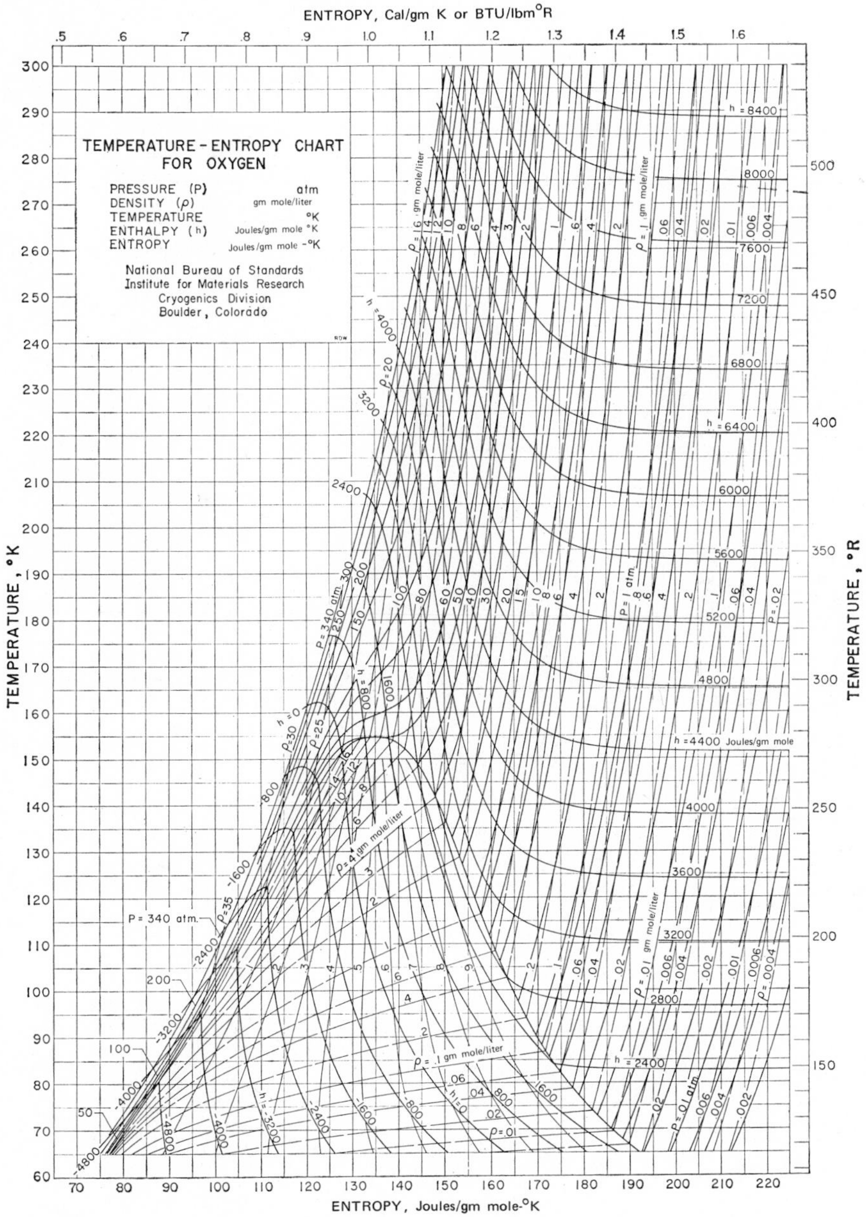

Fig. B.5 Temperature-entropy diagram for oxygen (Reprinted by permission from ASHRAE Handbook of Fundamentals 1967).

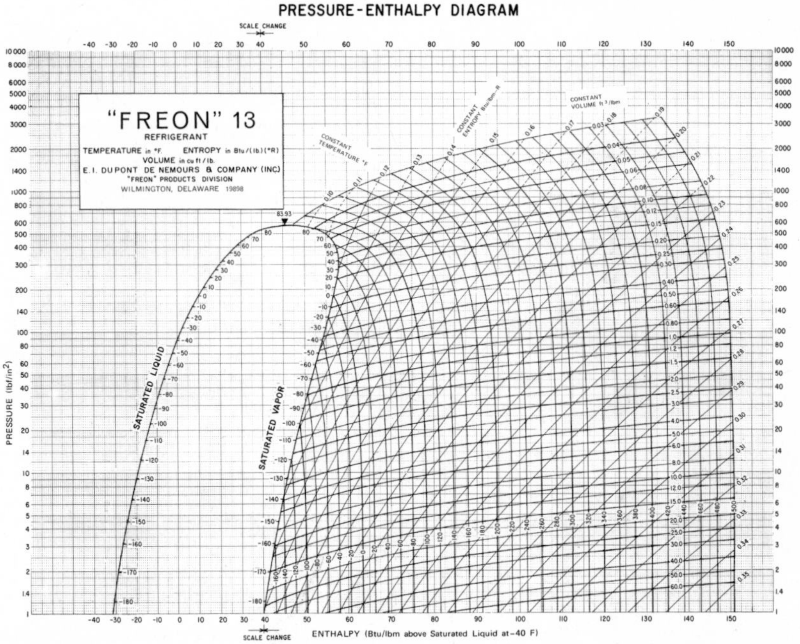

Fig. B.6 Pressure-enthalpy diagram for Freon-13 (Copyright 1960 by E. I. duPont de Nemours and Co., Reprinted by permission).

ASHRAE PSYCHROMETRIC CHART NO. 1

NORMAL TEMPERATURE

BAROMETRIC PRESSURE 29.921 INCHES OF MERCURY

COPYRIGHT 1963

AMERICAN SOCIETY OF HEATING, REFRIGERATING AND AIR-CONDITIONING ENGINEERS, INC.

ENTHALPY OF WATER-BTU PER LB DRY AIR

$$\frac{\text{SENSIBLE HEAT}}{\text{TOTAL HEAT}} = \frac{\Delta H_S}{\Delta H_T}$$

$$\frac{\text{ENTHALPY}}{\text{HUMIDITY RATIO}} = \frac{\Delta h}{\Delta W}$$

HUMIDITY RATIO (W) POUNDS MOISTURE PER POUND DRY AIR

WET BULB OR WATER TEMPERATURE F

SATURATION

ENTHALPY DEVIATION BTU PER POUND OF DRY AIR

ENTHALPY (h) BTU PER POUND OF DRY AIR

SATURATION TEMPERATURE F

VOLUME CU. FT. PER LB. DRY AIR

85 F WET BULB TEMPERATURE

10% RELATIVE HUMIDITY

DRY BULB TEMPERATURE — F

HUMIDITY RATIO (W) — POUNDS MOISTURE PER POUND DRY AIR

ENTHALPY (h) BTU PER POUND OF DRY AIR

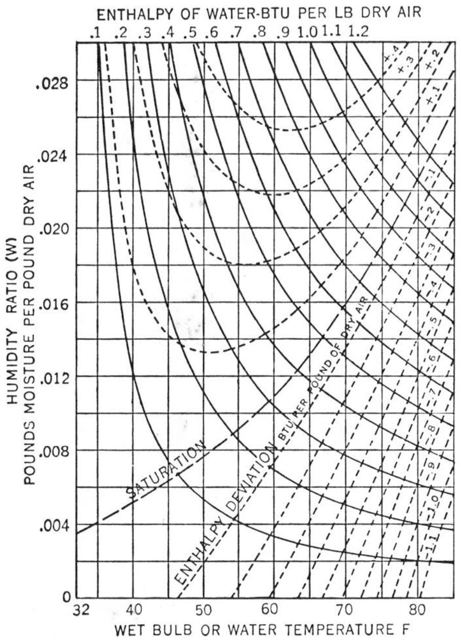

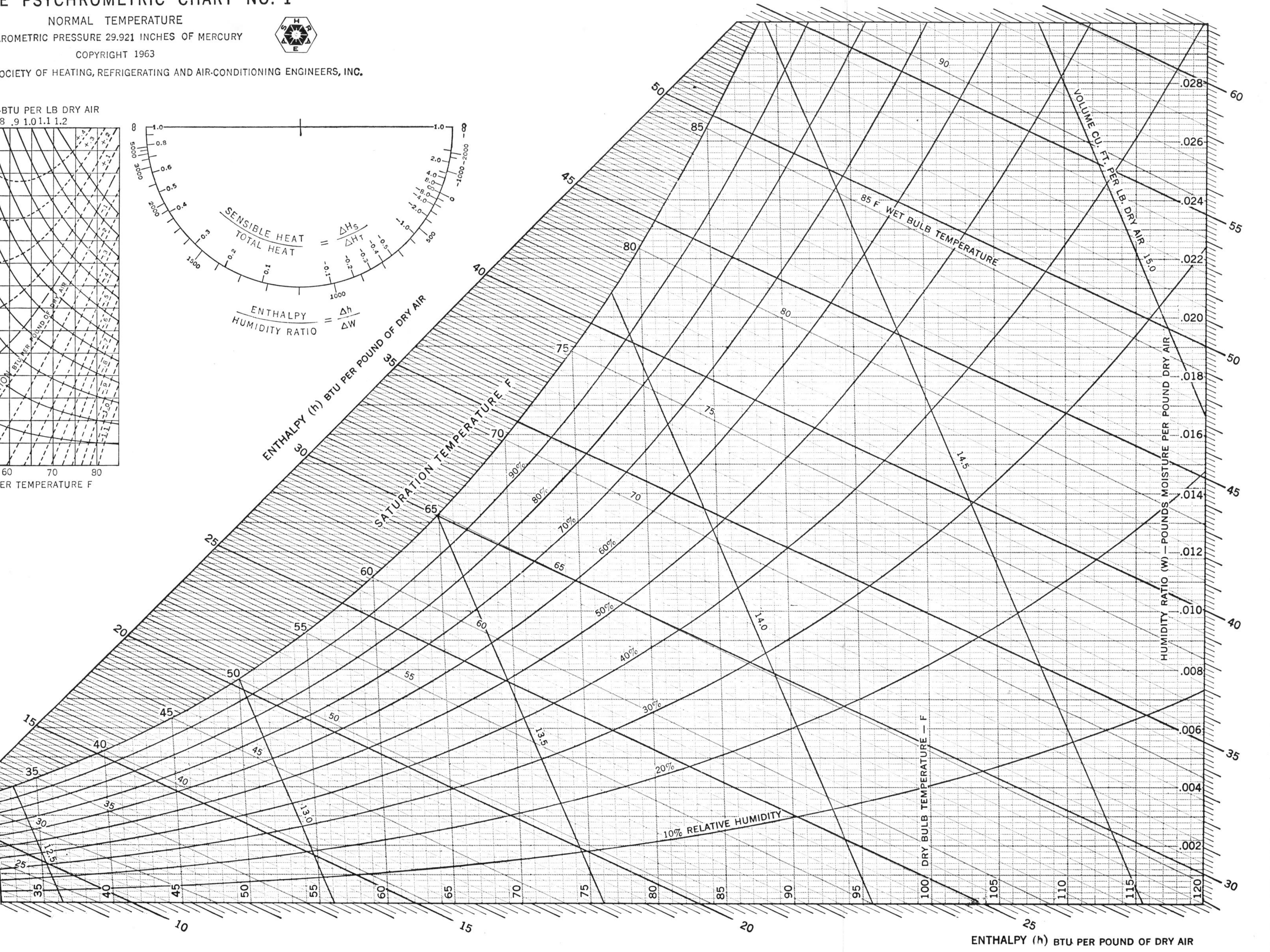

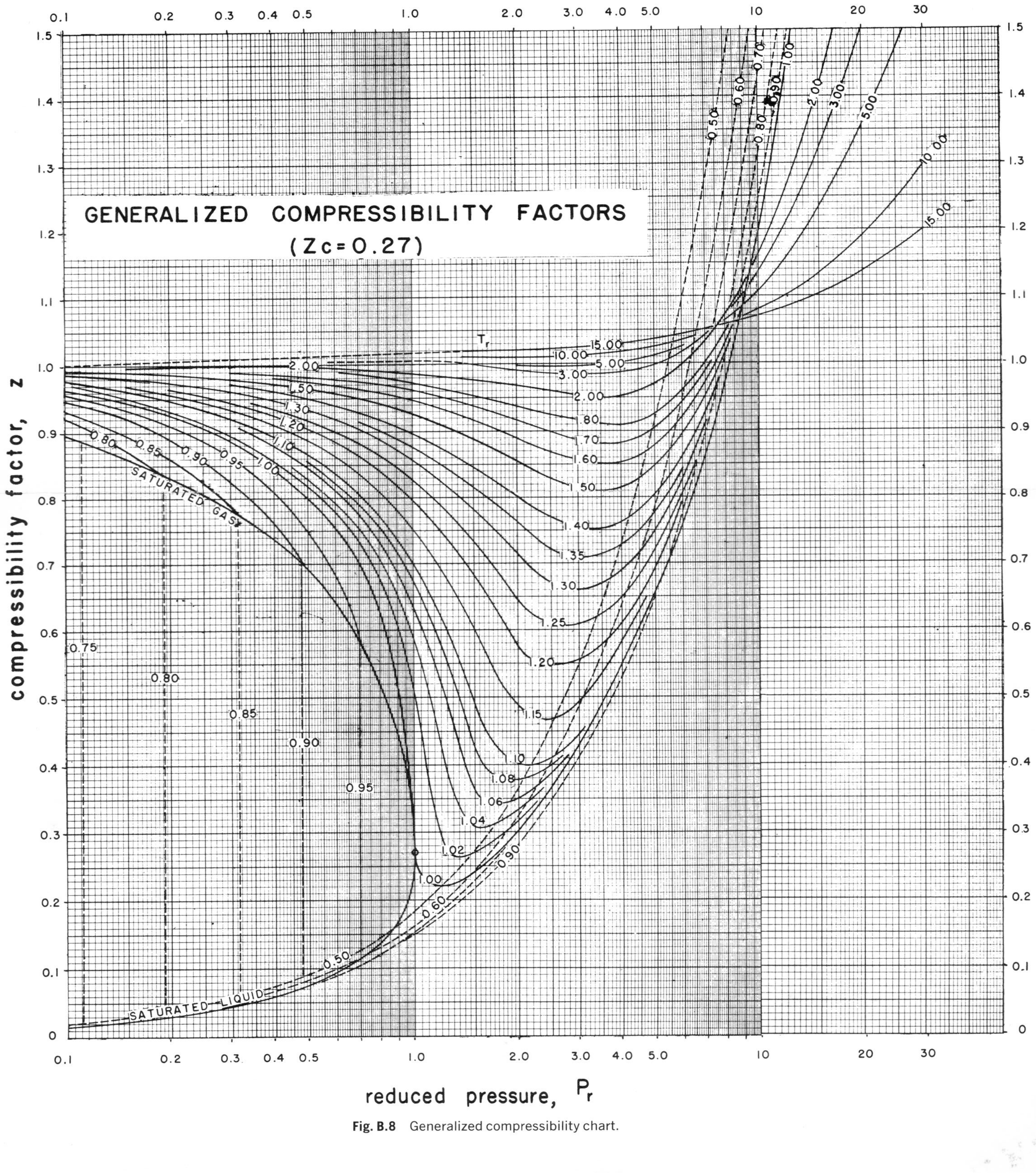

Fig. B.8 Generalized compressibility chart.

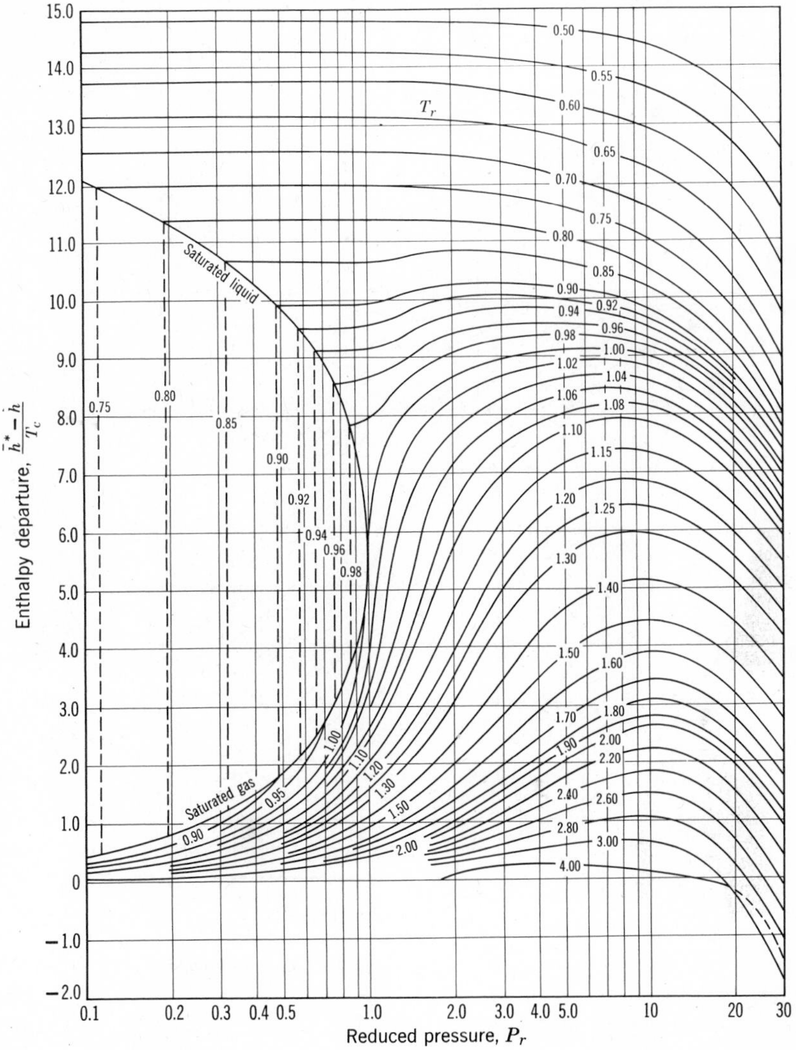

Fig. B.9 Generalized enthalpy correction chart (Based on Fig. B.8).

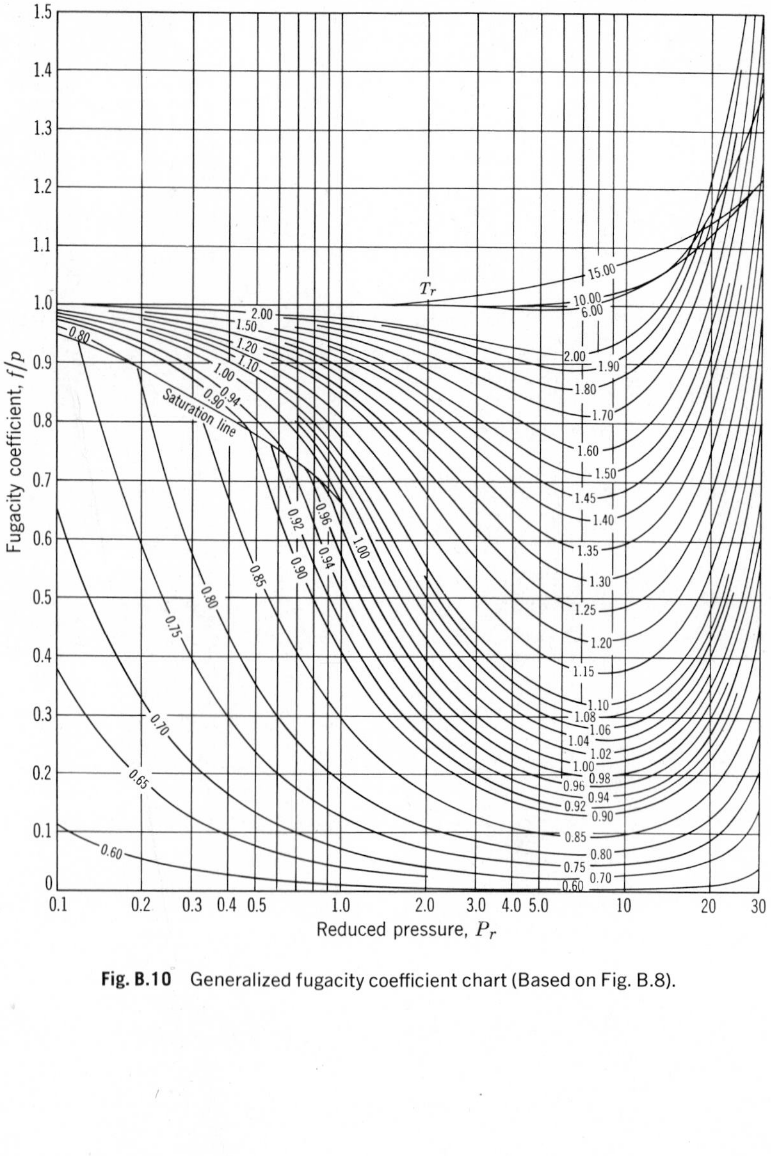

Fig. B.10 Generalized fugacity coefficient chart (Based on Fig. B.8).

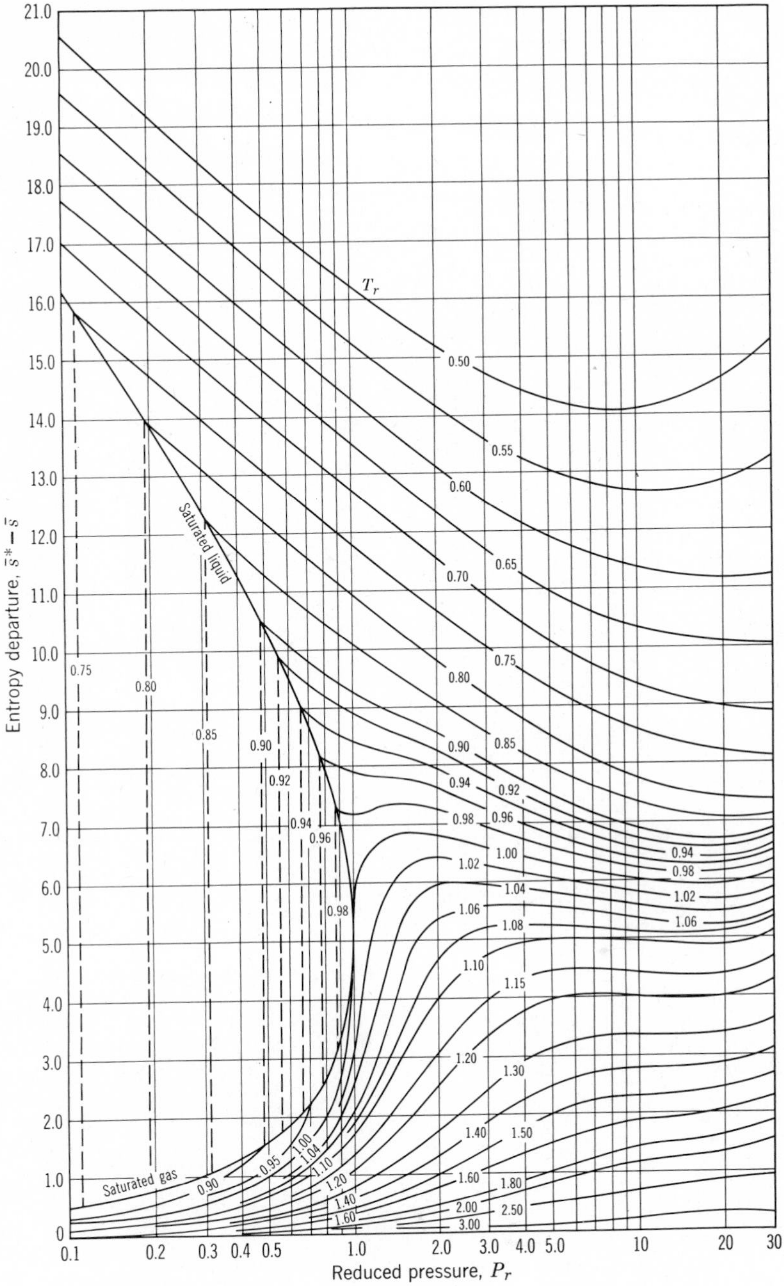

Fig. B.11 Generalized entropy correction chart (Based on Fig. B.8).

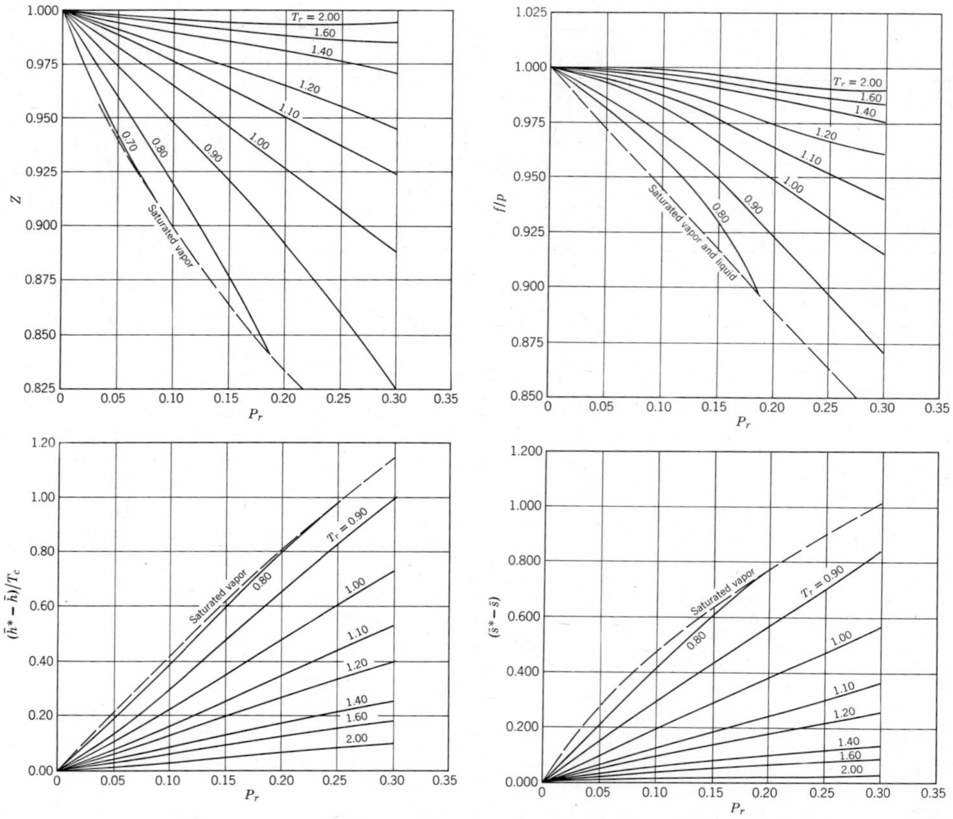

Fig. B.12 Generalized charts at low pressure.

Some Selected References

H. B. Callen, *Thermodynamics*, John Wiley and Sons, New York, 1960.

N. Davidson, *Statistical Mechanics*, McGraw-Hill Book Co., New York, 1962.

G. N. Hatsopoulos and J. H. Keenan, *Principles of General Thermodynamics*, John Wiley and Sons, New York, 1965.

O. A. Hougen, K. M. Watson, and R. A. Ragatz, *Chemical Process Principles, Part Two: Thermodynamics*, Second Edition, John Wiley and Sons, New York, 1959.

J. H. Keenan, *Thermodynamics*, John Wiley and Sons, New York, 1941.

J. Kestin, *A Course in Thermodynamics*, Blaisdell Publishing Co., Waltham, Mass., 1966.

J. E. Mayer and M. G. Mayer, *Statistical Mechanics*, John Wiley and Sons, New York, 1940.

E. F. Obsert, *Concepts of Thermodynamics*, McGraw-Hill Book Co., New York, 1960.

H. Reiss, *Methods of Thermodynamics*, Blaisdell Publishing Co., Waltham, Mass., 1965.

W. C. Reynolds, *Thermodynamics*, Second Edition, McGraw-Hill Book Co., New York, 1968.

R. E. Sonntag and G. J. Van Wylen, *Fundamentals of Statistical Thermodynamics*, John Wiley and Sons, New York, 1966.

G. J. Van Wylen and R. E. Sonntag, *Fundamentals of Classical Thermodynamics*, John Wiley and Sons, New York, 1965.

M. W. Zemansky and H. C. Van Ness, *Basic Engineering Thermodynamics*, McGraw-Hill Book Co., New York, 1966.

Answers to Selected Problems

2.3 302 lbf; 321.74 lbm
2.6 1440 lbm
2.9 55 lbf/in.^{2}; 3.74 atm
2.12 14.71 lbf/in.^{2}
2.15 16.6 lbf/in.^{2}

3.3 (*a*) 170 F
(*b*) 25.2 lbf/in.^{2}
3.6 19.3 lbf/in.^{2}; 1.67 ft^3
3.15 0.0208
3.21 3380 lbf/in.^{2}
3.24 0.467, 0.509, 0.024 lbm
3.27 16.9 lbm
3.30 24.1 lbf/in.^{2}

4.3 (*b*) 4680 ft lbf
(*c*) 2160 ft lbf
4.6 (*a*) 216 F
(*b*) 9030 ft lbf
4.9 (*a*) 7.07×10^7 ft lbf
(*b*) 247 ft^3
4.12 (*a*) 30,000 ft lbf
(*b*) 7%
4.15 (*a*) 991 lbf/in.^{2}
(*b*) 87.0 Btu

5.3 (*a*) 100,000 Btu
(*b*) 0
5.6 −502 Btu
5.9 15.8 Btu
5.12 1.6 Btu
5.15 −47.4 Btu
5.18 (*a*) 771 F
(*b*) 250 Btu/lbm
5.21 (*a*) 300 lbf/in.^{2}
(*b*) 948 F
(*c*) 212 Btu
(*d*) 2129 Btu
5.24 (*a*) 39.5 lbf/in.^{2}
(*b*) 6.4 Btu
(*c*) 8.9 Btu
5.27 (*a*) 889, 1248, 889, 0 Btu
(*b*) 889, 1248, 1248, 359 Btu
5.30 (*a*) 1250 Btu/lbm
(*b*) 248 Btu/lbm
(*c*) 203 Btu/lbm

6.3 16.8 HP
6.6 7.58 lbm/hr
6.9 0.955
6.12 183.7 Btu/lbm
6.15 (*a*) 1150 Btu/lbm; 0.976
(*b*) 24,800 HP
(*c*) 30,100 HP
(*d*) 0.0354
(*e*) 243.5 F
(*f*) 410.7×10^6 Btu/hr
(*g*) 0.26
6.18 (*a*) 20,200 lbm/hr
(*b*) 3.28×10^6 lbm/hr
6.21 (*a*) 79.1 F
(*b*) 80 F
6.24 0.15 lbm/min
6.27 6530 HP; 1750 HP
6.30 (*a*) 755 R
6.33 500 lbf/in.^{2}
6.36 (*a*) 1125 Btu
(*b*) 48,000 Btu
6,39 8700 Btu
6.42 (*a*) 0.045
(*b*) 0.844

7.6 6.0
7.9 0.754
7.12 (*a*) 2040 Btu/hr
(*b*) 100 F
7.15 1400 cal/mole; 0.0142

8.3 (*b*) 0.8543; 0.1655
(*c*) 165.4 Btu/lbm
8.6 −100.8 Btu/lbm
8.9 216 ft/sec
8.12 705 Btu; 450 F
8.15 (*a*) 4.07, 4.35, 14.37 Btu/R
8.21 2.83 lbm; 715 R
8.24 (*a*) 13.0 lbm
(*b*) 13.2 lbm
8.27 10,330 Btu
8.33 71.1 lbf/in.^{2}; 599 R

9.3 316,000 lbm/hr
9.6 47.5 lbf/in.^{2}; 472 F
9.9 230 F; 597 F; 0.618 lbm
9.12 (*a*) 0.149 Btu/lbm
(*b*) 100.15 F
9.15 178.7 HP
9.21 (*a*) 9.2 lbm; −1470 Btu
(*b*) 14.6 Btu/R
9.24 (*a*) 0.0923
(*b*) 0.25 Btu/lbm; 141.4 Btu/lbm
9.27 (*a*) 441 HP; 788 R
(*b*) 538 HP; 847 R
9.30 (*a*) 1.83; 0.9467
(*b*) 3.74; 0.9761
9.33 −68.4 Btu/lbm; 11.4 Btu/lbm
9.36 (*a*) 1.76 HP
(*b*) 1780 Btu/hr
9.39 (*a*) 1000 R; 500 Btu
(*b*) 1250 R; 222 Btu
9.42 (*a*) 291.8 Btu/lbm
(*b*) 167.1 Btu/lbm
(*c*) −1.8, 168.9, −69.4, 236.5, −115.2, 282.3 Btu/lbm

10.3 33.15%, 0.261; 34.3%, 0.177; 36.65%, 0.1047; 40.3%, 0.0283
10.6 36.5%; 380.6 Btu/lbm

10.12 (*a*) 5530 lbm/hr
(*b*) 4.0 HP
(*c*) 30.9%
10.21 (*a*) 5.2 Btu/lbm
(*b*) 1.183 tons
(*c*) 2.23
10.24 (*a*) 1850 lbf/in.^{2}; 8208 R; 56.6%; 320 lbf/in.^{2}
(*b*) 1472 lbf/in.^{2}; 6462 R; 48.1%; 273 lbf/in.^{2}
10.27 98 Btu/lbm; 400 Btu/lbm; 75.5%
10.30 60.5%
10.36 (*a*) 37.2 lbf/in.^{2}
(*b*) 110.7 Btu/lbm; 2300 lbm/hr
(*c*) 1524.7 R; 63.7%
10.39 41 lbf/in.^{2}; 2430 ft/sec
10.42 0.259

11.3 2.75 Btu/R
11.6 570 R; 0.177 Btu/lbm-R
11.9 (*a*) 549 Btu/mole
(*b*) −549 Btu/mole
11.12 36 F; 57.7 F
11.15 (*a*) 0.878
(*b*) 403 Btu/min
11.18 63.6 F; 0.00269 lbm
11.21 (*a*) 37 F; 0.03375
(*b*) −1970 Btu/hr
11.24 (*a*) 32.4 Btu/lbm
(*b*) 0.378
11.27 (*a*) 73.4 lbf/in.^{2}
(*b*) 0.52 gal/hr
(*c*) −55,500 Btu/hr

12.15 (*a*) 3800 ft/sec
(*b*) 4790 ft lbf
12.18 234 lbf/in.^{2}; 432 R; 0.323 ft^3
12.27 −44,750 Btu/min
12.30 43 F
12.33 (*a*) 16.3 lbm
(*b*) 1015 Btu
12.36 (*a*) 13.17 Btu/lbm
(*b*) 13.66 Btu/lbm
(*c*) 16.95 Btu/lbm
12.39 −41.7 Btu/lbm; −92.0 Btu/lbm
12.42 1.03%; −5.5%; −8.0%
12.45 2220 Btu

12.48 −3000 Btu; −2385 Btu
12.51 (*a*) 2.58 lbm
(*b*) 342 lbf/in.^{2}

13.3 (*a*) 1.105
(*b*) 1.413
13.6 5040 R
13.9 −560,350 Btu/mole
13.12 320%
13.15 2263 R
13.18 3250 R
13.21 (*a*) −10,640 Btu/lbm
(*b*) 4476 K
13.24 (*a*) 20,113 Btu/mole
(*b*) −545,515 Btu/mole
13.27 (*a*) −101,810 Btu/mole
(*b*) 1377 Btu/R
13.30 (*a*) 907,094 Btu/mole
(*b*) 898,876 Btu/mole
13.33 (*a*) −13,923 Btu/hr
(*b*) 13.8%

14.3 1122 lbf/in.^{2}; 0.786 CH_4
14.6 170 lbf/in.^{2}; 1920 Btu/mole
14.12 71.7% NH_3, 21.4% N_2, 6.9% H_2
14.15 0.1076
14.18 (*a*) 38.2
(*b*) 86.6% F, 13.4% F_2
(*c*) 74,940 Btu/min
14.21 (*a*) −37,833 Btu
(*b*) 40,620 Btu
14.24 41.9 Btu/R
14.27 8.9
14.30 46.38% O_2, 40.76% H_2O, 10.93% OH, 1.93% H_2
14.33 49.9% e^-, 48.35% Ar^+, 1.05% Ar, 0.7% Ar^{++}

15.3 (*a*) 0.91
(*b*) 4
15.6 0.72
15.9 12,600
15.12 10/243

16.3 (*a*) 4.00; 0.246
(*b*) 2114, 672, 214
16.9 1261, 220 m/sec; 3980, 695 m/sec
16.12 (*b*) 0.739

17.3 (*b*) $10^{3.95\times10^{24}}$
17.6 $10^{1.307\times10^{13}}$; 1.0×10^{-10} cal/mole-K
17.9 33.87 cal/mole-K (difference $\bar{R}\ln 4$)
17.15 1.6×10^{-21} cal/mole-K

18.6 7300 cm/sec; 1 cm/sec
18.9 (*a*) 3.44×10^{-29}, 3.28×10^{-27} erg
(*b*) 9.8×10^{-31}, 9.33×10^{-29} erg
18.15 (*a*) 2.42, 4.84, 7.26, 9.7, 12.1×10^{-14} erg
(*b*) 0.96, 1.92, 2.88, 3.84, 4.8×10^{-16} erg
18.18 (*a*) 1.79×10^{-11} erg
(*b*) 3.65×10^{-12} erg

19.6 2.79×10^{23}
19.9 27.43 cal/mole-K
19.24 7.434, 48.41 cal/mole-K; 8.912, 67.92 cal/mole-K
19.27 −6043 cal/mole; −10.37 cal/mole-K
19.30 44,320 cal/mole; 72.55 cal/mole-K
19.33 55.50 cal/mole-K; 9.81 cal/mole-K
19.39 2.48×10^{-9} atm
19.45 34.05 cal/mole-K
19.51 (*a*) 0.431
(*b*) 47.37% N^+; 47.37% e^-, 5.26% N

20.3 2.87 cal/mole-K; −8.6 cal/mole
20.6 56,270 cal
20.9 (*b*) 7×10^{-13} dyne/cm^2; 2093 cal/mole
20.12 (*b*) 9×10^{-12} erg; 2.14×10^5 atm
20.15 1.255, 0.0046 cal/mole-K; 5.89, 0.046 cal/mole-K
20.21 (*c*) 29.1 amp/cm^2

Index